STOCHASTIC HYDRAULICS 2000

PROCEEDINGS OF THE EIGHTH INTERNATIONAL SYMPOSIUM ON
STOCHASTIC HYDRAULICS / BEIJING / CHINA / 25 - 28 JULY 2000

Stochastic Hydraulics 2000

Edited by

Zhao-Yin Wang
*Department of Hydraulic Engineering, Tsinghua University &
International Research and Training Center on Erosion and Sedimentation,
Beijing, China*

Shi-Xiong Hu
*International Research and Training Center on Erosion and Sedimentation,
Beijing, China*

Taylor & Francis
Taylor & Francis Group

LONDON AND NEW YORK

Cover photo courtesy of Zhao-Yin Wang

The texts of the various papers in this volume were set individually by typists under the supervision of either each of the authors concerned or the editor.

Published by Taylor & Francis
2 Park Square, Milton Park, Abingdon, Oxon, OX14 4RN
270 Madison Ave, New York NY 10016

Transferred to Digital Printing 2007

ISBN 90 5809 166 X

Stochastic Hydraulics 2000, Wang & Hu (eds) © 2000 Taylor & Francis, ISBN 90 5809 166 X

Table of contents

Turbulence

Sediment transport

Flood estimation and control

Environmental hydraulics and global climate modeling

Wave and coastal processes

Probabilistic control of reservoir management

Stochastic modeling

Miscellaneous

Stochastic Hydraulics 2000, Wang & Hu (eds) © 2000 Taylor & Francis, ISBN 90 5809 166 X

Foreword

The 8-th IAHR Conference on Stochastic Hydraulics continues a tradition which was initiated in 1972 with the first conference in Pittsburgh, Pa. USA, by Ch.-L. Chiu, who for many years has been the chairman of the IAHR section on stochastic methods in hydraulics. The basic idea was to concentrate on stochastic aspects of sediment transport and turbulence – fields in which increases in basic understanding through stochastic methods were to be expected. Such methods were used in the fundamental work of G.I.Taylor for describing dispersion in continuous media and by H.A.Einstein in analyzing sediment transport. These were pioneering contributions to these fields. Later applications included work on water surface waves and, of course, on hydrology, such as stochastic simulation or extreme value analysis. These more traditional subjects have been the main topics in the past.

To me, one of the appeals of stochastic hydraulics has been the potential for a marriage of hydrology and hydraulics, with stochastic methods as important link. For many practical applications hydrology provides the input into hydraulic models. In traditional design the hydrologist determines a design discharge, or a design hydrograph, which a hydraulician, usually with little knowledge of hydrology, uses as design information as if it were a deterministic quantity, without regard to questions of uncertainty or accuracy. The increasing power of the computer and the availability of a larger data base has made it possible to address the problems of uncertainty of design information on the one hand, and the concept of design by reliability on the other hand. By using the whole range of possible hydrological inputs the engineer is capable to optimize his design not only for one special extreme condition, but for a whole range of likely extreme or in other ways important conditions. Through estimations of the uncertainty of the hydrological data base it also has become possible to obtain an estimate for the uncertainty of a design.

The applications of stochastic methods in the design by reliability include the better utilization of hydrological information, but its use goes beyond combining hydrology and hydraulics. With statistical methods one can answer the question of "how safe is a hydraulic system?" - a question which in many cases traditional design cannot answer, because usual design proceeds by cumulating safety factors for different loads, different components and materials. In this way it adds safety features of different kinds to ensure reliable performance of a hydraulic system. An example is the design of a dam, which is based on the assumption of a full reservoir, to which a design discharge corresponding to the flood protection function plus the super-elevation for the spillway, such as the elevation due to the PMF (probable maximum flood), plus a freeboard is added. All these are random variables, each with a very distinct and quite different probability function. If their combined effect is not properly determined by stochastic methods, then the failure probability for such a dam is not known. The statistically accurate method for such a determination is reliability theory, which from its development in structural engineering has been gradually transferred to applications in hydraulics. The work done by B.C.Yen and his students has been particular important in this context, as was strikingly demonstrated during the 4th IAHR Conference on Stochastic Hydraulics, which he organized. However, stochastic design by reliability only replaces one design criterion: the classical criterion of critical values which are not to be exceeded, by another, i.e. by a failure probability which is not to be exceeded. This criterion is more satisfactory, as it permits comparison of components of a design on an equal basis, however, it becomes part of a complete and fully consistent design theory only if the concept of risk minimization is introduced. This is one of the applications of stochastic hydraulics which I see as most promising. Although it is not a new method in principle, it has not been used as extensively as it should be: this remains as a challenge for hydraulic engineers, and for future research in applications of stochastic methods to hydraulic engineering.

New fields are evolving in hydraulics and hydraulic engineering as part of integrated water resources management, such as re-establishing the natural environment of rivers, detection and sanitation of ground-water, urban sewer system design, to name only a few. They all have design aspects which should be solved by means of stochastic methods. I thus see a continuing need for stochastic methods in hydraulics, and I am sure that the present conference will contribute to further successful development of methods for science and applications. In the name of IAHR and its council I would like to thank Prof. Z.Y.Wang and his colleagues for organizing this conference.

Erich J. Plate
April 2000

Stochastic Hydraulics 2000, Wang & Hu (eds) © 2000 Taylor & Francis, ISBN 90 5809 166 X

Acknowledgements

ORGANIZATION

International Research and Training Center on Erosion and Sedimentation
Tsinghua University

SPONSORS

International Association for Hydraulic Research
(IAHR)

National Natural Science Foundation of China
(NSFC)

United Nations Educational, Scientific and Cultural Organization
(UNESCO)

International Association for Hydrological Sciences (IHES)
Chinese Hydraulic Engineering Society
(CHES)

Water Resources Engineering Division of American Society of Civil Engineers
(WREDASCE)

German-Sino Unsteady Sediment Transport
(GESINUS)

Keynote lectures

Stochastic Hydraulics 2000, Wang & Hu (eds) © 2000 Taylor & Francis, ISBN 90 5809 166 X

Stochastic design – Has its time come?

Erich J. Plate
Universität Karlsruhe (TH), Germany

ABSTRACT: Stochastic design is a method of including the stochastic variability of loads, resistances, and consequences into design decision models. Instead of relying on safety factors or ad hoc assumptions, one replaces uncertainties by probability functions. It serves a number of important purposes not included in traditional design concepts. First of all, it is a method which permits to handle uncertainties of all kinds: natural variability, data uncertainties, model uncertainties, and parameter uncertainties. In a complex model, which transforms an input depending on many variables through a system with many parameters into a design output, it permits to study the transformation and combination of uncertainties in a logical and analytically consistent manner. Because of this ability, stochastic methods applied to a specific design problem can provide guidance for deciding how much effort one should spend in improving the accuracy, either of the theoretical basis, or of the construction. Stochastic design may aid in balancing the value of improvement in theory against the advantages gained by additional measurements and closer studies of all factors influencing the design.

The most appealing aspect of stochastic design, however, is that it is an important component of a complete theory of design, which optimizes a water resources system. By means of typical examples the advantages of stochastic design will be illustrated, first by looking at examples for the different design levels of stochastic design, and finally by presenting the general framework for an application within the optimization of a water resources project.

Key words: Stochastic design, Variability of loads, Safety factors, Uncertainties

1. INTRODUCTION

Hydraulic engineers and hydraulic engineering research had an important role in the development of water resources. Indeed, one could identify different ages by the contributions of hydraulic engineering. The classical modern age of hydraulic design culminated in the thirtieth of the 20-th century, when the large dams built in the USA added a glamour to the profession which it did not enjoy before. By making water available in every household and for every industrial purpose in sufficient quantity at all times engineers have contributed significantly to reaching the quality of life of today. Navigable rivers, water supply and hydro power were among the accomplishments which gave hydraulic engineers a very high reputation among all engineers. Today, the benefits of hydraulic engineering in these fields are taken for granted, and hydraulic engineering has become an "old" profession, whose development curve has become very flat: the basic knowledge needed for most tasks of hydraulic engineering has not changed much during the last few decades: large investments in research recently have yielded only small improvements in the traditional design of pipes and canals, dams and surge tanks, coastal defense and port structures, sewers and retention basins, to name a few typical structures. This became already apparent in the early 60-es, and I think that the "Handbook of Hydraulic Engineering" of Davis et. al or the book "Engineering Hydraulics" edited by Hunter Rouse and John McNown in the USA, or the chapters in Schleichers "Handbuch des Bauingenieurwesens" in Germany summarized the hydraulic knowledge of the classical age - almost as well as any book on hydraulics of today could do.

If the types of problems for hydraulic engineering had remained as they were in the 70-ies, a certain stagnation in hydraulic research would have been felt earlier. However, hydraulic engineering entered a second modern age, with the advent of environmental concern. Hydraulic design profited greatly from new research based on turbulent processes such as mixing and dispersion, involving not only neutral flows, but density stratified water bodies and pollutant transport in rivers, lakes and the ocean. New types of structures were developed : cooling water intakes for giant power plants, outfalls of sewage into the ocean, river and groundwater pollution control, renaturalisation of rivers. All these problems were approached in an essentially deterministic way, using what may be called the classical environmental approach to design.

We now find that today hydraulic design has entered another phase of development, which Prof. Mike Abbott calls the age of Hydroinformatics, which is based on the power of the computer. It had the first result of dramatically improving accuracy and speed of calculations for the classical design procedures, and to permit the numerical solution of partial differential equations for complex boundary conditions: the methods employed are more sophisticated, but generally speaking, the more powerful computers are mostly used to make a much more refined analysis for the traditional design practice.

The classical design methods of hydraulics, be they structure or environment oriented, are based on one-dimensional application of known principles - based on natural sciences, experiments and experience - to technical problems of varying degrees of complexity. They consist of more or less deterministic applications of fluid mechanical principles to complex practical problems. The design is subdivided into three steps. The first step is the determination of the loads (where loads can be either forces, such as the wave forces on a breakwater, or demands, such as the required storage of a reservoir for irrigation). Usually, the largest reasonable load was assumed. The second is the design of the structures capable to withstand the loads. The third is the recognition of uncertainties of the basic assumptions in this design concept. The uncertainties are taken care of by factors of safety specified on the basis of experience in codes and regulations which codify good practice. The characteristics of this approach is that the engineer who designs according to the standards is assumed to produce absolutely safe structures, - if a structure should fail under these design rules, it is either human error, or an act of god. By being employed for this type of design, the computer is a tool for doing what has always be done – only much faster and much more accurately.

But there is another aspect of the computerization of the design phase: the fact that a computer can handle large quantities of data. Not only data which are results of measurement, and which can be used to obtain empirical probability distributions of measured quantities, but also data which are artificially generated, which makes it possible to create dynamic models of the real world and to study their behavior by means of simulations. The ability of studying the response of a system under varying loads, under varying operation rules, under varying constraint has added a new dimension to hydraulic design. Through this capability, design of hydraulic structures and water resources system has entered a new age, which opens new avenues of research and enables engineers to become partners in a broader approach to project design, in which the interaction of a design with other aspects of a water resources development project is possible. In a formulation of a general cost benefit analysis, the application of data oriented procedures provide the means of giving weights to failures of a system. In this context, design is to be understood as a means of adjusting the system to have a failure probability which is lower or equal to what can be tolerated by the project owner. Instead of designing according to fixed design parameters, design is oriented on a probability of failure – which is the basis for stochastic design. Stochastic design thus is the building block, which the hydraulic engineer contributes to the design of water resources systems. In using this method, he can participate directly and interactively in the exciting development of water resources systems, giving a new perspective to an old profession.

2. THE LEVELS OF STOCHASTIC DESIGN

2.1 What is stochastic design?

Stochastic design in general replaces deterministic design information through probabilistic information. It takes account of the fact that no design is absolutely safe, every product or project may fail due to many causes: extreme loads, aging and corrosion, wanton destruction, misuse etc. In particular, it replaces deterministic design criteria by probabilistic criteria – such as replacing the design height of a dike for a river by the probability of failure of the dike. The structures used in stochastic design are traditional: stochastic design is only used to determine the dimensions of the structure. In this sense, by using stochastic design the engineer is only using different design criteria: the standard of stochastic design is based on probability: the probability of failure, or the risk, i.e. the expected value of the losses which can occur if a system fails. The concept of stochastic design is dating from the early part of this century, but its application to real world problems has only been possible with the advent of the computer, which is needed for solving non-linear problems, and which is also needed to handle large amounts of data from which to determine empirical probability distributions. As shall be shown by means of simple examples based on the experience of the writer, stochastic design has the following advantages:
- It provides a logical method for determining the loads as random processes derived from many basic processes, and covering a continuum of load conditions for the design process.
- It combines, for natural loads, the results of hydrology and of hydraulics into a consistent and interactive frame
- It provides a means to replace safety factors (which are a result of extensive experience of experts) through statistical estimates (which are based on measurements and probabilistic concepts and thus need less experience)
- It provides the basis for quantifying, in an objective manner, the risk which has to be used in a comprehensive theory of design, - a theory which is based on the optimization of a suitable objective function

These advantages should give stochastic design a very useful place in the tool box of the hydraulic engineer – in particular when he has to consider natural processes as loads for his designs. In using it, the power of a computer can be more fully realized in design. It is of course not a method to be used in every design case. Many of the classical problems of hydraulics can be handled in the traditional manner. But even where handbook knowledge suffices for a design, stochastic design methods should be used for establishing the codes: the empirical factors used in design codes should be based on statistics and should be given with their probability density distribution or at least with their standard deviation. It is a task for hydraulic research to provide such information for improvement of codes, which should be addressed to put the design standards on the basis of stochastic design.

Stochastic design has been developed to a large extent by structural engineers, but its main advantage can be realized as well or even better in studies in which the loads have to be evaluated as complex random processes, such as hydrologic processes of river flows, or wave forces. Examples from the work of the writer include dike safety (Plate, 1998) and the design of reservoirs for flood storage (Plate and Meon, 1988). But the extent of applications is not limited to loads and resistances, the method can also be applied in a planning environment. Loads can be water demands, as for example water demand from a reservoir (Plate, 1989). And in water quality problems, the load can be a mass of a pollutant, which has to be handled in water quality problems associated with sanitary engineering tasks (Plate, 1991, Schmitt-Heiderich and Plate, 1995). The application of stochastic design is not restricted to single failure causes: through the method of failure and event trees (Vrijling, 1989) it can handle, for example, compound effects and their joint failure probability.

2.2 Levels of stochastic design and examples

Depending on the stochastic information used for the design, different levels of design are identified, illustrated by means of the dike problem of Fig.1. In this case one has: a load s – which is the water level h_{des} in the canal, - a resistance r – i.e. the ability to withstand the load, which is the dam height h. Failure of the dike is assumed to occur when the height h exceeds the dike height h_{des}.

We distinguish between four different design levels, depending on the degree to which stochastic information is used (Plate and Duckstein, 1988).

- **Level I design:** the traditional approach, which is based on assuming that failure occurs if the load exceeds the deterministic resistance. The design is based on a single value obtained from an extreme value of the load
- **Level II design:** a design also based on failure probability, but which uses resistances and loads as random variables. It uses only first moments (mean values) and second moments (variances) of resistance and load.
- **Level III design:** a design which uses the joint probability density function of resistance and load and integrates over the failure region to obtain the failure probability of the system.
- **Level IV design:** risk based design which includes the jpdf from Level II or Level III and also the consequences of failure and can be used as input into a risk based benefit cost analysis.

Level I design: example. An application to the dike problem of Fig.1 shows that for the level I design, the exceedance probability P_E of h is needed, which is set equal to the failure probability P_F, i.e. it is assumed that:

$$P_F = P\{h > h_{des}\} = P_E(h_{des}) \tag{1}$$

Level II design: example. For level II design, h is a random variable with mean μ_h and variance σ_h^2, and the dike height h′ is a random variable with mean $\mu_{h'}$ and variance $\sigma_{h'}^2$. Level II is called design by FOSM (first order, second moment analysis). This method is simple in its application, but difficult to understand. Starting point is the definition of a safety margin z = r-s , whose probability density function has to be determined. Failure occurs if the safety margin z is equal to zero. If r und s are stochastic independent random variable with mean values μ_r and μ_s , and with variances σ_r^2 and σ_s^2 then z has a pdf (probability density function) with parameters:

$$\mu_Z = \mu_r - \mu_S \quad \text{and} \quad \sigma_Z^2 = \sigma_r^2 + \sigma_S^2 \tag{2}$$

From these, the safety index ß can be obtained:

$$\beta = \frac{\mu_r - \mu_S}{\sqrt{\sigma_r^2 + \sigma_S^2}} \tag{3}$$

If r and s are normally distributed random variables, then there exists the direct relationship of safety index to failure probability $P_F = \Phi(-\beta)$, where $\Phi(\eta)$ is the standardized normal distribution with standardized variable $\eta = (z - \mu_Z)/\sigma_Z$. Analogously, for r and s lognormally distributed, the FOSM method can be applied to the ratio r/s, which is a safety factor. However, if r and s are not normally distributed, then the FOSM method provides only an approximation. The level II method is well developed and permits to include the probability

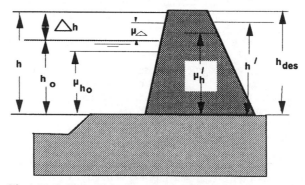

Fig.1: Definitions of the dimensions of a stochastic design problem

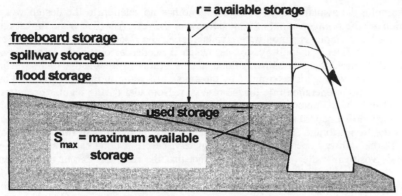

Fig.2 Definition for the problem of dam safety.

distributions of n – with n any number - basic variables $x_1, x_2,....x_n$ for the load, and m basic variables $x_1, x_2,....x_n$ for the resistance, i.e. for $s = g_1(x_1, x_2,....x_n)$ and $r = g_2(x_1, x_2,....x_n)$ into the design, which is exceedingly difficult to do for the case of the more general level III design. Therefore, many methods have been developed for fitting the normal distribution to the extreme end of arbitrary pdfs (Ang and Tang, 1984, Plate, 1992). The application of the FOSM method to dikes was discussed at length by Plate (1998). Vrijling, (1989) and Reeve, (1998) give examples for the application to coastal engineering works, such as sea dikes including many different kinds of failure mechanisms (Vrijling, 1989), which have to be combined using failure and event trees (Henley and Kumamoto, 1981).

The failure probability is associated with a time unit, usually a year. Therefore it is a quantity that changes if either s or r changes with time. For example, the resistance of dikes is a function of time, unless they are regularly inspected and repaired under a regime of strict maintenance. Also, the load may change with time. The methods for determining extreme value distributions for floods are well developed, they usually are empirical distributions to which a suitable pdf is fitted. Generally, the fitting is done under the assumption that the data are stationary, i.e. that the extreme values are taken from a time series extending over many years, whose statistical parameters, such as mean and variance, are not changing with time. For natural loads, the validity of this assumption is questioned. However, there is no statistical trick by means of which a non-stationary development can be predicted. An extrapolation of a trend into the future requires the identification of causes, i.e. requires deterministic models. Exceptions are such processes which vary gradually and have been observed over long time. Such a process is the rising of the sea level, which has been rising at the rate of about 16 cm/century, - but this may be increased through the increase of CO_2 in the atmosphere and the resulting green house effect. The frequency of extreme events, such as floods may also change due to climate variability, but the evidence for this is even more difficult to asses.

Level III design: example. The basis of level II design is the solution of the general failure integral for general relations between resistance and load:

$$P_F(D) = \iint_\Omega f_{rs}(r, s) \cdot dr \cdot ds \Big|_D \tag{4}$$

where Ω is the failure region $s > r$ in the two-dimensional probability space for r,s, and D is the design vector denoting the chosen design. If resistance and load are stochastically independent, then Eq.4 reduces to the Freudenthal integral:

$$P_F(D) = \int_0^\infty \left\{ \int_0^r f_s(s) \cdot ds \right\} \cdot f_r(r) \cdot dr \Big|_D \tag{5}$$

where the inner integral is the probability function $F_s(r, D)$

An example of a level III design is the analysis of the failure probability for a dam, which fails if the level in the reservoir exceeds the height of the dam. It is convenient to formulate this problem not in terms of reservoir level, but in terms of flood volume. Resistance is the available storage during flood, and load is the in-flowing flood volume. The different storages in the reservoir are indicated in Fig.2.

In this example, the available flood storage volume has no relation to the design values of storage, which are the storage volumes for the purpose of the reservoir (below the lowest dashed line in Fig.2) – for example irrigation water availability - , the flood storage, for the protection of the people down river, the spillway storage, which is needed to obtain the super-elevation required for safely discharging the spillway flood, and the storage available due to the freeboard. Usually, the available storage is determined by the amount of water in the reservoir, and has to be determined from the operation rule for the reservoir. Note that during regular operation the flood storage has to be kept empty all the time; the only time it will be filled is when a rainfall event had occurred that caused a flood large enough to fill both the unused part of the storage volume plus the flood storage. The flood has to be even bigger in order to fill the reservoir to the brim, with the spillway operating at design capacity, before a flood can occur which does harm to the downstream valley. Obviously, the fact that the reservoir can at most be empty implies that the random variable r (i.e. the resistance = available storage) is bounded from above by the maximum storage of the reservoir and from below by zero for a full reservoir. Interestingly, used reservoir storage and storage demand by the flood are independent random variables, the former determined by the operation rule of the reservoir, the latter determined through the hydrology of the catchment above the reservoir. Meon (Plate and Meon, 1987) has solved this problem for the actual case of the existing Prim reservoir in Southwest Germany, using the freeboard as decision variable.

Level IV design: The reservoir problem, and the problem of the dike design typically follow a scheme which is quite general, as is outlined in Fig.3 and which can be expanded to include level IV design. The starting information is a set of basic random variables, typically in the form of a hydrological time series, which have to be translated by means of a model (the transformation in Fig.3) into loads s. The design leads to dimensions D which combine into the resistance r. Both r and s have to be seen together and determine the jpdf $f_{rs}(r,s)$, which is used to calculate the failure probability according to Eq.4 or 5. However, $f_{rs}(r,s)$ also is needed for calculating the risk RI(t), which combines the consequences RC(t,r,s) (usually given as costs) at time t of a given combination r,s with the probability for the consequences to occur. RI(t) is the risk (usually in terms of cost) for year t. It is defined as an expected value:

$$RI(t) = \int_{s=0}^{\infty} \left\{ \int_{r=0}^{\infty} RC(t,r,s) \cdot U(t,r,s) \cdot f_{rs}(r,s,t) \cdot ds \right\} dr \qquad (6)$$

where RC(t,r,s) denotes the consequence cost (i.e. damage cost) which occur if the event: "combination r,s" occurs during year t. U(t,r.s) is the utility function, which expresses the attitude of the decision maker toward the risk: U(t) is larger than one for a risk averse decision maker, whereas a risk taking decision maker would accept a utility function smaller than 1. Note that the utility function may change with the magnitude of event: a decision maker may well take chances for a smaller risk, whereas he is risk avers towards large events – or vice versa. The utility function also has the function to express the uncertainty of risk estimation. As the risk is a random variable, which very likely has a symmetric pdf –such as the normal distribution - , the probability that the actual risk is larger than the expected value is 50 percent. A risk-shy decision maker may not be willing to base his decision on a 50 – 50 chance of being right and is more likely to assume a larger risk.

In many cases, the risk cost does not depend on r and s independently, but is a function of the magnitude by which s exceeds r, i.e. it is a function of the safety margin z. In this case, the procedure which has to be followed is to:
- Calculate the pdf for z
- Use this pdf to calculate the risk based on the risk cost function RC(z):

$$RI(t,D) = \int_{-\infty}^{0} RC(t,z) \cdot f_z(z) \cdot dz \Big|_D \qquad (7)$$

The calculation of the risk is the final stage of the design at Level IV. Although usually risks are expressed through risk costs, it is not necessary to use risk as a monetary quantity. Another form of risk estimation is expected loss of life due to the failure of a system, and in general, the risk cost function RC can be any type of loss.

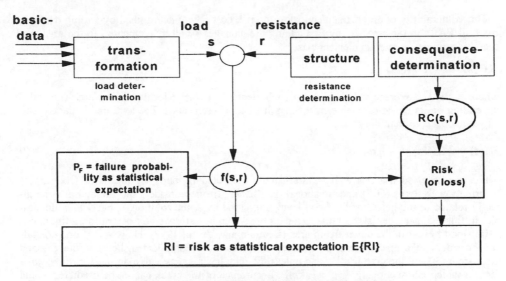

Fig.3: Schematic of a stochastic design, including risk

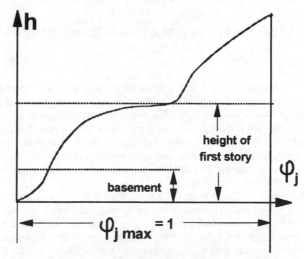

Fig.4: relative vulnerability of a two-story house

A typical example of a risk cost determination is the evaluation of a the expected loss due to floods. Starting point is the determination of the flooded area with associated exceedance probability, which result in flood damage to buildings and other elements EAR (= "elements at risk") located in the flooded area. The damage is to be quantified by means of an analysis of the potential damage to each of the EAR. Let the index j denote element j, and let the maximum damage which this element can suffer be k_j. (for example, k can be the replacement cost for a flood damaged building, or k = 1 can denote loss of life). Most of the time, not every element will suffer maximum losses, instead it will suffer an event s dependent damage percentage $\varphi_j(s)$ of k. $\varphi_j(s)$ is the relative vulnerability of the element j. It is found from a statistical analysis of observed damages. Therefore it can be interpreted as probability that an element of type j which is exposed to s suffers a damage k_j, i.e. $\varphi_j(s)$ is a conditional probability.

9

The vulnerability of an area is identical to the risk cost RC in determined by adding damages over all EARs in the area, i.e. by first adding the damages of all n_j elements j in the area, and then by summation over all element classes:

$$RC(s) = \sum_j n_j \cdot \overline{v}_j(s) \qquad (8)$$

where v_j is the average vulnerability of each element in class j located in the area (note that depending on its elevation each element may have a different $v(s)$). The total risk is the expected value of the vulnerability:

$$RI(t) = \int_0^\infty RC(s) \cdot f_s(s) \cdot ds \qquad (9)$$

The results of such a risk analysis are usually presented in maps, by means of Geographic Information Systems (GIS) and relational data banks. If one considers the flooded area as directly related to the height of the flood level, expressed by s, the solution of Eq. 9 is quite simple. In this manner, the risk cost due to a dam break can be estimated by determining the maximum flood height in the downstream area (Betamio and Viseu, 1997). However, if one considers the safety of a dike system, the solution becomes more complicated, because the flooded area occurs where the dike breaks, and unless the dike has a safety valve, in the form of a dike section which is intentionally kept weak, the occurrence of the break cannot be predicted – and one also has to be aware that the breach of a dike in one place may result in reducing the probability of a dike break occurring at other downstream places.

3. RISK COST AS DECISION VARIABLE IN WATER RESOURCES PLANNING

3.1 The planning environment

Although the determination of the failure probability is an important aspect of stochastic design, and an excellent basis for obtaining better safety factors in codes, the most promising aspect of stochastic design is the inclusion of the risk cost in water resources planning. During recent years, new criteria for the utilization of water resources have evolved based on the concept of sustainable development – requiring a development of water resources systems that meets the need of the present generation of people without compromising the ability of future generations to meet their needs. In particular, this implies that water resources development must proceed in conformity with the preservation of nature.

To illustrate this process, the case of integrated planning for flood safety and a healthy environment as part of a sustainable project is shown schematically in Fig.5 (adapted from A. Götz, Swiss Institute for Water Resources. Personal communication). The societal goal of sustainable development is converted into a set of objectives: objectives for the safety, and objectives for the preservation of natural functions. If an analysis of the existing situation is showing that the existing conditions meet the objectives, then the only action required is to keep it that way, i.e. to maintain the conditions and to prevent intrusion of external demands that could alter the situation to the negative – i.e., it might be necessary to set up legal barriers against abuse. If the existing situation does not meet the objective, a process is initiated for improving the situation. The next stage is the decision process for finding the alternate which does meet the objectives of the design. There are many cases when it is impossible to meet all requirements, in particular when constraints are set, which might be financial, social, or political. Then it is necessary to change the objectives, to make them more to conform to reality. In this manner, many well meaning nature preservation objectives had to be overruled, or protection objectives had to be set aside. Finally, the alternative selected is implemented, and the project is prepared.

3.2 Risk cost in a decision environment

The logically consistent translation of the decision process in Fig.5 is by means of an optimization formulation. The task is the determination of the technical or non-technical measure which optimizes a given objective function. This implies, that the measures are given monetary or other weights, which permit a comparison on equal terms. Consequently, all quantities have to be made dimensionally homogeneous, for example by the same monetary units. The total costs are combined into an objective function Z (Crouch and Wilson, 1982):

$$Z = \frac{1}{T} \sum_{t=1}^{T} \frac{1}{(1+i)^{tn}} \left(B(t) - C_A(t) - C_s(t) - RC(t) \right)_D \qquad (10)$$

where Z is the annual average net benefit. In Eq.10, T is the design life in number of years, t is the time index, t = 1,2,....T in number of years, i is the discount percentage, and D indicates the decision variable, i.e. it denotes the different solutions for the system, such as dike height for a dike system, or reservoir size for a flood retention reservoir, etc . However, D can also denote a set of entirely different alternatives, such as a non-technical solution vs. technical solutions. The solution to the optimization problem is found, in principle, by varying the decision D until a solution is found which minimizes Z.

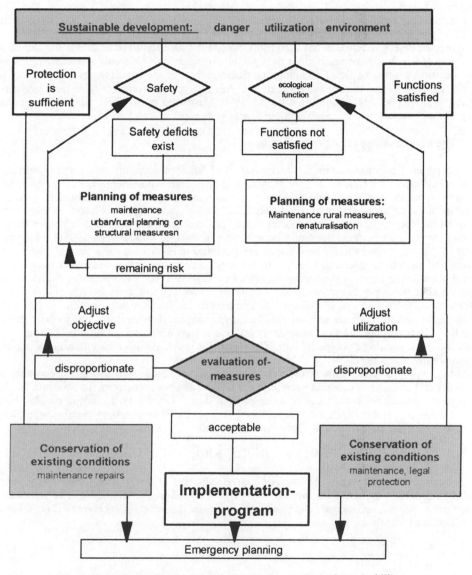

Fig.5: The concept of project planning under the paradigm of sustainability

The total benefits consist of the direct benefits B(t) obtained from the solution, which have to be balanced against the sum of the costs, where $C_A(t)$ is the cost for producing the optimum solution, $C_S(t)$ the social cost – i.e. the marginal costs of the losses which occur if the decision D is made minus the cost of maintaining the condition as is – including the cost of resettlement of persons or goods displaced by the project from their original location, or the difference in maintenance cost, which may reduce the benefits by sizeable amounts. Generally, the social costs are covered by others than the owner of the project, and careful evaluation of the benefit distribution is needed to make the decision not only optimum in total cost, but also equitable in that the people share the benefits, who have to carry the burden of the project. Finally, RC(t) is the risk cost, which also is a function of D.

Usually the objective function is constrained by social and other limitations. A solution cannot be optimum, when the price for an inexpensive solution is an unacceptable threat to human life – which cannot be expressed in monetary terms. Other constraints are imposed by ecology or nature protection, or by the preservation of historical buildings and landmarks. These constraints have to be considered, and frequently, they narrow the choices for a protection system down to a very small number of alternatives with only very few decision variables – such as the dike height, or the size of a reservoir.

When the objective function has been formulated, a decision has to be made for the choice of the project which is to be realized. In the hydraulic literature there are only a few examples for the decision process of Eq. 1 including the determination of the remaining risk, according to Eq.2. Examples for a systematic application of this concept in hydraulics have been given by R.A.Freeze and his co-workers (Freeze et al. 1990, Massmann et al. 1990), who used this concept for a systematic management of groundwater resources.

3.3 The uncertainty of the objective function

In this paper, I only touched on the determination of NB. The determination of the benefit and cost aspects shall not concern us here. It suffices to mention that a modern view of the costs and benefits is much more discerning than traditional economic evaluations and requires to give due consideration to sustainability of the project. Unfortunately, although there exists a vast literature on the subject of classical economic benefit and cost analysis, not much is known on the social cost as seen from the point of view of sustainability (see Loucks et.al, 1998 for a discussion of this issue). In a careful analysis, all the quantities in Eq. 1 have to be seen as function of times and have to be discounted to the present value: a process which permits to consider foreseeable variations, such as the limited benefit of a a large reservoir during the first years of operation after being put into commission. Obviously, the direct cost must include not only construction cost, but also cost of maintenance and repairs. Furthermore, all quantities in Eq. 1 including the discount rate are random variables, and influence the value of Z due to their uncertainties. However, this does not necessarily imply that the uncertainty of all quantities need be considered. It should be subject of careful analysis if it is actually necessary to account for the variability of all costs.

For example, the uncertainty of the risk has to be seen in comparison with the uncertainty of benefits and costs, as calculated by the formulas which have been presented. To illustrate this let NB be the annual net benefit due to a design D at time t. Let this be a random variable with mean value μ_{NB} and variance σ_{NB}^2. Also, let the annual risk RI be a random variable with mean μ_{RI} and variance σ_{RI}^2. Then Z is also a random variable:

$$Z = \frac{1}{T}\sum_{t=1}^{T}\frac{1}{(1+i)^t}\big[NB(t) - RI(t)\big] = a_{iT} \cdot [NB(t) - RI(t)] \tag{11}$$

The expression in the square bracket has a mean value $\mu = \mu_{NB} - \mu_{RI}$ and for stochastic independent quantities NB and RI a variance $\sigma^2 = \sigma_{NB}^2 + \sigma_{RI}^2$. Assume NB and RI to be stationary, i.e. their mean and variance are independent of time and the same for each year. Then Z has a coefficient of variation:

$$CV_Z = \frac{\sigma}{\mu} = \frac{\sqrt{\sigma_{NB}^2 + \sigma_{RI}^2}}{\mu_{NB} - \mu_{RI}} \tag{12}$$

This expression shows clearly that σ_{RI}^2 and σ_{NB}^2 must be of approximately the same magnitude in order for both of them to be effective. In most cases a closer analysis will show that one of the two will dominate, and then it is advantageous to concentrate ones efforts at improving the data base for the dominant quantity. Water resources engineers planning complex systems are well advised to make a close investigation of the uncertainty of their conclusions: Often, uncertainty considerations can make a solution near the optimum more acceptable.

4. CONCLUSION

Stochastic design as a new method is actually not so very new, but a widespread application of this technique has yet to be found. This is in part based on the tradition, and research engineers who have promoted stochastic design, therefore saw as the first purpose of stochastic design to put the standards used in hydraulic design on a better theoretical footing. Design by level I techniques is still the most prevailing in hydraulic engineering working with natural loads. This is fairly well justified in many cases, where the uncertainty of the resistance is small as compared to the uncertainty of the load. Level II and III techniques are useful in particular to determine the weak link in a design chain. When many random processes combine into one uncertain load or resistance, then stochastic design is a useful approach. Its greatest usefulness is in level IV problems, such as water resources systems.

When one considers advantages and limitations of stochastic design, one comes to the conclusion that stochastic design is an added and valuable tool contributing to better design. By replacing a multitude of design criteria by the criteria of permissible failure probability or minimum cost (or highest benefit), stochastic design greatly unifies design practice and uses the computer in a most productive way. Every engineer should become familiar with this technique and be able to employ it wherever it might lead to increased safety or increased cost savings.

REFERENCES:

Ang,A.H. and W.H.Tang 1984: "Probability concepts in engineering planning and design, Vol.2" J.Wiley, New York

Betamio de Almeida, A. and T.Viseu, 1997: Dams and safety managment at downstream valleys. Balkema, Rotterdam

Crouch, E.A.C. und R.Wilson 1982: Risk Benefit Analysis, Ballinger Publisher, Boston, Mass. USA

Freeze, R.A., J.Massmann, L.Smith,T.Sperling, and B.James 1990: Hydrogeological decision analysis: 1. A. framework, Groundwater, Bd. 28, S.738-766

Henley,E.J. and H.Kumamoto, 1981: "Reliability engineering and risk assessment" Prentice Hall Publishers, Englewood Cliffs, N.J.

Loucks, D.P. et.al 1998: Task Committee on Sustainability criteria, American Society if Civil Engineers, and Working Group UNESCO/IHPIV Project M-4.3 Sustainability criteria for water resources systems. ASCE, Reston, VA. USA

Plate, E.J., 1989: Reliability concepts in reservoir design. Nordic Hydrology, Vol. 20, 1989, pp. 231 – 248

Plate, E.J., Duckstein, L., 1988: Reliability-based design concepts in hydraulic engineering. In: Water Resources Bulletin, Vol. 24, 1988, No. 2, pp. 235 – 245

Plate, E.J., Meon, G., 1988: Stochastic aspects of dam safety analysis. In: Proceedings of the Japan Society of Civil Engineers; Hydraulic and Sanitary Engineering, No. 393/II-9, 1988, May, pp. 1 – 8

Plate, E.J., 1991: Probabilistic modelling of water quality in rivers. In: Water Resources Engineering Risk Assessment. Ed.: J. Ganoulis, Berlin: Springer-Verl. 1991, (NATO ASI Series G 29), pp. 137 – 166

Plate, E.J. 1992: "Statistik und angewandte Wahrscheinlichkeitslehre für Bauingenieure" (Statistics and applied probability theory for civil engineers) Ernst und Sohn, Berlin

Plate, E.J. 1998 Stochastic hydraulic modelling - a way to cope with uncertainty. in: K.P.Holz et.al. (ed.)1998: ADVANCES IN HYDRO - SCIENCES AND –ENGINEERING, Vol. III, Proceedings of the 3rd International Conference on Hydro – Science and – Engineering, Cottbus/Berlin, Germany (CD-ROM)

Reeve, D.E. 1998: Coastal flood risk assessment, , Journal of the Waterways, Port, Coastal, and Ocean Engineering, Am.Soc.Civil Eng., pp.219-228

Schmitt – Heiderich, P., E.J.Plate, 1995: River pollution from urban stormwater runoff. In: Statistical and Bayesian Methods in Hydrological Sciences, Paris, UNESCO, Vol.2, Chapter 2 pp.1-17

Vrijling, J.K. 1989: "Developments in the probabilistic design of flood defences in the Nederlands" Seminar on the reliability of hydraulic structures, Proceedings, XXIII Congress, International Association for Hydraulic Research, Ottawa, Canada, pp.88-138.

Stochastic Hydraulics 2000, Wang & Hu (eds) © 2000 Taylor & Francis, ISBN 90 5809 166 X

Numerical calculation of flow and sediment transport in rivers

W. Rodi
Institute for Hydromechanics, University of Karlsruhe, Germany

Abstract: The calculation of flow and sediment transport is one of the most important tasks in river engineering. The task is particularly difficult because of the many complex and interacting physical phenomena which need to be accounted for realistically in a model that has predictive power. Both 2D depth-average and 3D models were developed and tested in the author's group over the years which allow the calculation of flow, both suspended and bed-load sediment transport and the associated bed deformation for natural rivers. These two model types and their interrelation are described with their components such as the hydrodynamic model including the turbulence model, bed-load and suspended-load model and the numerical procedures for solving the model equations. Applications of the models to a classical laboratory situation and to stretches of the rivers Rhine and Elbe and the Three Gorges Reservoir Project in the Yantze River are presented and compared with experiments whenever possible.

Key words: Numerical calculation, 2D and 3D models, Sediment transport , River flow

1. INTRODUCTION

Flood protection, keeping rivers navigable, securing household and irrigation water supply and providing the conditions for hydropower generation are the prime tasks of river engineers. In fullfilling this task, not only economical and technical considerations are important but increasingly also ecological ones and the impact of any human measures on the environment. For planning such measures, the engineers need to be able to predict the consequences of these. In the past, such predictions were mainly based on integrated programmes of field and hydraulic model studies, but these are very costly and time consuming. With the rapid increase in computer power and the advancement of numerical methods, more and more computer models are used, and the prime task of such models is to predict the flow and sediment transport in rivers and the impact of human measures. This is a difficult taks because of the many complex and interacting phenomena involved, like irregular geometry which can vary with time causing complex flow patterns, turbulence, suspended and bed-load transport with deposition and erosion causing bed deformation.

Different levels of idealisations and empirical input have been used in computer modelling. 1D models are widely used in practice and can be applied to long river stretches. With these models, only cross-sectional averaged quantities are calculated so that details cannot be resolved and many influences require an empirical treatment. Depth-averaged 2D models resolve horizontal variations and provide many more details like the influence of changing cross-sections and irregular side boundaries. These models are based on the assumption of hydrostatic pressure distribution and cannot account for secondary motions which are particularly strong in bends. Today they are also used already for solving practical problems, albeit only for shorter river stretches. A number of models suitable for such applications are available and a review can be found in Minh Duc (1998). 3D models are the most powerful ones; they need not assume hydrostatic pressure and can calculate secondary motions, but they are also the most costly ones and are so far used only for calculating more local phenomena. Again several models, some fully 3D, some only partly, are available and a brief review is given in Wu et al (2000). The recently published model of Gessler et al (1999) should further be mentioned here.

In the author's research group both 2D and 3D models for calculating the flow and sediment transport in natural rivers with complex geometries have been developed over the years. Both models have much in common in terms of numerical solution procedure as well as turbulence and sediment-transport modelling. The paper gives a description of the main features of the two models and shows their interrelation. Test calculations for a classical laboratory situation as well as for various natural river situations are presented which show the performance and the potential of the two models and also their relative merits.

2. HYDRODYNAMIC MODELS

Since the concentration of suspended sediments is generally small and the bed-load layer thin, the assumption can be made that the flow development is not influenced directly by the presence of the sediments, i.e. the flow equations can be solved decoupled from the sediment-transport equations. However, the sediments may have an indirect effect on the flow in that they can deform the bed and change the geometry and bed roughness. This is accounted for by adjusting the flow geometry and in the calculation of the bed friction.

2.1 3D Model

With the above assumptions, the three-dimensional flow field in rivers is determined by the following Reynolds-averaged continuity and Navier-Stokes equations, written here in Cartesian coordinates (see Fig. 1).

$$\frac{\partial u_i}{\partial x_i} = 0 \tag{1}$$

$$\frac{\partial u_i}{\partial t} + \frac{\partial (u_i u_j)}{\partial x_j} = -\frac{1}{\rho}\frac{\partial p}{\partial x_i} + \frac{1}{\rho}\frac{\partial \tau_{ij}}{\partial x_i} + F_i \tag{2}$$

where u_i (i = 1, 2, 3) are the velocity components, F_i is the gravity force per unit volume, ρ the fluid density and p the pressure. The Einstein summation convention applies. The turbulent stresses τ_{ij}, which in the averaged equations represent the momentum exchange through the turbulent fluctuating motion, have to be calculated with a turbulence model. Most models used in hydraulics employ the eddy-viscosity relation

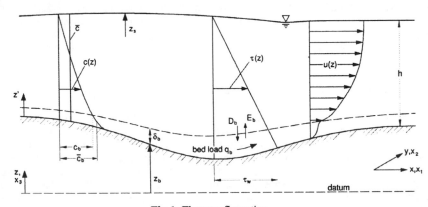

Fig.1: Flow configuration

16

$$\tau_{ij} = \rho v_t \left(\frac{\partial u_i}{\partial x_j} + \frac{\partial u_j}{\partial x_i} \right) - \frac{2}{3} \delta_{ij} k \tag{3}$$

introducing an artificial, turbulent or eddy viscosity v_t. For this, often a constant, suitably adjusted value is used. However, this does not do justice to complex flow patterns that evolve in rivers with irregular boundaries and hence the more universal k-ε turbulence model (see Rodi, 1993) is used here which allows to calculate the distribution of v_t over the flow field. This is achieved by relating v_t to two local turbulence parameters, namely the turbulent kinetic energy k and its dissipation rate ε, through $v_t = c_\mu k^2 / \varepsilon$ and by then determining the distribution of k and ε from the following model equations:

$$\frac{\partial k}{\partial t} + \frac{\partial (u_j k)}{\partial x_j} = \frac{\partial}{\partial x_j} \left(\frac{v_t}{\sigma_k} \frac{\partial k}{\partial x_j} \right) + G - \varepsilon \tag{4}$$

$$\frac{\partial \varepsilon}{\partial t} + \frac{\partial (u_j \varepsilon)}{\partial x_j} = \frac{\partial}{\partial x_j} \left(\frac{v_t}{\sigma_\varepsilon} \frac{\partial \varepsilon}{\partial x_j} \right) + (c_{\varepsilon 1} G - c_{\varepsilon 2} \varepsilon) \frac{\varepsilon}{k} \tag{5}$$

Here $G = v_t \left(\frac{\partial u_i}{\partial x_j} + \frac{\partial u_j}{\partial x_i} \right) \frac{\partial u_i}{\partial x_j}$ is the production of k through velocity gradients. The

standard values of the model coefficients as given by Rodi (1993) are used.

The boundary conditions for solving the 3D hydrodynamic model equations are described in detail in Wu et al (2000). At the free surface, basically the vertical gradients of the horizontal velocity components and k are specified as zero and ε is related to the surface value of k and to the water depth. The water level is calculated from a 2D Poisson equation for the surface height z_S which is obtained from the depth-averaged 2D momentum equations. At the river bed, the wall-function approach is used, relating the values of the horizontal velocity components, k and ε at the first grid point (placed outside the viscous sublayer and above any roughness elements) to the bed shear stress. For the velocity, a log law is used containing a roughness parameter which is calculated from the roughness height k_s. In situations with sand waves, k_s is related to the height and length of the sand waves through an empirical formula of van Rijn (1984).

2.2 2D Model

The 2D model calculates only depth-averaged quantities; the governing equations are obtained by integrating the 3D equations (1) and (2) over the water depth (i.e. from $z = z_b$ to z_s), yielding with the assumption of hydrostatic pressure:

$$\frac{\partial h}{\partial t} + \frac{\partial h \bar{u}_i}{\partial x_i} = 0 \tag{6}$$

$$\frac{\partial h \bar{u}_i}{\partial t} + \frac{\partial h \bar{u}_i \bar{u}_j}{\partial x_j} = -gh \frac{\partial z_s}{\partial x_i} + \frac{1}{\rho} \frac{\partial h \bar{\tau}_{ij}}{\partial x_j} + \frac{1}{\rho} \frac{\partial h D_{ij}}{\partial x_j} - \frac{1}{\rho} \tau_{b,i} \tag{7}$$

Here the quantities with an overbar are depth-averaged values (e.g. $\bar{u}_i = \frac{1}{h} \int_{z_b}^{z_s} u_i \, dz$) and i can

take the values 1 and 2 indicating the two horizontal directions x_1 (= x) and x_2 (= y). Depth-averaging introduces the bed-friction components $\tau_{b,i}$. These are determined with a quadratic friction law relating them to the square of the depth-averaged velocity via the friction coefficient c_f. This coefficient depends of course on the bed roughness, and for rough beds it is related to the Manning coefficient n via $c_f = n^2 g / h^{1/3}$. The depth-averaged turbulent stresses $\bar{\tau}_{ij}$ are

determined from a depth-averaged version of the k-ε turbulence model described in detail in Rodi (1993). Again an eddy-viscosity model is introduced and the eddy viscosity is related to $\bar{k}$ and $\bar{\varepsilon}$ characterising the depth-averaged state of the turbulence. These two quantities are then determined from 2D analogues of the k- and ε-equations (4) and (5). In these, in addition to turbulence production through horizontal velocity gradients, additional terms are introduced accounting for the production of turbulence due to vertical velocity gradients near the bed. The production terms $P_{kv} = c_k u_*^3/h$ and $P_{εv} = c_ε u_*^4/h^2$ are introduced respectively in the k- and ε-equations where u_* is the local bed-friction velocity and the coefficients c_k and $c_ε$ are related to the other turbulence model constants and to c_f but also to an additional empirical coefficient e^* which is a dimensionless diffusivity and could be determined, for example, from dye-spreading experiments. This parameter accounts for the bed-generated turbulence but also for other transport and mixing effects brought in through dispersion to be discussed next. The dispersion terms D_{ij} in (7) arise also from the depth-averaging and represent horizontal momentum transport due to vertical non-uniformities of the velocity components. Often these terms are neglected in the depth-averaged momentum equations (7), but in the 2D calculations to be reported later they were accounted for by an increased eddy viscosity obtained via calibration of the dimensionless eddy diffusivity e^* (see Minh Duc, 1998).

For cases where the water depth goes gradually to zero at the side boundaries of the river, boundary conditions need not be prescribed there but only at the inflow and outflow boundaries. In cases with vertical walls at the side boundary, the wall-function approach (see Rodi, 1993) is used.

3. SEDIMENT TRANSPORT MODEL

The overall sediment transport in rivers is governed by the following equation, which is the sediment mass-balance equation integrated over the water depth h (i.e. from $z = z_b$ to z_s, see Fig. 1):

$$(1 - p') \frac{\partial z_b}{\partial t} + \frac{\partial}{\partial t}(hC) + \frac{\partial(q_{Tx})}{\partial x} + \frac{\partial(q_{Ty})}{\partial y} = 0 \qquad (8)$$

Here z_b is the local bed level above datum and z_s the level of the free surface, p is the porosity of the bed material, C is the depth-averaged sediment concentration and q_{Tx} and q_{Ty} are the components of the total-load sediment transport in the horizontal x- and y-directions, respectively. The second term, which is the storage term, can be neglected for the quasi-steady flow conditions considered here (see e.g. van Rijn, 1987). The above equation relates the bed deformation $\partial z_b/\partial t$ to the spatial changes in the sediment-transport load. This load is subdivided into suspended load and bed load and hence the flow domain is subdivided into a bed-load layer with thickness δ_b and the suspended-load region above it with thickness h - δ_b (see Fig. 1). The exchange of sediment between the two layers is through deposition (downward sediment flux) at rate D_b and entrainment from the bed-load layer (upward flux) at rate E_b. In situations without bed-load transport, these quantities express the exchange between the suspended load and the bed itself. The net flux across the two layers or between suspended-sediment layer and the bed is hence D_b - E_b. In a 3D model the vertical variation of the sediment concentration c in the suspended-load layer is resolved with a 3D convection-diffusion equation and the sediment flux D_b - E_b at the lower boundary of this layer needs to be specified as boundary condition. In a 2D depth-averaged model, only the depth-averaged value $\bar{c}$ in the layer of thickness (h - δ_b) is calculated and in the 2D equation for this the exchange with the bed or bed-load layer, D_b - E_b, appears now as source or sink term. How the net flux D_b - E_b is determined will be described in the section on the suspended-load model. The bed-load layer is not resolved in the vertical direction but only the transport rate at a point x, y is calculated. This is a 2D process and hence the same bed-load transport model is used in both the 3D and the 2D model and will be discussed first.

3.1 Bed-Load Transport

The mass-balance equation for sediment transport within the bed-load layer can be written as (neglecting again the storage term):

18

$$(1 - p')\frac{\partial z_b}{\partial t} + D_b - E_b + \frac{\partial \alpha_{bx} q_b}{\partial x} + \frac{\partial \alpha_{by} q_b}{\partial y} = 0 \tag{9}$$

where $\alpha_{bx} q_b$ and $\alpha_{by} q_b$ are the components of the bed-load transport q_b in the x- and y-directions respectively. α_{bx} and α_{by} are the direction cosines; under the assumption that the bed load is in the direction of the bed shear stress these cosines are known from a three-dimensional flow calculation. Gravity-induced bed-load transport on transverse bed slopes can change these cosines and this can be accounted for through empirical formulae, e.g. the one due to Struiksma et al (1985). In general, and especially in cases with significant secondary flow, the direction of the bed shear stress does not correspond with the direction of the depth-averaged velocity. Hence, in 2D depth-averaged calculations, this difference needs to be accounted for through an additional model based on the assumption of a vertical velocity profile (see Minh Duc et al, 1998).

An equation for local bed-load transport q_b can be obtained from (9) when a model assumption is introduced for the bed deformation term. Following Wellington (1978), Philips and Sutherland (1989) and Tranh Tuc (1991) this is related to the difference between the actual bed-load transport and the bed-load transport q_{b^*} that would prevail under equilibrium conditions:

$$(1 - p')\frac{\partial}{\partial} \frac{z_b}{t} = \frac{1}{L_s}(q_b - q_{b^*}) \tag{10}$$

where L_s is a non-equilibrium adaptation length for bed-load transport which has to be prescribed empirically. Unfortunately, considerable uncertainty exists about this prescription, and a wide variety of different values have been adopted. Some discussion on this can be found in Wu et al (2000) where it is concluded that the value of L_s is not so important in near-equilibrium situations but that the development towards this state may be markedly influenced by this value. Further, L_s has also an influence on the numerical stability of the calculation which may be a determining factor when coarse grids are used. In most of the calculations presented in this paper, the empirical relation of van Rij (1987) given below has been used. In these calculations the bed-load transport q_b is determined from the governing equation which results from combining (9) and (10). The equilibrium bed-load transport q_{b^*} appearing in this equation also needs to be determined from an empirical relation. Here one of the many transport formulae available in the literature can be used; again in the calculations presented below, a formula due to van Rijn was employed.

3.2 Suspended-Load Transport

In the 3D model, the distribution of the sediment concentration in the suspended-load layer is governed by the following 3D convection-diffusion equation:

$$\frac{\partial c}{\partial t} + \frac{\partial}{\partial x_j}[(u_j - \omega_s \delta_{j3})c] = \frac{\partial}{\partial x_j}(\frac{v_t}{\sigma_c}\frac{\partial c}{\partial x_j}) \tag{11}$$

where c is the local sediment concentration, ω_s the settling velocity of the sediment, δ_{j3} the Kronecker delta with $j = 3$ indicating the vertical direction and σ_c the turbulent Schmidt number relating the turbulent diffusivity of the sediment to the eddy viscosity v_t. In the case of non-uniform particle size distribution, several concentration equations for different classes or particle sizes could be solved, but in the calculations presented below only one concentration equation for a representative particle size was used. Equation (11) is solved with the following boundary conditions: at the free surface the vertical sediment flux is set to zero while at the lower boundary of the suspended-sediment layer, the net flux D_b - E_b across this boundary has to be specified as mentioned above. The deposition rate at the boundary, which is located at $z' = \delta_b$, is $D_b = \omega_s c_b$ while for the entrainment rate E_b a model has to be introduced. Following van Rijn (1987) and Celik and Rodi (1988) it is assumed that the entrainment is equal to the one under equilibrium conditions (i.e. when $E_b = D_b$) so that $E_b = \omega_s c_{b^*}$ where c_{b^*} is the equilibrium concentration at $z' = \delta_b$. The sediment flux at the lower boundary of the suspended layer is therefore prescribed as

$$D_b - E_b = \omega_s(c_b - c_{b*})$$ (12)

For determining the equilibrium concentration c_b. again an empirical relation of van Rijn given below is used. The reference level $z' = \delta_b$ at which this concentration is to be determined is taken as $2d_{50}$ for flat beds and 2/3 - 1 of the height of roughness elements in the case of beds with such elements.

In the two-dimensional model, the depth-averaged concentration $\bar{c}$ is obtained by solving a 2D convection-diffusion equation which is obtained by integrating the 3D equation (11) over the depth. This equation reads

$$\frac{\partial h\bar{c}}{\partial t} + \frac{\partial h\bar{u}_j \bar{c}}{\partial x_j} = \frac{\partial}{\partial x_j}(h\frac{\bar{v}_t}{\sigma_c}\frac{\partial \bar{c}}{\partial x_j}) + E_b - D_b$$ (13)

where summation is to be taken only over j = 1,2. As was mentioned already, the exchange of sediment between the suspended-load layer and the bed or bed-load layer, the flux $D_b - E_b$, now appears as a source or sink term in this equation. Originally, a dispersion term appears in the depth-averaged concentration equation which is due to non-uniformities of the velocity and concentration profiles and represents a horizontal sediment tansport due to such non-uniformities which may be important particularly in cases with secondary flow. This effect is accounted for in the 2D model by increasing the eddy viscosity $\bar{v}_t$ and hence the diffusivity $\bar{v}_t/\sigma_c$ by using a suitably adjusted empirical parameter e* which enters through the extra production term $P_{\varepsilon v}$ in the $\bar{\varepsilon}$ -equation of the depth-averaged turbulence model. Details are decribed in Minh Duc (1998) and Minh Duc et al (1998). The depth-averaged eddy viscosity $\bar{v}_t$ in (13) is provided by the depth-average k-ε turbulence model. For calculating $D_b = \omega_s c_b$ in the sink term, a vertical profile assumption for c is necessary in order to relate the concentration c_b at the reference level $z' = \delta_b$ to the depth-averaged concentration $\bar{c}$. Here the well known equilibrium distribution (Rouse profile) is adopted.

3.3 Empirical Input

In the sediment transport models described above, the near-bed equilibrium concentration at reference level, c_{b*}, the equilibrium bed-load transport q_{b*} and the non-equilibrium adaptation lenght L_s have to be provided through empirical relations. Here the following relations proposed by van Rijn (1987) are used:

$$c_{b*} = 0.015\frac{d_{50}T^{15}}{\delta_b D_*^{03}}, \qquad q_{b*} = 0.053[\frac{\rho_s - \rho}{\rho}g]^{05}\frac{d_{50}^{15}T^{21}}{D_*^{03}}, \qquad L_s = 3d_{50}D_*^{06}T^{09}$$ (14)

where the particle-size diameter $D_* = d_{50}[(\rho_s - \rho)g/\rho v^2]^{1/3}$ and the non-dimensional excess bed shear stress $T = [(u_*')^2 - (u_{*cr})^2]/(u_{*cr})^2$. In these relations, d_{50} = median diameter of the bed material, ρ_s = density of this material, ρ = density of water, u' = effective bed shear velocity related to the grain and u_{*cr} = critical bed shear velocity for sediment motion given by the Shields diagram.

3.4 Calculation of Bed Deformation

Once the bed load and the suspended sediment concentration have been determined with the above models, the resulting change of the bed level z_b is calculated from equation (8), with the storage term (second term) neglected. The total-load components q_{Tx} and q_{Ty} are composed of the bed-load components q_{bx} and q_{by} and the suspended-load components q_{sx} and q_{sy}. Here $q_{bx} = \alpha_{bx}q_b$ and $q_{by} = \alpha_{by}q_b$ and the suspended-load components are determined from

$$q_{s,i} = \int_{\delta_b}^{z_s} u_i c \, dz' \ (3D \, model) = \ h\bar{u}_i \bar{c} \ (2D \, model)$$ (15)

4. NUMERICAL SOLUTION PROCEDURES

The partial differential equations for the mean flow, for the turbulence model and for the suspended-sediment concentration are solved with extended versions of the finite-volume codes FAST3D and FAST2D in the 3D and 2D model, respectively. FAST2D is a reduced 2D version of the three-dimensional code FAST3D which is described in detail in Zhu (1992) and also in Majumdar et al (1992). The 2D code is described in detail in Minh Duc (1998). Both finite-volume codes use non-staggered curvilinear numerical grids. In the 3D model the grid is adjusted to the changing free surface and bed boundaries as the solution proceeds in time; in the 2D model there is of course no vertical discretization and the variable depth enters as calculated quantity in the equations. The convective fluxes are discretized with the standard hybrid upwind/central differencing scheme and the pressure velocity coupling is achieved by using the SIMPLE pressure correction algorithm. The discretized algebraic equations are solved with the strongly implicit line relaxation procedure of Stone. The non-equilibrium bed-load transport equation (9) with (10) is solved by a 2D finite-volume procedure using second-order upwind differencing. Finally the bed-load changes are computed either with a Lax-Wendroff scheme or a predictor-corrector scheme. Further details can be found in Wu et al (2000) and Minh Duc (1998).

5. CALCULATION EXAMPLES

In this section, a number of calculation results obtained with the 2D and 3D models introduced above are presented. Both models were tested first for a number of relatively simple laboratory situations like curved and meandering open-channel flow, scour due to a jet entering into a rectangular basin, net entrainment from a loose bed and a suspended-load situation with net deposition (see Minh Duc, 1998, for the 2D model and Wu et al, 2000, for the 3D model).

5.1 180° Open Channel Bend with Movable Bed

As first application example, calculations obtained with both the 2D and 3D model are presented for a 180° open-channel bend situation studied in the laboratory by Odgaard and Bergs (1988). The channel was filled at the beginning of the experiment with a sand layer which had initially a flat surface. Both the water and the sediment it carried were recirculated through the flume and the return circuit until steady-state conditions were obtained. For these conditions the horizontal velocity components were measured as well as the bed profiles. The sediment moved through the channel mainly as bed load which was measured by a bed-load sampler. In both 2D (Minh Duc, 1998) and 3D (Wu et al, 2000) calculations the experimental development was simulated in that the calculations started with a flat sand bed and zero initial velocity and ran until the bed form and the flow field did not change any more. The experimental discharge and bed-load were prescribed at the inlet as well as uniform velocity. Both bed-load and suspended-load model were switched on in the calculations, but in agreement with the experimental observation, the main mode of transport was bed load (about 80% from the 3D model). The streamwise and lateral directions were discretized by 121 x 22 and 102 x 23 grid points in the 3D and 2D model, respectively, and in the former the vertical direction by 15 grid points. Further details can be found in the relevant publications.

Fig. 2 displays an example of secondary flow velocity vectors and streamwise velocity contours calculated by the 3D model at the section 45° in the bend, showing clearly that the bed has developed a closely triangular shape with a scour channel near the outer bank and a point bar near the inner bank. Fig. 3 (left) shows for both models the predicted contours of water depth in comparison with the measurements. For both model calculations, there is generally good agreement about the development of the scour channel at the outer bank and a point bar at the inner bank, but the deviations between calculations and measurements are larger for the 2D model, especially in the outflow part of the bend. This indicates that 3D effects, and especially the secondary motion, which cannot be accounted for properly in a 2D model, are of significance. Fig. 3 (right) shows the corresponding contours of the depth-averaged velocity in the bend. There is again fairly good agreement, with the maximum velocities appearing near the outer bank and the minimum ones near the inner bank, which is a consequence of the water depth being larger near the outer than near the inner bank. Again, the agreement is somewhat better for the 3D model.

5.2 Section of River Rhine with Groynes

A 3.6 km long section of the river Rhine in the region Vynen/Rees near the Dutch/German border was simulated with the 2D model by Wenka et al (1993) and later also by Minh Duc (1998). This section was studied by the Bundesanstalt für Wasserbau (BAW), Karlsruhe, in a laboratory model. The flow situation considered corresponds to a well documented situation from 1965 with a discharge of Q = 2300 m³/s. The river bed was covered by gravel and a Manning coefficient n = 0.023 s/m$^{1/3}$ was used. In the stretch considered, the Rhine has a 90° bend and 16 groynes distributed on the two banks which are not submerged. The numerical grid is shown in Fig. 4 (left) which has 280 points in the streamwise and 60 points in the lateral direction. The white blocked-out regions represent the 16 groynes. Fig. 4 (right) shows the calculated velocities in the complete domain and demonstrates that the complex flow field with recirculation zones past and between the groynes could be resolved well in the calculations. Fig. 5 provides a comparison of the calculated streamwise velocity profiles at selected river cross-sections (see Fig. 4) with measured depth-averaged velocities. The profiles show generally good agreement with the measurments, including the reverse-flow region at km 830.3. The changes from cross-section to cross-section are well reproduced.

5.3 Stretch of River Elbe

A stretch of the river Elbe (km 506,0 to 513,0) in the area of the former East/West German border was simulated by both 2D (Minh Duc et al, 1998) and 3D (Fang, 2000) models; also, both field measurements and laboratory experiments have been carried out by BAW. An impression of the geometry of the river in this region can be obtained from Fig. 6, which also shows the calculation domain (full lines) used in the 2D model with the flood plains included. For the discretization of this domain a curvilinear grid with 140 points in the lateral and 701 points in the streamwise direction was used. In the main channel, the grid was fairly fine and conforms to the geometry of more than 50 groynes on either bank of the river; on the flood plains coarser grid elements were used. The bed elevation at the initial state of the simulation was interpolated from the river and flood plain topography data provided by the BAW. In a first step, the model coefficients n and e* were calibrated by simulating steady-state medium-discharge (Q = 795 m³/s) and high-discharge situations (Q = 1557 m³/s). Fig. 6 shows the calculated velocity field for the high-water situation. In this case, some areas are flooded with the velocities there being very small (dark areas). The groynes are submerged at this discharge so that there is no complex flow pattern developing in their vicinity and they act merely as roughness elements. Fig. 7 shows a blown-up picture a the part of the calculation domain at the medium-discharge case. In this case, the groynes are only partly submerged and a complex flow field develops in their vicinity with recirculation regions. The figure demonstrates that this is resolved fairly well in the calculations. Fig. 8 compares calculated and measured velocity profiles for the medium and high-water situations at km 508.5 and km 510. Both field and laboratory measurements are included. It should be noted, however, that the measured velocities are surface velocities while the calculated ones are depth-averaged velocities. It can be seen that the calculated velocities agree well with the field and laboratory measurements in the main river

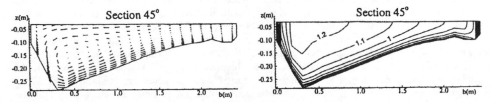

Fig.2: Secondary flow velocity vectors and normalised streamwise velocity contours u/U$_{in}$ at section 45° in 180° open channel bend as calculated by 3D model (Wu et al, 2000)

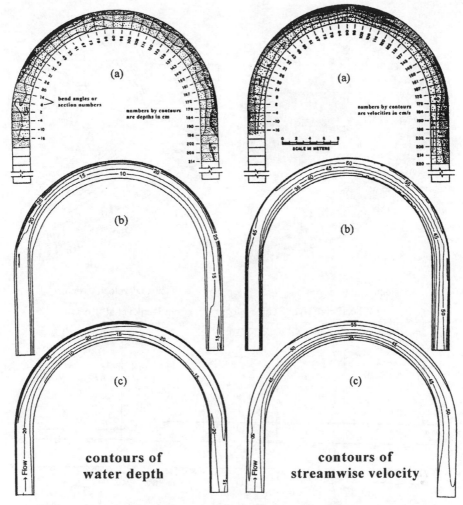

Fig.3: 180° open channel bend: (a) measured, (b) calculated by 3D model (Wu et al, 2000), (c) calculated by 2D model (Minh Duc, 1998)

For the high water, where flooding of the flood plains occurs, the laboratory data show higher velocities, especially in the flood plains. As there may be considerable differences between the surface and the depth-averaged velocity, part of this deviation may be due to these.

The 2D model was then applied to simulate a 26 day flood period (18/12/93 - 12/01/94), now including a sediment-transport calculation with both the bed-load and suspended-load model. The calculations were started from the bed obtained from the BAW topography data and the bed deformation during the days simulated was then calculated with the sediment-transport model. Concerning boundary conditions for the sediment-transport model, at the downstream boundary a zero gradient condition was applied, while at the other boundaries bed-load transport and suspended-load concentration for equilibrium conditions were used. The discharge was prescribed from a hydrograph varying from 800 m³/s to 1600 m³/s in the peak and then falling again to about 1400 m³/s, and the corresponding water elevation at the downstream boundary was prescribed. Fig. 9 shows calculated bed-load and suspended-load transport rates at the end of the simulation period. Unfortunately, there were no measurements available for comparison, but the result that the suspended load dominates over bed load is in accordance with the observations from the field data.

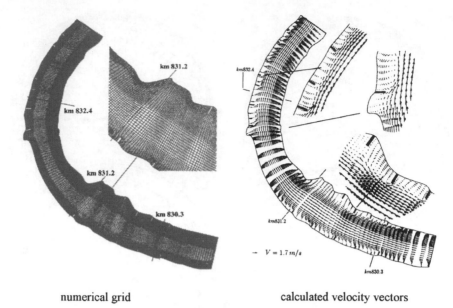

numerical grid   calculated velocity vectors

Fig.4: Section of river Rhine with groynes (from Wenka et al, 1993)

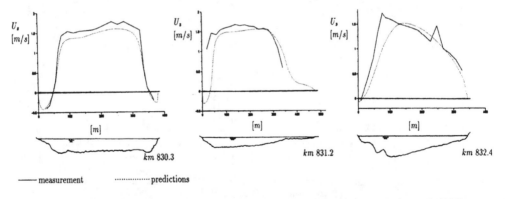

——— measurement   ·············· predictions

Fig.5: Section of river Rhine with groynes: velocity profiles (from Wenka et al, 1993)

The BAW Berlin carried out a laboratory experiment for the same Elbe stretch starting with a flat movable bed. The horizontal length scale was 1:110 and the vertical scale 1:40 and the channel cross-section was trapezoidal with vertical walls. At the beginning of the experiment, the channel was filled with a 10 cm thick layer of sand with a flat surface. The experiment was carried out for a low-water discharge of 12.54 l/s which corresponds to Q $\approx$350 m^3/s in the natural river. At the inlet, 1070 cm^3/min of model sediment with median diameter d$_{50}$ = 2.1 mm was fed in permanently and this was also the bed-load transport rate in the channel. The experiment was run for 6 hours and the bed level was measured at that time. However, a state close to equilibrium was established after 2 - 3 hours and the surface velocity was measured then at a number of cross-sections. Fang (2000) simulated the experiment just described with the 3D model, switching on only the bed-load sediment transport model because this was the only important mode in the experiment. The calculations started also with a flat sand bed and this was then allowed to deform according to the sediment transport calculated. The calculation domain can be seen from Fig. 10; it is restricted to the main river part as the flood plains were not flooded at the discharge simulated. For this part, the horizontal grid was taken over from Minh Duc (1998) and had also 701 points in the streamwise direction but only 62 points in the

24

lateral direction because the flood plains were excluded. The vertical direction was discretized by 25 grid points. The experimental sediment input was used as inflow condition for the bed load. In initial calculations the non-equilibrium adaptation length L_s was determined from van Rijn's formula (14) which yielded values considerably smaller than the grid size. These small values led to rather erratic behaviour and a decoupling of the results from the boundary condition of bed-load input. As was found also in other sediment-transport calculations using L_s, this parameter should be closer to the mesh size and hence it was adopted to be equal to the average grid size. However, further work is necessary to study the influence of this parameter on the calculations.

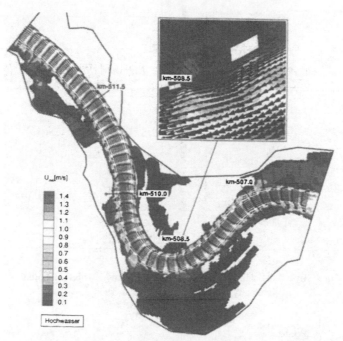

Fig.6: River Elbe: depth-average velocity field at high-water discharge
(Q = 1557 m3/s) calculated by 2D model (Minh Duc, 1998)

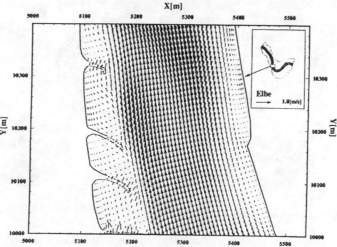

Fig.7: River Elbe: velocity vectors at medium discharge (Q = 795 m³/s)
calculated by 2D model (Minh Duc, 1998)

25

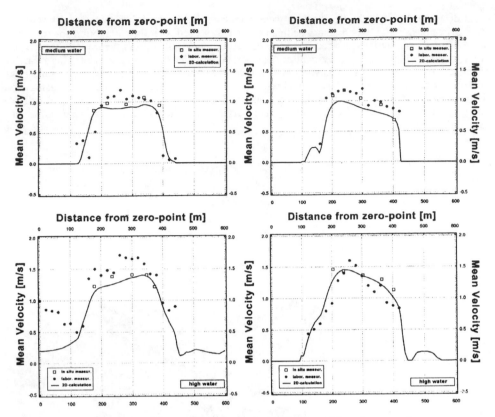

Fig.8: River Elbe: comparison of calculated (2D model) and measured velocity profiles (Minh Duc et al, 1998)

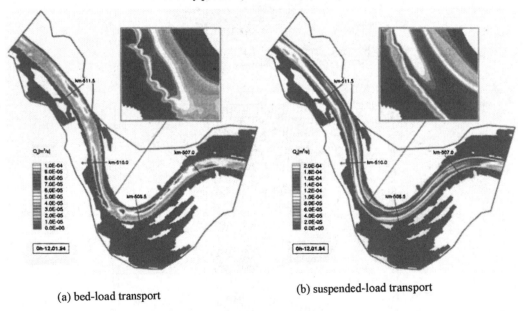

(a) bed-load transport

(b) suspended-load transport

Fig.9: River Elbe: sediment transport rates calculated by 2D model (Minh Duc et al, 1998)

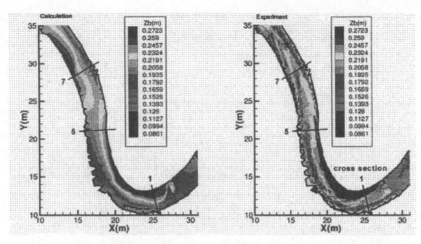

Fig.10: River Elbe: river bed morphology; comparison of 3D calculation and laboratory experiment (from Fang, 2000)

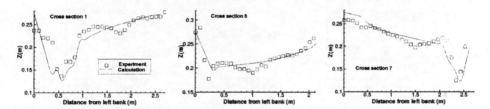

Fig.11: River Elbe: comparison of bed profiles calculated by 3D model and measured in the laboratory (from Fang, 2000)

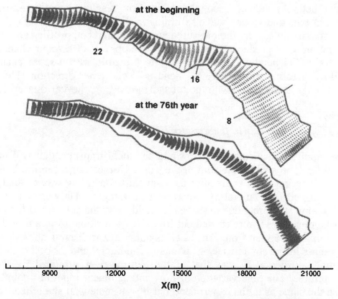

Fig.12: Three Gorges project reservoir in Yangtze River: surface velocity vectors calculated by 3D model (Fang and Rodi, 2000)

27

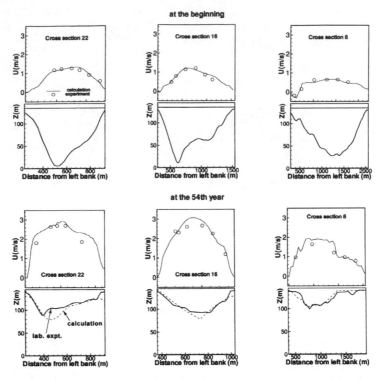

Fig.13: Three Gorges project reservoir in Yangtze River: comparison of surface velocity and bed profiles calculated by 3D model and measured in the laboratory (from Fang and Rodi, 2000)

Fig. 10 compares calculated and measured morphology of the channel bed after 6 hours starting from a flat bed. As can be seen, the calculation results generally agree with the experiment, with point bars and scour channels forming in the curved sections. Fig. 11 allows a more quantitative comparison as here the measured and calculated bed profiles are compared at 3 cross-sections. The model reproduces very well the development of a scour channel near the outer left bank in the initial bend and a change-over to a profile with a scour channel on the other side in the downstream region where the bend is in the other direction. The calculated surface velocities also agree fairly well with the measured ones at the various cross-sections (see Fang, 2000).

5.4 Three Gorges Project Reservoir in the Yangtze River

The 3D model was applied to calculate the flow and sediment transport that will occur in the reservoir which will be generated by the dam of the Three Gorges project in the Yangtze River from the year 2003. A period of 76 years after the dam starts operating was simulated and this period was also investiged in a laboratory experiment at Tsinghua University (1996). As the sediment transport is almost entirely due to suspended load, only the suspended-load model was used in these calculations. These were carried out for a 16.7 km reach upstream of the dam and the calculation domain can be seen from Fig. 12. This domain was discretized with a numerical grid that had 234 points in the streamwise, 42 points in the lateral and 22 points in the vertical direction. The initial bed geometry was that of the natural river provided by the National Yangtze River Committee. The time variation of flow and sediment input at the inflow section was basically taken the same as in the experiment, but it was somewhat smoothed by averaging over certain periods. Also, the experimental water depth at the downstream boundary (dam) was prescribed. Various time steps were used in the calculation: the shortest one for the calculation of the flow field with fixed bed geometry, an intermediate time step for the calculation of the

28

sediment transport and the bed deformation and the longest time interval for the periods of constant discharge and suspended-load input. Further details can be found in Fang and Rodi (2000).

Fig. 12 shows the calculated surface velocities at the beginning of the dam operation period and after 76 years of operation. At the beginning, when the natural bed still exists, the flow occupies the entire calculation domain and obviously is faster in the narrower upstream reach and has a lower velocity and a more complex pattern in the vicinity of the dam where the cross-section is considerably wider. After 76 years, much sediment has deposited on the sides and the river flows only in a fairly narrow cross-section similar to its behaviour before the dam was erected. The change in the bed and hence the flow cross-sections can be seen from Fig. 13 where the bed profiles are given for the year zero and for the 54th year, showing clearly the rise of the bed elevation due to the sediment deposition over the years. The figure compares the calculated bed profiles with those measured in the laboratory experiment and the agreement can be seen to be quite good. For the same cross-sections and times, the figure also compares the profiles of calculated and measured surface velocity, and again the agreement is good, even in the region near the left bank at cross-section 8 where negative velocities occur. Altogether, these results demonstrate that the model is able of simulating the flow and sediment transport processes in this practically important application example realistically and hence can be used as a tool in river engineering.

6. CONCLUSIONS

2D and 3D models developed in the author's group for calculating the flow and sediment transport in rivers have been presented. The models consist of a hydrodynamic module, which includes the k-ε model, and bed-load and suspended-load sediment transport modules with a model for calculating the interchange between them. The bed-load model is based on the mass-balance equation for non-equilibrium bed-load transport and the suspended load is calculated by solving the convection-diffusion equation for sediment concentration. Various empirical relations due to van Rijn (1987) appear in the sediment-transport models. The bed deformation is finally calculated from the depth-integrated sediment mass balance equation. The 2D model can calculate only depth-averaged quantities; it cannot simulate directly any secondary flow effects, but such effects are accounted for by adjusting the diffusivity coefficient e* in the depth-averaged k-ε model and through a quasi-3D flow approach in the sediment-transport module.

Applications of the models have been presented for a classical laboratory situation with simple geometry and for and various natural river situations with complex geometry. The capabilities of the models for calculating flow with complex geometry and bed-load and suspended-load transport were thereby tested. In general, the main flow behaviour was simulated well by both models, including complex patterns in the vicinity of structures such as groynes and a proper response to changes in cross-section. Of course, the 2D model does not provide any information on the secondary flow and there is a certain loss of accuracy when strong deviations occur between surface and depth-average velocity. Two laboratory experiments in curved channels were simulated in which the mobile bed was initially flat and then deformed mainly due to bed-load transport. This process could be reproduced quite well by both models, which predicted realistically the development of a scour channel near the outer bank and of a point bar near the inner bank. For the case which was simulated by both 2D and 3D models (180° bend with regular geometry), the 3D model yields somewhat better agreement with the experiments, and this superiority is due to the fact that it can account directly for the secondary motion. However, for most practical purposes the results of the 2D model would be good enough. There is still some uncertainty about the specification of the non-equilibrium adaptation length L_s appearing in the bed-load model; so far the comparison with data was for near-equilibrium situations where the value of this parameter is of little influence. Further tests should be carried out for really non-equilibrium situations and also for large-scale natural conditions. The suspended-load model involves fewer uncertainties, especially when it is run for cases where deposition is predominant. Application to the Three Gorges Project Reservoir has shown that the long-term development of the bed and the associated flow can be predicted quite well for realistic situations so that the model can be used as a predictive tool. The 3D calculations are, however, still quite expensive as they took hundreds of CPU hours on a powerful vector computer; for larger stretches such calculations cannot yet be carried out on the more widely available PC's or workstations. For the same horizontal resolution, 2D models require 10 to 20 times less computing time; for simulating larger areas they are still expensive, but with the rapidly increasing power of PC's and workstations they will soon become important and widely used tools for river engineers.

ACKNOWLEDGEMENTS

The author should like to thank Mr. B. Hentschel of the BAW Berlin/Karlsruhe for providing unpublished measurement data for the Elbe test case.

REFERENCES

Celik, I. and Rodi, W. (1988), „Modeling suspended sediment transport in non-equilibrium situations", ASCE J. of Hydraulic Eng., Vol. 114, pp. 1157-1191.

Fang, H.-W. (2000), ;;Three-dimensional calculations of flow and bed-load transport in the Elbe river", Rept. Institute for Hydomechanics No. 763, University of Karlsruhe, Germany.

Fang, H.-W. and Rodi, W. (2000), „Three-dimensional calculations of flow and suspended-sediment transport in the neighbourhood of the dam for the Three Gorges Project (TGP) Reservoir in the Yangtze River", Rept. Institute for Hydomechanics No. 762, University of Karlsruhe, Germany.

Gessler, D., Hall, B., Spasojevic, M., Holly, F., Pourtaheri, H., Raphelt, N. (1999), „Applicaton of 3D mobile bed, hydrodynamic model", J. Hydraulic Eng., ASCE, Vol. 125, pp. 737-749.

Majumdar, S., Rodi, W., Zhu, J. (1992), "Three-dimensional finite-volume method for incompressible flows with complex boundaries", J. Fluids Eng., Vol. 114, pp. 496-503.

Minh Duc, B. (1998), „Berechnung der StrÖung und des Sedimenttransports in FlÜsen mit einem tiefengemittelten numerischen Verfahren", Ph.D. Thesis, University of Karlsruhe.

Minh Duc, B., Wenka, Th., Rodi, W. (1998), „Depth-average numerical modelling of flow and sediment transport in the Elbe river", Proc. 3rd Int. Conf. on Hydroscience and Engineering, Cottbus/Berlin.

Odgaard, A.J. and Bergs, M.A. (1988), „Flow processes in a curved alluvial channel", Water Resources Research, Vol. 24, pp. 45-56.

Phillips, B.C. and Sutherland, A.J. (1989), „Spatial lag effects in bed load sediment transport", J. of Hydraulic Research, Vol. 27, No. 1, pp. 115-133.

Rodi, W. (1993), „Turbulence Models and their Application in Hydraulics", 3rd. ed., IAHR Monograph, Balkema, Rotterdam.

Struiksma, N., Olesen, K.W., Flokstra, C. and de Vriend, H.J. (1985), „Bed deformation in curved alluvial channels", J. Hydraulic Research, Vol. 23, pp. 57-79.

Tran Thuc (1991), „Two-dimensional morphological computations near hydraulic structures", Doctoral Dissertation, Asian Institute of Technology, Bangkok, Thailand.

Tsinghua University (1996), „The report on the sediment scale model experiment of the neighborhood of the dam for Three-Gorges-Project (TGP), Sediment Research Laboratory.

van Rijn, L.C. (1984), „Sediment transport, part III: bed forms and alluvial roughness", ASCE J. of Hydraulic Eng., Vol. 110, pp. 1733-1754.

van Rijn, L.C. (1987), „Mathematical modelling of morphological processes in the case of suspended sediment transport", Delft Hydraulics Communication No. 382.

Wellington, N.W. (1978), „A sediment-routing model for alluvial streams", M. Eng. Sc. Dissertation, University of Melbourne, Australia.

Wenka, Th., Rodi, W., Nestmann, F. (1993), "Depth-average calculations of flow in river reaches with flood control and sediment regulation structures", Proc. 25th IAHR Congress, Tokyo.

Wu, W., Rodi, W., Wenka, Th. (2000), „3D numerical modeling of flow and sediment transport in open channels", ASCE J. Hydraulic Eng., Vol. 126, pp. 4-15.

Zhu, J. (1992), „An introduction and guide to the computer program FAST3D", Report Institute for Hydromechanics, University of Karlsruhe, Germany.

Stochastic Hydraulics 2000, Wang & Hu (eds) © 2000 Taylor & Francis, ISBN 90 5809 166 X

Stochastic approach to free-surface turbulence

John S.Gulliver
St. Anthony Falls Laboratory, Department of Civil Engineering, University of Minnesota, Minn., USA

Joseph J.Orlins
Department of Civil and Environmental Engineering, Rowan University, Glassboro, N.J., USA

Aldo Tamburrino
Centro de Recursos Hidráulicos, Departamento de Ingenieria Civil, Universidad de Chile, Santiago, Chile

ABSTRACT: Free surface turbulence is important to the transfer of heat and mass across an air-liquid interface, and to the remote sensing capabilities for ship movement. Velocity measurements near a free surface, however, are complicated by limitations on probe resolution and fluctuations of the free surface itself. New approaches and techniques are used for the quantifications of free-surface turbulence. Free surface velocities were measured using an adaptation of streakline photography and particle image velocimetry (PIV) in a flume and an oscillating grid chamber. Quantitative analysis of the free-surface turbulence data requires a stochastic approach. However, auto-correlations revealed a pattern to the free-surface turbulence in an open-channel flow, on the scale of the outer dimensions of the flow field. These patterns are qualitatively related to turbulence patterns in the bulk of the flow. Turbulent frequency spectra in the oscillating grid chamber revealed the predominance of a -2 slope on a log-log plot. The spectra also have a transition to a -3 slope at higher frequencies, indicative of viscous dissipation.

Ramifications of these measurements for air-water mass transfer are also discussed.

Key words: Stochastic approach, Free surface turbulence, Fluctuating velocity, Turbulence spectrum

1. BACKGROUND

Turbulence on the surface of free liquid flows is of importance in the transfer of heat and mass across the liquid-atmospheric interface. Streams, water and wastewater treatment processes, and other natural and engineered systems rely on transfer of mass from one phase to another across such a surface. These transport processes are controlled by both gas- and liquid-film transfer coefficients. For slightly soluble compounds such as oxygen, the transfer rate is dominated by the liquid-side mass transfer coefficient (Gulliver, 1991). This liquid-film coefficient, in turn, is related to features of the flow, in particular the turbulence structure at the interface (Hanratty, 1991). In many cases (such as open channel flows free from wind shear), the turbulence is generated well below the liquid surface (*e.g.* at the stream bed or channel bottom) and propagates upwards towards the air-liquid interface.

Because of difficulties with the flexible nature of the boundary, free-surface turbulence measurements in an open-channel flow have been the subject of relatively few investigations. Utami and Ueno (1977) presented the evolution of the vorticity distribution on the water surface in a shallow flow, where near-wall structures could be affecting the free surface configuration. Komori *et al.* (1982) studied the velocity field close to the free surface, computing turbulence intensities, a budget of the turbulence kinetic energy and the wave number power spectra of the velocity fluctuations from their measurements. Rashidi and Banerjee (1988) presented vertical profiles approaching the free surface for several quantities of interest in turbulence, such as velocity, Reynolds stress, turbulence intensities, and energy dissipation rate. These three investigations focused on the near-wall turbulence generation at the channel bottom and its influence on the free-surface turbulence in a channel flow with a Reynolds number of up to 2500. Tamburrino and Gulliver (1994) presented measurements of free-surface turbulence that were at a higher bulk Reynolds number (Re $\leq$ 116,000) and indicated that the largest length scale of surface vorticity seems to be influenced by the outer variable of flow depth (5 cm $\leq$ h $\leq$ 16 cm) instead of the near-wall spacing of turbulent bursts and streaks.

Computational estimates of free-surface turbulence have also been hindered by the flexible boundary condition that the free surface presents. Lam and Banerjee (1992) and Handler *et al.* (1993) predict free-surface flow characteristics with a rigid, slip-free surface. Komori *et al.* (1993) predict free-surface turbulence with small-amplitude surface deformations. Pan and Banerjee (1995) extended the work of Lam and Banerjee to investigate the formation of free-surface vortices with the rigid, slip-free wall boundary condition. These computational studies suffer, however, from the restrictions of computational speed and space (Papavassiliou and Hanratty, 1997). The influence of boundary assumptions on the flow field is still unresolved for these computational techniques.

Another aspect of free-surface flows of interest is the gradient of vertical velocity relative to the free surface location. The vertical turbulent velocity and its gradient close to the free surface are important in determining the liquid-side mass and heat transfer coefficient at the interface. McCready, Vassiliadou and Hanratty (1986) began with the mass transport equation in the concentration boundary layer and developed the following expression for the liquid-side mass transfer coefficient, K_L:

$$K_L \sim \sqrt{D\beta'} \tag{1a}$$

or

$$St \sim \sqrt{Hn/Sc} \tag{1b}$$

where St is a Stanton number (K_L/U_*), Sc is the Schmidt number (ν/D), Hn will be called the Hanratty number ($\beta'\nu/U^2$), U_* is a shear velocity, ν is the kinematic viscosity of the liquid, D is the diffusivity of the compound in water, and β' is the root-mean-square gradient of the vertical velocity component at the free surface. At a given moment in time,

$$\beta = \frac{\partial w}{\partial z} \tag{2}$$

where w is the vertical component of velocity and z is the distance from the interface. The Hanratty number is so-called because of the work throughout the 1970's and 1980's by Hanratty and co-workers illustrating the importance of the vertical velocity gradient to interfacial mass transfer.

Equation (1a) bears a remarkable resemblance to the Danckwerts' (1951) surface renewal theory:

$$K_L = \sqrt{Dr} \tag{3}$$

where r is a mean surface renewal rate. The difficulty in applying Eq. (3) has been its conceptual nature, *i.e.* there is no definition of what constitutes a surface renewal eddy. Equation (1) is the first equation applied at a free surface that will enable characterization of surface renewal, or

$$r \sim \beta' \tag{4}$$

2. MEASUREMENTS

In this paper, we will present measurements of free-surface turbulence taken in two dramatically different laboratory test stands. Turbulence on the surface of an open-channel flow was investigated using a moving-bed flume, as described by Tamburrino (1994) and Tamburrino and

Gulliver (1995). In this specialized flume, turbulence is generated at the channel bottom by means of a conveyor belt; the turbulence propagating up through the water column to the air-liquid interface. This device provides the equivalent of a Lagrangian reference frame, while allowing stationary observation. In the second experimental setup, free-surface turbulence was measured in a closed tank, with turbulence generated by a vertically oscillating grid, as described by Orlins (1999). The turbulence, once again propagated up to the water surface.

For both experimental facilities, measurements of free-surface turbulence were made using photographic techniques. A form of streakline photography was used to capture spatial motions of small tracers floated on the surface of the flume, and Particle Image Velocimetry (PIV) was used to measure both spatial and temporal movements of tracers floated on the water surface in the grid-stirred tank.

Most of the experimental techniques used in these measurements of free-surface turbulence were fairly standard, and documented in Tamburrino (1994) and Orlins (1999). Only the aspects that are unique to free-surface turbulence measurements will be discussed here.

3. IMAGING TECHNIQUES

Traditionally, streakline photography and particle image velocimetry have been used to study fluid motions constrained within a relatively narrow plane in the flow field. For each of these techniques, a plane of light is generated with a narrow width (typically laser light), so that the plane location is known for the two-dimensional turbulence measurements. Tracer particles are placed in the flow, with photographic images or video used to determine components of the velocity vectors in the illuminated plane.

In our case, however, we did *not* want to determine velocity components in a fixed two-dimensional plane at or near the free surface. As described below, the fluid motions *on* the water surface are important for interfacial mass transfer. Since the water surface is mobile, we established a coordinate system with its vertical origin at the water surface. Thus, if the water surface moves vertically, the coordinate system moves vertically as well.

Most hydraulic engineers will recognize the simplicity that this introduces into the measurements, because they have thrown confetti on a water surface and measured temporal mean velocities through streakline imagery. To measure free-surface turbulence through streakline imagery, we employed small particles (~ 100 μm) and higher camera shutter speeds. Since the particles float on the water surface, no specialized planar light source (or light 'sheet') is required. The same simplification applies to PIV measurements of free-surface turbulence. No plane of light needs to be generated: in essence, all one must do is simply throw small particles on the free surface, videotape their motions, and process the data.

4. COMPUTATION OF VERTICAL VELOCITY GRADIENT AT THE FREE SURFACE

As customary in turbulent flow, the velocity vector, $\vec{v}$ is split in two components, a temporal mean value, $\vec{V}$ and a turbulent fluctuation, $\vec{v}'$:

$$\vec{v} = \vec{V} + \vec{v}' \tag{5}$$

Using the momentum and continuity equations, expanding in Taylor series and imposing proper boundary conditions, it can be shown (Tamburrino, 1994) that very close to the free surface:

$$u' = \alpha_1 \quad v' = \alpha_2 \quad w' = \beta z \tag{6}$$

where u', v' and w' are the fluctuating components in the horizontal directions and vertical direction, respectively, z is the distance below the water surface, and α_1, α_2, and β are functions of the spatial coordinates x, y and the time, t. β is easily determined from continuity:

$$\beta = -\left(\frac{\partial u}{\partial x} + \frac{\partial v}{\partial y} \right) \quad \text{at } z=0 \tag{7}$$

This is equivalent to the two-dimensional divergence at $z = 0$, or on the water surface.

5. FREE-SURFACE TURBULENCE IN AN OPEN-CHANNEL FLOW

Initial experiments were conducted in the moving-bed flume, as described above. Measured values of velocity on the free-surface of the open-channel flow are provided in Figs. 1 and 2. In these figures, the mean free surface velocity has been subtracted from the original data, making it easier to distinguish the vortices on the free surface. The vertically oriented vortices seen in Fig. 1 generally seem to form in longitudinal rows, with a spacing and scale that increases with depth. A qualitative assessment of the number of vortex rows is given in Table 1, which also provides some other relevant flow characteristics.

Table.1:.--Experimental conditions for flume experiments

Exp	H cm	U cm/s	u. cm/s	B/H	Re	Re.	Vortex Rows
61	12.65	18.30	0.89	6.0	22,884	1,113	2-4
81	9.36	19.07	0.96	8.1	17.163	864	3-4
103	7.12	66.44	2.78	10.7	45,486	1,903	5-6
121	6.07	15.04	0.83	12.5	8,532	471	4-8
122	6.00	41.53	2.02	12.7	23,289	1,133	4-8
123	5.96	55.60	2.60	12.8	30,970	1,448	4-8

2 cm/s

← Belt motion

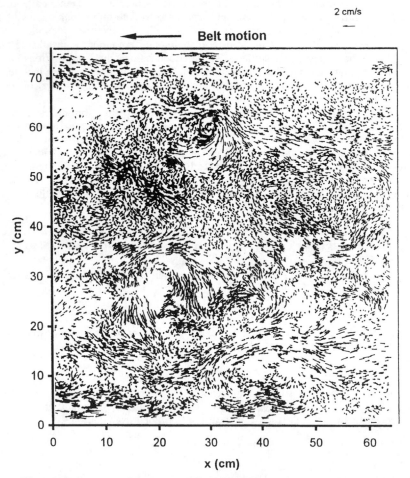

Fig.1: Velocity vector field for Experiment 61 with free surface mean velocity subtracted. H=12.63 cm, B/H=6.0, Ub=18.3 cm/s, Re=22,900, and Re.=1113.

34

4 cm/s

⟵ **Belt motion**

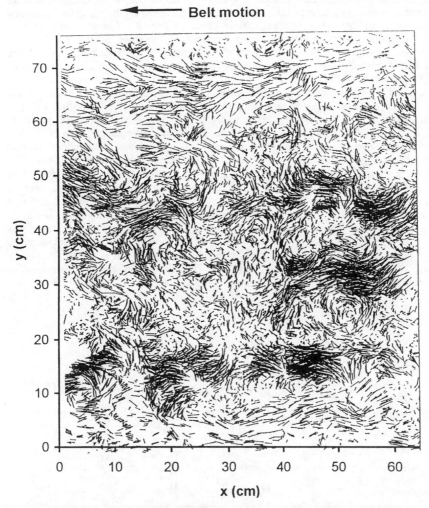

Fig.2: Velocity vector field for Experiment 103 with free surface mean velocity subtracted.
H=7.12 cm, B/H=10.7, Ub=66.47 cm/s, Re=45,500, and Re.=1900.

6. QUALITATIVE ANALYSIS OF TURBULENCE MEASUREMENTS

Figures 1 and 2 indicate the importance of depth on the scale of the larger surface vortices. The scale increases proportionally with increasing depth, regardless of the flow velocity or turbulence generated at the channel bottom, as indicated by shear velocity. The three surface flow patterns of Fig. 2 also indicate that the intensity of the surface turbulence *does* increase with the shear velocity. In an open-channel flow without an externally applied surface shear (*i.e.* without wind) the flow structure at the free surface can be qualitatively described as composed of alternating "strips" of vortices and zones with lower vorticity. Although Prandtl visualized this structure at the beginning of the last century, he did not explain its origin and referred to the configuration as "unpleasantly complicated" (Prandtl, 1926). It has recently been postulated that this free-surface pattern is a manifestation of the structure existing in the bulk of the flow, characterized by large streamwise vortices scaling with the flow depth (Gulliver and Halverson, 1987; Tamburrino and Gulliver, 1999). Tamburrino and Gulliver further state that these

35

streamwise vortices represent the largest scale of wall-generated turbulence in a confined boundary layer, stretched by the flow depth and velocity. They appear to be coherent through their juxtaposition with regard to each other. A conceptual sketch of the hypothesized temporal mean location of large streamwise vortex cells is given in Fig. 3. Unless constrained by sidewalls, frequent shifting will occur in the spanwise direction, as shown in Figs. 1 and 2. The upwelling motion induced by those vortices is associated with a vortex pattern seen on the free surface and the downwelling motion is associated with the regions of higher longitudinal velocities (Utami and Ueno, 1977).

The lateral location of surface vorticity, however, continually shifts as one moves in the longitudinal direction. The large streamwise vortices existing in the outer region of open channel flows are related to the variation of the shear velocity across the bottom of the flume (Nezu and Nakagawa, 1984; Tamburrino and Gulliver, 1992) and a similar variation of the free surface longitudinal velocity. The upwelling regions are associated with zones of smaller longitudinal velocities and eddies (Utami and Ueno, 1977; Imamoto and Ishigaki, 1984). The alternate bands of eddies and larger longitudinal velocity are easier to detect when there is a mechanism that avoids the shifting of the large streamwise vortices across the flume, such as interaction of the flow with the sand bed (Nezu et al., 1989) or alternating regions of different roughness on the bed (Müller, 1982). In our experiments, the bed was flat and behaved as hydrodynamically smooth. This allowed a free shifting of the large streamwise vortices across the flume, impeding the formation of well-defined rows of surface vorticity, interspaced by rows of large longitudinal velocities. The temporal mean velocity in the center of the flume, therefore, has little indication of secondary currents. This shifting of the large streamwise vortices can be inferred from plots of the longitudinal surface velocity.

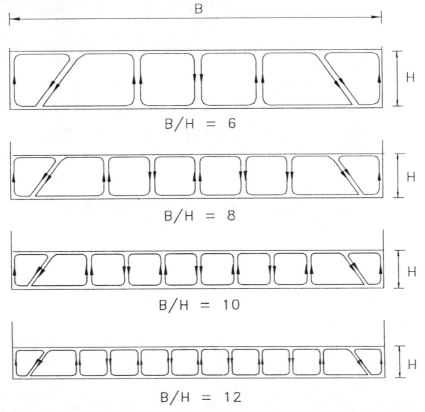

Fig.3: Sketch of the temporal mean location of cells defined by the large streamwise vortices for the range of nominal aspect ratios covered in the experiments. The bars over the water surface indicate the locations where a downwelling "streaks" (higher longitudinal velocity) would exist in the surface photographs.

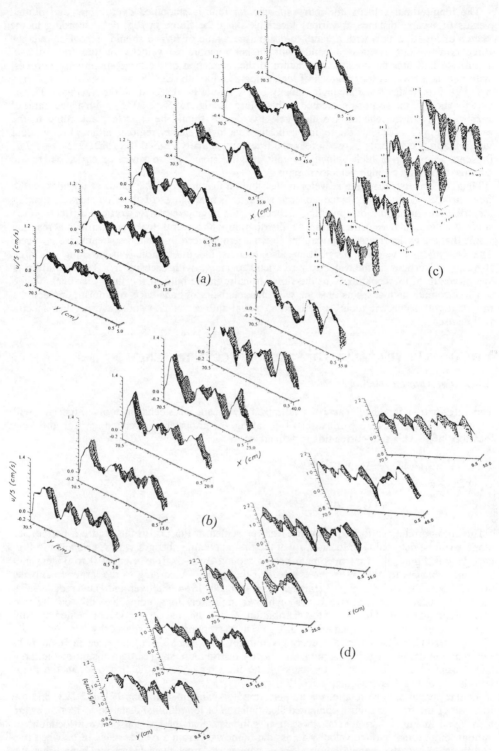

Fig.4: Smoothed longitudinal surface velocity for a) Experiment 61: B/H=6.0, b) Experiment 81: B/H-8.1, c) Experiment 103: B/H=10.7, and d) Experiment 122: B/H=12.7.

The longitudinal velocity for four experimental runs is smoothed over 1 cm^2 and plotted versus transverse distance at various locations along the flume in Fig. 4. According to the sketch of Fig. 3, a flow with a width-to-depth aspect ratio of $B/H=6$ should present on average three downwelling regions or bands with higher longitudinal velocity on the free surface. Figure 4a indicates the complicated nature of the interaction of the large streamwise vortices, with between two and four regions of high velocities. Fig. 4b, Exp. 81 with aspect ratio $B/H = 8.1$. Fig. 3 indicates that four high velocity bands should be detected on the average. This is true for most of the sections presented in this figure. In Fig. 4c (Exp. 103A), with aspect ratio of $B/H = 10.7$. An average of five high velocity regions should be expected, according to the model of Fig. 3. Most of the region indicates five high velocity regions, although individual interpretations could vary between four and 6 regions. Finally Fig. 4d (Exp. 122D, $B/H = 12.7$) presents a number of high velocity regions ranging from four to seven or eight, where the average should be six high velocity regions.

Thus, free surface longitudinal velocity distribution than to recognize bands of vortices on the free surface. Vertically-oriented vortices seem to "smear" most of the free surface, making identification of regions of upwelling induced by the large streamwise vortices difficult. Free surface longitudinal velocities present a distribution with several peaks and valleys across the flume that are easier to associate with the flow pattern induced by the large streamwise vortices. The longitudinal velocity distribution also reflects the interaction and merging of large streamwise vortices as the velocity peaks tend to merge with each other. This variation is symptomatic of the complexity of the flow structure in the bulk of the fluid. To make some sense of something that seems random, with characteristics of coherence, one must approach the raw data with stochastic tools. In this case, spatial auto-correlation and wave number spectra will be used.

7. QUANTITATIVE ANALYSIS OF FREE SURFACE TURBULENCE

7.1 Spatial Auto-correlation

The information in Fig. 4 can be synthesized by means of stochastic tools. First, we will compute the spatial auto-correlation of u′ along the transverse direction, y, for different locations of x. This auto-correlation is defined as:

$$\rho_u(x, \Delta y) = \frac{\int_0^B u'(x, y)u'(x, y + \Delta y)dy}{\int_0^B u'(x, y)^2 dy} \tag{8}$$

The auto-correlations for experiment 61 are presented in Fig. 5. In this figure a periodicity along y can be detected. The location Δy of the first maximum changes with x, as shown in Fig. 6a. In this figure, the first maximum occurs around $\Delta y/H=2$ from x equal 0 to 35 cm and suddenly changes to $\Delta y/H=3$ in the last half of the graph. According to the large streamwise vortex model illustrated in Fig. 3 (Imamoto and Ishigaki, 1984; Gulliver and Halverson, 1987), the mean distance between two bands of higher (or lower) longitudinal velocity on the free surface should be equal to $2H$. If the distribution of u on the free surface is represented by only one harmonic, its auto-correlation should present maximum values spaced by $2H$ ($\Delta Y/H=2$) This occurs in the first half of the record given in Fig. 6a. After 35 cm, however, there are only two maxima dominating the longitudinal velocity distribution, suggesting a fusion of the large streamwise vortices existing in the bulk of the fluid region with the first peak in the auto-correlation of Fig. 5 approaching $\Delta Y/H=3$.

Similar spatial auto-correlations were performed on experimental runs 121 and 123, and the location of the first maximum computed as a function of longitudinal distance. The results are also given in Fig. 6. In general, these results indicate that the spacing between locations of similar longitudinal surface velocity across the flume vary from a $\Delta Y/H$ values of between two and three. Thus, the spacing is, on average, somewhat larger than that of our hypothesis and indicated in Fig. 3. Our interpretation of this observation is that the largest turbulence scales in the flow are slightly above the flow depth, and the streamwise vortices often merge into a larger, flat vortex for a distinct period of time. This merging is reflected in the spatial auto-correlations of longitudinal surface velocity across the transverse direction in the flume.

The mean number of longitudinal streaks and vertically-oriented vortex streaks on the free surface can now be specified as

$$\text{Longitudinal Streaks} = \text{An Integer of } \left(\frac{B/H}{2 \to 3} \right) \tag{9}$$

$$\text{Vertical Vortex Streaks} = \text{An Integer of } \left(\frac{B/H}{2 \to 3} + 1 \right) \tag{10}$$

Equation (9) may be used to compute between 2 and 3 longitudinal streaks for an aspect ratio of 6 and between 4 and 6 longitudinal streaks for an aspect ratio of 12. These generally correspond to the observations of Figs. 5 and 6.

Visual observations indicate that the large streamwise vortices undergo frequent lateral shifts in location. These shifts are not always clean, and some merging or splitting of the streamwise vortices can occur. Figures 4 through 6 do indicate that there is a coherent structure in the flow that cannot be discerned from temporal mean velocity profiles. We have chosen to call this structure "large streamwise vortices" because they scale with the flow depth.

7.2 Gradient of the Vertical Velocity

As discussed previously, the liquid-film mass transfer coefficient is related to the two-dimensional divergence at the water surface. The divergence (*i.e.* the gradient of the vertical velocity) β, was computed from the interpolated values of u and v replacing the derivatives in Eq. (3) by finite differences. The β values, smoothed over 1 cm^2, are plotted versus longitudinal transverse distances for four representative experiments in Fig. 7. The spatial scales of β seem to be correlated on a smaller scale than longitudinal velocity, indicating that a direct correlation with large streamwise vortices, as hypothesized by Gulliver and Halverson (1989), does not exist. In fact, a detailed inspection of the data indicates that large β values do not occur at the center of the streamwise vortex upwelling regions but instead occur at the edges of these boils, where the gradient of inertial forces on the free surface are large. Thus, it appears to us that the intense horizontal velocity gradients of the smaller vortices created by the upwelling is responsible for the high values of β, instead of the upwelling itself.

8. FREE-SURFACE TURBULENCE GENERATED WITH AN OSCILLATING GRID

An oscillating grid is often utilized to approach horizontally homogeneous turbulence in a small chamber. The oscillating-grid chamber has been used as a more convenient substitute for flumes and tanks, especially when investigating the transport of potentially toxic substances (*Connolly et al.*, 1983; Valsaraj et al., 1997). The turbulence characteristics have been fairly well characterized (Hopfinger and Toly, 1976; Thompson and Turner, 1975), but the turbulence at the water surface has not previously been measured.

In the second set-up, experiments were conducted in a square chamber 50 cm on a side, with a horizontal 7x7 grid of 12 mm square bars spaced 62.5 mm apart. The grid was connected by a vertical shaft and eccentric drive to a variable-speed DC motor with a programmable speed controller. A stationary sleeve mounted to the lid of the tank surrounded the vertical drive shaft and projected below the water surface, to minimize surface waves caused by the motion of the shaft. The grid stroke length could be varied from 2 to 4 cm, and the vertical oscillation frequency maintained in the range 3 to 7 Hz. The operational characteristics and turbulence levels in the bulk of the tank fluid are described in Orlins (1999). The turbulence right on the water surface was measured using Particle Image Velocimetry (PIV), as shown schematically in Fig. 8. A commercial Hi-8 video camcorder was used to record the motions of polystyrene particles floating on the water surface for different turbulence conditions in the tank.

By recording the particle motions over relatively long times (1-5 minutes), both the spatial and temporal nature of the free-surface turbulence could be investigated.

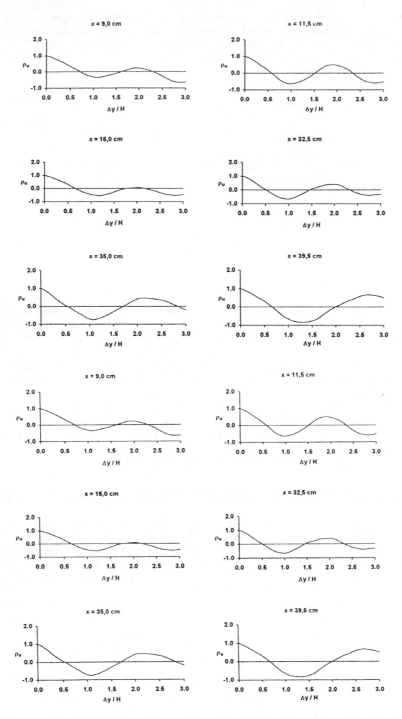

Fig.5: Spatial auto-correlation of longitudinal velocity (u) along the transverse direction (Δy) at different longitudinal locations, X. Experiment 61: B/H=6.0.

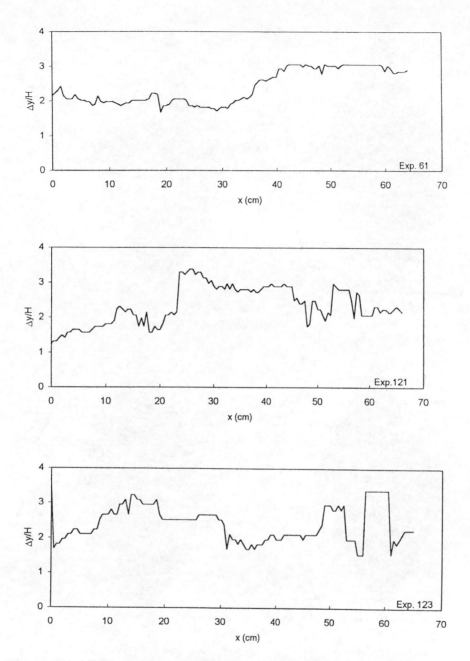

Fig.6: Location of the first bank of auto-correlation of u as a function of the longitudinal distance for Experiment 61 (top), Experiment 121 (middle), and Experiment 123 (bottom).

41

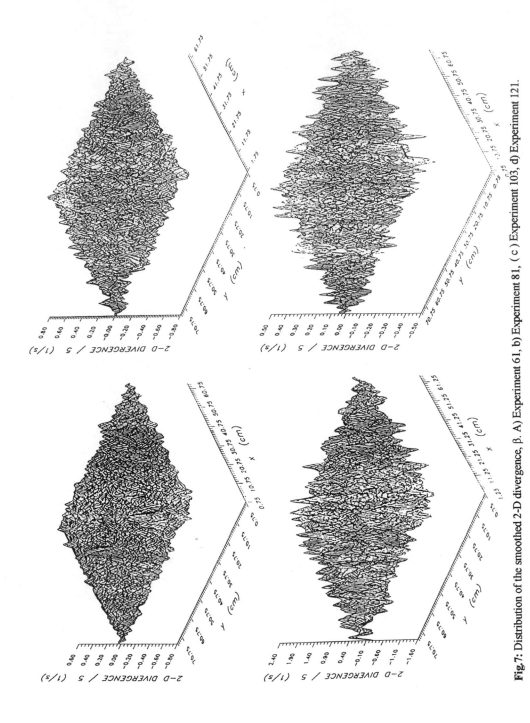

Fig.7: Distribution of the smoothed 2-D divergence, β. A) Experiment 61, b) Experiment 81, (c) Experiment 103, d) Experiment 121.

42

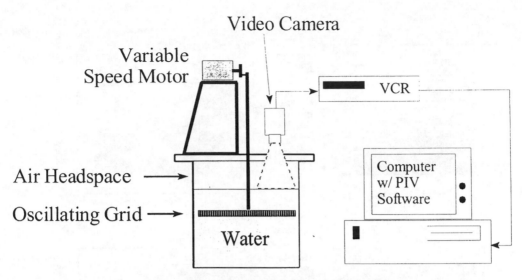

Fig.8: Experimental setup with the oscillating grid.

9. QUALITATIVE ANALYSIS OF TURBULENCE MEASUREMENTS

A sample of the results is shown in Fig. 9, which gives typical velocity, vorticity, and 2-D divergence for a grid oscillation frequency of 5 Hz. These measurements are shown for two instants, with a temporal difference of 0.1 sec.

The velocity field on the water surface was found to be well developed, changing gradually over time and space. Local velocities ranged from near-zero to over 20 centimeters per second, depending on the grid oscillation frequency and the location in the tank. For a grid oscillation frequency of 3 Hz, the peak surface velocity was about 8 cm/second. The average magnitude of the surface velocity increased at a grid oscillation frequency of 5 Hz, with regions of "hot spots" or intense rapid surface movement associated with upwelling of fluid from the bulk, as shown in Fig. 9. At a grid oscillation frequency of 7 Hz, the surface flow patterns were much more active, with strong circulation patterns over some image regions and areas of little surface movement in others.

Eddies form and disappear on the free surface and persist in stable forms for relatively long periods of time. The vorticity field evolves relatively slowly over time (tens of seconds), with spatial scales on the order of 2 cm for the smaller eddies and 10 cm for the larger ones.

There is little variation in vorticity over a time span of 0.1 seconds, as shown in Figures 9b and 9e. However, the vorticity field does undergo substantial temporal changes as the eddies appear, evolve, migrate, and disappear over longer time periods. It appears that eddies first form in the corner regions of the tank for small energy inputs (*i.e.* low oscillation frequencies). As the grid frequency is increased, these vortex structures increase in size and magnitude, and additional regions of high vorticity form near the sides and then near the center of the chamber. At the highest energy level investigated (grid frequency of 7 Hz), the regions of large vorticity magnitude appear evenly distributed over the water surface of the tank.

The two-dimensional divergence (Figs. 9c and 9f) has both spatial and temporal scales much smaller than the large-flow features such as vorticity. The length scale of the finer features is on the order of 1 cm. The temporal variations are rapid, with the divergence changing often, at frequencies approaching the image capture rate. The spatial distribution of divergence is qualitatively quite similar to that measured in open channel flows (Tamburrino, 1994, Kumar *et al.*, 1999)

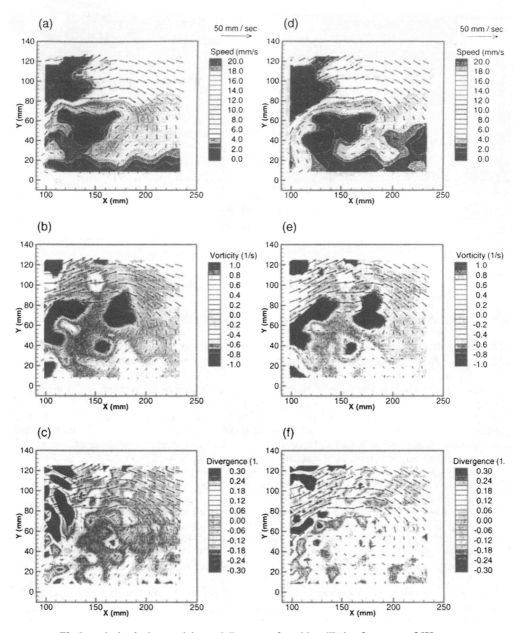

Fig.9: typical velocity, vorticity, and divergence for grid oscillation frequency of 5Hz.
(a) - (c): time = t; (d) - (f): time = t + 0.1 sec.

10. SPECTRAL ANALYSIS

Our stochastic tools were also applied to the turbulence of the oscillating grid chamber. For each test condition, PIV was used to measure surface velocities in eight sub-areas on the water surface of the tank. For each of the eight image analysis regions, time series data were extracted at nine locations from the measured flow field. Frequency spectra of the U-component of velocity and divergence were computed for each of these 72 locations, and then averaged to create an overall spectrum for each of the three grid oscillation frequencies tested.

44

The frequency spectra of free-surface velocity are given in Fig. 10. These curves seem to indicate a slope of –2 on a log-log plot in the frequency decade above that of the peak values. This is so consistent that we believe it may have some relevance for free-surface flows.

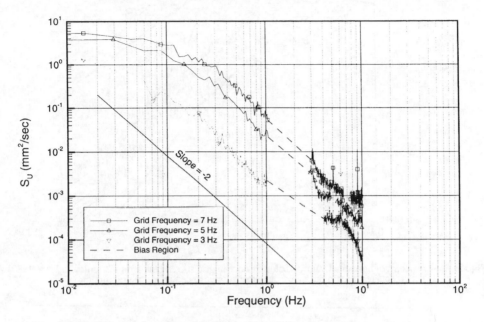

Fig.10: Edited frequency spectra of horizontal velocity component.

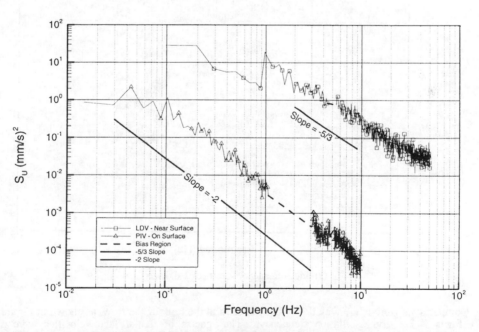

Fig.11: Frequency spectra of horizontal velocity in grid in grid-stirred tank. Top curve: measured with LDV, below water surface. Bottom curve: measured with PIV on water surface.

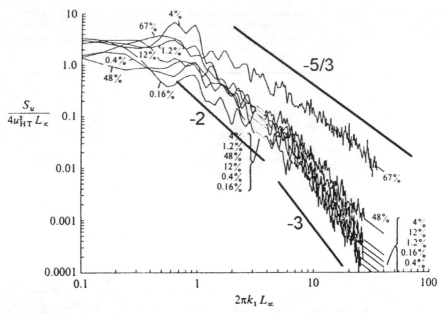

Fig.12: Power spectra of horizontal velocity below surface in grid stirred tank (from Brumley & Jirka, 1987). Percentages refer to distance from water surface to centerline of grid motion.

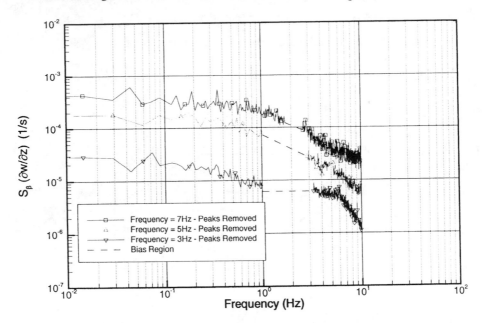

Fig.13: Edited frequency spectra of 2-d divergence.

Comparisons were made with the velocity spectra in the bulk of the flow, as shown in Figure 11 for a 5 Hz grid oscillation frequency. The spectra of free-surface velocities have a magnitude that is approximately a factor of $10^{3.5}$ lower than the spectra in the bulk flow. This seems to indicate that there is significant dissipation of turbulence as the free surface is approached. In addition, while the bulk flow spectra follow a slope of −5/3 between 1 and 40 Hz, corresponding to the inertial subrange, the spectra on the water surface follows a slope of −2 between 0.1 and 5 Hz.

46

It is interesting to make a comment regarding the slope of the velocity spectra. In three-dimensional turbulence, a $-5/3$ slope is representative of the inertial sub-range. However, this is not the case here because close to the free surface the motion is restricted in the vertical direction and the flow becomes more two-dimensional. For two-dimensional turbulence, the inertial sub-range is characterized by a proportionality to κ^{-3} (Batchelor, 1969). A dependency on $\kappa^{-5/3}$ has been detected in the productive sub-range (Farge, 1992). There is, however, little turbulence production at a free-surface with no wind, and we are well below the frequency of the inertial sub-range. Thus, these spectra represent neither the production of turbulence nor the inertial sub-range.

The -3 slope of the inertial subrange for free-surface turbulence seems to occur above a frequency of 5 Hz. There are few studies of free-surface turbulence to compare with these results. Brumley and Jirka (1987), however, used rotating hot-film probes to measure the turbulence below but *near* the free surface in a grid-stirred tank. In their work, a moving probe was used to measure a turbulence field that was varying both in space and time. Brumley and Jirka calculated spatial power spectra from their velocity measurements, as shown in Fig. 12. Their results have been normalized and are based on wave number, so it is difficult to make direct comparisons to the present study, but qualitative comparisons can be made.

In Fig. 12, the turbulence spectra in the bulk of the flow follow a $-5/3$ slope. As the free surface is approached, however, we can see the pattern of the free-surface velocity spectra develop with a slope of -3 at the higher frequencies and a slope of -2 at the lower frequencies, corresponding to what is seen in Fig. 11. Since there is no production of turbulence at the free-surface in these flows, we believe that the -2 slope is related to the stretching of turbulent eddies from 3-dimensional into 2-dimensional forms.

The spectra of 2-D divergence ($\partial w/\partial z$) were also computed, as shown in Fig. 13. The spectra follow a slope of -1 until a low-frequency plateau is reached, similar to the spectra of McCready *et al.* (1986).

11. CONCLUSIONS

Turbulence on the free surface of air-liquid flows is important to the transfer of heat and mass across the interface. Experiments in a flume and an oscillating grid chamber have allowed us to quantify some turbulence characteristics at the free surface, when the turbulence is generated from below the air-liquid interface.

The flow structure existing in the outer region of an open-channel flow can be inferred from its effects in the free surface longitudinal velocity. However, it was the application of stochastic analysis tools that allowed us to quantify the effect. A spatial auto-correlation of longitudinal velocity allowed quantification of the flow patterns in the flume. The location of these peaks and valleys shown in the velocity distribution tend to shift in their transverse position across the flume. Even the number of those peak changes, indicating the complex interaction and merging of the large streamwise vortices. This is clearly seen in sectional plots of the longitudinal velocity, or in the variation along the flume of the lag corresponding to the first peak of the spatial auto-correlation of longitudinal velocity. Thus, the structure at the free surface gives credence to the model of large streamwise vortices in the bulk of the fluid proposed by some researchers, where the pattern of alternating bands of higher and lower longitudinal velocity measured at the free surface separated a distance of roughly 1 to 1.5 times the flow depth corresponds to the zones of downwelling and upwelling motion induced by the vortices.

The lateral shifting, stretching, and squeezing of these large streamwise vortices over time results in their being one of the larger scales of turbulence in an open channel, with no temporal mean secondary velocity except near the side walls. We believe that the streamwise vortices are ubiquitous in a fully developed boundary layer when confined by a free surface. The intermittent nature of the outer part of an unconfined boundary layer (VanDyke, 1982) is simply manifested as streamwise vortices when these large eddies are confined.

Measurements made of free-surface turbulence in a oscillating grid chamber were made using Particle Image Velocimetry. The results were compared with prior studies of turbulence below the free surface in oscillating grid tanks. Turbulence intensity increases with increasing grid oscillation frequency, as do surface velocity, vorticity, and two-dimensional divergence. The velocity and calculated vorticity change slowly over time and space, while the divergence fluctuates more rapidly and over a smaller spatial scale. The velocity spectra at the water surface decays in the inertial range more rapidly than in the bulk, which is in agreement with measurements made by Brumley and Jirka (1987) just below the free surface. The predominant slope of the spectra was −2. At higher frequencies, the slope was −3, which corresponds to the dissipation range for free-surface turbulence.

The spatial variability of 2-D divergence in the grid-stirred tank is strikingly similar to that found from the measurements in a laboratory flume. Since both heat- and mass-transfer across the air-water interface are related to the spectrum of 2-D divergence, this suggests that the oscillating grid chamber may be appropriate as a direct analogue to open-channel flows for studying interfacial transport phenomena. Identification of the relationship between free-surface turbulence in an open channel flow and surface renewal rate in open-channel flows and stirred chambers will allow more detailed investigations of mass transfer in the future.

ACKNOWLEDGEMENTS

This research was funded in part by support from the National Science Foundation under Grants Nos. CES-8615279 and BES 9522171. Any opinions, findings and conclusions or recommendations expressed are those of the authors and do not reflect those of the National Science Foundation. The research was also supported by a doctoral dissertation fellowship and doctoral dissertation special grant provided by the Graduate School of the University of Minnesota.

REFERENCES

Batchelor, G.K., 1969. "Computation of the energy spectrum in homogeneous two-dimensional turbulence," *Physics of Fluids* 12(12), 11,233-239.

Brumley, B.H., and G.H. Jirka. 1987. "Near-surface turbulence in a grid-stirred tank," *J. Fluid Mech.* 183, 235-263.

Connolly, J.P., N.E. Armstrong, and R.W. Miksad. 1983. "Adsorption of Hydrophobic Pollutants in Estuaries," *J. of Environmental Engineering*, 109(1), 17-35.

Danckwerts, P.V., 1951. "Significance of liquid-film coefficients in gas adsorption," *Ind. Eng. Chem.*, 43(6), 1460-1467.

Farge, M., 1992. "The continuous wavelet transform of two-dimensional turbulent flows," *Wavelets and Their Applications*, MB. Ruskai, G.G. Beylkin, R. Coifman, I. Daubechies, S. Mallat, Y. Meyer, and L. Raphael (eds.) Jones and Bartlett Publishers, 275-302.

Gulliver, J.S. and M.J. Halverson, 1987. "Measurements of large streamwise vortices in an open-channel flow," *Water Resources Research*, 23(1), 115-123.

Gulliver, J.S. and M.J. Halverson, 1989. "Air-water gas transfer in open channels," *Water Resources Research*, 25(8), 1783-1793.

Gulliver, J. S. 1991. "Introduction to air-water mass transfer," *Air-water Mass Transfer*, J.S. Gulliver and S.C. Wilhelms (eds.), American Society of Civil Engineers, Washington, D.C.

Handler, R.A., T.F. Swean, R.I. Leighton, and J.D. Swearingen, 1993. "Length scales and the energy balance for turbulence near a free surface," *AIAA Journal*, 31(11), 1998-2007.

Hanratty, T. J. 1991. "Effect of gas flow on physical adsorption," *Air-water Mass Transfer*, J.S. Gulliver and S.C. Wilhelms (eds.), American Society of Civil Engineers, Washington, D.C.

Imamoto, H. and T. Ishigaki, 1984. "Visualization of longitudinal eddies in an open-channel flow," *Proc. Fourth Int'l. Symp. on Flow Vis.*, T Asanuma (ed.), Hemisphere Publ. Co.,333-337.

Komori, S. and H. Ueda, 1982. "Turbulence structure and transport mechanism at the free surface in an open channel flow, *Int'l. Jour. Heat and Mass Transfer*, 25(4), 513-521.

Komori, S., R. Nagaosa, Y. Murakami, S. Chiba, K. Ishii, and K. Kuwahara, 1993. "Direct numerical simulation of three-dimensional open channel flow with zero-shear gas-liquid interface," *Physics of Fluids A*, 5(1):115-127.

Kumar, S., R. Gupta, S. Banerjee, 1998. "An experimental investigation of the characteristics of free-surface turbulence in channel flow," *Physics of Fluids*, 10(2): 437-456.

Lam, K. and S. Banerjee, 1992. "Condition of streak formation in a bounded turbulent flow," *Physics of Fluids A*, 4(2):306-320.

McCready, M.A., E. Vassiliadou, and T.J. Hanratty, 1986. "Computer simulation of turbulent mass transfer at a mobile interface," *AIChE Jour.*, 32(7), 1108-1115.

Müller, A., 1982. "Secondary flow vortices: A structure in turbulent open channel flow," in *Structure of Turbulence in Heat and Mass Transfer*, S.P. Zaric (ed.), Hemisphere Publ. Corp, 451-560.

Nezu, I., and H. Nakagawa, 1984. "Cellular secondary currents in straight conduit," *J. Hydraulic Engineering*, ASCE, (110), 173-193.

Nezu, I., H. Nakagawa, M. Mitsunary, and N. Kawachima, 1989. "Measurements of cellular secondary currents and sand ribbons in fluvial open-channel flows," *Laser and Heat Wire/Film Velocimeters and Their Applications* II, Proc. Fourth Osaka Symp on Flow Measurement Tech., pp. 9-24.

Orlins, J. J., 1999. Free-surface turbulence and mass transfer in an oscillating-grid chamber, *Ph.D. Thesis*, University of Minnesota, Minneapolis, MN.

Papavassiliou, D.V. and T.J. Hanratty, 1997. "Interpretation of large-scale structures in a turbulent plane Couette flow," *Int'l. Jour. Heat and Fluid Flow*, 18:55-69.

Pan, Y. and S. Banerjee, 1995. "A numerical study of free-surface turbulence in channel flow, *Physics of Fluids A*, 7(7):1649-1664.

Prandtl, L., 1926. *Turbulent Flow*, Int'l. Congress for Applied Mech., Zurich. Also as NACA Technical Memo No. 435, 1927.

Rashidi, M. and S. Banerjee, 1988. "Turbulence structure in free-surface channel flows" *Physics of Fluids*, 31(9), 2491-2503.

Tamburrino, A., 1994. "Free-surface kinematics: Measurement and relation to the mass transfer coefficient in open-channel flow," *Ph.D. thesis*, University of Minnesota, Minneapolis, MN.

Tamburrino, A. and J.S. Gulliver, 1994. "Free-surface turbulence measurements in an open-channel flow," ASME, FED Vol. 181, *Free-Surface Turbulence*, E.P. Rood and J. Katz (eds.), 103-102.

Tamburrino, A. and J.S. Gulliver, 1999. "Large flow structures in a turbulent open-channel flow," *Jour. Hydraulic Research*, 37(3).

Utami, T. and T. Ueno, 1977. "Lagrangian and Eulerian measurement of large scale turbulence by flow visualization techniques," in *Flow Visualization*, in Proc. Int. Symp. on Flow Vis., T. Asanuma (ed.), Hemisphere Publ. Co., 221-226.

Stochastic Hydraulics 2000, Wang & Hu (eds) © 2000 Taylor & Francis, ISBN 90 5809 166 X

Hierarchy of hydraulic geometry relations

Vijay P. Singh
Department of Civil and Environmental Engineering, Louisiana State University, Baton Rouge, La., USA

ABSTRACT: Hydraulic geometry relations have been derived using a variety of hypotheses, including (a) regime theory, (b) empirical regression and correlation theory, (c) tractive force theory, (d) extremal theory (e) hydrodynamical theory, (f) minimum variance theory, (g) minimum energy dissipation rate theory, and (h). maximum entropy theory. These hypotheses lead to unique relations between channel form characteristics and discharge both at a station and downstream along a stream network in a homogeneous basin. However, if the hypotheses based on the minimum energy dissipation rate and the maximum entropy are combined, the result is a hierarchy of hydraulic geometry relations.

Key words: Extremal hypotheses, hydraulic geometry, maximum entropy, minimum energy dissipation, regime theory, stable channels, and tractive force.

1. INTRODUCTION

The term "hydraulic geometry" connotes the relationships between the mean stream channel form and discharge both at-a-station and downstream along a stream network in a hydrologically homogeneous basin. The channel form includes the mean cross-section geometry (width, depth, etc.), and the hydraulic variables include the mean slope, mean friction, and mean velocity for a given influx of water and sediment to the channel and the specified channel boundary conditions. Leopold and Maddock (1953) expressed the hydraulic geometry relationships for a channel in the form of power functions of discharge as

$$B = aQ^b, \quad d = cQ^f, \quad V = kQ^m \tag{1a}$$

where B is the channel width; d is the flow depth; V is the flow velocity; Q is the flow discharge; and a, b, c, f, k, and m are parameters. To equation (1a), also added are

$$n = NQ^p, \quad S = sQ^y \tag{1b}$$

where n is Manning's roughness factor, S is energy slope, and N, p, s, and y are parameters. Exponents b, f, m, p and y represent, respectively, the rate of change of the hydraulic variables B, d, V, n and S as Q changes; and coefficients a, c, k, N and s are scale factors that define the values of B, d, V, n and S when Q = 1.

The hydraulic variables, width, depth and velocity, satisfy the continuity equation:

$$Q = BdV \tag{2}$$

Therefore, the coefficients and exponents in equation (1a) satisfy:

$$ack = 1, \quad b + f + m = 1 \tag{3}$$

The at-a-site hydraulic geometry entails mean values over a certain period, such as a week, a month, a season, or a year. The concept of downstream hydraulic geometry involves spatial variation in channel process and form at a constant frequency of flow. The downstream hydraulic geometry involving channel process and form embodies two types of analyses, both of which are expressed as power functions of the form given by equations (1a, b). The first type of analysis is typified by the works of Leopold and Maddock (1953) and Wolman (1955) who formalized a set of relations, such as equations (1a, b), to relate the downstream changes in flow properties (mean width, mean depth, mean velocity, slope and friction) to mean discharge. This type of analysis describes regulation of flow adjustments by channel form in response to increases in discharge downstream, and has been applied to both at particular cross-sections and in the downstream direction.

The second type of analysis is a modification of the original hydraulic geometry concept and entails variation of channel geometry for a particular reference discharge downstream with a given frequency. Implied in this analysis is an assumption of an appropriate discharge that is the dominant flow controlling channel dimensions. For example, for perennial rivers in humid regions, the mean discharge or a discharge that approximates bankfull flow (Q_b), such as Q_2 or $Q_{2.33}$, is often used in equations (1a, b). This concept is similar to that embodied in the regime theory (Blench, 1957). It should, however, be noted that the coefficients and exponents are not constrained by the continuity equation when the selected discharge substantially differs from the bankfull flow.

The hydraulic geometry relations are of great practical value in prediction of channel deformation, design of stable canals and intakes, layout of river training works, design of river flow control works, design of irrigation schemes, design of river improvement works, and so on. Richards (1976) has reasoned that hydraulic exponents can be used to discriminate between different types of river sections. Because of the complexity of most natural rivers, several width-discharge curves can be defined up to bankfull stage.

The hydraulic geometry relations of equations (1a, b) are derived using a variety of hypotheses. Each hypothesis leads to unique relations between channel form parameters and discharge, and the relations corresponding to one hypothesis are not necessarily identical to those corresponding to another hypothesis. The objective of this paper is to review these hypotheses and discuss that by combining the hypotheses based on the principles of maximum entropy and minimum energy dissipation rate a hierarchy of hydraulic geometry relations is obtained. This hierarchy encomapsses the hydraulic geometry relations corresponding to other hypotheses.

2. CHARACTERISTICS OF HYDRAULIC GEOMETRY RELATIONS

The hydraulic geometry relations of equations (1a, b) possess certain characteristics a brief discussion of which is in order.

2.1 Basis of hydraulic geometry relations

The mean values of the hydraulic variables of equations (1a, b) are known to follow, according to Langbein (1964) and Yang, et al. (1981) among others, necessary hydraulic laws and the principle of the minimum energy dissipation rate. As a consequence, these mean values are functionally related and correspond to the equilibrium state of the channel. This state is regarded as the one corresponding to the maximum sediment transporting capacity. The implication is that an alluvial channel adjusts its width, depth, slope, velocity, and friction to achieve a stable condition in which it is capable of transporting a certain amount of water and sediment. Leopold and Maddock (1953) have stated that the average river system tends to develop in such a way as to produce an approximate equilibrium between the channel and the water and sediment it must transport.

2.2 Tendency for equilibrium state

Knighton (1978a) observed that at cross-sections undergoing systematic change, the potential for adjustment toward some form of quasi-equilibrium in the short term is related to flow regime and channel boundary conditions. Marked changes to channel form and associated hydraulic geometry can occur over a short period of time in the absence of exceptionally high flows and in a channel with high boundary resistance. This suggests that the approach to quasi-equilibrium or establishment of a new equilibrium position is relatively rapid.

Ponton (1972) found the hydraulic geometry of the Green and Birkenhood river basins in British Columbia, Canada, to significantly depart from the previous works, and attributed the departure to the recent glaciation in the area and the strong control which glacial features exercise on streams. He concluded that the equilibrium throughout the stream was not established. Many reaches within each system may have reached a quasi-equilibrium, as indicated by the at-a-station hydraulic geometry. However, these reaches were not yet adjusted to each other because of glacial features which separate them.

Heede (1972) investigated the influence of a forest on the hydraulic geometry of two mountain streams. He found that dynamic equilibrium was attained in the streams. Sanitation cuts (removal of dead and dying trees) would not be permissible where a stream was in dynamic equilibrium and bed material movement should be minimized.

2.3 Limitations of equilibrium assumption

Rhodes (1978) emphasized geometric adjustments to discharge downstream in response to environmental history, bed load and climate. The power function model has often been applied to small basins which in many cases have varying geology and climate, or human intervention has disturbed the long-term equilibrium situation. Thus, the equilibrium assumption can only address "at-a-station" geometry, as each cross-section of the channel adjusts to the discharge of water and sediment in a unique surface and subsurface environment. Furthermore, for discharges deviating in the extreme from the mean discharge, i.e., extremely low and high flows, even the "at-a-station" interpretation of hydraulic exponents and coefficients will be less than meaningful. This is because the influence of geology, soil, widening and narrowing floodplains, upstream bog and marsh environments, or network topology on hydraulic geometry is not properly understood at the full range of space-time scales.

2.4 Stability of hydraulic geometry relations

The relations of equation (1a, b) have been calibrated for a range of environments, using both field observations and laboratory simulations. Chong (1970) stated, without a firm basis, that rivers over varying environments behaved in a similar manner. However, hydraulic geometry relations are stable for rivers that have achieved "graded equilibrium." Parker (1979) has stated that the scale factors, a, c, and k, vary from locality to locality but the exponents, b, f, and m, exhibit a remarkable degree of consistency, and seem independent of location and only weakly dependent on channel type. Rhoads (1991) examined the factors that produce variations in hydraulic geometry parameters. He hypothesized that the parameters are functions of channel sediment characteristics and flood magnitude, and that the parameters vary continuously rather discretely. Phillips and Harlin (1984) found for a subalpine stream in a relatively homogeneous environment that hydraulic exponents were not stable over space. The interactions amongst channel form; discharge; and atmospheric, surface and subsurface environments in the system produce variables which even in the short run and at a station are neither consistently dependent nor independent. The exponents and coefficients of hydraulic geometry relations of equations (1a, b) vary from location to location on the same river and from river to river, as well as from high flow range to low flow range. This is because the influx of water and sediment and the boundary conditions that the river channel is subjected to vary from location to location as well as from river to river. This means that for a fixed influx of water and sediment a channel will exhibit a hierarchy of hydraulic geometry relations in response to the boundary conditions imposed on the channel. It is these boundary conditions that force the channel to adjust its allowable hydraulic variables. For example, if a river is leveed on both sides, then it cannot adjust its width and is, therefore, left to adjust other variables, such as depth, friction, slope, and velocity. Likewise, if a canal is lined, then it cannot adjust its friction. This aspect does not seem to have been fully explored in the literature.

2.5 Effect of river channel patterns

Chitale (1973) investigated into river channel patterns and found that the coefficients and exponents in equations (1a, b) depended on the type of the pattern. He classified channels into three groups: (1) single, (2) multi-thread, and (3) transitional channels. Single channels were further subdivided into (a) meandering, (b) straight, (c) and transitional between meandering and straight. Meandering channels were distinguished as regular or irregular and simple or compound. Multithread channels were divided into (a) braided and (b) branching-out channels. For small streams, including straight, shoaled and meandered, the slope varied inversely with 0.12 power of discharge, whereas the power of discharge changed to - 0.21 for critical straight line water surface slope. For braided channels, the power changed to - 0.44. The meander length changed with discharge raised to the power of 0.5, but it changed with the range of discharge used. Similarly, the surface width changed with discharge raised to the power of 0.42.

Knighton (1972) investigated changes in a reach morphology and hydraulic geometry. The processes of erosion and deposition interact with channel hydraulics. A braid may gradually be transformed into a meandering reach by a slight modification to the flow pattern and without any change in the independent variables. In braided reaches, large differences in flow behavior exist between juxtaposed channels, and streamflow may be concentrated in that channel that offers the least resistance.

2.6 Boundary conditions

The boundary conditions that a channel has to satisfy vary with the type of the channel. Based on their boundary conditions and hydrologic input, open channels can be classified (Yu and Wolman, 1987) into three groups: (a) channels with rigid boundaries and uniform flow, (b) canals with erodible beds and banks, and (c) natural rivers in equilibrium and regime. In the first case, there is only one flow depth above critical depth associated with uniform flow. If the discharge is known, the geometry is determined. In the second case, geometry, resistance, flow and sediment transport are interrelated. The channel geometry is determined by the laws of fluid flow, sediment transport and bank stability. Assuming there is no appreciable scour or silting in a channel if the boundary conditions do not change over a long-term average or the channel is in equilibrium, the regime theory is invoked in engineering practice. In the third case, the hydraulic geometry is determined by the stability of channel bank, availability of sediment (bed and bank material, and cohesive material) for transport, and vegetative cover, in addition to the magnitude and variability of flow. In analogy with regime in canals, the channel hydraulic geometry is assumed to be determined by discharge. Yu and Wolman (1987) investigated the effect on the channel hydraulic geometry of variable flow conditions characteristic of natural rivers.

3. HYPOTHESES FOR HYDRAULIC GEOMETRY RELATIONS

There have been various approaches for deriving functional relationships among the aforementioned hydraulic variables. These approaches are based on the following theories: (1) regime theory, (2) empirical regression and correlation theory, (3) tractive force theory, (4) extremal theory, (5) hydrodynamic theory, (6) minimum variance theory, (7) minimum energy disspation rate theory, and (8) maximum entropy theory. A short review of these theories and the resulting approaches is in order.

3.1 Regime theory

Lindley (1919) defined the regime concept as: The dimensions, width, depth and gradient of a channel to carry a given supply (of water) loaded with a given silt were all fixed by nature. Regime represents a long-term average of river form rather than an instantaneously variable state. That means stable or "in regime" channels do not change over a period of one or several water years. It then expresses the natural tendency of channels carrying sediment within alluvial boundaries to seek a dynamic equilibrium.

The regime theory defines a regime channel as a nonsilting, nonscouring equilibrium channel carrying its normal suspended load (of a certain kind and quantity). The theory implies a unique solution for a stable channel, at a given steady discharge, transporting a known concentration of solids in alluvium of given character. The regime concept was developed, based on analyses of data from stable canals on the Indian Subcontinent. Kennedy (1895), using data from stable canals, derived a relationship between mean velocity and depth of flow. Lacey (1930) proposed regime equations for wetted perimeter, hydraulic mean depth, and hydraulic gradient in terms of mean channel discharge and Lacey's factor. Blench (1957) derived hydraulic geometry relations which were similar to those of Lacey's but they distinguished between the influence of bed alluvium and bank material where these were dissimilar. Later, the regime concept was modified by including sediment concentration into geometric relationships. Lacey (1957-58) suggested a group of nondimensional equations which, if sufficient data were available, would define a stable channel for alluvium of any diameter and sediment

3.2 Empirical regression and correlation theory

The regression and correlation theory is empirical and is based on measurements of channel geometry and the corresponding water and sediment discharge. There is huge literature on this empirical approach. Exemplifying this approach are the regime approach illustrated by the equations of Lacey (1958), among others; and the power function relations pioneered by Leopold and Maddock (1953) and Wolman (1955). The power function relations are obtained either graphically or using regression analyses, and have been developed for both at-a-station geometry and downstream geometry. A short discussion of these relations is in order.

3.2.1 At-a-station relations: Since the classic work of Leopold and Maddock (1953) and Wolman (1955), studies on at-a-station hydraulic geometry have been voluminous. They have, however, focussed on five main issues: (1) validity of power relations, (2) stability of exponents in power relations, (3) variability of exponents, (4) extension of power relations to drainage basins, and (5) dependence of exponents on climatic and environmental factors.

3.2.1.1 Vailidity of power relations: Dury (1976) confirmed the validity of power function relations for hydraulic geometry using extended sets of data at the 1.58-year mean annual discharge. Chong (1970) stated that hydraulic geometry relations of equations (1a, b) were similar over varying environments. Thus, it seems that the regional generalizations proposed in the literature are acceptable for rivers that have achieved "graded-time" equilibrium (Phillips and Harlan, 1984). However, Park (1977) suggests that simple power functions are not the best way to describe hydraulic geometry. Richards (1976) has reasoned that since the depth and velocity are functions of roughness, when the rate of change in roughness is not uniform, the power function model for depth and velocity will not reflect the true hydraulic nature. He then proposed a model for describing the nonuniform variation of roughness in relation to similar nonuniform changes of depth and velocity with discharge.

Betson (1979) developed procedures for predicting variations in geomorphic relationships in the Cumberland Plateau in Appalachia for use in streawmflow and sediment routing model components of a planning-level strip mining hydrology model. He found that at high flows the width and area could be predicted with 25% accuracy about 2/3 of the time. These results were comparable to or better than the results that would normally be obtained from the reconnaissance field surveys used in streamflow routing for planning-level applications.

Knighton (1974) investigated the downstream and at-a-station variation in width-discharge relation and its implication for hydraulic geometry. The width of a channel with cohesive banks and no marked downstream variation in bank erodibility increased with discharge in the downstream direction and the rate of increase was principally a function of discharge. The channel width at a cross section was determined by flow exceeding the threshold of erosion and its magnitude increased regularly downstream with drainage area. The at-a-station rate of change was controlled by the bank material composition, especially silt-clay content. Thus, at a cross section, the rate of change of width can increase due to deposition of noncohesive sediment in the form of point bars and central islands, suggesting that the b exponent can be used to distinguish meander and braided reaches from straight reaches. The effect of this adjustment is to decrease the mean velocity range.

3.2.1.2 Stability of exponents: From an analysis of a subalpine stream in a relatively homogeneous environment, Phillips and Harlin (1984) found that hydraulic exponents were not stable over space. Knighton (1974) emphasized variations in exponents as opposed to mean values. Rhodes (1978) noted that the exponent values for high flow conditions can be vastly different than those for low flow conditions.

3.2.1.3 Variability of exponents: Kolberg and Howard (1995) examined the variable exponent model of hydraulic geometry for piedmont and midwestern streams. They analyzed active channel geometry and discharge relations using data from 318 alluvial channels in the Midwestern United States and 50 Piedmont sites. Their analysis showed that the discharge-width exponents were

55

distinguishable, depending on the variations in materials forming the bed and banks of alluvial channels. For example, highly cohesive channels (high silt and clay beds), gravel and cobble streams, and noncohesive sand stream channels had statistically distinguishable exponents for midwestern streams. The estimated width-discharge exponents for high silt and clay, gravel, or cobble bed channels deviated significantly from those for alluvial beds with 30% or greater sand content. However, no significant trend was apparent for estimated exponents among other sand-silt-clay channel categories. In case of Piedmont data, no significant departure of the discharge-width relations was apparent for different groups of stream types based on sediment categories. Both midwestern and piedmont data indicated that the width-discharge exponents ranged from 0.35 to 0.46 for groups of streams with width to depth ratios less than 45 range. For groups of streams with width to depth ratios greater than 45, the width-discharge exponents decreased to values below 0.15, suggesting a systematic variation in the exponents and a diminished influence of channel shape.

Rhoads (1991) has reasoned that variations among the three channel types are not discrete thresholds but can be viewed as continuous variations. These variations support the assertion of Howard (1980) that thresholds in the hydraulic geometry of alluvial sand and gravel streams exist. However, the implicit control of width-discharge relations by sediment characteristics was not strongly evident for the Virginia and North Carolina Piedmont streams investigated by Kolberg and Howard (1995). Despite the lack of strong relation between channel shape and sediment type for sand and gravel Piedmont streams, one can argue that certain hydraulic exponents could be characteristic of different climatic and environmental regimes.

3.2.1.4 Power relations for drainage basins: McConkey and Singh (1992) found for subbasins of the Sangamon and Vermilion rivers that the power function relationship had poor performance for low flows. They observed there that the variation in discharge was consistently dependent on drainage area and annual flow duration. Because of the existence of riffles and pools in natural streams, the hydraulic geometry relations are rendered less reliable, especially at low flows which typically create more critical conditions for fish species and various aquatic life forms. The power function relations were reasonable for watersheds greater than 100 square miles but may not be reliable for smaller drainage area streams.

3.2.2 Downstream relations: Allen et al. (1994) employed downstream channel geometry for use in planning level models for resource and impact assessment. For the data on channel dimensions obtained from the literature, they found that over a large variety of stream types and physiographic provinces the channel width and depth can be predicted with 86 percent efficiency.

3.2.2.1 Variation of channel width with discharge: Klein (1981) analyzed the variation of channel width with downstream discharge and found that the value b = 0.5 was a good average. The low b values normally occur for small basins (in lower flows) and for very big basins (in very high flows). Thus, the b = 0.5 value, being a good average, tends to smooth out deviations from the average. The value of b ranged from 0.2 to 0.89. Klein argued that the simple power function for hydraulic geometry was valid for small basins and that did not hold over a wide range of discharges.

3.2.2.2 Variation in channel velocity: Mackin (1963) has noted that in individual segments there are just as many such segments with a downstream decrease in velocity as there are with a downstream increase in velocity. Carlson (1969) found in Susquehanna River that the number of streams with a downstream velocity increase are balanced by an equal number of streams with either a constant velocity or a downstream decrease in velocity. The most common relationship on long segments of rivers is a nearly constant velocity; however, in many smaller streams velocity may increase or decrease downstream because of geological influences present at the mean annual discharge. Large rivers, such as the Mississippi, accommodate a downriver increase in depth, whereas lesser rivers generally accommodate the downriver increase in discharge principally through increase in width. Leopold (1953) has shown that the large scale floods which move large quantities of sediment have nearly constant downstream velocity.

3.2.2.3 Effect of stream size: Thornes (1970) analyzed the differences in the explained variances of the power function geometry relations for streams of different sizes-smaller as well as bigger-(e.g., Susia-Missu and Araguaia rivers in Brazil) undergoing significant changes. Smaller streams were unstable and out of phase with the steady state condition in the main stream. The difference was, therefore, between the minor stream and major tributary of the area, rather than between streams above and below a given discharge. Three possible explanations were advanced: (1) For the long term: Smaller channels undergo greater fluctuations from steady state conditions and are geomorphologically more active with strong slopes. (2) For the medium term: Small channels are substantially impacted by human activities, in the form of extensive land clearances. The width to depth ratios are substantially affected by suspended load. (3) For the short term: Seasonal changes affect most significantly the geometric relationships in smaller channels. In smaller streams, because of little or no baseflow, there is a marked difference between bankfull discharge and low discharge. This instability may reflect adjustments to long-term erosion of the upland areas, changes accommodating the impact of human activities on small basins or the effects of seasonal differences. The channel form adjusts to wet season.

3.2.2.4 Effect of land use: Lane and Foster (1980) analyzed adjustments of stream channels due to changes in discharge and channel characteristics resulting from changing land use. Their results showed that channel widths increased as discharge and hydraulic resistance increased and that narrower channels resulted from a larger critical stress. Therefore, the land use that caused these changes caused a readjustment in streams. For example, changes in the amount of sediment eroded from the channel boundary as the boundary adjusted to changing discharge reflected changes in land use.

3.3 Tractive force theory

The tractive force theory, developed by Lane (1937), is based on application of the principle of a limiting tractive force for a boundary of any given material for nonscouring condition. The concept of limiting tractive force can be applied to channel banks as well as bed, with proper allowance for the gravitational component of the stress of noncohesive materials with an inclined face. Evaluation of this limiting stress depends on experiments, and the Shields diagram for incipient motion is widely accepted to that end. It postulates a granular bed on the threshold of motion, and relates the cross-sectional geometry and weight of individual particles to the shearing force of the fluid. Thus, the design of a stable channel is based on the premise securing a distribution of the tractive force along the sides and bottom of channels such that the magnitude of this force at all points will be sufficiently large to prevent sediment deposits in unacceptable quantities and at the same time it will be small enough to prevent unacceptable scour. Then, limiting values of the tractive force are computed for various conditions that a channel must satisfy. This approach leads to a set of relations between hydraulic geometry parameters and discharge.

3.4 Extremal theory

A unique solution of slope, width and flow depth is needed for design of stable channels.
With reasonably reliable methods available for prediction of form and grain roughness, it is possible to predict flow velocity and depth with sufficient accuracy for a specified bed width. The bed slope for uniform flow in alluvial channels depends on the quantity and quality (i.e. size) of sediment being transported. Thus, sediment transport and friction equations are sufficient to predict the slope and flow depth for a specified water discharge, sediment load and bed width. To bring a closure to the problem of determining a unique solution, another equation is needed for determining the channel width. To that end, extremal theories are employed and they employ a criterion to minimize or maximize a key parameter as to yield a third equation. By analogy with the theory of least work, Chang (1980) hypothesized that the dependent variables, width, depth and slope, must have such values that the total rate at which work is done upon the water and sediment mixture by external forces must be minimum. In this way, with an assumed width, the depth and slope are computed for a given water discharge, sediment discharge and sediment size. The channel dimensions leading to the minimum slope are supposed to be the regime dimensions.

Another approach, which is variational in principle, was presented by White et al. (1982) in which it was assumed that an alluvial channel adjusts its slope and geometry to maximize its transport capacity. In other words, for a given discharge and slope, the width of a channel adjusts itself to give a maximum transport rate. The maximization of transport rate is equivalent to minimization of slope. Using the sediment transport formulas of Ackers and White (1973) and the friction relationships of White, et al. (1982), together with the principle of maximum sediment transporting capacity and a variational principle, White et al. (1982, 1986) derived regime relationships for width, depth and slope of a channel in equilibrium for a wide range of practical applications. Comparisons with laboratory data showed that predictions of depth were excellent, except for very large sand channels and for meandering laboratory channels, and the prediction of slopes showed scatter when compared with observations. Davies and Sutherland (1983) provided an alternative hypothesis based on the principle of maximum friction factor for determining river behavior and justified on the premise that it is compatible with the characteristics of turbulent flows and nonlinear processes.

3.5 Hydrodynamical theory

The hydrodynamic theory is based on the principles of conservation of mass and momentum for water and sediment. In practice, the continuity equation of water, a resistance equation, the sediment transport equation, and a morphological relation are invoked. These equations have four unknowns: width, depth, slope, and velocity. The morphological relation is derived in different ways, depending on the hypothesis to be employed. Four popular hypotheses are: (1) least mobility, (2) minimum stream power, (3) minimum variance, and (4) maximum efficiency principle.

Ackers (1964) summarized the results of experiments conducted at the Hydraulics Research Station at Wallingford, England, on small streams that achieved stability at discharges between 0.4 and 0.5 cfs. Empirical correlations of stream geometry with discharge were found to be consistent with those deduced by the combination of three physical relationships: (1) the resistance formula, (2) sediment transport formula, and (3) the ratio of width to depth. The usefulness of these relationships depends on the quality of the data and should be applied in situations similar to those for which the data were collected.

Applying the Manning equation and a well-known empirical formula for transport of total sediment load, Smith (1974) derived a downstream hydraulic geometry for steady-state channels, subject to three conditions: (1) Sediment mass is conserved during transport; (2) the channel has a form just sufficient to transport its total water discharge, and (3) the channel has a form just sufficient to transport its total sediment load. Three assumptions were made for deriving the form parameters of width, depth, velocity and slope: (a) The channel has a finite width, which is essential for specifying boundary conditions; (b) the channel is carved in noncohesive materials, with most of the sediment transport occurring close to the channel bed, which is approximately valid for a large class of channels; and (c) there is freedom to choose a time scale for which the channel has an essentially steady-state form. Smith's hydraulic geometry relations corresponded well with those of Leopold and Maddock (1953). The analytical approach developed by Smith (1974) is promising and insightful but surprisingly it has received little attention.

3.6 Theory of minimum variance

Employing the theory of minimum variance, Langbein (1964) derived power relationships between a hydraulic variable, such as velocity, depth, width, or slope and bankfull discharge. His fundamental postulate is that a change in stream power is accommodated by the channel change encompassing as equal a change of each component of power as possible, the components of power being velocity, depth, width, and slope. This condition of equal change is met when the sum of the variances of the components of power is a minimum. Langbein (1964) discussed three examples which are meaningful in discussion of hydraulic geometry relationships: (1) Response to changes in flow over a sand bed between the fixed walls of a circulating flume at constant discharge: This example has one degree of freedom, i.e., the liberty to change roughness. (2) The accommodation of a river channel, at a given cross-section, to changing discharge: This example also has one degree of freedom. (3) A river in a humid region has the liberty to adjust its profile, velocity, depth and width to accommodate the downstream increase in discharge: There are three degrees of freedom in this case. Although Kennedy et al. (1964) severely criticized Langbein's approach, the basic tenets are insightful indeed.

Knighton (1977b) presented an alternative derivation of the minimum variance hypothesis. His main argument was that the stream would approach or converge to a limit state or quasi-equilibrium state through a sequence of channel adjustments. By arguing in Euclidean space terms, the minimum variance hypothesis was shown to be a special case of the more general problem. Knighton showed how the variance hypothesis might be used to examine transitory states in natural streams and how a stream might approach a more probable state through adjustment of its channel form.

Williams (1978) examined the theory of minimum variance in regard to its validity and its ability to predict observed hydraulic exponents. He identified velocity, depth, width, bed shear stress, friction factor, slope (energy gradient), and stream power for use of the theory. If the slope is constant for a particular station, then only the first five of these variables need to be considered. To that end, he considered five cases: (1) cross-sections where both width and slope are approximately constant; (2) stations with cohesive but non-vertical banks where water surface slope is constant; (3) streams in which the slope at a station is constant but the entire flow boundary is loose and readily eroded; (4) stations having one firm and one loose bank; and (5) channels in loose, readily erodible material as in case (3) but with water surface slope varying with discharge.

Dozier (1976) examined the validity of the variance minimization principle by studying changes in the longitudinal distribution of the bed shear-stress components over a two-month period in two reaches, one meandering and one straight, of a rapidly eroding supraglacial stream. The coefficients of variation of bed shear stress, skin resistance, and form resistance decreased in both reaches. Except for skin resistance, the values were lower for the curved reach than for the straight reach.

Riley (1978) examined the role of the minimum variance theory in defining the regime characteristics of the lower Namoi-Gwydir drainage basin. The mean condition of a stream results from minimization of the variation of certain free and constrained variables within the channel system. He noted that the theory had a great value in predicting possible range of channel shapes, given a range of hydraulic conditions, but was of limited value in predicting a hydraulic condition with any degree of certainty and geometry.

3.7 Theory of minimum energy dissipation rate

Yang et al. (1981) employed the theory of minimum energy dissipation rate to derive hydraulic geometry relations. According to the theory, when a channel is equilibrium, its rate of energy dissipation is at its minimum and this minimum value depends on the boundary conditions the channel has to satisfy. Restricting their analysis to approximately rectangular channels, they were able to derive width and depth relations with discharge which agreed with laboratory experiments, and the exponents of these relations were within the range of variations of measured data from river gaging stations. The theoretical values of hydraulic exponents for channel depth and slope derived from the theory of minimum energy dissipation rate were 9/22 and –2/11, respectively. These values are very close, respectively, to the usually mentioned values of 0.41 for gaging stations in the United States and –1/6 for the regime formula used in India and Pakistan.

3.8 Theory of maximum entropy

Employing the principle of maximum entropy and the concept of unit stream power Deng and Zhang (1994) derived morphological equations. Their equations were based on the assumption that for a given discharge the flow depth and width were independent variables among five hydraulic variables. They did not advance any justification for this key assumption. Singh and Yang (2000) relaxed this assumption and derived a hierarchy of relations.

4. HIERARCH OF HYDRAULIC GEOMETRY RELATIONS

A channel responds to the influx of water and sediment coming from its watershed by the adjustment of stream power (SP). Indeed Yang (1972) found the SP to be the dominating factor in determination of the total sediment concentration. Cheema et al. (1997) determined stable width of an alluvial channel using the hypothesis that an alluvial channel attains a stable width when the rate of change of unit stream power with respect to its width is a minimum. This means that an alluvial channel with stable cross-section has the ability to vary its width at a minimum consumption of energy per unit width per unit time.

With the flow discharge expressed by Manning's equation, the stream power entails five hydraulic variables: Q, S, B, n and h. Three of these variables, including n, B, and h, are independent variables for a given discharge. Thus, it is hypothesized that for a given influx of discharge from the watershed the channel will adjust its stream power by adjusting these three independent variables. The spatial adjustment of stream power along a river involves the spatial rate of adjustment of friction, the spatial rate of adjustment of width, and the spatial rate adjustment of flow depth. Thus, the total spatial rate of adjustment of stream power can be interpreted as the proportion of adjustment of the stream power by friction, the proportion of adjustment of the stream power by channel width, and the proportion of adjustment of the stream power by flow depth. According to the principle of maximum entropy (Jaynes, 1957), any system in equilibrium state under steady constraints tends to maximize its entropy. When a river reaches a dynamic (or quasi-dynamic) equilibrium, the entropy should attain its maximum value. The principle of maximum entropy (POME) states that the entropy of a system is maximum when all probabilities are equal, i.e., the probability distribution is uniform or rectangular. Applying this principle to a river in its dynamic equilibrium, it can be hypothesized that the probability of the spatial adjustment of SP due to flow depth is the same as that due to flow width and also as that due to friction. This can be interpreted to mean that the self-adjustment of SP is equally shared among n, B, and h. This is supported by Williams (1978) who found from an analysis of data from 165 gaging stations that a channel adjusted all its hydraulic parameters (B, h, S, V) in response to changes in the influx of water and sediment and that self-adjustments were realized in an evenly distributed manner among factors. Thus, combining the theories of minimum energy disspiation rate and maximum entropy leads to a hierarchy of hydraulic geometry relations that encompass the relations derived using all other theories.

As emphasized by Langbein (1964) and Yang et al. (1981) amongst others, equations (1a, b) correspond to the case when the channel is in equilibrium. It is then hypothesized that corresponding to this state, when a channel adjusts its hydraulic variables, the adjustment is then shared equally among the hydraulic variables. However, in practice the channel is seldom in equilibrium state and this means that the adjustment among hydraulic variables will be unequal. It is not clear as to the exact proportion in which the adjustment will be shared among variables. Nevertheless, two points can be made. First, there will be a hierarchy of hydraulic geometry relations, depending on the adjustment of hydraulic variables. Second, the adjustment can explain the variability in the parameters (scale and exponents) of these relations. These two points have not been pursued fully in the literature.

5. CONCLUSIONS

The following conclusions are drawn from this discussion: (1) The application of the principles of the minimum energy dissipation rate and maximum entropy leads to a hierarchy of hydraulic geometry relations. These relations correspond to four different possibilities, depending on the way the spatial change in stream power is distributed among variables. (2) The exponent values are not fixed, rather they have ranges dictated by the value of the associated weighting factors. The exponent values vary continuously. (3) The exponent values derived here encompass the reported ranges. (4) The scale factors are not fixed; rather they vary with hydraulic variables.

REFERENCES

Ackers, P., 1964. Experiments on small streams in alluvium. Journal of the Hydraulics Division, Proc. ASCE, Vol. 90, No. HY4, pp. 1-37.

Ackers, P. and White, W. R., 1973. Sediment transport: New approach and analysis. Journal of the Hydraulics Division, ASCE, Vol. 99, No. HY11, pp. 2041-2060.

Allen, P. M., Arnold, J. G. and Byars, B. W., 1994. Downstream channel geometry for use of in planning-level models. Water Resources Bulletin, Vol. 30, No. 4, pp. 663-671.

Betson, R. P., 1979. A geomorphic model for use in streamflow routing. Water Resources Research, Vol. 15, No. 1, pp. 95-101

Blench, T., 1957. Regime Behavior of Canals and Rivers. 138 pp., Butterworths, London.

Carlston, C. W., 1969. Downstream variations in the hydraulic geometry of streams: special emphasis on mean velocity. American Journal of Science, Vol. 267, pp. 499-509.

Chang, H. H., 1980. Fluvial Processes in River Engineering. John Wiley and Sons, New York.

Cheema, M. N., Marino, M. A. and DeVries, J. J., 1997. Stable width of an alluvial channel. Journal of Irrigation and Drainage Engineering, Vol. 123, No. 1, pp. 55-61.

Chitale, S. V., 1973. Theories and relationships of river channel patterns. Journal of Hydrology, Vol. 19, pp. 285-308.

Chong, S. E., 1970. The width, depth and velocity of Sungei Kimla, Perak. Geographica, Vol. 6, pp. 72-93.

Davies, T. R. H. and Sutherland, A. J., 1983. Extremal hypotheses of river behavior. Water Resources Research, Vol. 19, No. 1, pp. 141-148.

Deng, Z. and Zhang, K., 1994. Morphologic equations based on the principle of maximum entropy. International Journal of Sediment Research, Vol. 9, pp. 31-46.

Dozier, J., 1976. An examination of the variance minimization tendencies of a supraglacial stream. Journal of Hydrology, Vol. 31, pp. 359-380.

Dury, G. H., 1976. Discharge prediction, present and former, from channel dimensions. Journal of Hydrology, Vol. 30, pp. 219-245.

Heede, B. D., 1972. Influences of a forest on the hydraulic geometry of two mountain streams. Water Resources Bulletin, Vol. 8, No. 3, pp. 523-530.

Howard, A. D., 1980. Thresholds in river regimes. In: Thresholds in Geomorphology, edited by D. R. Coates and J. D. Vitek, pp. 227-258, Allen and Unwin, Winchester, Massachusetts.

Jaynes, E.T., 1957. Information theory and statistical mechanics, I. Physical Review, Vol.106, pp. 620-630.

Kennedy, J. F., Richardson, P. D. and Sutera, S. P., 1964. Discussion of "Geometry of River Channels by W. B. Langbein." Journal of Hydraulics Division, ASCE, Vol. 90, No. HY2, 332-341.

Kennedy, R. G., 1895. On the prevention of silting in irrigation channels. Proceedings, Institution of Civil Engineers, Vol. 119.

Klein, M., 1981. Drainage area and the variation of channel geometry downstream. Earth Surface Processes and Landforms, Vol. 6, pp. 589-593.

Knighton, A. D., 1972. Changes in braided reach. Geological Society of America Bulletin, Vol. 83, pp. 3813-3922.

Knighton, A. D., 1974. Variation in width-discharge relation and some implications for hydraulic geometry. Geological Society of America Bulletin, Vol. 85, pp. 1069-1076.

Knighton, A. D., 1977a. Short-term changes in hydraulic geometry. In: Rivewr Channel Changes, edited by K. D. Gregory, John Wiley & Sons, New York, pp. 101-119.

Knighton, A. D., 1977b. Alternative derivation of the minimum variance hydpothesis. Eological Society of America Bulletin, Vol. 83, pp. 3813-3822.

Kolberg, F. J. and Howard, A. D., 1995. Active channel geometry and discharge relations of U. S. piedmont and midwestern streams: The variable exponent model revisited. Water Resources Research, Vol. 31, No. 9, pp. 2353-2365.

Lacey, G., 1930. Stable channels in alluvium. Proceedings of the Institution of Civil engineers. London, Vol.227, pp. 259-384.

Lacey, G., 1957-58. Flow in alluvial channels with sand mobile beds. Proceedings, Institution of Civil Engineers, London, Vol. 9, pp. 146-164.

Lane, E. W., 1937. Stable channels in erodible materials. Transactions, ASCE, Vol. 102, pp. 123-142.

Lane, L. J. and Foster, G. R., 1980. Modeling channel processes with changing land use. Proceedings, ASCE Symposium on Watershed Management, Vol. 1, pp. 200-214.

Langbein, W. B., 1964a. Geometry of river channels. Journal of the Hydraulics Division, ASCE, Vol. 90, No. HY2, pp. 301-311.

Langbein, W. B. and Leopold, L. B., 1964b. Quasi-equilibrium states in channel morphology. American Journal of Science, Vol. 262, pp. 782-792.

Leopold, L. H., 1953. Downstream change of velocity in rivers. American Journal of Science, Vol. 25, pp. 606-624.

Leopold, L. B. and Maddock, T. J., 1953. Hydraulic geometry of stream channels and some physiographic implications. U. S. Geological Survey Professional Paper 252, 55 p.

Lindley, E. S., 1919. Regime channels. Proceedings, Punjab Engineering Congress, Vol. 7, pp. 63.

Mackin, J. H., 1963. Rational and empirical methods of investigation in geology. In: The Fabric of Geology, edited by Albritton, C. C., Jr., Addison-Wesley Publishing Co, Reading Massachusetts, pp. 135-163.

McConkey, S. A. and Singh, K. P., 1992. Alternative approach to the formulation of basin hydraulic geometry equations. Water Resources Bulletin, Vol. 28, No. 2, pp. 305-312.

Park, C. C., 1977. World-wide variations in hydraulic geometry exponents of stream channels: An analysis and some observations. Journal of Hydrology. Vol. 33, pp. 133-146.

Parker, G., 1979. Hydraulic geometry of active gravel rivers. Journal of Hydraulic Division, Proc. ASCE, Vol. 105, No. HY9, pp. 1185-1201.

Phillips, P. J. and Harlin, J. M., 1984. Spatial dependency of hydraulic geometry exponents in a subalpine stream. Journal of Hydrology, Vol. 71, pp. 277-283.

Ponton, J. R., 1972. Hydraulic geometry in the Green and Birkenhead river basins, British Columbia. In: Mountain Geomorphology: Geomorphological Processes in the Canadian Zcordillera, edited by H. O. Slaymaker and H. J. McPherson, pp. 151-160, Tantalus Research Limited, Vancouver, Canada.

Rhoads, B. L., 1991. A continuously varying parameter model of downstream hydraulic geometry. Water Resources Research, Vol. 27, No. 8, pp. 1865-1872.

Rhodes, D. D., 1978. World wide variations in hydraulic geometry exponents of stream channels: an analysis and some observations-Comments. Journal of Hydrology, Vol. 33, pp. 133-146.

Richards, K. S., 1976. Complex width-discharge relations in natural river sections. Geological Society of America Bulletin, Vol. 87, pp. 199-206.

Riley, S. J., 1978. The role of minimum variance theory in defining the regime characteristics of the lower Namoi-Gwydir basin. Water Resources Bulletin, Vol. 14, pp. 1-11.

Smith, T. R., 1974. A derivation of the hydraulic geometry of steady-state channels from conservation principles and sediment transport laws. Journal of Geology, Vol. 82, pp. 98-104.

Thornes, J. B., 1970. The hydraulic geometry of stream channels in the Xingu-Araguaia headwaters. The Geographical Journals, Vol. 136, pp. 376-382.

White, W. R., Paris, E. and Bettess, R., 1982. River regime based on sediment transport concepts. Report IT 201, Hydraulics Research Station, Wallingford, England.

White, W. R., Bettes, R. and Wang, S. Q., 1986. A study on river regime. Hydraulics Research Station, Wallingford, England.

Williams, G. P., 1978. Hydraulic geometry of river cross-sections- Theory of minimum variance. Professional paper 1029, U.S. Geological Survey, Washington, D.C.

Wolman, M. G., 1955. The natural channel of Brandywine Creek, Pennsylvania. U. S. Geological Survey Professional Paper 271, Washington, D. C. noncohesive sands. U. S. Geological Survey Professional Paper 282-G, 183-210.

Yang, C. T., 1972. Unit stream power and sediment transport. Journal of the Hydraulics Division, ASCE, Vol. 98, No. HY10, pp. 1805-1826.

Yang, C. T., Song, C. C. and Woldenberg, M. T., 1981. Hydraulic geometry and minimum rate of energy dissipation. Water Resources Research, Vol. 17, pp. 877-896.

Yu, B. and Wolman, M. G., 1987. Some dynamic aspects of river geometry. Water Resources Research, Vol. 23, No. 3, pp. 501-509.

Stochastic Hydraulics 2000, Wang & Hu (eds)© 2000 Taylor & Francis, ISBN 90 5809 166 X

Stochastic theory of sediment motion

Han Qiwei & He Mingmin
China Institute of Water Resources and Hydropower Research, Beijing, China

ABSTRACT: This paper describes the development of stochastic theory of sediment motion in China, especially the authors work since the sixties, which contributes to establish a new field of sediment research. .In the first section the previous studies on three aspects of stochastic theory has been reviewed . The second section is about the mechanical and stochastic model of single particle motion, including the incipient condition, the distribution of incipient velocity, the incipient probability, the movement equations of three motion states: rolling, jumping and suspension, the related parameters , and the transition conditions and probabilities. In the third section the life distributions of four motion states, the distributions of number of exchange particles and exchange intensities are introduced. The forth session is concerned with the stochastic model of transport rate. This model provides the relation between the distribution and average value of transport rate and unifies the rules of transport rate of various movement . The fifth section summarizes the general dispersion modes of bed load, including constant or variable parameters ; point , line and degenerate surface sources and the case that motion time period taken into considered. The final section is about the application of the theory of exchange intensities and transport rates to the sediment engineering. The application field extends to the seven aspects: incipient velocity, nonequilibrium transportation of nonuniform suspended load, carrying capacity of nonuniform suspended load, nonequilibrium transport of bed load, sediment prevention in diversion works, ratio between suspended load and bed load and dry density.

Key word: Stochastic theory, Sediment motion, Incipient probability, Bed load, Exchange intensity

1. INTRODUCTION

The previous study on stochastic theory of sediment motion mainly includes three aspects. First is the stochastic dispersion of bed load; the second concerns the single particle motion and its parameters; and the third is about the average transport rate. Most of the studies are based on the mechanical analysis. Only some probabilities are adopted to express the related parameters.

The first aspect was started by Einstein (1937). Einstein considered a complicated problem, i.e., under an initial condition (moving or stationary), after a stochastic motion process in a certain time interval, what the distribution of a certain amount of particles along the flow will be. By adopting the probability approach, Einstein obtained a compound Poisson distribution. The study of Einstein shows that the probability approach has an advantage over mechanical method to reveal the mechanism and essence of sediment motion in some aspects. The reason is that by using the mechanical method if the average parameter of flow and sediment are provided only the determinate velocity and location of particle can be obtained but its distribution and probability are still unknown. Although Einstein is the pioneer in the study of stochastic theory, the parameters in his distributions, not being connected with the hydraulic and sand factors, can only be presented by the experiment. Unfortunately, the experiment technology was backward at that time. Besides, as compared with the transport rate of bed load, less attention was paid to the dispersion in sediment engineering. Therefore, the dispersion theory has not been further developed for a long time. In the fifties only Velikanov (1958) had carried out some research. Even Einstein himself concentrated on the average transport rate by the third approach as mentioned above to meet the requirement of engineering (1942). In the sixties, because of the development of radioactive tracer technique for observation of the sediment dispersion, and because of the requirement of environment study and deep study on mechanism of sediment motion, more attention was attracted to the study on dispersion .The main achievements have been presented by CrikmoreLean (1962), Hubbell-Sayre (1964), Todorovic (1967,1970,1975), Yang (1968), Yang-Syre (1971), Shen-Todorovic (1971) and Cheong-Shen (1976,1977). Moreover, Yano (1969) presented the empirical relation between the parameters of the dispersion model and the factors of hydraulics and sediment.

The main existing problems of the previous study on dispersion are: 1.All the models provided by the others are the same as the original one of Einstein. 2.The relation between both the step length and stationary time period and the parameters of hydraulics and sediment is very complicated and has not been revealed for a long time. The theoretical results can be obtained only through deep study on mechanism of sediment motion. 3.The application of dispersion to sediment engineering is very limited and it is very difficult to provide the transport rate of bed load based on the dispersion model.

Most of the studies concerning the second aspect are carried out by some Japanese researchers. Because of the fluctuation of flow and the variation of location of particle on bed surface, the practical motion is a stochastic process. Gonchanov (1938), Bagnold (1956,1973), Yalin (1963) and Dou (1964) studied the locus of particle motion only by the determinate method of mechanics. Tsuchiya (1969), Losinskee-Lubomilonova (1976),Nakagawa-Tsujimoto (1976) have studied the distributions of incipient jumping and jumping height and distance. Later Hu (1995) also presented these distributions based on the experiment. But, as these studies were limited in extent and were of an empirical characteristics, they did not establish a complete stochastic model of single particle motion in theory yet .

Einstein (1942,1950) is also the first to apply the probability method to study the average transport rate of bed load. In his study the expressions of probabilities of incipient motion and incipient jumping are introduced in theory to form his well-known formula of average transport rate of bed load. Einstein's achievement has attracted more attention and promoted further study in this field. Some experts verified his formula by experiment and then made some supplements and adjustment. Some other scientists have developed the formula in some degree. Velikanov (1955) introduced two types of probabilities, Dou (1964) suggested two types of probabilities and derived parameters of jumping by mechanical analysis. Paintal (1971) introduced the concept of particle location on bed surface and revised the expression of incipient probability and the derivation of transport rate. Strictly speaking, all these studies cannot be considered as a complete stochastic model , because except average transport rate of bed load , no other average transport rate and any distributions of transport rate are provided. In the sixties the stochastic theory has been developed by Han and He (1984). Their studies introduced a new field concerning the stochastic model of single particle, exchange intensity and transport rate, and extended to the case of non-uniform sediment, and thus formed a complete stochastic model. Moreover, a lot of their research has been applied to the problem of sediment engineering. This paper with the addition of Sun (1996) and some other studies in China will mainly describe the contribution of the authors.

2. MECHANICS AND STOCHASTIC CHARACTERISTICS OF SINGLRLE PARTICLE MOTION

By combining the mechanical and probability method, Han and He   (He-Han,1980,1981; Han-He,1984) established a complete stochastic model of single particle motion. First, the variables are defined as determinate and the mechanical analysis is carried out to derive the expressions of particle motion. Then the expressions are taken as the relation of random variables to form the stochastic model. Four motion states are supposed: stationary, rolling, jumping and suspension. The stationary sediment is called bed material, the rolling and jumping sediment is bed load and the suspened sediment is suspended load. The step motion and unit motion are introduced as the fundamental concept of stochastic model. All the mechanical parameters corresponding to the step motion and unit motion for various motion are derived. The step motion of rolling is defined as the motion between two adjacent lowest location of a moving particle, the jumping step and suspension step are the motion between two adjacent contact points of a moving particle with the bed surface . The unit motion consists of several steps of the same motions. At the end of a unit the particle transfers to the other motion state.

2.1 Basic random variables for single step of particle motion and their distribution

There are three random variables which determine the particle motion. V_b longitudinal bottom velocity, Δ,the location of particle on bed surface, and $R_{1,l}$ ($l=1,2,...n_l$), size distribution of sediment. V_b and $V_{b,y}$—the vertical component of bottom velocity have normal distribution, which is different from Einstein's suggestion that the lift force has a normal distribution. Some Chinese experts (Chen,1989;Sun,1991;Wan,1995; also disagreed Einstein's suggestion and revised his transport rate formula. Δ is defined by Han and He as the vertical distance between the lowest point of a particle and its contact point with the next downstream particle (Fig.1). It effects the maguitude of force and force arm acted on the particle and hence effects incipient velocity of particle. Δ has a uniform distribution. Its distribution density is

$$p(\Delta') = (1 - \Delta'_m)^{-1} \qquad (\Delta'_m \leq \Delta' \leq 1) \qquad (2.1)$$

in which, $\Delta' = \Delta/R$ is the relative location, R the radius of particle, Δ'_m=0.134, the minimum value of Δ'. For the nonuniform sediment the distribution of Δ' is expressed as the following. For the coarse size group, $D_l > \overline{D}/\Delta'_m=0.745\overline{D}$, the distribution function is

$$F(\Delta'_l) = \begin{cases} 0, & (if\ \Delta'_l < \dfrac{\overline{D}}{D_l}) \\ 1, & (otherwise) \end{cases} \qquad (2.2)$$

which is a point distribution . It means that all the location is smaller than Δ'_m.For the medium size group $\overline{D}/\Delta'_m \geq D_l \geq \overline{D}$, the distribution density is

$$p(\Delta'_l) = \begin{cases} (\dfrac{\overline{D}}{D_l} - \Delta'_m)^{-1} & (if\ \Delta'_m \leq \Delta'_l \leq \dfrac{\overline{D}}{D_l}) \\ 0 & (otherwise) \end{cases} \qquad (2.3)$$

For the fine size group, $D_l < \overline{D}$, then

$$p(\Delta'_l) = \begin{cases} (\dfrac{\overline{D}}{D_l} - \Delta'_m)^{-1} & (if\ \Delta'_m \leq \Delta'_l \leq 1) \\ 0 & (otherwise) \end{cases} \qquad (2.4)$$

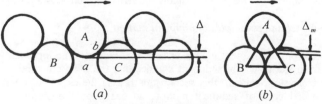

(a)   (b)

Fig.1: Location of Particle on Bed Surface

Han and He's definition of particle location on bed surface is simpler than that of Paintal, which has three parameters to determine the location. In reality, the location of particle related to the next downstream particle is the main affect on particle motion. Therefore, according to the definition proposed by Han and He, without any additional coefficients the mechanical parameters can be expressed as the function of Δ'. Liu-Cheng (1987) proposed the location of particle according to the field data:

$$e_i = \begin{cases} A(D_i - D_A) & \text{(if all the bed layer can be moved)} \\ A(D_i - D_c^*) & \text{(if part of the bed layer can be moved and } D_i \leq D_c^*) \\ A(D_c^* - D_A) & \text{(if part of the bed layer can be moved and } D_t > D_c^*) \end{cases} \tag{2.5}$$

in which

$$D_A = \begin{cases} \displaystyle\sum_{l=1}^{L_n} D_c R_{1,l} & \text{(if all the bed layer can be moved)} \\ \displaystyle\sum_{l=1}^{L_k} D_c R_{1,l} + \sum_{l=L_k+1}^{L_n} D_c^* R_{1,l} & \text{(if part of the bed layer can be moved)} \end{cases}$$

D_i is the size of particle being considered, D_c^* is the size of particle in incipient motion, L_K defined by $D_k \leq D_C^* < D_{k+1}$, A empirical constant with value from 0.1 to 0.4. For fine particle, its location is negative, otherwise it is possitive. But the distribution has not been presented by Liu. Sun (1996), suggested that instead of location, the force arm is a random variable and is the function of particle size and its height level y on bed surface, and the conditional distribution density of y is uniform

$$p(\xi_y|_{\xi_{D_l}} = D_l) = \frac{1}{0.7(D_K + D_l)} \quad (-0.2(D_K + D_l) \leq y \leq 0.5(D_K + D_l)) \tag{2.6}$$

and the combined probability density of ξ_{D_l} and y is

$$p(\xi_{D_y}, \xi_{D_l}) = \frac{R_{1\,l}}{0.7(D_k + D_l)} \tag{2.7}$$

in which D_l is the size of adjacent particle downstream particle D_K, $R_{1,l}$ is the size distribution. Sun then derived the force arms of drag force and lift force and gravity and their distribution and the expectation of force. However, when the incipient probability is calculated, the assumption that the force arms are random variables is not adopted, only its expectation is used.

The size of particles, especially of coarser particles, changes rapidly at different location of bed surface. At a certain location, the particle size is not determinate. Therefore, the particle size of nonuniform sediment is a random variable and its probability can be expressed by the size distribution

$$R_{1\,l} = P[D_l - \frac{(D_l - D_{l-1})}{2} \leq \xi_D < D_l + \frac{D_{l+1} - D_l}{2}], (l = 1,2\cdots n_l) \tag{2.8}$$

in which n_l is the number of size group. Usually the size distribution is a discrete variable.

2.2 Stochastic model and principles of incipient motion

The general concept of the incipient motion of sediment indicates the incipient rolling, jumping and suspension from stationary. In sediment engineering, the incipient motion only means the incipient rolling from stationary. In this section this is also the case will be studied. Incipient motion is the special case of sediment motion, so based on the study on mechanics and stochastic process of sediment motion described in section 2-3. the general results can be easily extended to incipient motion   (Han,1982;He-Han,1982;Han- He , 1984,1999).

2.2.1. Critical velocity of incipient motion

For uniform sediment the instant critical velocity of incipient motion is

$$V_{b.k_1} = \varphi(\Delta')\omega_1(D,H,\frac{t}{\delta_1}) \tag{2.9}$$

in which

$$\varphi(\Delta') = \sqrt{\frac{\sqrt{2\Delta' - \Delta'^2}}{(\frac{4}{3} - \Delta') + \frac{1}{4}(\frac{1}{3} + \sqrt{2\Delta' - \Delta'^2})}} \tag{2.10}$$

$$\omega_1(D,H,\frac{t}{\delta_1}) = [3.33(\frac{\gamma_s - \gamma}{\gamma})gD + 0.0465(3 - \frac{t}{\delta_1})(\frac{\delta_1^2}{t^2} - 1)\frac{\delta_1}{D} + 1.55 \times 10^{-7}(3 - \frac{t}{\delta_1})(1 - \frac{t}{\delta_1})\frac{H}{D}]^{\frac{1}{2}} \tag{2.11}$$

γ_s and γ are the specific weight of particle and water respectively, g the gravity acceleration , H water depth, $\delta_1 = 4 \times 10^{-7} m$ is the thickness of film water, and $t \leq \delta_1$ is the clearance between particles and t/δ_1 depends on the dry density of sediment.

2.2.2. Probability of incipient motion

The probability of incipient motions is

$$\varepsilon_1 = P[V_b > V_{b k_1}] = \int_{\Delta'_m}^{1} p(\Delta')d\Delta' \int_{\omega_1\varphi(\Delta')}^{\infty} p(V_b)dV_b \tag{2.12}$$

By using the distribution of particle location of non-uniform sediment, it is easy to obtain the total probability of incipient motion of non-uniform sediment:

$$\varepsilon_1 = \sum_{l=1}^{n} \varepsilon_1(D_l)P(D_l) = \sum_{l=1}^{n_l} P_{1l} \int_{\Delta'_m}^{\Delta'_M} p(\Delta'_l)d\Delta'_l \int_{\omega_1\varphi(\Delta'_l)}^{\infty} p(V_b)dV_b \tag{2.13}$$

The previous formulae only consist one or two random variables.   However, Han and He's result including three random variables, is strictly derived in theory without any additional assumption.

2.2.3. Distribution and expectation of incipient velocity

From the distribution of Δ' the distribution function and density of dimensionless incipient velocity $\tilde{V}_{b c} = V_{b c}/\omega_1 = \varphi(\Delta')$ can be derived. The calculated result of distribution density is shown in Fig.2. and has been verified by Han (1965). $\tilde{V}_{b.c}$ varies from 0.594 to 1.225. The expectation of $\tilde{V}_{b c}$ is 0.916. The expectation of incipient velocity of non-uniform sediment is the function of particle size and its numerical values are listed in.

Table.1: Expectation of incipient velocity of nonuniform sediment

$D_l/\overline{D}$	0.25	0.5	0.9	1.0	1.5	2.0	3.0	5.0	7.46	10
$\overline{V}_{b c}/\omega_{1 l}$	1.156	1.081	0.951	0.916	0.804	0.749	0.688	0.631	0.596	0.553

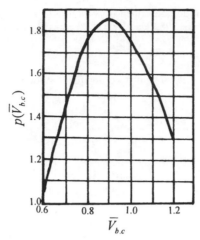

Fig.2: Distribution Desity of Instant Incipient Velocity

2.3 Single step motion of rolling

The step of rolling can be divided into two half steps. The first half is from the initial point to the highest point of the particle and the rolling particle is continuously in contact with the bed surface. The second half is from the highest point to the lowest point and the particle is free in contact with other particles on bed surface. The rolling equation of mass center θ of first half and the equation of longitudinal velocity $u_{2.x2}$ of second half have been derived by the authors. By solving the motion equations of rolling under different conditions, some important results can be obtained.

2.3.1. If the particle starts rolling from stationary, the critical velocity of incipient rolling from stationary is :

$$V_{b.k_1} = \varphi(\Delta')\omega_0 \tag{2.14}$$

where $\varphi(\Delta')$ is expressed in Eq. (2.10), and

$$\omega_0 = \sqrt{\frac{4}{3C_x}\frac{\rho_s - \rho}{\rho}gD} \tag{2.15}$$

Instead of ω_1, ω_0 is adopted. ω_0 is available for the coarser particles. In this case the cohesion force and additional pressure of film water disappeared. The solution of rolling equation also shows that once the particle starts rolling from stationary it will roll over $\theta = \pi/2$, since at $\theta = \pi/2$, $d\theta/dt > 0$.

2.3.2. The rolling equation demonstrates that if the rolling is uniform, then the longitudinal velocity of rolling $u_{2.x} = V_b - V_{b.k_1}$, which is the same as some previous results.

2.3.3. The velocity of first half step u_{2x1} can be solved form rolling equation. Then under the condition that at location $\theta = \pi/2$, $u_{2x1}=0$, the general critical velocity of incipient rolling can be derived:

$$V_{b.k_0} = V_{b.k_0}(\Delta', u_{2.x.0}, u_{2.x.1}\Big|_{\theta=\frac{\pi}{2}} = 0) \tag{2.16}$$

68

$V_{b\,k_0}$ is smaller than $V_{b\,k_1}$, since the velocity has an initial value u_{2x0}, and the cohesion force and additional pressure of film water has already disappeared. Therefore $V_{b\,k_0}$ is also the ceasing velocity.

2.3.4. By solving the rolling equation, the other rolling parameters, such as, x_{2x} step length, t_{20} time interval of incipient rolling, and U_{2x} average rolling velocity of a step, have been obtained. From the distribution of basic variables the distributions and expectations of all rolling parameters can also be derived. Moreover, from the transition conditions , the transition probability of rolling and the parameters of unit rolling are provided.

2.4 Single step motion of jumping

2.4.1. By analyzing the forces acting on the jumping particles, Han and He established the motion equation of jumping. The forces includes longitudinal drag force, lift force (existing from bed surface up to the height of particle size D), resistance of motion, gravity and additional mass force.

2.4.2. Han and He also reveal the essence of the force dispersing the particle. This force actually is the vertical conponent of collision force between moving particle and bed surface and is proportional to the square of particle velocity. For the rolling particle the collision force is

$$F_y = \frac{2\rho_s(1+K_0)}{3eK} \frac{\pi D^2}{4} \frac{\rho u_{2x}^2}{2} \sin 2\theta \tag{2.17}$$

which is not directly related with the flow velocity as shown by some other studies.

2.4.3. The velocity of the particle after collision with the bed surface is the initial velocity of jumping. Therefore, it is more direcly to use the initial velocity rather than the collision force. The velocity after collision can be expressed as

$$u_{3x0} = u_{1x}[1-(1+K_0)(2\Delta'-\Delta'^2)] + u_{y0}[(1+K_0)(1-\Delta')\sqrt{2\Delta'-\Delta'^2}]$$
$$u_{3y.0} = u_{1x}[1-(1+K_0)(1-\Delta')\sqrt{2\Delta'-\Delta'^2}] + u_{y0}[1-(1+K_0)(1-\Delta')] \tag{2.18}$$

in which K_0 is the recovery saturation coefficient. u_{1x} and u_{1y} the conponents of particle velocity before collision, and $i=2,3,4$ represent rolling, jumping and suspension state respectively.

2.4.4. By solving the motion equation of jumping the parameters of jumping have been obtained. Those are t_{30}, the time interval for particle leaving the bed layer; h_m, the maximum height of jumping; t_{31}, the time interval for particle lifting to the maximum height; x_3, the step length; and U_{3x}, the average longitudinal velocity. The distribution and expectation of these parameters have also been derived. These parameters are fundamental ones for exchange intensity and transport rate. Some of the parameters have been verified by the experiment data. For example, the jumping height is one to five times of particle size, and jumping distance is three to 150 times of particle size.

2.4.5. Condition and probability of incipient jumping Han and He assumed that the difference between jumping and rolling is the motion height larger or smaller than the particle size. In other words, when the motion particle reaches height D, if the particle velocity $u_{3.y.D}$ is higher than zero, then the particle is in jumping state. As a result, the necessary condition for particle jumping is

$$u_{3.y.D} \geq 0 \tag{2.19}$$

which can also be expressed as

69

$$V_b > \sqrt{V_1^2 - 1.297 \times 12 u_{3y0}^2} \qquad (2.20)$$

If the particle is from stationary to jumping, then $u_{3y0}=0$ and $V_b>V_1$, V_1 is the critical velocity of jumping. $V_b=V_1$ means the lift force is equal to gravity. If the particle is from rolling to jumping, in equation of u_{3y0} take $u_{2x}=V_{b.k_1}-V_{b.k}$ and $u_{2y}=0$ and change inequality (2.20) into equality, then V_b has two solutions $V_{b.k_2}^{(1)}$ and $V_{bk_2}^{(2)}$, which are the critical velocities of incipient jumping from rolling. The non-conditional expectation of $V_{bk_2}^{(2)}$ is $1.694\overline{V}_{b.k_2}$. $V_{bk_2}^{(2)}$ has small or negative value . While negative, the critical velocity should be zero. If the particle is from suspension to jumping, the vertical velocity is much smaller than the longitudinal velocity $u_{4.x}=V_b$, then the critical velocity of incipient jumping from suspension is

$$V_{b.k_2 4}^{(\Delta')} = \sqrt{\frac{12}{12+15.564 f_2(\Delta')}} \qquad (2.21)$$

in which

$$f(\Delta') = (1-\Delta')(1+K_0)\sqrt{2\Delta'-\Delta'^2} \qquad (2.22)$$

The expectation $\overline{V}_{b.k_2 4}$ is equal to $1.10\overline{V}_{b.k_1}$ .One more condition:

$$V_{by} < \omega \qquad (2.23)$$

is required for the sufficient condition of jumping. Inequality (2.23) means the jumping particle is the one which cannot transfer to suspension. From the conditions (2.20) and (2.23) the transition probabilities $\beta_{i.3}(\Delta')$ and $\beta_{i.3}$ can be derived. In some cases the numerical calculation is required to obtain the value of the critical velocities and transition probabilities.

2.5 Suspension

The motion of suspension is very complicated. In order to reveal the essence of suspension, Han and He (1984) have established a simplified model. It is assumed that when a particle is suspended from the bed surface, it will reach a certain height—called the suspension height, and then drop down to the bed surface. This motion process is called a single step of suspension. In a half step (suspended up or drop down) the bottom velocities V_b and V_{by} are supposed not being changed. The variation of fluctuation of vertical velocity along water depth is very small so that the vertical flow velocity V_y is assumed to be equal to bottom velocity V_{by}. Another suggestion is that the particle motion related to the flow is small, except in vertical direction , where the related motion is caused by the settling of particle. Then the suspension velocity of particle is

$$\begin{aligned} U_{4.x}(y) &= V(y) \\ U_{4.y}(y) &= V_{b.y} - \omega \end{aligned} \qquad (2.24)$$

The suspension height is a random variable and can be expressed by the distribution of sediment concentration along the water depth. According to this model, the parameters of suspension and their distribution have been derived, some basic principles can be revealed and some new concepts can be introduced.

2.5.1.Condition and probability of incipient suspension

For loosely tied sediment at bottom the suspended condition is $V_{by}>\omega$, But the fine stationary particles on bed surface, usually acted upon by the cohesion force and additional pressure of film water, are not loosened before suspension. Therefore, the loosened condition is required for starting suspension. The critical velocity for particle loosening is

$$V_{by k_3}^{(1,2)} = \frac{V_b}{8} \pm \sqrt{\frac{\omega_1^2}{4} - \frac{3}{64}\overline{V}_b^2} \qquad (2.25)$$

Then the probability from stationary to suspension is

$$\beta_{1.4} = P[V_{b\,y} > V^{(1)}{}_{b.y.K_3}] + P[\omega < V_{b\,y} \le V^{(2)}_{b\,y\,k_3}] = \varepsilon_{4.1} - \varepsilon_{42} + \varepsilon_4 \tag{2.26}$$

in which

$$\varepsilon_{4.1} = P[V_{b.y} > V^{(1)}_{b.y.K_3}]$$

$$\varepsilon_{4.2} = P[V_{b.y} \le V^{(2)}_{b.y.K_3}] \tag{2.27}$$

$$\varepsilon_4 = P[V_{b.y} > \omega]$$

2.5.2. Height of suspension

It is assumed that the conditional distribution along water depth of suspended particle is uniform:

$$F_{y|h}(y) = p[\xi_y = y | \xi_h = h] = \begin{cases} 1 & (h < y \le H) \\ y/H & (0 < y \le h) \\ 0 & (y = 0) \end{cases} \tag{2.28}$$

in which H is water depth and h suspension height. Then the distribution function $F_{\eta_h}(h)$ can be derived as:

$$F_{\eta_h}(h) = \frac{1}{H\overline{S}} + \int_y^H S(t)dt - \frac{hS(h)}{H\overline{S}} \tag{2.29}$$

in which $S(y)$ is the concentration distribution, and $\overline{S}$ is the average concentration along vertical direction. If the concentration has exponential distribution

$$S(y) = S_b \exp[-\frac{6\omega}{ku_*}(\frac{y}{H})]$$

then the relative suspension height is

$$\frac{h}{H} = 2[\frac{Ku_*}{6\omega} - \frac{e^{-\frac{6\omega}{ku_*}}}{1 - e^{-\frac{6\omega}{ku_*}}}] \tag{2.30}$$

from which, if $\omega/u_* = 0.001, 0.1, 0.5$ and 5, then $h/H = 0.997, 0.759, 0.133$, and 0.0133 respectively.

2.5.3. Distance and Average Velocity of Single Step of Suspension

Through a series of derivation, the expectation of single step distance of suspension is obtained:

$$\overline{x}_4 = \frac{\overline{V}}{\omega}[\frac{1}{\overline{u}_{4\,y.u}} + \frac{1}{|\overline{u}_{4\,y\,d}|}]\int_0^1\int_0^h p_h(h)V(t)dtdh \tag{2.31}$$

in which

$$\overline{u}_{4.y.u} = M[V_y - \omega | V_y > \omega] = \frac{u_*}{\sqrt{2\pi\varepsilon_4}}e^{-\frac{1}{2}(\frac{\omega}{u_*})^2} - \omega$$

$$\tag{2.32}$$

$$\overline{u}_{4\,y\,d} = M[V_y - \omega | V_y > \omega] = \frac{U_*}{\sqrt{2\pi}(1-\varepsilon_4)}e^{-\frac{1}{2}(\frac{\omega}{u_*})^2} + \omega$$

$\overline{u}_{4.y.\mu}$ and $\overline{u}_{4.y\,d}$ are the average vertical velocity of the half step of suspending up and down respectively. For fine particle, $\bar{u}_{4yu}/\omega$ is closed to $\bar{u}_{4yd}/\omega$. When ω/u_* is increase, both $\bar{u}_{4yu}/\omega$ and $\bar{u}_{4yd}/\omega$ are decreased, but the former decreases more rapidly than the later. When $\omega/u_*>5$, $\bar{u}_{4yu}/\omega$ approximates to zero, and $\bar{u}_{4yd}/\omega$ to one, This is the expected result.

3. STOCHASTIC CHARACTERISTICS OF SEDIMENT EXCHANANGE

The concept of sediment exchange was first introduced by Einstein (1942,1950). Based on the assumption that the amount of particles transferred from stationary on bed surface to bed load is equal to that from bed load to stationary, the formula of transport rate of bed load is derived by Einstein. This idea has been adopted in some researches latter. Han-He (1981,1984) studied all the transition conditions among different motion states and the transition probabilities, and proposed that there are four states of sediment movement, sixteen transition conditions and sixteen corresponding transition probabilities. And thus established a complete stochastic model of exchange and corresponding stochastic principles.

3.1 Transition probabilities with discrete time

The stochastic variables related to certain transition conditions of various states are not completely independent. Han and He established the matrix of transition probabilities for certain conditions. Under the most simplified condition that vertical bottom flow velocity $V_{b,y}$ is independent of longitudinal velocity V_b and before suspension the sediment is already loosely tied, then the matrix of simplified transition probabilities is derived as

$$\{\beta_{ij}\} = \begin{cases} (1-\varepsilon_1)(1-\varepsilon_4) & (\varepsilon_1-\varepsilon_2)(1-\varepsilon_4) & \varepsilon_2(1-\varepsilon_4) & \varepsilon_4 \\ (1-\varepsilon_0)(1-\varepsilon_4) & (\varepsilon_0-\varepsilon_2)(1-\varepsilon_4) & \varepsilon_2(1-\varepsilon_4) & \varepsilon_4 \\ (1-\varepsilon_0)(1-\varepsilon_4) & (\varepsilon_0-\varepsilon_2)(1-\varepsilon_4) & \varepsilon_2(1-\varepsilon_4) & \varepsilon_4 \\ (1-\varepsilon_0)(1-\varepsilon_4) & (\varepsilon_0-\varepsilon_2)(1-\varepsilon_4) & \varepsilon_2(1-\varepsilon_4) & \varepsilon_4 \end{cases} \tag{3.1}$$

If V_b and $V_{b,y}$ are linearly correlated and the sediment should be loosened before suspension then the transition probability turns to be more complicated. $\beta_{1.4}$ is expressed in Eq. (2.26) and the expressions of $\beta_{1.1}, \beta_{1.2}$ and $\beta_{1.3}$ are more complicated. Each has more than ten formulae. The other simplified assumptions are that the rolling and jumping particle considered as bed load and V_b and $V_{b,y}$ are linearly correlated, then the matrix of transition probabilities turns to be (Han - He,1999a)

$$\{\beta_{ij}\} = \begin{cases} (1-\varepsilon_1)(1-\beta) & \varepsilon_1(1-\beta) & \beta \\ (1-\varepsilon_0)(1-\varepsilon_4) & \varepsilon_0(1-\varepsilon_4) & \varepsilon_4 \\ (1-\varepsilon_0)(1-\varepsilon_4) & \varepsilon_0(1-\varepsilon_4) & \varepsilon_4 \end{cases} \tag{3.2}$$

in which $\beta = \beta_{14}$ is expressed by Eq. (2.26), $\varepsilon_0 = P[V_b < V_{bk_0}]$ is the probability without ceasing, V_{bk_0} is the flow velocity for ceasing. If no adhesion is considered among particles , then $V_{bk_0} = V_{bk_1}$ and $\varepsilon_0 = \varepsilon_1$.

Sun (1996) assumed that there are three motion states of particle and derived the transition matrix, and suggested that if particle transferred from stationary to suspension or reversely , it should be transferred to bed load first .

3.2 The life distribution of particle at rest and in incipient motion

It takes a certain time interval for the transition of motion state. For example, a moving particle will transfer to another state only after having finished the single step. Therefore, the transport process is a Markov process with discrete states and continuous time. The main results of sediment motion according to the Markov process will be described in the following.

Among n_o stationary particles on bed surface, the probability of incipient motion for K particles in time interval t, under some simplified conditions , can be derived as,

$$W_K(t) = e^{-\lambda_1 t} \frac{(\lambda_1 t)}{K!} \tag{3.3}$$

which is a Poisson distribution. Then the probability of none of the particles starting moving in time interval is

$$W_0(t) = e^{-\lambda_1 t} \tag{3.4}$$

The probability that at least one of the n_o particles starts moving in time t is the probability of incipient motion

$$F_{1.n_0}(t) = \sum_{k=1}^{\infty} e^{-\lambda_1 t} (\frac{\lambda_1 t}{k!})^k = 1 - e^{-\lambda_1 t} \tag{3.5}$$

which is also called the life distribution of n_o stationary particles Obviously, among n_o particles on bed surface, the average particle number start moving in time t is

$$\overline{K} = \lambda_1 t \tag{3.6}$$

where

$$\lambda_1 = n_0 (\frac{\beta_{12}}{t_{2.0}} + \frac{\beta_{13}}{t_{3.0}} + \frac{\beta_{14}}{t_{40}}) = n_0 \sum_{j\neq 1} \frac{\beta_{1j}}{t_{j.0}} = \sum_{j\neq 1} \lambda_{j1} \tag{3.7}$$

is called the intensity of incipient motion, in which

$$\lambda_{1j} = n_0 \frac{\beta_{1j}}{t_{j0}}, \qquad (j=2,3,4) \tag{3.8}$$

are the intensity of transition from rest to state j, in which t_{j0} is the incipient time from rest to state j, that is the time required for particle to leave the bed surface. For rolling particle, t_{20} is the time for a rolling step, and for jumping and suspended particle t_{30} and t_{40} are the time required for particle to be lifted up from bed surface for one particle size. All of t_{jo} have been presented by the motion equations of single particle. $1/\lambda_{1j}$ is the average incipient period of n_o particles, $i.e.$ in time interval $1/\lambda_{1j}$, in average there is one particle to start moving.

3.3 Life distribution of moving particle

Some researchers pointed out that the single moving step approximately has an exponential distribution. This has been proved in theory by Han and He under some simplified conditions. Han and He also proved that unit step has an exponential distribution. The distribution density for K_i particles in ith state is

$$P_{1.K_i}(x) = (1 - \sum_{j\neq i} \beta_{ij}) K_i \mu_i e^{-(1-\sum \beta_{ij}) K_i \mu_i x}$$

$$= (1 - \sum_{j\neq i} \beta_{ij}) K_i \mu_i U_i e^{-\beta_{ii} K_i \mu_i U_i t} \tag{3.9}$$

in which

$$\beta_{ii} = 1 - \sum_{j\neq i} \beta_{ij} \quad (i = 2,3,4; j = 1,2,3,4) \tag{3.10}$$

is the probability for particles continuously kept in ith state. The moving distance x is related to the moving time interval t, $x=U_i t$, in which U_i is the velocity for ith moving state and is a constant. Corresponding to Eq. (3.9), the probability of at least one of K_i particles in ith state changing its moving state is

73

$$F_{i,K_i}(t) = 1 - e^{-\lambda_i \cdot t} \tag{3.11}$$

in which

$$\lambda_i = (1 - \beta_{i,i})K_i\mu_iU_i = \sum_{\substack{j=1 \\ j\neq i}}^{4} \beta_{ij}K_i\mu_iU_i = \sum_{\substack{j=1 \\ j\neq i}}^{4} \lambda_{ij} \qquad (i = 2,3,4) \tag{3.12}$$

is transition intensity to state i, μ_i is the reciprocal of single step length, and $1/\lambda_i$ expresses the average period of one of K_i particles transferred from ith state to the other one.

$$\lambda_{ij} = \beta_{ij}K_i\mu_iU_i \qquad (i = 2,3,4; j = 1,2,3,4; j \neq i) \tag{3.13}$$

is the transition intensity from state i to state j. Eq. (3.11) is also the life distribution of K_i particles in state i.

3.4 Stochastic principles of motion state transition

Based on the study described above, Han and He obtained the conditional distribution of m_i particles which are among K_i particles transferred from ith state in time t:

$$W_{m_i,k_i}(m,t) = p[\xi_{m_i}(t) = m|\xi_{K_i} = K_i] = e^{-\lambda_i t}\frac{(\lambda_i t)^m}{m!} \qquad (i = 2,3,4) \tag{3.14}$$

The conditional expectation of particle transition in unit area is

$$\overline{m}_i(K_i) = \lambda_i t, \qquad (i = 2,3,4) \tag{3.15}$$

The non-conditional expectation is

$$\overline{m}_i = \overline{\lambda}_i t = \sum_{j\neq i} \overline{\lambda}_{ij}t = \sum_{j\neq i} m_{ij}, \qquad (i = 2,3,4) \tag{3.16}$$

in which

$$\overline{m}_{ij} = \overline{\lambda}_{ij}t = \beta_{ij}\overline{K}_i\mu_iU_it \tag{3.17}$$

is the average particle number transferred from state i to state j in unit area in unit time period The corresponding transition intensities are:

$$\overline{\lambda}_{ij} = \beta_{ij}\overline{K}_i\mu_iU_i \tag{3.18}$$

These twelve transition intensities, nine shown in Eq. (3.18) and three in eq (3.8) demonstrate the entire transition principle of particles in four moving states in quantity.

The derivation of transition intensities is based on the theory of stochastic process. It shows that the number of transition particles is a random variable and thus has its own distribution. In other words, the transition intensity is the expectation of the number of transition particles and the transition distribution is derived from life distribution. Instead of the stochastic process with the intuitive method the transition intensities can also be derived. But the number of transition particles can not be expressed as a random variable and its distribution can not be obtained either. All these study results can be extended to the case of nonuniform sediment. The transition intensity of ith group is expressed as

$$\lambda_{i,j,l} = \beta_{i,j,l}K_{i,l}\mu_{i,l}U_{i,l} \qquad (i = 2,3,4, j = 1,2,3,4, j \neq i, l_n = 1,2\cdots n_i) \tag{3.19}$$

$$\lambda_{1,j,l} = n_{1,l}\frac{\beta_{1,j,l}}{t_{j,l}} \qquad (j = 2,3,4; l_n = 1,2,\cdots n_i) \tag{3.20}$$

4 STOCHASTIC MODEL OF SEDIMENT TRANSPORT RATE

Einstein, followed by Velikanov, Dou, Rosinske and Paintal, used the probability method to study the formula of sediment transport rate. All of their studies are concerned with the average rate only, but none of the distribution or fluctuation of transport rate. Therefore,those cannot be considered as stochastic model. Han and He (He-Han,1980b; Han-He,1984) are the first to establish a complete stochastic model of transport rate by a unified theoretical approach, which includes the average value and the distribution of transport rate and the other random parameters.

4.1 Conditional and non-conditional probability and corresponding mathematical expectation

From the life distribution shown in Eq. (3.5) and   (3.11) , Han and He derived the probability and the expectation of number of transport particles . In a unit water column with the height of water depth and unit bottom area, the conditional distribution of rolling particle number $K_2 = K$ is

$$W_{K_2}[K|\xi_{K_3} = K_3, \xi_{K_4} = K_4] = \frac{\frac{1}{K!}(\frac{\lambda_{12} + \lambda_{32} + \lambda_{42}}{m_2})^K}{\sum_{l=0}^{N_2}\frac{1}{l!}(\frac{\lambda_{12} + \lambda_{32} + \lambda_{42}}{m_2})^l} \qquad (K = 0,1,2,\cdots N_2) \qquad (4.1)$$

The non-conditional distribution is

$$W_{K_2}[K] = \sum_{K_3=0}^{N_3}\sum_{K_4=0}^{N_4}W_{K_2}[K|\xi_{K_3} = K_3, \xi_{K_4} = K_4] \qquad (4.2)$$

in which K_3 and K_4 are the numbers of jumping particles and suspended particles respectively in the same volume, and N_2, N_3, N_4 the maximum number of rolling, jumping and suspended particles respectively. Eq (5.1) is an Erlang distribution. When $N_2 \to \infty$ it approaches a Poisson distribution. The conditional expectation of rolling particle number is

$$\overline{K}_2(K_3,K_4) = \frac{\lambda_{12} + \lambda_{32} + \lambda_{42}}{m_2}[1 - W_{K_2}(N_2)] \qquad (4.3)$$

and the non-conditional expectation is

$$\overline{K}_2 = \frac{\lambda_{1.2} + \overline{\lambda}_{3.2} + \overline{\lambda}_{4.2}}{m_2}[1 - W_{K_2}(N_2)] \qquad (4.4)$$

Similarly, the conditional distribution and expectation as well as non-conditional distribution and expectation of jumping   and suspended particles can be derived. From eqs (4.1) and   (4.2) and their similar equations for jumping and suspension, it can be seen that the non-conditional distributions of rolling, jumping and suspension particles are related to each other through a set of $N_2+N_3+N_4+3$ linear equations, by which these distributions can be solved. But this set of equations is too complicated. In reality, the approximate formulae of the distribution are adopted:

$$W_{K_i}(K) = \frac{\frac{1}{K!}(\frac{\lambda_{1.i} + \overline{\lambda}_{j.i} + \overline{\lambda}_{n.i}}{m_i})^K}{\sum_{l=0}^{N_i}\frac{1}{l!}(\frac{\lambda_{1.i} + \overline{\lambda}_{j.i} + \overline{\lambda}_{n.i}}{m_i})^l}, (K = 1,2\cdots K_i; i = 2,3,4; j = 2,3,4, j \neq i; n = 2,3,4, n \neq i, j) \qquad (4.5)$$

The non-conditional expectation is

$$\overline{K}_i = \frac{\lambda_{1.i} + \overline{\lambda}_{j.i} + \overline{\lambda}_{n.i}}{m_i}[1 - W_{K_i}(N_i)], (i = 2,3,4; j = 2,3,4, j \neq i; n = 2,3,4, n \neq i, j) \qquad (4.6)$$

in which

$$m_i = \sum_{\substack{j=1 \\ j \neq i}}^{4} \beta_{i,j} \mu_i U_i, \quad (i = 2,3,4) \tag{4.7}$$

The probability of maximum particle number W_{K_i} (N_i) is very small and thus can be neglected, then Eq. (4.1)、(4.5) can be replaced by Poisson distribution.

4.2 Average transport rate and its distribution under weak equilibrium as indicated by particle numbers

If W_{K_i} (N_i) is neglected, then from Eq. (4.6) , there is

$$\sum_{\substack{j=1 \\ j \neq i}}^{4} \overline{\lambda}_{j,i} = \sum_{\substack{j=1 \\ j \neq i}}^{4} \overline{\lambda}_{i,j} \tag{4.8}$$

From which it can be seen that under equilibrium transportation the particle number transferred from one to other three states is just equal to that from three other states. This condition is called weak equilibrium. From eq (4.8) the number of transport particle can be solved.

$$\overline{K}_i = \frac{\Delta_i}{\Delta_0} \frac{n_{10}}{\mu_i} \frac{1}{U_i} \quad (i = 2,3,4) \tag{4.9}$$

in which Δ_0 and Δ_i are the determinants of the coefficients of Eq. (4.8). K_i is the particle number in ith state in unit water column. Therefore ,the transport rate accounted in particle number should be

$$\overline{q}_{N_i} = U_i K_i = \frac{\Delta_i}{\Delta_0} \frac{n_{10}}{\mu_i} \quad (i = 2,3,4) \tag{4.10}$$

which has Poisson distribution as K_1

$$W_{N_i} = P[q_{N_i} = N] = e^{-\overline{q}_i} \frac{(\overline{q}_{N_i})^N}{N!} \tag{4.11}$$

4.3 Average transport rate and its distribution under strong equilibrium as indicated by particle numbers

The strong equilibrium means that

$$\overline{\lambda}_{j,i} = \overline{\lambda}_{i,j} \quad (i,j = 2,3,4; i \neq j) \tag{4.12}$$

i.e. the particle number transferred from state i to state j is equal to that from state j to state i. Under this condition the transport rate indicated in particle numbers can be derived as

$$\overline{q}_{N_i} = U_i \overline{K}_i = \frac{\beta_{1,i}}{\beta_{1,1}} \frac{n_{1,0}}{\mu_i t_{10}} \quad (i = 2,3,4) \tag{4.13}$$

It also has Poisson distribution as K_i

$$W_{N_i}(N) = e^{-\overline{q}_{N_i}} \frac{(\overline{q}_{N_i})^N}{N!} \tag{4.14}$$

4.4 Average transport rate indicated by weight

If the moving particles are not large in number, then the particle number on bed surface is

$$n_1 = \frac{m_0}{\frac{\pi}{4}D^2} \tag{4.15}$$

where $m_0 = 0.4$ is the coefficient of stationary density of sediment on bed surface. If the moving particles are large in number then

$$n_1 = \frac{P_1 m_0}{\frac{\pi}{4}D^2} \tag{4.16}$$

where P_1 is the probability of particle at rest on bed surface. Only the Eq. (4.16) is adopted by early researches. From eqs (4.10) and (4.13) the transport rate in unit width indicated by weight is expressed as

$$\bar{q}_i = \frac{2}{3}m_0\gamma_s DP_1 \frac{\Delta_i}{\Delta_0 \mu_i} \tag{4.17}$$

$$\bar{q}_i = \frac{2}{3}m_0\gamma_s DP_1 \frac{\beta_{1i}}{\beta_{i1}}\frac{1}{\mu_i t_{i0}} \tag{4.18}$$

Their distributions are also the Poisson ones.

From eq (4.18) the transport rate of jumping sediment is

$$\bar{q}_3 = \frac{2}{3}m_0\gamma_s DP_1 \frac{\beta_{1.3}}{\beta_{3.1}}\frac{\bar{l}_3}{t_{3.0}} \tag{4.19}$$

where the unit step length $\bar{l}_3$ is the reciprocal of μ_3, β_{13} is the probability from stationary to jumping and β_{31} is the probability from jumping to stationary . This formula summarizes the formulae of Einstein, Velikanov and Dou.

To sum up, Han and He, by adopting an approach combining the stochastic process with mechanical method established the stochastic model of transport rate, which is a complete theoretical model including the distribution and average value of transport rate of sediment in rolling, jumping and suspension conditions.

4.5 Transport rate of nonuniform sediment[1]

If the nonuniform sediment is divided into n_l size groups, then the total transport rate for weak equilibrium is

$$\bar{q}_i = \sum_{l=1}^{n_l} q_{il} = \sum_{l=1}^{n_l} R_{il}\bar{q}_i = \frac{2}{3}m_0\gamma_s \sum_{l=1}^{n_l} R_{il}D_l P_{11} \frac{\Delta_i(l)}{\Delta_0(l)\mu_i(l)} \tag{4.20}$$

and that for strong equilibrium

$$\bar{q}_i = \sum_{l=1}^{n_l} R_{il}\bar{q}_i = \frac{2}{3}m_0\gamma_s \sum_{l=1}^{n_l} R_{il}D_l \frac{P_{1l}\beta_{1il}}{\beta_{i11}\mu_{il}t_{i0l}}, (i = 2,3,4) \tag{4.21}$$

in which P_{1l} and R_{il} are the size distribution of bed material and transport rate respectively, subscript l indicates the parameters of lth size group. The distribution of transport rate of lth size group is also an Erlang one.

5 STOCHASTIC MODEL AND CHARACTERISTICS OF BED LOAD DISPERSION

Based on the previous studies, the dispersion model of bed load has been further developed by Han and He (1980,1984). A most general model has been established, and the various models introduced previously by others can be considered as its special cases.

77

5.1 Description of the general dispersion model

Suppose that X_0 is the original location of the particle at the initial instant $T_0, \xi_{t,i}$ the moving time period of the ith unit, and $\xi_{x,i}$ the distance of ith moving unit, then $\xi_t^{(n)}$, the time interval from instant T_0 to the end of nth rest time period and $\xi_\tau^{(n)}$, the time interval from T_0 to the end of time period of nth unit motion can be expressed as

$$\xi_t^{(n)} = \theta_{t0} + \xi_{\tau 1} + \xi_{t.1} + \xi_{\tau.2} + \xi_{t.2} + \cdots \xi_{\tau.n} + \xi_{t\,n} \quad (n \geq 0)$$

$$\xi_\tau^{(n)} = \theta_{t0} + \xi_{\tau 1} + \xi_{t1} + \xi_{\tau.2} + \xi_{t2} + \cdots \xi_{t\,n-1} + \xi_{\tau.n} \quad (n \geq 0)$$

$$(5.1)$$

in which

$$\theta = \begin{cases} 1 & (if\ particle\ at\ rest\ at\ T_0\ ) \\ 0 & (if\ particle\ in\ moving\ at\ T_0) \end{cases} \tag{5.2}$$

The moving distance of particle just after nth moving unit is

$$\xi_x^{(n)} = \begin{cases} \displaystyle\sum_{i=1}^{n} \xi_{x\,i} & (n \geq 1) \\ 0 & (n = 0) \end{cases} \tag{5.3}$$

There are three types of distribution of dispersion : deposition, motion and transport respectively. The deposition distribution is the distribution of particle at rest on the bed surface along X at instant T, the motion distribution is the distribution of particle moving on the bed surface along X at instant T, and the transport distribution is the distribution of particle passed through location X in time interval $[0, \ T]$. These three distributions can be expressed as

$$F_{1\,\theta T}(X) = P(B_{1.\theta}) = \int_0^{T_1} \int_0^{X_1} F_{1\,\theta T}(X|X_0,T_0) p_0(X_0,T_0) dX_0 dT_0 \tag{5.4}$$

$$F_{2\,\theta T}(X) = P(B_{2.\theta}) = \int_0^{T_1} \int_0^{X_1} F_{2\,\theta T}(X|X_0,T_0) p_0(X_0,T_0) dX_0 dT_0 \tag{5.5}$$

$$F_{3.\theta.X}(T) = P(B_{3.\theta}) = \int_0^{T_1} \int_0^{X_1} F_{3\,\theta T}(X|X_0,T_0) p_0(X_0,T_0) dX_0 dT_0 \tag{5.6}$$

in which the conditional distribution functions are:

$$F_{1\,\theta T}(X|X_0,T_0) = \sum_{n=0}^{\infty} p[\xi_\tau^{(n)} \leq T - T_0 < \xi_t^{(n)}, \xi_x^{(n)} \leq X - X_0] \tag{5.7}$$

$$F_{2\,\theta T}(X|X_0,T_0) = \sum_{n=0}^{\infty} p[\xi_t^{(n-1)} \leq T - T_0 < \xi_\tau^{(n)}, \xi_x^{(n)} \leq X - X_0] \tag{5.8}$$

$$F_{3.\theta Z}(X|X_0,T_0) = \sum_{n=0}^{\infty} p[\xi_\tau^{(n)} \leq T - T_0, \xi_x^{(n-1)} \leq X - X_0 < \xi_x^{(n)}] \tag{5.9}$$

and $p_0\ (X_0,T_0)$ $\ (0 \leq X_0 \leq X_1, 0 \leq T_0 \leq T_1)$ is the density of surface source. One of its special case is degenerate surface source, and its distribution is

$$p_0(X_0,T_0) = \begin{cases} P_1 \dfrac{p_{T_0}(T_0)}{\Delta X_0} & (0 \leq X_0 \leq \Delta X_0, 0 \leq T_0 \leq T_1) \\[3mm] P_2 \dfrac{p_{X_0}(X_0)}{\Delta T_0} & (0 \leq X_0 \leq \Delta X_1, 0 \leq T_0 \leq T\ ) \end{cases} \tag{5.10}$$

78

in which ΔX_0 and ΔT_0 are sufficient small and $P_1+P_2=1$. The degenerate surface source corresponds to the following two line sources. p_0 (X_0),the line source along X, is the distributions of particle moving or deposit on bed surface at the initial instant. Another line source p_0 (T_0) is the distribution of inflow sediment at inlet section. In practice the scouring by clear water downstream the dam site is just the case of initial line source $p_0(X_0)$.

The general model established by Han and He summarizes various dispersion distribution and initial source distribution and is an important development in this research field.

5.2 Dispersion of point source with constant parameters

If the motion time can be neglected, the single unit distance and rest time period have exponential distribution and their parameters are constant , the average unit distance is $1/\mu$ and the average rest time period is $1/\lambda$,then according to the results mentioned above, the four distributions with point source are derived.

The deposit distribution under initial conditions of stationary and motion has been provided by Einstein (1937), the transport distribution under initial condition of rest was derived by Hubbell and Sawyer (1964), The transport distribution under initial condition of motion was first introduced by the authors.

5.3 Dispersion of degenerate source with constant parameters and moving time neglected

Under the initial condition of particle at rest the deposit distribution is

$$F_{11T}(Z) = \frac{P_1}{\lambda T_1} e^{-\mu X} \sum_{n=0}^{\infty}\sum_{r=n}^{\infty} \frac{(\mu X)^r}{r!} \sum_{k=0}^{n} [e^{-\lambda(T-T')} \frac{\lambda^k (T-T')^k}{k!} - e^{-\lambda(T-T')} \frac{(\lambda T)^k}{k!}$$

$$+ \frac{P_2}{\mu x_2} e^{-\lambda T} \sum_{n=0}^{\infty} \frac{(\lambda T)^n}{n!} \sum_{r=n}^{\infty}\sum_{k=0}^{r} [e^{-\mu(X-X')} \frac{\mu^k (X-X')^k}{k!} - e^{-\mu X} \frac{(\mu X)}{k!} , (X>0, T>0)$$

(5.11)

in which

$$T' = \begin{cases} T & (if\ T < T_1) \\ T_1 & (if\ T \geq T_1) \end{cases} \qquad X' = \begin{cases} X & (if\ X < X_1) \\ X_1 & (if X \geq X_1) \end{cases}$$

(5.12)

$$P_1 = P[\xi_{X_0} = 0, 0 \leq \xi_{T_0} \leq T_1]$$

(5.13)

$$P_2 = P[0 \leq \xi_{X_0} \leq X_1, \xi_{T_0} = 0]$$

(5.14)

and $P_1+P_2=1$.When $P_1=0$, $P_2=1$, the deposit distribution under the initial condition of particles at rest is just the deposit distribution of particles scoured from bed surface by clear water downstream the density.

5.4 The distribution under more complicated conditions

Some more complicated dispersions introduced by Han and He, includes the distribution with variable parameters and neglected moving time and the distribution with constant parameters but with the moving time in consideration. The distribution of Shen-Todorovic (1971) is a special case of the former. The deposit distribution with the moving time in consideration is expressed as

$$F_{1.1T}(X|X_0,T_0) = e^{-\lambda(T-T_0)} + \sum_{n=0}^{\infty} \lambda^n \mu^n e^{-\lambda(T-T_0)} [\sum_{n=0}^{\infty} \frac{(-1)^n A_{r_n}}{(\lambda - \mu')^{2n-r_n}} \frac{(T-T_0)^{r_n}}{r_n!}$$

$$- \sum_{h=1}^{n} (-1)^n \frac{(X-X_0)^{n-k}}{(n-K)!} \sum_{r_k=0}^{n} \frac{A_{r_k}}{(\lambda - \mu')^{n-r_k+k}} e^{(\lambda-\mu')X} \frac{(T-T_0 - \frac{X-X_0}{U})^{r_k}}{r_k!}]$$

(5.15)

79

In which

$$A_{r_k} = \sum_{r_{k-1}=r_k}^{n} \cdots \sum_{r_2=r_3}^{n} (n+1-r_2), \qquad (k=3,\cdots\cdots n) \tag{5.16}$$

and $\mu' = U\mu$.

Sun (1996) obtained two distribution of dispersion, in which the moving time period is Considered.

6. APPLICATION OF EXCHANGE INTENSITY AND DISTRIBUTION OF TRANSPORT RATE

The theoretical results of exchange intensity and distribution of transport rate have been applied to sediment engineering to develop a series of new results.

6.1 Application to incipient velocity (Han,1982; Han-He,1984,1999b; He-Han,1982)

The low transport rate indicates the average transport rate at which incipient rolling occurs. It can be expressed as

$$\lambda_{q_b} = \frac{q_b}{\gamma_s D \omega_1} = \frac{2}{3} m_0 \varepsilon_1 \frac{U_2}{\omega_1} = f(\frac{\overline{V_b}}{\omega_1}) \tag{6.1}$$

in which q_b is the transport rate of unit width, γ_s the specific weight of sediment, D particle size, U_2 the average rolling velocity of particle, and $\overline{V_b}$ the average flow velocity at bottom.

The low transport rate is an important criterion for incipient motion. Han and He have demonstrated that the general principle of transport rate is still available to the particle in incipient motion, only the rate is very low and most particles are in rolling. As a result the incipient velocity is not a critical velocity to distinguish movement and stationary of particle. At this velocity a small amount of particles are always in moving. Therefore, the incipient velocity is an artificial standard. From the formula of low transport rate different critical standard of incipient motion expressed by average bottom velocity are able to be provided. According to a lot of experiment, the incipient standard expressed by dimensionless transport rate is about $\lambda_{q_{bc}} = 0.219 \times 10^{-3}$,which corresponds to the incipient bottom velocity $\overline{V}_{bc}/\omega_1 = 0.433$ and average vertical velocity

$$V_c = 0.116\varphi\omega \tag{6.2}$$

in which

$$\varphi = 6.5(\frac{H}{D})^{\frac{1}{4+\lg(\frac{H}{D})}}$$

Based on the stochastic model of incipient motion the authors have studied a series of theoretical and practical problems of incipient motion in the book "Incipient principle and Incipient Velocity of Sediment".

6.2 Application to nonequilibrium transportation of nonuniform sediment (Han,1979; Han-He,1997)

From the exchange intensity the bottom boundary condition of non-equilibrium transport equation of nonuniform sediment can be expressed as :

$$\varepsilon_y \frac{\partial S_l}{\partial y}\Big|_{y=0} + \omega_l S_{bl} = \alpha_l \omega_l (S_{bl} - S_{bl}^*) \quad (l=1,2\cdots N_l) \tag{6.3}$$

80

By using Eq. (6.3) and integrating the diffusion equation along the vertical direction the one-dimensional equation of non-equilibrium transport of nonuniform sediment is obtained

$$\frac{dS_l}{dx} = -\frac{\alpha_l \omega_l}{q}(P_{4l}S - P_{4l}^* S^*), \quad (l = 1,2,\cdots n_l) \tag{6.4}$$

in which, α_l is the recovery saturation coefficient

$$\alpha_l = (1-\varepsilon_{0l})(1-\varepsilon_{4l})[1+\frac{1}{\sqrt{2\pi}(1+\varepsilon_{4l})}(\frac{u_*}{\omega_l})e^{-\frac{1}{2}(\frac{\omega_l}{u_*})^2}] \tag{6.5}$$

ε_y the diffusion coefficient, S concentration, P_4 and P_4^* the size distribution of suspended load and carrying capacity respectively, ω the terminal velocity, q the flow discharge of unit width, and u_* the frictional velocity of flow, subscript b indicates the parameters at bottom and l the number of size group. Eq (6.5) is the theoretical result of α_l. Its values either larger than or less than one are in good agreement with the empirical values. It is thus a good answer to the controversy over different opinion on α.

6.3 Application to non-equilibrium transport of bed load   (Han-He,1984)

From exchange intensity the non-equilibrium transport equation of nonuniform bed load is derived:

$$\frac{d\overline{q}_{bl}}{d_x} = -[\sum_{l=1}^{l_m}\varepsilon_{4l}\mu_{bl}(R_{bl}\overline{q}_b - R_{bl}^{**}q_b^*) + \sum_{l=1}^{l_m}(1-\varepsilon_{0l})(1-\varepsilon_{4l})\mu_{bl}(R_{bl}\overline{q}_b - R_{bl}^*\overline{q}_b^*) \tag{6.6}$$

In which $\overline{q}_{bl}$ is transport rate of unit width, R_{bl} the size distribution of bed load, $\overline{q}_b^*, \overline{q}_b^{**}$ and R_{bl}^*, R_{bl}^{**} are the carrying capacities and size distributions of bed load when bed material is in equilibrium with bed load and suspended load respectively, and their formulae expressed in terms of exchange intensities have been obtained.   Eq (6.6) shows that in order to be in exchange equilibrium between bed load and both bed material and suspended load, $\overline{q}_{bl}^* = \overline{q}_{bl}^{**}$ and $R_{bl}^* = R_{bl}^{**}$ should be exist. But it is almost impossible for nonuniform sediment. Therefore, the bed load transport is always not in equilibrium. As a result, the transport equation based on the experiment in laboratory flume is not consistent with that   measured from field data.

6.4 Application to sediment prevention for water diversion works (Han - He, 1984)

The average height of suspension h_l for different size groupis is an important criterion for determining the water level for diversion   works to prevent the coarse particles flowing into diversion channel or hydraulic power station.

6.5 Application to measurement of bed load

Han - He (1982,1984) have studied the fluctuation in the measurement of transport rate of bed load from the distribution of transport rate of bed load. The variation coefficient is used to indicate the measurement error:

$$C_V = \frac{\sqrt{D[\xi_{q_b}]}}{\sqrt{M[\xi_{q_b}]}} = \sqrt{\frac{1}{m}\sum_{l=1}^{l_m}R_{bl}(\frac{R_{bl}}{R_{1l}}-1) + \frac{T_1\gamma_s D_0^3}{66t\overline{q}_b'}} = \sqrt{\frac{1}{m}\sum_{l=1}^{l_m}R_{bl}(\frac{R_{bl}}{R_{1l}}-1) + \frac{1}{w_n}} \tag{6.7}$$

in which R_{1l} is the size distribution of bed material , b the width of samplers, t the time interval for sampling, l_m the number of repeated sampling, $\overline{q}_b'$ the transport   rate of unit width of bed load entering into samplers, $\overline{W}_n$ the transport amount of bed load entering samplers , D_0 the equivalent size of nonuniform sediment

$$D_0^3 = \sum_{l=1}^{l_m} D_l^3 R_{1l} \tag{6.8}$$

The first item under the radical sign in Eq. (6.7) is the error caused by random size of bed material. The second term $1/\overline{W}_n$ is the error dependent on the measurement time, average transport rate and the width of samplers. It is obvious that the larger the equivalent particle number, the smaller the error. If the sediment is uniform or its size distribution is definite, then only the second item exists. Since the larger the time interval for sampling, the smaller the error, increase in the number of repeated sampling will reduce the error.

6.6 Application to carrying capacity of nonuniform sediment

From the strong equilibrium condition $\lambda_{14l} = \lambda_{41l}$, the carrying capacity of nonuniform suspended load can be expressed as

$$S*(l) = \frac{2}{3} m_0 \frac{\varepsilon_{4l}}{(1-\varepsilon_{0l})(1-\varepsilon_{4l})} \frac{L_{4l}\omega_l}{q} \frac{D_l}{t_{4.0l}\omega_l} \tag{6.9}$$

in which $L_{4l} = 1/\mu_{4l}$ is single step length. It can be seen that the carrying capacity is proportional to relative step length L_{4l} and ω_l / q and is inversely proportional to the relative time for incipient suspension. It is also related to the transition probabilities. The numerical calculation by Fang (1998) shows that eq (6.9) is in good agreement with the field data without any corrected coefficient. The total carrying capacity from eq (6.9) is

$$S* = \sum_{l=1}^{l_n} R_{1l} S*(l) \tag{6.10}$$

6.7 Application to ratio of bed load to suspended load   (Han - He, 1999a)

From the stochastic model of particle motion, Han and He have presented three state probabilities P_1, P_2, and P_3 for stationary , bed load and suspended load respectively. The ratio of bed load to suspended load is:

$$m(l) = \frac{q_b(l)}{q_s(l)} = \frac{\Delta_{ml}}{H} \frac{U_{2l}}{V} \frac{\lambda_{bc}}{\lambda_{sc}} \frac{P_3}{P_2} \tag{6.11}$$

in which q_b (l) and q_s (l) are the transport rate of bed load and suspended load respectively, Δ_{ml}/H the ratio between depth of bed load and suspended load   (water depth) respectively, U_{2l}/V the ratio between velocity of bed load and suspended load respectively, λ_{bc} and λ_{sc} the bottom concentration of bed load and average concentration of suspended load respectively. Eq (6.11) has been verified by the field data . This equation is only applicable to the sand bed load because only the sand bed load can be either transported as bed load or be suspended.

6.8 Dry specific density of nonuniform sediment   (Han-Wan,1981b; Han,1997)

By establishing dry-density model of random void-filling, Han and He introduced the contact probability and calculated the dry-density of nonuniform sediment.

If there are two size groups and $\gamma'_{s.1}$ and γ'_{s2} are the dry-density weight of coarse and fine groups respectively, then the dry-density of mixing sand is

$$\frac{1}{\gamma_s'} = P_1(\frac{Q}{\gamma_{s1}'} + \frac{1-Q}{\gamma_s}) + \frac{P_2}{\gamma_{s2}'} \qquad (6.12)$$

In which γ_s' is the dry-density weight of mixing sand and γ_s the specific weight of sand particle, P_1 and P_2 the size distribution of coarse and fine groups respectively. $Q = 1 - Q_2^{n+1}$ is the probability of voids among coarse particles which are not filled by fine particles and

$$Q_2 = \frac{P_2}{D_2} / (\frac{P_1}{D_1} + \frac{P_2}{D_2}) \qquad (6.13)$$

is the probability at which the coarse particles are in contact with fine particles, n the maximum number of fine particle layer, which is two   when D_2 is about ten times of D_1 . If $Q=1$, no fine particles will be filled in the voids of coarse particles, then

$$\frac{1}{\gamma_s'} = \frac{P_1}{\gamma_{s1}'} + \frac{P_2}{\gamma_{s2}'} \qquad (6.14)$$

which is also the equation suggested by Colby. If $Q=0$, the voids of coarse particles are fully filled by fine particles then the dry-density turns to be

$$\frac{1}{\gamma_s'} = \frac{P_1}{\gamma_s} + \frac{P_2}{\gamma_{s2}'} \qquad (6.15)$$

REFERENCES

Bagnold, R.A.1956. The flow of cohesionless grains in fluids, Phil. Trans . Roy. Soc. Ser. A, 249,964.

Bagnold, R.A.1973. The nature of salation and of bed-load transport in water, Proc. Royal, Soc. Ser A, 332,473.

Chen,Y.S. 1989. A Study on Einstein's formula of bed load transport Rate. J.Hyd. No.7.

Cheong, H.F. and Shen, H.W.1976. Stochastic characteristics of sediment motions, J. Hy. Div. ASCE,102 (7), 1035.

Cheong, H.F. and Shen, H.W.1977. On the motion of sand particles in an alluvial channel, Proc. 2nd Int . Symp. on Stochastic Hydraulics, IAHR, Hjorth, P., Jönsson, L. And Larson, P. ed . Lund Inst. of Technology, Univ. of Lund, 219.

Crikmore, M.J. and Lean, G.H. 1962. The measurement of sand transport by the time-integration method with radioactive tracers. Proc. Royal Society of London, A 270, 27.

Dou,G.K.1964. Theory of sediment movement. Nanjing Ist. of Water Resources Res.

Einstein, H.A.,1937. Bed load transport as a probability problem in sedimentation. Translated in English by Sayre, W.W. [Einstein].Shen, H.W. Ed. Appendix C, P.O.Box 606, Fort Collins, Colorado,1972.

Einstein, H.A.1942 Formulas for the transportation of bed load. Trans. ASCE. Vol. 117: 561.

Einstein, H.A.1950. The bed-load function for sediment transportation in open channel flows. Tech, Bulletin 1026.USDA Soil Conservation Service.

Fang, C.M., Han, Q.W., He, M.M. and Jia, X.L.1998. Numerical method of stochastic theory of carrying capacity of nonuniform sediment and verification. J.Hyd.No.2:199-210.

Gonchanov, B.H.1938.Sedement Movement. Lelinglad.

Han, Q.W.1965, Report of pebble experiment in field flume at Mangqi River in Sichuan. Inst . of Yangtze River.

Han, Q.W.1979. A study on nonequilibrium transport of nonuniform suspended load, Science Report. No.17.

Han,Q.W.1982. Incipient principles and incipient velocity of sediment. J.of Sediment Res. No.2.

Han,Q.W.1997. Distribution of dry-density of sedimention and its application. J. Sed. Res. No.2.10-16.

Han,Q.W. and He,M.M.1980. Stochastic model and statistical characteristics of bed load dispersion, Scientia Sinica, 23 (8), 1006.

Han, Q.W. and He , M.M. 1981. Stochastic principles of sediment exchange . J. Hyd, No.1: 10-22.

Han, Q.W. and He, M.M.1982. Bed load fluctuation: applications. J. of Hyd , Div . of ASCE , Vo.108, No. 2, 199-210.

Han., Q.W. and He, M.M.1984. Stochastic theory of sediment movement, China Science Press.

Han, Q.W. and He, M.M.1997. Two-dimensional non-equilibrium transport equation and its boundary conditions of nonuniform bed load, J.of Hyd. No.1. 1997.

Han, Q.W. and He, M.M.1999a. A study on sediment exchange at bottom layer and state probability and ratio between bed load and suspended load. J.Hy No.10.

Han,Q.W. and He M.M. 1999b. Incipient principles and incipient velocity of sediment. China Science Press.

Han, Q.W., Wang. Y.C. and Xing, X.L.1981. Initial dry density of sedimentation. J.of Sediment Res. No.1.

He M.M. and Han, Q.W. 1980a. Stochastic model of single particle movement, Proc.3rd Int. Symp. on Stochastic Hadraulics, IAHR, Tokyo, Japan, 289.

He, M.M.and Han,Q.W.1980b. Stochastic model and principles of transport rate, J.of Mechanics. No.3.

He,M.M. and Han, Q.W.1981. Mechanics and stochastic principles of single particle movement . J.of Mechanics, No.1 (Suplement).

He, M.M and Han , Q.W. 1982. Stochastic model of incipient sediment motion. J.Hyd, Div., ASCE, vol.108,No.2,211-224.

Hu, C.H., and Hui,Y.J.1995.Mechanics and stochastic principles of sediment flow mixture in open channel, China Science Press.

Hubbell, D.W. and Sayre, W.W.1964. Sand transport studies with radioactive tracers, J. Hy. Div. ASCE, 90 (3),39.

Liu,X.L and Cheng,Y.X., 1987. Transport rate of nonuniform bed load. J. Chendou Univ. of Sci. And Tech.

Losinskee, K.E. and Lubomilova, K.C.1976.Movement mechanism of river sediment. Proc. 4th National Symp on Hydrology.USSR.10-22.

Nakagawa, N. and Tsujimoto , T. 1976. On probabilistic characteristics of motion of individual sediment particles on stream beds, Proc. 2nd Int. Symp. On Stochastic Hydraulics, IAHR. 1976, Lund Inst. of Technology, Univ. of Lund, 219 .

Paintal, A.S.1971.A stochastic model of bed load transport. J.Hy. Res 9 (4),527.

Shen, H. W and Todorovic, P.1971. A general stochastic sediment transport model. Int . Symp. On Stochastic Hydraulics, C.L. Chiu ed , School of Engineering, Univ. Of Pittsburgh, 489.

Sun,Z.L.1991. A Discussion on Einstein's formula of bed load, J.Sed. Res. No.1.

Sun,Z.L. 1996. Stochastic theory of nonuniform sediment transport .Ph.D.Dissertationm, Wuhan Univ.of Water Res.and Elec.Pow.

Todorovic, P. 1967. A stochastic process of monotonous sample function, Mathematnykn Bechnk, 4 (19) CB.2, 149.

Todorovic, P. 1970. On some problems involving random number of random variables, The Ann. of Math. Statistics, 41 (3), 1059.

Todorovic, P. 1975. A stochastic model of dispersion of sediment particles released from a continuous source, Water Resources Research, 11 (6), 919.

Tsuchiya, Y., 1969. On the mechanics of salation of a spherical sand particle in a turbulent stream, IAHR 13th Cong, 191.

Velikanov, M.A. 1955.River dynamics. National Tech. Press. Vol 2.21.

Velikanov,M.A.,1958,Fluvial process, National Press of Math-Phsics.

Wan,S.Q. 1995. A study on rivision of Einstein's formula of transport rate of nonuniform bed load. J. Sediment Res. No.1. 44-55.

Yalin, M.S.1963. An expression for bed-load transportation, J. Hy. Div. ASCE, 89 (2).

Yang, C.T. 1968. Sand dispersion in laboratory flume, Dissertation, Colorado State Univ., Fort Collins, Colorado.

Yang, C.T. and Sayre, W.W.1971. Stochastic model for sand dispersion, J. Hy. Div. ASCE, 97 (2),265.

Yano, K., Tsuchiya, Y.and Michive, M.1969. Tracer studies on the movement of sand and gravel , IAHR 13th Cong., 121.

Stochastic Hydraulics 2000, Wang & Hu (eds) © 2000 Taylor & Francis, ISBN 90 5809 166 X

Suspension flow in open channels

W.H.Graf
Laboratoire des Recherches Hydrauliques, Ecole Polytechnique Fédérale, Lausanne, Switzerland

M.Cellino
Bonnard and Gardel Consulting, Lausanne, Switzerland

ABSTRACT: This paper reports on measurements of velocities and concentration profiles made with a sonar instrument, called Acoustic Particle Flux Profiler (APFP), in suspension flow over a mobile bed. From these measurements the turbulent and the sediment fluxes and subsequently the β-values were extracted. This allowed the determination of the concentration distribution over the flow depth by using the Rouse equation. In suspension flows over a mobile bed without bed forms the measured $\bar{\beta}$-values at *capacity* condition are smaller than unity, $\bar{\beta} < 1$. It could also be shown that for flows over a mobile bed with bed forms the $\bar{\beta}$-values are larger than unity, $\bar{\beta} > 1$.

Key words: Open-channel flow, Suspension flow, Concentration distribution, Diffusion coefficients

1.INTRODUCTION

The transport of suspended sediments in a steady uniform open-channel flow is calculated (see *Graf*, 1973, p.173) as:

$$q_{ss} = \int_a^h \bar{c}_s(y) \cdot \bar{u}(y) \cdot dy \qquad \left[\frac{m^3}{s \cdot m} \right] \tag{1}$$

with $c_s = c_s - c'_s$ as the mean suspended particle (volumetric) concentration, where c'_s is the fluctuating component; $\bar{u} = u - u'$ is the mean flow velocity; h is the flow depth and a is the distance from the bed, above which the sediment particles move in suspension. Thus, to compute the suspended load, q_{ss}, the vertical distribution of the velocity, $u(y)$, and of the concentration, $c_s(y)$, have to be known.

2.THE EXPERIMENTS

The measurements were performed in a channel, being 16.8 [m] long and 0.60 [m] wide, at the centerline of its cross section. The measuring section was located at 13[m] from the entrance of the channel; the flow depth was maintained at h=0.12[m]. Sediment particles were added to the flow till a layer, 2 [mm] thick, formed itself on the bottom of the channel; in this way it was assured that the sediment transport is always in full (saturation) capacity (see *Cellino*, 1998 and *Graf* et *Cellino*, 2000). Previously, flows at non capacity conditions were also investigated (see *Cellino* et *Graf*, 1999).

The instantaneous vertical concentration and velocity profiles, $c_s(y,t)$ and $u(y,t)$, were measured by a non-intrusive sonar instrument, the Acoustic Particle Flux Profiler (APFP), developed in our laboratory (see *Shen* et *Lemmin*, 1996). The vertical mean concentration profiles, $c_s(y)$, have been directly measured by the isokinetic suction method, which in turn was also used to calibrate the APFP instrument (see *Cellino*, 1998).

The point velocities, $u(y, t)$ and $v(y, t)$, and the corresponding concentrations, $c_s(y, t)$, are extracted with a frequency of 16[Hz], from a measuring volume, having a diameter of $\Phi \approx 13$ [mm] and a height of $\Delta d \approx 5$ [mm]. A special (tristatic) configuration of the APFP instrument permitted to measure the longitudinal, $u(y, t)$, and vertical, $v(y, t)$, velocity components, but not the concentration, $c_s(y, t)$, with a frequency of 39[Hz]. The velocities measured by the instrument appears to be the one of the water-sediment mixture (see *Shen* et *Lemmin*, 1996); the accuracy is generally better than 2[mm/s].

Eleven runs using sand I, $d_{50} = 0.135$ [mm], and six runs using sand II, $d_{50} = 0.230$ [mm], were performed (see *Cellino*, 1998, and *Graf* et *Cellino*, 2000). In all runs, the bed presented itself without bedforms. Additionally, suspension flows with bed forms were also studied (see *Cellino* et *Graf*, 2000).

3. VELOCITY DISTRIBUTION

The longitudinal mean velocity profile for a dilute suspension shall be given with the velocity-defect expression (see *Graf* et *Altinakar*, 1998, p.44), or :

$$\frac{\bar{u}_c - \bar{u}}{u_*} = -\frac{1}{\kappa} ln\left(\frac{y}{\delta}\right) + \frac{2\Pi}{\kappa} cos^2\left(\frac{\pi \cdot y}{2 \cdot \delta}\right) \tag{2}$$

where Π is Coles' wake-strength parameter and $y = \delta$ is the position of the maximum velocity, $\bar{u}_c$. The Karman constant was taken as $\kappa = 0.4$, being valid for clear-water and suspension flow (see *Coleman*, 1986, p.1382). Above equation is valid in the inner and outer region of the flow and is independent of the bed roughness, k_s.

By best-fitting the defect law, eq. 2, to the measured velocity profiles of the suspension flow (see Fig. 1) the wake-strength parameter, Π, could be obtained. (Note, that only data in the outer region, $y/\delta > 0.2$, have been used, since data in the inner region are probably of questionable quality). The resulting average $\overline{\Pi}$-values for sand I and II are $\overline{\Pi} \cong 0.50$ and $\overline{\Pi} \cong 0.24$, respectively. In either case they are larger than the clear-water value of $\overline{\Pi} \cong 0.1$ (see *Kironoto* et *Graf*, 1994, p. 336, and *Coleman*, 1981, p. 221). The tendency of the Π-values to increase with the concentration, itself parametrized by a Richardson number (see *Coleman*, 1981, p. 221), is also confirmed (see Fig. 2).

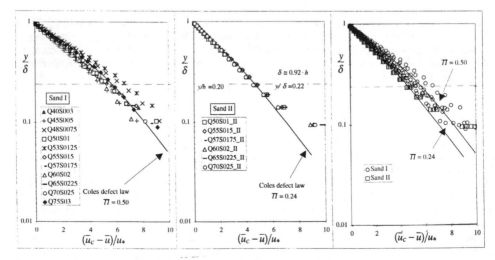

Fig.1: Longitudinal velocity profiles in defect form

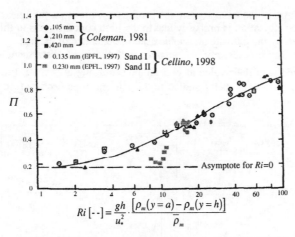

$$Ri\,[\,-\,-\,] = \frac{gh}{u_*^2}\cdot\frac{\left[\rho_m(y=a)-\rho_m(y=h)\right]}{\overline{\rho}_m}$$

Fig.2: Coles' wake strength parameter as a function of Richardson number

4.CONCENTRATION DISTRIBUTION

The vertical concentration profile for a dilute suspension shall be given with the Rouse equation (see *Graf* et *Altinakar*, 1998, p.388), or :

$$\frac{\overline{c}_s}{\overline{c}_{sa}} = \left(\frac{h-y}{y}\cdot\frac{a}{h-a}\right)^{z'} \tag{3}$$

with the Rouse number defined as: $z' = \dfrac{z}{\overline{\beta}} = \dfrac{v_{ss}}{\kappa u_*}\cdot\dfrac{1}{\overline{\beta}}$; the reference concentration, $\overline{c}_{sa}$, is

evaluated at $a = 0.05h$. A depth-averaged value of :

$$\overline{\beta} = \frac{1}{h-a}\int_a^h \beta(y)\cdot dy \tag{4}$$

enters the above equation to allow for the fact that momentum, ε_m, and sediments, ε_s, are not necessarily diffused in the same way; it is given by the ratio :

$$\beta(y) = \frac{\varepsilon_s(y)}{\varepsilon_m(y)} = \frac{\overline{c_s'v'}}{\partial\overline{c}_s/\partial y}\bigg/\frac{\overline{u'v'}}{\partial\overline{u}/\partial y} \tag{5}$$

being the respective diffusion coefficients. v_{ss} is the settling velocity and κ is the Karman constant; u_* is the shear velocity.

Typical dimensionless mean concentration profiles, $\overline{c}_s/\overline{c}_{sa}$, are presented in Fig. 3. Shown are the measured_with the suction method_data, which are compared to the Rouse equation, eq. 3, taking $\overline{\beta}$=1. The Rouse equation, eq. 3, with the assumptions, $\overline{\beta}$=1 or $\overline{\varepsilon}_s(y)\equiv\overline{\varepsilon}_m(y)$, does not very well predict the measured data. It is therefore, that using the existing data set a careful study of the terms in eq. 5 is necessary.

5.FLUX DISTRIBUTION

The data obtained with the APFP instrument rendered an unique possibility to study the influence of the turbulence on the velocity and concentration profiles in suspension flow.

The measured profiles of the Reynolds stress, $\overline{u'v'}$, for the suspension flows were evaluated for both sands. By extrapolating the Reynolds-stress profiles – measured with the APFP instrument – towards the bed, the shear velocity, $u_{*_r}\equiv u_*$, can be obtained. The dimensionless Reynolds-stress profiles, normalized using u_*, are presented in Fig. 4. Besides some scattering, these profiles retain their linear trend, implying that the flow is a uniform one; they are unaffected by the suspended sediments. In the inner region, the profiles deviate from the linear trend falling to zero rather rapidly. This deviation is probably due to the high concentration in this region, but was probably also affected by the parasite ultrasonic echoes, generated at the bed.

The measured sediment-flux profiles, $\overline{c_s'v'}$ — obtained with the APFP instrument — are shown in Fig. 5, being normalized with their value at y=a. Some scattering is present, but globally the profiles for both sands are of a shape, being similar to the concentration profiles (see Fig. 3).

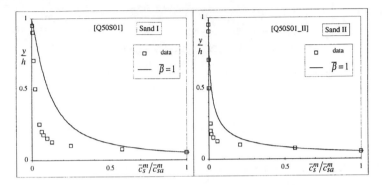

Fig.3: Dimensionless concentration profiles, measured data compared to the Rouse equation with $\beta = 1$

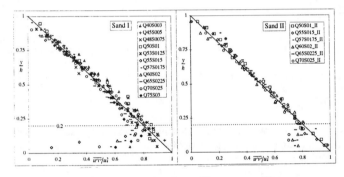

Fig4: Dimensionless Reynolds-stress profiles

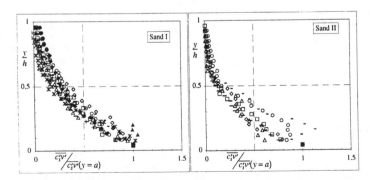

Fig.5: Dimensionless sediment-flux profiles

88

6.DIFFUSION COEFFICIENTS

The sediment-diffusion coefficient, $\varepsilon_s(y)$, as well as the momentum-diffusion coefficient, $\varepsilon_m(y)$, defined with eq. 5, can be evaluated from the measured data set, given with Fig. 3 and Fig. 5 as well as Fig. 1 and Fig. 4, respectively. This is shown in Fig. 6, where the dimensionless form of the experimental – obtained with the APFP instrument – sediment-diffusion coefficient, $\varepsilon_s(y)$, profiles (black points) are compared with the experimental (gray points) and the theoretical (full line) clear-water momentum-diffusion coefficient ,$\varepsilon_m(y)$, profiles. The shapes of the diffusion coefficient profiles of the different runs for both sands are rather similar. However, scattering is large and apparently the concentration and the sand size should still be considered.

Close to the bed, where the concentration is high, the momentum-diffusion coefficients are always larger than the sediment-diffusion coefficients for both sands. In the middle of the flow depth the difference is still very large, but in the upper part of the flow, where the concentration is small, the two diffusion coefficients become rather similar. Near the bed and near the water surface the momentum diffusion coefficient profiles (gray points) are similar to the theoretical (full line) clear-water one, while in the middle of the flow depth the difference is rather large. Despite the scatter, there is sufficient evidence, that the sediment-diffusion coefficients, $\varepsilon_s(y)$, are always smaller than the momentum-diffusion coefficients, $\varepsilon_m(y)$; the latter are almost always smaller than the theoretical value (see *Graf* et *Altinakar*, 1998, p. 387), postulated for clear-water flows.

7.EXPERIMENTAL $\bar{\beta}$-VALUES

The ratio of the sediment- and momentum-diffusion coefficient defines the $\beta(y)$-value, given with eq. 5. Some typical vertical distributions of the β_{APFP}-values – the index recalls that the data are taken with the APFP instrument – are plotted in Fig. 7. The β_{APFP}-values are close to zero at the bed and increase with the distance from the bed, up to $y/h < 0.5$. Above the mid-depth and towards the water surface, the β_{APFP}-values reach their maxima, but a clear tendency is not evident. Also shown are the depth-averaged values, $\overline{\beta_{APFP}}$, which are distinguishable smaller than unity, $\bar{\beta} = 1$, for all our experiments.

Earlier we have seen in Fig. 3, that the Rouse equation, eq. 3, with $\bar{\beta}=1$ is unable to predict well the measured data. In Fig. 8 it is evident that the same equation, eq. 3, with the depth-averaged value, $\bar{\beta}_{APFP}$, obtained from the measurements with the APFP instrument, renders a much better prediction of the vertical concentration distribution. As a conclusion it is rather obvious that the $\bar{\beta}$-values for suspension flow of small particles, should be taken to be $\bar{\beta} < 1$.

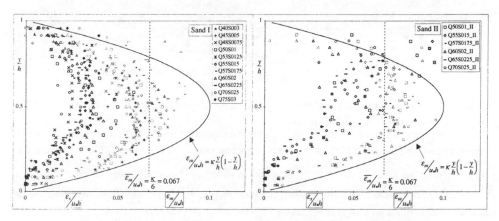

Fig.6: Dimensionless sediment- and momentum-diffusion coefficient profiles

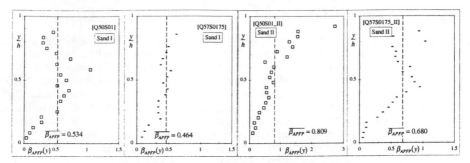

Fig.7: Vertical profiles of the β_{APFP}-values (sand I and sand II)

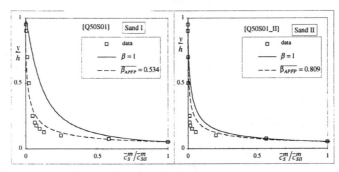

Fig.8: Dimensionless concentration profiles, measured data compared to the Rouse equation with β_{APFP}

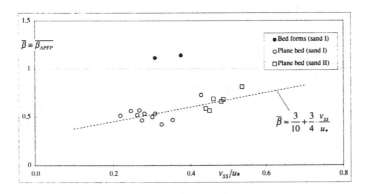

Fig.9: Correlation between the $\overline{\beta_{APFP}}$-values and the scaling parameters

This conclusion is in agreement with the fluid-mechanics arguments presented by *Hinze*, 1975, p. 466 and reviewed by *Graf*, 1984, p. 177.

An attempt shall be made to relate the depth-averaged $\overline{\beta}$-values to a parameter easily obtainable also in field situations. It has been suggested (see *Graf* et *Cellino*, 2000) to scale the $\overline{\beta}$-Values with the parameter v_{ss}/u_*; this is shown in Fig. 9. For both sands a tendency of the $\overline{\beta_{APFP}}$-Values to increase with this scaling parameters, v_{ss}/u_*, is evident. A linear relationship using a regression line was suggested (taking only the plane bed values), such as :

$$\overline{\beta} \equiv \overline{\beta_{APFP}} \approx \frac{3}{10} + \frac{3}{4} \cdot \frac{v_{ss}}{u_*} \qquad \left[R^2 = 0.762 \right] \qquad \text{for } 0.2 < v_{ss}/u_* < 0.6 \qquad (6)$$

Even if agreement is not excellent (see Fig. 9), this relation can be used to obtain an approximated $\overline{\beta}$ _value, if the settling velocity of the suspended sediments, v_{ss}, and the shear velocity of the flow, u_*, are known. For the sake of comparison the $\overline{\beta_{APFP}}$ _values measured over bed forms (see *Cellino* et *Graf*, 2000) are also shown; note that in this case $\overline{\beta_{APFP}} > 1$.

While the literature reports many data on suspension flow over plane bed, only few of those are for flows in capacity condition. After a carefully examination, only those data which satisfy capacity or near-capacity conditions have been evaluated and are plotted in Fig. 10. Beside the data of *Coleman*, 1986, the other ones seem to show the tendency presented in Fig. 9.

In the literature, there has also been reported evidence, that $\overline{\beta} \geq 1$ (see Graf, 1984, p. 177, and Rijn, 1984, p. 1621). Carefully study of this, points to the fact that the bed forms on the mobile bed might be responsible for these large $\overline{\beta}$-values. In a recent investigation (see Cellino et Graf, 2000) we could show, that the $\overline{\beta}$-values in presence of bed forms are always larger then the unity. Such large values are mainly due to the effect of a high turbulence region, generated by the bed-form crest, where the sediment diffusion is considerably enhanced and the momentum diffusion is suppressed, leading to an augmentation of the $\overline{\beta}$-values.

Suspension flows over beds with bed forms have been evaluated from the channel data of *Vanoni* and *Nomicos*, 1960, and *Coleman*, 1970 as well as for the river data of *Nordin* and *Dempster*, 1963 (see Fig. 11). Here is to be observed that the $\overline{\beta}$-values are always large, $\overline{\beta} > 1$, and have a tendency to increase with the parameter, v_{ss}/u_*. This corroborates with the conclusion drawn by *Rijn*, 1984, p. 1621, who suggested the functional relationship, $\overline{\beta} = 1 + 2(v_{ss}/u_*)^2$, also shown in Fig. 11.

The data of the literature and of this study are summarized in Fig. 11. Here it is particularly evident that for the suspension flows over bed without bed forms: $\overline{\beta} < 1$, while for the ones over bed with bed forms: $\overline{\beta} > 1$.

8. CONCLUSIONS

The important result of this study is a recommendation, that the Rouse equation, eq. 3, with an improved $\overline{\beta}$-value – itself to be estimated from Fig. 11 – can be used to establish the dimensionless concentration profile. For beds without bed forms the $\overline{\beta}$-values are $\overline{\beta} < 1$, while for beds with bed forms the $\overline{\beta}$-values are $\overline{\beta} > 1$. This recommendation is confirmed by analyzing many vertical concentration distributions presented in the literature (see Fig. 11).

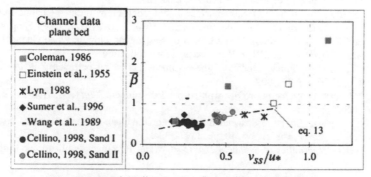

Fig.10: Data of the literature on flows without bed forms

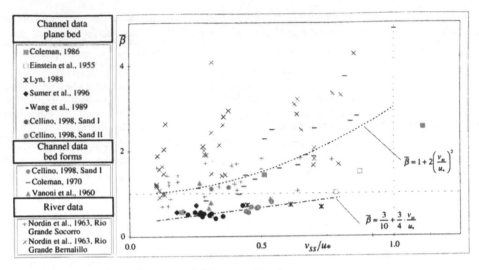

Fig.11: Summary of all data considered

The APFP instrument has been successfully employed to measure the instantaneous velocity and concentration profiles in suspension flows. In all, 11 runs for sand I and 6 for sand II, have been performed. From these measurements, the vertical distribution of the momentum-, $\varepsilon_m(y)$, and the sediment-diffusion, $\varepsilon_s(y)$, coefficient could be obtained. Despite the scatter, there is sufficient evidence, that the sediment-diffusion coefficients, $\varepsilon_s(y)$, are always smaller than the momentum, $\varepsilon_m(y)$, ones. Finally, the $\beta_{APFP}(y)$_values as well as the depth-averaged $\overline{\beta_{APFP}}$_values (see Fig. 7) could be calculated.

ACKNOWLEDGMENTS

This work was made possible by the support of the Swiss National Science Foundation, grant number 20-39495.93. The use of the APFP instrument was greatly facilitated by the help of D. Hurther, U. Lemmin, and C. Shen of our laboratory.

REFERENCES

CELLINO, M. (1998). "Experimental Study of Suspension Flow in Open Channel" *Doctoral dissertation No. 1824*, Ecole Polytechnique Fedeale, Lausanne.

CELLINO, M. et GRAF, W. H. (1999). Suspended sediment-laden Flow in Open Channels under non-Capacity and Capacity Condition." *Am. Soc. Civ. Eng., J. Hydr. Engr.*, vol. 125, N°. 5, pp. 455-462.

CELLINO, M. et GRAF, W. H. (2000). "Experiments on Suspension Flow in Open Channel with Bed Forms." *Int. Assoc. Hydr. Res., J. Hydr. Res.* Accepted for publication.

COLEMAN, N. L. (1970). "Flume Studies of the Sediment Transfer Coefficient." *Water Resources Research*, vol. 6, N.° 3, pp. 801-809.

COLEMAN, N. L. (1981)."Velocity profiles with suspended sediment." *Int. Ass. Hydr. Res., J. Hydr. Res.*, vol. 19, N° 3, pp. 211-229.

COLEMAN, N. L. (1986). "Effects of Suspended Sediment on the Open-Channel Velocity Distribution." *Water Resources Research*, vol. 22, N°. 10, pp. 1377-1384.

GRAF, W. H. (1984). *Hydraulics of Sediment Transport*, Water Resources Publications, Littleton, CO, USA.

GRAF, W. H. et ALTINAKAR, M. S. (1998). *Fluvial Hydraulics*, J. Wiley & Sons, Chichester, GB and New York, USA.

GRAF, W. H. et CELLINO, M. (2000). "Suspension Flows in Open Channels; Experimental Study." Submitted for publication.

HINZE, J. O. (1975). *Turbulence*, McGraw-Hill, New York, USA.

KIRONOTO, B. A. and GRAF, W. H. (1994). "Turbulence characteristics in rough uniform open-channel flow." *Proc. Instn Civ. Engrs, Wat., Marit. & Energy*, vol. 106, pp. 333-344.

NORDIN, C. F. and DEMPSTER, G. R. (1963). "Vertical Distribution of Velocity and Suspended Sediment Middle Rio Grande New Mexico.", *U.S. Geol. Survey; Professional Paper 462-B*, Washington, DC, USA.

RIJN, L. C. van (1984). "Sediment transport, part II: Suspended Load Transport." *Am. Soc. Civ. Eng., J. Hydr. Engr.*, vol. 110, N°. 11, pp. 1613-1641.

SHEN, W. and LEMMIN, U. (1996). "Ultrasonic measurements of suspended sediments. A concentration profiling system with attenuation compensation" *Meas. Sci. Tech.*, vol. 7, p. 1191-1194.

VANONI, V. A. and NOMICOS, G. N. (1960). "Resistance Properties of Sediment-Laden Streams.", *Trans., Am. Soc. Civ. Eng.*, vol. 125/I, pp. 1140-1175.

Stochastic methods in open channel hydraulics

Stochastic Hydraulics 2000, Wang & Hu (eds) © 2000 Taylor & Francis, ISBN 90 5809 166 X

Modeling open channels using moments of the Boltzmann equation

J.Q. Deng & M.S. Ghidaoui
Department of Civil Engineering, Hong Kong University of Science and Technology, Kowloon, China

Abstract: The two dimensional statistical mechanics equation of Boltzmann for 1 and 2-D shallow waters is formulated. It is shown that the 2-D unsteady depth average shallow water equations are derivable from the zero and first moments of the Boltzmann equation. The relationship between the statistical equation of Boltzmann and the shallow water equations is exploited in the formulation of an accurate, efficient and robust algorithms for (i) a 2-D depth averaged shallow water equations, where turbulent stresses are neglected, and (ii) a 2-D Large Eddy Simulation (LES) for open channel flows. The results obtained by the model which ignores turbulence stresses compare well with measured data from dam break laboratory experiments. In addition, the results of the LES model compare well with measured data from compound channel laboratory experiments.

Key words: Shallow water, Boltzmann equation, Unsteady flows

1. STATISTICAL MECHANICS EQUATION OF BOLTZMANN FOR 1 AND 2-D SHALLOW WATER FLOWS

According to the molecular theory of matter, an infinitesimal volume of a fluid contains an enormous number of molecules (or particles) which are moving randomly with different speeds in all directions. That is, the velocity vector $\vec{c} = (c_x,\ c_y,\ c_z)$ of a given particle is a random variable. Let $p = p(\vec{c}, \vec{x}, t)$ be the probability density function in $\vec{c}$, where $\vec{x} = (x, y, z) =$ position and $t =$ time. Therefore, the probability that a given particle has a speed between $\vec{c}$ and $\vec{c} + d\vec{c}$ is $p\,dc_x dc_y dc_z$. The Boltzmann theory provides the equation which governs the change of this probability with $(\vec{c}, \vec{x}, t)$.

This section provides a heuristic derivation of the Boltzmann equation for the case of a two dimensional unsteady depth-averaged shallow water flows. For ease of presentation, the Boltzmann equation in 1-D unsteady shallow water flows is first derived. In this case, the independent variables are (x, t, c). Note, while the probability density function p depends on (x, t, c), the water depth in the channel h only depends on (x, t). The product of fluid density and the probability density function, ρp, defines the density distribution of particles in (x, t, c). Referring to the control volume (C.V.) in figure 1, the instantaneous mass per unit width of water particles, located between x and $x + \Delta x$ and whose speeds are between c and $c + \Delta c$, is $\rho h p \Delta x \Delta c$. Applying the principle of mass conservation to the control volume shown in Figure 1 gives:

<table>
<tr><td>The change of the total mass of particles inside the C.V. during Δt</td><td>=</td><td>Net flux of mass from the C.V. along x during Δt due to convection</td></tr>
</table>

<table>
<tr><td>+</td><td>Net flux of mass from the C.V. along c during Δt due to external forces</td></tr>
</table>

<table>
<tr><td>+</td><td>The change of mass within the C.V. during Δt due to particle interaction</td></tr>
</table>

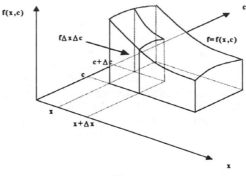

Fig. 1

To obtain the Boltzmann equation, all the terms in the above mass balance equation needs to be quantified. This is accomplished below.

The change of the total mass inside the C.V. during Δt is:

$$[\rho hp\Delta c\Delta x]_{t+\Delta t} - [\rho hp\Delta c\Delta x]_t = ([hp]_{t+\Delta t} - [hp]_t)\rho\Delta c\Delta x \tag{1}$$

Note, the incompressibility assumption has been invoked in equation (1).

The mass of influx per unit width at x during Δt of all the particles (i.e., regardless of their speed) is $[\rho hc\Delta t]_x$. However, of interest here is the mass flux of the particles with speeds between c and $c+\Delta c$. This mass flux is determined by the multiplying the total mass flux by the probability of the particles' speed being between c and $c+\Delta c$ which result in the following: $[\rho hc\Delta tp\Delta c]_x$. Similarly, the mass out-flux at $x+\Delta x$ is $[\rho hc\Delta tp\Delta c]_{x+\Delta x}$. Therefore, the net mass flux from the C.V. along x during Δt of those particles whose speeds are in the range $[c, c+\Delta c]$ is:

$$[\rho hpc\Delta c\Delta t]_x - [\rho hpc\Delta c\Delta t]_{x+\Delta x} = ([hpc]_x - [hpc]_{x+\Delta x})\rho\Delta c\Delta t \tag{2}$$

Now, the net flux of expected mass from the C.V. due to external force need to be quantified. According to Newton Second Law, external forces cause particles to accelerate or decelerate, depending on the direction of the external forces with respect to the motion of particles. Therefore, external forces may result in some particles whose initial speeds at time t are outside the range $[c, c+\Delta c]$ to acquire speeds within the range $[c, c+\Delta c]$ during the time interval $[t, t+\Delta t]$. Conversely, external forces may result in some particles whose initial speeds at time t are within the range $[c, c+\Delta c]$ to acquire speeds outside the range $[c, c+\Delta c]$ during the time interval $[t, t+\Delta t]$. Let F_x denote the net external force per unit mass (i.e., acceleration or deceleration) along x. Clearly, the rate at which particles whose initial speeds is below c will acquire a speed that exceeds c is proportional to the particle's acceleration during Δt. In fact, the mass influx across the control surface $C = c$ (see Figure 1) of the particles located between x and $x+\Delta x$ during Δt due to F_x is $[\rho h F_x p \Delta t \Delta x]_c$. Similarly, the mass out-flux across the control surface $C = c + \Delta c$ of the particles located between x and $x+\Delta x$ during Δt due to F_x is $[\rho h F_x p \Delta t \Delta x]_{c+\Delta c}$. Therefore, the net mass flux from the C.V. due to external forces is:

$$[\rho h F_x p \Delta t \Delta x]_c - [\rho h F_x p \Delta t \Delta x]_{c+\Delta c} = \left([h F_x p]_c - [h F_x p]_{c+\Delta c}\right)\rho \Delta t \Delta x \tag{3}$$

Particles in a liquid are closely packed and interact mainly through their molecular bonding. In the other hand, particles in a gas usually lie outside each others force field and thus mainly interact through collision. Particle interaction, whether by collision or through molecular bonding, can changes the velocity vector of each of the interacting particles. The changes in velocity vectors induced by particle interaction will result in a decrease in the mass of particles with speeds in the range $[c, c+\Delta c]$ if particles with speeds that belong to the range $[c, c+\Delta c]$ before the interaction acquire speeds that are outside the range $[c, c+\Delta c]$ after the interaction. Conversely, the changes in velocity vectors induced by particle interaction will result in an increase in the mass of particles with speeds in the range $[c, c+\Delta c]$ if particles with speeds that are outside the range $[c, c+\Delta c]$ before the interaction acquire speeds that are within the range $[c, c+\Delta c]$ after the interaction. Let $Q(p,p)\rho$ define the rate of increase of particles' density in the speed range $[c, c+\Delta c]$ due to molecular interaction. Therefore, the net mass flux into the C.V. due to interactions among particles is:

$$Q(p,p)\rho h \Delta x \Delta t \Delta c = Q(f,f)\rho \Delta x \Delta t \Delta c \tag{4}$$

where $f=hp=$ non-equilibrium distribution function. During particle interaction, there is a net transfer of energy and momentum from particles with high momentum and energy to particles with low momentum and energy. Therefore, molecular interaction reduces the gradients of momentum and energy in the flow fluid. If there are no external inputs of energy and momentum, molecular interaction will eventually result in a global equilibrium where the gradients of momentum and energy are zero throughout the flow field. However, if the flow is continually subjected to external inputs of energy and momentum, molecular interaction reduces but cannot totally cancel the gradients of energy and momentum. The instantaneous value of the local gradients of flow properties depend on the relative magnitude between the rate at which molecular interaction reduces these gradients and the rate at which external inputs replenish these gradients.

The fact that molecular interaction reduces gradients of momentum and energy implies a movement of flow properties towards equilibrium and an increase in the flow's thermodynamic entropy. Bhatnagar-Gross-Krook (BGK) proposed that the local rate at which the flow moves towards equilibrium at time t is proportional to the difference between the local non-equilibrium probability distribution of particles speeds, $p(t)$, and the local equilibrium probability of distribution, $p_e(t)$. Mathematically, the BGK model is as follows:

$$Q(f,f) = \frac{q - f}{\tau} \tag{5}$$

where $q(t)=$local equilibrium distribution whose form is derived later in the paper and $\tau=$interaction (relaxation) time which is a measure of the time scale of molecular interaction. Therefore, $1/\tau$ =frequency of interaction between particles. Combining equations (4) and (5) and inserting the result along with equations (3), (4) and (5) into the principal of mass conservation and dividing by $\rho \Delta x \Delta t \Delta c$ gives:

$$\frac{[hp]_{t+\Delta t} - [hp]_t}{\Delta t} = \frac{[hpc]_x - [hpc]_{x+\Delta x}}{\Delta x} + \frac{[hF_x p]_c - [hF_x p]_{c+\Delta c}}{\Delta c} + \frac{q - f}{\tau} \tag{6}$$

Taking the limit as $(\Delta x, \Delta t, \Delta c) \to (0,0,0)$ gives:

$$\frac{\partial f}{\partial t} = -c\frac{\partial f}{\partial x} - \frac{\partial F_x f}{\partial c} + \frac{q - f}{\tau} \tag{7}$$

Equation (7) is the BGK Boltzmann equation for 1-D unsteady depth averaged open channel flows. The spatial symmetry in the laws of physics and the fact that mass is a scalar makes the extension of equation (7) to 2-D unsteady depth averaged open channel flows straightforward and results the following:

$$\frac{\partial f}{\partial t} + c_x \frac{\partial f}{\partial x} + c_y \frac{\partial f}{\partial y} + \frac{\partial F_x f}{\partial c_x} + \frac{\partial F_y f}{\partial c_y} = \frac{q - f}{\tau} \tag{8}$$

If gravitation and friction are the only external forces, then $F_x = g(S_{0x} - S_{fx})$ and $F_y = g(S_{0y} - S_{fy})$, where g = gravitational acceleration; S_{0x}, S_{0y} = channel bed slope components in x and y directions, respectively and S_{fx}, S_{fy} = frictional slopes in x and y directions, respectively. Since these forces are independent of particle speeds, equation (8) becomes as follows:

$$\frac{\partial f}{\partial t} + c_x \frac{\partial f}{\partial x} + c_y \frac{\partial f}{\partial y} + g(S_{0x} - S_{fx})\frac{\partial f}{\partial c_x} + g(S_{0y} - S_{fy})\frac{\partial f}{\partial c_y} = \frac{q - f}{\tau} \tag{9}$$

Of course, other forces such wind shear or Coriolis forces can be included in equation (8). In addition, forces that may depend on particle speeds can also be easily handled.

2. RELATIONSHIPS BETWEEN THE MACROSCOPIC AND THE MICROSCOPIC PROPERTIES IN 2-D SHALLOW WATERS

In order to link equation (9) with the shallow water equations, the relationships between the microscopic variables of and the macroscopic variables are derived in this section. For example, the relationship between the water depth and p is as follows:

$$\int_{-\infty}^{\infty} \int_{-\infty}^{\infty} h(x,y,t)p(c_x,c_y,x,y,t)dc_x dc_y = \int_{-\infty}^{\infty} \int_{-\infty}^{\infty} f(c_x,c_y,x,y,t)dc_x dc_y = h(x,y,t) = h \tag{10}$$

Realizing that the macroscopic flowrate per unit width along x at (x,y,t), hu, is simply the mean value of the flowrate of particles along x at (x,y,t), leads to:

$$\int_{-\infty}^{\infty} \int_{-\infty}^{\infty} c_x f(c_x,c_y,x,y,t)dc_x dc_y = h(x,y,t)u(x,y,t) = hu \tag{11}$$

where u = x-component of the depth averaged velocity in the channel. Similarly, the macroscopic flowrate per unit width along y at (x,y,t), hv, is simply the mean value of the flowrate of particles along y at (x,y,t), leads to::

$$\int_{-\infty}^{\infty} \int_{-\infty}^{\infty} c_y f(c_x,c_y,x,y,t)dc_x dc_y = h(x,y,t)v(x,y,t) = hv \tag{12}$$

where v = y-component of the depth averaged velocity in the channel. Particle velocity can be decomposed as follows: $c_x = u + c_x'$ and $c_y = v + c_y'$, where c_x' and c_y' are the velocity fluctuations of particles with respect to the macroscopic flow velocity in the channel u and v, respectively. Therefore, the stress tensor, Γ_{ij}, is as follows (Vincenti 1965):

$$h(x,y,t)\Gamma_{ij} = \int_{-\infty}^{\infty} \int_{-\infty}^{\infty} c_x' c_y' f(c_x,c_y,x,y,t)dc_x dc_y \tag{13}$$

where $i = x, y$ and $j = x, y$. When $i = j$, equation (13) gives the normal stresses (i.e., pressure). In open channel, when the hydrostatic pressure distribution assumption is invoked, the average normal stress in a cross section becomes: $\rho g h / 2$. Using equation (13) and the hydrostatic pressure distribution assumption, the mean momentum flux of all particles along x at (x,y,t) is:

100

$$\int_{-\infty}^{\infty}\int_{-\infty}^{\infty}c_x^{\,2}f(c_x,c_y,x,y,t)dc_xdc_y = h(x,y,t)u^2(x,y,t)+gh^2(x,y,t)/2 \tag{14}$$

Similarly, the mean momentum flux of all particles along y at *(x,y,t) is:*

$$\int_{-\infty}^{\infty}\int_{-\infty}^{\infty}c_y^{\,2}f(c_x,c_y,x,y,t)dc_xdc_y = h(x,y,t)v^2(x,y,t)+gh^2(x,y,t)/2 \tag{15}$$

When $i \ne j$, equation (13) gives the shear stresses. That is, shear stresses result from the correlation between of c_x' and c_y'. These shear stresses vanish when the flow is in local equilibrium. To explain, when the flow is in local equilibrium, equation (5) states that the rate at which particle interaction changes the distribution of particles' speeds is identically zero. That is, for every interaction that, say, increases the speed of particles in a certain range R, there is, at the same time, an inverse interaction that decreases the speed of particles within R in such way that complete cancellation is achieved. Therefore, for every interaction which cause a shear of $c_x'c_y'$ at *(x,y)* at time t, there is another interaction which cause a shear of $-c_x'c_y'$ at *(x,y)* at time t. As a result, when the flow is in local equilibrium, of c_x' and c_y' are uncorrelated and $\Gamma_{xy}=0$. Thus,

$$\int_{-\infty}^{\infty}\int_{-\infty}^{\infty}c_xc_yq(c_x,c_y,x,y,t)dc_xdc_y = h(x,y,t)u(x,y,t)v(x,y,t) \tag{16}$$

The entropy is given by (Ghidaoui et al. 1999, Deng 2000, Jou et al., Vincenti 1965)

$$S(x,y,t) = -\int_{-\infty}^{\infty}\int_{-\infty}^{\infty}f\ln(Af)dc_xdc_y \tag{17}$$

where A is a normalizing coefficient which will be determined subsequently.

3. LOCAL EQUILIBRIUM DISTRIBUTION

According to the H-theorem (Boltzmann 1873), the entropy function given by equation (17) is maximum when the flow is in a state of local equilibrium. Using Lagrange multipliers, the maximization of S must subject to the constraints given by equations (10), (11), (13), (14) and (14) gives (Ghidaoui et al. 1999, Deng 2000):

$$q(x,y,c_x,c_y,t) = \frac{1}{\pi g}e^{-\frac{(c_x-u)^2+(c_y-v)^2}{gh}} = \left(\frac{1}{\sqrt{\pi gh}}e^{-\frac{(c_x-u)^2}{gh}}\right)\left(\frac{1}{\sqrt{\pi gh}}e^{-\frac{(c_y-v)^2}{gh}}\right)h \tag{18}$$

where the normalizing coefficient is found to be $A = e\pi g$ (see Ghidaoui et al. 1999, Deng 2000). Note that $p_e = q/h$ is a two-dimensional (bivariate) Gaussian distribution with mean u and v and standard deviation $(gh/2)^{\frac{1}{2}}$. The fact that the equilibrium distribution is Gaussian can be understood using the central limit theorem. To explain, at local equilibrium, the flow in a channel is a result of a random motion of a large number of particles, where of c_x' and c_y' and, thus, $c_x = u + c_x'$ and $c_y = v + c_y'$ are uncorrelated. Therefore, the equilibrium flow meets the condition of the central limit theorem (see Benjamin and Cornell 1970; Ang and Tang 1975) and thus it is not surprising the distribution is in this case Gaussian.

4. OPEN CHANNEL BOLTZMANN EQUATION AND ENTROPY

Multiplying the BGK Boltzmann equation (9) by $ln(A f)$ and integrating gives expression for the rate of entropy production, σ (The detailed derivation can be found in Ghidaoui et al. 1999 and Deng 2000):

$$\sigma = -\frac{1}{\tau} \int_{-\infty}^{\infty}\int_{-\infty}^{\infty} (q-f)\ln\frac{q}{f} dc_x dc_y \qquad (19)$$

It is clear from equation (19) that

$$\sigma \geq 0 \quad \text{for all } f \text{ and } \sigma = 0 \text{ only if } q = f \qquad (20)$$

This proves that the BGK Boltzmann equation always satisfies the entropy condition.

5. COMPATIBILITY CONDITIONS

The classical physical laws state that mass, momentum and total energy of particles are invariant during particle interaction (Jou et al. 1996). As a consequence,

$$\int_{-\infty}^{\infty}\int_{-\infty}^{\infty} c_x c_y \frac{q(c_x,c_y,x,y,t)-f(c_x,c_y,x,y,t)}{\tau}\begin{pmatrix}1\\ \vec{c}\\ c^2/2\end{pmatrix} dc_x dc_y = 0 \qquad (21)$$

There are three equations imbedded in (21): the first is the mass invariant condition, the second is the momentum invariant condition and the third is the total energy invariant condition. These three conditions are often referred to as compatibility conditions. The energy invariant condition along with the entropy equation (20), state that, during particle interaction, while the total energy is conserved, the conversion of mechanical energy into heat is irreversible.

Thus far, (i) the Boltzmann equation for 2-D unsteady, depth averaged shallow water equation has been formulated, (ii) the form of the equilibrium distribution has been derived, (iii) the BGK interaction term has been provided, (iv) the relationships between the microscopic and macroscopic variables in 2-D channel flow has been determined, (v) the compatibility equations has been given and (vi) the fact that the Boltzmann model satisfies the second law of thermodynamics has been proven. However, if the Boltzmann equation is to be meaningful, we need to prove that equation (9) is consistent with the shallow water equations. This proof is provided below.

6. MOMENTS OF THE STATISTICAL BOLTZMANN EQUATION UNDER LOCAL EQUILIBRIUM STATE: CLASSICAL SHALLOW WATER EQUATIONS

This section shows that under the local equilibrium state (i.e., $q = f$), the zeroth and the first moments of the BGK Boltzmann equation give the classical shallow water equations.

6.1 Zeroth Moment of BGK Boltzmann Equation under Local Equilibrium State: Continuity Equation

Taking the zero-th moment of equation (9) and setting $f=q$ gives

$$\int_{-\infty}^{\infty}\int_{-\infty}^{\infty}\left\{\frac{\partial q}{\partial t}+c_x\frac{\partial q}{\partial x}+c_y\frac{\partial q}{\partial y}+\frac{\partial F_x q}{\partial c_x}+\frac{\partial F_y q}{\partial c_y}\right\}dc_x dc_y = 0 \qquad (22)$$

Invoking equations (10), (11) and (12) and realizing that the fourth and fifth terms in equation (22) are zero because the equilibrium (Guassian) distribution and all its derivatives are zero $c_x = \pm \infty$ and $c_y = \pm \infty$ gives:

$$\frac{\partial h}{\partial t} + \frac{\partial (uh)}{\partial x} + \frac{\partial (vh)}{\partial y} = 0 \tag{23}$$

which is the continuity equation for 2-D unsteady depth averaged shallow water flows.

6.2 First Moments of BGK Boltzmann Equation under Local Equilibrium State: Momentum Equations

To derive the macroscopic x-momentum equation under local equilibrium state, take the first moment of equation (9) (i.e., multiply this equation by c_x and integrate the result) gives

$$\int_{-\infty}^{\infty} \int_{-\infty}^{\infty} c_x \left\{ \frac{\partial q}{\partial t} + c_x \frac{\partial q}{\partial x} + c_y \frac{\partial q}{\partial y} + \frac{\partial F_x q}{\partial c_x} + \frac{\partial F_y q}{\partial c_y} \right\} dc_x dc_y = 0 \tag{24}$$

Using relationships (11) through (16) and realizing that the fourth and fifth terms in (24) are zero because the equilibrium (Guassian) distribution and all its derivatives are zero $c_x = \pm \infty$ and $c_y = \pm \infty$ gives:

$$\frac{\partial (uh)}{\partial t} + \frac{\partial (u^2 h + \frac{gh^2}{2})}{\partial x} + \frac{\partial (uvh)}{\partial y} + gh(S_{fx} - S_{0x}) = 0 \tag{25}$$

Similarly, the first moment of local equilibrium Boltzmann equation with respect to c_y gives the following macroscopic y-momentum equation:

$$\frac{\partial (vh)}{\partial t} + \frac{\partial (uvh)}{\partial x} + \frac{\partial (v^2 h + \frac{gh^2}{2})}{\partial y} + gh(S_{fy} - S_{0y}) = 0 \tag{26}$$

Discussion: Equations (23), (25) and (26) are exactly the 2-D unsteady shallow water equations. These equations were obtained as moments of the Boltzmann equation under the assumption that the flow is in local equilibrium. With the local equilibrium assumption, equation (20) is identically zero. Therefore, implicit in the shallow water equations is that assumption that the entropy production is zero. The assumption of local equilibrium is valid when local velocity and depth gradient are not significant. Across jumps, bores and expansion waves, flow gradients are large, entropy production is far from zero and the flow departs significantly from local equilibrium. The zero-entropy assumption in the shallow water equations has important implications in numerical modeling. To explain, Jha et al. (1995) and Alcrudo et al. (1993) found that models based on the classical shallow water equations can produce solutions that satisfy mass and momentum but are not physical (i.e., do not satisfy entropy and energy). This is not surprising since the mass and momentum (i.e., shallow water equations) are satisfied regardless of whether flow is from subcritical to supercritical or vice versa. In fact, in regions where there is flow transition from subcritical to supercritical, the shallow water equation produce non-unique solutions. This non-uniqueness is due to the fact that the classical shallow water equation are based on the assumption of zero-entropy production. To remove the non-uniqueness, numerical modelers add artificial terms which are often referred to as entropy fixes (Lax 1972, LeVeque 1992, Alcrudo et al. 1993, Jha et al. 1995). However, these entropy fixes luck physical basis and thus may not always be successful. For example, Jha et al. (1995) found that the entropy fix suggested by Alcrudo et al. (1993) could not avoid the problem of expansion waves. In this regard, Jha et al. [1995], pg. 881) wrote: *"These vertical drops in the water-surface profile are due to violation of the entropy-inequality condition, which is obviously not remedied by using a constant value of Δ ..as suggested by Alcrudo et al. (1993)".* Note, in dam break computation, the expansion shock usually manifests itself as a vertical drop in the water-surface profile at the location of the dam.

The non-equilibrium statistical Boltzmann equation satisfies the second law of thermodynamics by ensuring that the entropy production is always positive (see equation 20). Therefore, building a model based on the non-equilibrium Boltzmann equation removes the need for non-physical entropy fixes. However, prior to building an open channel model based on the non-equilbrium distribution, it is essential to check whether or the moments of the BGK Boltzmann equations are consistent with the macroscopic mass and momentum equations.

7. MOMENTS OF THE STATISTICAL BOLTZMANN EQUATION UNDER NON-LOCAL EQUILIBRIUM STATE : SHALLOW WATER EQUATIONS

Using the Chapman-Enskog expansion and carrying the zero and first moment of the BGK Boltzmann equation leads to (Ghidaoui et al. 1999, Deng 2000):

$$\frac{\partial h}{\partial t} + \frac{\partial (uh)}{\partial x} + \frac{\partial (vh)}{\partial y} = 0 \tag{27}$$

$$\frac{\partial (uh)}{\partial t} + \frac{\partial (u^2 h + \frac{gh^2}{2})}{\partial x} + \frac{\partial (uvh)}{\partial y} + gh(S_{fx} - S_{0x})$$

$$= \frac{\partial}{\partial x}\left[\tau \frac{gh^2}{2}\left(\frac{\partial u}{\partial x} - \frac{\partial v}{\partial y} \right) \right] + \frac{\partial}{\partial y}\left[\tau \frac{gh^2}{2}\left(\frac{\partial u}{\partial y} + \frac{\partial v}{\partial x} \right) \right] \tag{28}$$

$$\frac{\partial (vh)}{\partial t} + \frac{\partial (v^2 h + \frac{gh^2}{2})}{\partial y} + \frac{\partial (uvh)}{\partial x} + gh(S_{fy} - S_{0y})$$

$$= \frac{\partial}{\partial x}\left[\tau \frac{gh^2}{2}\left(\frac{\partial u}{\partial y} + \frac{\partial v}{\partial x} \right) \right] + \frac{\partial}{\partial y}\left[\tau \frac{gh^2}{2}\left(\frac{\partial v}{\partial y} - \frac{\partial u}{\partial x} \right) \right] \tag{29}$$

Setting $\tau = v / gh$, the right-hand-side terms in (28) and (29) become the viscous terms for a Newtonian fluid. On the other hand, as $\tau \to 0$, the right-hand side terms approach zero and the BGK Boltzmann equations collapse to the shallow water equations. Therefore, the solution of the classical shallow water equations can be obtained as the limit of equations (27,28,29) as $\tau \to 0$, the solution of the viscous shallow water equations can be obtained when $\tau = v / gh$ and the solution of depth averaged turbulent flows in shallow waters can be obtained by relating τ to eddy viscosity. , equations (27,28,29) become the depth averaged turbulent model for shallow waters. This limit of equations (27,28,29) as $\tau \to 0$ produces solutions that are unique and physically realizable (satisfies entropy) (see Lax 1972) and thus eliminates the need for entropy fixes (Ghidaoui et al. 1999, Deng 2000).

8. BOLTZMANN BASED MODELS FOR OPEN CHANNELS: BACKGROUND

The above analysis shows that the moments of the differential form of the BGK Boltzmann equation (9) produces the differential equations of shallow waters (27,28,29). Therefore, if a stable and consistent discretization is applied to equation (9), the moments of the difference form of the BGK Boltzmann equation (9) produce the difference equations of shallow waters (27,28,29) . The Boltzmann scheme differs from classical schemes in the following respects. The classical approaches entail the discretization of a system of non-linear partial differential equations of shallow is required. However, the Boltzmann approach entail the descretization of a single linear, first order partial differential equation (i.e., equation 9). The algorithm for shallow water equations is obtained by taking moments of the difference equation of Boltzmann. The moments of the difference equation of Boltzmann are performed analytically.

The Boltzmann scheme provides a unified numerical modeling approach. To explain, the generic form of BGK Boltzmann equation used to recover, say, the 2-D Navier-Stokes equation is identical to that used to recover the shallow water equations or the 2-D mass transport equation. The fact that problems which appear different at the macroscopic level are, in fact, similar at the mesoscopic level is best illustrated by the following quote from Feynmann (see Rothmann and Zaleski 1997, pg.4): ``We have noticed in nature that the behavior of fluids depends very little on the nature of the individual particles in that fluid. For example, the flow of sand is very similar to the flow of water or the flow of ball bearings." Given that the generic form of the BGK Boltzmann equation from which, say, the 2-D Navier-Stokes equation is recovered is the same as that from which the shallow water or 2-D mass transport are recovered, to name a few only, a BGK difference equation constructed for one application can be adapted for other applications. In fact, the difference form of the BGK Boltzmann equation used in the formulation of a numerical scheme for shock waves in gas dynamics (Xu et al. 1995 and 1996) is identical to that used in the formulation of numerical scheme in shallow waters (Ghidaoui et al. 1999, Deng 2000) and mass transport (Deng 2000).

Examples where Boltzmann based models have been applied include shock waves in compressible flows (e.g., Reitz 1981, Chu 1965, Xu et al. 1995 and 1996), multicomponent and multiphase flows (e.g., Gunstenson et al. 1991, Xu et al. 1997a, He et al. 1998), flows in complex geometries (e.g., Rothman 1988, Chen and Doolen 1998), turbulent flows (e.g., Chen et al. 1992, Martinez et al. 1994), low Mach number flows (e.g., Su et al. 1998), heat transfer and reaction diffusion flows (e.g., Qian, 1993, Xu, 1997b), and open channel flows (e.g., Deng and Ghidaoui 1998 and Ghidaoui et al. 1999). These applications revealed a number of advantages of Boltzmann based schemes. For instance, numerical models based on the collisional Boltzmann theory are found to satisfy the entropy condition; thus, precluding the emergence of physically non-realizable solutions (Xu et al. 1995; Ghidaoui et al. 1999). In addition, Boltzmann based schemes have been noted for the ease by which they can be extended to multi-dimensional cases and for their local characters which makes them ideal for implementation in parallel computers (Chen and Doolen 1998; Kumar et al. 1999). Furthermore, Boltzmann based techniques were found to be well suited for problems with complex geometry and boundary conditions (Reitz 1981; Frisch et al. 1986; Chen and Doolen 1998; Inamuru et al. 1999). Furthermore, using a Boltzmann based scheme, Su et al. (1998) showed that incompressible flow solutions can be obtained in the limit as Mach number tends to zero; thus, avoiding the tedious and difficult solution of the Poisson equation for the pressure field which arises when one solves the incompressible flow equations. The fact that the Boltzmann equation has a simple form and that this equation is kinetic in nature makes the incorporation of additional physics straightforward (Chen and Doolen 1998; He et al. 1998). The diffusion and viscous terms that appear as 2nd derivative terms in macroscopic modeling are represented by a simple algebraic difference term in mesoscopic modeling. This eliminates the need for separate treatment of the advection and diffusion terms.

9. BOLTZMANN BASED MODELS FOR OPEN CHANNELS: APPLICATIONS

2-D Dam Break: Figures 2 through 5 plots the results of the BGK Boltzmann model together with measured data from two-dimensional dam-break laboratory experiments carried by Bellos et al. (1991) and Soulis (1992). Figure 2: S_{0x}=0.01; n=0.012; at the dam. Figure 3: S_{0x}=0.01; n=0.012; 5 m downstream of the dam. Figure 4: S_{0x}=0.01; n=0.012; 8.5 m upstream of the dam.

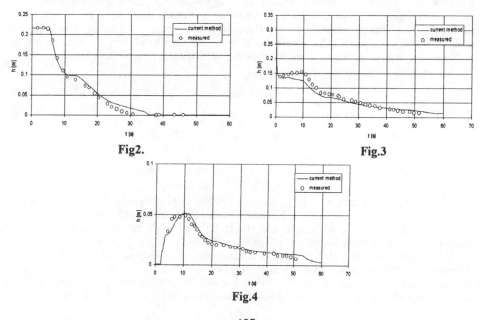

Fig2.   Fig.3

Fig.4

105

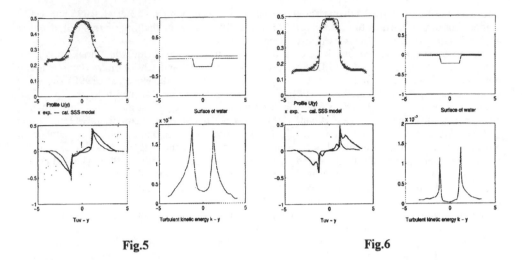

Fig.5   **Fig.6**

Compound Channel Flows: Figures 5 and 6 plots the results of a Boltzmann based non-eddy viscosity sub-grid scale model LES model for open channels along with measured data from Shiono and Knight (1991). Figure 5 is for $D_r=0.242$ and figure 5 is for $D_r=0.0111$, where D_r = relative depth in the channel of the compound channel defined as the ratio of the water depth in the flood plain to that in the main channel. Each of the figures display the transverse velocity profile, the channel shape, the transverse turbulent stresses and the instantaneous production of turbulent kinetic energy.

10.CONCLUSION

The two dimensional statistical mechanics equation of Boltzmann for 1 and 2-D shallow waters is formulated. It is shown that the 2-D unsteady depth average shallow water equations are derivable from the zero and first moments of the Boltzmann equation. The relationship between the statistical equation of Boltzmann and the shallow water equations is exploited in the formulation of an accurate, efficient and robust algorithms for (i) a 2-D depth averaged shallow water equations, where turbulent stresses are neglected, and (ii) a 2-D Large Eddy Simulation (LES) for open channel flows. The results obtained by the model which ignores turbulence stresses compare well with measured data from dam break laboratory experiments. In addition, the results of the LES model compare well with measured data from compound channel laboratory experiments.

ACKNOWLEDGEMENT

The financial support by the Research Grant Council of Hong Kong is gratefully acknowledged.

REFERENCES

Alcrudo, F. and Garcia-Navarro, P. (1993). "A high-resolution Godunov-type scheme in finite volumes for the 2D shallow water equations". Int. J. Numer. Methods in Fluids, 16, 489-505.

Alcrudo, F., Garcia-Navarro, P. and Saviron, J.M. (1992). "Flux difference splitting for 1D open channel flow equations". Int. J. Numer. Methods in Fluids, 14, 1009-1018.

Bellos, C.V., Soulis, J.V. and Sakkas, J.G. (1991). "Computation of two-dimensional dam-break-induced flows". Adv. Water Resources, 14, 31-41.

Bhatnagar, P.L., Gross, E.P. and Krook, M. (1954). A model for collision processes in gases, I. Small amplitude processes in charged and neutral one-component system, Phys. Rev., 94, 511-525.

Cercignani, C. (1988). The Boltzmann equation and its applications. Springer-Verlag. New York.

Chapman, S. and Cowling, T.G. (1990). The mathematical theory of non-uniform gases. Cambridge University Press. New York.

Chen, S. and Doolen, G.D. (1998). "Lattice Boltzmann Method for Fluid Flows." Ann. Rev. Fluid Mech. 30, 329-364.

Chen, H., Chen S., and Matthaeus W.H. (1992). "Recovery of the Navier-Stokes equations using a lattice-gas Boltzmann Method." Phys. Rev. A. 45, R5339-42.

Chu, C.K. (1965). Phys. Fluids, 8, 12.

Deng J.Q. and Ghidaoui M.S. (1999). Modeling Open-Channel Flow and Contaminant Transport Based on kinetic Theory. J. of Progress of Theoretical Physics (Supplement). No. 138, 2000 (in print).

Einfeldt, B. Munz, C.D. Roe, P.L. and Sjogreen, B. (1991). "On Godunov-type methods near low densities". J. Comput. Phys., 92, 273-295.

Glaister, P. (1988). "Approximate Riemann solutions of the shallow water equations". J. of Hydr. Res. 26(3), 293-306.

Ghidaoui, M.S. and Kolyshkin, A.A. (1999). Linear stability analysis of lateral motions in compound open channels. J. Hydr. Engrg., ASCE, 125(8), 871-880.

Ghidaoui, M.S., Deng, J.Q. Xu, K. & Gray, W.G. (1999). A Boltzmann based model for open channel flows. Submitted to Int. J. for Num. Methods in Fluids.

He, X., Shen, X. and Doolen, G.D. (1998). "Discrete Boltzmann equation model for nonideal gases." Phys. Rev. 57(1), R13-R16.

Jha, A.K., Akiyama, J. and Ura, M. (1995). ``First and second order flux difference splitting schemes for dam-break problem." J. Hydr. Engrg., ASCE, 121(12), 877-844.

Jou, D., Casas-Vazquez, J. and Lebon, G. (1996). "Extended irreversible thermodynamics." Springer-Verlag Berlin Heidelberg New York, N.Y.

Katopodes, N. (1984). A dissipative Galerkin scheme for open-channel folw. J. Hydraulic Engrg. ASCE. 110 (4), 450-466.

Lax, P.D. (1972). "Hyperbolic Systems of Conservation Laws and their Mathematical Theory of Shock Waves". Capital City Press, Montpelier, Vermont, USA.

LeVeque, R.J. (1992). Numerical methods for conservation laws. Brekhauser Verlag. Boston.

Pullin, D.I. (1980). "Direct simulation methods for compressible inviscid ideal gas flow." J. Comput. Phys., 34, 231-244.

Qian, Y. H. (1993). Simulating thermohydrodynamics with lattice BGK models. J. Sci. Comp. 8:231-41.

Quirk, J.J. (1994). "A contribution to the great Riemann solver debate". Int. J. Numer. Methods in Fluids, 18, 555-574.

Reitz, R.D. (1981). "One dimensional compressible gas dynamics calculations using the Boltzmann equations." J. Comput. Phys., 42, 108-123.

Rothman, D.H. (1988). Cellular-automaton fluids: a model for flow in porous media. Geophys. 53, 509-18.

Roberts, T.W. (1990). "The behavior of flux difference splitting schemes near slowly moving shock waves". J. Comput. Phys., 90, 141-160.

Shah, K.B. & Ferziger, J.H. (1995). Stimulate small scale SGS model and its application to channel flow. Center for turbulence research annual Research Brefs.

Shan, X. and Doolen, G. (1995). "Multicomponent lattice-Boltzmann model with interparticle interaction." J. Stat. Phys. 81, 379-93.

Shiono, K. & Knight, D.W. (1991). Turbulent open-channel flows with variable depth across the channel. J. Fluid Mechanics , 222, 617-646

Sod, G.A. (1985). "Numerical Methods in Fluid Dynamics." Cambridge University Press, New York, USA.

Soulis, J.V. (1992). "Computation of two-dimensional dam-break flood flows." Int. J. for Num. Methods in Fluids., 14, 631-664.

Su, M., Xu, K. and Ghidaoui, M.S. (1998). "Low speed flow simulation by the Gas-kinetic scheme". J. Comput. Phys., 150, 17-39.

Vincenti, G.H. and Kruger, Jr. C.H. (1965). "Introduction to Physical Gas Dynamics." Krieger Publishing Co., Malabar, Florida, USA.

Xu, K. (1998). Gas-Kinetic schemes for unsteady compressible flow simulation. Lecture series 1998-03, von Karman Institute for fluid dynamics, Feb. 23-27, 1998.

Xu, K. (1997a). "BGK-based scheme for multicomponent flow calculations." J. Comput. Phys., 134, 122-133.

Xu, K. (1997b). "A gas-kinetic scheme for the Euler equations with heat transfer." Proceedings of the international symposium on computational fluid dynamics. Beijing. 1997. 247-252.

Xu, K., Kim, C., Martinelli, L. and Jameson, A. (1996). "BGK-based schemes for the simulation of compressible flow". IJCFD. 7, 213-235.

Xu, K., Martinelli, L. and Jameson, A. (1995). "Gas-kinetic finite volume methods, flux-vector splitting and artificial diffusion." J. Comput. Phys., 120, 48-65.

Xu, K. and Prendergast, K. H. (1994). "Numerical Navier-Stokes solutions from gas-kinetic theory." J. Comput. Phys., 114, 9-17.

Stochastic Hydraulics 2000, Wang & Hu (eds) © 2000 Taylor & Francis, ISBN 90 5809 166 X

Stochastic modelling of river morphology: A case study

Hanneke van der Klis
Subfaculty of Civil Engineering, Delft University of Technology, c/o WL/Delft Hydraulics, Netherlands

ABSTRACT It is important to develop stochastic methods to predict river morphological processes. In this paper we presented considerations on the stochastic prediction of scour in a straight alluvial channel due to a constriction. Of all uncertainties involved, only the uncertainty in the future river discharge was taken into account. River morphological processes are non-linear, which makes it difficult to apply stochastic methods. Therefore, we used a Monte Carlo method, despite the large computation time it takes. The results showed an uncertainty in the predicted bed level of the same order as the mean bed level changes. The variations in time of the predicted statistical properties depended on the ratio between the duration of the discharge peak and the morphological time scale. The use of linearisation and constant discharges in stochastic modelling were discussed, which could probably be used to estimate statistical properties of morphological changes in a simplified way.

Key words: Stochastic modelling, River morphology, Monte Carlo simulations, Constriction, Scour

1. INTRODUCTION

One of the important research areas in river engineering is morphology, in which one examines the adaptation of the river bed to a variety of natural changes and human interferences. The ability to predict morphological changes is of considerable importance for river engineering subjects, as flooding, fairways for shipping, and the foundation stability of river works.

A river morphological prediction model is subject to several uncertainties:

- uncertainties in the model itself, due to a lack of knowledge of the morphological processes, or to discarding phenomena which are thought to be of minor influence;
- inaccuracies due to a limited availability of site-specific measured data, which have to be entered into the model;
- inherent uncertainties in the time-evolution of essential inputs, such as the future river discharge.

With the description of these uncertain factors another kind of uncertainty is introduced, namely the statistical uncertainty. The choice of a certain probability function and its parameters depends on a priori ideas and preferences of the modeller, and on the data used.

Despite these uncertain factors, predictions so far have mostly been made with deterministic models. This means that the model output represents only one realisation of a stochastic process, and thus a more or less arbitrary choice out of a range of possible occurrences. Strictly speaking, this should not be called a prediction.

Literature about stochastic methods that can be used in the prediction of river morphological processes is very scarce, if existent at all. Apparently, little is known about the reliability of morphological predictions. This paper presents the first results of a study in which we want to develop ideas about this subject. We discuss a one-dimensional case which concerns the morphological changes due to a constriction in a straight alluvial channel. Of all uncertainties involved, we only took the uncertainty in the river discharge into account. River morphological processes are non-linear, which makes it difficult to apply stochastic methods. Therefore, we used a Monte Carlo method, despite the large computation time it takes. The results showed an uncertainty in the predicted bed level of the same order of magnitude as the mean bed level changes. The variations in time of the predicted statistical properties depended on the ratio between the duration of a discharge peak and the morphological time scale. We shortly discussed the use of linearisation and constant discharges in stochastic modelling, which could probably lead to a simplified estimation of the statistical properties.

2. SCHEMATISATION AND METHOD

2.1. Schematisation

The constricted channel and the parameters describing it are shown in Figure 1, where h the water depth, B the channel width, u the flow velocity, C the hydraulic roughness coefficient and i the bed slope. The indices 0 and c refer to the equilibrium state outside and within the constricted reach, respectively. The channel contains fixed vertical banks.

We defined the boundary conditions as follows,
- the hydraulic condition at the downstream boundary is a rating curve, which specifies the water level as a function of the discharge following the equilibrium equations below (Eqs. 1-2). The location of the boundary is such that it does not significantly influence the morphology in and near the constricted reach;
- the hydraulic condition at the upstream boundary is a discharge hydrograph;
- the morphological condition at the upstream boundary is an imposed bottom level.

In case of a constant river discharge, the channel bed will approach an equilibrium state, described by the following equations (Jansen et al., 1994),

$$Q = Buh \tag{1}$$

$$u = C\sqrt{hi} \tag{2}$$

$$s = au^b \tag{3}$$

$$S = sB \tag{4}$$

with Q the discharge, s the bulk volume of sediment transported per unit of time and width and S the total amount of sediment transported per unit of time through a channel cross section. Equations (1-4) hold for each channel reach, where Q and S are constant and B, h, u and i differ per reach. We used the Engelund-Hansen formula (Engelund and Hansen, 1967) to describe the sediment transport, which means that $b=5$ in Equation (3).

It follows that the direct effect of the constriction is a higher flow velocity, and hence erosion of the channel bed in the constricted reach (Eqs. 1-4). In the first period of the constriction the major part of this erosion is realised. The celerity of this process is characterised by the morphological time scale which follows from the analytical solution of a similar case obtained by Klaassen and Struiksma (1988),

$$T_{morf} = \frac{L_c}{c_{b,c}} \tag{5}$$

with L_c the length of the constriction and $c_{b,c} = b\,s_0/h_0$ the celerity of the bottom waves.

In the equilibrium state the water depth in the constriction is larger and the bed slope is gentler than before the constriction had been put in place:

$$h_{e,c} = \left(\frac{B_c}{B_0}\right)^{-\frac{b-1}{b}} h_0 \tag{6}$$

$$i_{e,c} = \left(\frac{B_c}{B_0}\right)^{\frac{b-3}{b}} i_0 \tag{7}$$

110

in which the index e refers to the equilibrium state (Jansen et al., 1994). The bottom level upstream of the constriction is slightly lowered, because of the reduced bed slope in the constricted reach.

In case of a variable river discharge an equilibrium as described above will never be reached. With each discharge peak new bottom waves will be initiated, namely, an erosion wave at the beginning of the constriction and a sedimentation wave downstream of it. This time dependent process can be described by four equations, notably the continuity equation,

$$\frac{\partial Buh}{\partial x} = 0 \tag{8}$$

the momentum equation,

$$u\frac{\partial u}{\partial x} + g\frac{\partial(h + z_b)}{\partial x} + \frac{gQ^2}{B^2 C^2 h^3} = 0 \tag{9}$$

an equation describing the changes in bottom level,

$$\frac{\partial z_b}{\partial t} + \frac{\partial s}{\partial x} = 0 \tag{10}$$

and a sediment transport equation like Equation (3). In this equations z_b is the bottom level.

To illustrate the morphological effect of a variable discharge, Figure 2 shows the effects of a one-year discharge hydrograph containing one peak, starting from an equilibrium based on a constant discharge. The changes in bed level with respect to the unconstricted channel bed are shown. These are the results of a simulation with the river morphological model Sobek (WL|Delft Hydraulics, 1997), solving the one-dimensional equations mentioned above (Eqs. 8-10). The bottom waves initiated by the discharge peak were shown (Fig. 2). During low discharge these bottom waves travel downstream and reduce in height. Under the influence of a long term discharge hydrograph this process is repeated constantly. The channel bed tends to a dynamical equilibrium, in which the bottom level fluctuates around a mean value, while this mean value does not change in time anymore.

An important parameter is the ratio between the duration of the discharge peak and the morphological time scale discussed before (Eqn. 5). A large ratio implies erosion of the whole constricted reach during a discharge peak. Figure 2 illustrates the effect of a small ratio, in which case only part of the constricted reach is eroded during one discharge peak.

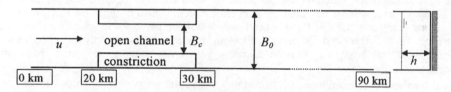

Fig.1: Overview and cross section of channel with constriction
(B_0=200m, B_c=160m, C=50m$^{1/2}$/s, i=9e-5).

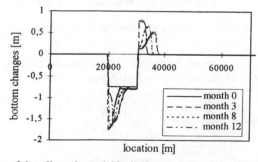

Fig.2: Illustration of the effect of a variable discharge on the channel bed, with a discharge peak during the first three months and a constant low discharge during the rest of the year.

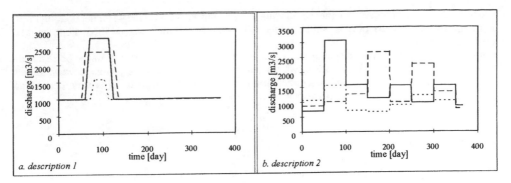

Fig.3: Examples of randomly drawn one-year hydrographs following descriptions 1 and 2.

2.2. Monte Carlo Simulations

Basically, Monte Carlo simulations consist of a large number of deterministic simulations where the input time series are constructed on the basis of a number of randomly generated parameters with a prescribed probability distribution. The set of results from these simulations is used to determine the statistical properties of the predictions. We used the previously mentioned model Sobek to perform the deterministic simulations.

We applied two statistical descriptions of the discharge hydrograph in separate simulations:
1. A statistical description as illustrated in Figure 3a. The duration of the yearly peak was assumed to be uniformly distributed between zero and half a year. The magnitude of the peak was normally distributed, with a mean of 2000 m³/s and a standard deviation of 500 m³/s.
2. A description as illustrated in Figure 3b, based on 50 years of daily realisations of the discharge through the Dutch river Waal. We divided a hydrological year into successive parts of 50 days, where in each part the magnitude of the discharge was statistically described by a lognormal distribution. This statistical description corresponded reasonably well with the data. In this description no attention was paid to possible correlations between the realised discharges in each part of 50 days.

The first description is based on test computations, from which it appeared that the periods of high discharge have a large influence on short-term morphological changes (in this case in the order of one year), and that the effects depend considerably on the actual magnitude and duration of the peak. On the other hand, the periods of low and moderate discharge only influence the long term morphological behaviour, and the discharge variations during these periods have little influence at this time scale.

Based on these two descriptions of discharge hydrographs we performed three Monte Carlo simulations, each containing 1000 Sobek simulations:

A. A discharge hydrograph of type 1, and a relatively large morphological time scale (T_{morf} in the order of 20 months);
B. A discharge hydrograph of type 1, and a relatively small morphological time scale (T_{morf} in the order of 4 months);
C. A discharge hydrograph of type 2, and a relatively large morphological time scale (T_{morf} in the order of 20 months).

The results of the Sobek simulations were used to calculate the following statistical properties of the morphological effects on the bed level, at each arbitrary moment and location,

$$\hat{\mu}(x,t) = \sum_{i=1}^{n} \frac{\Delta z_{b,i}(x,t)}{n} = \Delta \bar{z}_b(x,t) \tag{11}$$

$$\hat{\sigma}(x,t) = \sqrt{\sum_{i=1}^{n} \frac{\left(\Delta z_{b,i}(x,t) - \Delta \bar{z}_b(x,t)\right)^2}{n}} \tag{12}$$

112

with $\Delta z_{b,i}(x,t)$ the change in bed level at location x and time t following simulation i, $\hat{\mu}(x,t)$ the estimate for the mean change in bed level at location x and t, and $\hat{\sigma}(x,t)$ the estimate for the standard deviation of the change in bed level at location x and time t.

Furthermore, the 90%-reliability interval was estimated by calculating the 0.05- and 0.95-quantiles from the cumulative probability function, obtained by ordering the Monte Carlo results in descending order.

3. RESULTS

3.1. Morphological predictions

The results of Monte Carlo simulation A showed that in the constricted reach the reliability interval is of the same order of magnitude as the mean changes in bottom level (Fig. 5). Furthermore, these results showed that the statistical results do not change considerably with time. Comparison of Figure 5a and 5b showed that only within the first few kilometres of the constriction the effect of the discharge peaks is visible, from the erosion of the channel bed. This is further illustrated by the frequency distributions in Figure 4. Namely, in the upstream part (km 22) the frequency distribution changes considerably in time, whereas in the downstream part (km 26) the frequency distribution seems rather time-independent. This difference can be explained by the small ratio between the average duration of the discharge peak and the morphological time scale T_{morf}. As mentioned previously, this implies that only the first part of the constricted reach is directly affected by a discharge peak.

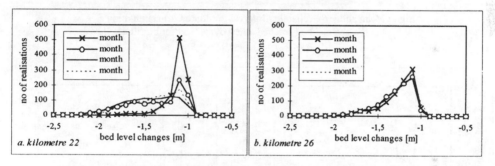

Fig.4: Frequency distributions of bed level changes within the constricted reach (simulation A), at four moments within a hydrological year after a new dynamical equilibrium had been reached.

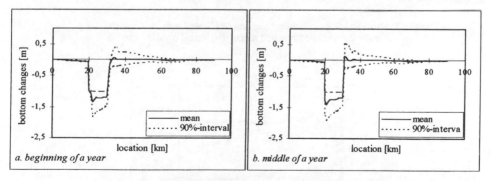

Fig.5: Mean and 90%-reliability interval of Monte Carlo simulation A, after the channel bed reached a new dynamical equilibrium.

113

The results of Monte Carlo simulation B showed the influence of a large ratio between the average duration of the discharge peaks and the morphological time scale (Fig. 6). In this case the main part of the constricted reach was directly affected by the discharge peaks. The effect of the larger ratio is also evident within the reach downstream of the constriction. The local reduction of the reliability interval a few kilometres downstream of the constriction (Fig. 6a) is due to the quick response of the channel bed to the varying discharge, in which the peaks occur during the same period every year.

The results of Monte Carlo simulation C (Fig. 7) showed a striking similarity with those of simulation A. Also in this case the statistical results did not change considerably with time. Due to the small ratio between the average duration of the discharge peaks and the morphological time scale, only the first kilometres of the constricted reach were affected by the discharge peaks.

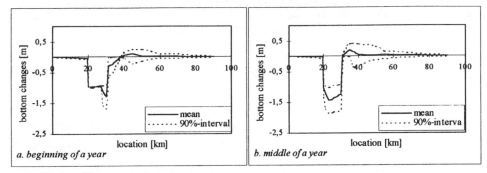

Fig.6: Mean and 90%-reliability interval of Monte Carlo simulation B, after the channel bed reached a new dynamical equilibrium.

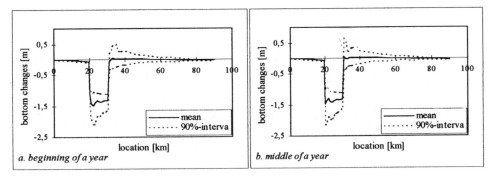

Fig.7: Mean and 90%-reliability interval of Monte Carlo simulation C, after the channel bed reached a new dynamical equilibrium.

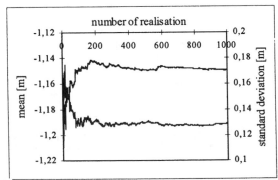

Fig.8: Illustration of the convergence behaviour of the mean (lower line) and the standard deviation (upper line) of changes in the bottom level, averaged over the constricted reach (simulation A).

114

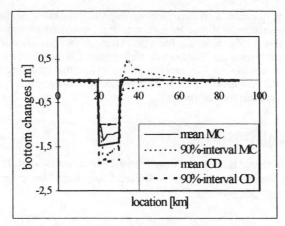

Fig.9: Comparison between mean and 90%-reliability interval based on constant discharges (CD), and Monte Carlo simulation A (MC).

3.2. Convergence

The convergence behaviour of the mean and the standard deviation of the bed level changes in simulation A showed that in this case at least 200 realisations were required to justify some confidence in the convergence of the results (Fig. 8).

4. DISCUSSION

4.1. Morphological predictions

The results of the Monte Carlo simulations (Figs. 5-7) showed a significant uncertainty in the predicted bed level changes, due to the uncertainty in the channel discharge. This illustrated the importance of the development of stochastic prediction models for river morphology.

The variation in time of the predicted statistical properties depended on the ratio between the duration of the discharge peaks and the morphological time scale. The results from Monte Carlo simulations A and C (Figs. 5 and 7) seemed to imply that in case of a small ratio, the quantiles of the bed level changes are not considerably affected by the number of discharge peaks per year and the moments on which they occur. In case of a large ratio, where the bed level reacts quickly on variations in the discharge (Fig. 6), we expect that the quantiles are much more affected by these aspects of the discharge hydrograph.

4.2. Convergence

The convergence behaviour of the mean and standard deviation in a Monte Carlo simulation (Fig. 8) illustrated the main disadvantage of this method, namely the large number of realisations needed to estimate the statistical properties accurately. Since this implies too much computation time to be practicable, it is necessary to find other stochastic prediction methods for river morphological processes. This is the subject of further research. In this paper, we pay attention to two simplifications often used in river morphological research, because they might lead to simplified stochastic methods.

Some relatively simple stochastic techniques are based on the linearisation of physical models. Also in the field of river morphology linearised models are regularly used (e.g. Klaassen and Struiksma, 1988). An important advantage of these models is the qualitative insight they give in the non-linear processes. However, river morphological processes are highly non-linear. Therefore, quantitative studies with linearised models are only suitable for small disturbances of a river. In the example of this paper, this means that a linearised model only leads to satisfactory results if the constriction is strongly limited.

Another interesting question is, if it is possible to predict quantiles of the bed level changes with constant-discharge computations. In a first attempt, we tried to approach the results of Monte Carlo simulation A (Fig. 5) by applying the corresponding quantiles of the peak discharge distribution as constant discharges. Comparison showed that the results of the constant discharge computations were promising for the erosion peaks within the constricted reach, where the reliability interval in the downstream part of the constriction was overestimated (Fig. 9). It seems worthwhile to investigate this method further.

ACKNOWLEDGEMENTS

The author wishes to thank Huib J. de Vriend from Delft University of Technology, Gerrit J. Klaassen from Netherlands Centre of River Research and Erik Mosselman from WL | Delft Hydraulics for their help and comments.

REFERENCES

Engelund, F. and E. Hansen, 1967. A monograph on sediment transport in alluvial streams. Teknisk Forlag, Copenhagen.

Jansen, P.Ph., et al., 1994. *Principles of River Engineering: The non-tidal alluvial river*. D.U.M.. Delft, fascimile edition. Original edition Pitman, London, 1979.

Klaassen, G.J. and N. Struiksma, 1988. On constriction scour in braided rivers. In *International conference on fluvial hydraulics*.

WL|Delft Hydraulics, june 1997. *Sobek, User Manuel, Version 1.18*.

Stochastic Hydraulics 2000, Wang & Hu (eds) © 2000 Taylor & Francis, ISBN 90 5809 166 X

Fractal and chaotic characteristics of alluvial rivers

Deng Zhi-Qiang
Department of Civil Engineering, Shihezi University, Xinjiang, China

Vijay P. Singh
Department of Civil and Environmental Engineering, Louisiana State University, Baton Rouge, La., USA

ABSTRACT: Using the fractal and chaotic theory the paper analyzes the geometrical characteristics and dynamic property of alluvial rivers. A new method is presented to determine the self-affine scaling exponents of rivers based on channel morphologic equations. The results show (1) alluvial rivers exhibit self-affinity, characterized by three self-affine scaling exponents $v_B \approx 0.464$, $v_H \approx 0.321$, and $v_L = 0.674$, in channel morphology, and (2) stable alluvial rivers display self-similarity, characterized by a fractal dimension $D \approx 1.20$, in channel planform. The meandering river pattern with fractal dimension $D \approx 1.20$ is an attractor for alluvial rivers, and they tend to be chaotic when channel sinuosity $S > 1.6$ or $S < 1.4$. A channel pattern with fractal dimension D approaching 1.20 should be maintained for alluvial rivers.

Key words: Alluvial rivers, Fractal, Chaotic characteristics

1. INTRODUCTION

The morphology and dynamics of alluvial rivers have been extensively studied over the past century by the use of river-mechanics, experimental, empirical and computer-simulation approaches. Although a significant progress has been made, the morphologic irregularity and dynamic complexity hinder the sound understanding and management of alluvial rivers. In the last three decades two new and closely related sciences known as "fractal" and "chaos" have given up deep insights into previously intractable inherently complex, nonlinear, natural phenomena and impacted upon most of traditional subjects. Fractals provide a mathematical framework for treatment of irregular, ostensibly complex shapes that display self-similarity or self-affinity, whereas chaotic dynamics underlies the fractal behavior of a nonlinear system. Technically, there is no connection between the fields of dynamic systems and fractal geometry (Devaney, 1993). Dynamics is the study of objects in motion such as iterative processes; fractals are geometric objects that are static images. However, ever since fractal geometry was introduced (Tsonis, 1992), it has become indivisible part of the theory of dynamical systems and chaos as most chaotic regions for dynamic systems are fractals. Hence, in order to completely understand dynamic properties of alluvial rivers, it is essential to understand the geometric structure of river fractals. There are a number of different definitions of fractals that are currently in use. Mandelbrot defined a fractal as a subset of R^n which is self-similar and whose fractal dimension exceeds its topological dimension (Devaney, 1993). Fractals can also be simply regarded as structures that possess a similarity dimension that is not an integer, and the dimension is the fractal dimension. The irregular planform of rivers invites description by fractal geometry.

Fractal investigations of alluvial rivers were started by Mandelbrot (Rodriguez-Iturbe and Rinaldo, 1997), who proposed the first simple models. Proceeding from self-similarity considerations, Mandelbrot concluded that river channels must have fractal properties because of Hack's empirical relation:

$$L_r \sim A^{06} \tag{1}$$

where L_r is the river length and A is the catchment area. If the relation were perfectly consistent dimensionally, the exponent in it would be 0.5; the deviation from dimensional consistency suggests fractal behavior. Writing Eq.(1) in the form

$$L_r^{1/D} \sim A^{0.5} \tag{2}$$

Comparing Eq.(1) with Eq.(2) Mandelbrot indicated that the main river length should be viewed as a fractal measure with dimension D = 1.2. Following the investigation into the world wide rivers, Japanese investigators found that fractal dimension of main rivers varies from 1.1 to 1.3 (Zhang, 1997). Nikora's finding (1991) supports the above result. Nikora et al (1996) summarized main conclusions concerning fractal structure of individual river channels and indicated that individual channels may exhibit self-similar pattern, or self-affine pattern, or self-similarity at small scales and self-affinity at large scales. It means that there is no common point of view, even on the simplest fractal properties. Moreover, it is not clear what D =1.2 and D = 1.1 - 1.3 imply, namely, the fractal behavior and the dynamic mechanism responsible for the fractal characteristics of alluvial rivers need to be further clarified.

The objective of this paper is (1) to investigate the fractal and dynamic characteristics of alluvial rivers by using the fractal and chaotic theories and field data, (2) to establish a connection between the geometric characteristics and the dynamic property of alluvial rivers, and (3) to make a suggestion on sound river management.

2. FRACTAL CHARACTERISTICS OF ALLUVIAL RIVERS

Alluvial rivers are considered to exhibit two distinct fractal characteristics: self-affinity and self-similarity (Sapozhnikov and Foufoula-Georgiou, 1996). The self-similarity is used to describes the plane pattern or channel pattern of an individual watercourse, while the self-affinity to describe the sectional profile or channel morphology. Objects showing the same spatial scaling in all directions are called self-similar fractals and can be characterized by their fractal dimension D. In a more general case, scaling properties are different in different directions. Such anisotropically scaled objects are called self-affine fractals and are characterized by more than one fractal exponents which properly reflect scaling in each direction.

2.1 Self-affinity in Alluvial Rivers

Self-affine behavior of alluvial rivers is attributed to gravity which makes the rivers scale differently in the direction of the main river slope and in the perpendicular direction (Sapozhnikov and Foufoula-Georgiou, 1996). In terms of fractal dimension a self-affine fractal is characterized not by one but rather by two fractal exponents $v_x = 0.72 - 0.74$ and $v_y = 0.51 - 0.52$, the x axis being oriented along the river and the y axis in the perpendicular direction. This conclusion implies that channel width and flow depth, both in the perpendicular direction, should change with same scale or rate along a river. Obviously, this conclusion is in contradiction with almost all existing hydraulic geometry equations summarized by Yalin (1992, p120). Moreover, the fractal exponents v_x and v_y are determined by use of the logarithmic correlation integral. The method involves the use of a least squares estimation technique and thus is empirical. The self-affinity of a river, therefore, needs to be further investigated by a new method.

Owing to the different scaling in the channel width B direction, the flow depth H direction and flow direction L in the fluvial processes, the self-affinity of a river should thus be characterized by three fractal exponents v_B, v_H and v_L, which reflect scaling of a river course in the three directions respectively. As mentioned earlier, the self-affine property reflects the different scaling in channel morphology which was widely investigated and described by various morphologic equations. Therefore, it should be a reasonable way to determine the fractal exponents v_B, v_H and v_L by means of the morphologic equations. Based on the principle of maximum entropy, following channel morphological equations were derived to describe the variation of channel width and flow depth with channel dominant discharge (Deng and Zhang, 1994; Deng and Singh, 1999).

$$B \propto Q^{13/28} \tag{3}$$

$$H \propto Q^{9/28} \tag{4}$$

Eqs.(3) and (4) clearly indicate that channel width and depth vary along a river with different exponents or scales, that is the typical characteristic of the self-affinity. The self-affine scaling exponents v_B and v_H can be therefore determined from Eqs.(3) and (4). That is, $v_B = 13/28 \approx 0.464$, $v_H = 9/28 \approx 0.321$. Using the approximate relationship $v_H \approx v_L - \theta (v_L + v_B)$, proposed by Nikora et al (1996), it is determined that $v_L = 0.674$ ($\theta = 0.31$). It follows from the results that the Hurst exponents $H_{LB} = v_B/v_L \approx 0.69$, $H_{HB} = v_H/v_B \approx 0.69$. These values agree well with that of rivers in various regions of the World (Nikora et al, 1996; Sapozhnikov and Foufoula-Georgiou,1996). It means that the channel morphology of alluvial rivers exhibits self-affinity rather than self-similarity and is characterized by self-affine scaling exponents $v_B \approx 0.464$, $v_H \approx 0.321$, and $v_L = 0.674$. It should be pointed out that the theoretical scaling exponents $v_B \approx 0.464$, $v_H \approx 0.321$ are applicable to all alluvial rivers, including braided rivers, meandering rivers and straight rivers, whereas the fractal exponents $v_x = 0.72 - 0.74$ and $v_y = 0.51 - 0.52$ are derived only from braided rivers and considered to be applicable to such rivers. Due to the dominant influence of channel width B and flow depth H, compared to channel length L, on a river, the two scaling exponents v_B and v_H are far important than the exponent v_L or v_x. Therefore, the two theoretical scaling exponents $v_B \approx 0.464$, $v_H \approx 0.321$ are accurate enough to describe the self-affine characteristics of alluvial rivers.

2.2 Self-similarity in Alluvial Rivers

The self-similarity is characterized by a fractal dimension D. Several methods have been proposed to calculate the fractal dimension (Zhang, 1997). The box-counting method is one of the most widely used due to the relative ease of mathematical calculations involved. It basically consists of drawing successively larger boxes and counting the number of boxes that cover the object, as shown in Fig.1.

Based on the box-counting technique, the scaling number N is proportional to the scale size r raised to the power -D (Tsonis, 1992), where D is the fractal dimension of the object, i.e.,

$$N(r) \propto r^{-D} \tag{5}$$

D has been theoretically defined by different expressions (Zhang, 1997), among which Hausdorff dimension is widely used and expressed as

$$D = \lim_{r \to 0} \frac{\ln N(r)}{\ln(1/r)} \tag{6}$$

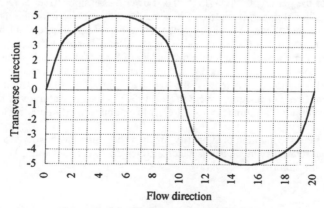

Fig 1: Box-counting technique for estimating the fractal dimension

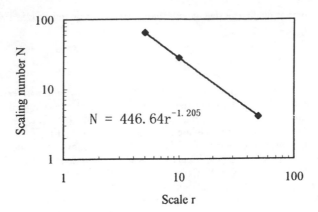

Fig 2: Fractal dimension of stable alluvial rivers

Eqs.(5) and (6) lead to a way of calculating D for an alternative semi-circular river path as such a meander loop can best describe the stable and equilibrium river pattern (Deng and Singh, 1999): Cover the circular flow loop with a two dimensional grid of size r_1 and count the number $N(r_1)$ of the boxes that contain a piece of the loop. Repeat for various sizes $r_2, r_3, \ldots, r_n$. Plot the logarithm of $N(r_1)$, $N(r_2)$, ..., $N(r_n)$ versus the logarithm of $r_1, r_2, \ldots, r_n$. If the resulting curve has a linear part over a wide range of scales, then the slope of that linear part is an estimate of the fractal dimension, as indicated in figure 2.

Following the above procedure, a correlation equation between the scaling number N and the scale size r is obtained for the alternative semi-circular river path.

$$N = 446.64r^{-1.205} \tag{7}$$

Comparing Eq.(5) and Eq.(7) yields the fractal dimension of the most stable river pattern D = 1.205, equaling Mandelbrot's empirical value of D ≈ 1.20. It means that the meandering rivers with fractal dimension D ≈ 1.2 possess the property of scale invariance and they are therefore self-similar. Owing to the preference of natural rivers for D ≈ 1.2, it appears that alluvial rivers are attracted to the fractal dimension D ≈ 1.20, corresponding to the channel sinuosity S ≈ 1.57.

Alluvial rivers exhibit the self-similarity in two ways: (1) same river strives to maintain self-similarity with time and along its channel; (2) different rivers tend to form same channel pattern or to maintain the similarity among them.

The fluvial processes of the Lower Mississippi River can best illustrate the first kind of similarity. Figure 3 shows the variation of the mean channel sinuosity of 24 reaches on the Lower Mississippi River for the period of 1765 - 1915. The Lower Mississippi River is regarded as one of the world's great alluvial rivers. It is affected by a number of variables in addition to hydrology and hydraulics. These could include tributary influences and geologic controls, such as bedrock faulting and uplift or subsidence (Schumm and Winkley, 1994). Figure 3 demonstrates that the river strives to maintain a channel pattern with a sinuosity of S ≈1.6 although it abruptly alters its pattern from meandering to straight or braided and vice versa. Similar phenomenon can be found in other alluvial rivers. It should be noted that the (same) magnitude S ≈ 1.4 - 1.6 is exhibited by substantially different freely meandering rivers (Yalin, 1992). It means that alluvial rivers tend to form or maintain the similar planform (S ≈ 1.4 - 1.6) or to exhibit the self-similarity longitudinally in a statistical sense.

As the evidence of the second kind of similarity, freely meandering rivers strive to maintain the following well known relations among the meander wavelength λ, channel width B, center radius R of curvature, and amplitude T_m.

$$R = 3B \tag{8}$$
$$\lambda = 12B \tag{9}$$
$$T_m = 4.3B \tag{10}$$

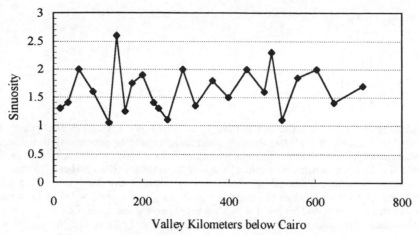

Fig 3: Variation of channel sinuosity of the Lower Mississippi River valley kilometers

These relations correspond to the channel sinuosity S ≈ 1.57 (Deng and Singh, 1999) or fractal dimension D ≈ 1.2 and imply that alluvial rivers tend to maintain the similar channel pattern with D ≈ 1.2. It is necessary to note that the planform meeting Eqs.(8-10) occurs in alluvial rivers with the highest frequency than other planforms, but it does not occur in all alluvial rivers.

To examine the long term behavior of freely meandering rivers, Stolum (1998) used a deterministic continuum model of meandering rivers for extensive simulations of free meandering motion. The simulation outcomes are consistent with a dynamic state of self-organized criticality, which has the following characteristic behavior: (1) stationary mean sinuosity of the final state; (2) the same final state is reached from any initial conditions; and (3) formation of a spatiotemporal fractal structure. Although Stolum did not obtain a quantitative measure for the final state, his conclusion is consistent qualitatively with our result.

It can be concluded that alluvial rivers have a tendency toward the planform with a fractal dimension D ≈ 1.2, but the dynamic mechanism responsible for the geometric characteristic is still not clear.

3. DYNAMIC CHARACTERISTICS OF ALLUVIAL RIVERS

Lyapunov exponents are of interests in the study of dynamic systems in order to characterize quantitatively the average rates of convergence or divergence of nearby trajectories in phase space, and, therefore, they measure how predictable or unpredictable the system is. Since they can be computed either from a mathematical model or from experimental data, they are widely used for the classification of attractors (Szemplinska-Stupnicka et al, 1991). Negative Lyapunov exponents signal periodic orbits; constant Lyapunov exponents imply that nearby trajectories do not diverge or converge; whilst at least one positive exponent indicates a chaotic orbit and the divergence of initially neighboring trajectories. Tsonis (1992) proposed the following definition of Lyapunov exponents :

$$\lambda_i = \lim_{T \to \infty} \frac{1}{T} \ln\left[\frac{P_i(T)}{P_i(0)}\right] \tag{11}$$

in which P_i can take channel width B, flow depth H, and channel length L for a river system.

Using 77 sets of field data from 42 alluvial rivers in different continents, a relationship between channel width-depth ratio $B^{1/2}/H$ and channel sinuosity S is given in Figure 4. The figure demonstrates that for channel sinuosity S < 1.4 or so, channel width to depth ratio increases remarkably with channel sinuosity S, i.e., $\lambda_B > 0$; and for S > 1.6 or so channel length L increases suddenly as channel sinuosity S increases, i.e., $\lambda_L > 0$. Based on the Eq.(11), λ_B and λ_L are

121

determined as follows :

$$\lambda_B = \lim_{T \to \infty} \frac{1}{T} \ln\left[\frac{B(T)}{B(0)}\right] > 0 \qquad\qquad \text{for S} < 1.4 \qquad\qquad (12)$$

$$\lambda_L = \lim_{T \to \infty} \frac{1}{T} \ln\left[\frac{L(T)}{L(0)}\right] > 0 \qquad\qquad \text{for S} > 1.6 \qquad\qquad (13)$$

It appears that $S \approx 1.4 - 1.6$ is a critical pattern for alluvial rivers. Kondratiev et al (Yalin, 1992) also demonstrated such a critical state through the meander expansion process. With the increase of channel sinuosity S from 1 to 4, the angular expansion speed $d\theta_0 / dt$ first increases and then decreases and achieves a maximum or critical value at the sinuosity $S \approx 1.6$, here $\theta_0 =$ the initial deflection angle. Similarly, Yalin (1992) found that for channel sinuosity $S < 1.4$ to 1.6 the expansion speed of river meander loops increases rapidly, i.e., $\lambda_B > 0$, for $S \approx 1.4$ to 1.6 it reaches its maximum value, $\lambda_B \approx 0$, and for $S > 1.4$ to 1.6 it progressively decreases, i.e., $\lambda_B < 0$, but $\lambda_L > 0$. As mentioned earlier, alluvial rivers strive to maintain a channel pattern with sinuosity $S \approx 1.6$. As a result, $\lambda_B \approx \lambda_L \approx \lambda_H \approx 0$ for $S \approx 1.6$ in a statistical sense. It means that the spectrum of the Lyapunov exponents is (+, 0, -) in case of fractal dimension $D < 1.2$ (S<1.57) due to the increase of river width, and the spectrum (+, 0, 0) occurs when $D > 1.2$ due to the increase of channel length, and (0, 0, 0) occurs when $D \approx 1.2$.

It follows from above discussion that the channel pattern with a fractal dimension $D \approx 1.2$ corresponds to the spectrum of the Lyapunov exponents (0, 0, 0), the only spectrum of no positive and critical Lyapunov exponents, and therefore is an attractor of the nonlinear dynamic river system in a statistical sense. The tendency of a dynamic system toward the spectrum of minimum Lyapunov exponents is the driving force inducing alluvial rivers to maintain the planform with a fractal dimension $D \approx 1.2$, while the spectrum of no positive minimum Lyapunov exponents represents the most stable state of the dynamic system.

4.CONCLUSIONS

Alluvial rivers exhibit self-affinity, characterized by three self-affine scaling exponents $v_B \approx 0.464$, $v_H \approx 0.321$, and $v_L = 0.674$, in channel morphology.

Stable alluvial rivers display self-similarity, characterized by a fractal dimension $D \approx 1.20$, in channel planform.

Alluvial rivers have an attractor, characterized by the meandering channel pattern with fractal dimension $D \approx 1.20$, and tend to be chaotic when $S > 1.6$ or $S < 1.4$, respectively.

The tendency of a dynamic system toward the spectrum of minimum Lyapunov exponents is the driving force inducing alluvial rivers to maintain the planform with a fractal dimension $D \approx 1.2$.

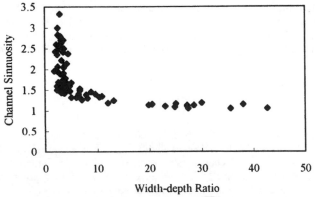

Fig 4: Variation of width-depth ratio $\sqrt{B}/H$ with channel sinuosity S

The channel pattern with fractal dimension D approaching 1.20 should be maintained for alluvial rivers. The closer the channel planform to the pattern $D \approx 1.20$, the more stable the river is.

REFERENCES

Devaney, R. L. (1993), *A first course in chaotic dynamical systems: theory and experiment*, Addison-Wesley Publishing Company, U.S.A.

Deng, Z.Q., and Zhang, K.Q. (1994). "Morphologic equations based on the principle of maximum entropy." International Journal of Sediment Research, 9(1), 31-46.

Deng, Z.Q., and Singh, V.P. (1999). "Mechanism and conditions for change in channel pattern." Journal of Hydraulic Research, IAHR, 37(4), 465-478.

Nikora, V.I. (1991), Fractal structures of river plan forms, Water Resources Research, 27 (6), 1327-1333..

Nikora, V.I., Ibbitt, R, and Shankar, U. (1996), On channel network fractal properties: a case of study of the Hutt River basin, New Zealand, Water Resources Research, 32 (11), 3375-3384.

Rodriguez-Iturbe, I., and Rinaldo, A. (1997), *Fractal river basins: chance and self-organization*, Cambridge University Press, New York.

Sapozhnikov, V.B., and Foufoula-Georgiou, E. (1996), Self-affinity in braided rivers, Water Resources Research, 32 (5), 1429-1439.

Stolum, H.-H.(1998), Planform geometry and dynamics of meandering rivers, Bulletin of the Geological Society of America, 110(11), 1485-1498.

Schumm, S.A., and Winkley, B.R. (1994), *The variability of large alluvial rivers*, ASCE Press.

Szemplinska-Stupnicka, W., Ippt-Pan, W., and Troger, H. (1991), *Engineering applications of dynamics of chaos*, Springer-Verlag Wien-New York.

Tsonis, A. A. (1992), *Chaos: from theory to applications*, Plenum Press, New York.

Yalin, M.S. (1992), *River mechanics*, Pergamon Press.

Zhang, J. Z. (1997), *Fractals* (in Chinese), Tsinghua University Press, Beijing, China.

Stochastic Hydraulics 2000, Wang & Hu (eds) © 2000 Taylor & Francis, ISBN 90 5809 166 X

A stochastic process in synthesizing streamflow data

Edmond D.H.Cheng
University of Hawaii at Manoa, Honolulu, USA

ABSTRACT: In order to provide a better method of generating synthetic streamflow data, a stochastic model of simulating long-term streamflow record is being developed. Basically, this method uses limited historical streamflow data to establish Markov transition probabilities at an intended project site. These probabilities will be the guide for producing synthesized streamflows of a desired duration. An application of this stochastic model is presented.

Key words: Streamflow, Markov process, Time series, Synthesizing.

1. INTRODUCTION

There are numerous models for use in generating synthetic data. (Matalas and Jacobs, 1964; Fiering, 1967; Scheidegger, 1970; Fiering and Jackson, 1971; Mejia, et al., 1972; and Yakowitz, 1979). In water resources engineering, the linear auto-regressive models were considered by far the most popular methods for generating long-term monthly streamflow records of planing and management purposes (Fiering, 1967; Fiering and Jackson, 1971). In these approaches, the serial correlation coefficients of historical flows are the key elements in the recursive simulation formula. However, these serial correlation coefficients have to exceed the conventionally acceptable minimum values for meeting the required statistical significance. Unfortunately, in many cases, not all the serial correlation coefficients can meet this condition. As a consequence, the simulated streamflows would be less desirable. To improve this situation, and to generate streamflow data for a shorter time unit, a stochastic computer simulation model for generating synthetic daily streamflow records is thus being developed. Basically, this method uses historical streamflow data to establish Markov transition probabilities at an intended project site. These probabilities will be the guide for producing synthesized streamflows for a desired period.

The basic strategy underlying this proposed Markov stochastic model is to have the time series of daily streamflows generated in parts: for those streamflows associated with well-behaved climates, in which extraordinary high flows or floods will not be expected to occur, and for those floods belonging to tropical cyclones or some other non-tropical cyclones. In other words, consider these to be generated time series of daily streamflows as a time-dependent system, and this system is decomposed into a set of components for which operating rules of these components may be designed. Under these operational rules, a system of streamflow data will be constructed. An application of this stochastic model is demonstrated.

2 SIMULATION MODEL

Many meteorological phenomena that are considered random in their occurrences also exhibit various degrees of periodicity and persistence. In general, the daily streamflows change with time of day but largely recur from season to season or year to year. It is because of this known fact, the basic simulation strategy adopted in this paper is aimed at the prevailing phenomenon of seasonal periodicity of streamflow patterns, i.e., for a given season, daily streamflows will be simulated on a seasonal basis. In a well-behaved climate, persistence in streamflow records is expected, i.e., the magnitude of daily streamflow at any given day is dependent on that of the previous day or days. It is these two fundamental characteristics of streamflows, viz., persistence and periodicity, which formed the basis of the stochastic simulation model.

Two major assumptions are made in this paper: (1) the correlation between two adjacent daily streamflows depends only on the time interval between them, and (2) the degree of expected persistence between successive streamflows does not depend on the level of the streamflows. The following steps are necessary to construct the model.

2.1 State divisions

The first step is to divide the entire range of observed streamflows into a finite member of states. This task shall be performed with reference to the probability histogram, which is derived from the observed streamflow data at a site. The state divisions should be made in principle of even distribution of measured streamflow values among the states; therefore, state intervals will result in various non-equal sub-ranges of streamflows.

2.2 Distribution functions

The second step in the model is the probability density functions (PDF) and the cumulative distribution functions (CDF) of streamflows in various states. In this paper, three types of PDF are utilized to fit a streamflow histogram, viz., uniform, linear, and some extreme functions. If relative uniform streamflows were observed within a state or states, constant values were assigned to the state of states. Likewise, if linear streamflow variations were observed in a state, a linear PDF is assigned to that state. The extreme PDF is exclusively reserved for the last state in order to take care of high flows or floods.

2.3 Transition probability matrices

In transition probability, p_{ij} is defined as the probability of a streamflow in state j which will occur in the next day, given that a streamflow in state i has occurred in this day.

For a flow field of m finite states, p_{ij} is actually a conditional transition probability of streamflow Q_t going from state i at day t to streamflow Q_{t+1} going from state i at day t to streamflow Q_{t+1} of state j at day t+1, or

$$P_{ij} = P(Q_{t+1} = j | Q_t = i) \tag{1}$$

with m states determined, an mxm transition probability matrix PM, can be determined as

$$PM = [P_{ij}] \quad \text{for } i,j = 1,2,\ldots,m.$$

in which, p_{ij} have the following properties:

$$\sum_{j=1}^{m} P_{ij} = 1, \quad \text{for } i = 1,2,\ldots,m.$$

and

$$P_{ij} \geq 0, \quad \text{for all i and j.}$$

In the process of determining a transition probability matrix, first, a mxm tally matrix, TM, is computed from historical records as

$$TM = [f_{ij}], \quad \text{for } i,j = 1,2,\ldots,m.$$

where f_{ij} = the number of transitions of Q_t going from state i at day t to Q_{t+1} of state j at the next day within a time period under consideration. Then the transition probability, p_{ij}, can be estimated from the tally matrix as follows:

126

$$P_{ij} = f_{ij} / \sum_{j=1}^{m} f_{ij}$$

(2)

for i,j=1,2,...,m.

In this paper, variation of the mean daily streamflows is accounted for by grouping consecutive days with similar streamflow trends into a number of seasons for a year. Let the number of seasons in a year be S, then there will be S transition probability matrices in the simulation process.

A typical tally matrix for a given season s can be expressed as

$$TM(s) = [f_{ij}^{s}]$$

Therefore, by means of Equation (2), the transition probability matrix will be

$$PM(s) = [p_{ij}^{s}]$$

(3)

where s = 1,2,...,S.

2.4 Simulation procedures

Some sequential steps of generating daily streamflow data points at a given project site are briefly described as follows:

1. Divide the historical streamflow data into subsets.
2. Define S, number of seasons in a year.
3. Calculate PDFs and CDFs of the historical data.
4. By means of Equation (3), compute PM (s), the transition probability matrix of seasons s. A total number of S transition probability matrices will be obtained.
5. Determine the state of the succeeding day's streamflow. For any given streamflow in state i of this current day (with specified season s), the succeeding day's streamflow state interval "k" can be determined.
6. Determine the value of the succeeding day's streamflow. With the state k determined in Step 5. the simulated mean streamflow for the succeeding day can thus be obtained.

Repeat steps 5 and 6 until a desired period of simulation is attained.

3 MARKOV PROPERTY AND STATIONARY TESTS

A test of Markov property, the existence of dependency between two adjacent hourly wind speeds, must be carried out. Let the null hypothesis be that the successive events are independent of each other; then the purpose of this test is intended to reject the null hypothesis. For a specified season s, the test statistic is (Anderson and Goodman, 1957):

$$\alpha_1 = -2\ln \lambda_1 = 2\sum_{i,j}^{m} f_{ij}^{s} \ln(p_{ij}^{s} / p_{j}^{s})$$

(4)

where
- λ_1 = the likelihood ratio
- m = number of states
- $f_{ij}^{s}(t)$ =frequency tally for the transition from state i to state j in the t-th subinterval of period r in season s.
- p_{j}^{s} = marginal probabilities for state j; and

$$p_{j}^{s} = \sum_{i}^{m}(f_{ij}^{s}) / \sum_{i,j}^{m}(f_{ij}^{s}).$$

In Equation (4), α_1 is distributed asymptotically as a chi-square distribution with $(m-1)^2$ degrees of freedom, and it is computed for each of the S transition probability matrices. These values are then compared with the table value of chi-square distribution for a specified significance level and degrees of freedom.

Furthermore, the simulation process is only applicable to a stationary time series. Consequently, a test of stationarity of the historical streamflow is necessary. When the null hypothesis of stationarity is satisfied, then

$$p_{ij}^s(t) = p_{ij}^s$$

for t = 1,2,...,T

is distributed asymptotically as a chi-square distribution with (T-1) [m(m-1)] degrees of freedom. The test statistic is:

$$\alpha_2 = -2\ln\lambda_2 = 2\sum_t^T\sum_{i,j}^m f_{ij}^s(t)\ln\frac{p_{ij}^s(t)}{p_{ij}^s}$$

(5)

where λ_2 = the likelihood ratio.

If the null hypothesis of stationarity is to be accepted, then the calculated α_2 values must be less than the table values of chi-square distribution for the specified level of significance and degrees of freedom.

4. ILLUSTRATION

The simulation model based on the described procedures is applied to the streamflow records at Kalihi Stream on the Island of Oahu, Hawaii. In this illustration, six seasons in a year (two adjacent months form a season) were considered. Therefore, by using Equation (3), six transition probability matrices were calculated.

Based on the 70 years (1915-1984) of mean daily streamflow records at Kalihi Stream, five simulation runs were made. Each run generated 100 years of mean daily streamflows. Historical record periods for the five runs are: three 10-year (1915-1924, 1925-1934, and 1945-1954), one 15-year (1960-19740) and all 70 years (1915-1984). Record periods for the first four runs were arbitrarily chosen.

The streamflow duration functions for the historical as well as the 100-year simulation records are plotted in Figure 1. As shown in this figure, the duration curves derived from the simulated records are closely matched with those duration curves for the historical records, which implies that the characteristics of the historical mean daily streamflows at Kalihi Stream were adequately represented.

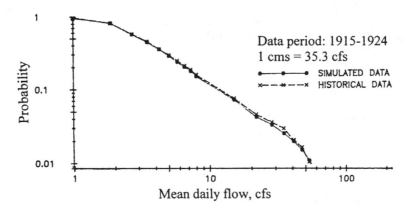

Fig.1: Duration functions of historical and simulated 100-year mean daily streamflows at Kalihi Stream on Oahu, Hawaii, USA
(to be continued)

128

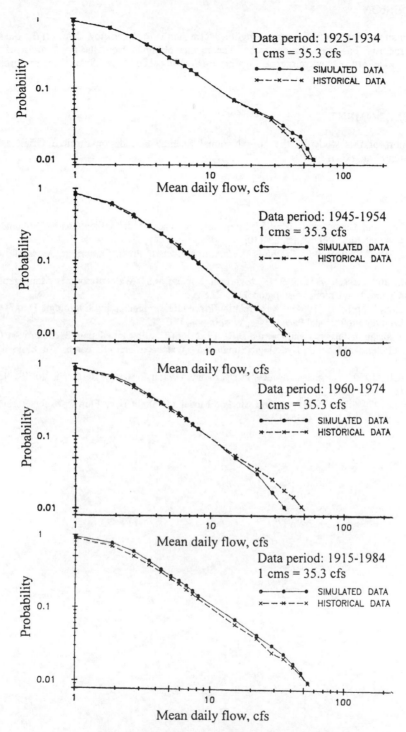

Fig.1: Duration functions of historical and simulated 100-year mean daily streamflows at Kalihi Stream on Oahu, Hawaii, USA

5. CONCLUSION

A stochastic simulation model for generating long-term synthetic streamflow data, on the basis of short-term records, has been demonstrated. The results obtained from the application of this preliminary model are very encouraging. Further research is needed to reinforce these preliminary findings.

ACKNOWLEDGMENT

Partial support of this study by the U.S. National Science Foundation through Grant BCS-912224 is greatly acknowledged.

REFERENCES

Anderson, T.W., and Goodman, L.A., 1957; "statistical Inference about Markov Chains," Annals of Mathematical Statistics, Vol. 28, pp. 89-110.

Fiering, M.B., 1967: "Streamflow Synthesis," Harvard University Press, Cambridge, Massachusetts, U.S.A.

Fiering, M.B., and Jackson, B. B., 1971: "Synthetic Streamflows," Water Resources Monograph 1, American Geophysical Union, Washington, D.C., U.S.A.

Matalas, N.C., and Jacobs, B., 1964: "A Correlation Procedure for augmenting Hydrologic Data," U.S. Geological Survey Professional Paper 434-e, Washington, D.C., U.S.A.

Mejia, J.M., Rodriguez-Iturbe, i., and Dawdy, D.R., 1972: "Streamflow simulation, 2, the broken Line Process as a Potential Model for Hydrologic Simulation," Water Resources Research, Vol. 8, no. 4, pp. 931-941.

Sceidegger, A.E., 1970: "Stochastic Models in Hydrology," Water Resources Research, Vol. 6, No. 3, pp. 750-755.

Yakowitz, s.J., 1979: " A Nonparametric Markov Model for Daily river Flow," Water Resources Research, Vol. 15, No. 5, pp. 1035-1043.

Stochastic Hydraulics 2000, Wang & Hu (eds) © 2000 Taylor & Francis, ISBN 90 5809 166 X

The effect of the bend on the discharge coefficient of a triangular side-weir

A.Coşar & H.Ağaçcioğlu & S.Üç
Faculty of Civil Engineering, Yildiz Technical University, Istanbul, Turkey

ABSTRACT: A side weir, also used as a lateral intake structure is a free overflow weir set into the side of a river or channels which allows to spill a part of water over side. Side-weirs are generally used in channel systems to be able to control water levels during floods, as storm overflows from urban sewage systems, and as head regulators of distributors. There are a large number of studies for side-weirs, which are mostly on straight channels. The same, however, can not be mentioned for side-weirs on curved channels. The present experimental study focuses on the investigation of discharge coefficient of the apex angle 120° of sharp crested triangular side-weirs placed on various locations along a 180° bend. The triangular side-weir discharge coefficient along the bend was found to depend on the upstream Froude number and the overflowed length of the side-weir.

Key words: Discharge capacity; lateral intake; overflow; spill; apex angle.

1. INTRODUCTION

Side weirs are generally used in channel systems to be able to control water levels during floods, as storm overflows from urban sewage systems, and as head regulators of distributors. Although many investigators have been interested for the rectangular side-weir flow characteristic, the first rational approach was made De Marchi (1934). However, a good description on the variation of the side weir coefficient is still not available. The discharge over the side weir per unit length, q, is assumed as:

$$q = - \frac{dQ_s}{d_s} = C_d \sqrt{2g} \left[h - p \right]^{3/2} \tag{1}$$

Where q $\left(= - \frac{dQ_s}{d_s} \right)$ is the discharge per unit width over a rectangular side weir, g is acceleration due to gravity, p is height of the side weir, h is depth of flow and C_d is discharge coefficient.

As a result of dimensional analysis, C_d is dependent on the dimensionless parameters, which is indicated below (Subramanya and Awasthy, 1972).

$$C_d = f \left\{ Fr_1 = \frac{V_1}{\sqrt{gh_1}}, \; L/b, \; L/h_1, \; p/h_1 \right\} \tag{2}$$

Where V_1 is the mean velocity of flow at the upstream section of side-weir in the main channel, L is width of the side weir, b is width of the main channel and h_1 is depth of flow on the upstream end of side weirs in the main channel axis.

The general differential equation for flow along a side-weir as given by Chow (1959), and modified for a rectangular, prismatic, horizontal, frictionless channel is

$$\frac{dh}{ds} = \frac{Qh(-\frac{dQ}{ds})}{gb^2h^3 - Q^2} \tag{3}$$

in which s is the distance from the beginning of the side-weir, h is depth of the flow at a section located at s, dh/ds is water surface slope with respect to the bed, Q is discharge in the channel at section s.

As it can be seen the water level rises from the upstream end of the side-weir towards the downstream end of the side-weir in the main channel according to this equation for subcritical flow conditions in the main channel. However, Subramanya and Awasthy (1972) and El-Khashab (1975) have pointed out that the water level drops slightly at the upstream end of the weir. This has been attributed to the side-weir entrance effect at the upstream end. This effect doesn't extend to the centreline of the main channel and it is only near the weir crest. Then a rapid rise in the water level from the upstream end of the side-weir towards the mid-span of the crest and the rate of rise decreases substantially. The change in water level is not noticeable for nearly the last third of the weir length, and the water surface is nearly horizontal.

This behaviour of the water surface level agrees with the explanation of the effect of secondary flow due to lateral flow near the downstream end of the side-weir. The secondary flow that is created by the lateral flow (El-Khashab, 1975) causes strong disturbances of the water surface despite subcritical flow conditions at the last third of weir section in the main channel. In subcritical flow, Subramanya and Awasthy (Subramanya and Awasthy, 1972) and El-Khashab (El-Khashab, 1975) observed the separation zone and the reverse flow at the downstream end of the side-weir (see Figure 2). According to these researchers, the location and size of the separation zone and the reverse flow area were found to depend on Froude number at the upstream side of weir in the main channel and also on the length of the side-weir.

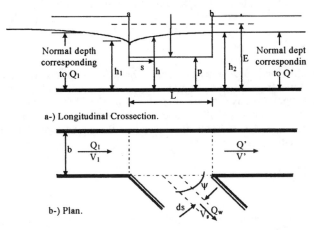

Fig.1: Definition sketch of subcritical flow over rectangular side-weir.

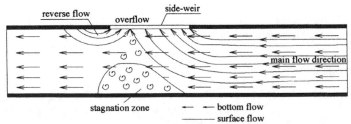

Fig.2: Definition sketch of flow structures along the side-weir.

132

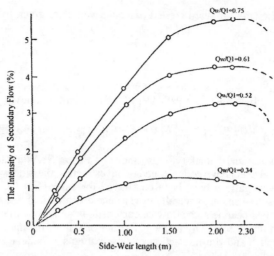

Fig.3: The change of the intensity of secondary flow created by lateral flow (1975).

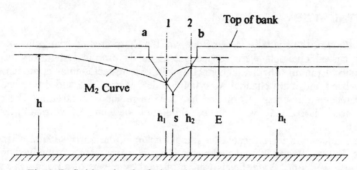

Fig.4: Definition sketch of triangular sharp crested side weir flow.

When the upstream Froude number (Fr_1) increases, the stagnation zone and the reverse flow area move towards the downstream end of the side-weir. The reverse flow area diminishes at higher upstream Froude number.

In order to delineate the magnitude and effect of the spiral flow for different curves under varying conditions of flow, Shukry (1950) has used a term called the strength of a spiral flow (or intensity of secondary flow). This term is defined as the percentage ratio of the mean kinetic energy of the lateral motion to the total kinetic energy of flow at a given cross section. El-Khashab (1975) was also used the same definition of the meaning of the intensity of secondary flow created by lateral flow. Figure 3 shows the variation of the strength or intensity of secondary flow along the side-weir with the ratio of Q_w/Q_1 for subcritical flow. As it is seen in Figure 3, the strength of secondary flow increases along the weir crest length and also is affected by the ratio of Q_w/Q_1. However, the secondary flow being a three-dimensional and in practice turbulent phenomenon, makes almost impossible analytically to solve the problem for the side-weir flow even in the straight channel.

Kumar and Pathak (1987) mentioned to dimensionless parameters for triangular side-weirs as follows:

$$C_d = f \{Fr_1, \theta, p/h_1\} \tag{4}$$

In which θ is the apex angle of triangular side weir (Figure 4).

Authors reported that the effect of p/h_1 on C_d was found to be insignificant for sharp crested triangular side weirs. Subramanya and Awasthy (1972) have also mentioned a similar finding about the negligible effect of p/h_1 on C_d for rectangular side weirs. Then Equation 4 reduces as follows:

$$C_d = f \{Fr_1, \theta\}$$

133

Thus, the discharge coefficients for sharp crested triangular weirs are given by the authors as follow:

$$Q_s = 0.5908 C_d \sqrt{2g} \tan\frac{\theta}{2}(h-p)^{5/2} \tag{5}$$

For sharp crested triangular side weirs having 60^0, $90°$ and $120°$ apex angles, side weir coefficient, C_d was obtained by Kumar and Pathak (1987) are given the following,

$$C_d = \left[0.811 - 0.321\tan\frac{\theta}{2} + 0.129\tan^2\frac{\theta}{2}\right] - \left[0.695 - 0.638\tan\frac{\theta}{2} + 0.150\tan^2\frac{\theta}{2}\right]Fr_1 \tag{6}$$

C_d versus Fr_1 by given Kumar and Pathak (98) also shows in Figure 5. As it can be seen from the literature, most of researchers have concentrated on investigating the side weirs on straight channels. Only minor investigations have been considered for side weirs located on curved channels. However, such side weirs are generally used as intake structure in the river bends.

When the flow is in curved channels, the flow characteristic is rather complicated due to the helicoidal flow pattern. In this case, the hydraulic behaviour of side weirs and curved channels should be evaluated together and the interaction of flow patterns of side weirs and curved channels must be explained.

2.EXPERIMENTAL STUDY

An experimental study of the spatially varied flow over the sharp crested triangular side weirs on a rectangular curved channels were carried out at the Hydraulic Laboratory of Yildiz Technical University, Istanbul. The main channel was 14 m long rectangular approach section and 180° channel bend and main channel were 0.4m wide, 0.4 m. deep and 0.001 bed slope. The bend was of rectangular cross section and had 2.95 m radius to centreline. The channel consisted of a smooth horizontal well-finished aluminium bed and of vertical plexi-glass sidewalls (Figure 6.).

Triangular shape side weirs were located on straight channel, the entry section of the bend and the outer side of 30°, 60° and 90° channel bend. The side weirs were sharp edged and fully aerated on the overflowed side. Water levels were measured by means of a point gage with an accuracy of the order of ± 0.1 mm. The depth of flow at the centreline of main channel was measured at the apex section. For a given side-weir, flow depth in the main channel was adjusted by a tailgate which is located at the end of the main channel.

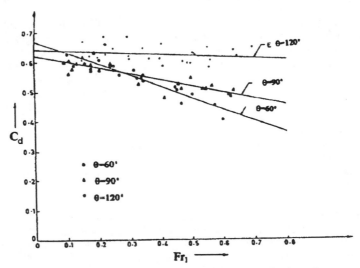

Fig.5:C_d versus Fr_1 given by Kumar-Pathak at straight channel.

134

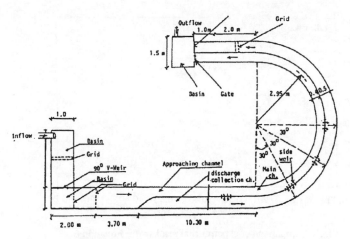

Fig.6: Schematic view of experimental set-up

The experiments were repeated for various combinations of depths and discharge. The details of weirs tested and ranges of various flow parameters are given in Table 1. Flow rates (Q_s) over side weirs were obtained by measuring the discharges at the downstream end of the collection channel. The discharge coefficients were found by Equation 5.

3. RESULTS AND DISCUSSION

In this study, experiments were first carried out on the main straight channel and obtained discharge coefficients were compared with Kumar-Pathak's (1987) results. Figure 7 shows that the relationship between C_d and Fr_1 with their results when $\theta = 120°$. Three different weir crest heights (12, 16 and 20 cm) studied and indicated together with the same curve. Side-weir discharge coefficients were obtained by measuring of the water surface level that is taken account as the upstream end of the weir in the centreline of the main channel. The discharge coefficient were computed by using Equation 5 and given in Figure 7. The difference between the present experimental results and Kumar-Pathak's results (from Equation 6) is due to the overflow length of the side-weir. As shown in Figure 5, the data given by Kumar-Pathak shows so much scatters. Scatter of the data is attributed neglecting the influence of L/b (dimensionless weir length) and p/h_1 (dimensionless weir crest height) parameters. However, when the overflow length (L or L/b) increases the magnitude of secondary flow to the direction of side-weir increases. Due to that reason the overflow rates and discharge coefficients increase (Agaccioglu, 1995). For the same nape thickness, the overflow length of the side-weir for the apex angle $\theta = 120°$ is longer then the other side-weirs which have the apex angle with $\theta = 60°$ and $\theta = 90°$.

Experiments were also carried out at the entrance section of the bend to compare the results of the straight channel. Figure 8 shows that the variation between the C_d and Fr_1 at the entrance section of the bend. As shown in Figure 8, the differences between the experimental results of the entrance section and the straight channel results are due to the early spiral motion. Francis and Asfari also mentioned that the secondary flow created by the bend starts before the upstream end of the bend and it develops its full strength in the bend. The intensity of the secondary flow also depends on Froude number and it is directly proportional by that. That's why the discharge coefficient obtained in the entry section of the bend is larger than the discharge coefficient obtained at the straight channel. Therefore, the differences between two curves increase as the increasing values of Froude number.

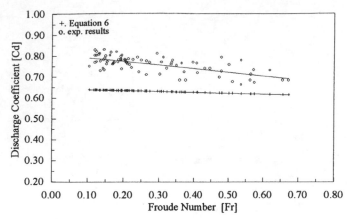

Fig.7: Comparison of experimental results with Equation 6

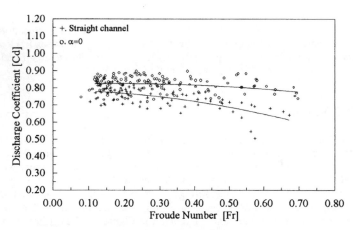

Fig.8: C_d versus Fr_1 for $\theta=120°$ on the entry section of the bend ($\alpha=0°$)

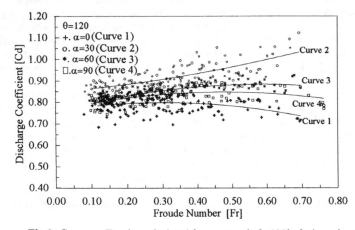

Fig.9: C_d versus Fr_1 along the bend for apex angle $\theta=120°$ of triangular side-weir.

Table.1: The details of weir and flow parameters.

Type of weir	θ	p (m.)	Q_1 (lt/s)	Q_w (lt/s)	Fr_1
Sharp crested	120°	0.12,0.16,0.20	9.56 - 73.03	0.32 - 16.0	0.065 - 0.6

At the last period of the study, experiments were carried out on the curved path of the channel for each 30° angle along the bend. Figure 9 shows that the relation between C_d and Fr_1 along the bend when the apex angle is $\theta=120°$.

As it can be shown in Figure 9 shows that the discharge coefficients along the bend is close to each other when Froude number is small. However, the divergence of these curves for large Froude number becomes significant due to an increasing intensity of the secondary flow. That means that the intensity of the secondary flow increases with an increasing of Froude number along the bend. On the other hand, the largest discharge coefficient were obtained at 30° section of the bend because of the biggest intensity of the secondary is formed that section of the bend.

4.CONCLUSIONS

In this study, the experiments carried out that Froude number is between 0.1 and 0.6 in the bend for apex angle 120° of triangular sharp crested side-weirs. The following results have been obtained:

1) The discharge coefficient obtained by the experimental results is bigger than the discharge coefficient given by Equation 6 due to the effect of the overflowed length of side weir for apex angle 120°.

2) The effect of the bend flow is started to notice before the upstream sections of the bend. In this case, the discharge coefficient obtained on the straight channel is smaller than the discharge coefficient obtained on the entry section of the bend.

3) The discharge coefficient in the bend is larger then the discharge coefficient obtained at the straight channel for the whole flow conditions due to the extra secondary flow created by bend flow.

ACKNOWLEDGMENT

Financial support of this work by the Research Foundation of Yıldız Technical University (Project 94-B-05-01-01) is gratefully acknowledged.

REFERENCES

Agaccioglu, H. (1995), Investigation of side - weir flow on curved channel. Ph. D. Thesis, Yildiz Technical University, Istanbul, Turkey. Chow, V. T. (1959). Open channel hydraulic., Mc Graw Hill, Chap. 16, 439 - 460.

De Marchi, G. (1934). " Saggio di teoria de funzionamente degli stramazzi laterali.", L'Energia Elettrica, Milano, Vol. 11.,pp. 849 - 860.

El-Khashab, A. M. M. (1975)." Hydraulics of flow over side-weirs.", Ph.D. Thesis, presented to the University of Southampton, England.

Kumar, C. P. and Pathak, S. K. (1987). Triangular Side Weirs., Journal of Irrigation and Drainage Engineering, ASCE, Vol.113, No.1, pp. 98 - 105

Shukry, A. (1950). " Flow around bends in open flume.", J. Hydr. Engrg., ASCE, 115, 751 - 759.

Subramanya, K., and Awasthy, S. C. (1972). Spatially varied flow over side-weirs. J. Hydr. Engrg., ASCE, 98(1), 1 - 10.

Stochastic Hydraulics 2000, Wang & Hu (eds) © 2000 Taylor & Francis, ISBN 90 5809 166 X

Comparison of sediment transport rates between symmetric and asymmetric straight compound channels

S. Atabay
School of Civil Engineering, University of Birmingham, UK

G. Seckin
Department of Civil Engineering, University of Cukurova, Turkey

ABSTRACT: Experiments have been carried out at a small laboratory scale to investigate sediment transport rates in compound channels comprising of one rectangular mobile main channel composed of uniform sand with a d_{35} of around 0.80mm, and one or two symmetrically disposed smooth floodplains. Results are presented for straight compound channels and include sediment transport rates and some bed form analysis. The results show that the cross-sectional shape of the channels affects the transport rate and bed forms. For all experiments a typical longitudinal bed form of dune was obtained.

Key words: Overbank flow, Compound channel, And stage-discharge relationships, Sediment transport rate

1. INTRODUCTION

One of the main objectives of a river engineer is to know how the flow structure affects the sediment transport rate. The determination of the sediment transport rate in many river studies is important because it governs the river bed behaviour.

Numerous experimental studies have been carried out on compound channels in laboratories in recent years. For example, studies on the UK Flood Channel Facility (FCF), supported by the UK Engineering and Physical Sciences Research Council (EPSRC), have made a significant impact on our understanding of morphological process in rivers, turbulent flow structure and overbank flow mechanisms (Knight, 1999). In this paper some results on straight compound channels having symmetric and asymmetric cross-sections are compared based on experiments undertaken at the University of Birmingham.

When a river flows out of bank significant dissimilar velocity distributions occur in the main channel and floodplains. Therefore in order to calculate the stage-discharge and sediment-load-discharge relationships in compound channels, the momentum transfer between the main river channel and the floodplains must be considered. Standard 1-dimensional methods such as the 'Coherence (COH)' method (Ackers, 1993), the single channel method (SCM), and the divided channel method (DCM), are generally used to calculate the stage-discharge and sediment load-discharge relationships in such channels. Shiono and Knight (1991) developed a 2-dimensional method that takes into account bed generated turbulence, lateral shear, and secondary flows to account for the interaction between the main channel and the floodplains (Shiono & Knight, 1991; Knight & Abril, 1996). Numerous sediment transport equations from laboratory flume and field data exist, and the Ackers & White (1973) approach is adopted herein.

2. EXPERIMENTAL PROCEDURES

The experiments were performed at the University of Birmingham in a non-tilting 22m long flume with a length of 18m. The flume was 1213mm wide, 400mm deep and configured into a two stage channel with a 398mm wide rectangular mobile main channel and two 407.3mm wide smooth floodplains at an elevation h=50mm above the mobile bed, as shown in Fig.1. A uniform sand size of d_{35}=0.80mm was used in the main central river section and recirculated in the flume for every test discharge via a slurry pump and 50mm pipeline, which itself was monitored by an Electro-magnetic flow meter.

At the end of the flume a series of three adjustable tailgates were located to produce uniform flow conditions in the 18m-test length. Water surface profiles were measured directly using pointer gauges. Preliminary experiments, with sediment recirculating in the 50mm pipeline, and bed forms allowed to develop within the main channel, allowed the tailgates to be calibrated so that the mean water surface slope and the mean longitudinal bed slope could be set equal to the valley slope of the floodplains, fixed at 2.204×10^{-3} for any required test discharge (Atabay & Knight, 1999).

Measurements were made of sediment transport rate for inbank and overbank flows with asymmetric and symmetric floodplains. Sediment samples were collected manually at the upstream side of the channel over 15-20 minute intervals. At the conclusion of each experiment one lateral and three longitudinal bed profiles were measured in detail every 10cm down the flume with an automatic HR touch sensitive bed profiler, one of them on the central line of the channel and others 10 cm apart from either side of the central line. Velocity measurements were made at 0.4 of the local depth using a Novar Nixon miniature propeller current meter with a propeller diameter of 13mm. Local boundary shear stress on the floodplains were measured with a 4.77mm diameter Preston tube. Further details of the apparatus and experimental procedures may be found in Ayyoubzadeh (1997)

3. EXPERIMENTAL RESULTS

In order to obtain a direct comparison between the effect of interaction and no interaction of the floodplain flows, a series of inbank flow tests had to be performed by isolating both floodplains from the main channel. These inbank flow tests, as well as asymmetric overbank flow tests, were performed by Ayyoubzadeh (1997). In this study asymmetric overbank flow tests have been repeated in order to check on the accuracy and reliability of the experimental procedures. Comparison of stage-discharge relationships between symmetric and asymmetric compound channels is shown in Fig. 2.

The presence of a mobile bed in the main channel caused large bed forms that made the setting up of uniform flow difficult, as well as the measurements of the bed load. Therefore quasi-steady uniform flow had to be established in order to obtain accurate data for the sediment concentration at each discharge. Figure 3 shows a comparison of sediment transport rates, [Gs (in mg/s)] for both channels.

For each experiment the longitudinal bed form results were normalised by the mean average bed level to identify the lateral variation in the bed forms more easily. After normalising the longitudinal bed profiles, the standard deviation, SD, was calculated to compare with other experiments. Some results of normalised bed form profiles at the centre of the main channel for the asymmetric and symmetric compound channels are shown in Figure 4. A detailed elevation of typical dune bed forms is shown in Figure 5 for the asymmetric compound channel with a discharge of 27.0l/s. The comparison of bed form index, σ (standard deviation of bed forms heights) versus Manning's n relationships between the symmetric and asymmetric compound channels is shown in Figure 6. This type of plot helps to understand the effect of the bed form index on the resistance.

4. DISCUSSION OF RESULTS

In the literature here are not many experimental data published for compound channels with the same geometry and asymmetric & symmetric floodplains. This study is therefore a small contribution to providing high quality data on stage-discharge and sediment transport rate-discharge relationships in order to improve our understanding concerning the effects of cross sectional shape.

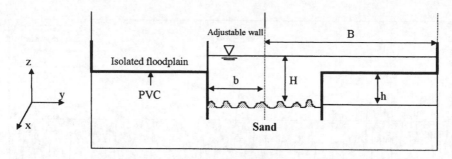

Fig.1: Schematic cross section of the 18m flume at the University of Birmingham

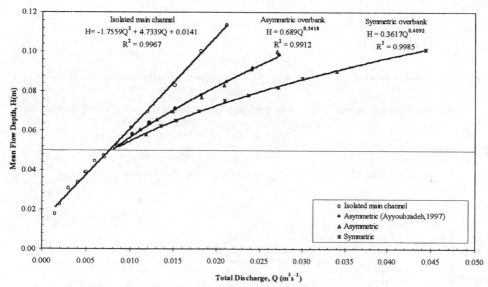

Fig.2 : Comparison of the stage-discharge relationships between the asymmetric and symmetric compound channels

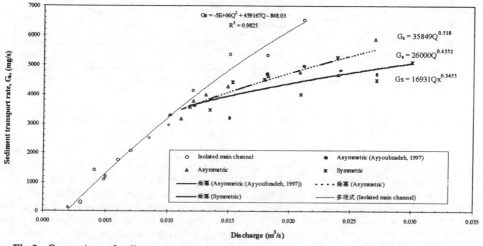

Fig.3:. Comparison of sediment transport rate-discharge relationships between the asymmetric and symmetric compound channels

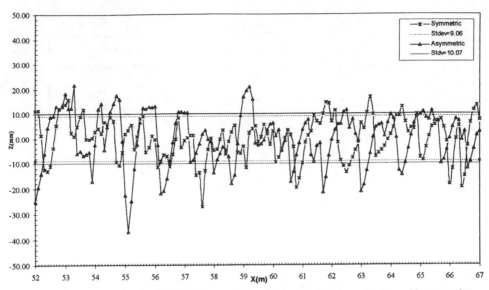

Fig.4:. Comparison of normolised bed form profiles at the central of the main channel between the asymmetric and symmetric compound channels (Q=24l/s)

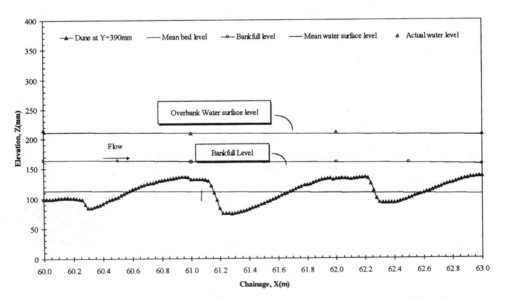

Fig.5: A detail elevation of typical dune bed forms found in asymmetric compound channel, test OB121196, overbank flow, Q=27.0ls-1 (Ayyoubzadeh, 1997).

Figure 2 shows the stage-discharge relationships for the asymmetric and symmetric floodplains as well as isolated mobile main channel. The best fit lines through the data and simple equations were derived in the form $H=\alpha Q^{\beta}$ for overbank flow cases and in the form $H=\alpha Q^{2}+\beta Q+C$ for inbank flow case, where α, β and C are constants, shown in Figure 2 for each experimental configuration, H is the flow depth in m and Q is the total discharge in m^{3}/s. For the asymmetric and symmetric overbank flow tests when the discharge increase above the bankfull level (H=0.05m) the stage does not increase as rapidly as it would with an isolated main channel. This is due to the sudden increase in hydraulic radius as the flow is out of bank.

142

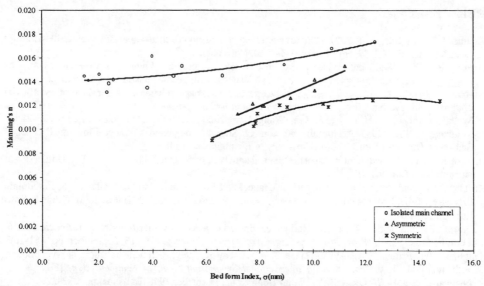

Fig.6: Comparison of bed form index-Manning`s n relationships between the symmetric and asymmetric compound channels

The equations of best fit shown in Figure 3 provide the simplest possible empirical flow discharge-sediment transport rate prediction formulae for overbank & inbank flows. For high discharges the sediment transport rates (in mg/s) in the symmetric channel are lower than those in the asymmetric channel. This might be expected since the stage will be lower in the channel with two floodplains, allowing for more lateral shear and retardation of main river flow (Knight et. al., 1999).

For all experiments, typical longitudinal dune bed form were present (Figure 4 & 5). Figure 6 shows that for a given bed form index, Manning`s n for the inbank flow is always higher than Manning`s n for overbank flows and the symmetric channel values are always lower than the asymmetric channel values ($n_{iso}>n_{asy}>n_{sym}$). This might possibly be because when flow is out of bank the reduction in resistance due to the smooth floodplain surface and discontinuity of cross section for the symmetric compound channel is always higher than the reduction in resistance for the asymmetric compound channel. This reduction in resistance was clearly shown by Atabay & Knight, (1999).

5. CONCLUSIONS

1. This study has produced some high quality data on stage-discharge and sediment transport rate-discharge relationships for mathematical modelers.
2. Because of the temporal variation in bed load rate on duned beds in compound channels, care needs to be taken over the measurement of basic data in order to establish mean values.
3. This study indicates that the sediment transport rates are affected by overbank flows and makes a small contribution to our understanding of the interaction effects and cross-section shape effects.
4. All the experimental results except two data point show that for a given discharge sediment transport rate for asymmetric compound channel is always higher than that for the symmetric compound channel.

REFERENCES

Ackers, P., and White, W. R., 1973, "Sediment transport: new approach and analysis," Journal Hydraulics Division, ASCE, Vol. 99, HY11, pp. 2041-2060, November.

Ackers, P., 1993, "Sediment transport in open channels: Ackers and White update," *Proc. Instn. Civ. Engrs Wat., Marit. & Energy*, 101, pp. 247-249, Water Board, Technical Note 619, December.

Atabay S., Knight D. W., 1999, Stage discharge and resistance relationships for laboratory alluvial channels with overbank flow, *Proc. Seventh International Symposium on River Sedimentation*, [Eds. A. W. Jayawardena, J. H. W. Lee & Z. Y. Wang] Hong Kong, December 1998, Balkema, pp. 223-229.

Ayyoubzadeh, S.A., 1997, "Hydraulic Aspects of Straight-Compound Channel Flow & Bed Load Sediment Transport," PhD Thesis, University of Birmingham, April.

Brown, F.A., 1997, "Sediment transport in river channels at high stage," PhD Thesis, The University of Birmingham, England, UK

Knight, D.W., and Shiono, K., 1996, "River channel and floodplain hydraulics," Chapter 5, Floodplain Processes. Edited by Malcom G. Anderson, Des E. Walling and Paul D. Bates, John Wiley & Sons Ltd., pp.139-181.

Knight, D.W., and Abril, J. B., 1996, "Refined calibration of a depth-averaged model for turbulent flow in a compound channel," *Proc. Instn Civ. Engrs Wat. Marit. & Energy*, No. 118, September, pp. 151-159

Knight D. W., Brown F. A., Ayyoubzadeh S. A., Atabay S., 1999, Sediment transport in river models with overbank flow, *Proc. Seventh International Symposium on River Sedimentation*, [Eds. A. W. Jayawardena, J. H. W. Lee & Z. Y. Wang] Hong Kong, December 1998, Balkema, pp. 223-229.

Knight D. W., 1999, Flow mechanisms and sediment transport in compound channels, *The International Journal of Sediment Research*, IRTCES, Beijing, Vol. 14, No 2, pp. 217-236

Stochastic Hydraulics 2000, Wang & Hu (eds) © 2000 Taylor & Francis, ISBN 90 5809 166 X

Effect of improper dredging of open channels

M.B.A. Saad
Hydraulic Research Institute, Delta Barrage, Egypt

A.F. Ahmed
Channel Maintenance Research Institute, Delta Barrage, Egypt

ABSTRACT: There are more than 32,000 km of canals and 16,000 km of open drains which form the irrigation and drainage networks respectively in Egypt. These channels are periodically dredged due to sedimentation and aquatic weed growth. In order to increase the conveyance efficiency of these canals and to ensure optimum water management, dredging is essential. In the past, clearing obstacles and aquatic weeds as well as reshaping canals to maintain its designed cross sections were manually executed. As modern and efficient equipment were invented, dredging was mechanically applied. In some cases, improper mechanical dredging led to sever bank erosion that caused deposition of eroded materials at certain locations in the open channel systems. One adverse impact of bank erosion is increasing the channel width that led to lower the water surface level and consequently more discharge is needed to meet the required water levels in the case of irrigation canals. In the case of open drains the decrease of flow depth enhanced the growth of several types of aquatic weeds which obstructed the flow currents.

This paper presents the effect of improper mechanical dredging on the geometry of the channel sections as well as its hydraulic characteristics. Two examples are presented to compare the hydraulic characteristics before and after several years of improper mechanical dredging. The paper revealed that improper mechanical dredging of canals and drains lead to dramatic changes in the channel cross section geometry and consequently significant changes in flow characteristics.

Key words: Open channels, Dredging, Field survey, Bank erosion, Channel maintenance

1. INTRODUCTION

The irrigation and drainage networks in Egypt mainly consist of unlined earth channels to distribute and collect water. The lengths of the irrigation and drainage networks are about 33,000 and 16,000 km respectively. As a result of High Aswan Dam construction in the late sixties, most of the annual transported sediment is deposited in Lake Nasser upstream the dam. Therefore, the flowing water in the river and irrigation network became sediment free which permitted deeper sunlight percolation through the water. Consequently, rapid infestation of several types of aquatic weeds took place within shallow depths. Moreover, drainage and seepage water from cultivated lands which contain fertilizer materials enhanced aquatic weed growth in most irrigation canals and open drains.

These caused lower flow velocities, higher water surface level at upper reaches, increase in hydraulic roughness and reduce the conveyance waterway capacity.

In this paper, field measurements including main hydraulic parameters were conducted for one main irrigation canal, and one open drain. Also in order to reveal the effect of improper dredging, comparison between the design and measured hydraulic parameters for both mentioned waterways was carried out.

2. CANAL AND DRAIN DESCRIPTION

The Suez Irrigation Canal, as shown in Fig.1, branches from El-Ismailia Canal through two head regulators and extends parallel to the Suez Canal from North to South on the West Side. The canal is 90.50 Km long with bed width ranging from (11.0 to 13.0) m and water depth ranging between (1.7 and 3.5) m. The average bed slope is about 3cm/km and the side slopes are 2:1. The present total served area of the canal is about 35,000 Feddans (15,000 Hectares) while a future plan was set to reclaim an extra 125,000 Feddans (52,000 Hectares). In addition, to satisfy industrial and drinking requirements, more than 24 water intakes are located along both sides of the canal. The most important one is the main potable plant of Suez City that is located at 800 m upstream of the canal tail end. For that reason, the water surface level at its intake structure should be kept at a certain value to avoid any operational disturbance.

The West Nubaria Drain was constructed in 1986 to serve about 287,000 Feddans (120,000 Hectares) of recently reclaimed lands located west of the Delta. The drain as shown in Fig.2 is 86.0 Km long, and the maximum allowable discharge is 63.0 m^3/s. The maximum flow depth is 4.8 m*s while the bed and water surface slopes are ranging from (3 to 5) cm/km. The design bed width ranges between 5.0 m*s at the upstream entrance located at km 86.00 and 18.0 m*s at the outlet which discharges its water by gravity to the Mediterranean Sea.

3. FIELD MEASUREMENTS

Concerning Suez Irrigation Canal, bed profile and cross section at 18 locations covering the entire canal length was measured. The velocity distribution at nine cross-sections and water surface levels along the canal were measured. Bed and water samples were collected at the measured cross sections. Table 1 shows the measured hydraulic parameters at nine cross-sections. In this Table B is the bed width, Y is the average flow depth, T is the top width, S is the water surface slope, A being the cross section area, P is the wetted perimeter, R is the hydraulic radius, and n is the Manning roughness coefficient.

Concerning West Nubaria Drain, 18 cross-sections located at 4 km apart were measured as well as the percentage of submerged weed infestation was determined. Bed, water and aquatic weed samples were collected at each of the measured cross sections. The velocity distribution at three cross sections and water surface slopes along the drain were measured. Moreover, locations of the infested areas by different types of weed were determined. Table 2 shows the measured hydraulic parameters.

4. RESULTS AND ANALYSIS

Data obtained for Suez Irrigation Canal revealed the existence of some obstacles and heavy weed infestation within some reaches along the canal. Consequently, the water level was higher than the design values at the infested reaches that impair the water requirement upstream the Suez City main potable plant. As the field measurements of Suez Irrigation Canal were conducted while the passing discharge was less than the design value, the corresponding design parameters to the measured discharge were computed as shown in Table 1. In this table (V) is the average flow velocity in m/s and (Q) is the discharge in m^3/s.

The comparison between the mean value of the design and existing parameters (Table 1) along the canal is illustrated in Fig. 3 which reveals the following results

B*	Y*	T*	A*	P*	R*	V**	N*
18.4 %	26 %	8 %	82 %	27 %	26 %	37 %	113 %

Note: * means increase;   ** means decrease

146

In the above Table (*) means increase while (**) means decrease. Accordingly, it can be deduced from the above results that the improper dredging of the canal led to major changes compared to the design parameters. These changes raised the water levels in some upstream reaches of the canal and reduced its conveyance capacity which result unsufficient water level at the intake of Suez potable water which is lant located at its end.

Table.1: Design and Measured Hydraulic   Parameters of Suez Irrigation Canal

C.S.No.	Location (Km)	B_d (m)	Y_d (m)	T_d (m)	S_d (cm/Km)	A_d (m²)	P_d (m)	R_d (m)	n_d	B_m (m)	Y_m (m)	T_m (m)	A_m (m²)	P_m (m)	R_m (m)	V_m (m/s)	Q_m (m³/s)	n_m
1	0.20	13.0	2.05	21.20	3.0	35.055	22.168	1.581	0.025	14.0	1.2	28.0	28.80	27.961	1.03	0.358	10.30	0.016
2	13.00	13.0	1.75	20.00	3.0	28.875	20.826	1.386	0.025	12.0	2.0	32.0	39.50	25.987	1.52	0.192	7.58	0.038
3	14.00	13.0	1.56	19.24	3.0	25.147	19.977	1.259	0.025	14.0	1.8	36.0	54.85	44.593	1.23	0.117	6.40	0.054
4	16.10	13.0	2.10	21.40	3.0	36.120	22.391	1.613	0.025	12.0	2.7	32.0	60.45	31.984	1.89	0.178	10.73	0.047
5	31.25	12.0	2.05	20.20	3.0	33.005	21.168	1.559	0.025	12.0	2.6	20.0	40.00	20.000	2.00	0.242	9.67	0.038
6	49.50	12.0	1.85	19.40	3.0	29.045	20.273	1.433	0.025	15.0	2.2	31.0	54.20	30.971	1.75	0.141	7.62	0.065
7	68.00	11.0	1.40	16.60	3.0	19.320	17.261	1.119	0.025	15.0	1.8	27.0	48.70	27.056	1.80	0.094	4.60	0.099
8	75.00	11.0	1.30	16.20	3.0	17.680	16.814	1.052	0.025	11.0	2.0	21.0	30.00	20.979	1.43	0.137	4.10	0.037
9	83.00	11.0	1.05	15.20	3.0	13.755	15.696	0.876	0.025	12.0	2.0	24.0	41.40	23.931	1.73	0.066	2.75	0.087

Table 2: Design and Measured Hydraulic Parameters for West   Nubaria   Drain

C.S.No.	Location (Km)	B_d (m)	Y_d (m)	T_d (m)	S_d (cm/Km)	A_d (m²)	P_d (m)	R_d (m)	n_d	B_m (m)	Y_m (m)	T_m (m)	S_m (cm/Km)	A_m (m²)	P_m (m)	R_m (m)	V_m (m/s)	Q_m (m³/s)	n_m
1	0.30	18.00	1.46	20.91	4.00	26.19	22.12	1.18	0.03	22.50	3.25	38.75	4.00	73.90	34.25	2.16	0.095	7.04	0.11
2	4.00	18.00	1.38	23.52	4.00	24.84	24.17	1.03	0.03	26.25	2.75	31.75	4.00	59.90	32.50	2.84	0.118	7.04	0.11
3	8.00	18.00	1.02	22.08	5.00	18.36	22.56	0.81	0.01	19.00	3.25	32.00	6.00	69.80	29.25	2.39	0.198	13.80	0.07
4	12.00	18.00	1.21	22.84	5.00	21.78	23.41	0.93	0.02	19.00	2.50	29.00	8.00	59.60	29.25	2.04	0.157	9.36	0.09
5	16.00	18.00	1.61	24.44	5.00	28.98	25.20	1.15	0.03	20.00	2.75	31.55	6.14	64.60	31.00	2.08	0.159	10.29	0.08
6	20.00	18.00	1.87	25.48	5.00	33.66	26.36	1.28	0.03	19.00	2.75	30.55	11.20	59.60	29.25	2.04	0.226	13.48	0.08
7	24.00	18.00	1.91	25.26	5.00	34.38	26.20	1.31	0.03	16.00	2.75	27.00	7.06	79.80	32.50	2.46	0.173	13.78	0.09
8	28.00	18.00	1.95	23.85	5.00	35.10	25.03	1.40	0.04	16.00	2.50	21.00	15.00	51.70	29.00	1.78	0.198	10.24	0.09
9	32.00	15.00	1.98	20.94	5.00	29.70	22.14	1.34	0.04	15.00	2.60	19.68	6.80	61.60	31.00	1.99	0.144	8.85	0.09
10	36.00	15.00	1.95	21.63	5.00	29.25	22.69	1.29	0.04	15.00	2.60	20.20	2.50	56.40	28.00	2.01	0.157	8.85	0.05
11	40.00	15.00	2.35	22.05	5.00	35.25	23.47	1.50	0.06	15.00	3.38	21.76	4.60	81.00	32.75	2.47	0.099	8.00	0.13
12	44.00	15.00	2.10	21.72	5.00	31.50	22.92	1.37	0.05	15.00	3.38	21.76	5.00	63.80	30.25	2.11	0.125	8.00	0.09
13	48.00	15.00	2.38	21.66	5.00	35.70	23.19	1.54	0.06	16.00	3.30	25.90	5.81	63.70	28.75	2.22	0.126	8.00	0.10
14	52.00	15.00	2.15	21.02	5.00	32.25	22.40	1.44	0.05	16.00	3.00	22.00	5.55	71.90	30.00	2.40	0.111	8.00	0.12
15	56.00	13.25	1.63	18.14	5.00	21.60	19.13	1.13	0.03	14.00	2.40	21.20	4.59	58.60	30.00	1.95	0.128	7.50	0.08
16	60.00	13.25	1.92	19.02	5.00	25.48	20.18	1.26	0.04	14.00	2.40	21.20	0.60	55.50	28.50	1.95	0.135	7.50	0.03
17	64.00	13.25	1.77	20.33	5.00	23.45	21.17	1.11	0.04	14.50	2.70	19.90	5.45	60.30	27.75	2.17	0.116	7.00	0.11
18	68.00	5.00	1.15	8.45	5.00	5.75	9.15	0.63	0.01	5.50	2.40	12.70	5.00	21.90	14.00	1.56	0.228	5.00	0.04

147

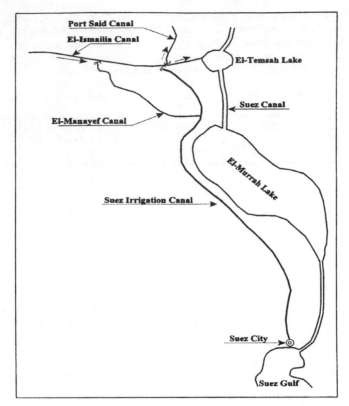

Fig.1: Layout of Suez Canal

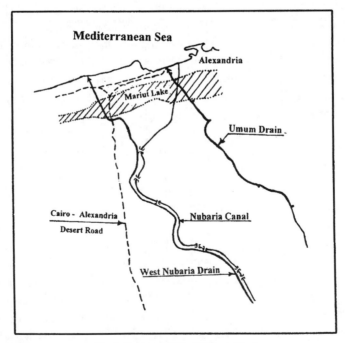

Fig.2: Layout of West Nobaria Drain

148

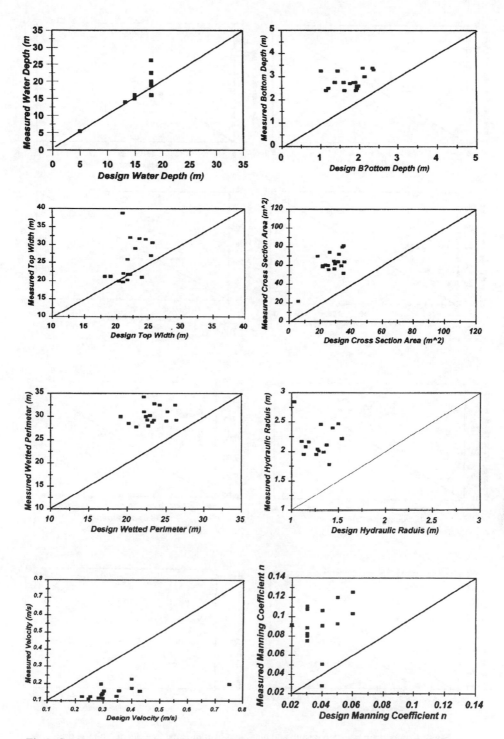

Fig.3: Comparison between design and measured hydraulic parameters of Suez Irrigation Canal

149

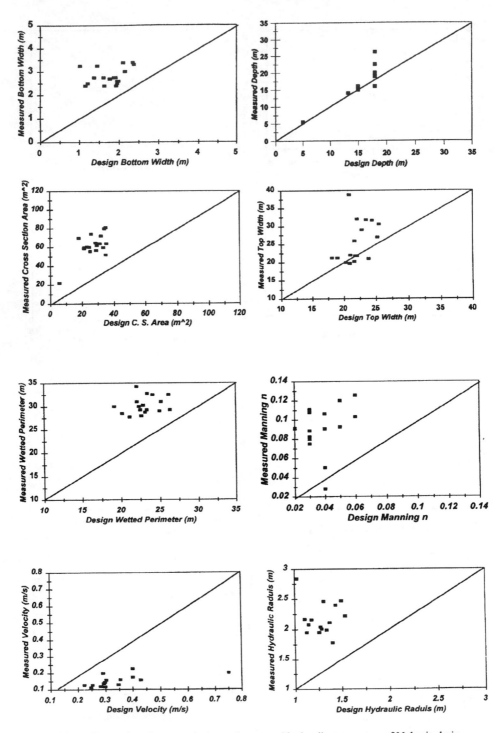

Fig.4: Comparison between design and measured hydraulic parameters of Nobaria drain

On the other hand, field survey and collected data for hydraulic parameters and weed infestation at West Nubaria Drain revealed that the total drain length could be divided into two reaches. Upstream reach starts from Km 86.00 and extends to Km 27.00, while downstream reach extends up to the Mediterranean Sea. The upstream reach is generally characterized by high weed infestation, low flow velocities, the average discharge is less than 50% of the design value, the water surface slope is less than the design value and the flow depth is almost 1.0 m higher than the design value. On the other hand, the downstream reach is clear of weed infestation that is obviously due to saltwater intrusion from the Mediterranean. Also at the downstream reach, the mean flow velocity is much higher, and the water surface slope is steeper than the design value. Although, bank erosion was noticed along some scattered areas, deposition only took place within the upstream reach, while erosion recorded within the downstream reach.

Figure (4) shows a camparison between measured and designed hydraulic parameters of the dain.

Table (2) shows the measured data of the drain.The comparison between the design and existing hydraulic parameters along the drain is also shown in Table 2. The comparison reveals the following results :

B*	Y*	T*	A*	P*	R*	V**	N*
7 %	68 %	16.8 %	140 %	33 %	85 %	56 %	184 %

Note: * means increase; ** means decrease.

In the above Table (*) means increase while (**) means decrease. Accordingly, one may conclude that the West Nubaria Drain is highly affected by improper dredging.

6. CONCLUSIONS AND RECOMMENDATIONS

The comparison between the existing and design parameters for the canal and the drain revealed a serious effect of the improper dredging on both of them. The analysis also shows a significant change in hydraulic parameters, which led to a deviation from the designed characteristics and consequently reduced the conveyance efficiency of the canal and the drain. The improper dredging resulted in the following:
- Increase in the water surface levels especially within the upstream reaches;
- Remarkable increase in Manning roughness coefficient;
- Reduction in the conveyance efficiency of the waterway; and
- Inefficient waterway performance for either water distribution or collection.

Based on the results obtained the following recommendations are stated :-
- The design parameters of the main channels should be maintained to restore the conveyance efficiency.
- Canals and drains should be kept free of weed infestation which leads to increase the roughness coefficient and consequently reduces the conveyance efficiency.
- Comprehensive plan, regular maintenance and weed control of irrigation canals and drains should be developed.
- Developing proper mechanical equipment which permit channel dredging with minimum violation of the hydraulic characteristics is desired.

REFERENCES

Kamel, Mahmoud, 1997, "Hydraulic Characteristics of Suez Irrigation Canal Problem and Provisions for the Future", Hydraulics Research Institute, Delta Barrage, Egypt.

Seif, H. M., 1998, "Aquatic Weed Problem in West Nubaria Drain", Channel Maintenance Research Institute, Delta Barrage, Egypt.

Turbulence

Stochastic Hydraulics 2000, Wang & Hu (eds) © 2000 Taylor & Francis, ISBN 90 5809 166 X

On development of inhomogenous turbulent scalar transport

A. Chanda & J. Das
Indian Statistical Institute, Calcutta, India

ABSTRACT: Transport of scalar in a turbulent flow adjacent to a solid wall has been studied in this paper. Transport starts in a tranquil state in presence of a source (sink). It effects an anomalous distribution of scalar concentration. It is assumed that no tangible density difference exists. Scalar is transported by its gradient only. It is argued that a turbulent velocity field is created by this anomalous distribution of scalar concentration and vice versa. We assume homogeneity in the plane parallel to the boundary. It is argued that this model may replicate many actual cases of turbulent transport to reasonable approximation. A numerical study has been included.

Key words: Turbulent scalar transport, Velocity field, Homogenous and inhomogeneous turbulence

1. INTRODUCTION

Lot of research works have been carried out on turbulent advection i.e.transport of scalar e.g. temperature and humidity. A good number of papers are available in journals on this aspect of turbulence. In this context the works of Batchelor (1959), Brost et al (1978), Kistler et.al. (vide Hinze (1959)), Launder et al (1975), Tavoularis et al (1981), Wyngaard (1975), Wyngaard et al (1971) and (1974), Orszag et al (1972), Castaing et al (1989), Chilla et al (1993) and (1993), Iada et al (1995) and Kasagi et al (1995) are worth mentioning.

In the present paper turbulent transport of scalar in a system with a solid boundary on one side is studied. The system also includes a source (or a sink). The turbulent field is inhomogenous in the direction perpendicular to the solid boundary. Homogeneity in the other two orthogonal directions, parallel with the boundary, has, however, been assumed. This is different from the case of complete homogeneity, studied earlier extensively. It may be showed that if we consider overall homogeneity, velocity field virtually becomes nonexistant. But in the present case because of inhomogeneity in one diection the effect of the velocity field on the transport phenomenon is explicitly present. This may be regarded as a replica of turbulent transport of humudity, thermal energy, pollutants etc in the atmospheric boundary layer.

2. GOVERNING EQUATION

We start with the standard three dimensional advection equation.

$$\frac{\partial c}{\partial t} + \overline{U_k}\frac{\partial c}{\partial x_k} + \frac{\partial \overline{C}}{\partial x_k}u_k + \frac{\partial u_k c}{\partial x_k} - \frac{\overline{u_k c}}{\partial x_k} = \lambda c + D\frac{\partial^2 c}{\partial x_k \partial x_k} \tag{1}$$

Here $\overline{C}$ and $\overline{U}_K$ are respectively mean scalar concentration and velocity at time t and position x. Fluctuations are respectively given by c and u_k. The first term on the right side of the equation stands for source (sink). Positive (negative) λ is the rate of generation (decay) of scalar concentration from (to) source (sink). D is diffusion coefficient. The source (sink) in the present case has been taken linear.

Now multiplying Eq.1 with $(n\, c^{n-1})$ and then averaging we may get

$$\frac{\partial \overline{c^n}}{\partial t} + \overline{U}_k \frac{\partial \overline{c^n}}{\partial x_k} + n\frac{\partial \overline{C}}{\partial x_k}\overline{u_k c^{n-1}} + \frac{\partial \overline{u_k c^n}}{\partial x_k} - n\frac{\partial \overline{u_k c}}{\partial x_k}\overline{c^{n-1}} =$$

$$n\lambda\overline{c^n} + D\frac{\partial^2 \overline{c^n}}{\partial x_k \partial x_k} - n(n-1)D\overline{c^{n-2}(\frac{\partial c}{\partial x_k})^2} \tag{2}$$

3. A STEADY STATE EQUATION OF SCALAR VARIANCE

In the present case we would try to visualise a liquid system with a source or sink of a particular scalar. It may be material one like a solute lump or physical one like heating or cooling source. Initially when the source or the sink is just activitated there is a localised concentration of scalar. As time passes scalar spreads through the liquid body. This is possible by diffusional and gradient of scalar concentration as well. With the spreading of scalar a movement is set up in the liquid body. In this way turbulence sets in by the gradient of scalar concentration and diffusional process. On the other hand turbulence activates mixing and effects transport of scalar.

The two processes are mutually dependant. This process of mixing and turbulence goes on till scalar concentration is almost uniform. The present study is limited to one such situation of development of turbulence vis-a-vis turbulent transfer of scalar. We may observe that in this case the mean flow may be taken equal to zero. In addition to this we assume that scalar gradient and scalar fluctuation are poorly correlated.

On the basis of the above assumptions we get the following steady state form from Eq.2.

$$n\frac{\partial \overline{C}}{\partial x_k}\overline{u_3 c^{n-1}} + \frac{\partial \overline{u_3 c^n}}{\partial x_3} - n\frac{\partial \overline{u_3 c}}{\partial x_3}\overline{c^{n-1}} = n\lambda\overline{c^n} + D\frac{\partial^2 \overline{c^n}}{\partial x_3 \partial x_3} - n(n-1)D\overline{c^{n-2}(\frac{\partial c}{\partial x_3})^2} \tag{3}$$

Here the suffix 3 stands for component normal to the boundary.

Again going back to our transfer system we would like to observe that gradient mechanism of production of fluctuation of a scalar concentration is related to diffusional transfer of velocity fluctuation and mean scalar gradient. This has led Sekundov (1978) to suggest a relation between scalar fluctuation and velocity fluctuation. Taking cue from the conjecture of Sekundov we suggest the following relation.

$$c = -\tau u_3 \frac{\partial \overline{C}}{\partial x_3} \tag{4}$$

where τ is the Lagrangian time scale.

Using Sekundov's conjecture in Eq.3 we get an equation as below.

$$-\frac{n}{\tau}\overline{c^n} - \frac{1}{\tau\frac{\partial \overline{C}}{\partial x_3}}\frac{\partial \overline{c^{n+1}}}{\partial x_3} + \frac{\frac{\partial^2 \overline{C}}{\partial x_3}}{\tau(\frac{\partial \overline{C}}{\partial x_3})^2}\overline{c^{n+1}} + \frac{n}{\tau\frac{\partial \overline{C}}{\partial x_3}}\frac{\partial \overline{c^2}}{\partial x_3}\overline{c^{n-1}} - \frac{n\frac{\partial^2 \overline{C}}{\partial x_3}^2}{\tau(\frac{\partial \overline{C}}{\partial x_3})^2}\overline{c^2}\overline{c^{n-1}}$$

$$= n\lambda\overline{c^n} + D\frac{\partial^2 \overline{c^n}}{\partial x_3^2} - n(n-1)\chi_c\overline{c^{n-2}} \tag{5}$$

Here $\chi_c = (\overline{\dfrac{\partial c}{\partial x_3}})^2$ is the rate of scalar dissipation.

At this stage we make a simplifying assumption that mean temperature grdient is a linear function of height. Eq. 5 is then reduced to the following form.

$$-\frac{n}{\tau}\overline{c^n} - \frac{1}{\tau\dfrac{\partial \overline{C}}{\partial x_3}}\frac{\partial \overline{c^{n+1}}}{\partial x_3} + \frac{n}{\tau\dfrac{\partial \overline{C}}{\partial x_3}}\frac{\partial \overline{c^2}}{\partial x_3}\overline{c^{n-1}} = n\lambda\overline{c^n} + D\frac{\partial^2 \overline{c^n}}{\partial x_3^{\,2}} - n(n-1)\chi_c\overline{c^{n-2}}$$

(6)

For n=2 we have ultimately the following form of steady state equation.

$$-\frac{2}{\tau}\overline{c^2} - \frac{1}{\tau\dfrac{\partial \overline{C}}{\partial x_3}}\frac{\partial \overline{c^3}}{\partial x_3} = 2\lambda\overline{c^2} + D\frac{\partial^2 \overline{c^2}}{\partial x_3^{\,2}}$$

(7)

4. THE CASE OF A NON-GAUSSIAN SCALAR FIELD

It is difficult to find τ directly from the field data. Continuous measurement of velocity and temperature at different locations and at different times following a fluid spec moving in a turbulent flow is not a feasible proposition. So we have to calculate Lagrangian time scale from Eulerian data, realised from the experiment. Methods of calculation have been suggested by Phillip (1978), Chanda (1978), Corrsin (1959) and Saffman (1963) and (1963). Shlien et al (1974) have opined that from theoretical view-point it is not possible to have a consistent relation between Eulerian and Lagrangian scales. But they have given some comparative analysis of the relations obtained by different authors. We, however, for our present purpose use the relation of Saffman (1963), which is given as

$$\tau = 0.4\frac{L}{u'}$$

(8)

where L and u' are respectively Eulerian length scale and rms velocity fluctuation in the vertical direction.

On the basis of these assumptions, so far discussed, Chanda et al (1999) have analysed a case of convection in this type of turbulence but without source (sink) and got the steady state solution for temperature variance when the temperature field is Gaussian. In the present case, however, we would analyse a non-Gaussian case with constant skewness factor S, given by

$\overline{c^3} = S(\overline{c^2})^{\frac{3}{2}}$. Using this skewness factor and further nondimensionalising the equation for n=2 with the help of the following relation

$$\overline{c^2} = f\overline{c^2}\big|_0, \quad \overline{C} = \Theta\sqrt{\overline{c^2}}\big|_0, \quad x_3 = z\sqrt{Dt}$$

(9)

we get the steady state equation for non-dimensional scalar variance as below.

$$f'' + af^{\frac{1}{2}}f' + bf = 0$$

(10)

where *prime* indicates differentiation w.r.t. z. Here

$$a = 3.75\frac{S}{\dfrac{\partial \Theta}{\partial z}}\frac{u'L}{D}, \quad b = 2(\frac{L^2\lambda}{D}) + 5(\frac{u'L}{D})$$

(11)

5. A NUMERICAL SOLUTION

Eq.10 is a nonlinear equation. Direct analytical solution may be difficult to get. In the present case we have done a numerical solution after discretising the equation. The result of the numerical solution has been exhibited in the figure. We may see that in equations 11 there are two nondimensional numbers. The first one $\dfrac{u'L}{D}$ may be taken as a Schmidt number for the scalar, whereas the second number $\dfrac{L^2\lambda}{D}$ may be linked with the rate of reaction of the source (or sink). While doing the numerical calculation we have more or less arbitrarily chosen the values of these nondimensional numbers. Two curves have been presented one with a positive rate of reaction and the other without any reaction vis-à-vis any source (or sink) but keeping the other terms unchanged. The effect of linear source or sink is thus exhibited. The values, considered for the numerical calculation and then plotting the graphs are as below.

$$S = 0.1, \quad \frac{\partial \Theta}{\partial z} = -1.0, \quad \frac{u'}{LD} = 1.0, \quad \frac{L^2\lambda}{D} = 1.75 \tag{12}$$

The result of the numerical calculation will vary from case to case. Our obective here is to present the model and to show theoretically how turbulent transport of scalar is likely to take place in inhomogenous turbulence in the boundary layer complicated by the presence of a source (or sink).

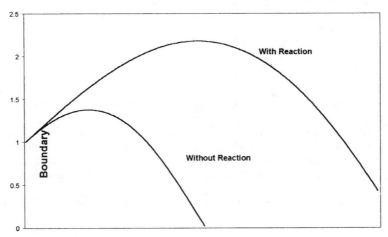

Nondim Scalar Variance vs Distance from Boundary

ACKNOWLEDGEMENT

The authors express their thanks to the Department of Science and Technology, the Government of India for financial assistant. The authors also express their gratitude to the Director, Indian Statistical Institute, Calcutta for providing infrastructural facilities.

REFERENCES

G. K. Batchelor, "Small scale variation of convected quantities like temperature in turbulent fluid. Part 1. General discussion and the case of the small conductivity", *J. Fluid Mech., vol. 5, pp 113-133,1959.*

R. Brost and J. C. Wyngaard, "A model study of the stably stratified planetary boundary layer", *J. Atmos. Sci.}, vol. 35, pp. 1427-1440, 1978.*

J. O. Hinze, *Turbulence. McGraw-Hill Book Co., 1959.*

B. E. Launder, G. J. Reece and W. Rodi, "Progress in the development of a Reynold's stress turbulent closure", *J. Fluid Mech.},vol. 68, pp 537-566, 975.*

S. Tavoularis and S. Corrsin, "Experiment in nearly homogenous turbulent shear flow with a uniform mean temperature gradient", *J. Fluid Mech., vol 104, pp 311-347, 1981.*

J. C. Wyngaard,"Modelling the planetary boundary layer - extension to the stable case", *Boundary layer Met., vol 9, pp 441-60, 1975.*

J. C. Wyngaard and O. R. Cote, "The budgets of turbulent kinetic energy and temperature variance in the atmospheric surface layer", *J. Atmos. Sci., vol 28, pp 190-201, 171.*

J. C. Wyngaard and O. R. Cote, "The evolution of a convective planetary layer - a higher order closure model", *Boundary Layer Met., vol 7, pp 284-308, 1974.*

S. A. Orszag and G. S. Patterson, "Numerical simulation of three dimensional homogenous isotropic turbulence", *Phys. Rev. Lett., vol 28, pp 76-79, 1972.*

B. Castaing, G. Gunaratne, F. Heslot, L. Kadanoff, A. Libacher, S. Thomae, X. Wu, S. Zaleski and G. Zanetti, "Scaling of hard thermal turbulence in Raleigh-Benard convection", *J. Fluid Mech., vol 204, pp 1-30, 1989.*

F. Chilla, S. Ciliberto, C. Innocenti and E. Pampaloni, "Spectra of local and average scalar fields in turbulence", *Europhysics Letters, vol 22, no. 1, pp 23-28, 1993.*

F. Chilla, S. Ciliberto and C. Innocenti, "Thermal boundary layer in turbulent thermal convection", *Europhysics Letters, vol 22, no. 9, pp 681-687, 1993.*

O. Iada and N. Kasagi, "Direct numerical simulation of unstably stratified turbulent channel flow", *Proc. 4th ASME-JSME Thermal Engng Joint Conf., Hawaii, vol 1, pp 417-424, 1995.*

N. Kasagi and N. Shikazono, "Contribution of direct numerical solution to understanding and modelling turbulent transport", *Proc. R. Soc., London, A, vol 451, pp 257-292, 1995.*

A. Sekundov, "A phenomenological model and experimental study of turbulence in the presence of density fluctuations", *Fluid Mechanics, Soviet Research, vol 7, no. 2, pp 11-18, 1978.*

J. Phillip, "Relation between Eulerian and Lagrangian statistics", *Phys. Fluids, vol 10 supple, no. 2, pp S67-S71, 1967.*

Amitabha Chanda, "Some aspects of turbulent medium studied from statistical idea", *Ph. D. Thesis, Calcutta University, Calcutta, India, 1978.*

S. Corrsin, "On outline of some topics in homogenous turbulence", *J. Geophys. Res., vol 64, pp 2134-2150, 1959.*

P. Saffman, "On the fine scale structure of vector field convected by a turbulent fluid", *J. Fluid Mech., vol 16, pp 545-572, 1963.*

P. Saffman, "An approximate calculation of the Lagrangian autocorrelation coefficient for stationary homogenous turbulence", *Appl. Sci. Res., vol A11, p 245, 1963.*

D. Shlien and S. Corrsin, "A measurement of Lagrangian velocity autocorrelation in approximately isotropic turbulence", *J. Fluid Mech, vol 62, pp 255-271, 1974.*

Amitabha Chanda and J. Das, "Inhomogenous vertical turbulent convection in the planetary boundary layer", *Mechanics Research Communications, vol 26, no.3, pp 371-377, 1999.*

Stochastic Hydraulics 2000, Wang & Hu (eds) © 2000 Taylor & Francis, ISBN 90 5809 166 X

Measuring of turbulence intensity interference on the wake flow of circular cylinder

Bao-Shi Shiau
Department of Harbor and River Engineering, National Taiwan Ocean University, Keelung, Taiwan, China

ABSTRACT: Measurement of turbulence intensity interference on the wake flow of a circular cylinder was conducted in the wind tunnel. Different turbulence intensity (3.2%, 8%) with the same mean velocity of incoming flows were created. This is to separate the turbulence intensity variations from the mean flow field. The separation of turbulence intensity from conventional mixing of the mean velocity and turbulence intensity has a merit to get a clear picture of turbulence intensity effect. The Reynolds number composed of the free incoming flow stream velocity and cylinder diameter is about 4,500. Effects of the turbulence intensity of incoming flow on the development of the mean velocity profile of wake flow and the vortex shedding behind the circular cylinder are presented.

Keywords: Turbulence intensity, Circular cylinder, Wake flow

1 INTRODUCTION

Circular cylinders are commonly used as the important component of hydraulic engineering structures (such as bridge piers, pipelines under water, superstructures offshore platform, chimney, etc.). Of these hydraulic engineering structures, the superstructures of offshore platform expose to wind loads. The effect of turbulence intensity of wind on the exposed offshore platform is an important design consideration. Also, the turbulent flow past cylinder is of practical interest and basic common phenomenon.

For many years, studies on the wake flow behind the circular cylinder had been made. Nashioka and Sato (1974) had measured the velocity distribution in the wake of a circular cylinder at Reynolds number 10~80. Cantwell and Coles (1983) had investigated the near wake flow for a uniform laminar flow past a circular cylinder at the Reynolds number of 1.4×10^5. Hussain and Hayakawa (1987) used hot wire rake to measure the organized flow structure behind the circular cylinder at Reynolds number 1.3×10^4. On reviewing these typical experimental studies, it is found that incoming flow on the circular cylinder is smooth flow (i.e. laminar flow, or very low turbulence intensity). In fact, the actual incoming wind on the superstructures of offshore platform is turbulent flow. The turbulent incoming flow may vary turbulence intensity. Up to now, the published literature shows that little attentions have been received on the turbulence intensity of incoming flow and relevant data are rarely available. Therefore in this paper, we measure and investigate the turbulence intensity interference on the wake flow of circular cylinder.

2 EXPERIMENTAL WORK

The empty wind tunnel turbulence intensity is less than 0.45 % at the free stream velocity of 5 m/s. Various turbulence intensity (turbulence intensity 3.2%, 8%) with keeping the same mean velocity of incoming flows were obtained by setting different sizes of grids on the entrance of the test section.

The circular cylinder has a diameter of 0.76 cm with 20cm long. The blockage ratio in wind tunnel is 3.8 %. According to the experimental results of Maskell (1965), Saathoff & Melbourne, and Biggs, they indicated that the blockage effect is of no significance when blockage ratio is less than 5.6 %(Seathoff & Melbourne, 1987,Biggs, 1954). The Reynolds number based on the free stream velocity of incoming flow and cylinder diameter was about 4,500. Fig.1 is the schematic diagram of the experimental arrangement.

The TSI IFA-300 constant temperature anemometer in conjunction with a KANOMAX 0252R-T5 cross-type hot wire was employed to measure the turbulent signals in the wake of circular cylinder.

3 RESULTS AND DISCUSSION

Two turbulence intensity cases (turbulence intensity 3.2%, 8%) with created and it keeps the same mean velocity of incoming flows for all cases. The reason for doing so is to separate the turbulence intensity variations from the mean flow field. The separation of turbulence intensity from conventional mixing of the mean velocity and turbulence intensity has a merit to get a clear picture of turbulence intensity effect on the wake of circular cylinder.

The wake mean velocity profile development and integral length scale downstream of the cylinder at different turbulence intensity of incoming flows are presented and discussed.

3.1 Mean velocity profile development

The mean velocity deficit of wake flow is defined as: $U_s = U_f - U$; U_f : the free stream velocity; U : mean velocity. Mean velocity at the wake center is U_c. The maximum mean velocity deficit, $(U_s)_{max}$ occurs at the wake center. The distance from where one half of maximum mean velocity deficit occurred to the center is defined as half width of the wake, b. Fig. 2 shows the mean velocity deficit distributions at different downstream distances for smooth flow. Here smooth flow represented the incoming flow with very low turbulence intensity (i.e. less than 0.45 %). Fig.3 and Fig.4 are the mean velocity deficit distributions at different downstream distances for incoming flow with turbulence intensity of 3.2 % and 8 %, respectively. Sato and Okada [7] had proposed the following equation to describe the wake flow mean velocity distribution.

$$\frac{U_f - U}{U_f - U_c} = \exp[-\ln 2(\frac{Z}{b})^2]$$ (1)

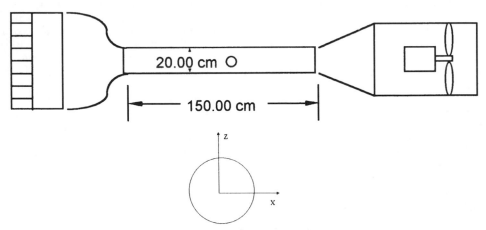

Fig1: Schematic diagram of the experimental arrangement

As comparing the present measurement results with Sato and Okada's equation, it is found that the mean velocity deficit distributions of the present measurements for three cases are all close to the equation.

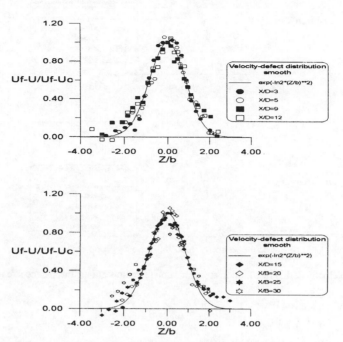

Fig2: Mean velocity deficit distribution at various downstream stations; smooth flow

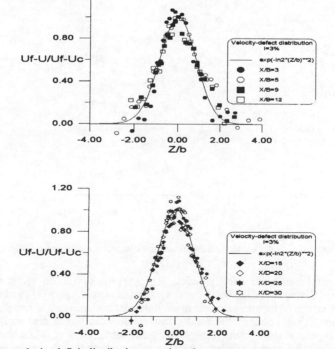

Fig3: Mean velocity deficit distribution at various downstream stations; turbulence intensity 3.2 %

163

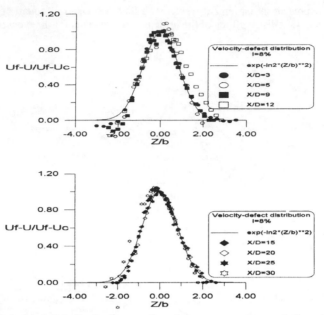

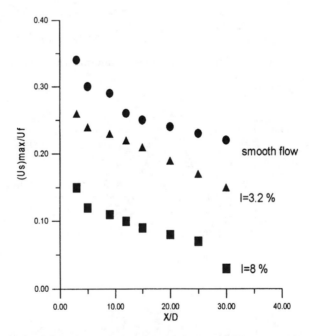

Fig4: Mean velocity deficit distribution at various downstream stations; turbulence intensity 8 %

Fig5: Maximum velocity deficit decays along the downstream distance for different turbulence intensity

Fig.5 shows the maximum mean velocity deficit, $(U_s)_{max}$ as function of downstream distance for different turbulence intensity of incoming flow cases. Nakamura & Ohya [8] , and Nakamura et al. [9] studied the effects of turbulence on the mean flow past square rods and two dimensional rectangular cylinders. They obtained the results that small scale turbulence accelerates the growth rate of the separated shear layers through enhanced entrainment of fluids from the near wake and external flow. So, as the turbulence intensity of incoming flow increases, it enhances the entrainment of wake flow. Therefore, $(U_y)_{max}$ becomes smaller as the turbulence intensity of incoming flow increases.

3.2 Integral length scale

The variation of integral length scale of wake flow, L_x for different turbulence intensity along downstream distance is presented in Fig. 6. It is shown that integral length scale of wake flow becomes larger along the downstream distance. For increasing turbulence intensity of incoming flow, the integral length scale of wake flow decreases.

3.3 Strouhal number

The vortex shedding phenomenon occurred in the near wake of cylinder. We measured the vortex shedding frequency, f by using the hot wire at the location behind cylinder $X/D=5$, $Z/D=0.5$. The measured vortex shedding frequency is scaled to a non-dimensional parameter, Strouhal number $St = \dfrac{fD}{U_f}$. Here D is diameter of cylinder, and U_f is the mean velocity of incoming flow. The present measurement of Strouhal number for smooth flow (very low turbulence intensity (less than 0.45%) of incoming flow) is 0.228. For the case of turbulence intensity 3.2 % of incoming flow, the measured turbulence intensity is 0.219. This value is smaller than that of the smooth flow case. Shih et al. [10] measured flow past rough circular cylinder at larger Reynolds number, $10^5 \leq Re \leq 10^7$. They obtained rougher cylinder producing the lower Strouhal number. In general, rougher cylinder creates larger turbulence. The trend of Shih et al.'s result consists with the present measurement.

4. CONCLUSION

Measurement of turbulence intensity interference on the wake flow of a circular cylinder was conducted in the wind tunnel. Different turbulence intensity (3.2%, 8%) with keeping the same

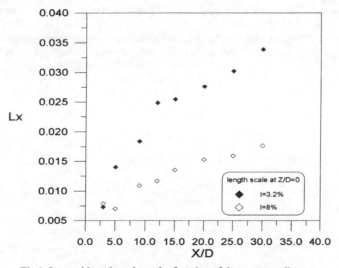

Fig6: Integral length scale as the function of downstream distance

165

mean velocity of incoming flows were created. This is to separate the turbulence intensity variations from the mean flow field. The separation of turbulence intensity from conventional mixing of the mean velocity and turbulence intensity has a merit to get a clear picture of turbulence intensity effect. Measurement results are summerized as follows:

The present measurement of mean velocity profiles of wake flow is good in agreement with equation proposed by Sato and Okada.

The maximum velocity deficit becomes smaller as the turbulence intensity of incoming flow increases.

For reducing turbulence intensity of incoming flow, the integral length scale of wake flow increases.

As increasing the turbulence intensity of incoming flow, the Strouhal number decreased.

ACKNOWLEDGEMENT

Mr. Cheng-huan Lee's help during the experiment is deeply appreciated.

REFERENCES

Biggs, J.M., "Wind Loads on Truss Bridges," Journal of Structural Engineering, ASCE, Vol.119, p879, (1954)

Cantwell, B. and Coles, D., "An experimental study of entrainment and transport in the turbulent near wake of a circular cylinder, " Journal of Fluid Mechanics, Vol.136, pp.321-374 (1983)

Hussain, A.K.M.F., and Hayakawa, M., "Evolution of large-scale organized motion in the turbulent plane wake," Journal of Fluid Mechanics, Vol. 180, pp. 193-229 (1987)

Maskell, E.C., "A theory of the blockage effects on bluff bodies and stall wings in a closed wind tunnel," Aerodynamic Research Council, R&M 3400, (1965)

Nakamura, Y., and Ohya, Y., "The effects of turbulence on the mean flow past two dimensional rectangular cylinder," Journal of Fluid Mechanics, Vol.140, pp.255-273, (1984)

Nakamura, Y., Ohya, Y., and Ozono, S., "The effects of turbulence on bluff body mean flow," Journal of Wind Engineering and Industrial Aerodynamics, Vol.28, pp.251-259, (1988)

Nishoka M., and Sato. H., "Measurement of velocity distribution in the wake of a circular cylinder," Journal of Fluid Mechanics, Vol.65, pp.97-112 (1974)

Saathoff, P.J., and Melbourne, W.H., "Free stream turbulence and wind tunnel blockage effects on streamwise surface pressures," Journal of Wind Engineering and Industrial Aerodynamics, Vol.26, pp.353-370, (1987)

Sato, H. and Okada O., "The stability and transition of an axisymmetric wake," Journal of Fluid Mechanics, Vol.26, pp.237-253, (1966)

Shih, W.C.L., Wang, C., Coles, D., and Roshko, A., "Experiments on flow past rough circular cylinder at large Reynolds number," Journal of Wind Engineering and Industrial Aerodynamics, Vol.49, pp.351-368, (1993)

Stochastic Hydraulics 2000, Wang & Hu (eds) © 2000 Taylor & Francis, ISBN 90 5809 166 X

Decay of turbulence generated by oscillating grid

Nian-Sheng Cheng & Adrian Wing-Keung Law
School of Civil and Structural Engineering, Nanyang Technological University, Singapore

ABSTRACT: Turbulence generated by a vertically oscillating grid in a water tank was investigated using the DPIV technique. The statistical turbulence characteristics computed based on the experimental data obtained agree well with the two main findings reported in the literature, i.e., the turbulence decays following the power law and the integral length scale increases linearly with the distance from the grid. The flow structure near the grid was particularly observed in details utilising the advantage of the planar measurements. It was found that the velocity fluctuations in the region near the grid vary depending on the grid geometry. The velocity fluctuations immediately over the bar position are significantly different from those over the grid openings. Turbulence with the highest intensity occurs above the intersection of the square bars that constitute the grid. The results imply that shear flow clearly exists near the grid and the homogeneity of the turbulence can only be achieved at a distance from the grid greater than about three mesh sizes. In addition, it was found that the number of velocity vector maps from the data sampling should be large enough to ensure the accuracy of the measured velocity fluctuations.

Keywords: DPIV, Flow measurement, Grid turbulence, Homogeneity, Turbulence decay, Reynolds stress, And integral scale.

1. INTRODUCTION

Turbulence generated by an oscillating grid in a water tank has been used in various laboratory experiments. One typical kind of experiments is related to the studies of the interfacial mixing in stratified fluids that widely exists in the ocean, the atmosphere and lakes (Thompson and Turner 1975; Hopfinger and Toly 1976). Similar experiments have also been performed to study the free-surface affected turbulence and the gas transfer processes at the gas/water interface (Brumley and Jirka 1987). In comparison with the turbulence in boundary layers and open channels, the turbulence generated by the oscillating grid is considered to be theoretically simpler because it is characterised by zero-mean velocity and homogeneity. However, such grid-induced turbulence is still complex since it is obviously affected by the shape and size of the grid, and the stroke and frequency of oscillation. Fortunately, most of the grids used in the previous studies are rather similar which allows a direct comparison of the results obtained in these studies. Even so, a number of questions regarding the characteristics of this mechanically driven turbulence remain to be answered.

The early measurements of the grid-induced turbulence were performed with a hot-film probe mounted on a spindle, which rotates on a plane parallel to the grid plane (Thompson and Turner 1975; Hopfinger and Toly 1976). These measurements directly yielded spatially averaged velocities. Thompson and Turner (1975) demonstrated that for a grid made of square bars, the integral length scale of the oscillating grid turbulence increased linearly while the r.m.s.

horizontal velocity decayed with the distance from the grid to the power of -1.5. Their results also showed that the power had to be much less negative than -1.5 to describe the decay of the turbulent fluctuation in the region close to the grid. Hopfinger and Toly's (1976) study confirmed the linear dependence of the turbulence length scale on the distance from the grid, as found by Thompson and Turner. For the r.m.s. horizontal velocity, they proposed the following empirical expression:

$$u = CfS^{1.5}M^{0.5}z^{-n}$$ (1)

where C = 0.25, n = 1, u = r.m.s. horizontal velocity, f = frequency, S = stroke, M = mesh size defined as the distance between the centres of two neighbouring openings, and z = distance from a virtual origin near the mid-position of the grid. Hopfinger and Toly thus concluded that the turbulent kinetic energy, k, decays with the distance from the grid according to a power law, $k \propto z^{-2}$, and that the turbulent Reynolds number remains approximately constant during the decay.

Besides the measurements with the hot-film probe described above, the grid turbulence has also been studied using other techniques such as LDV (e.g. McDougal 1979; De Silva and Fernando 1992), ADV (Brunk et al. 1996), and PIV (Lyn 1997). McDougal (1979) found that the turbulence generated with 1 cm stroke was far from homogeneous in the horizontal plane. His results also showed that the cutoff value of the oscillating frequency was approximately 7 Hz, above which a large circulating motion occurred and the r.m.s. horizontal velocity was no longer linearly related to frequency. Brunk et al. (1996) used five horizontally oscillating grids to simulate hydrodynamic and chemical processes associated with pollutant fate in aquatic environments. Despite the very different grid arrangement, their results on the kinetic energy of the turbulence are still close to those obtained earlier by Hopfinger and Toly (1976). Lyn (1997) measured the oscillating-grid flow using the PIV technique and reported that the spatially averaged turbulence characteristics agreed with the previous work, while the time-averaged results at a single section implied strong and persistent large-scale motion. The accuracy of his statistical results may however be notably affected by the small number of images taken which only ranged from 25 to 36. The effect of the sampling duration and frequency of the PIV measurement on the statistical results of the turbulence characteristics will be addressed further in this paper.

In this study, the Digital Particle Image Velocimetry (DPIV) technique is employed to study the turbulence induced by an oscillating-grid. The spatial decay of the turbulence generated and its horizontal homogeneity are re-examined using the new experimental data. The flow information obtained is expected to be helpful in further studies on particle dispersion in the confined turbulent environment.

2. EXPERIMENTAL APPARATUS AND PROCEDURES

2.1 Oscillating Grid System

Fig. 1(a) shows a schematic representation of the adopted turbulence-generating system which was very similar to those used by most of the previous researchers. The water tank was made of glass and had a dimension of 500×500 mm in cross section and 1000 mm in height. Its transparent walls and bottom allowed the projection of a laser sheet across the flow field and the associated image capturing with digital cameras. The tank was supported by a platform, of which the height was adjustable. A grid made of square bars of 10×10 mm had a mesh size of 50 mm, giving a solidity of about 36%. A solidity lower than 40% was considered to be capable of effectively avoiding strong inhomogeneity and secondary circulation. The grid was hanged vertically at the middle of the water depth of 600 mm by four steel bars with 5-mm diameter, which were connected to a speed-controlled motor.

To obtain a homogenous turbulent field, the grid arrangement was further improved by choosing its end condition in such a way that the walls acted as planes of symmetry. This is consistent with the suggestion given by E and Hopfinger (1986), who stated that the turbulence is strongly affected by the grid geometry near the wall. In addition, the gap between the glass wall and the ends of the grid bars was set as small as 2~3 mm to further avoid the influence of the end condition. The grid system was oscillated vertically with a stroke of 40 mm from 1 to 4 Hz. Since the oscillating grid turbulence is sensitive to the initial conditions similar to the ordinary grid turbulence in wind tunnels, measurements were taken at least twenty minutes after the start of the oscillation.

2.2 Digital Particle Image Velocimetry (DPIV)

The horizontal and vertical velocities were measured using the Digital Particle Image Velocimetry (DPIV) technique, which can be conveniently employed to obtain planar flow distribution in the area to be sampled. The planar capability is an important advantage compared to the point-based velocimetry techniques like LDA and ADV. The DPIV technique has emerged as the digital counterpart of the conventional laser speckle velocimetry and the film-based Particle Image Velocimetry (PIV) techniques. It enables two-dimensional digital images to be captured, viewed and processed in real time. In this study, a Dantec DPIV system was used that includes a dual-cavity frequency-induced Q-switched pulsed mini Nd:Yag laser with an energy level of 25 mJ per pulse, a Kodak Megaplus ES1.0 CCD camera with a resolution of 1008x1018 pixels, and the PIV 2000 Processor. This system has been successfully implemented in combination with Planar Laser Induced Fluorescence (PLIF) by Law and Wang (1998; 1999) to study the mixing processes induced by buoyant jets. Its performance was verified with good agreement between the experimental results obtained and the existing point-based measurements.

The typical processes involved in the DPIV data acquisition consist of seeding, illumination, recording and data analysis. Polyamid particles with a nominal diameter of 50 μm were used as flow field tracers. Seeding density was adjusted to ensure that at least about 5 particles are present within the interrogation area. The two laser beams from the dual pulsed laser were aligned such that the two vertical light sheets illuminated the same spatial area. The pulse duration was about 7 ns, which was much smaller than the time scale of the flow so that the velocity measured can be considered to be instantaneous. The pulse interval was optimised to achieve a maximum displacement of about 8 pixels, i.e., 25% of the side of the interrogation area. The light sheet had a typical thickness of 3 mm and a divergence angle of 32°.

Images with a size of approximately 10×10 cm were taken at different locations in the flow field. A plan view of the locations is shown in Fig. 1(b). The camera, of which the axis was perpendicular to the laser sheet, was adjusted with slight defocusing such that the size of the seeding particle was generally greater than 3 pixels. This was necessary to enhance the accuracy of the sub-pixel interpolation. The camera, having an intensity resolution of 8 bits, was equipped with a CCD chip of 9.1×9.2 mm. The sampling rate was set at 5 Hz, which was directly limited by the maximum readout frequency for the double images being approximately 7 Hz. Cross-correlation analysis was performed to derive velocity vectors from the images. The interrogation area was selected as 32×32 pixels with an overlap of 25%. The overlap does not increase the resolution of DPIV but does provide more spatial information especially at the edge of the interrogation area for the cross-correlation process.

The sampling duration of the experiments was chosen to be 100 s, i.e., 500 images were taken at each location. This duration was determined based on preliminary tests to explore the effect of the number of images on the statistical results of the velocity fluctuations. An example of the horizontal velocity taken at three different elevations at Section B is presented in Fig. 2, where the number of images used in the averaging calculations ranges from 1 to 500. It shows that the statistical results of the velocity fluctuation are not consistent for small samples and only approach asymptotic values when the sample exceeds approximately 400 images. This observation implies that a low accuracy may be associated with the statistical results in Lyn (1997) that were obtained based on only 25 to 36 images.

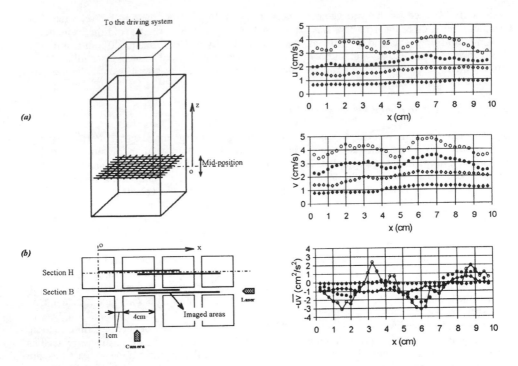

Fig. 1: Experimental Setup. (a) Water Tank; (b) Plan View of Locations Imaged

Fig. 2: Distributions of Velocity Fluctuations and Reynolds Shear Stress.

3. RESULTS

As shown in Fig. 1(b), two typical sections parallel with each other, Sections B and H, located within one quarter of the grid plane, were selected for the velocity measurements. Section B was aligned with the bar location, while Section H was located throughout the grid openings. It was expected that Section B, in comparison with Section H, might have different flow conditions because vortices are directly shed from the bar when the grid is oscillated. Each section was composed of four imaging areas with 50% overlap in the horizontal direction, of which the two lower areas were chosen just above the top position of the oscillating grid and the upper areas were set approximately 4 cm below the water surface. Verification tests using different water depths were performed to confirm that the water surface effect on the velocity fluctuation is insignificant at the selected imaged areas.

Fig. 3 shows the distributions of the temporally averaged values of the velocity fluctuations and the Reynolds shear stress at different elevations measured at Section B. Significant horizontal variations of the measured velocity fluctuations can be observed at a small distance from the grid, say, at z = 5 cm. With increasing distance from the grid, the turbulence generated becomes more homogeneous in the horizontal plane and the variations in the velocity fluctuations reduce. Furthermore, it can be seen in Figs. 3(a) and 3(b) that the velocity fluctuations for Section B are enhanced at the locations immediately above the bar intersections, i.e., x = 2~3 cm and 7~8 cm. This is probably due to the interactions of the vortices shed from the bars, which are aligned in the two directions perpendicular to each other.Correspondingly, the distribution of the Reynolds shear stress clearly indicates that the turbulence can only be considered to be shear-free given a distance sufficiently far away from the grid, as shown in Fig. 3(c). In the region near the grid, the Reynolds shear stress undulates around the zero value. Smaller variations near the grid are observed for the measurements taken at Section H than Section B.

170

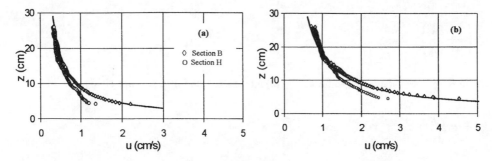

Fig. 3: Comparison of the Decay of Spatially-Averaged Horizontal Velocity Fluctuation at Two Sections with Empirical Formula Eq. (1). (a) f = 2 Hz; (b) f = 4 Hz.

The various distributions of the flow characteristics at the different elevations shown in Fig. 3 indicate that the mechanism of the turbulence generation is comparable to the qualitative observations previously reported by Hopfinger and Toly (1976) and Nokes (1988). Generally two kinds of flow, namely jets and wakes, can be identified with the grid oscillation. The jets are formed throughout the grid openings and the wakes are created behind the bars. Whether the jets or wakes are dominant depend on the grid geometry. For a grid with a high solidity like a plate with small holes, jets are important in generating the turbulence away from the grid. In comparison, for a grid with a low solidity like the 36% used in this study, the turbulence is formed primarily by the interaction of the wakes. This is the reason why the near-grid turbulence appears with the high intensity and high shear at Section B instead of Section H. The jets or wakes interact with each other, and the turbulence diffuses away from the grid. Shear-free turbulent motion with little or no mean horizontal variation finally occurs at a sufficiently large distance from the grid.

To compare the present results with the empirical scaling law for the turbulence decay as shown in Eq. (1), the measurements of the velocity fluctuations were further averaged spatially at each elevation for the two sections. The results of the horizontal velocity fluctuation are presented in Fig. 5 for f = 4 Hz, where the data are also compared to the empirical formula in Eq. (1). Fig. 5 shows that the near-grid turbulence at Section B is stronger than that at Section H. This difference between the two sections becomes insignificant when the distance from the grid is greater than about three mesh sizes, i.e. z = 15 cm. It is also interesting to note that Eq. (1) seems to represent an upper bound of the horizontal velocity fluctuations, which is close to the measurements taken at Section B.

Due to the advantage of the DPIV technique, the planar velocity measurements were taken simultaneously which enable an evaluation of the spatial correlation of the velocity fluctuations over the imaged area. The integral length scales can thus be computed using the longitudinal and transverse correlation coefficients, respectively. The longitudinal and transverse correlation coefficients were first evaluated for each imaged area and then an ensemble averaging was made using all the 500 images. The results indicate that the correlation coefficient for small distance lag increases with increasing the distance from the grid. The computed integral length scale is almost proportional to the distance from the grid. The coefficient of proportionality derived from the longitudinal correlation coefficient ranges from 0.1 to 0.2, while that derived from the transverse correlation coefficient varies from 0.06 to 0.17. These values are close to the one given by Hopfinger and Toly (1976), who reported that the integral length scale is approximately equal to 10% of the distance from the virtual origin of the grid.

With the observation that the length scale is proportional to the distance from the grid, the eddy viscosity can thus be expressed in the form (Rodi, 1993)

$$v_t = c'_\mu \beta \sqrt{k} z \qquad (2)$$

where v_t = eddy viscosity, c_μ' = empirical constant, β = ratio of the length scale to the distance from the grid, and z = vertical distance from the grid mid-plane. On the other hand, the k-equation reads

$$\frac{\partial}{\partial z}\left(\frac{v_t}{\sigma_k}\frac{\partial k}{\partial z}\right) = c_D \frac{k^{\frac{3}{2}}}{\beta z} \qquad (3)$$

where σ_k, c_D = constants. Substituting Eq. (2) into Eq. (3), and then solving yields

$$k^{\frac{3}{2}} = c_1 z^{\sqrt{c}} + c_2 z^{-\sqrt{c}} \qquad (4)$$

where c_1, c_2 = constants and $c = (1.5 c_D \sigma_k) / (c'_\mu \beta^2)$. Considering the boundary condition that k approaches zero at a large distance from the grid, c_1 should be equal to zero and therefore Eq. (4) simplifies to

$$k = c_2^{\frac{2}{3}} z^{-\frac{2\sqrt{c}}{3}} \qquad (5)$$

Eq. (5) shows that the kinetic energy of the turbulence and thus the velocity fluctuation decay with increasing distance following the power law, which is consistent with Eq. (1) in spite of the unknown constants included.

5. CONCLUSIONS

The DPIV technique was employed to study the distribution of the turbulence generated in a water tank by a vertically oscillating grid. The experimental results confirm the previous findings obtained based on accurate point-based measurement techniques. In addition, the present study quantitatively reveals the flow structure in the region near the grid. It was found that the structure is closely related to the grid geometry. The intensity of the near-grid turbulence is clearly enhanced in the region at the bar intersections. Shear flow obviously exists near the grid and the turbulence generated can only be considered to be horizontally homogeneous when the distance from the grid mid-plane is greater than approximately three mesh sizes.

REFERENCES

Brumley, B. H. and Jirka, G. H. (1987). "Near-surface turbulence in a grid-stirred tank." J. Fluid Mech.. 183, 235-263.

Brunk, B., Weber-Shirk, M., Jensen, A., Jirka, G., and Lion, L. W. (1996). "Modeling natural hydrodynamic systems with a differential-turbulence column." J. Hydr. Engrg., ASCE, 122(7), 373-380.

De Silva, I. P. D., and Fernando, H. J. S. (1992). "Some aspects of mixing in a stratified turbulent patch." J. Fluid Mech., 240, 601-625.

E, X. and Hopfinger, E. J. (1986). "On mixing across an interface in stably stratified fluid." J. Fluid Mech., 166, 227-244.

Hopfinger, E.J. and Toly, J.A. (1976) "Spatially decaying turbulence and its relation to Mixing across density interfaces." J. Fluid Mech., 87, part 1, 155-175.

Law, A.W.K. and Wang, H.W. (1998). "Simultaneous Velocity and Concentration Measurements of Buoyant Jet Discharges using Combined DPIV and PLIF." Proc. 2nd International Symposium on Environmental Hydraulics, Hong Kong, 211-216.

Law, A.W.K. and Wang, H.W. (1999). "Simultaneous Velocity and Scalar Measurements of Axisymetric Plume using Combined DPIV and PLIF." Proc. PIV Workshop, Santa Barbara, USA, 445-450.

Stochastic Hydraulics 2000, Wang & Hu (eds) © 2000 Taylor & Francis, ISBN 90 5809 166 X

Simulation of three-dimensional coherent and turbulent motions around a rectangular pile in an open channel flow

Chi-Wai Li & Pengzhi Lin
Department of Civil and Structural Engineering, Hong Kong Polytechnic University, Kowloon, China

ABSTRACT : In this paper, a three-dimensional Large-Eddy- Simulation (LES) model is used to study the flow motions around a rectangular pile exposing to a turbulent channel flow. The flow pattern around the pile is analyzed, with the emphasis on the vortex shedding characteristics that are directly related to the forces acting on the structure. Attention is also drawn to the time-averaged vortical motion in the vertical plane, which is the unique feature due to the presence of free surface in the channel flow. The resolved coherent and turbulent structures behind the pile are separated from the mean flow with the aid of energy density spectrum analysis. Different mesh systems have been used to simulate the same problem for the mesh convergence test. The results are also used to provide the guideline of choosing the optimal mesh size in the LES model.

Key words: LES model, Channel flow, Pile, Coherent motion, Turbulent motion

1. INTRODUCTION

Structures in rivers, estuaries, or coastal regions are subject to the long period flow action such as river flows or tides. The stability analysis of structures in these regions relies on the accurate prediction of flow and turbulence structures around the objects. Using the commercial model FLOW-3D that employs the k-ε turbulence model to treat turbulence fluctuation, Richardson and Panchang (1998) studied the scour problem near a bridge pile. Unfortunately, the turbulence effect on the pile was not adequately addressed in their study.

Compared with the k-ε model, the Large Eddy Simulation (LES) model is distinct in the way it handles the turbulence averaging process. In the last decade, it has been receiving increasing attention from researchers all over the world due to its capability of providing large scale chaotic flow pattern directly with moderate computational cost. For example, Shah and Ferziger (1997) used a LES model to study wind passing a cubic obstacle. Rodi (1997) made comparisons between the LES models and other turbulence models for the same problem. It is noted that no free surface is present in the above studies. On the other hand, Hodges and Street (1999) proposed a LES model for turbulent free surface flows. Li and Wang (1999) also developed a LES model to study the turbulence characteristics of a free surface shallow water flow.

The study of a free surface flow interacting with solid bodies using a LES model, however. has rarely been reported. In this study, a three-dimensional numerical model using σ-coordinate transformation (Lin and Li, 2000) will be employed to investigate the flow structures around a rectangular pile in a turbulent channel flow. In the following sections, a brief introduction of the mathematical formulation is provided, followed by the discussions of numerical results. The forces on the structure, the flow and turbulence pattern around the structure, and the distinct flow pattern induced by the presence of free surface will be analyzed in detail. Finally, a concise conclusion will be provided based on the current study.

2. MATHEMATICAL FORMULATION

In the LES model, the governing equations for the spatially averaged mean flow are first derived from the original Navier-Stokes equations as follows,

$$\frac{\partial \overline{u_i}}{\partial x_i} = 0 \tag{1}$$

$$\frac{\partial \overline{u_i}}{\partial t} + \overline{u_j}\frac{\partial \overline{u_i}}{\partial x_j} = -\frac{1}{\rho}\frac{\partial \overline{p}}{\partial x_i} + \frac{\partial}{\partial x_j}(\tau_{ij} + R_{ij}) \tag{2}$$

where $\overline{u_i}$ is the spatially averaged velocity in the i-th component, $\overline{p}$ the averaged pressure, τ_{ij} the stress by molecular viscous effect, and R_{ij} the stress due to the averaging process. We shall discuss in more detail about the averaging process and its implication to the numerical results when the calculated mean flow and turbulence are discussed.

The term R_{ij} is modeled by the subgrid scale (SGS) model as follows:

$$R_{ij} = 2(L_s^2 \sqrt{2S_{ij}S_{ij}})S_{ij} \tag{3}$$

where S_{ij} is the strain rate of the mean flow and L_S is the characteristics length scale which equals $C_s(\Delta_1\Delta_2\Delta_3)^{1/3}$ with $C_s=0.15$.

The numerical model solves equations (1) and (2) using the operator splitting method (Li and Yu, 1996). The σ-coordinate transformation is used to map the irregular physical domain into a cube where the free surface boundary condition could be precisely applied. The details of numerical implementation are referred to Lin and Li (2000).

3. PROBLEM SETUP

In this study, a rectangular pile with the dimension of 10cm × 10cm is deployed in the computational domain that measures 3.0m × 1.0m on horizontal plane. The center of the pile is 1.0m away from the left boundary and in the middle of the y-direction. The still water depth is 0.1m. The incoming channel flow has a uniform velocity of 0.22m/s. No artificial velocity fluctuation is introduced from the inflow boundary. The small perturbation that is necessary for the generation of turbulence will be automatically induced by the numerical round-off errors. The Reynolds number (Re) in this case is estimated as $Re=UD/\nu=2.2\times10^4$, where U is the velocity of incoming flow and D is the width of the pile. The value of Re in this study is the same as that in the experimental study by Lyn et al. (1995), based on which Shah and Ferziger (1997) and Rodi (1997) validated their numerical models. A non-uniform reference mesh system is proposed that has a total number of 130×80 grids on the horizontal plane with the finest grid $\Delta x=\Delta y=0.005$m arranged near the pile (see Figure 1). Totally 20 uniform grids are used in vertical direction. A constant $\Delta t=0.001$s is used that ensures the numerical stability during the entire simulation. The computation is carried up to t=50s. The radiation boundary condition is used on the downstream side and the rest of lateral boundaries are assumed to be the solid walls.

Four more mesh systems are used for the purposes of mesh convergence test and better understanding of resolvable turbulence structures with the LES model. Three of them have larger grid size and one has smaller grid. The total number of meshes for the testing cases are 33×24×6 (A), 65×40×10 (B), 104×64×16 (C), 130×80×20 (reference case; R), and 156×96×24 (E), respectively. The corresponding finest mesh sizes near the pile, Δx_f, are 0.0167m, 0.0100m, 0.00625m, 0.00500m, and 0.00417m.

4. DISCUSSION

4.1 Vortex shedding

It is well known that when Re is high enough, most of flows passing a bluff body will become turbulent and vortices will be formed at the separation points that are generally located at the sharp corners if they exist. If the flow is periodic and the oscillation frequency is high, the vortices might be confined near the separation points and no vortices shedding will occur. This is often happens if the flow is mainly induced by the wave and the Keulegan-Carpenter (KC) number, $KC=UT/D$, is low, where T is the wave period. When KC is large or approaches infinity (e.g., channel flows), vortices shedding will occur. Normally, the shedding is periodic and the shedding frequency f depends upon both the incoming flow and the size and shape of the solid body. A non-dimensional parameter that is related to this frequency is defined as, $St=fD/U$, which is commonly referred to as Strouhal number.

In this study, since the pile is exposed to the channel flow at Re=22,000, it is expected that the flow around the pile will experience the states from the laminar flow to fully developed turbulent flow. This process has been shown in Figure 2 in terms of vorticity field around the pile for the reference mesh system (without otherwise indicated, the following discussions are all based on the reference mesh system). It is clearly seen that at the initial stage (t=5s), the flow is basically laminar and symmetric about the centerline of the pile in y-direction. As the time progresses, the flow starts to become chaotic and experience the transition from laminar flow to turbulent flow (t=10s). The vortex shedding occurs in the later time (t=15s) and eventually reach the nearly stable state (t=20s).

The vortices shedding behind the pile is periodic. In this study, the shedding period is found to be about 3.2s. This renders that St=0.142, which is close to the values obtained from the experiment and most of numerical computations for the case without free surface (Rodi, 1997). Therefore, it is safe to conclude that the presence of the free surface will not affect the vortices shedding pattern significantly, at least on the horizontal plane. We shall, however, demonstrate later that the velocity field in the vertical plane does exhibit distinct feature when free surface exists. Figure 3 shows the vorticity field between t=36.5s to t=50.0s, during which about one vortex shedding cycle is completed. It is seen that the vortices are originated from the front corners of the pile and they are carried over the back corners and eventually form the periodic shedding pattern that is about symmetric within one shedding period. Near the back corner, there is a small region where the reverse flow is induced and thus the vorticity has the opposite sign to its neighborhood (e.g., t=46.5s or t=48.0s). This is also the feature similar to the case without free surface (Rodi, 1997).

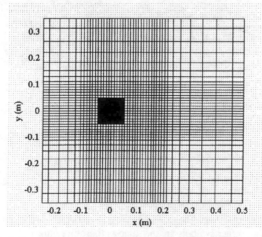

Fig.1: Mesh arrangement near the pile for the reference mesh system; lines are plotted every two grid nodes for easier visibility.

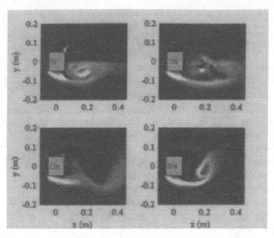

Fig.2: Vorticity field on the middle elevation around the pile that changes from laminar flow to nearly stable turbulent flow; white color represents positive vorticity and black negative with the vorticity ranging from −10 1/s to 10 1/s.

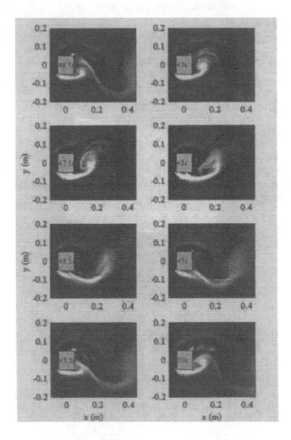

Fig.3: Vorticity field on the middle elevation around the pile at stable state; other symbols are the same as those in figure 2.

176

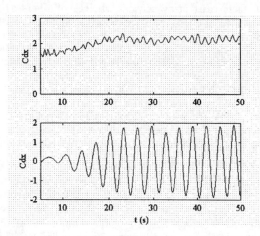

Fig 4: Time histories of calculated drag force coefficients in x- and y-directions.

4.2 Force Coefficients

The vortex shedding alters the pressure field around the pile and thus causes the forces acting on the pile to exhibit oscillating features. Figure 4 shows the calculated drag coefficients in both x- and y-directions. It is observed that the magnitudes of both Cd_x and Cd_y increase from $t=5s$ (laminar flow) to $t=20s$ (nearly stable turbulent flow). After $t=25s$, the stable periodic vortices shedding is formed and so do the forces on the pile. The force in the y-direction has the period of 3.2s, which is the same as the vortex shedding period. The force in the x-direction has the period that is half of that in the y-direction. The variation between any two shedding cycles is caused by the turbulent fluctuations in the flow. Based on the data between $t=25s$ to $t=50s$, the important force coefficients are calculated as,

Table 1: calculated force coefficients

$<Cd_x>$	$<Cd_x'^2>^{1/2}$	$<Cd_y'^2>^{1/2}$
2.173	0.093	1.247

These values are rather close to what have been obtained from other numerical computations for the case without free surface (Rodi, 1997). In the following paragraph, however, we shall demonstrate that these values are also the function of mesh size, at least when the mesh size is not adequate to resolve the energy containing flow motions.

4.3 Analysis of Coherent and Turbulent Flow Motion

Behind the pile, there coexist temporally averaged, periodic coherent, and turbulent motions. By looking into this issue, we are able to gain more insight about the relationship between the coherent structures and turbulent fluctuations with respect to certain mesh size in the LES model. It is well known that for any turbulent flow with periodic coherent structures, the instantaneous velocity u_i can be decomposed into three parts,

$$u_i = <u_{it}> + <u_{ip}> + u_i' \tag{4}$$

where $<u_{it}>$ is the time-mean velocity, $<u_{ip}>$ is the periodic coherent velocity, and u_i' is the turbulent velocity fluctuation. The time-mean velocity in this case is a constant over time but varies in space.

In the LES model, the spatial filter is first applied to u_i as follows,

$$u_i = \overline{u_i} + u_i'' \tag{5}$$

where u_i'' is the residual velocity. It is noted that in the above formulation, the spatial filter is not

177

precisely defined. In fact, there exist many sensible choices of filters as long as the basic normalization condition for the filter is satisfied. Corresponding to different filters, the characteristics of residual velocity vary and so do the resolvable stresses R_{ij}. Therefore, rigorously speaking, with respect to a specific filter that has special properties to remove high wave number modes, a unique closure model should be proposed and different interpretation of results should be pursued. Such approach, however, is tedious to apply to the actual engineering computation. In practice, the vagueness of the filter type is normally retained with the compromise that the cutoff wave number can not be clearly and accurately defined. For example, if the mesh size Δx, which is often regarded as the filter size, is used, the cutoff wave number can only be roughly estimated as $k=2\pi/\Delta x$. However, it is possible that some higher wave modes are actually retained during the computation and some lower wave number modes are filtered.

Substituting (4) into (5), we shall find that the spatially averaged velocity $\overline{u_i}$, which is the solution from the LES model, includes three components as well. The residual velocity also has three components. Ideally, the mesh size should be such selected that it is the largest grid (for economic consideration) resolving the smallest energy containing coherent and turbulent flow motion (for model accuracy consideration). This is possible for high Re turbulent flow when the energy containing eddy motion is far separated from the viscous dissipation motion in terms of characteristic length scale. Under this circumstance, the so-called inertial subrange exists within which the turbulence energy spectrum has a universal form (Tennekes and Lumley, 1972) and can be well modeled by simple eddy viscosity model. This situation is illustrated in Figure 5.

In practice, the ideal cutoff wave number is not always accessible due to the high computational expense. Most of practical grid size may have to fall into the overlapping region between the coherent motion and turbulent motion. This is especially true when the turbulence is mainly induced by the coherent motion rather than the time-mean flow motion, where the energy spectrum of turbulence and coherent motion might be coupled together. Therefore, the choice of any affordable grid size will cause the loss of certain coherent motion and thus affect the calculation of resolvable velocity field and force coefficients. We shall see this more clearly later when the mesh convergence test is performed.

In Figure 6, the time history of velocities at the middle elevation of $(x,y)=(0.2m, 0m)$ is plotted together with their corresponding energy spectra. The sampling frequency in the time history plots is 100 Hz. Similar to the force coefficients, it is observed that the velocities reach stable state after $t=25$s. The energy spectra are obtained from the data between $t=25$s to $t=50$s. It is noted that the time-mean quantities, which have the peak value at $f=0$ Hz, are not shown in the log-log plot of energy spectra. Thus, the spectra in Figure 6 mainly represent the resolved coherent flow motion and energy containing turbulent motion. It is not difficult to identify from the energy spectra plots that the peak values for u and v are 0.64 Hz and 0.32 Hz, respectively, agree with the oscillation frequencies of the coherent motion observed in the time history plots.

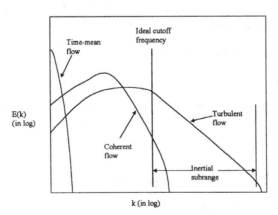

Fig.5: Illustration of selection of ideal cutoff wave number.

178

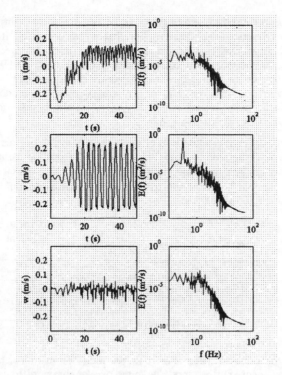

Fig.6: Time histories of velocities at the middle elevation of $(x,y)=(0.2m, 0m)$ [left column] and their corresponding energy spectra calculated from data between $t=25s$ to $t=50s$.

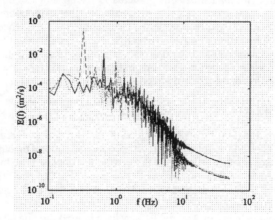

Fig.7: Energy spectra for u (solid), v (dashed), and w (dotted) in Figure 6.

For the vertical velocity w, the motion is rather chaotic, indicating that there is no obvious organized flow motion in this direction or the coherent motion is inseparable from the turbulent motion. Accordingly, in the energy spectrum, no dominant peak value can be identified. However, the vertical velocity fluctuation has nearly the same order of magnitude as those in the horizontal plane, implying that the instantaneous flow field is indeed three-dimensional.

When we plot out the energy spectra for three velocity components together (see Figure 7), we can observe that in the low frequency domain ($f<1.0$ Hz), where the coherent motion

179

dominates over turbulent motion, the energy spectra for $\bar{u}$, $\bar{v}$, and $\bar{w}$ look different with their peak values located at different places. This implies that in the organized flow motion, the boundary effect is still strong that makes the flow different in three directions from the statistical point of view. As the frequency further increases (1.0 Hz$<f<$10.0 Hz), the turbulent motion gradually prevails and the energy spectrum becomes more and more similar in three directions in statistical sense. Beyond $f>$10 Hz, the grid size used in LES model can no longer resolve the flow motion accurately. This roughly agrees with the estimation of the cutoff frequency at this point, $f_c=<\overline{u_t}>/\Delta x\sim0.1/0.005\sim20$ Hz.

Based on the energy spectrum, it is possible to estimate the magnitude of time-mean velocity, coherent motion, and turbulent motion by doing the numerical integration using chosen cutoff frequencies. If we assume that the time-mean motions are confined within the frequency 0.0 Hz $\leq f < 0.2$ Hz, coherent motion 0.2 Hz $\leq f < 1.0$ Hz, and turbulent motion 1.0 Hz $\leq f$, the important values are summarized as follows:

Table 2: Calculated time-mean velocity and fluctuation intensity due to coherent and turbulent motions.

Magnitude (m/s)	Time-mean	Coherent motion	Turbulent motion
$\bar{u}$	0.106	0.034	0.015
$\bar{v}$	0.011	0.180	0.014
$\bar{w}$	0.013	0.017	0.022

It is noted, however, the values in Table 2 are only the rough estimation, which varies with the selection of cutoff frequency. Furthermore, due to the overlapping of different types of flow motion, the complete separation of coherent structure from turbulence is nearly impossible. Nevertheless, Table 2 can still provide us the useful information for assessing the relative importance of each flow mechanism at this particular location. For example, it is ready to see from Table 2 that the time-mean velocity dominates in the x-direction while the periodic coherent motion dominates in y-direction. In the z-direction, the turbulence motion prevails or at least is of the same importance as the coherent motion. It is also found that in the resolved scale, the turbulence is still going through the transition from two-dimensional to three-dimensional because the turbulence fluctuations in the horizontal plane are nearly the same but they are different from that in the vertical direction.

4.4 Time-Mean Velocity Fields

When the flow is confined in the vertical direction, the time-mean vertical motion is zero, though there exist three-dimensional coherent motion and instantaneous turbulent motion. However, when the free surface is present, there will be nonzero time-mean vertical motion induced by the difference of free surface displacement in front of and behind the pile. It is observed from Figure 8 that the strongest vortical motion occurs near the middle elevation right behind the pile.

The non-zero vertical velocity on the middle elevation behind the pile alters the velocity field on the horizontal plane nearby (see Figure 9). It is noted that right behind the pile, the value of $\overline{u_t}$ is positive. The positive $\overline{u_t}$ persists regardless of the choice of averaging time, though the details of flow pattern vary slightly with the use of different averaging duration. This feature is different from that in the case no free surface is present, where the returning vortices extend all the way to the pile (Rodi, 1997). It is also noted that behind the pile some asymmetries are observed. This is probably caused by the presence of relatively strong turbulence there, which has not been completely removed by the time averaging process.

Near the bottom or free surface, the vertical motion is negligible. The velocity fields there then exhibit the similar feature to that observed in the case without free surface (Figure 10).

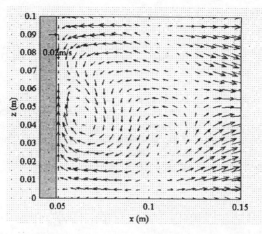

Fig.8: Time-mean (between t=25s to t=50s) velocity field at the centerline of the pile in the y-direction behind the structure.

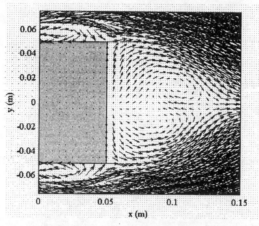

Fig.9: Time-mean velocity field on the middle elevation around the pile.

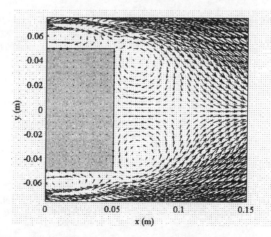

Fig.10: Time-mean velocity field above the bottom around the pile.

4.4 Mesh Convergence Test

In this section, we shall perform some mesh convergence test, which hopefully can provide us the information of how to determine the optimal mesh system in the LES model with the tradeoff of numerical accuracy and computational efficiency. As mentioned before, totally five mesh systems have been used to simulate the same problem. In general, the finer the grid size is, the more detailed flow structure could be obtained. In this particular case, since the forces are closely related to the coherent structure, the accurate prediction of drag force coefficient relies on the adequate resolution of coherent structure. It is found that when the mesh size is too coarse (e.g., mesh system A), the vortices are elongated up to the outflow boundary. No vortices shedding is observed and the flow is basically laminar and symmetric about y-coordinate. This is obviously incorrect when compared with the laboratory data.

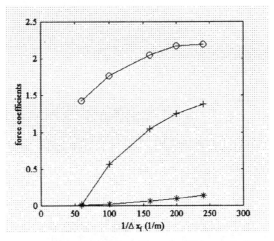

Fig.11: Convergence test in terms of force coefficients; circle represents $<Cd_x>$, star $<Cd_x'^2>^{1/2}$, and plus sign $<Cd_y'^2>^{1/2}$.

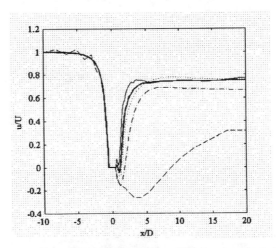

Fig.12: Time-mean u at the middle elevation and centerline of pile in y-direction using mesh systems A (dashed), B (dash-dotted), C (dotted), R (thick solid), and E (solid).

182

To find out what is the sufficient mesh resolution to this problem, we first plot out the calculated force coefficients as the function of inverse of finest mesh size Δx_f (Figure 11). It is seen that at the reference mesh system, the solution already approaches the relatively stable stage in terms of the mean drag force coefficient in the x-direction. For the force coefficients in terms of fluctuation intensities in the x- and y-directions, the further increase of mesh resolution should provide more accurate solution. It is expected that the solution to all force coefficients may become completely stable when $1/\Delta x_f \sim 500$ 1/m from the extrapolation of Figure 11. However, the computational cost of such resolution is too high at this moment with our computational tool, A Pentium III 600 MHz PC.

Another convergence test is based on the time-mean $\bar{u}$ at the centerline of pile in y-direction and on the middle elevation. The velocity is plotted in Figure 12. It is observed that except for the coarsest mesh system A, all other mesh systems generate almost identical smooth curve in front of the pile. Behind the pile, most of mesh systems predict the recirculating flow induced by the vortex shedding. However, the coarse mesh system (e.g., mesh system B) tends to predict longer recirculation length. Finer mesh systems (C, R, and E) capture the forwarding flow right behind the back face of the pile (a small upward kink in Figure 12), which is induced by the vortical motion on the vertical plane (see Figure 8). This is also consistent with the velocity field plot in Figure 9.

5. CONCLUSION

In this study, a three-dimensional LES model is used to investigate the problem of a rectangular pile exposing to a turbulent channel flow. The vortices shedding pattern and calculated drag force coefficients are found to be similar to those in the case without free surface. The unique time-mean vortical field in the vertical plane, however, is observed, which might be induced by the presence of free surface. The strongest vertical vortical motion occurs near the middle elevation right behind the pile, which consequently affects the velocity field on the horizontal plane nearby. The coherent and turbulent flow motions resolved by the LES model are analyzed by using the energy spectra. It is found that the time-mean motion and coherent structure dominates on the horizontal plane. In the vertical direction, the turbulent field is strongly coupled with the coherent structure with nearly the same order of magnitude. The mesh convergence test shows that at least 20×20 meshes should be used to resolve the rectangular pile to predict the reasonable force coefficients and vortex shedding pattern. The mesh system that is too coarse is not able to capture the periodic coherent structure and thus generate incorrect force coefficients.

REFERENCE

Hodges, B. R. and Street, R. L.: On simulation of turbulent nonlinear free-surface flows, *J. Comp. Phys.*, Vol. 151, (1999), pp. 425-457.

Li, C. W. and Yu, T. S.: Numerical investigation of turbulent shallow recirculating flows by a quasi-three-dimensional k-e model, *Int. J. Num. Met. Fluids*, Vol. 23, (1996), pp. 485-501.

Li, C. W. and Wang, J. H.: Large eddy simulation of free surface shallow water flow, accepted by *Int. J. Num. Met. Fluids*, (1999).

Lin, P. and Li, C. W.: A σ-coordinate three-dimensional numerical model for surface wave propagation, submitted to *Int. J. Num. Met. Fluids*, (2000).

Lyn, D. A., Einav, S., Rodi, W., and Park, J.-H. : A laser-Doppler velocimetry study of ensemble-averaged characteristics of turbulent near wake of a square cylinder, *J. Fluid Mech.*, Vol. 304, (1995). pp. 285-319.

Richardson, J. E. and Panchang, V. G.: Three-dimensional simulation of scour-induced flow at bridge piers, *J. Hydraul. Engng.*, ASCE, Vol. 124, No. 5, (1998), 530-540.

Rodi, W.: Comparison of LES and RANS calculations of the flow around bluff bodies, *J. Wind Engng. and Indust. Aerodyn.*, Vol. 69-71, (1997), pp. 55-75.

Shah, K. B. and Ferziger, J. H.: A fluid mechanicians view of wind engineering: Large eddy simulation of flow past a cubic obstacle, *J. Wind Engng. and Indust. Aerodyn.*, Vol. 67 & 68, (1997), pp. 211-224.

Tennekes, H. and Lumley, J. L.: A first course in turbulence, (1972), Cambridge, Mass., MIT Press.

Stochastic Hydraulics 2000, Wang & Hu (eds) © 2000 Taylor & Francis, ISBN 90 5809 166 X

An experimental investigation of side overflow under live bed conditions

H.Ağaçcioğlu & A.Coşar
Faculty of Civil Engineering, Yildiz Technical University, Istanbul, Turkey

ABSTRACT: Side weirs have been used extensively in irrigation, land drainage, flood protection, and urban drainage works. Side-weir located on the fixed bed conditions have attracted considerable interest and research effort. The same thing however is not true for side-weirs located on the straight channels under the live bed conditions. The present study is aimed that the effect of the side overflow under the live bed conditions to the bed profile has been experimentally investigated. The secondary flow that is created by lateral flow causes very strong disturbance to the bottom topography in the main channel. The stagnation zone was observed along the inner bank and reverse flow occurred due to the stagnation zone. This study shows that the effect of the lateral flow to the bed profile depends on the approach velocity in the main channel, the dimensionless side-weir length and the size of the bed material.

Key words: Lateral flow, Incipient motion, Bed profile, And migration.

1. INTRODUCTION

Side-weirs, also known as lateral weirs, are widely used in irrigation, land drainage, and urban sewage systems by flow diversion or intake devices. Many investigators have been interested in the weir discharge coefficient with the main channel's upstream Froude number by using De Marchi's equations for stable channel. Probably the first rational approach was made by De Marchi (1934). However, there is no sufficient investigation for lateral flow under the live bed conditions. The purpose of this paper is to report on the variation of the bed profiles along the weirs in the main channel under the live bed conditions.

The discharge over the side-weir per unit length, q, is assumed to be

$$q = -\frac{dQ}{ds} = C_d \sqrt{2g}(h-p)^{3/2} \tag{1}$$

where s is the distance from the beginning of the side-weir, q or (-dQ/ds) is the discharge per unit width over a rectangular side-weir, g is acceleration due to gravity, p is height of the side-weir, h is depth of flow measured from the channel bottom along the channel centerline, and C_d is the discharge coefficient, which is defined by dimensionless analysis as indicated below;

$$C_d = f\{ \ F_1 = V_1/\sqrt{gh_1}, L/b, L/h_1, p/h_1 \ \} \tag{2}$$

where V_1 is mean velocity of flow of the upstream section of side-weirs in the main channel, L is the length of the side-weir, b is the width of the main channel and h_1 is the upstream depth of the side-weirs (see Figure 1).

The general differential equation for flow along a side-weir as given by Chow (1959), and modified for a rectangular, prismatic, horizontal, frictionless channel is

$$\frac{dh}{ds} = \frac{Qh(-\dfrac{dQ}{ds})}{gb^2h^3 - Q^2} \qquad (3)$$

in which s is the distance from the beginning of the side-weir, h is depth of the flow at a section located at s, dh/ds is water surface slope with respect to the bed, Q is discharge in the channel at section s, -dQ/ds is discharge per unit length spilling over the side-weir.

As it can be seen in Figure 1, the water level rises from the upstream end of the side-weir towards the downstream end of the side-weir in the main channel in accordance with the above equation for subcritical flow conditions in the main channel. However, Subramanya and Awasthy (1972) and El-Khashab (1975) have pointed out that the water level drops slightly at the upstream end of the weir. This has been attributed to the side-weir entrance effect at the upstream end. This effect doesn't extend to the centerline of the main channel and it is only near the weir crest. Then a rapid rise in the water level from the upstream end of the side-weir towards the mid-span of the crest and the rate of rise decreases substantially. The change in water level is not noticeable for nearly the last one third of the weir length, and the water surface is nearly horizontal.

This behavior of the water surface level agrees with the explanation of the effect of secondary flow due to lateral flow near the downstream end of the side-weir. The secondary flow that is created by the lateral flow (El-Khashab 1975) causes strong disturbances of the water surface despite subcritical flow conditions at the last one third of weir section in the main channel. In subcritical flow, Subramanya and Awasthy (Chow, 1959) and El-Khashab (1975) observed a separation zone and the reverse flow at the downstream end of the side-weir (see Figure 2). According to these researchers, the location and size of the separation zone and the reverse flow area were found to depend on Froude number at the upstream side of weir in the main channel and also on the length of the side-weir. When the upstream Froude number (Fr_1) is increased, the stagnation zone and reverse flow area move towards the downstream end of the side-weir. The reverse flow area diminishes at higher upstream Froude numbers.

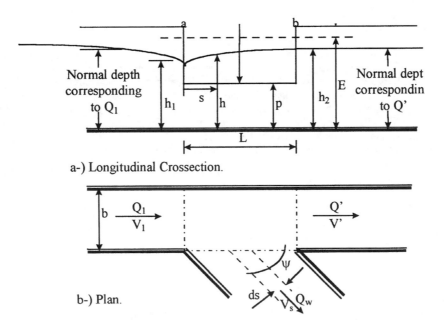

a-) Longitudinal Crossection.

b-) Plan.

Fig.1:. Definition sketch of subcritical flow over rectangular side-weir.

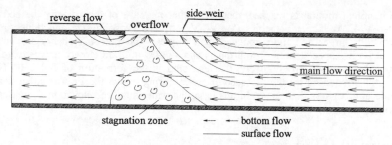

Fig.2:. Definition sketch of flow structures along the side-weir.

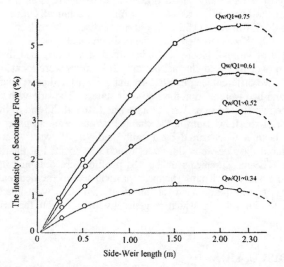

Fig.3: The change of the intensity of secondary flow created by lateral flow [4].

In order to delineate the magnitude and effect of the spiral flow for different curves under varying conditions of flow, Shukry (1950) has used a term called the strength of a spiral flow (or intensity of secondary flow). This term is defined as the percentage ratio of the mean kinetic energy of the lateral motion to the total kinetic energy of flow at a given cross section. El-Khashab (1975) was also used the same definition of the meaning of the intensity of secondary flow created by lateral flow. Figure 3 shows the variation of the strength or intensity of secondary flow along the side-weir with the ratio of Q_w/Q_1 for subcritical flow. As it is seen from in Figure 3, the strength of secondary flow increases along the weir crest length and also is affected by the ratio of Q_w/Q_1. However, the secondary flow being a three-dimensional and in practice turbulent phenomenon, makes almost impossible analytically to solve the problem for the side-weir flow even in the straight channel.

2. FORMATION OF BED FORMS

Flow conditions with small discharge under steady turbulent uniform flow over a uniform non-cohesive bed material are similar to those over a fixed bed. If the discharge is gradually increased, the magnitude of the resultant effect of disturbing forces also increases. After a critical condition is reached (V_{cr}, to critical velocity), the equilibrium of the particles is diminished and the random motion of individual particles is observed on the bed. This condition is called as incipient motion or threshold of the sediment particles. In general, observation of the incipient motion is very difficult in nature because of the highly random and non-uniform properties of sediment and flow.

After the incipient motion is reached, then on further increase of the velocity the bed develops ripples ($V_1>V_{cr}$). At higher velocities ($V_2>V_1>V_{cr}$) ripples grow in size and dunes may form. Dunes are both larger and more rounded than ripples. The bed level decreases or migrates downstream with respect to time. Both dunes and ripples migrate slowly downstream. As this phenomenon takes place progressively, the next step is the formation of flat bed. The water surface profile influenced by the presence of dunes during which the flow is subcritical with a flow Froude number less than unity. In this case, the water surfaces elevation decreases and the velocity increases above the dune crest.

3. EXPERIMENTAL SETUP

An experimental study of the spatially varied flow over the side-weirs in a rectangular channel has been carried out at the Hydraulic Laboratory of Yildiz Technical University, Istanbul. The channel was 14-m length, 0.4-m wide, 0.55-m height and a discharge collection channel, which is 0.5 m. wide, and 0.55 m. high was situated along the main channel.

The channel consisted of vertical plexi-glass sidewalls and a quartz bed that the thickness of quartz layers is 0.15 m over the fixed channel bottom. Side-weirs were located at the mid of the main straight channel. Two different lengths (0.25-m and 0.50-m) and crest height (0.17 m) of the weir over the quartz sand bed were used at the side-weir station. The side-weirs were sharp edged and fully aerated on the overflowed side. A sluice gate was fitted at the end of the main channel in order to enable control of the flow depth in the channel. Water was supplied from an overhead tank with a constant head and measured by a calibrated $90°$ V weir. A calibrated standard rectangular weir measured the side-weirs flow rate, which was located at the downstream end of the collection channel. The measurements of the bed profile were obtained using point gauges capable of reading to the nearest 0.1-mm. The mean particle size of the bed material is 1.28 mm. The critical velocity (V_{cr}) used quartz sand for initiation of motion was obtained as 0.5 m/sec. Experiment were run at the velocity which is generally bigger than the critical velocity for initiation of motion.

4. RESULTS AND DISCUSSION

In the study, the change of the bed profiles in the main channel along the side-weirs are carried out for the side-weirs is L=0.25 m, p=0.17 m and L=0.50 m, p=0.17 m. In the experiments, velocities were generally chosen higher velocity than the critical velocity which can be started to incipient motion of the sediment particles. The created flow conditions are summarized in Table 1 and Table 2. Figure 4 and Figure 5 are also shown the obtained bed profiles for chosen side-weir dimensions along the weir crest for the different flow conditions in the main channel. As shown in Figure 4, the flow conditions on the upstream of side-weirs in the main channel were taken account to be able to incipient motion to the gravel sand bed (i.e., the approach flow velocity higher than the critical velocity) and ripples or dunes formation are formed in the main channel. The deposition in front of the side-weir crest are not exactly observed and ripples or dunes coming from the upstream of the weir along the main channel can be passed to the downstream of the weir section due to small weir length. When the side-weir length is small, the discharge over the side-weir is small. Therefore, the velocity in the main channel on the downstream of the side-weirs section is higher than the critical velocity and ripples or dunes are passed to the downstream of the weir. This situation can be absolutely seen in Table 1.

Table.1:. Experimental program for the weir L=0.25 m and p=0.17 m

Test run	Q_1 (lt/s)	Q_w (lt/s)	$Q_r=Q/Q_w$ (lt/s)	V_1 (m/s)	V' (m/s)	Fr_1	h_1 (cm)
1	49.62	4.78	9.63	0.56	0.51	0.38	5.17
2	54.37	10.20	18.76	0.53	0.43	0.34	8.44
3	60.14	3.98	6.62	0.69	0.64	0.47	4.80

Table.2: Experimental program for the weir L=0.50 m and p=0.17 m

Test run	Q_1 (lt/s)	Q_w (lt/s)	$Q_r=Q/Q_w$ (lt/s)	V_1 (m/s)	$V^{'}$ (m/s)	Fr_1	h_1 (cm)
1	43.50	17.41	40.01	0.45	0.27	0.301	6.72
2	57.48	36.01	62.65	0.51	0.19	0.311	10.94
3	62.05	27.74	44.71	0.59	0.33	0.366	9.35

In Figure 5, the deposition or accumulation in the different levels or migration in front of the last one third of weir length can be observed. As seen Run 1 in Table 2, the flow velocity in the main channel is almost equal to the critical velocity. But, the sand materials transported by rolling-traction mode from the upstream of the side-weirs section of the main channel are carried toward the downstream of the weir. On the other hand, for Run 2 and Run 3, the flow velocity in the main channel is higher than the critical velocity and ripples or dunes are passed from the upstream to the direction of the downstream of the weir. However, for this weir, the flow velocity on the downstream of the weir is smaller than the critical velocity due to large overflow rate. When the side-weir length is large, the discharge along the side-weir increases. Thus, the velocity in the main channel to the direction of downstream end of the weir is smaller than the critical velocity due to the large overflow rate. So, the depositions in front of the last one third of the side-weir crest are appeared and ripples or dunes coming from the upstream can not be passed to the downstream of the side-weir section. The materials transported from the upstream are thrown over the weirs to the collection channel. The scour holes are formed at the downstream of the weir in the main channel due to the separation zone and the reverse flow area. This situation causes the accumulation at the different sections in front of the weir crest. As also mentioned above, the location and size of the separation zone and the reverse flow area depend on the weir length. If the weir length is small, the flow rates over the weir increase.

CONCLUSION

In the presented study, the experimental work was carried out under the live bed conditions in the main channel. The following results were obtained:
1- Lateral flow affects to the bed topography profiles along the weir crest in the main channel after the velocity in the main channel was reached to the critical velocity to be able to start to incipient motion of the bed material.

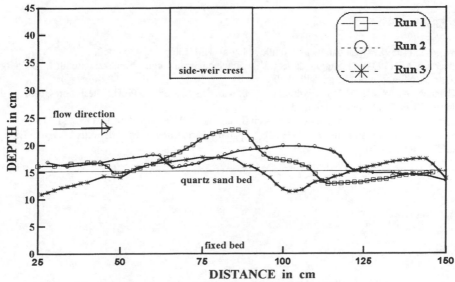

Fig.4: Bed topography profiles for L=0.25 m and p=0.17 m

189

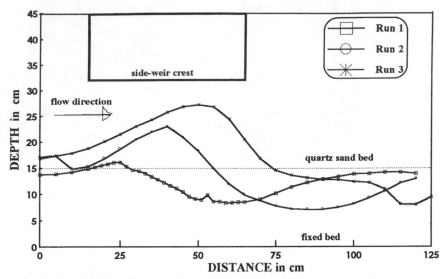

Fig.5: Bed topography profiles for L=0.50 m and p=0.17 m.

2- The deposition or accumulation in front of the last third of the side weirs are dependent on the velocity at the downstream end of the side-weir in the main channel. If the velocity at the downstream end of the weir in the main channel is smaller than the critical velocity which can be started to incipient motion of the bed material, then or the deposition in the different levels are formed at the different places of the side-weir crest length. On the other hand, the materials transported from the upstream are thrown over the weirs to the collection channel.

ACKNOWLEDGMENT

Financial support of this work by the Research Foundation of Yıldız Technical University (Project 98-A-05-01-03) is gratefully acknowledged.

REFERENCES

Chow, V. T. (1959)." Open channel hydraulic., Mcgraw Hill, Chap. 16, 439 - 460.

De Marchi, G. (1934). " Saggio di teoria de funzionamente degli stramazzi laterali.", L'Energia Elettrica, Milano, Vol 11., pp. 849 - 860.

El-Khashab, A. M. M. (1975)." Hydraulics of flow over side-weirs.", Ph.D. Thesis, presented to the University of Southampton, England.

Shukry, A. (1950). " Flow around bends in open flume.", J. Hydr. Engrg., ASCE, 115, 751 - 759.

Subramanya, K., and Awasthy, S. C. (1972). " Spatially varied flow over side-weirs."J. Hydr. Engrg., ASCE, 98(1), 1 - 10.

Stochastic Hydraulics 2000, Wang & Hu (eds) © 2000 Taylor & Francis, ISBN 90 5809 166 X

Structure of turbulent flow in scour holes downstream of submerged jets

S.L.Liriano & R.A.Day
Water Engineering Research Group, University of Hertfordshire, Hatfield, UK

ABSTRACT: Scouring downstream of circular jets is a common problem and some research has been conducted to assess the extent of scouring that can be expected under particular flow conditions. However, little attention has been directed towards developing an understanding of the flow structures leading to scouring downstream of jets. Research has been carried out at the University of Hertfordshire using facilities at H.R. Wallingford to explore scour downstream of jets and study the flow fields in and surrounding the scour hole, the preliminary findings are presented here. The results indicate that the initial scour hole development is a result of the jet impacting on the bed and transporting sediment out of the scour hole. Further growth in scour depth is caused by sweep events in the region of maximum scour depth. Over the dune region high contributions to the Reynolds stress are observed in all four quadrants with a dominance in quadrant 1 corresponding to outward interaction events. These events have been observed for open-channel flow over dunes downstream of the point of re-attachment and may be responsible for further sediment transport.

Key words: Jets, Turbulent flow structure, ADV

1. INTRODUCTION AND BACKGROUND

High exit velocities from jets downstream of structures such as culverts and outfalls in general leads to scouring and can result in damage to the ooutlet structure and flooding and damage to the surrounding infrastructure. Investigations have been carried out to measure the extent of scouring downstream of jets (Lim, 1995, Abt et al. 1984, Breusers and Raudkivi 1991) however very few researchers have considered flow structure in and around the scour hole. Johnston and Halliwell (1986) measured centreline velocity profiles downstream of a two-dimensional jet without any bed forms present and found different flow patterns for different tailwater depths. Ali and Lim (1986) measured centreline velocity profiles in scour holes downstream of two-dimensional and three-dimensional outlets at various stages of development and observed that whilst there were areas of reverse flow downstream of two-dimensional jets reverse flow was not observed downstream of three-dimensional jets. Schoppman (1975) measured streamwise turbulence intensity in scour holes downstream of a wedge and reported that turbulence intensity decreased as the scour hole developed and that a peak in the turbulence intensity occurred at the point of flow reattachment.

Measurements and observations of flow structures in scour holes formed around bridge piers by Melville (1975) and Melville and Raudkivi (1977) has increased understanding of sediment transport in this region. Such measurements in scour holes downstream of outlets have not been reported previously. This paper presents a series of experiments using scour holes fixed at different stages of development in order to investigate the magnitude of the mean velocities, turbulence intensities, Reynolds stresses and the bursting structure in scour holes.

2. EXPERIMENTAL FACILITIES AND METHODOLOGY

The 25m long, 2.4m wide, 1.2m deep High Discharge Flume at H.R Wallingford was used in order to carry out the work on a large-scale model culvert. A pipe 6m long, with an internal diameter of 311mm was placed upstream of a tailwall, see fig 1. Downstream of the tailwall the channel was filled with uniformly graded gravel, $D_{50}=9mm$, up to the pipe invert level. A flow rate of $0.1272m^3/s$ and a tailwater depth of $1.0d_0$ were used to create a scour hole downstream of the outlet. Scour holes were fixed during the development stages in order to observe changes in the flow structure as the scour hole developed. The bed was fixed by spraying a cement grout over the surface when the required scour depth had been achieved. The turbulent flow structure for the centreline of the beds fixed at 0%, 50% and 99% of the equilibrium scour depth measured under identical flow conditions are reported here.

Point velocity measurements were made using a downward facing three-component Acoustic Doppler Velocimeter (ADV). This instrument has the advantages of measuring three velocity components simultaneously and being suitably robust to be used downstream of the outlet. Additionally only minimal disturbance is made to the flow as the sampling volume is located 5cm beneath the instrument and no calibration is required.

To obtain mean velocities and other parameters required for the turbulence analysis the data collected was processed using the WinADV post-processing package that allows the results to be imported into an Excel spreadsheet. WinADV is available as freeware via the Internet from the Water Resources Research Laboratory, Colorado. For the analysis of the coherent structures ExploreADV, an additional post-processing package capable of high-order mathematical analysis was used. This software can be purchased from Nortek-AS.

The ADV data requirements were two fold. First, detailed near bed data was required to make an analysis of the near-bed bursting structures and Reynolds stresses. Second overall flow field information in the vicinity of the scour hole was necessary to develop a general understanding of the nature of the flow structures. The region *close to the bed* was considered to be the section to a height of 150mm above the bed and 10, 120 second readings were taken in this region. The closest reliable reading to the bed was 5mm and reading were taken at 5mm intervals up to 25mm and then 25mm intervals up to 150mm. The less detailed data followed on from the near-bed data by continuing sampling at 25mm intervals for 30 seconds starting at 175mm from the bed up to the closest point to the free surface the ADV could obtain a reliable reading. Measurements were made at 200mm intervals in the downstream direction starting from 100mm from the outlet and continuing until the dune crest was reached.

At the start of each test the sediment bed was screeded level and to the same height as the pipe invert. Covers were placed over the bed to prevent bed disturbance whilst the channel filled. The required flow rate and tailwater depth were established and a settling time of 1 hour was waited for steady conditions to be achieved, after which time the covers were removed. Scour then took place at the outlet until the required percentage of the final scour depth was attained at which point the flow was stopped and the channel drained. The scour hole profile was measured and fixed. The channel was then refilled and the same flow rate and tailwater depth set as created the scour hole and velocity data collected. Upon the completion of the velocity data collection the fixed bed was discarded.

3. RESULTS AND DISCUSSION

3.1 Mean velocity

Figure 2 shows the centreline mean velocity vectors at progressive stages of scour hole development. For the 0% developed case a high downstream velocity is observed as would be expected at the pipe outlet with the vertical velocity almost insignificant by comparison. Close to the bed the fluid immediately adjacent to the outlet appears to be entrained by the faster moving fluid in the jet centre, this can be observed up to 2 pipe diameters from the outlet. After a distance of 4.5 pipe diameters from the outlet the jet has moved away from the outlet centreline. This occurred for all the flat bed tests and has also been noted by Bohan (1970) and Blaisdell and Anderson (1988).

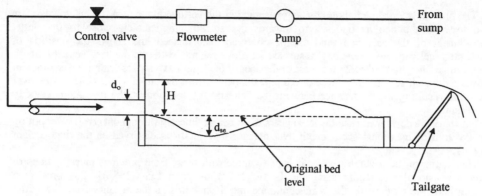

Fig.1: Schematic diagram of experimental set-up.

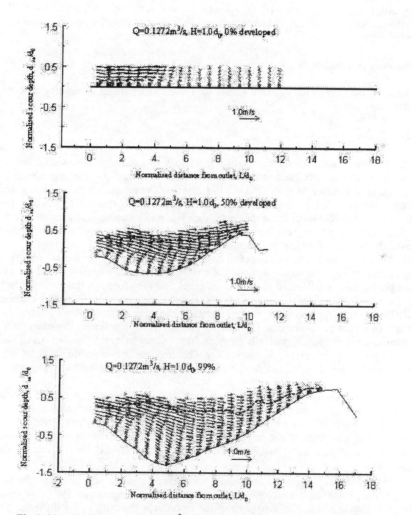

Fig.2: Vector plots for Q=0.1272m³/s, H=1.0d$_p$ at different stages of development

The 50% developed plot shows the jet expanding in the vertical plane as it travels downstream and towards the bed as opposed to towards the surface. Immediately close to the outlet entrainment of the near bed fluid can be observed. The dashed line shows the position of maximum velocity and indicates a surface jet flow regime with the highest velocity at an approximate height of $y^+=0.25$ for most of the scour hole length. This rises with a lower gradient than the dune slope after a distance of 6.0 pipe diameters from the outlet as the flow passes over the dune resulting in the distance between the bed and the location of the maximum velocity decreasing.

For the 99% developed case the jet appears to be spread linearly throughout the flow depth by the position of maximum scour depth. No increase in velocity is observed as the dune height increases.

The results of the mean downstream velocity agree with those from previous investigations of three-dimensional jets (Rajaratnam and Berry (1977), Ali and Lim (1986)). The prevailing pattern is of the jet entering the channel and moving downwards as the distance from the outlet increases, attaching to the bed and moving over the dune. The results do not indicate any recirculating flow close to the outlet in the near-bed region prior to the point of deepest scour as noted by Ali and Neyshaboury (1991), for two-dimensional flow. Ali and Lim (1986) also reported this to be a significant difference between their results from two-dimensional and three-dimensional tests.

3.2 Turbulence intensities

Figure 3 shows the turbulence intensity values for the different stages of development that have been calculated from:

$$\frac{\sqrt{(u')^2}}{U_0}, \frac{\sqrt{(v')^2}}{U_0}, \frac{\sqrt{(w')^2}}{U_0} \tag{2}$$

for the downstream, cross-stream and vertical directions respectively. For the flat bed case the profiles show a peak in turbulence intensity in the downstream and cross-stream directions close to the bed up to a distance of $y^+=1.5$. After this distance the maximum value is observed to be further from the bed and beyond 6 pipe diameters from the outlet the turbulence intensities decrease as the jet curves away from the centreline. The maximum near bed turbulence intensity for all three components is observed approximately 4 pipe diameters from the outlet. This coincides with the location of the maximum scour depth at the fully developed stage.

For the 50% developed case two peaks in each turbulence intensity profile are observed up to the point of maximum scour depth. A section of uniform turbulence intensity is observed between these two peaks, which is located inside the jet core, and therefore the flow is not being acted upon by the surrounding fluid, hence this has a low turbulence intensity. When the distance from the outlet increases to greater than 7 pipe diameters the maximum turbulence intensity in the downstream direction is observed to be close to the bed at all locations. The magnitude of the near bed turbulence intensities gradually increase from the outlet to the point of maximum scour and then a more rapid increase is noted. The absolute maximum value is observed in the downstream direction approximately 8 pipe diameters from the outlet, half way along the rise of the dune.

For the fully developed flow profiles the initial patterns are the same, as can be seen in fig 3. Peak values in turbulence intensity in the downstream direction are observable close to the bed but with a 45% reduction from the 50% developed case. Unlike the developing bed case no peak is observed in the magnitude of the turbulence intensities along the dune and at this location the magnitude of the downstream turbulence intensity is also reduced by 45% from the 50% developed case.

During the developing stages the maximum turbulence intensity was observed on the rising slope of the dune, which agrees with the results presented by Schoppmann (1975) for scour holes downstream of wedges. The measurements of the turbulence intensity in the different directions show the turbulent flow to be non-isotropic with u'≅1.6v'≅1.8w' for the fully developed bed. Bennett and Best (1993) observed the streamwise turbulence intensity for the flow behind two-dimensional dunes to be three times larger than the vertical turbulence intensity and Nelson et al.

(1993) found that the vertical intensity decreased relative to the horizontal turbulence intensity and that this was most prominent as the slope of the dune rose.

3.3 Reynolds stresses

Reynolds stress profiles at each stage of development are presented in figure 4. The Reynolds stress values presented here have been calculated from:

$$\tau_{uw} = \frac{-\overline{u'w'}}{U_0^2}, \quad \tau_{uv} = \frac{-\overline{u'v'}}{U_0^2}, \quad \tau_{vw} = \frac{-\overline{v'w'}}{U_0^2} \tag{3}$$

From the measurements taken over the flat bed it can be observed that the largest Reynolds stresses occur close to the outlet and close to the bed. For a distance of greater than two pipe diameters from the outlet the profiles show a small magnitude close to the bed increasing as y^+ increases.

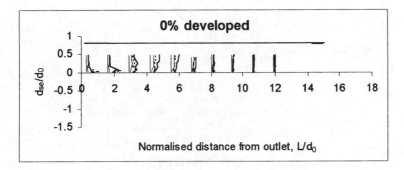

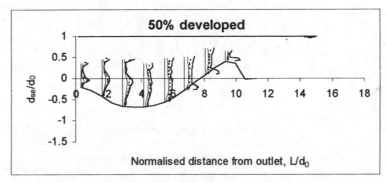

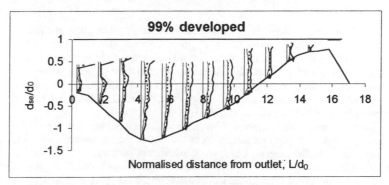

Fig.3: Turbulence intensities at different stages of development.
——— TI_u, ---- TI_v, ——— TI_w

195

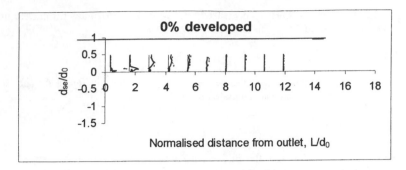

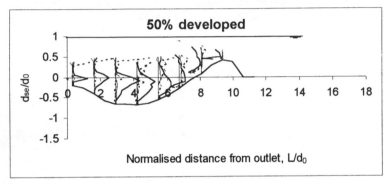

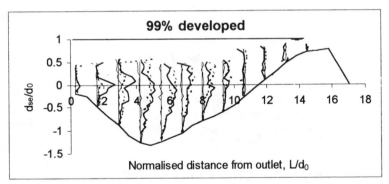

Fig.4: Reynolds stresses at different stages of development.

———— τ_{vw} ----- τ_{uv} ———— τ_{uw}

For the 50% developed case large positive Reynolds stresses are observed up to and just beyond the point of maximum scour depth. Positive Reynolds stresses have a retarding effect on the flow. Close to the bed in this region τ_{uw} has a small magnitude that increases to a maximum just below the original bed level and then decreases. On the dune slope negative Reynolds stresses are observed close to the bed at a distance greater than 6.5 pipe diameters from the outlet, and then throughout the Reynolds stress profile once this distance increases to 8 pipe diameters. The largest Reynolds stress is observed 15-20mm from the bed.

The 99% developed case shows a 77% lower τ_{uw} values than that observed at the 50% developed stage. Up to 8 pipe diameters from the outlet the maximum value of the Reynolds stress is again close to the original bed level, a significant distance from the bed at this stage of development. Close to the bed the Reynolds stress approaches zero. This very small near bed Reynolds stress is observed throughout the length of the scour hole at this stage of development with no negative Reynolds stresses observed and no peaks close to the dune.

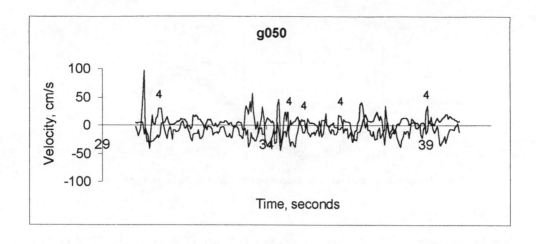

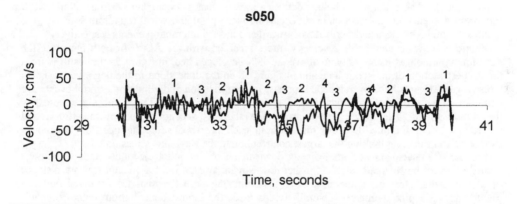

Fig.5: Examples of traces in the scour hole (g050) and in the dune region (s050) illustrating different events at each location. Each figure is a 10 second sample of a 120 second trace.

The Reynolds stress profiles revealed a peak at the edge of the jet. The peaks in the Reynolds stress observed just under the original bed level are attributed to the outer boundary of the jet where a high shear zone is present as observed in the turbulence intensity results. The largest magnitude of τ_{uw} in the dune region of the scour holes downstream of the culvert outlet was observed to be close to the bed, however the sign of the Reynolds stress was negative which is confirmed by the findings of Nelson et al. (1993).

Negative Reynolds stresses occur infrequently and are generally associated with areas of negative or reverse flow where the Reynolds stress is positive with respect to the local flow but negative with respect to the main body of the flow. However, the region of negative Reynolds stress discussed here is not in an area identified as having a reverse flow component and it was recorded consistently throughout the stages of development. It is considered here that these areas of negative τ_{uw} may be the result of some kind of vortex structure occurring at this location in the flow. This is verified by observations by Nezu and Nakagawa (1993) of negative Reynolds stresses occurring at the head and tips of hairpin vortices and McLean et al. (1996) who observed negative or virtually zero Reynolds stresses near the bed downstream of steps.

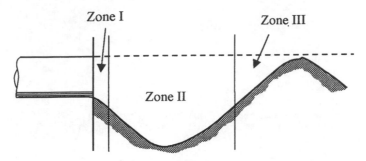

Fig.6: Scour hole divided into zones based on the results of the bursting analysis.

3.4 Bursting analysis

Utilising the ExploreADV software, the quadrant technique (Wallace et al., 1972, Lu and Willmarth, 1973) has been used to identify bursting trends within scour holes downstream of pipe culvert outlets. The quadrant technique defines a sweep event as occurring when $u'>0$ and $w'<0$ (quadrant 4 event) and an ejection event occurring when $u'<0$ and $w'>0$ (quadrant 2 event), the quadrant 1 and 3 events are described as outward and inward interaction events respectively.

Figure 5 shows examples of 2 velocity traces recorded with the ADV. In both instances the sampling volume was located 50mm above the 99% developed bed, in figure 5a the result is from the deepest section of the scour hole and in figure 5b on the dune slope in the negative Reynolds stress region. Events have been labelled for ease of identification and quadrant 2 and 4 events are dominant in the scour hole and quadrants 1 and 3 in the dune region. These traces are typical of the results found. The traces shown are just a short 10-second period of the 120-second recording. Using ExploreADV the results from each near-bed sampling position have been examined and the event making the largest contribution to the Reynolds stress noted.

For the 0% developed bed immediately downstream of the outlet ejections made the largest contribution to the Reynolds stress. Further downstream sweeps were dominant up to a distance of 3.5 pipe diameters with interaction events dominating once the distance from the outlet is greater than 3.5 pipe diameters. Overall sweeps made the largest contribution to the Reynolds stress with 44% of all positions indicating sweeps made the largest contribution to the Reynolds stress.

For the scoured beds, based on the trends observed from the bursting analysis the scour hole can be divided into 3 sections as shown in figure 6. The first section is very close to the outlet, which is then followed by a large section covering the area up to approximately the mid-point of the dune slope and thirdly an area including the remainder of the dune up to the crest.

Ejections are observed close to the outlet for all stages of development but over a smaller length of the channel as the scour hole increases in size. For the 50% developed bed ejections are dominant at 25% of the sampling points, mainly clustered at the scour hole outlet and in section 3 very close to the bed, reducing to 16% for the 99% developed bed with the reduction being due to ejections in section 3 becoming less dominant. Fewer ejection events are also observed close to the outlet as the scour hole enlarges because the measurements are taken relative to the bed and therefore the distance from the jet to the measuring volume increases as the scour depth increases.

In the second section the results show that for much of the scour hole, for all tailwater depths and at all stages of development, sweeps dominate which is typical of flow over rough beds, Raupach (1981). Sweeps dominate up to the region where the Reynolds stress become negative and the percentage of sweep events dominating remains consistently around 50% irrespective of the stage of development.

In the third section the inward and outward interaction events are the most significant. The change in event type from sweep to interaction events takes place at progressive locations downstream as the scour hole develops. This changeover occurs at approximately 7.4 pipe diameters from the outlet for the 50% developed case and 10.6 pipe diameters from the outlet for the 99% developed case, reflecting the increase in scour length. As the scour hole develops the proportion of quadrant 1 and quadrant 3 events changes. For the 50% developed bed quadrant 3 events are dominant for 14% of the locations with 11% of the sampling points showing that quadrant 1 makes the largest contribution to the Reynolds stress. For the 99% developed bed quadrant 1 makes the largest contribution to the Reynolds stress at 30% of locations and the quadrant 3 events only dominate at 3% of locations. The presence of these events coincides with high shear values in the flow.

4. CONCLUSIONS

The analysis of the mean velocity and turbulence intensity has shown that the near-bed velocities decrease as the scour hole develops, as one would expect. As the scour hole develops a critical velocity is reached which is no longer capable of transporting sediment. Areas of high turbulence intensity and low Reynolds stresses have been observed close to the dune crest in addition to a large number of inward and outward interaction events being present in this area of the scour hole. This is very similar to observations made downstream of dunes and steps and it is suggested here that the flow structure in the upstream of the dune crest may be comparable to that downstream of steps and dunes.

ACKNOWLEDGEMENTS

This paper presents some of the work conducted on an EPSRC research grant entitled 'Structure of the turbulent flow field at culvert outlets', GR/L15296. The authors would like to thank the technical staff both at the University of Hertfordshire and HR Wallingford for their time and assistance in setting up the experimental tests.

REFERENCES

Abt SR, Kloberdanz RL and Mendoza C (1984). "Unified culvert scour determination" J. of Hyd Eng. ASCE. Vol 110. Oct 1984 pp1475-9

Ali KHM and Lim SY (1986). "Local scour caused by submerged wall jets". Proc Inst Civ. Engrs. Vol 81 pt 2. Dec 1986 pp607-645.

Ali KHM and Neyshaboury AAS (1991). "Localised scour downstream of a deeply submerged horizontal jet". Proc Inst Civ Engrs. Vol 91 pt 2. March 1991 pp1-18.

Bennett SJ and Best JL (1993). "Structure of turbulence over two-dimensional dunes." Sediment transport mechanisms in coastal environments and rivers. Belorgey M, Rajona RD and Sleath JFA. Euromech 310 Sept 1993. World Scientific Pubs. ISBN 981 02 1854 0

Blaisdell FW and Anderson CL (1988). "A comprehensive generalised study of scour at cantilevered pipe outlets." J. of Hyd Res. Vol 26 pt4. April 1988 pp357-76.

Bohan JP (1970). "Erosion and riprap requirements at culvert and storm-drain outlets." US Army Engrs Waterways Experiment Station, Vicksburg, Mississippi, Report no. H-70-2. Jan 1970.

Breusers HNC and Raudkivi (1991). "Scouring. Hydraulic structures design manual, volume 2." Balkema, Rotterdam (1991) pp109-116.

Johnston AJ and Halliwell (1986). "Jet behaviour in shallow receiving water." J. of the Instn of Civ Enginrs Pt2 Vol.81 pp549-658.

Lim SY (1995). "Scour below unsubmerged full flowing culvert outlets." Proc Inst Civ Engrs Wat, Marit and Energy. Vol 112 pt2. June (1985) pp 136 - 149.

Lu SS and Willmarth WW (1973). "Measurements of the structure of the Reynolds stress in a turbulent boundary layer." J. of Fluid Mech Vol.60 pp481-511.

McLean SR, Nelson JM, Shreve RL (1996). "Flow-sediment interactions in separating flows over bed-forms." From Coherent flow structures in open channels, Ashworth PJ, Bennett SJ, Best JL and McLelland SJ eds. Wiley pubs 1996 ISBN 0 471 95723 2

Melville BW (1975). "Local scour at bridge sites." University of Auckland, NZ, School of Engineering Rep. No 117.

Melville BW and Raudkivi AJ (1977). "Flow characteristics in local scour at bridge piers." J. of Hyd. Res. Vol.15 pp373-380.

Nelson JM, McLean SR and Wolfe SR (1993). "Mean flow and turbulence fields over two-dimensional bed forms." Water Res. Res. Vol 29 pt 12 December 1993 pp3935 – 3953.

Nezu I and Nakagawa H (1993). "Turbulence in open-channel flows." IAHR Monograph, AA Balkema, Rotterdam, 1993. ISBN 90 5410 1180

Rajaratnam N and Berry B (1977). "Erosion by circular turbulent wall jets." J. of Hyd Res. IAHR Vol.15 pt3. Mar 1977. pp277-289.

Raupach MR (1981). "Conditional statistics of Reynolds stress in rough-wall and smooth-wall turbulent boundary layers." J. of Fluid Mech. Vol 108, 1981 pp363-382

Schoppmann B (1975). "The mechanics of flow and transport of a progressive scour hole." 16th IAHR Congress, Sao Paulo, Brazil 27th July - 1st August 1975. Volume 2 pp189 - 195.

Wallace JW, Ecklemann H and Brodkey RS (1972). "The wall region in turbulent shear flow." J. of Fluid Mech., Vol54 pp39-48.

Stochastic Hydraulics 2000, Wang & Hu (eds) © 2000 Taylor & Francis, ISBN 90 5809 166 X

On local scour around spur dikes

Liu Qingquan, Chan Li & Li Jiachun
Institute of Mechanics, Chinese Academy of Sciences, Beijing, China

ABSTRACT: Estimation of possible maximal scour depth around the head of spur dikes is an important step in the design of spur dikes foundation. This paper briefly analyses the characteristics of flow pattern and the complex eddy structure near the head of the spur dike. The strong rotational flow exhibits high shear stress resulting in intense local scour. Staring from the sediment continuity equation and the eddy's pick-up action on the sediment, a semi-empirical method for unsteady flow and stratified bed case has been established to show the time evolution of the local scour around the spur dike head. The method proposed in the present paper has been applied to simulate the local scour process of S1 and S2 spur dikes of the waterway project in the Yangtze River estuary. The computed and measured depths of scour holes agree fairly well with each other exhibiting its satisfactory prediction capability.

Key Words: spur dike, vortex structure, local scour, scour hole, Pickup function

1. INTRODUCTION

Scour at the foundation of spur dikes often causes failure. So it is an important step in spur dike design to accurately evaluate the local scour at the head. The scientific theory of structural design of spur dikes is highly advanced, but a unified theory for estimating scour depth at spur dikes head is still in an infant stage, mainly duo to the complex nature of the scour problems. Many scholars, such as Mukhamedov(1971), Garde(1961), Gill(1972), Tyagi(1973) etc., investigated the local scour of the spur dike head in early period. These treatments of the scour problem usually start with the statement that the depth of scour depends on variables which characterize fluid flow, bed material in the stream (grading, particle size and shape, alluvial or cohesive), the geometry of the spur dike (the height of the head, the side angle of the spur dike), and end up with an empirical formula based on the experimental results and observations. With these formulas, only the final scour depth can be obtained under the circumstance of steady flow and homogeneous river bed, but the time evolution of scour can not be shown. For this reason, these formulas can not be applied to unsteady flows and stratified bed cases. As a matter of fact, the flow is usually unsteady in many cases, such as flooding water, tide flows in estuaries, and the sediment composition of the river bed varies greatly along the depth.

In order to consider the influence of unsteady flow and to describe the process of local scour, the variation of scour depth in unsteady flow has to be studied. The objective of the present study is to develop a semiempirical method for determining the process of scour hole around the head of spur dikes. The current method is based on the sediment continuity equation, the sediment pickup rate and the similarity of local scour as well. Finally, applying this method in waterway project of the Yangtze River estuary, the process of local scour near the heads of S1 and S2 spur dikes has been successfully simulated.

2. MECHANISM OF LOCAL SCOUR AROUND THE SPUR DIKE HEAD

The local scour of a spur dike head is the direct consequence of flow resuspension action against the river bed. Generally speaking, after building up spur dike engineering in the waterway, the fluid velocity increases with restrained river cross section, and then the general scour occurs. What's more, the flow will separate at the head of spur dike, forming complex 3-D structure and resulting in intense local scour near the head.

Firstly, because of the block action of the spur dike, the flow close to the spur dike will concentrate and the kinetic energy will increase in the forward side of spur dike. On the other hand, the approaching flow goes to zero at the upstream face of the spur dike, part of the kinetic energy is transformed into pressure energy. Because of non-even vertical distribution of the horizontal velocity component (larger above and smaller below) and the pressure (smaller above and larger below), a separation plane will occur slightly below the streamline with maximal approaching velocity. The flow above the separation plane turns to the free surface resulting in local rise of the surface level; The flow below turns to bottom, forms a downflow and creates a vortex flow. The steady vortex flow near the bed will drastically pickup the sediment from bed and thus form a scour hole in the foreward side of the head of the spur dike.

Secondly, the vortex flow at the bottom will combine the concentrated longitudinal flow coming from the bottom of upstream near the spur dike head, which forms a spiral flow and goes downward. Under the intense action of the spiral flow, the bottom sediments around the spur dike head are continually scoured and carried downward. Then the scour hole grows deeper and bigger.

Thirdly, during the process of the flow around the spur dike, the separate flow will drives a rear vortex at the back of the spur dike head. So the complicated 3-D structure of the flow around the spur dike head and its pickup action on the sediment are intensified.

It is show that the complex vortex flow around the spur dike head (see fig.1) exerts important influence on the scour and sediment transport, and turns out the major driving force for intense local scour.

With the scour hole growing deeper and larger, the capacity of flow carrying sediment decreases gradually. When the rate of scour and the sediment deposition balance, the scour process stops, and a steady local scour hole is formed.

Lots of experiments had shown that the local scour of the spur dike head often occurs at the forward side of the head at initial stage, and a little scour hole takes shape, which develops rapidly with the increase of fluid velocity. When the scour hole develops to a certain degree, the rate of scour will decrease gradually till dynamic equilibrium. Generally, the scour hole is largely formed in a short time in the initial stage, and the scour quantity is very little for a long time during the later period. The local scour can be classified into two types according to the movement of bed load. If the velocity of approaching flow is less than the critical one U_c of the sediment there is no sediment transport and no sediment supply into the scour hole from upstream, this is known as clear-water scour condition, in which the scour hole develops rapidly. If approaching flow velocity is greater than the critical one, sediment is carried into the scour hole from upstream, the development of the scour hole will slow down and its depth fluctuates in response to the passage of bed features. This is the live-bed scour condition. Therefore, the scour depth of clear-water scour is generally a little larger than the depth of live-bed scour.

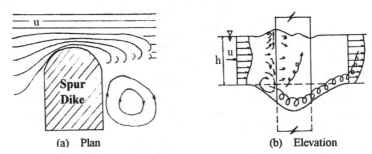

(a) Plan   (b) Elevation

Fig.1: Diagrammatic vortex movement around the head of spur dike

3. PRINCIPLES AND MODELING

Starting from the sediment continuity equation, A.Melih Yanmaz and H.Dogan Altmbilek (1992) established the correlation between scour hole volume and scour depth and the relation between volume change rate and entrainment rate, and computed effectively the local scour process around the pier. Because of the similarity between the spur dike and pier local scours, these principles can also be applied to simplify and compute the local scour around the spur dike head reasonably.

3.1 Local Scour Model

Based on the principle of mass conservation, the sediment continuity equation for the local scour hole near the head of a spur dike can be expressed as:

$$\frac{dV}{dt} = Q_{so} - Q_{si} \tag{1}$$

where, dV/dt is the change rate of the scour hole volume V with respect to time t; Q_{so} is the volume rate of sediment carried out of the scour hole by the flow. Q_{si} is the volume rate of sediment carried into the scour hole.

Ignoring the upstream sediment supply, we only consider the clear-water case, ie. $Q_{si} = 0$, the sediment continuity equation of the scour hole can be simplified as:

$$\frac{dV}{dt} = Q_{so} \tag{2}$$

The rate of sediment transport out of the scour hole, Q_{so} can be obtained by utilizing the sediment entrainment rate, ie. sediment pickup function. So the relationship between the volume rate of sediment carried out of scour hole and sediment entrainment rate looks like:

$$Q_{so} = E \times A \tag{3}$$

where, E is the sediment entrainment rate; A is the area of scoured bed (scour hole).

There are lots of the sediment pickup studies by previous investigators such as LeFeuvre (1965,1970), Van Rijn(1984), Cao Zhixian(1996). According to the fact that sediment entrainment is the direct result of turbulence bursting near the river bed, Cao obtained the sediment entrainment rate (ie. pickup rate) based on the turbulence bursting research.

$$E = E_n \cdot \sqrt{sgd} = P \cdot d^{1.5} \cdot (\frac{F}{F_c} - 1) \cdot F \cdot \sqrt{sgd} \tag{4}$$

where, E_n is the dimensionless sediment pickup rate; F is the Shields parameter, $F = U_*^2/(sgd)^{0.5}$; F_c is the critical Shields parameter, $Fc = U_{*c}^2/(sgd)^{0.5}$; U_* is the friction velocity of fluid; U_{*c} is the critical friction velocity of fluid; $s = \rho_s/\rho_f - 1$, ρ_s is the sediment density, ρ_f is the density of water; d is the diameter of sediment particles; g is the gravitational acceleration; $P = \lambda S_v(sg)^{0.5}/(\mu T_B^+)$, λ is the even area of turbulence bursting on unit bed area. $\lambda = 0.02$ according to experimental data; S_v is the sediment concentration in volume; μ is the kinematic viscosity of water; T_B^+ is the dimensionless average period of turbulence bursting, $T_B^+ = 100$ according to experimental results.

3.2 Similarity of the Local Scour

According to experiments, various local scour, especially the local scour of the same type, display good similarity in scour process and geometric shape of the scour hole. This similarity means that: 1) The geometry shape of the same kind scour hole is generally analogous; 2) A scour hole keeps its geometry analogy during the developing process. That is to say, its geometry shape maintains analogous in the whole process of scouring, though the depth and region of scour hole vary constantly with time.

Observations show that the local scour hole of the spur dike head is asymmetrically similar to a banana (as shown in Fig.2). The incline angle of the scour hole tends to be a bit less than the natural angle of repose for sediment. Generally, the inclination of upstream is larger than the downstream's. During the scouring process, the whole scour hole grows larger and deeper, but the angle of upstream and downstream approximately keeps constant. The deeper, the larger the region of the scour hole is. What's more, both depth and region of the scour hole are the functions of time and change as time goes on. The deepest of the scour hole locates nearby the head of the spur dike and deviates from the upstream.

Consequently, we suppose that the geometric shape of scour hole remains similar during the whole local scour process, that is, the scour hole grows deeper and bigger in a similar shape as time elapses. To express the volume of the scour hole explicitly, we combine the results of experiments and observations, and simplify the shape of local hole on the spur dike head as a part of concentric circles as shown in Fig. 3.

3.3 The Expression of the scour hole

To determine the rate of change of side inclination of the scour hole, a three-dimensional coordinate system is established in Fig.4. We suppose that the upstream and downstream side angles to be θ_1 and θ_2, respectively. The width of the spur dike head is b, and the depth of scour hole is h_s, so the expression for the volume of the scour hole can be derived as follows:

$$V = \frac{1}{2}\left[k_1(\pi\int_0^{h_s} x_1^2 dz - \pi(\frac{b}{2})^2 h_s)\right] + \frac{1}{2}\left[k_2(\pi\int_0^{h_s} x_2^2 dz - \pi(\frac{b}{2})^2 h_s)\right] \tag{5}$$

k_1, k_2 are the proportional coefficients of scour hole's upstream and downstream scoured area and the semi-circle area, and turn out 0.5 to 0.8 respectively.

Representing x in terms of y as given in Fig.4,

$$x_1 = \frac{z}{\tan\theta_1} + \frac{b}{2} \tag{6}$$

$$x_2 = \frac{z}{\tan\theta_2} + \frac{b}{2} \tag{7}$$

Substitute them into the formula above and integrating,

$$V = \frac{\pi}{2}\left[k_1(\frac{h_s^3}{3\tan^2\theta_1} + \frac{h_s^2 b}{2\tan\theta_1}) + k_2(\frac{h_s^3}{3\tan^2\theta_2} + \frac{h_s^2 b}{2\tan\theta_2})\right] \tag{8}$$

Time derivative of (8) is

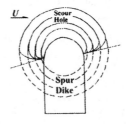

Fig.2: Sketch of scour hole around the head of spur dike

Fig.3: Simplified shape of scour hole

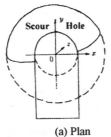

(a) Plan

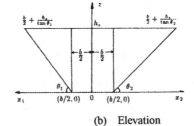

(b) Elevation

Fig.4: Geometric description of scour hole and sketch of coordinate system

$$\frac{dV}{dt} = \frac{\pi}{2}\left[k_1\left(\frac{h_s^2}{\tan^2\theta_1} + \frac{h_s b}{\tan\theta_1}\right) + k_2\left(\frac{h_s^2}{\tan^2\theta_2} + \frac{h_s b}{\tan\theta_2}\right)\right]\frac{dh_s}{dt} \tag{9}$$

In the same way, the actual scoured area of the scour hole can be obtained:

$$A = \frac{\pi}{2}\left[k_1\left(\frac{h_s^2}{\tan^2\theta_1} + \frac{h_s b}{\tan\theta_1}\right) + k_2\left(\frac{h_s^2}{\tan^2\theta_2} + \frac{h_s b}{\tan\theta_2}\right)\right] \tag{10}$$

3.4 The Method and Steps

For combining equations (2),(3), (4), (9) and (10) to compute the developing process of the scour hole nearby the spur dike head, the shear velocity U_* of the flow and the critical shear velocity U_{*c} of sediment have to be known.

Assuming the velocity vertical distribution of the approaching flow close to the spur dike head still accord with the logarithmic law, ie:

$$\frac{U}{U_*} = \frac{1}{\kappa}\ln\frac{y}{y_0} \tag{11}$$

Integrating Eq.(11), the average velocity can be obtained as follow:

$$\frac{U_{av}}{U_*} = \frac{1}{\kappa}\left(\ln\frac{h}{y_0} - 1\right) \tag{12}$$

where, U is the velocity at the height of y when the bottom of the flow is supposed as zero, U_{av} is the average velocity; h is the height of the water surface, ie. The depth of water; κ is the Karman constant, y_0 is the theoretical height where the velocity is zero.

After y_0 is determined, the formula mentioned above can be utilized to compute U_* and U_{*c} by using of the concept of viscous sublayer. In the viscous sublayer,

$$\tau = \mu\frac{du}{dy} \tag{13}$$

where, τ is the shear stress of the flow, and $\mu = \rho\nu$ is the kinematic viscosity coefficient.

Assuming τ as constant, the shear force at the bottom is $\tau_b = \rho U_*^2$. After integrating we obtain:

$$\frac{U}{U_*} = \frac{U_* y}{\nu} \tag{14}$$

The thickness of viscous sublayer is generally taken as $y = 11.6\nu/U_*$, so at the top of viscous sublayer,

$$\frac{U}{U_*} = 11.6 \tag{15}$$

Using the logarithmic distribution of velocity, the flow velocity is expressed as

$$U = \frac{U_*}{k}\ln\frac{11.6\nu}{U_* y_0} \tag{16}$$

Hence we have:

$$\frac{11.6\nu}{U_* y_0} = e^{11.6k} \tag{17}$$

ie.

$$y_0 = 0.112\nu/U_* \tag{18}$$

Substitute it into formula (17),

$$U_{av} = \frac{U_*}{\kappa}\left(\ln\frac{hU_*}{0.112} - 1\right) \tag{19}$$

So U_* can be worked out numerically by iteration

$$U_* = \frac{U_{av}k}{\dfrac{hU_*}{0.112} - 1} \tag{20}$$

Choosing an appropriate sediment critical velocity formula, the critical shear velocity U_{*c} of the sediment can be just worked out through critical shear stress in the same way as deriving U_* from U_{av}. For example, the critical velocity derived out by Dou Guoren is expressed like this:

$$U_c = 0.74 \lg(11\frac{h}{k_s})(\frac{\rho_s - \rho_f}{\rho_f} gd + 0.19\frac{gh\delta + \varepsilon_k}{d})^{\frac{1}{2}} \tag{21}$$

where, κ_s is the roughness, $\varepsilon = 2.56 cm^3/s^2$, and $\delta = 0.213 \times 10^{-4}$ cm, according to experiments.

When the scour hole grows to a certain depth, an anti-circulating flow with the horizontal axis perpendicular to flow direction will occur. The intensity of scour will increase because of the pickup action of complex vortex movement. In this time, the direction of the flow neighboring the bottom will be reversal to the direction of the upper flow (or the even flow). The direction of the flow in viscous sublayer is also reverse to the direction of the upper flow. The shear stress at the bottom is negative, ie. $\tau_b = -\rho U_*^2$, so, in the viscous bottom,

$$\frac{U}{U_*} = -\frac{U_* y}{v} \tag{22}$$

Assuming that the thickness of viscous bottom is still $y = 11.6$ v/U_*. At the top of viscous bottom, there is

$$\frac{U}{U_*} = -11.6 \tag{23}$$

By using of the logarithmic distribution law of velocity, the velocity is as follows:

$$U = \frac{U_*}{k} \ln\frac{11.6v}{U_* y_0} \tag{24}$$

So

$$y_0 = 1201v/U_* \tag{25}$$

Now, the formula of U_* can be expressed as

$$U_* = \frac{U_{av}k}{\ln\frac{hU_*}{1201v} - 1} \tag{26}$$

So far, the friction velocity U_* can be worked out by iteration formula (20) and (26). In the same way, the critical shear velocity U_{*c} of the sediment can be also computed from formula (21). Furthmore, combining the equations (2), (3), (4), (9) and (10), the developing process of scour hole around the spur dike head can be simulated.

3. CASE STUDY

Taking the spur dikes S1 and S2 in the waterway project of the Yangtze River estuary as an example, the method mentioned above is applied to simulating the local scour process of these two spur dikes respectively.

For the tide with the discharge of 30000m³/s under unsteady condition, its tidal level and velocity processes are shown in Fig 5. The sediment composition of the river's bed is stratified as the actual geological section structure (as Tab. 1 shown).

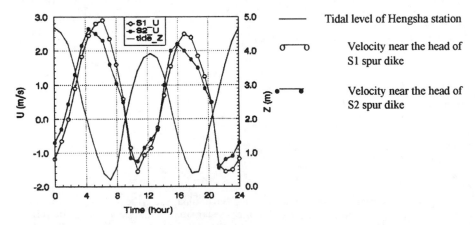

Fig.5: Tidal level of 3000 m³/s tide and velocities of the head of S1、 S2 spur dikes

206

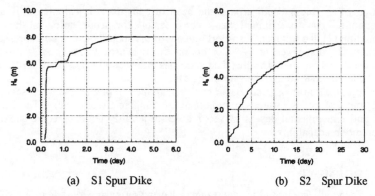

(a)   S1 Spur Dike   (b)   S2   Spur Dike

Fig.6: Computed results of the depth of scour holes with time

According to the properties of sediment, the angle of repose is about 30 degrees in the Yangtze river estuary area. Based on experimental data of model testing at Hohai University and Nanjing Water Science Academy, the side angle of scour hole close to the head of spur dike, S_1 (the upstream side) and S_2 (of the downstream side) can be selected approximately as 25 degrees and 20 degrees respectively. With the parameters of flow and sediment, the local scour of spur dikes S1 and S2 can be simulated. The results are shown in Fig 6.

Comparing the calculated maximum scour depth of spur dikes S1 and S2 with the physical modeling at Hohai University and Nanjing Water Science Academy, we can find out that the computational results accord with the modeling test results perfectly.

Fig. 6 shows the simulated results of the varying depth of the scour hole and the developing process of the local scour around the spur dike head. That is, the scour hole develops very quickly in initial stage. Subsequently, the scour velocity slows down as it develops to a certain degree. Finally, it will maintain a slight scour for a long time prior to reaching the equilibrium scour depth. The results accord well with the actual developing process of scour hole in physical law, showing this method's satisfactory prediction capability.

4. CONCLUSIONS

The water-restrain action of the spur dike results in increasing approaching velocity around the head of spur dike and leads to strong intense local scour, which is mainly caused by vortex movement and its pick up role. That's to say, such rotational flow around the spur dike head is the major driving force of local scour.

Table.1: Properties of sediment particles at the head of spur dikes (S1 & S2)

S1 spur dike	Depth of bed (m)	0.0-3,5	3.5-4.8	4.8-7.4	>7.4
	Sediment d_{50} (mm)	0.093	0.046	0.020	0.005
S2 spur dike	Depth of bed (m)	0.0-1.0	1.0-1.4	1.4-4.6	>4.6
	Sediment d_{50} (mm)	0.046	0.030	0.020	0.005

Table.2: Comparison of experimental and computed scour results

	Computed results	Hohai Unv.	Nanjing Water Science Academy
Depth of scour hole (m) S1 spur dike	8.00	7.60	9.50
Depth of scour hole (m) S2 spur dike	6.02	4.60	6.50

207

The variation of the scour depth around the head of spur dikes is calculated by solving a differential equation due to sediment mass conservation for scour hole. In the sediment continuity equation (1), the sediment transport rate out of the scour hole can be expressed by the sediment pickup function given in (4). Combining the two aspects above, a semi-empirical method for unsteady flow and stratified bed case has been established to describe the time evolution of the local scour around the spur dike head.

The method proposed in the present paper has been applied to predict the local scour process of S1 and S2 spur dikes of the waterway project in the Yangtze River estuary. The comparison of computed and experimental depths of scour holes in a fairly good agreement exhibits its satisfactory prediction capability.

ACKNOWLEDGEMENT

The project is financially supported by the Yangtze River Waterway Construction Co.Ltd. The authors are most grateful to professor Y. X. Yan, Hohai University, for helpful discussions and collaboration. Professor Yan also provided us with useful data in the computation.

REFERENCES

Altinbilek,H.D. & Basmac,Y.(1980). Localized scour below submerged vertical gates. Proc. Conf. on Computer and Physical Modelling in Hydraulic Engineering, New York, N.Y.,39-50.

A. J. Raudkivi(1986). Functional Trends of Scour at Bridge Piers, J. Hydr. Eng. Div., ASCE, 112(1), 1-13.

Cao Zhixian(1996), Turbulent bursting-based sediment entrainment function. Journal of Hydraulic Engineering, 1996(5), 18-21. (in Chinese)

Garde,R.J., Subramanya,K. & Nambudripad,K.D.(1961), Study of Scour around Spur Dikes. Journal of Hydraulic Division,ASCE, Vol.81,No.Hy6.

Gill,M.A.(1972), Erosion of Sand Beds around Spur Dikes. Journal of Hydraulic Division, ASCE, Vol.98,No.Hy9.

Lefeuvre,A.R.(1965). Sediment Transport Functions with Special Emphasis on Localized Scour, Thesis presented to the GeorgiaInst. of Tech., at Atlanta, Georgia, in partial fulfillment of the requirements for the degree of Doctor of Philosophy.

LeFeuvre,A.R., Altinbilek,H.D. & Carstens,M.R.(1970). Sediment Pickup Function. J. Hydr. Eng. Div., ASCE,96(10),2051-2063.

Mukhamedov,A.M., Abaurapov,R.R., Irmukhamedov,KH.A., Urkinbaev,R.(1971), Study of Local Scour and Kinematic Structure Of Flow around Solid and through Spur-Dikes, Proc. of 14th Congress of IAHR.

Subhasish DEY(1997), Local scour at Piers, Part I: A Review of Developments of Research, International Journal of Sediment Research}, 12(2), 23-46.

Tyagi,A.K.(1973), Modeling of Local Scour around Spur Dikes in Streams, International Symposium on River Mechanics, Vol.1.

Van Rijn,~L.~C.(1984), Sediment Pickup Functions. J. Hydr. Eng. Div., ASCE, 110(10), 1494-1502.

Yanmaz,A.M., Altmbilek,H.D.(1991). Study of Time-Dependent Local Scour Around Bridge Piers. Journal of Hydraulic Engineering,ASCE,Vol.117,No.10,1991.10,1247-1268.

Stochastic Hydraulics 2000, Wang & Hu (eds) © 2000 Taylor & Francis, ISBN 90 5809 166 X

Statistical characteristics of pressure fluctuations downstream of deflectors

M. R. Kavianpour
Civil Engineering Department, K.N. Toosi University of Technology, Tehran, Iran

ABSTRACT: The present investigation pertains to the pressure fluctuations and its statistical characteristics downstream of plates and ramps. The normal plate which typifies a leaf gate, is used for flow regulation in high pressure conduits in hydraulic structures. The ramp is also used in spillway of high dams to introduce air into the flow to reduce the tendency for cavitation to damage the surface.

Measurements were performed downstream of the deflectors, on the side wall and beneath the recirculating zone and their probability distributions were obtained and compared with the normal distribution. Although extensive measurements have been carried out for a wide range of conditions, only typical results are presented herein to describe the features of the results. More detailed results can be found in Kavianpour 1997, Kavianpour & Alimohamady 1999, and Nakhaei Abbolabady & Kavianpour 1999. It is hope that this information will give an insight into view of the pressure field and its characteristics downstream of deflectors.

Key words: Pressure fluctuation, Deflector, Reattaching flow, Probability distribution

1. INTRODUCTION

Over the years considerable attention has been given to reattaching flows since they arise in many situations of engineering importance. Some of the flows in which reattaching occurs are those over a normal wall, through a partially closed gate valve, past upstream and downstream-facing steps, and in sudden changes in conduit sections. Major problems encountered in such applications are cavitation, structural vibration and fatigue failure.

Recent investigations show that pressure fluctuations can give an insight into the structural behavior, vibration, and cavitation possibilities. It has been established long ago that pressure fluctuations could be responsible for cavitation to occur sometimes even when the mean pressure is well above the vapor pressure (Lopardo et al 1987). The fluctuating pressures may cause vibrations, fatigue of materials and large instantaneous pressures close to the vapor pressure of liquid. Damage due to pressure fluctuations has been serious at a number of structures, such as, spillways, stilling basin and reattaching flows (Narayanan & Reynolds 1972, Kavianpour 1997). Therefore, a knowledge of the turbulence pressure fluctuations is neccessary in many areas of practical flows (Narayanan 1984). However, owing to difficulties in making reliable measurements, there is little information available on pressure fluctuations and its characteristics. In this paper experiments were carried out to examine the wall pressure fluctuations and their statistical characteristics downstream of the normal plates and ramps.

2. EXPERIMENTAL SETUP

A set of experiments were performed in a closed conduit at Water Research Center of the Ministry of Energy, and Water and Watershed Management Research Center of the Ministry of

Construction the Jahad, Tehran, Iran. Experiments were performed in a duct to avoid the effect of self aeration from the upper surface of water on the flow downstream of deflector. Figure 1 shows a diagram of experimental set up. The main section was made of perspex to visualize the flow. The sections were joined carefully to avoid zones of local separation. Water was supplied from a constant head tank through pipes to the test section. A contraction was fixed at the entrance to the duct to ensure a uniform flow in the test section.

The main section contained several tappings on the side wall and along the center line of the bed of the duct to fix the pressure transducer. The minimum longitudinal distance between the tappings was set equal to 15mm and 20mm inside and outside the recirculating zone, respectively. The minimum vertical distance between the tappings on the side wall was 15mm. The first row was set at 5mm from the bed. The transducer was kept flush with the surface of the duct to avoid flow separation which would affect the measurements of pressure.

Water was supplied by a constant head tank and the quantity of flow was controlled by valve. The measurements of fluctuating pressures were made by a pressure transducer. Its maximum operating pressure ranges to 1bar. The electrical signals of pressure transducer were passed through a variable gain amplifier, and then to an analog to digital converter (Microlink) having a 12 bit ±5 volts range, for both slow and fast sampling rates. The pressure signals were collected using a personal computer and a standard data collection computer program. An option was implemented in the program to select the sampling rate which controlled the number of data and the time interval. Initial tests with three different sampling rates, namely 50, 100 and 150 samples/sec showed that a sampling rate of 100 samples/sec was appropriate. Comparison of results with those obtained with a sampling rate of 150 samples/sec revealed an error not in excess of 3% (Kavianpour 1997; Pirooz et al. 1999).

The flow separates after the plates and ramps and forms a recirculating zone. In this zone measurements of mean and fluctuating pressure were carried out for different plates and ramps having the slopes of 5° and 10°. The relative height of plate/ ramp to the height of duct (H/D) varied from 0.1 to 0.8.

3. PROBABILITY ASSESSMENT

Because of the random nature of the pressure fluctuations in the regions of recirculation, information regarding the statistical characteristics of pressure fluctuations were also studied in detail. Therefore, the data were analyzed to determine the intensity of pressure fluctuations with different probably occurrence and the results were compared together. To obtain the probability distributions of a sample of time depended $p(t)$ data, first the maximum and minimum pressures are computed. Therefore, the probability that the instantaneous pressure assumes a value within the range $(p, p + \Delta p)$ may be equal to t_p / T, where t_p is the amount of time that pressure falls within the range $(p, p + \Delta p)$ and T is the total time in which the data were collected. In equation form, the probability that the instantaneous value p(t) is less than or equal to some value p is equal to the integral of the probability density function from minus infinity to p. This function is known as the probability distribution function. So the probability that pressure falls inside a range (p, p_{min}) is obtained by:

$$prob[p_{min} \langle p(t) \langle p] = \int_{p_{min}}^{p} p_d(p).dp \tag{1}$$

where, p_d(d) is the probability density function within the range of $(p, p + \Delta p)$. Based on this equation, a computer program was written to calculate the probability distribution of pressure fluctuations, measured by pressure transducer.

4. RESULTS AND DISCUSSION

Experiments were performed with different height ratios (height of deflector to depth of water/duct). The data were analyzed to determine the intensity of pressure fluctuations with different probably occurrence and compared together. The root mean square (r.m.s.) values of pressure fluctuations is expressed non-dimensionally as $C'p = \sqrt{\overline{p^2}} / (0.5 \rho U_{vc}^2)$, where U_{vc} is

210

the velocity at the ideal vena-contracta. The values of U_{vc} are calculated using the results of Narayanan (1970). Measurements beneath the region of recirculation was restricted to those on the centerline of the duct along the x-direction. Figure 1 display the longitudinal variations of maximum C'p beneath the region of recirculating zone and on the side wall. The figures show that the maximum intensity of pressure fluctuations occurs upstream of the recirculating point at 0.7 x/x_r 0.95. The peak values of C'p on the side wall and beneath the region of recirculating are shown as a function of H/D in Figure 3. The distribution of C'p shows that the maximum intensity occurs on the side wall which is comparable to the results of other investigators (Narayanan 1972).

The probability distributions of the results were determined for all sets of data. Significant deviations from the Gaussian distribution were observed. The most important deviation occurred for the larger negative values of the distributions. Figures 4 and 5 show the typical results of maximum, minimum, and different probability occurrence of pressure fluctuations beneath the recirculating zone and on the side wall respectively. It was observed that the distribution of pressure fluctuation in most parts of the regions of recirculating zone have large negative pressures than predicted based on a normal distribution. This is associated with a high negative skewness which are shown in Figures 6 and 7. The figures show a negative skewness associated with high intensity of pressure fluctuations shown in figures 3 and 4.

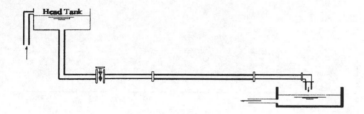

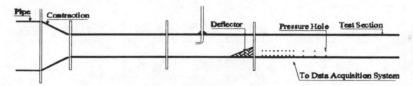

Fig.1: Schematic view of experimental set up

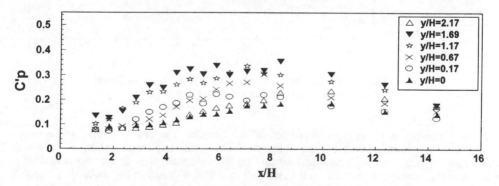

Fig.2: Variation of C'p on the side wall (H/D=0.3)

211

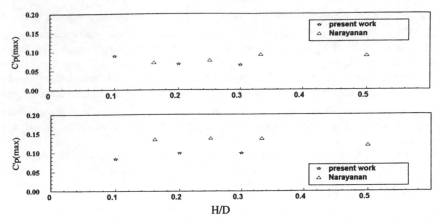

Fig.3: Variation of C'p$_{max}$ with height ratio beneath the recirculation zone and on the side wall

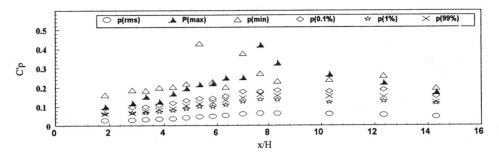

Fig 4: Variation of C'p$_{max}$, C'p$_{min,}$rms values of C'p,and C'p with different probability occurrence beneath the recirculation zone (H/D=0.3)

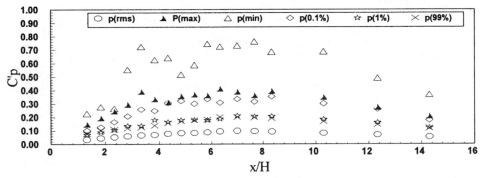

Fig.5: Variation of C'p$_{max}$, C'p$_{min,}$rms values of C'p,and C'p with different probability occurrence on the side wall (H/D=0.3).

The experiments were also performed for different ramps. The results shown a relatively similar behavior with respect to plates. Figure 8 shows the maximum values of pressure fluctuations beneath the recirculating zone and on the side wall. The figure shows that the maximum fluctuations of pressure on the side wall increases with increasing the angle of ramp. For the small angles considered in the present investigation (θ=5° and 10°), C'p$_{max}$ changes only very slightly but its value is much larger for normal plates (θ=90°). Figures 9 and 10 display the variations of maximum pressure fluctuation beneath the recirculation zone and on the side

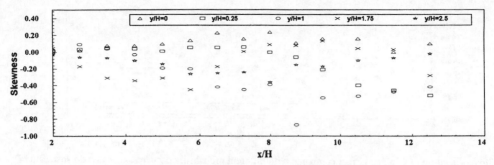

Fig 6: Variation of Skewness on the side wall (H/D=0.2)

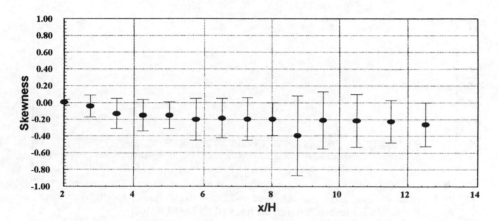

Fig 7: variation of skewness downstream of the plate(H/D=0.2)

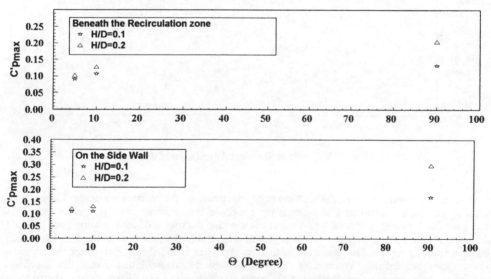

Fig 8: Variation of Maximum C'p with angle of ramp on the side wall and beneath the recirculation zone

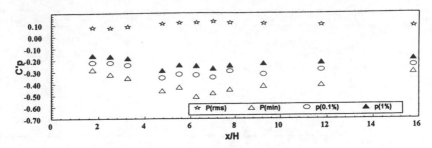

Fig9: Variation of C'p and C'p$_{min}$ compared with different probability occurrence of C'p downstream of ramp beneath the recirculation zone (H/D=0.2 θ=10°)

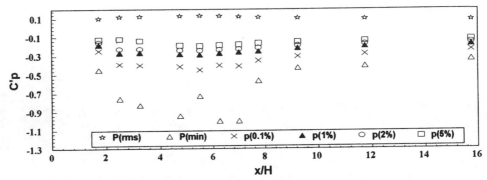

Fig 10: Variation of C'p$_{max}$ and C'p$_{min}$ compared with C'p with different probability occurrence downstream of ramp on the side wall (H/D=0.2 θ=10°)

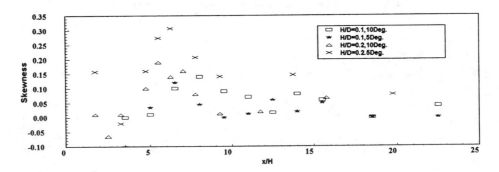

Fig.11: Variation of Skewness downstream of the ramp.

wall. The maximum pressure fluctuations occur upstream of the reattaching point. The ratio of C'p$_{max}$ on the side wall to that beneath the region of recirculating zone is smaller than that observed for flow past plates. It was observed that within the region of recirculating zone the data depart from the normal distribution. This was also related to the skewness of the data which can be seen from figure 11. The figures show that the skewness in the shear layer where the pressure fluctuations are intense, is negative. Downstream of the recirculating zones the skewness increases again and remains constant in the transverse direction. The distribution of skewness is similar to that of C'p beneath the region of recirculating zone.

REFERENCES

Kavianpour, M. R., 1997, "The Reattaching Flow Downstream of Deflectors, Including the Effect of Air Injection", A thesis submitted to The University of Manchester Institute of Science and Technology for the degree of Doctor of Philosophy.

Kavianpour M. R., and Alimohamady A., 1999, "Pressure Fluctuations Downstream of Aerators", 2nd Iranian Conference on Hydraulics, University of Science and Technology, Tehran, Iran.

Lopardo R. A., Lio J. C., and Henning R. E., 1987, "modeling techniques for preventing cavitation in structures submerged in hydraulic jumps", 22nd Congress IAHR, Lausanne, Switzer land, PP. 177-182,

Nakhaei Abbolabady M., and Kavianpour M. R., 1999, "Hydrodynamic Pressure Downstream a Leaf Gate", 2nd Iranian Conference on Hydraulics, University of Science and Technology, Tehran, Iran.

Narayanan R., 1970, "The reattachment flow downstream of a leaf gate", A Thesis submitted to the Department of Mechanical Engineering, Brunel University, in partial fulfillment of the requirements for the degree of Doctor of Philosophy.

Narayanan R., 1984, "The role of pressure fluctuations in hydraulic modeling", Symposium on Scale Effects in Modeling Hydraulic Structures, Esslingen, Germany, PP. 1.12-1 to 1.12-6.

Narayanan R., and Reynolds A. J., 1972, "Pressure fluctuations in Reattaching flow", Journal of Hydraulic Division, ASCE, Vol. 94, No. 6, March, PP. 1383-1398.

Pirooz B., Kavianpour M. R., and Montazeri Namin M., 1999, "pressure Fluctuation in Hydraulic Jump-Study the Effects of Inflow Conditions and Tail Water", 2nd Iranian Conference on Hydraulics, University of Science and Technology, Tehran, Iran.

Stochastic Hydraulics 2000, Wang & Hu (eds) © 2000 Taylor & Francis, ISBN 90 5809 166 X

Use of a submerged wall in controlling sedimentation at El Nasr canal intake structure

S. Mansour & M. B. A. Saad
Hydraulics Research Institute, Delta Barrage, Egypt

ABSTRACT: Right angle canal intakes suffer from sedimentation problems and unequal distribution of flow velocities through the intake openings (vents). Two main methods are used to overcome the sedimentation problem. The first one is periodically dredging the accumulated sediment from the intake while the second is to improve the approach flow conditions in the source canal to bypass and minimise or exclude the entering sediment to the intake of the branch canal. El Nasr Canal, one of the irrigation canals in Egypt, which irrigates an area of around 40,000 hectares, takes its water from El Nobaria canal. The canal has a right angle intake, which suffers from sedimentation problem. El Nasr Canal authority used to dredge the accumulated sediment from the downstream channel regularly. This process proved to be very expensive. Therefore, searching for another more sound solution becomes necessary. The present paper aims at finding the optimum solution to enhance the flow pattern and to bypass the sediment at the intake area. A physical model of scale 1:25 was constructed at the Hydraulics Research Institute, Egypt, to test the different alternatives that may be used to solve this problem. The sediment transport was studied by applying the flow velocity measurements in the well-known sediment transport equations. These measurements were also used to check the approach flow pattern in addition to track the surface current by floats and dye. The obtained results have shown that the use of a bottom wall has improved the flow distribution through the intake vents and that the construction of a sediment trap in the main channel upstream of the intake structure prevents the sediment from entering the canal.

Key words: Sedimentation problem, Flow pattern, Bypass, Physical model, Sediment transport, Bottom wall, Intake structure

1. INTRODUCTION

El Nasr canal irrigates the new reclaimed land in the northern west part of Egypt. The intake structure of this canal is located on the left bank of El Nobaria canal at a right angle orientation. Due to this orientation a severe sedimentation problem is characterised at the intake location and at the approach channel downstream it. As the in-curve vents attract less water than the outer vents then the sedimentation is deposited in the inner curve. Figure (1) shows a schematic definition of the three dimensional flow properties at the site of the intake. This problem hinders the delivery of water to El Nasr canal. In addition, the sediment particles that are sucked through the pumping station, which is situated on the canal, cause abrasion to its vanes. Upon the request of El Nasr Canal Authority, the Hydraulics Research Institute was assigned to carry out field and laboratory studies to find a solution for this problem.

A scale model was constructed at the northern experimental hall of the Institute, and different solutions were attempted. These solutions may be listed as follows:

1-   Use of a submerged wall 10.5m upstream the beginning of the concrete raft foundation (fig. 2.1).
2-   Use of an inclined submerged wall right at the beginning of the concrete raft foundation (fig. 2.2).
3-   Use of bottom vanes (fig. 2.3).

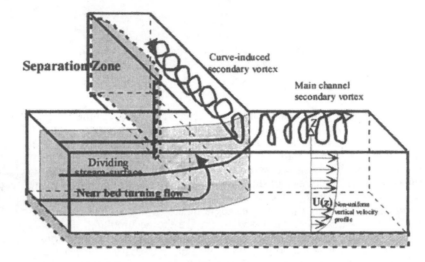

Fig.1: Three-Dimensional Flow Pattern in a 90-Degree Rectangular Diversion

Sectional Elevation of the Intake Structure with Bottom Wall

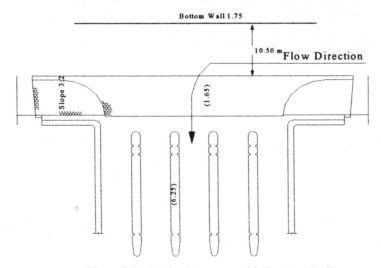

Plan of the Intake Structure with Bottom Wall
Fig.2.1: Bottom Wall 10.50 m Upstream the Raft Foundation

218

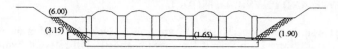

Sectional Elevation of the Intake Structure with Bottom Wall

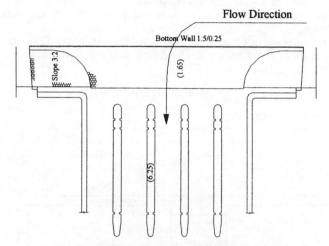

Plan of the Intake Structure with Bottom Wall

Fig.2.2: Bottom Wall (Option 2)

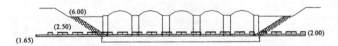

Sectional Elevation of the Intake Structure with Bottom Vanes

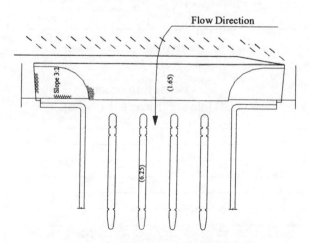

Plan of the Intake Structure with Bottom Vanes

Fig.2.3: Bottom Vanes

219

2. THE PROTOTYPE AND PHYSICAL MODEL

A detailed hydrographic survey was carried out for 2 km from El Nobaria Canal at the intake of El Nasr canal area, also it covers 2 km from it (approach channel). All relevant hydrological data were collected, such as maximum and minimum water levels and discharges in both canals. Bed and suspended sediment were collected. Based on the results of the field observations, an undistorted physical model of a scale 1:25 was constructed.

The scale was chosen to fit the available space and to ensure sufficient water depth for hydraulic measurements.

Figure No.3 shows the schematic plan of the model and the controlling elements.

Based on Froud similarity conditions, the different flow characteristics are presented in Table (1):

Table.1: Model scale

	Scale	Units
Length Scale	25	m
Time Scale	5	s
Velocity Scale	5	m/s
Discharge Scale	3125	m^3/s

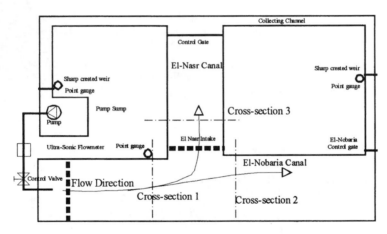

Fig.3: Schematic plan of the physical model

Velocity Distribution
60 m US Intake Structure (Nobaria)

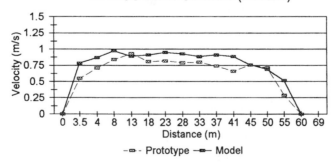

Fig.4: Velocity Distribution at Cross-Sections No. 1

Velocity Distribution
40 m DS Intake Structure (Nobaria)

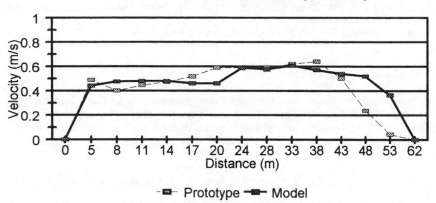

−▫− Prototype −■− Model

Fig.5: Velocity Distribution at Cross-Sections No. 2

Velocity Distribution
60 m DS Intake Structure (EL-Nasr)

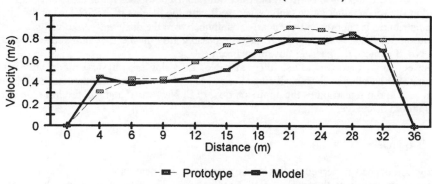

−▫− Prototype −■− Model

Fig.6: Velocity Distribution at Cross-Sections No. 3

Velocity Inside the Intake Structure
Flow Velocity Comparison

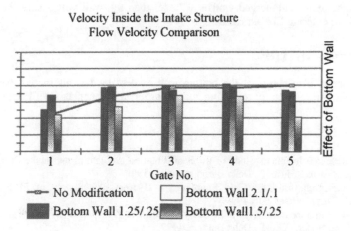

−▫− No Modification ☐ Bottom Wall 2.1/.1
▨ Bottom Wall 1.25/.25 ▨ Bottom Wall 1.5/.25

Fig.7: Velocity Comparison Between the different Options

To calibrate the model, the velocity distribution at three cross-sections was measured in the prototype and in the model for a given discharge. The water surface slope was controlled at the model and compared with that in the prototype. Figures 4, 5 and 6 show a good agreement between model and prototype results.

Therefore, it is possible to say that the model simulates the prototype.

3. TEST PROCEDURE AND RESULTS

The tests are based on flow velocity, and sediment tracing measurements. These measurements give an idea about the flow pattern and how it could be enhanced. To improve the flow distribution through the intake vents, several submerged wall configurations were tested in order to find the best dimensions of the wall. Table (2) shows the carried out tests:

Table.2: Test Programme

	Wall Height at Left Bank (m)	Wall Height at Right Bank (m)
Test No.1	0.25	1.25
Test No.2	0.25	1.5
Test No.3	0.1	2.1

Figure 7 shows a comparison between the three wall configurations. From these alternatives it is clear that the second alternative (submerged wall with the heights of 0.25/1.5m) is the best one as it redistributes the approach flow nearly equally between the different intake vents.

While alternative No.1 has hardly increased the passing flow in the first and second vents, alternative No. 3 reduced the passing flow dramatically.

The most effective solution was to use alternative No. 2. Using the physical model, the best configuration of the wall was chosen. It stops the bed load, which is the predominant sediment load in this case and it redistributes the approach flow to El Nasr canal and lets the left side two vents attract more discharge.

4. CONCLUSIONS

The construction of the submerged wall with the specified alignment as in alternative No.2 will enhance the flow pattern at the intake of El Nasr canal. The flow will be nearly equally distributed between the gate openings. The sediment load moves in a layer of a maximum height of 0.1 m as the most of which consists of bed load as given by the hydrographic survey. As the height of the submerged wall is 0.25 m, then the wall will assure the elimination of sediment from entering El Nasr canal.

5. RECOMMENDATION

It is recommended to construct the bottom wall as described in alternative No. 2 in order to reach an effective solution for the sedimentation problem at the intake of El Nasr canal.

REFERENCES

Hydraulics Research Institute, "Study of the Sedimentation Problem in the Intake Structure of En Nasr Canal and How to Reduce it", Delta Barrage, August 1999.

Hydraulics Research Institute, "Field Study and a Hydrographic Survey of El Nasr Canal Intake Structure", Delta Barrage, July 1999.

Hydraulics Research Institute, "Hydraulic Study of Sediment Transport at the Intake Structure of the Nasr Canal off the Nobaria Canal", Delta Barrage, 1987.

Iowa Institute of Hydraulic Research, "Enhanced Performance and Reliability of Water Intakes for Generating Stations", Iowa, 1992.

V.A. Vanoni, ASCE, "Sedimentation Engineering", New York, 1975.

Hendrickson, M.A. Drain, A.M Asce, R.H. Hotchkiss and M. Asce, "Hydraulic Model Investigation of Submerged Vanes for the Intake Structure at Fort Calhon Station", Water Resources, San Antonyo, Texas, 1995.

Stochastic Hydraulics 2000, Wang & Hu (eds) © 2000 Taylor & Francis, ISBN 90 5809 166 X

Characteristic values of suspended sediments sizes in turbulent streams

Alex Adesman
Parsons Brinckerhoff Incorporated, N.Y., USA

ABSTRACT: For mathematical modeling of streams and for channel processes prognosis we need to know characteristic values of suspended sediments sizes. In offered work an analysis of vertical turbulent pulsation's impact on solid particles was used to get an appropriate dependence for maximal sediments size. Regarding the distribution of turbulent oscillation standard in a stream, as well as bed sediments saltation parameters, related equation was obtained. The results of it solution are in good correspondence with experimental data. These results, being combined with lognormal size distribution regularity (rubbing and crushing theory), allow to get an expression for averaged (mean) size of suspended sediments. The obtained dependencies could be used for sediment transport mathematical modeling and forecast.

Key Words: Suspended sediments; Maximal diameter; Mean diameter; Fall velocity; Turbulent pulsation.

1. INTRODUCTION

The sizes of suspended sediments, which could be transported by channel flow, depend on following factors:
1. size grade distribution of bed ground (in the case of non-cohesive ground);
2. turbulent flow's ability to suspend solid particles.

It is obvious, that certain limit of maximal sediment size should exist. This limit could be considered as first characteristic value, and it is determined by the last of factors listed above. First factor determines another characteristic value – mean (averaged) size of suspended sediments. Thus, two main characteristic values are – maximal size (maximal fall velocity) and averaged size (averaged fall velocity) of sediments.

At present time we have the theory of rubbing process of alluvial grounds forming (Kolmogorov.A N ,1962), which could allow us to model grading of monogene bed ground and use it for mean size determination.

As to the maximal (limiting) size (Shen H.W., 1971), existing empirical formulae of different authors recommend values, differing from each other in up to two times

2. MAXIMAL SIZE

At present time the maximal fall velocity of suspended sediments, which accords to its maximal size, is determined by empirical relation:

$$w_m = Au_* = A\sqrt{gHi} \tag{1}$$

where w_m - is the maximal fall velocity;

u_* - is the friction velocity of a flow;

A - is empirical coefficient;

H - is the stream depth;

i - is a slope;

g - is the gravitational acceleration.

The values of empirical coefficient A, recommended by different authors, are sufficiently different, altering in the diapason from 0.3 to 0.6 (Abalianz,S.KH.,1973). This divergence is too considerable for the calculations.

To solve this problem in analytical way, consider the vertical movement of a solid particle in a fluid:

$$m\ddot{y} - F_g + F_a - F_r = 0 \tag{2}$$

where F_g - is the gravitational force;

F_a - is Archimedes force;

F_r - is the force of a resistance (drag force);

m - is the mass of a particle;

y - is the vertical coordinate, directed up
and having the origin at the flow's bed.

Hence, in a supposition of a square law of the resistance, we get a differential equation:

$$m\ddot{y} - mg + V\rho g - \rho\xi S\frac{1}{2}(\dot{y} - u)|\dot{y} - u| = 0 \tag{3}$$

where ρ - is the density of a fluid;

V - is the volume of the particle;

S - is the area of particle cross-section;

ξ - is the drag coefficient of the particle;

u - is the vertical velocity of the fluid.

Using the relation between the mass, volume and the density of solid particle, we could write:

$$\rho_0\ddot{y} - \rho_0 g + \rho g - \frac{1}{2}\rho\frac{S}{V}(\dot{y} - u)|\dot{y} - u| = 0 \tag{4}$$

where ρ_0 - is a the density of the solid particle.

Hence, assuming that the particles could be modeled with somewhat equivalent spherical particle, we have differential equation of vertical movement of this solid particle:

$$\ddot{y} - g\left(1 - \frac{\rho}{\rho_0}\right) - \frac{3}{4}\xi\frac{\rho}{\rho_0}\frac{(\dot{y} - u)|\dot{y} - u|}{D} = 0 \tag{5}$$

where D - is the equivalent diameter.

This expression ties the acceleration $\ddot{y}$ of the particle with its velocity $\dot{y}$.

The extremal case of a movement is its absence, i.e. case of:

$$\begin{cases} \dot{y} = 0 \\ \ddot{y} = 0 \end{cases} \tag{6}$$

When the values (6) are exceeded, it takes place the ascending, and when they are not reached it takes place the descending of this particle.

Thus, the top size of a sediment, suspended in a given point of a stream, is determined by the substitution of conditions (6) to the equation (5):

$$\frac{3}{4}\xi\frac{\rho}{\rho_0}\frac{|u|u}{D} = g\left(1 - \frac{\rho}{\rho_0}\right) \tag{7}$$

Hence we have the necessary condition of a limiting equilibrium:

$$|u| = \sqrt{\frac{4}{3}\frac{gD}{\xi}\left(\frac{\rho_0}{\rho} - 1\right)} \tag{8}$$

In the case of uniform and straight-forward stationary turbulent flow with $Fr < 1$ the velocity u is a vertical component of the turbulent oscillations. Averaging the absolute values of this vertical pulsation, we have:

226

$$|\overline{v'}| = \sqrt{\frac{4}{3}\frac{gD}{\xi}\left(\frac{\rho_0}{\rho}-1\right)} \tag{9}$$

where $|\overline{v'}|$ - is the averaged absolute value of the vertical turbulent oscillations.

Supposing, that the distribution of these oscillations is Gaussian (Khodzinskaya,A.1988), we could use the expression:

$$\sigma(y) = \sqrt{\frac{2}{\pi}}|\overline{v'}| \tag{10}$$

where $\sigma(y)$ - is the standard of vertical velocity oscillations. Taking into consideration that the right part of equation (9) is the maximal fall velocity of a solid particle with the diameter D, we could write down the following condition:

$$w = \sqrt{\frac{\pi}{2}}\sigma(y) \tag{11}$$

where w - is a fall velocity of a particle, having an equilibrium in the certain point.

So far as suspended particles move with the chaotic trajectories in the whole space of a flow, it is obvious, that maximal of the all possible values of w is determined by the minimal value of σ, impacting the suspended sediments in the flow:

$$\max(w) = \min\left[\sqrt{\frac{\pi}{2}}\sigma(y)\right], \; y \in [h, H] \tag{12}$$

where h - is a low bound of suspending of particles

The distribution of vertical pulsation standard along the depth could be modeled with the following dependence (Adesman,A.1987):

$$\begin{cases} \sigma = 0.75u_* \left(\frac{y}{H}\right)^n e^{k\left(1-\frac{y}{H}\right)} \\ n = 0.95\sqrt{\lambda} = 0.95\sqrt{\frac{8g}{C}} \\ k = \left(\frac{\kappa n u_m}{u_*}\right)^{\frac{1}{n}} \approx 5n \end{cases} \tag{13}$$

where λ - is hydraulic friction coefficient ;
   C - is Chezy coefficient;
   u_m - is the maximal velocity of a flow;
   $\kappa = 0.4$ - is Von Karman's constant.

Formulae (13) demonstrate that the least values of σ are located near the bed and they are decreasing toward the bottom.

The admitted model of the process of suspending is as following: the saltating bed-load particles "jump up" due to the collisions with other ones as well as another causes; at the top of their conventionally parabolic trajectories those of them, which have an appropriate size, could be picked up by the vertical pulsation and suspended. Thus, the more height of this jump (salto), the more chance is that the particle will be suspended, because of increase of vertical velocity values.

Consequently, the maximal (limiting) fall velocity of the suspension is determined by the pulsation value at the top of related particles jumps.

The height of saltating particles jump could be determined by the approximate empirical dependence (Khodzinskaya,A.,1988):

$$\frac{z}{D} \approx 4.54\frac{u_*}{w} \tag{14}$$

where z - is the height of the jump.

227

According to the admitted model, this height could be considered as the parameter h in (12).

The dependencies (13) show that the standard in this point has the least value in the area (12). Substituting the expression (14) into (13), we have:

$$\min(\sigma) = 0.75 u_* \left(4.54 \frac{u_* \, D}{w \; H} \right)^n e^{k\left(1 - 4.54 \frac{u_* D}{w H}\right)} \tag{15}$$

Hence, according to (12), the maximal fall velocity of a particle, which could be suspended by the flow, is:

$$w_m = 0.94 u_* \left(4.54 \frac{u_* \, D_m}{w_m \; H} \right)^n e^{k\left(1 - 4.54 \frac{u_* D_m}{w_m H}\right)} \tag{16}$$

where D_m - is the maximal diameter of the particle

To simplify the expression (16) we could use the fact that in real streams the following relation takes place:

$$\frac{h}{H} \ll 1 \tag{17}$$

In this case the equation (16) could be written in approximate form:

$$\frac{w_m}{u_*} = 0.94 \left(4.54 \frac{u_* \, D_m}{w_m \; H} \right)^n e^k \tag{18}$$

Maximal fall velocity is connected with the maximal diameter by the relation:

$$w_m = \sqrt{\frac{4}{3} \frac{g D_m}{\xi_m} \left(\frac{\rho_0}{\rho} - 1 \right)} \tag{19}$$

where ξ_m - is the resistance coefficient of maximal sized particle.

Applying to (18) the formulae (1) and (19), we obtain the equation, determining the maximal fall velocity of suspended sediments:

$$\frac{w_m}{u_*} = 0.93 \left(505.4 \frac{\xi_m i}{\rho_0/\rho - 1} \right)^{\frac{n}{1-n}} \tag{20}$$

The value of frontal resistance, consisted in the equation (20), depends on the regimes of solid particle movement in a fluid (laminar, turbulent or transitional ones). In general these regimes could be represented by Reynolds number:

$$Re = \frac{D_m w_m}{\nu} \tag{21}$$

where ν - is the cinematic viscosity of a fluid.

An empirical function of the drag coefficient, presented in [6], could be used to close the equation (20):

$$\xi_m = \begin{cases} \dfrac{24}{Re}, & Re \le 1; \\[2mm] \dfrac{24}{\sqrt{Re}}, & 1 \le Re \le 400 \\[2mm] 1.2, & Re \ge 400 \end{cases} \tag{22}$$

Thus, the expressions (19), (20) and (22) compose the closed system of equations, allowing to determine in numerical way the values of maximal fall velocity and, accordingly, the maximal diameter of suspended sediments. Therefore coefficient A in the formula (1) proves to be a variable, depending on the general parameters of the flow and bed. It alters in the diapason, corresponding to experimental data, considered above.

228

3. MEAN SIZE AND GRADING

The expressions (19)-(22) characterize just an ability of a flow to suspend certain fractions of bed ground. This means that all the particles with a size smaller than maximal one, could be suspended.

Thus, we could evaluate the supremum of a multitude of suspended sediments values. As to the composition of this suspension, it is determined by the grain-size distribution of the bottom ground.

Suppose, that bed ground grains could be unlimitedly small. It is obvious, that in this case the mentioned multitude is limited by zero from below.

If the maximal size of bed ground particles is smaller then D_m, the suspension grading is coincides with the bed ground grading. Otherwise, the suspended sediments sizes multitude's upper bound is D_m, and related grading is identical to the grain-size distribution in the interval between 0 and D_m.

An alluvial ground forming is the result of stuff rubbing processes. Consequently, Kolmogorov theory of crushing (Kolmogorov A.N.,1962) could be applied to the grain analysis.

According to (Kolmogorov A.N.,1962) when the ground forming has a monogene nature the distribution of probability's density of the certain diameter could be described with a lognormal distribution:

$$P = \frac{1}{\sqrt{2\pi}\delta} \exp\left(-\frac{\ln^2 \frac{D}{D_e}}{2\delta^2} \right) \tag{23}$$

where P - is the density of probability;

   D - is the diameter of particles;

   D_e - is the mathematical expectation (mean value) of bed ground grains sizes;

   δ - is the least square of a random quantity.

The parameter δ could be determined as:

$$\delta = \frac{1}{P_e \sqrt{2\pi}} \tag{24}$$

where P_e - is the maximal value of P, corresponding to D_e.

Uniting the solutions (20) and (23), we could see that the distribution of suspended sediments sizes differs from the bed grains ones by the restrictions:

$$\begin{cases} 0 \leq D \leq D_m \\ \ln \frac{D}{D_e} \in \left[-\infty, \ln \frac{D_m}{D_e} \right] \end{cases} \tag{25}$$

Mean diameter of a mixture of particles could be determined as:

$$D_a = \frac{\sum_i N_i D_i}{\sum_i N_i} \tag{26}$$

where j - is a number of the certain fraction;

   N_j - is a quantity of the particles of certain fraction;

   D_j - is the diameter of the particles of certain fraction;

   D_a - is the mean diameter.

Converting to the continuous distribution of the type of (23), we could write:

$$D_a = \frac{\int_0^{D_m} DPdD}{\int_0^{D_m} PdD} \qquad (27)$$

Submitting (23), we have:

$$D_a = \frac{\int_0^{D_m} D\exp\left(-\frac{\ln^2\frac{D}{D_e}}{2\delta^2}\right)dD}{\int_0^{D_m} \exp\left(-\frac{\ln^2\frac{D}{D_e}}{2\delta^2}\right)dD} \qquad (28)$$

To present this expression in convenient form, we could change the variable accordingly for numerator and denominator:

$$\begin{cases} t_1(D) = \frac{1}{\delta}\ln\frac{D}{D_a} - 2\delta \\ t_2(D) = \frac{1}{\delta}\ln\frac{D}{D_a} - \delta \end{cases} \qquad (29)$$

where t_1 and t_2 - are the new variables.

Hence, after the transformations we have:

$$D_a = D_e e^{1.58^2} \frac{\text{erf}[t_1(D_m)]}{\text{erf}[t_2(D_m)]} \qquad (30)$$

where erf - is the function of errors.

Thus, the mean value of sediment size is a function of the maximal size, determined by the flow properties, as well as the mathematical expectation and the least square of the fractions sizes, which are the alluvial bed ground geological properties. The related value of mean fall velocity could be figured out in the same way as (19).

4. CONCLUSION

The dependencies (19) and (30) allow to calculate the characteristic sizes of suspended sediments. The input data for these calculations are the stream parameters (slope, depth, hydraulic friction coefficient), solid particles parameters (density, resistance coefficients) and the bed ground grading parameters (prevalent fraction size and comparative quantity).

The restriction for applying the dependencies for maximal size is the moderate values of a slope and small value of Froude number; the dependencies for mean size determination is restricted with the case of monogene descending of the bed ground. Besides that, the common condition for both the group of dependencies is the presence of non-cohesive alluvial bed ground. Under the listed conditions these dependencies could be used for the prognosis and calculations.

REFERENCES

Abalianz, S. Kh,1973.: Steady and transitional processes in artificial channels, Hydrometizdat,
Adesman, A,1987. Cinematic characteristics of the flat turbulent flows in canals, Thesis, MCEI.
Hinze I. O,1952. Turbulence, Hafner Publishing.
Khodzinskaya, A,1988. Bed-load motion and evaluation of channel deformations, Thesis, VNIIGIM.

Kolmogorov A. N,1962. About a logarithmically normal regularity of the solid particles sizes distribution while carving, Papers of SU Academy of Sciense, vol. XXXI.

Schlichting, H,1955. Boundary layer theory,McGraw-Hill.

Shen H.W.,1971. Wash load and bed load, River Mechanics, H.W. Shen,.

Sediment transport

Stochastic Hydraulics 2000, Wang & Hu (eds) © 2000 Taylor & Francis, ISBN 90 5809 166 X

Invited lecture: Stochastic and deterministic sediment transport model concepts

Helmut M. Habersack
Institute for Water Management, Hydrology and Hydraulic Engineering, Universität für Bodenkultur, Vienna, Austria

ABSTRACT: The aim of this paper is to discuss and compare stochastic and deterministic sediment transport model concepts. Concepts that exist in the literature are analysed with respect to initiation of motion, transport path, transport rates, river morphology and sediment budget/catchment-wide aspects. It is shown that depending on the processes and scales either stochastic or deterministic concepts have been dominating in the past. Due to a lack of input data and a complex interaction of various processes the dominance of deterministic concepts is given for small scales whereas large (e.g. catchment-wide) scales were often modelled by the use of stochastic approaches. In the paper it is concluded, and several examples show this, that there is a recent tendency for combining stochastic and deterministic model concepts in order to improve the quality of the simulations and predictions.

Key words: Sediment transport, River morphology, Stochastic modelling, Deterministic modelling

1. INTRODUCTION

Bed load transport is commonly regarded as a deterministic phenomenon, but bed load transport rate time series exhibit pronounced short-term, temporal variability that is often associated with bed form migration (Gomez et al., 1989). According to Gomez and Phillips, 1999, > 80 % of the variability can be explained by the sequential passage of bed forms as modeled by Hamamori´s logarithmic cumulative distribution function and the remainder may be attributable to deterministic uncertainty (chaos)

Fig. 1 shows such a temporal variation of bedload transport for the river Drau in Austria. It is obvious that between the minimum and maximum transport rate there is a four-fold difference. Generally used deterministic models are based on mean flow velocities and mean transport rates that don´t regard stochastic elements in the modelling procedure.

Whereas in the past either pure deterministic or stochastic models were used recently a combination of these two concepts gains increasing interest.

The aims of this paper are:
- Description of stochastic and deterministic model concepts
- Analysis of various concepts concerning
    - initiation of motion
    - transport path
    - transport rates
    - river morphology
    - sediment budgets and catchment-wide aspects
- Comparison of stochastic and deterministic models in relation to various scales

A selection of relevant models given in the literature forms the basis of this paper and examples of their application illustrate their performance.

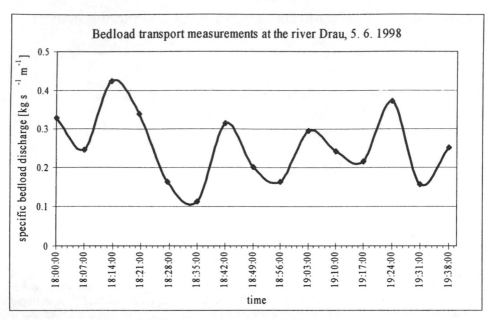

Fig.1: Temporal variability of bedload transport at the river Drau

2. INITIATION OF MOTION

Deterministic relationships have been established between the critical value of the displacing forces (e.g. flow velocity or tractive forces) and the parameters of the bed load and the stream (Stelczer, 1981). The most famous equation for determining the critical values for initiation of motion was derived by Shields (1936):

$$Fr^{*\prime} = Fr_c^* = \frac{\mu h_{cr} I}{\rho' d_{ch}}$$

(1)

Deviation of laboratory and field investigations showed that no single critical value can explain the initiation of motion and Zanke (1982, 1989) argued that the Shields diagram had to be modified by the introduction of the risk of movement R for which he derived the following empirical relation

$$R = (10(Fr^* / Fr_c^*)^{-9} + 1)^{-1}$$

(2)

where ρ_w, ρ_s = fluid, sediment density, τ_{cr} = critical shear stress, g = gravitational acceleration, $d_{ch, m}$ = characteristic grain diameters, Q, Q_s = discharge, sediment discharge, h_{cr} = the critical water depth, S = the slope, $k_{st,r}$ = the Strickler and grain roughness values, Fr_c^* = critical Froude number, v_{cr} = critical flow velocity, $\rho' = (\rho_s - \rho_w)/\rho_w$, v = kinematic viscostiy. Fig. 2 shows that an increase in R gives higher critical Shields-values. Therefore instead of one critical value a dependency of the initiation of motion on the risk of movement is assumed. In general the risk of grains to be moved is depending on the density distribution of the velocities (or of another value describing the resistence against movement) and the density distribution for critical velocities. Using measurement data at the river Drau in Austria (Habersack, 1997a, Fig. 2) it can be seen that for the hydraulic conditions at which initiation of motion had occured in nature according to the original Shields diagram neither for the subsurface material nor for the transported bedload motion should have taken place. Taking into consideration the risk for movement subsurface material is initiated at R-values less than 1% whereas the grain size similar to the transported material shows that the risk is between 1 and 10 % to be eroded. One reason for this behavior of initiation at low critical Froude numbers must be the influence of turbulence and related velocity fluctuations as well as the combination of drag and lift forces.

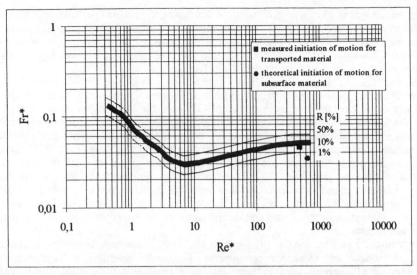

Fig.2: Shields diagram including the risk for movement R according to
Zanke (1989) and data for the river Drau in Austria

In stochastic model concepts no single critical value for initiation of motion is assumed. Instead, a probability of incipient motion is used. Einstein, 1950, noted, that a particle is eroded from the bed when the probability of the dynamic lift force F_L is larger than its weight (under water).

Sun et al., 1997 stated the probability that force moments causing particle motion exceed those keeping particles at rest to be the probability of incipient motion and Sun & Donahue derived the following equation for describing the probability of the fractional incipient motion for nonuniform sediment

$$\alpha_k = 1 - \frac{1}{\sqrt{2\pi}} \int_{-2.7(\sqrt{0.0822\psi_k}+1)}^{2.7(\sqrt{0.0822\psi_k}-1)} e^{-0.5x^2} dx \tag{3}$$

where ψ_k as intensity of flow

$$\psi_k = \theta_k^{-1} = \frac{(\rho_s - \rho)g(D_k D_m)^{0.5}}{\tau_0' \sigma_g^{0.25}} \tag{4}$$

where $(D_k/D_m=0.5\sigma_g^{0.25}$ reflects the shelter-exposure effect: σ_g and D_m = geometric deviation and mean diameter of bed material, D_k = average diameter for the kth size fraction; and τ_0 = bed shear stress with respect to grain roughness. If we apply equation 3 to the data of the river Drau the probability of the fractional incipient motion is 0.8.

3. TRANSPORT PATH

The use of step lenghts and rest periods in order to model the transport path of bedload was initiated by Einstein, 1937. Later the stochastic basis for sediment transport was further developed by Hubbell and Sayre, 1964, Hung and Shen, 1971, Nakagawa and Tsujimoto, 1980, Paintal, 1970, Shen and Cheong, 1980, Yang and Sayre, 1971 and Lisle et al., 1998. Basically the combination of the distribution of step lenghts and rest periods leads to the so-called gamma exponential model, which is mainly used in order to calculate the spatial probability density through time

$$f_t(x) = \lambda_1 e^{-(\lambda_1 x + \lambda_2 t)} \sum_{n=1}^{\infty} \frac{(\lambda_1 x)^{n \cdot r-1}}{\Gamma(nr)} \frac{(\lambda_2 t)^n}{n!} \tag{5}$$

where λ_2, which acts as a scale parameter, = the number of rest periods per time unit [1/minute], λ_1 (the inverse of the mean step length) = scale parameter [1/meter], $\Gamma(\ )$= gamma function, and r = shape parameter of $\Gamma(\ )$, x = displacement of the particle from the origin, t = time taken during this displacement, and n = number of single step length and rest periods. Fig. 3 shows the basic assumption for the stochastic model by Einstein.

In order to relate the stochastic model to a physical basis Busskamp, 1984 tried to determine the shape and scale parameters for the individual distributions. In distinction to the Einstein like model Lisle et al.,1998, developed a probability concept for soil erosion where the duration spent by the particle in motion cannot be neglected. The elementary theory of Markov processes implies that the rest and motion intervals R_i, M_i are independent and exponentially distributed random variables with means h^{-1} and k^{-1}. It is interesting to note that Lisle et al., 1998, show that a suitable averaging of the stochastic particle motions gives rise to the deterministic erosion differential of Hairsine and Rose, 1991. There the three parameters u, h and k are the particle velocity u, which is related to the overland flow velocity (for soil erosion), h, which is related to the rate of erosion and k, which is related to the deposition rate or particle settling velocity. In this way a combination of stochastic and deterministic model concepts improves the sediment transport simulation.

Recently discrete particle models (McEwan et al., 2000) combine a deterministic concept of bed particle motion with stochastic components. The model generates a three dimensional random packing of discrete, spherical particles and it includes a three dimensional rebound. These models allow to study e.g. the importance of relative roughness on the mean step length.

4. TRANSPORT RATES

According to Gomez and Church, 1989, four principal approaches have emerged to the design of bed load transport formulae, based upon:
- bed shear stress (e.g. du Boys, 1879)
- stream discharge (e.g. Schoklitsch, 1934)
- stream power (e.g. Bagnold, 1980)
- stochastic functions for sediment movement (e.g., Einstein, 1950).

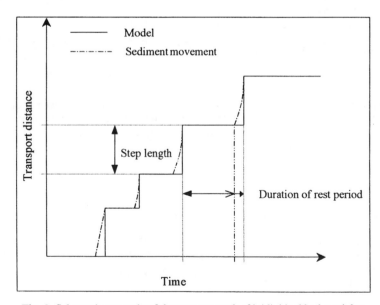

Fig. 3: Schematic example of the transport path of inidividual bed particles

In view of the contents of this paper we can devide the model concepts for calculating transport rates into deterministic and stochastic approaches. The deterministic formulas include critical values for the initiation of motion (e.g. τ_c, v_c, ω_c) and compare the actual shear stresses, flow velocities or stream power with these critical values. Einstein, 1937, 1950 was the first to introduce a probabilistic concept to sediment transport rates, where the bed load equations can be written as follows:

$$p = 1 - \frac{1}{\pi^{0.5}} \int_{-B_*\psi - 1/\eta_0}^{B_*\psi - 1/\eta_0} e^{-t^2} dt = \frac{A_*\phi}{1 + A_*\phi} \qquad (6)$$

using intensity values for the bedload and water discharges:

$$\Phi = \frac{q_s}{\gamma_s} \sqrt{\frac{\rho}{\rho_s - \rho} \frac{1}{gd^3}} \qquad (7)$$

$$\Psi = \frac{\rho_s - \rho}{\rho} \frac{d}{SR_b'} \qquad (8)$$

where p = probability for motion; $A_* = 43.5$; $B_* = 0.143$; $\eta_0 = 0.5$; t is a dummy variable of integration; Φ = intensity of bedload discharge; Ψ = flow intensity; ρ_s = density of sediment, g = acceleration due to gravity, R_b' = hydraulic radius with respect to bed and granular boundary. As basis for the derivation of the bedload equations Einstein formulated the following statistical laws:

1. The probability of a given sediment particle being moved by the flow from the bed surface depends on the particle's size, shape and weight and on the flow pattern near the bed, but not on its previous history;
2. The particle moves if the instantaneous hydrodynamic lift force overcomes the particle weight,
3. Once in motion, the probability of the particle's being redeposited is equal in all points of the bed where the local flow would not immediately remove the particle again,
4. The average distance traveled by any bed-load particle between consecutive points of deposition in the bed is a constant for any particle and is independant of the flow condition, the rate of transport and the bed composition. For the sediment grain of average sphericity this distance may be assumed to be 100 grain-diameters,
5. The motion of bed particles by saltation may be neglected
6. The disturbance of the bed surface by moving sediment-particles my be neglected in water.

In addition to the statistical formulation of the problem Einstein added some deterministically oriented aspects, which are related to some constants; like the roughness diameter k_s was assumed to be constant and equal to the D_{65} of the substrate, an average lift force ($p_L = c_L s_f u^2/2$, with c_L, s_f as constants, u as flow velocity), the standard deviation of the pressure fluctuations due to turbulence being 0.364 of the average lift. Sun & Donahue, 2000, argue that as result of many coefficients (A^*, B^*, η_0, ε and Y) the formula can rarely be precise or reliable. Nevertheless especially for weak transport rates where deterministic formulas don´t give any transport, the stochastic approach of Einstein gives bedload discharge (Habersack, 1998, Fig. 4). Sun and Donahue, 2000, statistically derived the following bedload formula for any fraction of nonuniform sediment

$$\phi_k = \frac{P_{0k}\alpha_k}{A\psi_k^{0.5+B}(1 - \alpha_k)} \qquad (9)$$

with $A = C/5m_0$ = comprehensive coefficient, A and B are constants determined from measurements of bedload transport, α_k = probability of incipient motion, P_{ok} = percentages of size D_k in the bed materials and ψ_k as intensity of flow. Thereby a stochastic model of sediment exchange and the probability distribution of fractional bedload transport rates was used. Furthermore relations for the probability of fractional incipient motion (equation 3) and for the average velocity of particle motion were introduced in the formula. Fig. 4 shows that the stochastic formula of Einstein gives reasonable values for subsurface grain sizes whereas the deterministic Meyer-Peter,Müller equation mostly doesn´t give any transport due to the high critical shear stress.

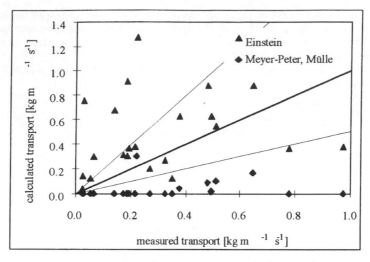

Fig. 4: Comparison of Einstein, Meyer-Peter, Müller formulas for
subsurface material at the river Drau, Austria

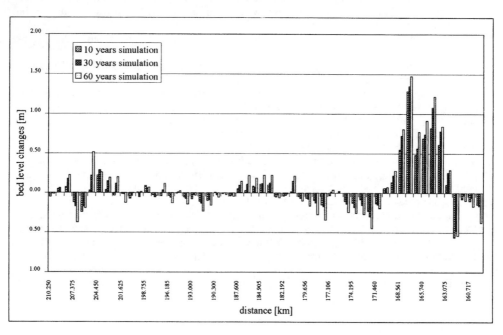

Fig. 5: Result of a 1-D model at the river Salzach in Austria (1-km average bed level changes)

5. RIVER MORPHOLOGY

Analysis of river morphology is performed mainly with 1-D sediment transport models at the
sectional scale (up to 100 km), 2-D models at the local scale and especially for suspended load
modelling with 3-D models. An example for a 1-D deterministic sediment transport model
result is given in Fig. 5 for the upper reach of the river Salzach in Austria.

Based on a calibration on observed bed level changes the future morphological changes are
demonstrated with respect to a time frame of 60 years. Although in general no significant bed

level changes had been measured over the last 10 years on average it can be seen that a significant aggradation occurs in various locations (around km 168-163) already after 30 years. This causes also an increase in the water level, leading to problems with flood protection. At other places minor degradation can be found in Fig 5.

The simulation of morphological characteristics of braided rivers cannot be done sufficiently by 1-D models. Enggrob and Tjerry, 1999, used a two-dimensional numerical, deterministic model to simulate the braided pattern of the Brahmaputra-Jamuna River in Bangladesh. Beside the hydrodynamics, where the direction of the bed shear stress in a curved flow plays a crucial role, and sediment transport, where a depth-averaged equation for the concentration of suspended sediment was used, the morphological changes due to bed level changes were found by solving the sediment continuity equation:

$$(1-n)\frac{\partial z}{\partial t}+\frac{\partial S_x}{\partial x}+\frac{\partial S_y}{\partial y}=\Delta S_e \tag{10}$$

where S_x = total sediment transport in x-direction, S_y = total sediment transport in y-direction, n = sediment porosity, z = bed level, t = time, (x,y) = Cartesian co-ordinate system, ΔS_e = lateral sediment supply from bank erosion. The bank erosion rate was described as:

$$E_b = -\alpha_b\frac{\partial z}{\partial t}+\beta_b\frac{S}{h}+\gamma_b \tag{11}$$

where E_b = bank erosion rate in m/s, z = local bed level near the bank, S = near bank sediment transport, h = local water depth near the bank, $\alpha_b, \beta_b, \gamma_b$ = calibration coefficients specified in the model.

Enggrob and Tjerry, 1999, applied the model for short-term and long-term predictions. No real calibration and verification could be made but even for the long-term predictions the simulated river in the upstream part looks remarkably like the real Jamuna river. In order to model the real morphology of a braided river in a deterministic way the basic physics of individual bed particle motion, its interaction with bed form developments and large scale morphological changes caused by the dynamics of convergent and divergent zones of braided rivers will have to be understood before such a modelling will be successful in predicting nature.

In comparison to such a classic deterministic concept cellular automata models have found a wide application in the physical sciences (Sapozhnikov et al., 1998) and such a model was proposed by Murray and Paola (1994, 1997) for braided streams. Beginning from upstream cells water is distributed to the downstream cells and sediment is also transported from one cell to three neighbouring cells by some rules like

$$Q_{su}=K(Q_iS_i+Q_iC)^m \tag{12}$$

$$Q_{su}=K(Q_iS_i+\varepsilon\sum_j Q_iS_i)^m \tag{13}$$

where K = a constant and Q_i and S_i = discharges and slopes from the cell in question to downstream neighbour i. C and ε = constant. The analysis of the modelled rivers (including the Brahmaputra) showed that they eventually develop into a state exhibiting anisotropic spatial scaling (self-affinity, Sapozhnikov et al., 1998). The presence of spatial scaling in natural braided rivers was interpreted by Sapozhnikov and Foufoula-Georgiou, 1996, as a strong indication that the same physical mechanisms are responsible for the formation of braided patterns at different spatial scales, from the scale of the smallest channel to the scale of the braid plain width (Sapozhnikov et al., 1998). It can be stated that the mentioned models are able to reproduce characteristic features of braided rivers but no simulation of the type of the river specifically can be achieved or a real prediction of the future morphology can be made. Beside the lack of fundamental knowledge especially of large size particle transport and the interactions with a dynamic morphology and the relatively simple model concepts so far, stochastic components have not yet been really implemented into such models, which could treat the somewhat chaotically behaviour (e.g. caused by influences of vegetation, sediment input from tribuaties etc.).

6. SEDIMENT BUDGETS AND CATCHMENT-WIDE ASPECTS

The larger the scale the more stochastic elements enter the calculation of sediment transport (Habersack, 1999). Deterministic model concepts are of limited value because for whole catchments not all of the necessary input and calibration data are available and cannot be monitored directly. Furthermore especially in mountainous environments bedload is influenced by the geometry, hydraulics but also sediment supply, caused by side erosion, shallow landslides etc.. The random occurence of sediment supply and the variation of bed load discharge rates during comparable floods lead to an increasing need for stochastic approaches. Whereas suspended sediment yields can be measured or calculated on the basis of regressions of measured mean annual yields against hydrologic, physiographic and climatic parameters, Griffith, 1980, presented a method for stochastic estimation of bed load yield from a given catchment, where the main points are:

- the occurence of floods is Poisson distributed,
- bed load yields from floods for a given period may be modelled by a shifted exponential distribution:

$$T_m = e^{\lambda(g_m - g_0)} \tag{14}$$

where T_m is the return period of a certain flood that produced the sediment yield g_m, g_0 is a location parameter, λ is the parameter of the cdf.

- because of the exponential distribution is defined by a single parameter, one value of bed load yield for a single flood is required for the definition of its expected value. Griffiths, 1980, suggests to calculate this yield by use of a bed load transport formula, regional flood frequency relations and input data from the reference catchment,
- the expected number of floods is multiplied by the expected bed load yield per flood to give a prediction of bed load yield for a specified period.

Although the model concept of Griffith is calibrated for a stream in New Zealand the principle may be used also for other catchments. Another example for a catchment-wide, stochastically oriented model is dealing with the simulation of slope instabilities and shallow landslides, being important as sediment (especially bedload) sources. A variety of model concepts exists with respect to modelling these processes (e.g. Montgomery & Dietrich, 1989, Dietrich et al. 1992, Montgomery & Dietrich 1994). Based on the theory of Montgomery & Dietrich 1994 Pack et al. 1998 developed a model for the simulation of shallow landslides („Stability INdex MAPping"). This slope stability model contains a stability factor

$$FS = \frac{C + \cos\theta[1 - wr]\tan\phi}{\sin\theta} \tag{15}$$

where $w = h_w / h$, $r = \rho_w / \rho_s$ (16)

$$C = (C_r + C_s)/(h\rho_s g) \tag{17}$$

where C_r = root cohesion [N/m²], C_s = soil cohesion [N/m²], θ = slope, ρ_s = density of the soil [kg/m³], ρ_w = density of water [kg/m³], g = acceleration due to gravity with 9,81 m/s², h = slope parallel soil thickness [m], h_w = water depth [m] und ϕ = internal friction angle [-]. In combination with a topographical wetness index

$$w = \min(\frac{Ra}{T\sin\theta}, 1) \tag{18}$$

where R = stationary recharged discharge [m/h], a = relation between contributing area and contour line length, T = soil transmissivity [m²/h], the following dimensionless safety factor and stability index is derived

$$SI = FS_{\min} \frac{C_1 + \cos\theta[1 - \min(x_2 \frac{a}{\sin\theta}, 1)r]\tan\phi_1}{\sin\theta} \tag{19}$$

with C_1 and $\tan\phi_1$ as lowest values and x_2 (x=R/T) as largest value. Where the safety factor is less than 1 a certain probability for failure exists. Fig. 6 shows a not calibrated example for the Soelk catchment in Styria/Austria.

With the help of such results the catchment can be devided into zones with high and low risk for slope instabilities. At a smaller scale the distinguished high risk areas can be investigated in more detail and as more input and calibration data are available there (e.g. one slope) deterministic sediment transport models may be applied.

7. SUMMARY AND CONCLUSIONS

In this paper stochastic and deterministic concepts for sediment transport modelling were discussed. For the initiation of motion, which is sensitive to all modelling efforts, it could be shown that an extension of the Shields diagram allows to incorporate a risk assumption. This kind of combination between a stochastic and deterministic concept is also given for the Einstein's bedload transport formula which has been recently again the basis for new developments concerning bedload transport equations. Although transport paths have originally been modelled mainly on a stochastic basis, a coupling with physical interpretations of the used parameters is aimed and examples are shown.

In the field of river morphology generally up to now the main focus was based on deterministic concepts, but these could not predict the exact shape of the river bed, especially e.g. for braided rivers. Due to the lack of data at the catchment-wide scale often stochastic approaches have been used which allow an analysis of the whole area with respect to risks for sediment input and transport. After defining these high risk zones deterministic approaches can be used to model transport processes in detail.

The main conclusion of this paper is that at various scales and for specific processes either stochastic or deterministic approaches were dominating in the past but due to the deficiences of using only one concept a combination of them will be aimed in the future.

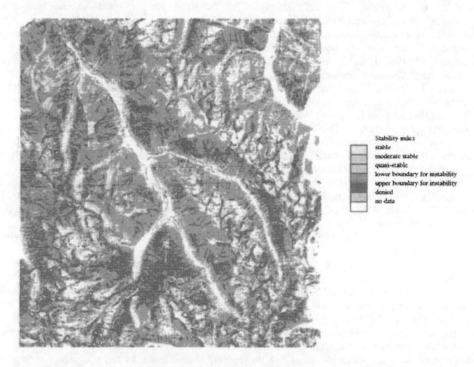

Fig. 6: Distribution of the slope stability index SI in the catchment of the Soelk-valley in Austria (dark areas tend to instability, modified after Habersack, 2000)

ACKNOWLEDGEMENT

The author wants to thank the Austrian Ministry for agriculture and forestry, the provincial governments of Carinthia, Styria and Salzburg for financing various projects. Furthermore some results were derived within the EU-project WARMICE ENV4-CT98-0989.

REFERENCES

Bagnold, R. A. (1980). "An empirical correlation of bedload transport rates in flumes and natural rivers." Proc. R. Soc. London, Ser. A., 372, 453-473.

Dietrich, W.E., Wilson, C.J., Montgomery, D.R., McKean, J., Bauer, R. (1992). "Erosion thresholds and land surface morphology" Geology, Vol. 20; 675-679.

du BOYS, M. P. (1879). "Etudes du régime et l'action exercé par les eaux sur un lit à fond de graviere indéfiniment affouiable." Annales des Ponts et Chaussées, 5(18), 141-195.

Einstein HA. (1937) "The bed load transport as probability problem" Mitteilung der Versuchsanstalt fuer Wasserbau an der Eidgenössischen Technischen Hochschule. Zürich; 110.

Einstein HA. (1950) "The bed load function for sediment transport in open channel flows" United State Dept. of Agriculture, Washington D.C., Technical Bulletin 1026: September.

Enggrob, H. G. & Tjerry, S. (1999) "Simulation of morphological characteristics of a braided river" I.A.H.R Symposium on River, Coastal and Estuarine Morphodynamics (proceedings), Vol.1 585-594.

Gomez, B. and Phillips, J. D. (1999), "Deterministic uncertainty in bedload transport." J. Hydr. Engrg, 125(3), 305-308.

Gomez, B., and Church, M., (1989). "Assessment of bed load sediment transport formulae for gravel bed rivers." Water Resour. Res. 25(6), 1161 -1186.

Griffiths, G. A. (1980) "Stochastic estimation of bed load yield in pool-and-riffle mountain streams" Water Resources Research, Vol. 16(5); 931-937.

Habersack, H. (1997a). "Raum-zeitliche Variabilitäten im Geschiebehaushalt und dessen Beeinflussung am Beispiel der Drau", Wr. Mitteilungen Vol. 144; S. 295.

Habersack, H. (1997b). "Catchment-wide, sectional and local aspects in sediment transport modelling and monitoring", Journal of Sediment Research. Vol. 12, No.3; 120-130.

Habersack, H. (1998). "Numerical sediment transport models-theoretical and practical aspects", IAHS - Publ. Nr. 249, "International Symposium on Modelling Soil Erosion, Sediment Transport and Closely Related Hydrological Processes", Wien; 299-308.

Habersack, H. (1999). "The river scaling concept (RSC) – a basis for ecological assessments" Hydrobiologia (in print).

Habersack, H. (2000). "Radio-tracking gravel particles in a large braided rivers in New Zealand – a field test of the stochastic bedload transport theory by Einstein" Hydrological Processes (in print).

Hairsine, P. & Rose, C. (1991) "Modeling water erosion due to overland flow using physical principles" Soil Sci. Soc. Am. J., Vol. 55(2), 320-324.

Hubbell DW, Sayre WW. (1964) "Sand transport studies with radioactive tracers" J. of the Hydraulic Division 90: 39-68.

Hung CS, Shen HW. (1979) "Statistical analysis of sediment motions on dunes" J. of the Hydraulics Division 105: 213-227.

Lisle, I.G., Rose, C. W., Hogarth, W.L., Hairsine, P.B., Sander, G.C, Parlange, J.-Y. (1998) "Stochastic sediment transport in soil erosion" J. of Hydrology, Vol. 204, 217-230.

McEwan, I., Habersack, H.M., Heald, J. (2000) "Discrete particle modeling and active tracers: new techniques for studying sediment transport as a Langrangian phenomenan" to appear in: Gravel bed rivers 2000, New Zealand.

Montgomery, D.R. & Dietrich, W. E. (1989). "Source areas, drainage density and channel initiation" Water Resources Research, Vol. 25(8); 1907-1918.

Montgomery, D.R. & Dietrich, W. E. (1994). "A physically based model for the topographic control on shallow landsliding" Water Resources Research, Vol. 30(4); 1153-1171.

Murray, A. B. & Paola, C. (1996) "A new quantitative test of geomorphic models applied to a model of braided streams" Water Resources Research, Vol. 32(8); 2579-2587.

Murray, A. B. & Paola, C. (1997) "Properties of a cellular braided stream model" Earth Surf. Processes Landforms, Vol. 22, 1001-1025.

Nakagawa H, Tsujimoto T, Nakano S. (1982) "Characteristics of sediment motion for respective grain sizes of sand mixtures" Bull. Disaster Prev. Res. Inst., Kyoto Univ. 32(1).

Pack, R. T., Tarboton, D. G., Goodwin, C.N. (1998) "A stability index approach to terrain stability hazard mapping", report and user's manual, Utah.

Paintal AS. (1971) A stochastic model of bed load transport. J. of Hydraulic Res. 9: 527-554.

Sapozhnikov, V. B. & Foufoula-Georgiou, E. (1997) "Experimental evidence of dynamic scaling and indications of self-organized criticality in braided rivers" Water Resources Research, Vol. 33; 1983-1991.

Sapozhnikov, V. B., Murray, A. B., Paola, C. & Foufoula-Georgiou, E. (1998) "Validation of braided-stream models: spatial state-space plots, self-affine scaling, and island shapes" Water Resources Research, Vol. 34(9); 2353-2364.

Schoklitsch, A. (1934). "Der Geschiebetrieb und die Geschiebefracht." Wasserkraft Wasserwirtschaft, 39(4), 1-7.

Shen HW, Cheong FH. (1980) "Stochastic sediment bed load models", In Application of Stochastic Processes to Alluvial Phenomean. Shen HW, Kikkawa H (eds). Water Resources Publications: Littleton, Colorado; 10/1-10/19.

Shields, A. (1936) "Anwendung Änlichkeitsmechanik und der Turbulenzforschung auf die Geschiebebewegung", Mitt. der Preussichen Versuchsanstalt für Wasserbau und Schiffbau, Berlin.

Stelczer K. (1981). Bed load transport: theory and practice. Water Resources Publications: Littleton, Colorado.

Sun, Z. & Donahue, J. (2000) "Statistically derived bedload formula for any fraction of nonuniform sediment" J. of Hydraulic Engineering, Vol. 126(2), 105-111.

Yang CT, Sayre WW. (1971) "Stochastic model for sand dispersion". J. of the Hydraulics Division 97: 265-288.

Zanke, U. (1989) "The beginning of sediment transport as a probability-problem" Proceeding of the International Symposium on Sediment Transport Modelling, New Orleans.

Stochastic Hydraulics 2000, Wang & Hu (eds) © 2000 Taylor & Francis, ISBN 90 5809 166 X

Stochastic nature of sediment transportation

Zhao-Yin Wang
Department of Hydraulic Engineering, Tsinghua University and International Research and Training Center on Erosion and Sedimentation, Beijing, China

Onyx W.H.Wai
Department of Civil and Structural Engineering, Hong Kong Polytechnic University, Kowloon, China

ABSTRACT: The paper studies the stochastic nature of sediment transport in rivers. Sediment transport formulas are valid only for prediction of average sediment transportation over long reaches of rivers and long periods of time. However, flow in rives is unsteady and non-uniform and failures of hydraulic works may occur because of instantaneous and local behaviour of sediment movement. An experimental study demonstrates fluctuation of the bed load transportation, which is so obvious that the maximum instantaneous transport rate is higher than 3 times of the average and the minimum is less than 5% of the average. In the upper Yangtze River, broad and shallow sections are between narrow and deep sections. Different from the normal law sediment transportation in broad sections exhibits high transport rate in low flow and low rate in high flow. Q_b is approximately inversely proportional to Q. Gravel bed load moves along a so called conveying belt and there is almost no bed load movement outside of the belt. The phenomenon was investigated in a scale model and the results shows the conveying belt is relatively fixed and is of only 10% of the total width of the channel bed.

Key words: Stochastic nature, Sediment movement, Bed load, Fluctuation of the bed load

1. INTRODUCTION

Sediment transportation exhibits stochastic nature, which is significant for river-bed deformation and construction of hydroelectric projects. A lot of formulas have been presented to relate sediment discharge to hydraulic and hydrologic factors. These can be expressed in the general form

$$Q_b = f(Q, J, U, B, h, d, \omega, \gamma_s, \gamma, \upsilon, \ldots) \tag{1}$$

in which Q_b and Q are sediment discharge and water discharge , respectively ; J and U are surface slope and average velocity of the flow; B and h are average width and depth; d is diameter of sediment , ω is fall velocity, γ_s and γ are the specific weights of sediment and water, respectively; υ is viscosity (Wang et al., 1993).

The formulas are valid only for prediction of average sediment discharges over long reaches of natural rivers and over long periods of time. However, flow in rives, especially during flood season , is unsteady and non-uniform along its course. Failures of hydraulic projects involving wharf, quay berths, terminal facilities, navigating channels, river bank protection, ship locks, diversion works, dams, reservoirs, and power stations may occur because of sediment movement, deposition, and erosion at a special place and a particular time. The instantaneous and local behaviour of sediment is unpredictable because so many varying factors and the wide variety of local conditions affect the movement of sediment. For example, the sediment discharge is quite different for rising and receding flood flows, even at the same water discharge; the deposition and erosion at a particular place are affected by the deposition and erosion that had occurred there previously. The characteristics of sediment movement are quite different for a broad and shallow section than it is for a deep and narrow section. River bends and the water and sediment entering from tributaries can have a major effect on sediment movement, siltation, and erosion, and random factors working in combination can also affect sediment transport.

2. UNCERTAINTY OF SEDIMENT DISCHARGE

Sediment discharge is unevenly distributed during a year. About 80%-99% of the annual load occurs on one season or in one half of the year, even though water flow in the same period is only 50-70% of the annual runoff. The bed load discharge in the Yangtze River is an example; the average annual sand bed load (d_{50}=0.23mm) measured at Yichang Station is 3.6×10^6 tons and the annual gravel bed load (d_{50}=23mm) is 4.3×10^4 tons. 98.1% of the sand bed load and 99.8% of the gravel bed load are transported in the period from May to October, and 70.4% of the sand bed load 86.9% of the gravel bed gravel bed load are transported in the shorter period from July to September.

The high concentration of rainfall in the flood season and the uneven distribution of the incoming sediment, of course, cause the uneven distribution of sediment discharge. To make the problem more complicated, sediment discharge can be quite different for the same conditions of water flow. From 675 sets of data measured over 9 years, Chen and Hu (1985) made a statistical analysis and found that the maximum rate of bed load transport can be more than 100 times higher than the minimum even if the flow is far greater than the initiation conditions. Thus prediction of sediment discharge is uncertain.

An intense fluctuation of bed load transport was measured at Yichang Station of the Yangtze River in 1973. Some 55-60 samples with duration of 0.5-3 minutes were taken in only a few hours, while the flow discharge remained constant. From the measured data (Huang, 1984), analysis is conduced and yields the results presented in Table 1. The maximum fluctuation of bed load discharge could be as high as 800% of the mean. But the positive fluctuations were much stronger than the negative ones.

Table.1: Statistical properties of bed load transportation in the Yangtze River (Yichang, 1973)

N0.	discharge(m^3/s)	Times of sampling	During of each sampling (min)	Average rate of bed load G_b(g/s)	Maximum G_{bmax} (g/s)	Minimum G_{bmin} (g/s)	Intensity of fluctuation G_{bmax}/G_b
1	21400	55	3	1.54	12.6	0	8.18
2	30100	59	3	40.0	268	2.33	6.64
3	19400	60	0.5	359	1190	54.7	3.16

No.	Mean square deviation (g/s)	Positive deviation Times	%	Negative deviation Times	%	Maximum relative deviation(%) Positive Negative	Remarks
1	2.14	17	30.9	38	69.1	718.1 100	Gravel
2	41.5	20	33.9	39	66.1	570 94.2	Gravel
3	271	22	36.7	38	63.2	232 84.5	Sand

An experiment was conducted in a tilting flume 15 m long and 50 cm wide of China Institute of Water Resources and Hydro-Power Research by the author to study the unsteady sediment transport (Wang et al., 1993). A bed load feeder with a hopper and conveying belt installed at the entrance of the flume supplies bed load material at given rate. Crushed coal of specific weight 1.48 g/cm^3 and median diameter 3 mm was used as the bed load. The bed load transport rate was measured with a sampler at the downstream end of the flume at a frequency 1 measurement/minute. The sediment transportation was fluctuating although the flow discharge was stable and the average sediment transportation was in equilibrium. Fig.1 shows the fluctuation of the bed load transportation at water flow rate 47.5 liters/second and energy slope 0.245%. The fluctuation of the bed load transportation is so obvious that the maximum transport rate is higher than 3 times of the average and the minimum is less than 5% of the average. The fluctuation is also obvious periodical, which reflects the effect of sand waves. The average length of sand waves was 0.8 m and the height of the waves was 3 cm. The speed of sand waves was measured at about 0.1m/min. Therefore, the major frequency of the transport rate fluctuation is about f=0.125/min. The same story occurs in the nature. For calibration of the sedimentation model of the Three Gorges Reservoir, we conducted continuous sediment measurement at Chongqing reaches and concluded that bed load discharge in rivers fluctuates intensively even if water flows remains stable.

Bed load moves on the surface of the bed and continually exchanges material with the bed. The rate of bed load transport is, of course, highly dependent on the carrying capacity of the flow. But incoming bed load also affects the local bed load discharge. This effect is often the result of bed armoring that impedes the replenishment of the bed load by scouring from the bed if the incoming bed load is deficient. Measurement of the bed load transportation and flow velocity at Yichang Station from April 6 to December 23, 1974 demonstrates that before August 10, the bed load discharge-flow velocity relation was as usual. A major flood (61600m^3/s) occurred on August 13 and then the incoming bed load was less than before. The rate of bed load transport was only 20% of that before August 10 at the same average velocity (Huang,1984).

Water surface slopes, and therefore velocities, are larger during the rising period of a flood and smaller during the receding rising period than it is during the receding period of a flood, at the same water discharge. Fig.2 shows the empirical curves of bed load discharge Q_b versus Q for rising and receding periods of floods. These curves were based on long-term records of bed load discharge at Tsintan Gauging Station on the Yangtze River.

3. NON-UNIFORM DISTRIBUTION OF SEDIMENT MOVEMENT

Natural rivers generally consist of broad shallow sections between regular sections or deep and narrow sections. The upper reach of the Yangtze River is like a long lotus root with many gorges forming the joints of the root . The profile of the channel bed in the upper reach of the Yangtze River consists of a series of irregular waves (Fig.3). Flow depths in deep and narrow sections are about 60-80 m larger than those in broad and shallow sections. The deepest point on a gorge near Yichang City has elevation as 40m below sea level. In a narrow and deep section, the average velocity and water surface slope J increase monotonically with increasing flow discharge. In a broad and shallow section, however, the variations of velocity and slope are complicated. They increase with increasing discharge if the discharge is less than 10000m^3/s, and then decrease with further increases of the discharge because the downstream narrow section rises the flow stage. Especially in the flood season, the water stage and width increase sharply, whereas the velocity reduces significantly. Therefore, the bed load transportation in the broad and shallow sections decreases with discharge, e.g. the Fengjie Station is in a broad and shallow section, flow velocity and water surface slope are smaller in the flood season than in low flow season (Hui et al., 1985).

The variation of hydraulic factors in different seasons and at different sections endows each river with its own pattern of sediment transport. In the flood season, gravel bed load is transported through narrow and deep sections. As bed load particles move into broad and shallow sections, they stop moving and accumulate there because the drag force of the flow acting on them is much smaller than that in narrow and deep sections. During the receding part of the flood cycle, in broad and shallow sections, both water stage and surface width are smaller, the area of cross-section of flow is much smaller, and the slope and velocity are larger. Bed load particles that accumulated there move again. In this intermittent way, bed load in the upper reach of the Yangtze River moves from narrow sections into broad sections in the flood season and from broad sections into narrow sections during the low-flow season. Consequently, the annual distance of travel for the gravel bed load in the Yangtze River depends not only on hydraulic factors but also on the distribution of broad sections and narrow sections along its course.

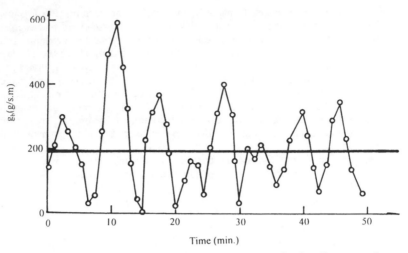

Fig.1: Fluctuation of bed load (crashed coal) transportation in a flume experiment
with water discharge 95 liters/s.m and energy slope 0.245%

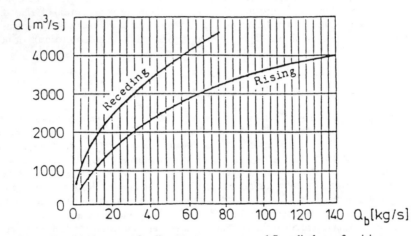

Fig.2: Empirical curves of sediment transport rate and flow discharge for rising
and receding periods of flood (Tsintan Station, Yangtze River)

The suspended load discharge Q_s roughly followed the variation of the flow discharge, but the bed load discharge Q_b varied in quite a different way. The peaks of gravel bed load occurred in the low-flow season. Statistics shows that the monthly discharge of gravel bed load averaged over many years demonstrates that over 80% of bed load is transported during the flood season (July-September) in regular sections (Zhuto, Wanxian, and Tsintan Stations), while in a broad section (Fengjie Station), only 13.4% is transported in this period. About 60% of the annual bed load is transported in December and January at Fengjie Station, coincidentally the period of lowest flow for the Yangtze River. Fig.4 gives the relation of bed load discharge Q_b to flow discharge Q measured at Fengjie Station in 1973-75. Q_b is approximately inversely proportional to Q. Such a phenomenon is difficult to be incorporated into the general law of sediment transport.

The intermittent motion of gravel bed load in the Yangtze Rive produces a long gravel dune. It differs from ordinary sand dunes in its formation, length, and frequency of occurrence in space, as determined by the flood peaks and the distribution of broad sections and narrow sections along the course. The height of a gravel dune can reach 3-5m in a shallow sections and over 10m in a deep section, and the length of the dune is generally 1000-3000m.

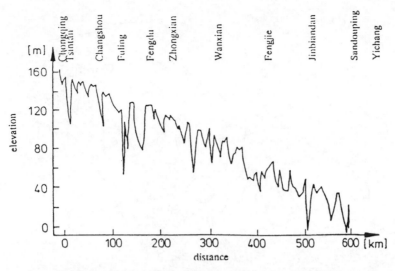

Fig.3 Bed profile of the upper Yangtze River from Chongqing to Yichang (Three Gorges Dam site)

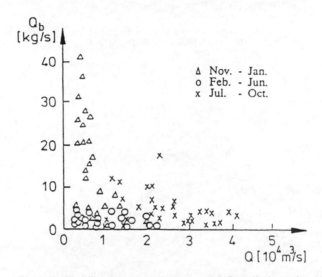

Fig.4: Bed load transport rate decreases with flow discharge in a broad section
(Fengjie Station) of the upper Yangtze River

Fig.5 shows an example of a gravel dune in a reach of the Yangtze River near Jiulongpo Harbour, Chongqing City, which is the most important freight harbour in the upper Yangtze River. The harbour is located in a broad and shallow section and at the convex bank of a large bend. The river is braided in this region, and water flows in the left channel in the low-flow season; in the flood season, water overflows in the whole valley. They pass through the upstream gorge section in the flood period and accumulate in the left channel and form a long dune. The dune moves into the gorge section downstream as the flood recedes.

The rate of bed load transport is unevenly distributed within the cross-sections of a river channel. Gravel bed load in the upper reach of the Yangtze River moves along a fixed narrow belt that is called a 'bed load conveying belt'. The width of the conveying belt is only 10-50% of the channel bed. Rarely occurs any bed load transported outside of the belt. In a regular and straight channel, the belt is near the main stream. At bends, however, the conveying belt is generally near the convex bank and can be far from the main stream.

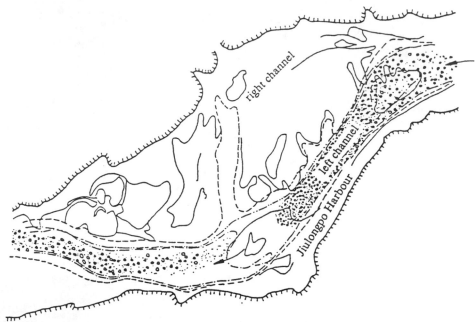

Fig.5: Gravel dune in the Jiulongpo Harbor

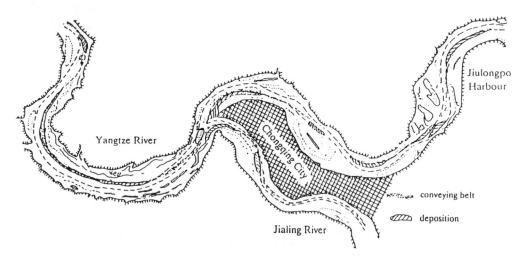

Fig.6: Conveying belt of gravel bed load in the Chongqing Reach of the Yangtze River

Gravel bed load movement and the phenomenon of the conveying belt were investigated in a scale model of the Chongqing reach of the Yangtze Rive in China Institute of Water Resources and Hydroelectric Power Research. Fig.6 shows the conveying belt in this reach. The total prototype length of this reach of the Yangtze River is 33km. The Jialing River-a main tributary of the Yangtze Rive- also shows such a belt. The location of the belt in the cross-sections is more or less fixed although it moves a little with variations of flow discharge.

The model test showed that, as gravel passes through a gorge, it does not fall into and then climb out of the bottom of the deep pools, but it moves on the convex bank wall instead. The conveying belt is at a bed elevation for which the submerged weight of gravel bed load is just balanced by the lift force of the circulating current. Many pools occur in the upper reach the Yangtze River, some of them have depths of more than 60m. The pool maintain a stable depth in

252

spite of the large amount of gravel bed load that passes through these reaches annually, mainly because the gravel moves only near the convex bank wall and never accumulates or is scoured at the bottom of the pools.

The rate of bed load transport per unit width was measured in the scale model test. Fig.7 shows one measured result. The cross-section is near the confluence (see Fig.6) and its width is about 700 m. No bed load moves in most of the section, but the rate of bed load transport is very high along the conveying belt, a part of the river that is only about 100 m wide.

The phenomenon of a bed load conveying belt is important. A power plant diverted cooling water from the Yangtze River by means of a mushroom-shaped intake in the river bed. The location of the intake happened to be near the bed load conveying belt, and it was often clogged by gravel. An understanding of the law of bed load motion and the location of the conveying belt in this reach allowed engineers to place a new intake some distance from the belt. As a consequence, the clogging was greatly reduced.

5 CONCLUSIONS

Sediment transport is a stochastic process. An experimental study demonstrates fluctuation of the bed load transportation, which is so obvious that the maximum instantaneous transport rate is higher than 3 times of the average and the minimum is less than 5% of the average. In the upper Yangtze River, broad and shallow sections are between narrow and deep sections. Different from the normal law sediment transportation in broad sections exhibits high transport rate in low flow and low rate in high flow. Q_b is approximately inversely proportional to Q. Gravel bed load moves along a so called conveying belt and there is almost no bed load movement outside of the belt. The phenomenon was investigated in a scale model and the results shows the conveying belt is relatively fixed and is of only 10% of the total width of the channel bed. Understanding of the location of the conveying belt has helped engineers to replace cooling water intake distant from the belt avoiding clogging.

ACKNOWLEDGMENT

The study was supported by the National Natural Science Foundation and Ministry of Water Resources of China (No.59890200).

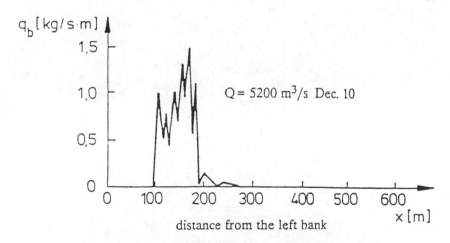

Fig.7: Distribution of bed load transport rate q_b in a cross section of the Chongqing Reach of the Yangtze River

REFERENCES

Chen J. and Hu C., 1985, Analysis of the rate and annual yield of gravel bed load in Dujiangyan Reaches of the Minjiang River, Chinese Journal of Sediment Research, No.4 (in Chinese).

Huang G., 1984, Analysis of bed load transport in Yichang Reaches of the Yangtze River, Assembly of Papers on Sediment Problems in Gezhouba Project, Water Conservation Commission of the Yangtze River (in Chinese).

Hui Yujia, Wang Guixian et al., 1985, Model study of sediment problem in the Yangtze River Gezhouba Project, Tsinghua University Academic Report (in Chinese).

Wang Zhaoyin, Kron Wolfgang, Lu Xiuzhen and Plate Erich, 1993, Flume experiments on sediment transport, erosion and siltation of a channel bed in unsteady and non-uniform flow conditions, in Contributions to Non-Stationary Sediment Transport (Kron Ed), Institute foer Hydrologie und Wasserwirtschaft, Universitaet Karlsruhe, B1-B82.

Wang Zhaoyin, Kron Wolfgang, Lu Xiuzhen and Xu Mingquan, 1993, Uncertainty of sediment discharge, siltation and erosion in rivers and its implications for hydraulic works, in Contributions to Non-Stationary Sediment Transport (Kron Ed), Institute foer Hydrologie und Wasserwirtschaft, Universitaet Karlsruhe, D1-D38.

Stochastic Hydraulics 2000, Wang & Hu (eds) © 2000 Taylor & Francis, ISBN 90 5809 166 X

Comparison of diffusion based approaches to sediment transport with stochastic interpretation

R.N.Bhattacharya
Indiana University, Bloomington, Ind., USA

D.C.Dalal
Indian Institute of Technology, Guwahati, India

J.K.Ghosh & B.S.Mazumder
Indian Statistical Institute, Calcutta, India

ABSTRACT: The Rouse equation is presented as the steady state distribution of mass for sand grains of different sizes under a Markovian diffusion process governed by the turbulent diffusion. The Kolmogorov forward equation is employed with reflection at a lower boundary set at a height equal to k_s, the roughness of the bed. To avoid difficulties of prediction with the Rouse equation, an equation analogous to the Rouse and a new model based on a combination of reflection and jump at the lower boundary are proposed. The jump is used to model the sediment transport due to the bursting phenomenon near the bed. The results of calculations show the new model with advantages over the previous two models. Predictions from the three models are compared with experimental data.

Key words: Sediment transport, Suspension, Markovian diffusion, Jump boundary condition, Flume experiments.

1. INTRODUCTION

We compare three diffusion based approaches to sediment transport and provide a stochastic interpretation. Only the flow parameters and the distribution of sand grains in the bed are used. The first two approaches have been published by Ghosh & Mazumder (1981) and Ghosh *et al.* (1986). Only the third approach is presented in this paper in details. These different methods were applied successfully to data arising from experiments conducted in recirculating flumes at Uppsala and Calcutta.

In the first approach, the Rouse equation for suspension

$$S_y(\phi) = S_a(\phi)\left(\frac{d-y}{y} \cdot \frac{a}{d-a}\right)^{\frac{c(\phi)}{\kappa u_*}} \tag{1}$$

is generally used to get results at a height above the reference level at $y = a$, which is well above the bed. Here $\phi = -log_2 D$ (*D in mm*), $S_y(\phi)$ is concentration of sediment particle of size ϕ at height y above the bed, d is the depth of fluid, $c(\phi)$ settling velocity of a particle of size ϕ, u_* shear velocity at the bed and κ von Karman constant (0.4). The Rouse equation becomes undefined at $a = 0$, i.e., if we take the reference level at the bed. So we proposed application of the Rouse equation to predict the grain-size distribution in suspension using the grain-size distribution in bed as the distribution at reference level $a = k_s$, where k_s is the roughness of the bed and may be taken as D_{65} (the representative grain diameter of the bed material at which 65% of the mixture, by weight, is finer, Einstein, 1950).

Specifically, for a grain size ϕ,

$$S_y(\phi) = \text{constant}.S_{k_s}(\phi)\left(\frac{d-y}{y} \cdot \frac{k_s}{d-k_s}\right)^{\frac{c(\phi)}{\kappa u_*}} \tag{2}$$

To eliminate the *constant* we work with

$$S'_y(\phi) = \frac{S_y(\phi)}{\sum S_y(\phi)} = S'_{k_s}(\phi)\left(\frac{d-y}{y} \cdot \frac{k_s}{d-k_s}\right)^{\frac{c(\phi)}{\kappa u_*}} \tag{3}$$

where $S'_{k_s}(\phi) = p(\phi)$ is the proportion of grain size ($\phi - \frac{1}{2} \leq size \leq \phi + \frac{1}{2}$) in the bed, which is known. Our calculation indicated that the accuracy of this crude method was comparable to the commonly used more sophisticated method of Einstein in which the Rouse equation is coupled with a separate equation for bed layer.

A stochastic interpretation of this is provided in Ghosh and Mazumder (1981). It is well known that the Rouse equation (1) is a steady state solution corresponding to the turbulent diffusion equation

$$\frac{\partial S_y(\phi,t)}{\partial t} = \frac{\partial}{\partial y}[c(\phi)S_y(\phi,t)] + \frac{\partial}{\partial y}\left[\in(y)\frac{\partial S_y(\phi,t)}{\partial y}\right] \tag{4}$$

with the boundary conditions

$$\in(y)\frac{\partial S_y}{\partial y} + c(\phi)S_y = 0 \text{ at } y = k_s \tag{5}$$

where $\in(y) = \kappa u_* \cdot y(1-y/d)$, is sediment diffusion coefficient.

The equation (4) may be interpreted as the Kolmogorov forward equation of a Markovian stochastic diffusion process with a reflecting boundary at height k_s (vide eqn. (8) of Ghosh & Mazumder, 1981). Under this model, one imagines at the micro level an individual grain size ϕ undergoing a random vertical displacement with mean $\mu(y)$ and variance $\sigma^2(y)$ per unit time, where

$$\left.\begin{array}{l}\mu(y) = \in'(y) - c(\phi)\\[2mm]\sigma^2(y) = 2\in(y)\end{array}\right\} \tag{6}$$

and the prime denotes differentiation with respect to y.

If the particle gets to the lower boundary, it gets reflected. It was observed in Ghosh & Mazumder (1981) that the upper boundary at the surface is inaccessible in the terminology of stochastic diffusion process. This means the upper boundary is like infinity. It can not be reached by a particle and hence no boundary condition is needed at the upper boundary. Mathematically, one verifies inaccessibility by checking

$$\int_{k_s}^{d} - \exp\left\{-\int_{k_s}^{y}\frac{2\mu(\xi)}{\sigma^2(\xi)}d\xi\right\}\left[\int_{k_s}^{y}\frac{2}{\sigma^2(\xi)}\exp\left\{\int_{k_s}^{\xi}\frac{2\mu(\eta)}{\sigma^2(\eta)}d\eta\right\}d\xi\right]dy = \infty \tag{7}$$

If d is accessible, i.e., the integral above is finite, then a boundary condition must be specified at d as well.

In the experiment reported in Ghosh et al. (1986), some material of known size distribution was injected in suspension, and deposits collected after sufficient time had passed for flow to attain steady state. An attempt was made to use the Rouse equation backwards by taking the reference level at the injection height at $y=a$, where S_a is considered as the amount of injected material and S_{k_s} is the deposited material. The backward use of Rouse equation yielded poor estimates of S_{k_s}. It was equally bad for predicting observed S_y of small y. This may be explained as follows. The amount of material at the reference level was small and small changes in the amount of material mean large relative changes, leading to unstable estimates of S_y for y near the bed.

To overcome these difficulties, we adopted the second approach (Ghosh et al. 1986). The reflecting boundary condition was changed in a somewhat adhoc way to

$$\in(y)\frac{\partial S_y}{\partial y} + (1-\alpha)c(\phi)S_y = 0 \quad \text{at } y = k_s \tag{8}$$

where $(1-\alpha)$ is the probability that a particle which comes in contact with the bed will be lifted, and solved a steady state version of turbulent diffusion equation (4) subject to (8). This

256

led to the following modification of the Rouse equation

$$S(y) = \frac{e^{-\frac{c(\phi)k_1}{u_*}}}{c(\phi)/u_*}\left[\left(\frac{d-y}{y}\right)^{\frac{c(\phi)}{\kappa u_*}} + \frac{\alpha}{1-\alpha}\left(\frac{d-k_s}{k_s}\right)^{\frac{c(\phi)}{\kappa u_*}}\right] \tag{9}$$

where $e^{-\frac{c(\phi)k_1}{u_*}}$ is given by the equation (21) of Ghosh *et al.* (1986). This equation is then used in a backward way as explained above. This method is reported in Ghosh *et al.* (1986). As indicated there, the fit of observed and predicted deposits of different grain sizes was much better than on the basis of Rouse equation.

However, a critical enamination of the above approach revealed two problems. First, if one uses eqn. (9) to predict S_y for y near the surface with $a < y$, then S_y does not tend to zero as y tends to surface height, contrary to what is observed. Secondly, using the theory of stochastic diffusion processes, it can be shown that no stochastic diffusion subject to (4) and (8) can attain a steady state.

It became clear that what one needs are boundary conditions that allow more mass near the bed than reflection and for which there does exist a steady state. From a physical point of view it was conjectured that a mathematical picture of the flow should contain a component corresponding to the bursting phenomenon near the bed. To explain this we introduce below what is called a jump boundary condition in the theory of Markov processes.

2. DERIVATION OF NEW MODEL

We consider the one-dimensional diffusion with infinitesimal generator

$$A = \frac{1}{2}\sigma^2(y)\frac{d^2}{dy^2} + \mu(y)\frac{d}{dy} \tag{10}$$

where y represents the co-ordinate in the vertical direction, and $\mu(y)$ and $\sigma^2(y)$ are the drift and diffusion coefficient and have been explicitly defined earlier in eqn. (6). We consider only a one-dimensional diffusion since turbulence affects only the displacement in the vertical direction, i.e., perpendicular to the direction of flow.

We subject the diffusion condition to reflection and jump at $y = k_s$ i.e., the domain of A consists of all twice differentiable functions (endowed with Supnorm) having a finite limit y↑d, satisfying the boundary condition

$$\in_1 f'(k_s) + \gamma \int_{k_s}^d (f(y) - f(k_s))\psi(y)dy = 0 \tag{11}$$

where $\in_1 = \kappa u_* k_s(1 - k_s/d) > 0$, $\gamma = \alpha\, c(\phi)$ and ψ is the jump probability density function. One must have Af is continuous on $[k_s,\, d]$ with a finite limit as y↑d. For details on A, see Mandl (1968).

If the second component in the boundary condition is dropped we get the reflecting boundary condition for the above backward diffusion equation at $y = k_s$

$$f'(k_s) = 0 \tag{12}$$

which we have seen earlier. In this case the forward equation (4) may be obtained by applying the adjoint A^T of A to the steady state density $\pi(y) = S(y)$ and the condition (12) transforms to the boundary condition (5) for the forward equation.

Under the more general boundary condition (11), one may use the very general results of Mandl (1968) or a somewhat simpler alternative treatment suitable for (11) as given in Bhattacharya and Ghosh (unpublished). We sketch a few essential steps from the latter treatment. The conditions given in Bhattacharya and Ghosh, namely, $\sigma^2(y)$ is strictly positive and continuously differentiable on $[k_s,\, d)$ and $\mu(y)$ is continuous on the same interval, are satisfied by our $\mu(y)$ and $\sigma^2(y)$.

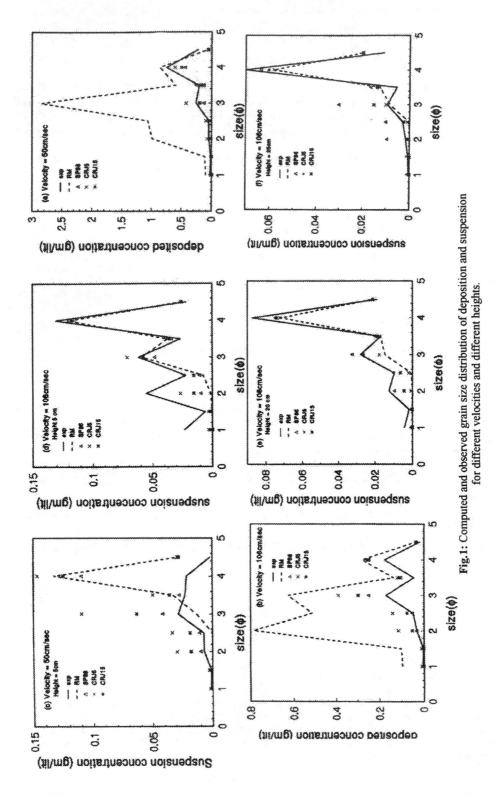

Fig.1: Computed and observed grain size distribution of deposition and suspension for different velocities and different heights.

Table 1: Grain size distribution of deposition and suspension for two values of maximum velocity.

Height (cm.)	Size (ϕ)	Velocity 50 cm./sec.					Velocity 106 cm./sec.				
		Exp.	RM	SP86	CRJ5	CRJ15	Exp.	RM	SP86	CRJ5	CRJ15
0.0250	1.0000	0.0226	0.0960	0.0011	0.0062	0.0020	0.0045	0.0958	0.0015	0.0076	0.0027
	1.5000	0.0119	0.1060	0.0013	0.0071	0.0024	0.0030	0.1027	0.0023	0.0100	0.0039
	2.0000	0.0441	0.9879	0.0141	0.0705	0.0252	0.0314	0.7860	0.0329	0.1151	0.0517
	2.5000	0.0444	1.0601	0.0200	0.0876	0.0340	0.0489	0.5126	0.0536	0.1405	0.0766
	3.0000	0.2540	2.8472	0.1291	0.4190	0.1971	0.1707	0.6278	0.2506	0.3908	0.2991
	3.5000	0.2054	0.5932	0.1287	0.2545	0.1669	0.0454	0.1370	0.1048	0.1190	0.1104
	4.0000	0.7456	0.8621	0.4355	0.6024	0.4961	0.1770	0.2795	0.2487	0.2620	0.2540
	4.5000	0.2180	0.0583	0.0526	0.0550	0.0535	0.0184	0.0355	0.0346	0.0350	0.0347
5.0000	1.0000	0.0012	0.0000	0.0010	0.0018	0.0019	0.0230	0.0000	0.0001	0.0028	0.0017
	1.5000	0.0010	0.0000	0.0011	0.0027	0.0020	0.0056	0.0000	0.0011	0.0032	0.0018
	2.0000	0.0069	0.0000	0.0099	0.0293	0.0177	0.0552	0.0014	0.0097	0.0265	0.0152
	2.5000	0.0071	0.0001	0.0109	0.0332	0.0185	0.0218	0.0075	0.0107	0.0239	0.0152
	3.0000	0.0280	0.0130	0.0415	0.1096	0.0634	0.0618	0.0590	0.0483	0.0715	0.0577
	3.5000	0.0225	0.0324	0.0274	0.0495	0.0355	0.0262	0.0373	0.0341	0.0382	0.0359
	4.0000	0.0213	0.1326	0.1098	0.1466	0.1250	0.1306	0.1199	0.1152	0.1203	0.1177
	4.5000	0.0016	0.0289	0.0279	0.0291	0.0284	0.0212	0.0258	0.0256	0.0258	0.0257
20.0000	1.0000	0.0024	0.0000	0.0010	0.0000	0.0000	0.0042	0.0000	0.0010	0.0000	0.0001
	1.5000	0.0008	0.0000	0.0010	0.0000	0.0000	0.0018	0.0000	0.0010	0.0000	0.0003
	2.0000	0.0077	0.0000	0.0099	0.0000	0.0013	0.0124	0.0000	0.0096	0.0008	0.0045
	2.5000	0.0073	0.0000	0.0109	0.0002	0.0031	0.0096	0.0007	0.0101	0.0021	0.0064
	3.0000	0.0199	0.0006	0.0411	0.0052	0.0223	0.0278	0.0146	0.0325	0.0176	0.0264
	3.5000	0.0036	0.0059	0.0226	0.0090	0.0170	0.0166	0.0170	0.0197	0.0175	0.0189
	4.0000	0.0237	0.0434	0.0699	0.0480	0.0614	0.0874	0.0718	0.0750	0.0723	0.0741
	4.5000	0.0071	0.0189	0.0195	0.0199	0.0194	0.0188	0.0212	0.0213	0.0212	0.0213
25.0000	1.0000	0.0004	0.0000	0.0010	0.0000	0.0000	0.0004	0.0000	0.0010	0.0000	0.0000
	1.5000	0.0006	0.0000	0.0011	0.0000	0.0000	0.0006	0.0000	0.0011	0.0000	0.0000
	2.0000	0.0008	0.0000	0.0099	0.0000	0.0001	0.0016	0.0000	0.0096	0.0002	0.0011
	2.5000	0.0008	0.0000	0.0109	0.0000	0.0004	0.0028	0.0002	0.0101	0.0008	0.0024
	3.0000	0.0018	0.0002	0.0411	0.0015	0.0065	0.0092	0.0082	0.0303	0.0100	0.0150
	3.5000	0.0018	0.0030	0.0221	0.0045	0.0085	0.0052	0.0124	0.0164	0.0127	0.0137
	4.0000	0.0069	0.0276	0.0628	0.0305	0.0390	0.0700	0.0583	0.0637	0.0583	0.0602
	4.5000	0.0037	0.0159	0.0170	0.0160	0.0163	0.0106	0.0195	0.0198	0.0196	0.0197

RM- computed by Rouse equation where constant has been calculated from the total mass. SP86- computd by Ghosh *et a.'s* model (see eqn. 22). CRJ5- computed by the proposed method (eqn. 20) where $\psi(y) = 1$ for $y \in [4.5, 5, 5]$, otherwise 0. CRJ15- computed values correspond to the height $y \in [4.5, 5, 5]$. All concentration values shown in the table are in gm/litre.

Let p(t; y, dη) denote the transition probability measure of a diffusing particle starting at y and ending up at η after t units of time. Let $\{T_t : t > 0\}$ be the corresponding semi group generated by A. Thus.

$$(T_t f)(y) = \int_{k_s}^{d} f(\eta) p(t; y, d\eta) \quad t > 0 \tag{13}$$

and $T_0 f = f$. Then

$$Af = \lim_{t \downarrow 0} \frac{T_t f - f}{t} = \left[\frac{d}{dt} T_t f \right]_{t=o} \tag{14}$$

A probability measure $\pi(dy)$ is steady state (or stationary or invariant) if

$$\int (T_t f)(y)\pi(dy) = \int f(y)\pi(dy) \quad t > 0 \tag{15}$$

for all f in the domain of A

Since $\dfrac{1}{h}(T_h f - f)$ converges uniformly to $(Af)(y)$ as $h \downarrow 0$, one can interchange the order of integration and differentiation to get

$$\int_{k_s}^{d}(Af)(y)\pi(dy) = 0,\, f \in \text{domain of } A. \tag{16}$$

Conversely, it can be shown a π satisfying(16) is the steady state distribution. Under our conditions, it can be shown that π has a smooth density $\pi(y)$ on $[k_s, d)$. Then eqn. (16) becomes

$$\int_{k_s}^{d}\left[\frac{1}{2}\sigma^2(y)f''(y) + \mu(y)f'(y)\right]\pi(y)dy = 0 \tag{17}$$

Integrating by parts using the boundary condition (11) repeatedly and using the fact that (17) must hold for all f in domain of A, one gets, for example,

$$\frac{d^2}{dy^2}\left(\frac{1}{2}\sigma^2(y)\pi(y)\right) - \frac{d}{dy}(\mu(y)\pi(y)) +$$

$$\gamma\left[\left(\frac{1}{2}\sigma^2(y)\pi(y)\right)_{k_s}^{d} - \mu(k_s)\pi(k_s)\right]\psi(y) = 0 \tag{18}$$

which takes on the role of a new diffusion equation in the presence of jump boundary conditions. The boundary condition on π takes the form

$$-\tfrac{1}{2}\sigma^2(k_s)\pi(k_s) + \in_1\left((\tfrac{1}{2}\sigma^2(y)\pi(y))_{k_s}^{d} - \mu(k_s)\pi(k_s)\right) = 0 \tag{19}$$

Integrating (18) and (19) one gets

$$\pi(y) = \frac{2}{\sigma^2(y)J(y)}\left\{(\tfrac{1}{2}\sigma^2(\eta)\pi(\eta))_{k_s}^{d} - \mu(k_s)\pi(k_s)\right\} \times$$

$$\times \left\{\in_1 + \gamma\int_{k_s}^{y}\left[\int_z^d \psi(\eta)d\eta\right]J(z)dz\right\} \tag{20}$$

where

$$J(y) = \exp\{-\int_{k_s}^{y} 2\mu(\eta)/\sigma^2(\eta)d\eta\} \tag{21}$$

The above treatment leading to (20) is valid for fairly general μ and σ. We now specialise to μ and σ as defined in (6). Using the somewhat curious fact

$$2c(\phi)\int_{k_s}^{y} J(\eta)d\eta = \sigma^2(y)J(y) - \sigma^2(k_s)J(k_s) \tag{22}$$

and setting $\psi(y) = 0$ if $y < y_1$ or $y > y_2$, one can verify that the solution (20) reduces essentially to the Rouse equation for $y > y_2$ and to the (1986) model of Ghosh et al. for $y < y_1$. So the equation (20) does provide an unification of our two models.

For numerical calculation we take ψ to be the uniform density $(y_2 - y_1)^{-1}$ between y_1 and y_2. The limits y_1 and y_2 are chosen on the basis of some pictures of trajectories of vortices near the bed.

3. COMPARISON WITH EXPERIMENTAL DATA

The experiments are conducted in a closed circuit laboratory flume (Ghosh et al. 1986; Sengupta et al. 1991) designed at the Indian Statistical Institute (ISI) Calcutta. Both experimental and the recirculating channels of the flume are of the same dimensions (10m X 50cm X 50cm). The experimental walls of the flume are made of perspex windows for a length of 8m, affording clear view of the sediment movements. Two non-clogging type of centrifugal pumps providing the flow are located outside the main body of the flume. The intake and outlet pipes are freely suspended from an overhead structure to allow tilting of the flume. Both the outlet pipes (Pumps - 1,2) are fitted with by-pass pipes and valves, so that by adjusting these valves in the outlet and the bypass pipes, the flow can be set at any desired speed up to 1.30 m/sec. for a water depth of 30 cm. A honeycomb cage fixed at the upstream bend of the experimental channel ensures smooth, vortex-free flow of water through the experimental channel. The vertical velocity distribution in this flume closely follows the logarithmic law. A certain amount of material was injected at a particular velocity in front of the outlet pipe in the flume and it was ensured that the material was in suspension. After the reduction of the flow velocity deposits were collected from the bed with the help of syphon tube. Details of the experimental findings are available in Ghosh *et al.* (1986).

Numerical comparisons between the three models - Rouse eqn. (3), model (9) of Ghosh *et al.* and the reflection-cum-jump model (20) are given in Table-1for different velocities, different heights y at which observation is made for different grain sizes. The experimental data are taken from Ghosh *et al.* (1986). The constant $\pi(k_s)$ appearing in the new model is chosen to take care of the total circulating mass in the flume. A graphical presentation for two velocities and different heights of data is given in Fig. 1(a-f).

The rigorously derived new model captures features of both the adhoc model of Ghosh *et al.* (1986) and the Rouse model based on diffusion with reflection, but none seems to be uniformly best. The Rouse model does not do as well as the other two.

ACKNOWLEDGEMENT

We thank Professor Supriya Sengupta for introducing us to the subject and for sharing with us many illuminating insights.

REFERENCES

Einstein H. A., The bed-load function for sediment transportation in open channel flows. Tech. Bull. US Dept. Agric., No. 1028 Soil Conserv. Serv., Washington, DC, (1950).

Mandl P., Analytical treatment of one-dimensional Markov process, Springer-Verlag, New York (1968)

Ghosh J. K. and B.S. Mazumder, Size distribution of suspended particles- Unimodality, symmetry and lognormality. Statistical Distributions in Scientific Work, (edited by C. Taillie et al. and published by D. Reidel Publishing Company), 6, pp. 21-32 (1981).

Ghosh J. K., B.S. Mazumder, M.R. Saha and Supriya Sengupta, Deposition of sand by suspension currents: Experimental and theoretical studies. Journal of Sedimentary Petrology, 56, 57-66 (1986).

Sengupta S., J. K. Ghosh and B. S. Mazumder, Experimental - theoretial approach to interpretation of grain size frequency distributions. In: Principles, Methods and Applications of Particle Size Analysis, (Ed. By J.P.M., Syvitski), Cambridge University Press, Cambridge, 264-279 (1991).

Bhattacharya Rabi and Jayanta Ghosh, Steady state distributions of one dimensional diffusions under general boundary conditions. (unpublished).

Stochastic Hydraulics 2000, Wang & Hu (eds) © 2000 Taylor & Francis, ISBN 90 5809 166 X

Coefficients in the sediment model of depth-averaged and moment equation

Guo Qingchao & Jin Yee-Chung
University of Regina, Sask., Canada

ABSTRACT: A sediment transport model using depth-averaged and moment equations has been established by Guo and Jin (1999a). However, it is impossible for sediment transport models to avoid any coefficients in the calculation. Like many other models, this model also contains coefficients. The objective of this paper is to find some semi-theoretical formulas to estimate these coefficients in order that, in practical application, the uncertainty in selecting coefficients may be avoided. Verification shows that the coefficients estimated from mathematical formulas give an excellent simulation of the channel bed deformation. The simulated profiles of velocity and concentration from the assumed and Rouse's equation agree with the measured considerably.

Key words: Sediment model ,Moment equation, Channel bed deformation

1.INTRODUCTION

The model using depth-averaged and moment equations (Guo and Jin 1999a) contains three coefficients. They are adjustment coefficient α, sediment-carrying capacity coefficient k and exponent m. The purpose of this paper is to find some guidelines to estimate these coefficients. In addition, a one-dimensional model is only concerned with the depth-averaged variables, such as velocity and concentration, without considering the distributions along the depth. This is determined by the characteristics of a one-dimensional model. To improve this kind of shortcoming, in Guo and Jin's model, the distributions of velocity and suspended sediment concentration along the flow depth are assumed to be linear. The assumed concentration distribution is very coarse. To get a better distribution, Rouse's equation is proposed. By establishing an expression for the adjustment coefficient α, the difficulty to calculate the reference concentration is resolved. In the later part of this paper, a good agreement between the calculated and measured concentration distribution will be shown.

2.BASIC EQUATIONS

The depth-averaged and moment equations for unsteady flow and suspended sediment transport with hydrostatic pressure distribution in a rectangular channel can be written as:

$$\frac{\partial h}{\partial t} + \frac{\partial h u_0}{\partial x} = 0 \tag{1}$$

$$\frac{\partial h u_0}{\partial t} + \frac{\partial h u_0 u_0}{\partial x} = -\frac{1}{3}\frac{\partial h u_1^2}{\partial x} - gh\frac{\partial(h+z_b)}{\partial x} - \frac{\tau_b}{\rho} \tag{2}$$

$$\frac{\partial hu_1}{\partial t} + \frac{\partial hu_0 u_1}{\partial x} = -hu_1 \frac{\partial u_0}{\partial x} + \frac{3}{\rho}(\tau_b - 2\bar{\tau}) \tag{3}$$

$$\frac{\partial hs_0}{\partial t} + \frac{\partial hu_0 s_0}{\partial x} = -\frac{\partial has_1}{\partial t} - \alpha\omega(s_0 - s_*) \tag{4}$$

$$\frac{\partial hs_1}{\partial t} + \frac{\partial hu_0 s_1}{\partial x} = hu_1 \frac{\partial(s_0 + as_1)}{\partial x} + 6\omega(s_0 + as_1) - 3\alpha\omega(s_0 - s_*) - \frac{12}{h}\varepsilon_z s_1 \tag{5}$$

Where h is the flow depth; t is time; x is the distance from the channel inlet; g is the gravitational acceleration; z_b is the bed elevation; τ_b is the bed shear stress; $\tau_b/\rho = gu_0^2/C^2$; C is Chezy coefficient; ρ is the water density; u_0 and s_0 are the depth-averaged velocity and concentration, respectively; u_1 and s_1 are equal to the half of the differences of velocities and concentrations near the channel bed and at the water surface, respectively; τ is a coefficient; $\bar{\tau}$ is the depth-averaged shear stress and $\bar{\tau}/\rho \approx 4\lambda^2 h^2 |u_1|/u_1$, λ is the mixing length coefficient; a is a coefficient equal to $u_1/3u_0$; α is the adjustment coefficient; ω is the settling velocity; s_* is the sediment-carrying capacity; and ε_z is the dispersion coefficient which is given by $\kappa u_* h/6$ and u_* is the shear-stress velocity.

The distributions of velocity u and concentration s along the depth relative, η, are

$$u = u_o + (2\eta - 1)u_1 \tag{6}$$

$$s = s_o + (1 - 2\eta + a)s_1 \tag{7}$$

Sediment transport results in the change of channel elevation is the primary concern here. To simulate the bed variation, one more equation to describe the deformation of channel bed is needed. The bed elevation equation reads as:

$$\gamma' \frac{\partial z_b}{\partial t} = \alpha\omega(s_0 - s_*) \tag{8}$$

Where γ' is the dry specific weight of bed materials.

There are six variables, h, u_0, u_1, s_0, s_1, and z_b, in this six-equation model. Thus the equations are closed. However, there are two important parameters, coefficient α and sediment-carrying capacity s_*, to be determined. They will be discussed in the following sections.

3. COEFFICIENTS

The model contains a total of three coefficients; adjust coefficient α, sediment-carrying capacity coefficient k and exponent m. Coefficients k and m are involved in the formula of sediment-carrying capacity.

3.1 Sediment-carrying capacity coefficient and exponent

Sediment-carrying capacity is the ability for flow to transport sediment without any deposition and erosion. For a given sediment composition, a certain flow can only carry a certain quantity of sediment without net deposition and erosion. This quantity is called the sediment-carrying capacity and the corresponding flow is said to be under saturated conditions.

Based on the basic relationship between work, energy expenditure of a stream and the quantity of the sediment transport by the stream, Bagnold (1966) obtained an expression to calculate sediment-carrying capacity:

$$s_* = \frac{B_r \gamma}{\dfrac{\gamma_s - \gamma}{\gamma_s}} \frac{1}{C^2} \frac{u_0^3}{h\omega} \tag{9}$$

Where γ and γ_s are the specific weights of clear water and of sediment, respectively. According to laboratory data, Bagnold (1966) gave a value of 0.01 for the coefficient B_r. For natural rivers, Rubey (1933) suggested that the coefficient B_r in equation (9) should be around 0.025 instead of

0.01. Large number of field data collected by Wuhan (1980) has shown that a similar relation between s_* and $u_0^3/h\omega$ is found to be:

$$s_* = k(\frac{u_0^3}{h\omega})^m \tag{10}$$

Where, k and m are two coefficients to be determined. The unit for s_* is kg/m^3 when $h=m$, $u_0 =m/s$, and $\omega = m/s$. Han (1980) used many field data from natural rivers and proposed values of 0.03 and 0.92 for k and m, respectively. However, these values are not always reliable.

Due to the difficulty in obtaining the values for coefficients k and m, the application of sediment transport models is limited. To solve this problem, it is suggested that coefficient k to be calculated from Bagnold's equation (9) if the equilibrium concentration is known:

$$k = \frac{B_r\gamma}{\frac{\gamma_s - \gamma}{\gamma_s}}\frac{1}{C^2} \tag{11}$$

Usually it is not difficult to get an equilibrium condition either in practical channels or in laboratory flumes. Once the coefficient k has been found, the coefficient m may be estimated from equation (10) as long as the equilibrium concentration is obtained. Slight variations in k and m values are acceptable since these values are usually obtained under the assumption of an equilibrium condition.

3.2 Adjustment coefficient α

Theoretically, coefficient α is the ratio of sediment concentration near the bed to the depth-averaged concentration under equilibrium condition. Based on the theoretical definition and the suspended sediment distribution presented by Rouse (1937), Guo and Jin (1999b) gave an expression to estimate the adjustment coefficient α:

$$\alpha = \frac{s_a}{s_0} = \frac{(\frac{1}{\eta_a} - 1)^z}{\left[(1 + \frac{\sqrt{g}}{\kappa C})\int_{\eta_a}^1 (\frac{1}{\eta} - 1)^z d\eta + \frac{\sqrt{g}}{\kappa C}\int_{\eta_a}^1 (\frac{1}{\eta} - 1)^z \ln\eta d\eta\right]} \tag{12}$$

Where the exponent z is equal to $\omega/\beta\kappa u_*$ and η_a is the reference relative depth. In this equation, β describes the difference between the diffusion of a discrete sediment particle and the diffusion of a fluid particle. Chien (1954) analyzed concentration profiles measured both in the laboratory and the field and stated that in some cases (especially for $\omega/u_*>0.4$), there is a clear indication of $\beta>1.0$. Rijn(1984) further gave an expression of $\beta=1+2(\omega/u_*)^2$ for $0.1<\omega/u_*<0.707$ Value of ω/u_* that is out of the range, $\beta=1.0$ for $\omega/u_* <0.1$ and $\beta = 2.0$ for $\omega/u_* >0.707$.

Currently, there isn't an appropriate method in the determination of the reference relative depth η_a. In order to estimate α, a value of about 0.01 for η_a is suggested based on experience. It should be noted that the reference concentration s_a can be obtained easily from equation (10) as long as the average concentration s_0 is determined. Therefore the concentration profile can be computed by using Rouse's equation:

$$s = s_a\left((\frac{1}{\eta} - 1)/(\frac{1}{\eta_a} - 1)\right)^z \tag{13}$$

4. MODEL VERIFICATION

The model was used to simulate migration of sediment within a trench. The experiment was carried out at Delft Hydraulics Laboratory (Galappati and Vreugdenhil, 1985). A flume of dimensions length = 30 m, depth = 0.7 m, and width = 0.5 m was used. To maintain equilibrium conditions in the section upstream of the trench, the suspended sediment concentration was about 0.145 kg/m^3. The entrance flow was uniform with a fully developed sediment concentration profile and a depth of 0.39 m and a mean velocity of 0.51 m/s. The mean diameter of sediment was 0.16 mm and the representative particle fall velocity was 0.013 m/s.

According to flow conditions and sediment composition, the values of C, u_*, β, and z are calculated to be 40.6, 0.04, 1.21, and 0.67 respectively. Applying equation (12) and selecting a reference depth η_a of 0.01, the value of α is calculated to be 15.5. Coefficient k was found to be 0.0097 from equation (11). The coefficient m is 0.83 using equation (10) when the equilibrium concentration is about 0.145 kg/m^3. In the numerical simulation, the actual values for coefficients α, k, and m were 18.0, 0.0088, and 0.85, respectively.

The computed and observed bed profiles for the trench with slopes 1:10 are presented in Figure 1. The results show very good agreement for both situations at 7.5 and 15 hours. This means that the model can not only simulate the final bed variation but also simulate the process of the bed deformation.

Figure 2 and 3 show the measured and computed velocity and sediment concentration profiles at initial time for the different reaches of the trench with slopes 1:3. Although the assumed distributions were considerably coarse, the trends for both velocity and concentration generally show a reasonable agreement with the measured data. Only the simulated profiles in the middle deep reach displayed some difference with the measured. This may result from the sharp change of the channel bed elevation that causes complex flow structure which can not be simulated by a one-dimensional model.

The profiles of sediment concentration from Rouse's equation are also presented in the Figure 3. Rouse's distribution can be obtained by the following: the reference concentration s_a can be calculated from equation (12) when s_0 and coefficient α are known; then the distribution can be computed by equation (13). From Figure 3, it can be seen that the concentration distributions between the measured and Rouse's equation match quite well. This match expresses the validity of the method suggested by this paper to estimate the three coefficients.

5.CONCLUSIONS

Some semi-theoretical formulas for estimating the coefficients in the model of depth-averaged and moment equations are found. They provide a guideline for selecting the values of coefficients.

Verification shows that the coefficients estimated from the formulas give an excellent simulation of the channel bed deformation. Additionally, the simulated profiles of velocity and concentration reasonably agree with the measured.

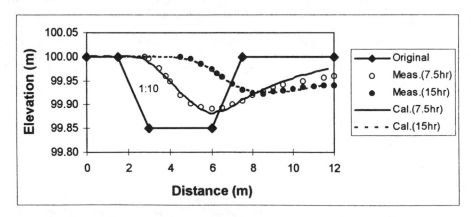

Fig.1: Comparison between the calculated and measured results at 7.5 and 15 hours

266

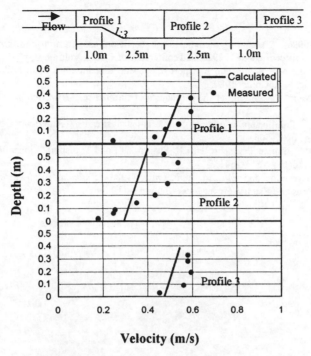

Fig.2: Distributions of calculated and measured velocity.

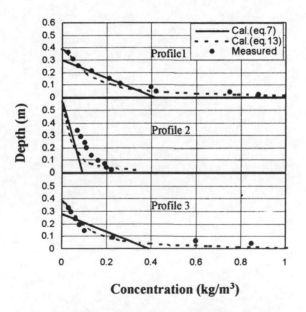

Fig.3: Distributions of suspended sediment concentration

Due to the complexity of sediment transport, it is difficult to obtain the accurate coefficient values. Hence, little modification in the values obtained from formulas is acceptable. In addition, the assumed sediment concentration profile may result in an unreasonable distribution near the water surface, but it does not significantly affect the depth-averaged concentration and the bed variation.

REFERENCE

Guo, Q. and Jin, Y., (1999b). "Estimating the adjustment coefficient used in non -equilibrium sediment transport modeling." Proceedings of 1999 annual conference of CSCE, Volume II, p217-226.

Han, Q. . "A study on the non-equilibrium transport of suspended sediment." *1st Int. Symp. On River Sedimentation*, Beijing, P. R. China, 1980.

Rouse, H., (1937). "Modern conceptions of the mechanics of turbulence," *Transaction, Amer. Geophy. Union*.

Rubey, W. W., (1933). "Equilibrium conditions in debris-laden streams." American Geophys. Union Trans., 14th Ann. Mtg., p497-505.

Rijn, L. C. van, (1984). "Sediment Transport, Part II: Suspended Load Transport." *Journal of Hydraulic Engineering, ASCE*, 110(11), 1613-1641.

Wuhan University of Hydraulic and Electric Engineering, (1980). *Sediment Engineering In Alluvial Rivers*, Water Resources and Electric Power Press, Beijing (In Chinese).

Stochastic Hydraulics 2000, Wang & Hu (eds) © 2000 Taylor & Francis, ISBN 90 5809 166 X

Network model for evaluation of sediment discharge in river basin

Mikio Kuroki & Tadaoki Itakura
Hokkaido University, Sapporo, Japan

ABSTRACT: A model to evaluate the sediment discharge and the stable longitudinal profiles of river channels was derived by taking the drainage channel networks of the basin into consideration. This model enables us to evaluate annual mean values of flow discharge and sediment discharge, mean diameter of bed materials, width of channels and flow depth at any point over river basin. This model requires values of bed elevation along the channel networks and flow parameters i.e. the flow discharge at outer links of the networks.

The model was tested by field data from the Kanayama dam basins in Hokkaido, JAPAN. The class of rocks in the deposited sediment in the dam, the capping of basin and the estimated sediment transport rate were compared each other. In the Kanayama dam basin, while the volcanic rock occupies almost the half of the basin area, it occupies only 20 % in the field data and 30 % in the model analysis. On the contrary, the metamorphic rock occupies only 15 % in basin area, but occupy 30 % in field data and 35 % in model analysis. This analysis shows that the proposed model could evaluate fairly well the mean sediment production and transport in the river basin.

Key word: Sediment production, Lithological classification of sediment, Network model, Stable channel

1. INTRODUCTION

Accurate evaluations of sediment discharge at any point in river basins are basic and important not only for river engineering but also for harbor and coastal engineering. The management of sediment from mountainous area to the sea region is becoming more and more important in Japan. However, in spite of its significance, it isn't cleared well yet. Although it includes many political problems, even in the engineering point of view it is not so easy problem. Even if the hydraulic data of flow and the algebraic data of channels are fully collected, it will be not easy to evaluate the rate and properties of sediment. This is partly because the complexity of sediment movement phenomena in channels and the vast variation of flow discharge. Unfortunately these data does not collected in many rivers in Japan. These are the reasons why the new concepts are needed to construct for rough estimation of sediment without hydraulic and algebraic data. The daily or monthly amount of sediment does not need but the annual amount of sediment is required for these porpoises in the first stage.

The authors (Kuroki & Itakura & Yamamoto, 1996) developed a theory of stable longitudinal profile of river channels. In this theory, critical flow discharge was introduced and the averaged profiles of river channels i.e. bed elevation, depth of flow, width of channel, mean diameter of bed sediment and the sediment discharge could became evaluated. If the channels could assume to be in stable or equilibrium state, only giving the longitudinal profiles of bed elevation to the theory, the residual profiles of channel could be evaluated.

The critical flow discharge, which is constant at a point under consideration, was introduced to analyse the property of geometry of river channels by many researchers. The one of authors (Kuroki & Kishi,1984) shows that the regime criteria of sediment bars are possible by the annual mean discharge. The sediment bars determine the properties of channels and the sediment transport in rivers. Therefore the annual mean discharge was introduced to the concepts of stable or equilibrium channels and are evaluated the annual mean sediment discharge.

2. BASIC EQUATIONS

The basic equations are as follows; the first one is the equation of motion for flow in one dimension;

$$\frac{1}{2g}\frac{d}{dx}\left(\frac{Q}{Bh}\right) + \frac{dh}{dx} = I_B - I_e \tag{1}$$

here Q; flow discharge, h; flow depth, B; width of channel, I_B; slop of river bed, i.e.; energy slop, g; acceleration due to gravity, x; distance from upstream. The second equation gives the resistance to flow where the flat bed is assumed in the present analysis;

$$\frac{(Q/Bh)}{\sqrt{ghI_e}} = 6.9\left(\frac{h}{d}\right)^{1/6} \tag{2}$$

where d; mean diameter of bed materials. The third equation is the continuity equation of sediment in equilibrium state;

$$\frac{dQ_B}{dx} = q_{out} \tag{3}$$

where Q_B; sediment discharge, q_{out}; sediment feed from outside of channel. The forth equation gives the sediment discharge in unit width;

$$q_B = 8\sqrt{sgd^3}\left(\tau_* - \tau_{*c}\right)^{3/2} \tag{4}$$

q_B; sediment discharge in unit width, s; specific weight of sediment in the water, τ_*; normalized bed shear stress, τ_{*c}; (=0.05) critical bed shear stress. The last equation gives the normalized bed shear stress in the equilibrium cross section;

$$\tau_*\left(=\frac{hI_e}{sd}\right) = \beta\tau_{*c} \tag{5}$$

Ikeda et. al. (1986) gives the value of β as approximately constant as $\beta = 1.23$.

After some abbreviations, the following simple but important relationship could derive.

$$\frac{dz}{dx} = C\left(\frac{Q}{Q_B}\right)^{-\frac{6}{7}} \tag{6}$$

where C; constant and are evaluated by known conditions. Eq.(6) are similar to the so called regime theory that is derived empirically while this was derived theoretically. If the longitudinal distributions of flow and sediment discharge, the longitudinal profile of river bed elevation in stable state could be estimated. Kuroki, Itakura and Yamamoto (1996) tested the Eq.(6) using the data of rivers in Hokkaido, JAPAN.

The additional relationships for the longitudinal change of flow depth, width of channels and mean diameter of bed materials could derive if the longitudinal distributions of flow and sediment discharge.

3. NETWORK MODEL

If the river channels could be assumed in equilibrium or stable state, inverse adaptation of the proposed theory enables to estimate the longitudinal change of river channel properties only by knowing the distributions of bed elevations. A model to evaluate the sediment discharge and the stable longitudinal profiles of river channels was derived by taking the drainage channel networks of the basin into consideration. This model enables us to evaluate annual mean values of flow discharge and sediment discharge, mean diameter of bed materials, width of channels and flow depth at any point over river basin. This model requires values of bed elevation along the channel networks and flow parameters i.e. the flow discharge at outer links of the networks.

The river networks were composed by using the GIS data and the basic unit area is about 1km x 1km, the unit channel which exists in the unit areas is about 1km long. The longitudinal distributions of flow and sediment discharges in unit channel were assumed as follows;

$$Q(\xi) = Q_u + q\xi \tag{7}$$

$$Q_B(\xi) = Q_{Bu} + q_s\xi \tag{8}$$

where $Q(\xi)$, $Q_B(\xi)$; flow and sediment discharge at arbitrary position in unit channel, Q_u, Q_{Bu}; flow and sediment discharges at the upper end of unit channel, q, q_s; rate of change, ξ; normalized distance from the upper end of unit channel.

Consider the simple networks as shown in Fig.1 as an example. The flow discharges, for example, at the upstream and downstream end of No. 1, 2 and 3 channels which has magnitude m=1 are expressed as follows;

$$[Q_u]_{(j)} \qquad \text{at upstream end of channel No.j,} \tag{9.a}$$

$$[Q_D]_{(j)} = [Q_u]_{(j)} + q_{(j)} \qquad \text{at downstream end of channel No.j.} \tag{9.b}$$

where the subscripts U and D denotes the upstream end and downstream end respectively, the subscript (j) denotes the number of Channel. For No.4 channel of magnitude m=2 are;

$$[Q_u]_{(4)} = [Q_D]_{(1)} + [Q_D]_{(2)} \qquad \text{at the upstream,} \tag{10.a}$$

$$[Q_D]_{(4)} = [Q_u]_{(4)} + q_{(4)} \qquad \text{at the downstream.} \tag{10.b}$$

For No.5 channel of magnitude m=3 are as a same manner;

$$[Q_u]_{(5)} = [Q_D]_{(1)} + [Q_D]_{(4)} \qquad \text{at the upstream,} \tag{11.a}$$

$$[Q_D]_{(5)} = [Q_u]_{(5)} + q_{(5)} \qquad \text{at the downstream.} \tag{11.b}$$

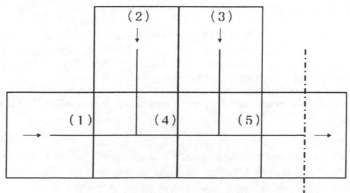

Fig.1: Schematic diagram of channel network.

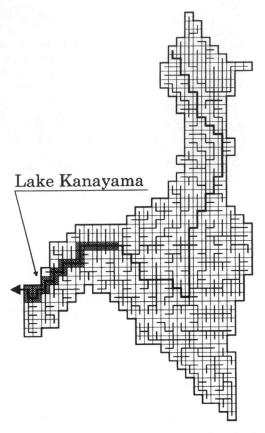

Lake Kanayama

Fig.2: Channel network in the Kanayama dam basin.

While the Eqs.(8)~(11) had been shown the flow discharges, the sediment discharges could be expressed as a same manner.

The values $[Q_u]_{j)}$ of the channel of m=1, and $q_{(j)}$ of all of the channels must be evaluated. In the present model it assumed that the value $[Q_u]_{j)}$ of all outer links are same for the simplicity of the analysis. The values of sediment discharge at upstream end of outer links $[Q_{B,u}]_{j)}$ could be evaluated from Eq.(6) because the longitudinal profiles of bed elevation were known.

4. APPLICATION OF MODEL

The model was tested by field data from the Kanayama dam basins in Hokkaido, JAPAN.
Fig.2 shows the networks of channels in the test basin. The total drainage area is 469km^2, the numbers of unit areas are 498 and therefore the unit area is 0.942km^2. The numbers of outer links are 190 and the numbers of unit channels are 497.

The values $q_{(j)}$ of all the unit channels were evaluated by using the longitudinal profile of bed elevation and the flow discharge at the inlet point of dam, sumQ, were evaluated as follows;

$$\text{sumQ}=111.5\text{x}[Q_{IU}]_{1)} \tag{12}$$

Using the data of flow discharge at the dam point, the mean annual maximum discharge had been evaluated sumQ=151.8 m^3/sec as a critical discharge for the analysis. Therefore the value of $[Q_u]_{1)}$ becomes to 0.137 m^3/sec. Using this value, the values of sediment discharge, flow depth, width of channels, and mean diameter of bed material at any point in the basin could be evaluated .

The flow discharge along the main streams is shown in Fig.3. Although the linear increase of flow discharge in the unit channels were assumed as shown in Eq.(7), the longitudinal distribution of flow discharge fits to the exponential distributions. This is because the conjunction of flow discharge by channel networks, and shows the validity of the assumption of Eq.(7). Fig.4 shows the longitudinal change of sediment discharge along the main streams. This distribution also fits to the exponential one.

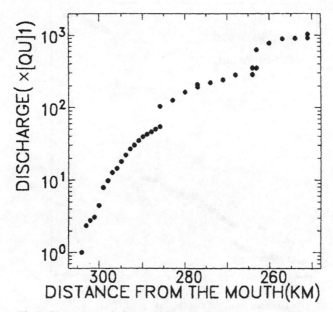

Fig.3: The estimated change of flow discharge along main channel.

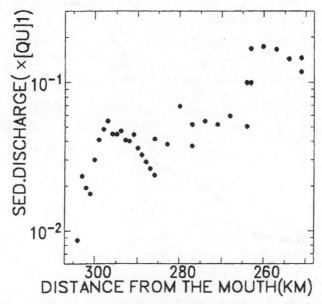

Fig.4: The estimated change of sediment discharge along main channel.

273

The width of channels along the main streams were compared in Fig.5. The white circles indicate of measured channel width and the small black dots indicated the evaluated value from the present model. The evaluated change along the main stream represents the observed distributions very well. Fig.6 shows the comparison of mean diameter of bed materials. This figure also represents that the evaluated change of diameters agree to the observed distribution fairly well.

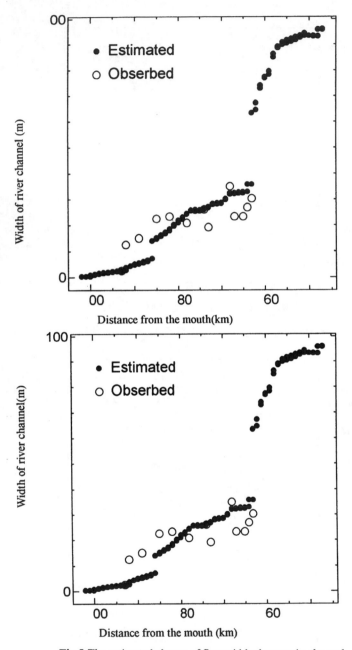

Fig.5: The estimated change of flow width along main channel.

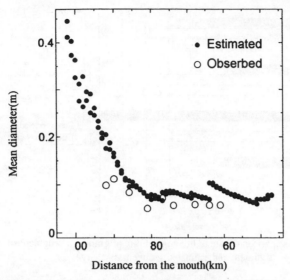

Fig.6: The estimated change of mean diameter of bed materials alongmain channel.

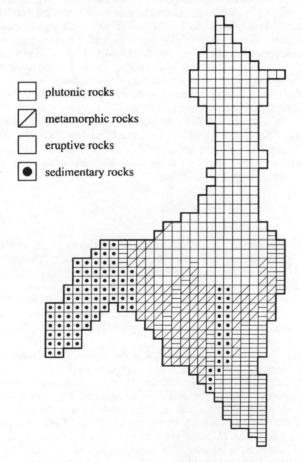

Fig.7: The aerial distributions of lithological classes of grand capping over the Kanayama dam basin.

275

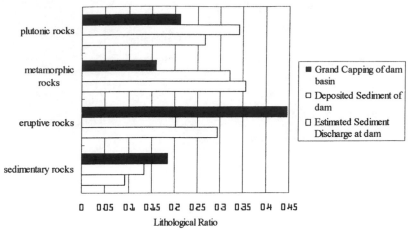

Fig.8: The comparison of rate in lithological rate of grand capping deposited
sediment and estimated sediment discharge.

The sediment discharge were checked by the transport rate of sediment in each class of rocks in the basin. The ground capping of basin varies lithologicaly and the sediment production also varies with each lithoid classes. Fig.7 shows the aerial distribution of lithological classes of the grand capping over the Kanayama dam basin. The lithoid information of ground capping over the basin introduced to the model, and the rate of sediment transport in each lithoid classes was evaluated. In the other hand the amount of the deposited materials in the dam reservoir in the each lithoid classes were analyzed and compared with the results of model analysis. Fig.8 shows the comparison among the aerial rate of ground capping, the observed volumetric rate of deposited sediment and the sediment discharge from the present analysis.

In the Kanayama Dam basin, while the volcanic rock occupies almost the half of the basin area, it occupies only 20 % in the field data and 30 % in the model analysis.   On the contrary, the metamorphic rock occupies only 15 % in basin area, but occupy 30 % in field data and 35 % in model analysis. This comparison shows that the proposed model could evaluate fairly well the mean sediment production and transport in the river basin.

5. CONCLUSION

(1) The network model was proposed by inverse application of the theory of the stable channels. Therefore the river basins under analysis was assumed in the stable state. The present theory consists with the constant, critical flow discharge. The mean annual maximum flow discharge was adopted to the analysis with some preliminary analysis.

(2) The proposed model required only the data of bed elevations in the basin area which easily obtained from the GIS data or the usual topographical maps, and the mean annual maximum flow discharge at least one position in the basin. The property reduces the difficulty of the data collection in the preceding movable bed analysis.

(3) This model could evaluate the channel properties that is the sediment discharge, flow depth, width of channel and mean diameters of bed materials at any point in the basin.

(4) The channel properties, except the flow depth, that could evaluate from the model checked by the field data in the Kanayama Dam basin. The width of channels and the mean diameter of bed material were compared directly with field data, and the agreements were fairly well respectively.

(5) The sediment discharge was checked indirectly by the transport rate of sediment in each class of rocks in the basin. The aerial rate of grand capping does not adapt to the field data of volumetric rate of deposited sediment in the dam. The estimated rate of sediment transport from the model adapts to the field data well.

ACKNOWLEDGMENT

The river channel Networks used in the present analysis were from the "Ishikari River Area Landscape Intelligences" from the Hokkaido River Disaster Prevention Research Center. The author thanks to the permission of the use of IRALI data.

REFERENCES

Kuroki, M., T. Itakura and T. Yamamoto: "Stable longitudinal profile of river channel", Journal of Hydroscience and Hydraulic Engineering, J. S. C. E. 14(1) : 37-45 ,1996.

Kuroki, M. and T. Kishi: "Regime criteria on bars and braids in alluvial straight channels", Proc. of JSCE, No.342, pp.87-96, 1984.

Ikeda, S., G.Parker, M. Chiyoda and Y. Kimura: "Stable channel cross-section of straight gravel rivers with actively transported bed materials", Proc. of JSCE, No.375/ II -6, pp.117-126, 1986.

Stochastic Hydraulics 2000, Wang & Hu (eds) © 2000 Taylor & Francis, ISBN 90 5809 166 X

Study on the relationship between runoff and sediment yield in rainfall erosion processes in the watershed on Loess Plateau

Zhanbin Li, Wenfeng Ding & Lingzhou Cui
Institute of Soil and Water Conservation, Chinese Academy of Sciences and Ministry of Water Resources, Yangling, China

Bing Shen
Xi' an University of Technology, China

ABSTRACT: The data of runoff and sediment yield in the watersheds on the Loess Plateau are used to analyze the characteristics of sediment yield in the single rainfall erosion event. According to mechanics of rainfall sediment yield processes in the watershed, the relationships between runoff discharge and sediment discharge of watershed have been built. Rainfall even-distribution coefficient and the antecedent affecting precipitation are taken as the principal influential factors are introduced to build the transference relationship from runoff to sediment concentration in the process of rainfall erosion sediment yield. The relationships can be used to calculate and predict the sediment yield in single rainfall event, it can also be used in the studies on soil and water loss control in watersheds.

Key words: Sediment yield by storm, Spatial distribution of rainfall , Antecedent precipitation Transference relationship from runoff to sediment

1. INTRODUCTION

The Loess Plateau, with the erosive area of 430000 km^2, is the most severe soil erosion area throughout the world. The Yellow River carries 1.6 billion tons of sediment loads is famous in the world due to the soil erosion on the Loess Plateau, which rushes the first in the water. The heavy soil erosion and sediment yield is caused by 2-3 storm events in a year. On the average, the sediment yield, which caused by storm events, accounts for about 95% in annual sediment yield in a watershed. Even more, the sediment yield, which caused by single storm event, account for about 80% in annual sediment yield. Because of the differences in area of the watersheds, soil characteristics, rainfall characteristics, canopy ratio, landforms and land pattern etc., there are different relationships between flow discharge and sediment concentration in outlet of the watersheds. In fact, the relationship between runoff and sediment in a watershed really reflect the characteristics of water erosion and sediment yield in the watershed. Therefore the relationship is useful for predicting the sediment yield in single storm event in the watershed.

2. METHODOLOGY

The hydrological data in Kuyehe watershed on Loess Plateau are used to study the relationship between runoff and sediment yield. According to the law of sediment movement, the sediment discharge is mainly related to runoff discharge, i.e.

$$Q_S = KQ^n \tag{1}$$

Where Q_S is the sediment discharge (kg/m^3); K and n is the parameter.

Generally, in the rising stage of a flood, the sediment concentration increases with the increasing of flow discharge. In the receding stage, the sediment concentration varies in a widely at the same flow discharge and decreases with the decreasing of flow discharge. No matter how much the gully area and difference in underlying characteristic is on the watershed, there exist a critical flow discharge in the relationship between the sediment discharge and flow discharge. The sediment concentration corresponding with the critical flow discharge is called critical sediment concentration. In the high sediment concentration condition, the characteristics can be divided into two flow regions by the critical sediment concentration. One is homogeneous flow region in high sediment concentration. Another is the non-homogeneous.

When the flow discharge is large than critical value, the flow is in the homogeneous flow region

$$Q_S = KQ^n, n \geq 1 \tag{2}$$

When the flow discharge is less than critical value, the flow is in the non-homogeneous flow region

$$Q_S = KQ^n, n \geq 1 \tag{3}$$

According to the previous research results (Cao et al, 1994), different watershed area has different relationships of Q~Q_s, table 1 shows the situations.

In the non-homogenous flow region, there are various $Q_S = KQ^n$ for different watershed. i. e. the parameter n , k is different in the formulae.

Table 1: The relationships in the watersheds

Name	Area $A(km^2)$	Q_k (m^3/s)	Q_s (kg/s)	S_k (kg/m^3)	Q_k/A	K	n
Tuanyuangou	0.49	0.065	30	460	0.1320	2920.0	1.74
Sanchuankou	21.00	1.400	700	500	0.067	418.0	1.88
Caoping	187.00	5.000	2700	560	0.0270	23.0	3.00
Dalihe	389.00	13.000	7500	577	0.0330	0.4	4.10

In the homogenous flow region, all the watersheds has same $Q_S=KQ^n$, Where k=550, n=1.06. The total stream power E_c of flow is large or equal to friction resistance power E_s in the interface of water and soil， i.e. $E_c \geq E_s$

$$E_S = \frac{A}{h} \tau_\beta V \tag{4}$$

$$E_C = \gamma_m AVJ \tag{5}$$

Then

$$\gamma_m hJ = \beta \tau_\beta \tag{6}$$

In which, $\beta \geq 1.0$, J is slope gradient, τ_β is the limit shear stress, γ_m is the capacity weight of muddy water, V is the average flow discharge, h is the flow depth and A is the area of water section.

$$\tau_\beta = 0.615 \times 10^4 \frac{S^5}{d_{50}^3} \tag{7}$$

In high τ_β range,
In which, S is the sediment concentration.
Substituting the τ_β in Eq.(6) with Eq. (7)

$$S = K_p (\gamma_m hJ)^{1/5} d_{50}^{3/5} \tag{8}$$

Where k_p is constant, k_p=5.725.
For the sallow flow,

$$h = CQ^m \tag{9}$$

280

Generally, m=0.3~0.4. We adopt m=0.3. Substitute m with 0.3 into equation (9) and both sides of the equation (9) are multiplied Q, we obtain that

$$Q_S = SQ = K'_p (\gamma_m J)^{1/5} d_{50}^{3/5} Q^{1\,06} \tag{10}$$

Where $K'_p = K_p C^{1/5}$

When J=1~3%, d_{50}=0.04~0.055, we can obtain

$$Q_S = KQ^{1\,06} \tag{11}$$

A large number of observed data and research results indicated that sediment yield processes by rainfall are related to the rainfall processes, temporal and spatial distribution of rainfall, antecedent precipitation, antecedent accumulation of dispersed materials, the topography, canopy on the ground surface, and the mankind activities etc. We use the P_a as the antecedent precipitation

$$P_{a,t+1} = K(P_{a,t} + P_t) \tag{12}$$

where $P_{a,t}$, $P_{a,t+1}$ is respectively the antecedent precipitation in time interval t and t+1 (mm); K is dissipation coefficient of soil moisture in watershed.

Because of extreme uneven spatial distribution of rainfall in the watershed, the sediment discharge processes are frequently different from the event to event with the similar storm conditions in the watersheds. Sometimes the small flow discharge is with high sediment concentration and sometimes the large flow discharge is with low sediment concentration. In order to quantitatively probe into the effect of spatial variation of rainfall upon the sediment yield in the watershed, the coefficient of the spatial distribution uniformity of rainfall, α, is introduced. Based on the precipitation in each gauge station in a single rainfall event and the critical precipitation p_c (postulation or trial method), the area weight of precipitation over p_c in watershed is obtained. The spatial even coefficient of this rainfall event, α, is equal the area weight. For series rainfall events, we can obtain α series in same p_c

3. RESULTS AND DISCUSSION

3.1 The relationship between flow discharge and sediment concentration

In the processes of the rainfall erosion, the erosion and sediment transportation by runoff play a very important role in sediment yield in watershed. Therefore, the discharge hydrograph in the outlet of watershed is the decisive factor to analyze the rainfall erosion processes.

By analyzing the rainfall runoff processes in Wangdaohenta station of Kuyehe watershed, it can be seen that the sediment concentration is proportional to the flow discharge, and the relationship in rising stage of discharge hydrograph is evidently different from that in receding stage. We selected 12 rainfall events with the single flood peak in Wangdaohenta station (1977~1989), and five points in rising and receding stage in hydrograph respectively to analyze the relationships. The results both in rising and receding stage are respectively shown as follows:

$$S = 15.00 Q^{0\,380} \qquad (R=0.9532) \tag{13}$$

$$S = 14.42 Q^{0\,400} \qquad (R=0.9515) \tag{14}$$

Where S is the sediment concentration (kg/m³), Q is the runoff discharge (m³/s).

From the Equations (13) and (14), it can be seen that the sediment concentration is proportional to runoff discharge. With the same discharge, the sediment concentration in receding stage is higher than in increasing stage. The calculation results of sediment concentration and sediment discharge of 12 rainfall events (1977~1989) were listed in table 2.

Table.2: Calculation results of sediment peak and sediment yield by Eq.(13)~(14)

flood No.	observed sediment (10^4T)	calculated sediment (10^4T)	relative error (%)	observed peak sediment concentration (kg/m^3)	calculated peak sediment concentration (kg/m^3)	relative error (%)
770722	11	5	45.4	574	219	61.8
770802	608	1059	74.2	337	643	90.8
770820	210	387	84.3	339	645	90.2
780831	4537	4268	5.9	1290	4468	9.4
790721	237	153	35.4	860	3198	62.9
790807	978	1246	27.4	621	772	24.3
800705	30	14	53.3	494	169	65.7
810721	86	73	15.1	349	306	12.2
820708	316	197	37.6	746	481	35.5
820804	445	298	33.1	1100	557	49.3
840730	2141	1578	26.3	1170	807	31.0
890721	2914	2168	25.6	1360	969	28.7

3.2 The relationship between sediment concentration and antecedent soil moisture

The antecedent soil moisture in the watershed directly affects the wind erosion, runoff erosion and the sediment transportation by water there. Meanwhile, the selling-shrinking property of loess is closely related to soil moisture. In winter, the soil moisture is a primary influencing factor of frozen-thaw erosion. The relationships between the sediment concentration and the runoff discharge, the antecedent precipitation in rising stage and in receding stage are respectively given as follows:

$$S = 17.46Q^{0.475}P_a^{-0.656} \qquad (R=0.9635) \tag{15}$$

$$S = 22.13Q^{0.467}P_a^{-0.127} \qquad (R=0.9424) \tag{16}$$

where P_a is the antecedent precipitation in the watershed (mm), which can be calculated by hydrological method .

From Equations (15) and (16), it can be seen that the sediment concentration in outflow is inversely proportional to P_a, i.e. with the same runoff discharge, the less the P_a is, the higher the sediment concentration is.

3.3 The effects of spatial distribution of rainfall in the watershed

We use the Thiessen polygon method to divide the area weight of each rainfall gauge in the watershed above the Wangdaohenta station. The different P_c levels are used to determine the α values. Then we get series α values. For same P_c level, the value of α in each rainfall event and corresponding Q, S values are used to analyze the function S = f (Q,α). The relationships both in rising and receding stage are respectively established as follows:

$$S = 8.91Q^{0.519}\alpha^{-0.528} \tag{17}$$

Where $P_c \geq 40$mm, multiple correlated coefficient R =0.9660

$$S = 7.08 Q^{0.515} \alpha^{-0.519} \tag{18}$$

Where $P_c >= 40$mm, multiple correlated coefficient R=0.9606

3.4 The comprehensive effects of Q、 P_a and α factors on S

From the analysis above, it can be seen that three factors of Q、 P_a、 α affect the processes of sediment yield in the watershed, and they influence mutually. For revealing the comprehensive effects of these factors on the sediment yield in the watershed, the comprehensive method of analysis and calculation have been used, and a synthetic equation has been adopted as follows:
In rising stage:

$$S = 12.88 Q^{0.514} \alpha^{-0.417} P_a^{-0.169} \tag{19}$$

Where $P_c >= 40$mm, R=0.9660
In receding stage:

$$S = 14.45 Q^{0.499} \alpha^{-0.648} P_a^{-0.299} \tag{20}$$

Where $P_c >= 40$mm, multiple correlated coefficient R=0.9649

The observed and calculation results of sediment yield and peak sediment concentrations are shown in table 3.

Table 3: Calculation results of sediment peak and sediment yield by S=f (Q, p$_a$,α)

flood No.	observed sediment (10^4T)	calculated sediment (10^4T)	relative error (%)	observed peak sediment discharge (kg/m^3)	calculated peak sediment discharge (kg/m^3)	relative error (%)
780516	413	357	13.6	571	663	16.1
780807	541	478	11.6	403	519	28.7
790810	1207	917	24.1	385	325	15.6
810701	547	419	23.4	1330	985	25.9
810725	86	97	12.8	370	411	11.2
840510	214	271	26.7	1380	1765	27.9
850708	799	438	36.1	1390	522	62.5
850805	6224	2701	56.6	1330	519	61.0
870701	85	136	60.1	396	651	64.4
870823	61	76	24.6	324	408	25.8
880712	1133	957	15.5	1630	1161	28.7

From Equations (19) and (20), it can be seen that in the comprehensive transference relationship from runoff to sediment concentration, the sediment concentration S is approximately proportional to $Q^{0.5}$ and inversely proportional to p_a and α. This result is the same as the previous one which shows that there exist the effect of p_a、 α on the processes of rainfall erosion sediment yield in the watershed.

3.5 Comparison of the transference relationships from runoff to sediment concentration

Since the transference relationships from runoff to sediment concentration are empirical equations, they can be used to predict the sediment yield by rainfall erosion. However, these equations must be calibrated with the observed data of sediment yield by storm erosion in the watershed. Taking

Table 4: The eligibility ratios of calculation sediment yield by various Eqs.

Equation form	Sediment amount (simulation)	Sediment peak (simulation)	Sediment amount (forecast)	Sediment peak (forecast)
$S=f(Q)$	41.7%	33.3%		
$S=f(Q,P_{a,})$	66.7%	58.3%		
$S=f(Q,\alpha)$	75.0%	58.3%	54.6%	54.6%
$S=f(Q,P_a,\alpha)$	83.3%	66.7%	72.7%	72.7%

the relative error of less than 30% as the standard, the calibration results are shown in table 4.

From table 4, it can be seen that the simulated results of sediment peak or the amount of sediment by equation $S = f(Q)$ are evidently inferior to the results by other equations, which indicate that the effects of p_a, α factor on the transference relationships from runoff to sediment concentration are evidently. The precision of simulating and predicting by $S = f(Q, P_a, \alpha)$ is evidently higher than the calculated results by other equations. It indicate that the transference relationship from runoff to sediment concentration with the p_a and α factors considered is superior to the others without or with one of p_a, α factor considered.

4. CONCLUSIONS

1. Not only the sediment discharge related to runoff discharge, but also the antecedent precipitation in Kuyehe watershed. The less the p_a is, the higher the sediment concentration is in the outflow.
2. The effect of spatial distribution of storm upon sediment yield in the watershed is evident. The quantitative relationships, which include the parameter α of rainfall even coefficient, show that the bigger the α value is, the lower the sediment concentration is in outflow.
3. The relationship between sediment concentration flow discharge, and antecedent precipitation, rainfall even coefficient is useful for sediment yield prediction in watershed.

ACKNOWLEDGEMENT

The work reported here was funded by the (Research Granted by China National Science Found And Ministry of Water Resources No.59890200) and State Key Lab of Soil Erosion and Dryland Farming on the Loess Plateau of China and this financial support is gratefully acknowledged.

REFERENCE

Fan Rongsheng, Li Zhanbin, 1994. Characteristic analysis of storm flood sediment in watershed of Kuyehe river. J. of Sediment Research, Vol.3.

Tang Keli et al, 1993. The changes of erosion and runoff, sediment in the Watershed of the Yellow River. Beijing, China Science and Technology Press.

Chen Yongzong, 1988. The proceeding of sediment source and the mechanism of erosion sediment yield in the Yellow River. Beijing, Weather Press.

Cao Ruxuan, Qian Shanqi, 1994, Behavior of Sediment Transportation in the Gullies of the Gullied-hilly Loess Areas, ACTA Conservationis Soli et Aquae Sinica Vol.2 No.4

Stochastic Hydraulics 2000, Wang & Hu (eds)© 2000 Taylor & Francis, ISBN 90 5809 166 X

Scour simulation with stochastic sediment transport model

Ali.A.Salehi Neyshabouri
Tarbait Modarres University, Tehran, Iran

ABSTRACT: In this paper the effects of using stochastic sediment transport model in numerical simulation of scour process is shown. This approach is useful where the impingement of flow to the bed occurs and theoretically the mean shear stress at the bed is zero. In such a case deterministic formula for sediment transport yields zero sediment transport and therefore no scouring, which is in contradiction to physical observations. Using stochastic formula which uses distribution of both the flow shear stress and critical shear stress for bed material results in an improvement in prediction of scour process in such a zone.

Keywords: Stochastic model, Scour, Turbulence.jet, Sediment transport

1. INTRODUCTION

In recent years, numerical simulation of scour process has been studied. Usually, governing hydrodynamic equations (continuity , momentum and transport equations for turbulence parameters) , along with sediment transport equation and continuity equation for sediment at the bed are solved numerically.

Ushijima et al (1992) considered scouring by warmed jet in 2-D and 3-D cases. Using finite difference method, they solved unsteady continuity and momentum equations along with the k-ε turbulence closure model. To solve the continuity equation for sediment at the bed, total bed load consisting of suspended and bed load were computed. Convection–diffusion equation for suspended load concentration was solved numerically and empirical relation proposed by Ashida (1978) was used to compute bed load rate. Sediment concentration at the bed was calculated based on the Lane and Kalinske (1941) relation. The results were satisfactory, however there were some problems in the form of undulations in the bed profiles which were not observed in the experimental results. To solve these problems, they did another work more recently (Ushijima et al, 1996). In this work they used an Arbitrary Lagrangian-Eulerian approach in which three-dimensional body-fitted coordinates are properly generated for the sand bed profile unsteadily deformed by the flow.

Hoffman and Booij (1993) presented a model for the flow in a trench based on the solution of 2-D Reynolds equation and convection-diffusion equation. Bed load and suspended loads are computed by the stochastic method of Van Rijn (1986). The results were satisfactory, even it was necessary to calibrate some parameters in the sediment transport formula for different flow zones.

Olsen (1993) simulated the scour process around a cylinder. Sediment concentration for the bed elements was calculated with Van Rijn's deterministic formula. Based on continuity for the bed elements, erosion and deposition were calculated. Comparing the numerical results with experimental data showed fairly good agreement.

In the present paper it is shown that using deterministic sediment transport relation leads to some problem in the vicinity of impinging point of a free falling jet. To solve this problem it was found appropriate to use a stochastic approach for sediment concentration calculation near the bed.

2. GOVERNING EQUATIONS

The main equations governing fluid flow are the Reynolds equations, i.e.:

$$U_j \partial U_i / \partial x_j = -1/\rho \partial (P\delta_{ij})/\partial x_j - \partial(\overline{u_i u_j})/\partial x_j \tag{1}$$

where U_i is time averaged velocity, P is dynamic pressure and $-\rho \overline{u_i u_j}$ is turbulent shear stress. The turbulent stresses are usually modeled by Bousinesque relation (Rodi, 1993) as follows:

$$-\rho \overline{u_i u_j} = \rho v_T (\partial U_i / \partial x_j + \partial U_j / \partial x_i) - 2/3\delta_{ij} \tag{2}$$

The turbulent eddy viscosity (v_T) is modeled with k-ε turbulence model. In this model the eddy-viscosity is obtained as follows:

$$v_T = c_\mu k^2 / \varepsilon \tag{3}$$

in which k is the turbulent kinetic energy and ε is the dissipation rate of k. The modeled equation for k and ε are :

$$U_j \partial \varepsilon / \partial x_j = \partial(v_T / \sigma_\varepsilon \partial \varepsilon / \partial x_j)/\partial x_j + C_{\varepsilon 1} \varepsilon / k P_k - C_{\varepsilon 2} \varepsilon^2 / k \tag{4}$$

$$U_j \partial k / \partial x_j = \partial(v_T / \sigma_k \partial k / \partial x_j)/\partial x_j + P_k - \varepsilon \tag{5}$$

where P_k is production of k and is defined as :

$$P_k = v_T \frac{\partial U_i}{\partial x_j}(\frac{\partial U_j}{\partial x_i} + \frac{\partial U_i}{\partial x_j}) \tag{6}$$

The partial differential equation for the sediment flow can be written as a convection-diffusion equation. For non-transient flow this becomes:

$$U_j \frac{\partial c}{\partial x_j} + w_s(\frac{\partial c}{\partial x_z}) = \frac{\partial}{\partial x_j}(\Gamma \frac{\partial c}{\partial x_j}) \tag{7}$$

here C is the concentration, U_j is velocity, w_s is the fall velocity of the sediment particles, and Γ is the diffusion coefficient which is proportional to the eddy-viscosity.

3. BOUNDARY CONDITIONS

For outflow boundary, zero-gradient boundary condition was used and velocities were set equal to the values in the closest elements to the outflow. At the water surface, zero-gradient conditions were used for velocity components parallel to the free surface and ε, while k, sediment concentration and velocity component perpendicular to free surface were set to zero.

For the bed, it was assumed that the center of the wall element was within the log-layer of the wall and therefore the logarithmic distribution for the velocity was used as follows:

$$\frac{U}{u_*} = 1/\kappa \ln \frac{Eyu_*}{v} \tag{8}$$

in the above equation, E=9.0 was used.

At the wall, based on the assumption of a balance between production and dissipation of k near the wall, ε for the closest elements to the wall is calculated from the following equation:

$$\varepsilon = \frac{C_\mu^{3/4}}{\kappa \delta_{np}} k_p^{3/2} \tag{9}$$

in which δ_{np} is the distance from the wall to the center of the bed element.

The sediment concentration near the bed is a very important boundary condition for the sediment calculation.. In the present study two different formulae for the bed concentration proposed by Van Rijn (1987), namely deterministic and stochastic formulae are used. In deterministic formula, C_a is calculated as follows:

286

$$C_a = 0.015 \frac{d_{50}}{a} \frac{T^{1.5}}{D_*^{0.3}}$$ (10)

in which:

$$T = \frac{\tau - \tau_c}{\tau_c}$$

$$D_* = d_{50} \left[\frac{(\rho_s - \rho_w)g}{v^2} \right]^{1/3}$$

where a is the distance from the bed to the center of the element closest to the bed.

Since in turbulent flow, actual shear stress oscillates around mean value, therefore to compute sediment transport based on the difference between flow shear stress and critical shear stress of the bed material, the stochastic behavior should be considered. To do this, the following stochastic formula was proposed by Van Rijn (1987) to calculate C_a

$$C_a = 0.03 \frac{d_{50}}{a} \frac{E_1(1.5) + E_2(1.5)}{D_*^{0.3}}$$ (11)

in which :

$$E_1(\gamma) = \int_0^\infty \left[T(\tau_0') \right]^\gamma P(\tau_0') d\tau_0'$$

$$E_2(\gamma) = \int_{-\infty}^0 \left[T(\tau_0') \right]^\gamma P(\tau_0') d\tau_0'$$

$$P(\tau_0') = \frac{1}{\sqrt{2\pi}\sigma_t} \exp\left[-\left[\frac{\tau_0' - \overline{\tau_0'}}{\sqrt{2}\sigma_t} \right] \right]$$

$$\overline{\tau_0'} = \mu\tau_0 = \mu\rho u_*^2 \qquad (\mu=1.0)$$

$$T(\tau_0') = \frac{-\tau_0' + \tau_2}{\tau_c} \qquad -\infty \langle \tau_0' \langle \tau_2$$

$$T(\tau_0') = 0 \qquad \tau_2 \langle \tau_0' \langle \tau_1$$

$$T(\tau_0') = \frac{\tau_0' - \tau_1}{\tau_c} \qquad \tau_2 \langle \tau_0' \langle \infty$$

here σ_t is standard deviation which should be determined based on the location and physical characteristics of the flow.

4.NUMERICAL SOLUTION

The partial differential equation for the flow and sediment concentration are integrated over the element using a finite volume technique. The resulting equation for any variable φ (velocity components , sediment concentration, k or ε) is written as (Patankar, 1980):

$$a_P \phi_P = \sum_{nb} a_{nb} \phi_{nb} + source$$ (12)

The notation nb indicates neighboring elements.

The difference schemes calculate the a_{nb} and a_P coefficients based on the convective and orthogonal diffusive fluxes in the equation. The non-orthogonal fluxes are discretized separately and put in the source term. Also the rest of the terms are calculated separately and are considered in the source term. Two different convective scheme namely, power-law scheme (POW) and second-order upwind scheme (SOU) are used. For velocity-pressure coupling, SIMPLE algorithm was followed. Since non-staggered grid was used in this study, the Rhie and Chow (1983) interpolation method was used to avoid instabilities in the calculation of the velocities and pressure.

5.APPLICATION OF THE MODEL

To check the effects of the sediment transport formula on the scour prediction, the above mentioned model was applied to the case of a free falling inclined jet. The details of the flow field considered is as follows: (fig. 1)

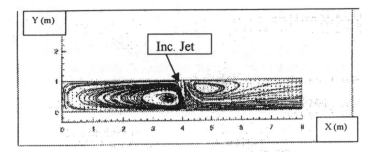

Fig 1: Flow field of an inclined free falling jet

Original flow field dimensions : depth = 1 m, length = 8 m
Inlet velocity in x direction (u_{in}): 0.5 m/s
Inlet velocity in y direction (v_{in}): 2.0 m/s
Width of the incoming jet : 0.2 m
Sediment density (ρ_s): 2650 kg/m^3
Sediment diameter : 2 mm
Sediment porosity : 40%
Grid no. : 12*62

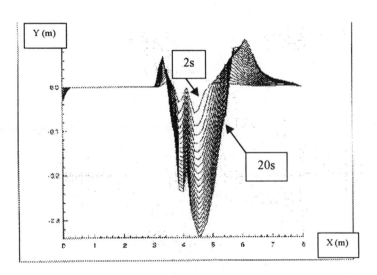

Fig. 2: Scour hole development during first 20 s.(using deterministic sediment transport formula)

The result of application of the present model using deterministic sediment transport formula of Van Rijn (Eq. 10), is shown in fig. 2 for duration of 20 sec. from the beginning of the scouring process. It can be seen that around impingement point of the jet, there is a pronounced less eroded area which is unreasonable. The reason for this is mainly due to the fact that

288

deterministic approach relates the sediment transport to the mean shear stress at the bed which at the impingement point is zero.

To solve the above mentioned problem, the stochastic formula of Van Rijn (Eq. 11) was used in the current application. The result for time= 20 sec. after beginning of scouring process is shown in fig. 3. It can be seen that a smooth scour hole along the channel bed, even near the impingement point is predicted. This example shows the benefit of using stochastic approach wherever, there is similar situation, such as scouring process caused by an offset jet.

6. CONCLUSION

It was shown in this paper that using deterministic approach for sediment transport causes the scouring prediction by numerical models to be in error around impingement zone where mean shear stress is zero. Using stochastic approach, on the other hand overcomes this difficulty and reasonable results could be achieved.

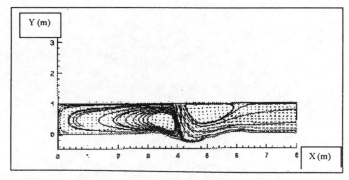

Fig.3: Flow field after 20s. from the beginning of scouring process

REFERENCES

K. Ashida, T. Kashashi, and T. Mizuyama, J. Japan Soc. Erosion Control Engng., Vol. 107, 1978.
Hoffmans, G.J.C.M., and R. Booij, "Two dimensional mathematical modeling of local scour holes", J. of Hyd. Res. I.H.A.R., Vol. 31, No. 5, 1993.
E. W. Lane and A. A. Kalinske, Trans. Am. Geophys., Union 22, 603, 1941.
Olsen, N. R. B. and Melaaen, M.C., "Three-dimensional calculation of scour around a cylinder", ASCE J. of Hydraulic Engng., Vol. 119, No. 9, Sept. 1993.
Patankar, S.V., "Numerical heat transfer and fluid flow", McGraw-Hill Book Company, New York, 1980.
Rhie, C.M., and Chow, W.L., "Numerical study of the turbulent flow past an airfoil with trailing edge separation", AIAA Journal, Vol. 21, No. 11, 1983.
Rodi, W., "Turbulence models and their application in hydraulics", IAHR, Delft,
Netherlands, 1993.
Van Rijn, L. C., "Mathematical modeling of morphological processes in the case of suspended sediment transport", Ph.D. Thesis, Delft University of Technology,1987.
Ushijima B., Shimitzu, T., Sakasi, A., Takizawa, Y., "Prediction method for local scour by warmed cooling water jets", J. Hyd. Engng., A.S.C.E., Vol. 118, No. 8,1992.

Stochastic Hydraulics 2000, Wang & Hu (eds) © 2000 Taylor & Francis, ISBN 90 5809 166 X

Grain size distribution – A probabilistic model for Usri River sediment in India

Barendra Purkait
Map and Cartography Division, Geological Survey of India, Calcutta, India

B.S. Mazumder
Flume Laboratory, Indian Statistical Institute, Calcutta, India

ABSTRACT: The present paper deals with the grain size distribution pattern of a modern river Usri, India. Sand samples were collected from the foresets of crossbeddings of specific size ranging in thickness from 23 to 40cm, developed on the point bars. Four point bars were chosen almost at a regular interval from the source to the mouth of the river Usri. The Usri is an ephemeral river of about 90 km in length and carries mainly the pebbly sands during monsoon discharge period. Grain size distribution pattern within point bar and between point bars were studied in the light of log-normal, log-hyperbolic and log-skew-Laplace distributions. Thus the best fit statistical model was developed to describe the longitudinal change of grain sizes vis-a-vis the sorting pattern with distance of transport.

Key words: River, Point bar, Grain-size distribution, Lognormal, Loghyperbolic.

1. INTRODUCTION

The patterns of size distribution have been studied by many investigators in different sedimentary environments. Bagnold and Barndorff-Nielsen (1980), Ghosh and Mazumder (1981), Barndorff-Nielsen *et al.*(1982), Christiansen *et al.* (1984), Wyrwoll and Smith (1985), Christiansen and Hartmann (1991) and others provide a coherent statistical approach on grain size distribution and suggested the loghyperbolic distribution of grain size data from a variety of sedimentary enviroments. Fieller and Flenley (1992) investigated the use of loghyperbolic and log-skew-Laplace models to distinguish different depositional environments and suggested the latter model to provide a simple and useful method for the classification of sand sediments. Kothyari (1995) analysed the data on bed material size distributions and found that the method of power transformation would normalize these mixtures. Sengupta *et al.* (1999) observed that lognormal grain size distribution in suspension once generated in a close-circuit flume at an appropriate combination of height and flow velocity is sustained till the flow condition is drastically altered. Lanzoni and Tubino (1999) noted in an experimental flume that the different mobility of individual grain size fractions within the mixture not only modifies the sediment transport capacity but also induces a longitudinal and transverse pattern of sorting.

The purpose of the present paper is to study the grain size distribution of the Usri river sediments in the light of lognormal, loghyperbolic and log-skew-Laplace distributions. Grain-size distribution patterns were critically examined within point bar and between point bars. The result of such studies might be helpful in palaeohydraulic interpretation of ancient fluvial deposits.

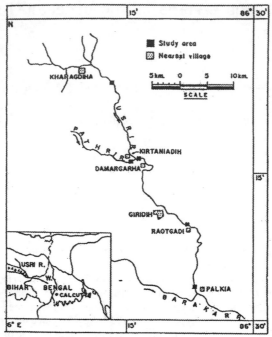

Fig1: Location map

2. STUDY AREA

The Usri is an ephemeral river of about 90 km in length situated in the state of Bihar, India (Fig.1). Several small streamlets at the source areas, in and around the villages Charghara, Kharagdiha etc. originated by headwar erosion during the monsoon season (June-August) feed the river Usri. The source area exhibits bad land topography with table like highs of highly erodible pebbly and sandy soils. The sediments at the source of the Usri are mainly pebbly sands derived from those easily erodible bad land topographic highs. The river almost maintains a uniform gradient of 1 : 66 except in the lower about 8.5 km reaches, i,e. near the confluence with the river Barakar where the gradient abruptly changes to 1 : 15 (between Raotgadi and Palkia villlages). During the winter and summer months (October-March), the river dries up almost completely exposing the channel and point bars. The maximum monsoon discharge was 36.62 m^3/sec.

3. DATA COLLECTION

Four point bars were chosen almost at a regular interval from the source (Kharagdiha) to the mouth (Palkia) of the river Usri where it meets the river Barakar. One sample (2/86) was collected from the stream bed of a feeder streamlet of the Usri about 7 km upstream of Kharagdiha near the village Charghara. Three samples from each of the point bars were collected from the upstream, intermediate and downstream parts of the bars. Samples were collected all along the foreset thickness of crossbeddings ranging in thickness from 23 to 40 cm. Mechanical analysis was done using A. S. T. M. sieves of 1/2 ϕ interval. For hyperbolic and Laplace distributions, the weight frequency (%) were converted to log scale as grain diameter was measured in ϕ scale ($\phi = -log_2 d$, where d is the grain diameter in mm).

Table 1: Estimated parameters of model distributions

Loghyperbolic				Log-skew-Laplace			Sample No.	Location
γ_1	γ_2	μ	δ	α	β	μ		
1.912	4.693	1.823	0.984	0.997	0.54	1.57	U-87	Kharagdiha (IM)
3.497	3.355	1.995	0.787	1.13	0.15	2.98	U-83	Kharagdiha (DS)
1.284	5.029	2.415	0.935	1.366	0.18	2.87	U-3	Damargarha (US)
2.762	3.948	1.947	2.806				U-8	Damargarha (IM)
2.652	3.612	2.099	2.856	0.905	0.423	2.46	U-12	Damargarha (DS)
1.599	2.611	1.341	1.712	2.476	0.136	3.499	U-35	Raotgadi (US)
4.231	3.06	1.482	0.276				U-31	Raotgadi (IM)
4.957	4.694	2.561	4.585	0.467	0.489	2.63	U-25	Raotgadi (DS)
2.165	2.624	1.068	2.897	1.049	0.602	1.36	U-70	Palkia (US)
1.712	2.502	0.801	0.901				U-72	Palkia(IM)
3.038	2.517	1.211	4.044	0.643	0.572	1.37	U-80	Palkia (DS)

Note: US = Upstream part of bar, DS = Downstream part of bar, IM = Intermediate part of bar

4. STATISTICAL MODELLING

The grain size frequency distributions provide different distributions like normal (Gaussian), hyperbolic and skew-Laplace. The density function for normal distribution is

$$g(x; \mu, \sigma) = \frac{1}{\sigma\sqrt{2\pi}} \exp\left\{ -\frac{1}{2}\left(\frac{x-\mu}{\sigma}\right)^2 \right\} \tag{1}$$

where $\sigma > 0$ represents standard deviation, and μ = mean size of the distribution.

The loghyperbolic distribution was introduced by Barndorff-Nielsen (1977) to describe the mass-size distribution of eolian sand. It is defined by its log probability function being a hyperbola whereas for normal distribution it is a parabola. Four parameters are utilized to describe the loghyperbolic distribution. These are γ_1, γ_2, μ and δ, where γ_1 and γ_2 represent the slopes of the left (coarser grade) and right (finer grade) asymptotes respectively, μ is the abscissa of the intersecting point of these two asymptotes, the ordinate v equals the logarithm of the norming constant, which is adjusted to agree with the observed frequency, and the scale parameter δ (>0) can be expressed as

$$\delta = \frac{2(\mu^* - \mu)\sqrt{\gamma_1\gamma_2}}{\gamma_1 - \gamma_2} \tag{2}$$

where μ^* is the observed mode of the distributions. In a symmetric distribution where $\gamma_1 = \gamma_2$, μ^* also equals μ. In such a situation the distribution may be described as lognormal.

Following Barndorff-Nielsen and denoting the size variable by ϕ and density function $S(\phi)$, the equation of the loghyperbolic distribution is of the form (Barndorff-Nielsen, 1977)

$$Log_e S(\phi) = v - \tfrac{1}{2}(\gamma_1 + \gamma_2)\sqrt{\delta^2 + (\phi - \mu)^2} + \tfrac{1}{2}(\gamma_1 - \gamma_2)(\phi - \mu) \tag{3}$$

However, computational difficulties in fitting this distribution make the use of a simplified version, the log-skew-Laplace distribution (Fieller et al., 1984). Log-skew-Laplace distribution can be written as

$$g(x; \alpha, \beta, \mu) = \begin{cases} (\alpha + \beta)^{-1} \exp\{(x - \mu)/\alpha\} & for\ x \le \mu \\ (\alpha + \beta)^{-1} \exp\{(\mu - x)/\beta\} & for\ x > \mu \end{cases} \tag{4}$$

where x indicates observed variable and α, β, μ are parameters to be estimated, in which α = slope of left asymptote (coarser fractions), β = slope of right asymptote (finer fractions) and μ = point of intersection of two asymptotes.

Distribution pattern was critically examined within and between point bars in the light of lognormal, loghyperbolic and log-skew-Laplace distributions. Initially, however, at the source (Charghara), the grain size distribution shows a polymodal distribution. Estimated parameters of different distributions are shown in table 1.

5.SIZE DISTRIBUTION WITHIN AND BETWEEN POINT BARS

The following four point bars named as-Kharagdiha, Damargarha, Raotgadi and Palkia after the nearest village were studied.

i) Kharagdiha

After about 7 km of transport from Charghara, at the upstream part of the Kharagdiha point bar, bed load sample U-92 was collected from the cross bedding of thickness (T) 40 cm. The grain size shows a polymodal distribution having an irregular pattern in log-log plots (Fig.2). For sample U-83 (T = 30 cm) at the downstream part of this bar, the distribution tends to be symmetric and hyperbolic with $\gamma_1 = 3.497$ and $\gamma_2 = 3.355$. But the sample (U-87, T = 25cm) at the intermediate part of the bar between U-92 and U-83, shows a very asymmetric distribution with $\gamma_1 = 1.912$ and $\gamma_2 = 4.693$. The log-skew-Laplace distribution does not fit well. The log probability plots for both samples in respect of log hyperbolic and log Laplace distributions show more than one linear segments indicating lack of lognormality.

ii) Damargarha

The bar is about 25 km downstream of the Kharagdiha point bar. The upstream end sample (U-3, T=23 cm) of the bar shows a very asymmetric grain size distribution pattern having $\gamma_1 = 1.284$, $\alpha = 1.366$ and $\gamma_2 = 5.029$, $\beta = 0.18$ whereas towards downstream part for U-12 (T=40 cm), the distribution pattern tends to become more symmetric ($\gamma_1 = 2.652$, $\alpha = 0.905$ and $\gamma_2 = 3.612$, $\beta = 0.423$). In the intermetiate part of the bar for U-8, $\gamma_1 = 2.762$ and $\gamma_2 = 3.948$ (Fig. 2), log probability plots for loghyperbolic and log-skew-Laplace distributions are studied for comparison. For U-3, at the upstream part of the bar, loghyperbolic distribution fits better than the log-skew-Laplace distribution. However, at the downstream part, sample U-12 shows a linear trend indicating the generation of lognormality.

iii) Raotgadi

The bar is about 24 km downstream of the Damargaha point bar. Sample U-35 (T = 30 cm) from the upstream part of the bar shows asymmetric distributions ($\gamma_1 = 1.599$, $\alpha = 2.476$ and $\gamma_2 = 2.611$, $\beta = 0.136$),whereas while on transport further downstream it takes again the symmetric pattern. At the downstream end for U-25, the grain size distribution pattern shows almost a perfect symmetry $\gamma_1 = 4.957$, $\alpha = 0.467$ and $\gamma_2 = 4.694$, $\beta = 0.489$, Fig. 2) with graphic skewness $Sk_1 = 0.031$ and kurtosis $K_G = 0.918$ (Folk and Ward, 1957). It indicates that after about 56km of transport from Charghara, the perfect lognormality attains. This is also shown by a single linear segment in logprobability plots in fig.3. Here for U-35, $Sk_1 = 0.004$, $K_G = 1.003$. Log-skew-Laplace distribution does not show such a single linear segment.

iv) Palkia

The bar is about 16.5 km downstream of the Raotgadi point bar. Three samples (U-70, U-72 and U-80, Fig. 2) were collected from its upstream to downstream parts with crossbedding thickness ranging from 30 to 35 cm. Sample U-70 from the upstream end of the bar exhibits a symmetric hyperbolic distribution $\gamma_1 = 2.165$ and $\gamma_2 = 2.624$ having a linear trend in logprobability plots from grain size range of 0.5ϕ to finer ones but not for the all plots Log-Laplace distribution also does not fit well and shows the slope values $\alpha = 1.049$ and $\beta = 0.602$. At the downstream part for the sample U-80, lognormal distributions pattern is also disturbed as is revealed from the deviation of straight line fit in logprobability plot.

294

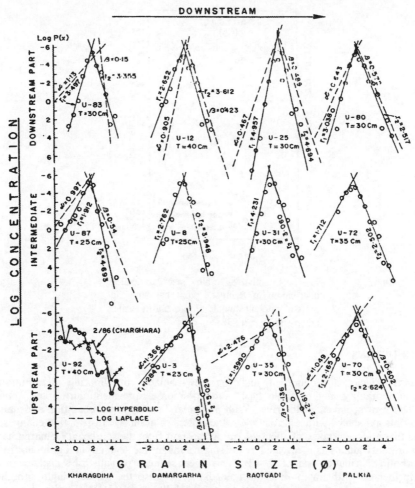

Fig2: Log-log plots of hyperbolic and Laplace distributions of the Usri River point bar samples (source to mouth)

6. BETWEEN POINT BARS

After examining the grain size distribution patterns of the point bars from the source to the mouth of the river Usri, it can be said that at the source area (Charghara), the grain size distribution is polymodal in nature. Bedforms are poorly developed and sediments are mainly pebbly sands. After a distance of about 7 km of transport, larger bedforms with crude crossbedding foresets develop as are observed within Kharagdiha point bar. The grain sorting continues within point bar and between point bars also. From Damargarha point bar after a distance of about 32 km from Charghara, lognormality attains as is shown by the single linear segment in logprobability plots. This lognormality prevails further downstream of about 24 km section of the river all along till this distribution reaches Raotgadi point bar at about 56 km from the source. At Raotgadi, the distribution becomes perfectly symmetric both in loghyperbolic and log-skew-Laplace distributions (For U-25, γ_1 = 4.957, α= 0.467, γ_2 = 4.694, β= 0.489, mean = 2.561ϕ, mode = 2.686ϕ). Further downstream, however, at Palkia point bar, the lognormality is disturbed and again the loghyperbolic distribution becomes asymmetric. This deviation from straight line fits may be attributed to the different hydrodynamic conditions of the river as well as the nature of the bed materials.

295

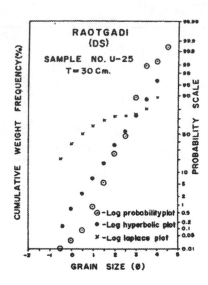

Fig3: Cumulative grain size frequency distribution of Raotgadi point bar showing plots of diferent models on probability scale, down-stream part (DS)

7.CONCLUSIONS

As already discussed, lognormal and log-skew-Laplace are limiting distributions of loghyperbolic family, not a single type of distribution is prevalent during the course of transportation of sediment in the river. With particular flow dynamic condition and nature of bed materials associated with bed roughness, the pattern of grain size distribution changes. It is observed that after a distance of about 24 km of transport from the source, at Damargarha point bar, lognormality attains as is revealed by the single straight line segment in logprobability plots for hyperbolic distribution (sample U-12). This linear segment prevails for a distance of about 24 km further downstream to Raotgadi point bar. Here also the perfect single straight line segment is reflected in loghyperbolic plots on logprobability scale (U-25). Further downstream, however, at Palkia point bar, the lognormal distribution becomes asymmetric. The deviation from lognormality may be attributed to the different hydrodynamic conditions of the river as well as the nature of the bed materials. Both the slope and discharge at Palkia area reduces as compared to its upstream area at Raotgadi bar.

ACKNOWLEDGMENTS

The authors are grateful to Prof. J. K.Ghosh of Indian Statistical Institute, Calcutta for his fruitful suggestions and encouragement; and to Prof. Supriya Sengupta for his valuable suggestions in the field. The first author is also thankful to the Director General, Geological Survey of India, for permitting him to work as a Visiting Scientist at I. S. I.

REFERENCES

Bagnold. R. A. & Barndorff-Nielsen, O. (1980) The pattern of natural size distributions.*Sedimentology*, **27**, 199-207.

Barndorff-Nielsen, O. (1977) Exponentially decreasing distributions for the logarithm of particle size. *Proceedings of the Royal Society of London*, **A 353**, 401-419.

Barndorff-Nielsen, O. & Blaesild, P. (1981) Hyperbolic distributions and ramifications:Contribution to theory and application. In : *Statistical Distributions in Scientific Work,* **V.4,**(Ed. C. Taillie, *et al.*) pp. 19-44, D. Reidel publishing comp., The Netherlands.

Barndorff-Nielsen, O., Dalsgaard, K., Halgreen, C., Kuhlman, H., Moller, J. T. & Schou, G.(1982) Variations in particle size distribution over a small dune. *Sedimentology,* **29,** 53-65.

Christiansen, C., Blaesild, F., and Dalsgaard, K. (1984) Re-interpreting "Segmented" grain size curves. *Geological Magazine,* **121,** 47-51.

Christiansen, C., & Hartmann, D. (1991) The hyperbolic distribution. In: *Principles, Methodsand Application of Particle Size Analysis,* (ed. J. P. M. Syvitski), Cambridge University Press.pp. 237-248.

Fieller, N. R. J., Gilbertson, D. D., & Olbricht, W. (1984) A new method for environmental analysis of particle size distribution data from shoreline sediments. *Nature,* **311,** 648-651.

Fieller, N. R. J. & Flenley, E. C. (1992) Statistics of particle size data. *Appl. Statistics*; **41,** 127-146.Folk, R. L. & Ward, W. C. (1957) Brazos river bar: a study in the significance of grain size parameters. *Jour. Sedim. Petrol,* **27,** 3-26.

Ghosh, J. K. & Mazumder, B. S. (1981) Size distribution of suspended particles - unimodality,symmetry and log-normality. In : *Statistical Distribution in Scientific Work,* **V.6** (Ed. by C. Taillie *et al.*), pp. 21-32. D. Reidel Publishing comp., The Netherlands.

Kothyari, U. C. (1995) Frequency distribution of river bed materials. *Sedimentology,* **42,** 283-291.

Lanzoni, S. & Tubino, M. (1999) Grain sorting and bar instability. *J. Fluid Mech.* **393,** 149-147.

Sengupta, S., Das, S. S., & Maji, A. K. (1999) Sediments transportation and sorting processes in streams. *INSA,* **65A,** No. 2, 167-206.

Wyrwoll, K. H. & Smyth, G. K. (1985) On using the log-hyperbolic distribution to describe the textural characteristics of eolian sediments. *Jour. Sedim. Petrol.,* **55,** 471-478.

Stochastic Hydraulics 2000, Wang & Hu (eds) © 2000 Taylor & Francis, ISBN 90 5809 166 X

Error analysis of bed and suspended load measurements in the Nile river

S. Abdel-Fattah
Hydraulics Research Institute, National Water Research Center, Ministry of Water Resources and Irrigation, Egypt

ABSTRACT : Error analysis was carried out for a set of measured sediment transport data in order to determine the accuracy of these measurements. These measurements were conducted on the Nile River in Egypt at a site located downstream the High Aswan Dam by 288 km. The cross section was divided into six stations. The analysis was performed for the bed and suspended load transport rates for each measured station and for the overall cross section as well.

The results showed good accuracy for the measured data. The overall relative error of the bed load transport was 29% for the cross section. The overall relative error of the suspended load transport was 9% for the cross section. These accuracy values are due to the variability of the transport processes along the bed form length.

Key words : Bed load suspended load, Measured sediment transport, Bed forms, Overall relative error.

1. INTRODUCTION

The location of the measuring cross section was selected in a straight and stable reach in order to avoid non-steady bed conditions during the measurements. Echo sounding for the cross section profile was performed. The total width of the cross-section was 578 m. The cross section profile was subdivided into six measuring stations, based on statistical error analysis for regular and irregular cross sections (Gaweesh and Van Rijn, 1994). Four longitudinal echo sounding profiles over 100 m length were conducted at 150, 200, 300 and 400 m from the left bank, respectively, to determine the mean bed form dimensions. The mean length and height of the bed form were found to be 22 m and 0.8 m, respectively.

The measurements of bed load, suspended load, and velocity profiles were conducted at the six measuring stations. At each station, measurements were performed at five locations (L1, L2, L3, L4 and L5) distributed along the length of the longitudinal section which is equal to the mean bed form length. Figure (1) shows the layout of the measuring stations and locations.

2. ERROR ANALYSIS

The accuracy of the cross section integrated transport depends on the relative error of the transport per station, width of each station across the river and the interpolation error. This interpolation error is related to the representation of a smooth curve by means of a series of line segments.

$$Q_b = \sum_{i=1}^{m} (q_{b,i} \Delta w_i) + \delta \tag{1}$$

where:

Q_b = cross section integrated bed load transport (kg/s);
$q_{b,i}$ = bed load transport at station i (kg/s/m);
Δw_i = width of station i (m);
m   = number of stations;
δ.   = interpolation error (δ -0 for m -∞).

The same equation is applied for the suspended load transport as follows:

$$Q_s = \sum_{i=1}^{m} (q_{s,i} \Delta w_i) + \delta \qquad (2)$$

where:

Q_s = cross section integrated suspended load transport (kg/s);
$q_{s,i}$ = suspended load transport at station i (kg/s/m).

The relative error of the transport rate per station, bed load or suspended load, can be computed from the variation of the transport rates per location.

Relative error = $(\Phi / (n-1)^{0.5}) / 0$ \qquad (3)

where:

Φ = standard deviation of the mean;
n   = number of locations at each station;
0   = mean value of transport rate at each station.

Tables (1) and (2) show the measured data and the computations of error analysis of the bed and suspended load transport rates, respectively. For example, in Table (1) and station 1, the mean bed load transport rate has a value of 0.0178 (kg/m/s) and the standard deviation is 0.0124. The relative error of the mean bed load transport is 0.34 or 34% which means that the measured mean bed load transport rate differs not more than ±34% from the actual mean bed load transport rate.

Because of the variability of the transport processes along the bed form length, the measurements were performed at five locations distributed along it. The bed load transport is maximum near the bed form crest and minimum near the bed form trough. These variations of the measured transports at the different locations are the reason for the big values of the relative error of the bed load transport for each station.

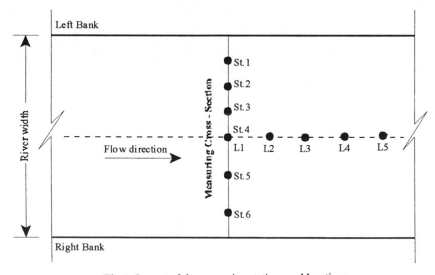

Fig.1: Layout of the measuring stations and locations.

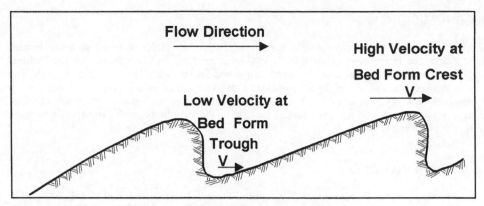

Fig.2: Sketch of bed forms in alluvial channels.

Table.1:

	Bed Load Transport Rate (kg/sm)					
	Station 1	Station 2	Station 3	Station 4	Station 5	Station 6
Location 1	0.0058	0.0162	0.0079	0.0025	0.0002	0.0003
Location 2	0.0106	0.0083	0.0041	0.0019	0.0000	0.0005
Location 3	0.0368	0.0112	0.0159	0.0014	0.0002	0.0019
Location 4	0.0281	0.0108	0.0037	0.0011	0.0002	0.0003
Location 5	0.0075	0.0136	0.0031	0.0006	0.0001	0.0002
Mean	0.0178	0.0120	0.0069	0.0015	0.0001	0.0006
St. Dev.	0.0124	0.0027	0.0048	0.0006	0.0006	0.0006
Rel. Err.	0.3440	0.1105	0.3452	0.2146	0.2683	0.4945

Table.2:

	Suspended Load Transport (kg/sm)					
	Station 1	Station 2	Station 3	Station 4	Station 5	Station 6
Location 1	0.0371	0.0366	0.0154	0.0047	0.0027	0.0035
Location 2	0.0403	0.0328	0.0110	0.0046	0.0030	0.0031
Location 3	0.0217	0.0390	0.0091	0.0050	0.0035	0.0031
Location 4	0.0287	0.0223	0.0067	0.0074	0.0028	0.0037
Location 5	0.0401	0.0347	0.0075	0.0063	0.0028	0.0037
Mean	0.0336	0.0331	0.0099	0.0056	0.0030	0.0034
St. Dev.	0.0073	0.0058	0.0031	0.0011	0.0004	0.0003
Rel. Err.	0.1084	0.0872	0.1560	0.0970	0.0485	0.0400

The relative errors of the six stations for the bed load transport rates were 0.34, 0.11, 0.34, 0.21, 0.26 and 0.49, respectively. The relative errors of the six stations for the suspended load transport rates were 0.11, 0.09, 0.15, 0.10, 0.05 and 0.04, respectively. The average relative error over the cross section for bed and suspended load transport rates are 0.29 and 0.09, respectively.

301

3. CONCLUSIONS

Sediment transport measurements were carried out in the Nile River in Egypt at a site located downstream the High Aswan Dam by 288 km. Error analysis were conducted for the bed and suspended load transports at each measured station and for the overall cross section as well. The results showed good accuracy for the measured data. The overall relative error of cross section for the bed load transport was 29%. The overall relative error of cross section for suspended load transport was 9%. These accuracy values are due to the variability of the transport processes along the bed form length.

ACKNOWLEDGMENTS

Prof. Dr. M. B. Saad, Director of the Hydraulics Research Institute and Prof. Dr. F. M. Abdel-Aal, Irrigation & Hydraulics Department, Faculty of Engineering, Cairo University, are gratefully acknowledged for their support and reviewing this work.

REFERENCES

Abdel-Fattah, S. (1997a). "Field measurements of sediment load transport in the Nile River at Quena", Technical Report, HRI, Delta Barrage, Egypt.

Bagnold, R.A. (1966). "An Approach to the Sediment Transport Problem from General Physics", Geological Survey Prof. Paper 422-1, Washington.

Gaweesh, M.T.K., and Van Rijn, L.C. (1994). "Bed Load Sampling in Sand Bed Rivers", Journal of Hydraulic Engineering, ASCE, Vol. 120, No. 12.

Van Rijn, L. C. (1993b)."Principles of sediment transport in rivers, estuaries, coastal seas and oceans", Aqua Publications, Amsterdam, The Netherlands.

Van Rijn, L.C., (1984). "Sediment Transport, Part II: Suspended Load Transport", Journal of Hydraulic Engineering, ASCE, Vol.110, No. 11.

Van Rijn, L.C., and Gaweesh, M.T.K. (1992). "A new total sediment load sampler", Journal of Hydraulic Engineering, ASCE, Vol.118, No. 11.

Stochastic Hydraulics 2000, Wang & Hu (eds) © 2000 Taylor & Francis, ISBN 90 5809 166 X

Flushing sediment deposits by water discharges at Suzhouhe River Tidal Barrier

Sun Jichao, Shen Zhigang & He Yousheng
Shanghai Jiao Tong University, China

Liu Songping
Shanghai Suzhouhe River Tidal Barrier Supervise Bureau, China

ABSTRACT Deposition below many tidal barriers, located in river mouths with large areas of mud deposits, impairs seriously the operation of barriers and releasing of river flood and waterlogging. Due to the complexity of real conditions, theoretical calculations cannot be carried out systematically and precisely. Field experiments were made at Suzhouhe River tidal barrier, a distinct tidal barrier with 17 sluice gates located in Shanghai China, for different water levels and operation modes.

Brief description is made in this paper on experimental results on the deposition patterns, on water levels, and on operation modes for flushing sediment deposits by water discharges. In the experiments, the 17 sluice gates were organized into different groups, and three different grouping modes are attempted in order to find the most effective one. It can be concluded that river sedimentation near the tidal barrier can be greatly mitigated by making use of the sluice gates rationally. The experimental results indicate that compared with the 3-4-gate-grouping mode and the 5-6-gate-grouping mode, the mix-grouping mode with the biggest flush-deposit ratio of 0.86 is the most effective mode and can keep the sluice gates under good working condition. It can also be concluded that a better scouring effect can be obtained for lower water levels at the cost of larger resedimentation.

Key words: Tidal barrier, Gate operation, Resiltation, Flushing sediment

1. GENERAL INTRODUCTION OF SUZHOUHE RIVER AND THE TIDAL BARRIER

Suzhouhe River, which originates from Tai Lake and ends at Huangpu River, is a mid-tidal river. Due to the influence of downstream Huangpu River and upstream flow from Tai Lake, the tide is irregular semi-diurnal tide, the mean runoff is $22.0\,m^3$/s, and the sediment's motion is quite complex here.

The tidal barrier, which is also called Wusong Road Tidal Bridge, is located at the mouth of Suzhouhe River in Shanghai. It is one of the most important constructions in Shanghai, and its main function is to impede tide as high as 5.90 meters during storm surge and mitigate the traffic problem on the Waibaidu Bridge. The bevel formed by the barrier and the river is about 74 degree, and the net width of the barrier is 62.084 meters. What makes this barrier different from other normal barriers is that it has 17 sluice gates instead of one, with each gate 3.60m in width and 9.47m in height. The gates can impede the tide level as high as 5.90m, and will stand the difference of water level as high as 3.30m. In order to prevent tide to overflow low land areas, the sluice gates will be closed during storm surges when tide level will reach 4.70m. Due to this operation mode and the influence of Huangpu River's tide and Tai Lake's water flow, the mean flow velocity here will be often below 0.5m/s, which leads to serious deposition below the barrier. The average sediment depth here is about 0.8-1.0 m, and this causes much trouble to the barrier's operation.

To reduce the sedimentation, in this paper efforts were made to seek for some regularities for reference, based on the experience gained in field experiments carried out by Shanghai Suzhouhe River Tidal Barrier Supervise Bureau.

2. METHODS AND PLAN OF FIELD EXPERIMENT

2.1 *Methods of Experiment*

The field experiment was carried out at the tidal barrier.

a.   Investigation of Morphology of the riverbed below the barrier.
In order to know the result of each flushing mode, the morphology of the riverbed below the barrier was observed three times every experiment.
1)The first time was just before the flushing experiment, which aims at knowing the situation of the deposit before flushing.
2)The second time was just after the flushing experiment, in order to observe the situation of the deposit just after flushing and calculate the clearance of deposit after flushing.
3)The third time was about half to one month after flushing, in order to know the resediments during this half to one month's time.
b.   Velocity observation
The velocity observation at the barrier includes surface observation and bottom observation, which were implemented simultaneously during flushing experiment. The upstream and downstream observation cross section were chosen 45 m and 40 m away from the barrier, respectively. Based on the velocity observation, the flushing area of different water heads and sluice gate operation modes can be analyzed separately. The observation was carried out in the follow ways:
1)Observation took place about 5 min after the gates were opened (the discharging flow will be steady at that time).
2)Surface observation point was 50 cm below the water surface, and bottom observation point was 50 cm above the riverbed.
3)All the observations were done simultaneously in the observation boats.

2.2 *Plan of experiment*

According to the fact that the flushing results were directly influenced by the flow velocity distribution and diffusion velocity of the flow behind the barrier, which were determined by sluice gate operation modes, in the present experiment, three kinds of sluice gate operation modes were chosen:
1)Mode A, 5-6-gate-grouping mode; the 17 gates were organized into 3 groups: 12-17, 8-12, 1-6 (each number stands for a gate).
2)Mode B, 3-4-gate-grouping mode; the 17 gates were divided into the following 6 groups: 17-14, 14-12, 10-8, 8-6 and 16, 4-6 and 16, 1-4 and 16.
3)Mode C, mix-grouping mode; in this case, the 17 gates were divided into the following 7 groups: 16-17, 16-14, 14-12, 12-10, 10-8, 8-6, 6-1.
In order to mitigate the flow confusion when sluice gates were opened, even gates were opened 25 degree and odd gates 15 degree, based on the experience got from field experiments, and each group's flushing time is 10 min. The particular grouping mode and time are shown in Table 3.

Table.1: Plan of Experiment

Date	Water Level(m)	Gate Grouping Mode	Type of Water Level
97.05.23	3.43	Mode A	Low
97.06.23	3.62	Mode B	Mid
97.11.03	3.79	Mode C	Mid

Table 1 shows the plan of experiment, and we can compare the flushing results of different flushing modes by using table 1 and table 3.

2.3 Water level

Table.2: Water Level Before And After Flushing Experiment (m)

Date	Gate Grouping Mode	Before Flushing	After Flushing	Highest Water Level (downstream)	Lowest Water Level (downstream)
5.23	Mode A	2.96	3.17	3.41	3.21
6.23	Mode B	3.17	3.40	3.64	3.35
11.3	Mode C	3.36	3.50	3.79	3.56

Table 2 shows Suzhouhe River's water level before and after each flushing experiment. Each flushing experiment and its water level's observation were completed in the same day. Table 3 shows the back flow situation of different flushing modes:

Table.3: Back Flow Situation of Each Flushing Mode

Date	Gate Grouping Mode	Operation Order NO.	Gate NO.	Operation Time (min)	Back Flow Situation
		1	12-17	20	No back flow
5.23	Mode A	2	8-12	20	No back flow
		3	1-6	20	To south bank
		1	14-17	10	No back flow
		2	12-14	10	To center
		3	8-10	10	To center
6.23	Mode B	4	6-8,16	10	To center and south bank
		5	4-6,16	10	To south bank
		6	1-4,16	10	To center and south bank
		1	16-17	10	No back flow
		2	14-16	10	No back flow
		3	12-14	10	No back flow
11.3	Mode C	4	10-12	10	To south bank
		5	8-10	10	To center and north bank
		6	6-8	10	To south bank
		7	1-6	10	To south bank

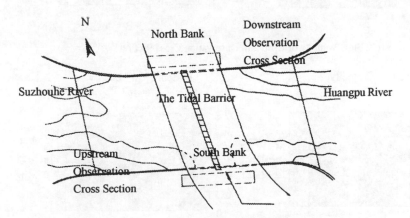

Fig.1: Sketch map of the Tidal Barrier

3. ANALYSIS OF THE EXPERIMENT'S RESULTS

Based on flow state, flow velocity, and resedimentation, each flushing mode's effects are analyzed.

3.1 Fflow state

Due to the 74-degree interacting angle formed by the barrier and river, and the fact that the opened sluice gates' width is relatively small compared with the river's width during the flushing experiment, different extents of back flow are easy to be formed here, which will cause serious bed scouring, sedimentation, and influence the flushing effect evidently.

In order to compare objectively, the gates were all opened from south bank to north bank, and there was no time interval when shifting the group.

From table 3, it is obvious that 5 groups of mode B and 4 groups of mode C have back flows. Mode B's back flow is the most serious one of all the three modes. Serious back flow can bring great possibility of sedimentation, that is, larger resedimentation, which is also proved by latter analysis.

3.2 Flow velocity:

Figure 2 and 3 show the observed surface and bottom flow velocity of the upstream and downstream observation cross section: m/s

From Figure 2 and 3, we can see that: On the upstream cross section mode C has the biggest average surface flow velocity, and mode A has the smallest one; as for average bottom flow velocity, mode B and A are both bigger than mode C. On the downstream cross section, mode C has the biggest surface flow velocity, and all the three modes have the same smallest flow velocity; as for bottom flow velocity, mode C has the biggest one and mode B has the smallest one. It is evident that upstream average surface and bottom flow velocities are all bigger than those of downstream. Based on the above analysis, it can be concluded that:

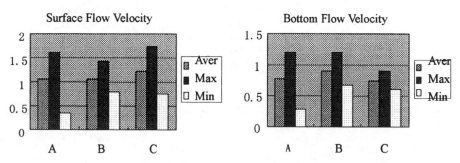

Fig2: Flow Velocity of Upstream Observation Cross Section

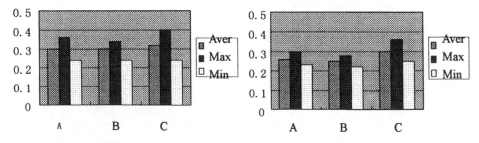

Fig.3: Flow Velocity of Downstream Observation Cross Section

306

1) Upstream flow velocity is bigger than downstream flow velocity. Downstream flow velocities are all below the sand's initial motion velocity of 0.5 to 0.7 m/s, while most of the upstream flow velocities are bigger than 0.5 m/s, which makes it difficult for the downstream sand to be washed away and results in serious deposition in the downstream.

2) Surface flow velocity is bigger than bottom flow velocity.

3) Generally, mode C's flow velocity is bigger than the other two's, and mode A has the smallest velocity.

Thus, it shows that to adopt mode C can be more likely to reduce the sediments and get a better effect on channel bed scouring.

3.3 Resedimentation:

Back flow induced by the flushing experiment forms the resedimentation in barrier's area, and the data provided by the following table show the channel bed resediments under different operation modes.

Table.4: Channel Bed Resediments of Each Grouping Mode

Grouping Mode	Date	Observation Date	Time Interval (day)	Average Bed Surface Elevation (m)	Difference of Bed Surface Elevation (m)	Scouring Effect (m^3)			
						Flushed Deposits	Resediment Deposits	Average Resediment Deposits (per day)	Unit Bed Area Sediment Reduction
Mode A	5.23	6.23	31	-2.10	22	748	1265	41	0.09
Mode B	6.23	7.6	13	-2.06	7	173	403	31	0.04
Mode C	11.3	12.2	29	-2.09	7	345	403	14	-0.01

In table 4, the observation of bed surface elevation is based on sea level and the negative sediment reduction stands for the decreasement of bed surface elevation; on the other hand, positive sediment reduction stands for the increasement of bed surface elevation.

Table 4 shows that the sediment reduction induced by mode A is the biggest of all, 748 m^3, and mode B is the smallest, 174 m^3. As for resediments, adopting mode A caused the biggest resediment deposits, 1265 m^3. To consider the fact that the scouring effect can be influenced by both flushed deposits and resediment deposits, we can give the flush-deposit ratio K to evaluate the scouring effect objectively:

K= flushed deposits/resediment deposits

The calculated flush-deposit ratio K under different grouping modes are shown in Table 5.

Table.5: Calculated Flush-Deposit Ratio K

Grouping Mode	Mode A	Mode B	Mode C
K	0.59	0.43	0.86

It is easy for us to find out that adopting mode C is the best choice. Although the time interval of mode B is only 13 days, not that consistent with the other two, mode B's effect can't match mode C's due to the comparison of bed surface elevation difference, resediment deposits, and everyday average resediment deposits, especially everyday average resediments.

Based on the comparison of flow states and flow velocity, it's obvious that the scouring effect is greatly correlated with back flow, and flow velocity.

Mode B has the strongest back flow, and mode A has the weakest. However, when considering velocity, mode A, which has the smallest velocity, can also result in serious resedimentation, just as mode B. Therefore, adopting mode C is most advantageous to scouring sediments.

From table 2 and table 4, we can also find out that mode A's water head, 2.96m, is greatly lower than the other two, 3.17m and 3.36m. Accordingly, adopting mode A can scour more sediments than the other two can do. Hence, it is easy to flush sediments when water head is low.

4. CONCLUSION:

A. Flushing under low water heads can get a better scouring effect.
B. Adopting mode C is more advantageous to scouring sediments.
C. Experiment shows that sluice gates can be used to flush sediments if adopt right grouping mode and under right water heads, although downstream scouring effect may not be good due to the fact that sluice gates can be only opened to one direction.

ACKNOWLEDGMENTS

The research is supported by Shanghai Suzhouhe River Tidal Barrier Supervise Bureau.

REFERENCES

Gang Wang, Baoliang Zhu, Qingsong & Ziran Bian (1999), "Impounding tidal flow for flushing sediment deposits below tidal barriers", River Sedimentation, Eds.Jayawardena, J.Lee & Z.Y.Wang, ISBN 90 5809 0345

Zhang Xiangfeng & Lian Daren (1999), "Fluvial processes of the Haihe River Estuary", River Sedimentation, Eds.Jayawardena, J.Lee & Z.Y.Wang,, ISBN 90 5809 0345

Ning Chien & Zhaohui Wan(1999),"Mechanics of Sediment Transport", American Society of Civil Engineers, ISBN 0-7844-0400-3

Stochastic Hydraulics 2000, Wang & Hu (eds) © 2000 Taylor & Francis, ISBN 90 5809 166 X

Numerical simulation of river-bed variations: Sensitivity analyses with respect of the initial hydraulic roughness

D. Termini
Dipartimento di Ingegneria Idraulica ed Applicazioni Ambientali, Università di Palermo, Italy

ABSTRACT: Two elements affect the efficacy of the numerical models of mobile-bed evolution: the prediction of the hydraulic roughness, related with both sediment distribution on the bed and flow depth, and the prediction of the hydraulic sorting, when the distribution of sediments on the bed is not uniform. In this work the bed profile variation is simulated, by using a numerical model proposed in previous works, in function of the variation of the friction factor, that reflects the coarsening of median sediment diameter, due to armouring process, and the variation of flow conditions. The variation of grains-size distribution of the bed material is predicted by using a probabilistic approach and the median sediment diameter and the dimensions of bed forms are estimated at each time step. The paper is devoted to show, by means of computer-based numerical experiments, the sensitivity of the numerical results with respect to the initial hydraulic roughness used. The choice of the numerical space and time steps is discussed and the relation with the initial roughness coefficient is reported.

Key words : Hydraulic sorting, Hydraulic roughness, Bed variation computation

1. INTRODUCTION

The computation of river-bed variations can be affected by some difficulties. First of all, the knowledge of the initial hydraulic roughness, due to bed material and to small-scale bed forms, and its variation during the process. On the other hand, with non uniform sediment mixtures the analyses of the sediment-transport mechanism is much more complicated. Another difficulty is due to the choice of the numerical discretization, used to solve the basic system of equations, that allows to obtain reliable results.

Many sediment transport theories have been developed in order to analyze the sediment transport mechanism. Some authors, supposing that the thickness of the active layer is not dependent on flow characteristics [Komura et al.,1967; Chin et al. 1994], use a deterministic approach. Other authors use a probabilistic approach, taking into account that the transport phenomenon is a stochastic process [Hsu and Holly, 1992; Cheng and Chiew, 1996]. In this work it is supposed that the probability to stay on the bed of each grain depends on its "mobility" and its "availability" to be transported by the flow or to form the active layer [Van Rijn, 1984]. Thus, it depends on the hydraulic conditions of the flow and on sediment properties. Both of them vary during the process and interact each other. In fact, the availability of each grain to be transported depends on its sheltering by larger grains and its exposure to the flow [Shen and Lu, 1983]. The variation of the grain size distribution, due to armoring process, produces the variation of the median sediment diameter and the variation of the bed roughness. The bed roughness affects the hydraulic flow conditions and the bed perturbations celerity, to which the numerical discretization of the solving equations system is related. In order to obtain stable results, in fact, the numerical time and space steps are limited by the Courant condition.

A one-dimensional numerical model, proposed in previous works [Termini et al., 1997; Termini and Bonvissuto, 1999(b)] is used to show the sensitivity of the results with respect the initial hydraulic roughness. The probability of each grain to stay on the bed, and thus the new grain-size distribution, is estimated at each time step. The analyses and the results presented herein suppose bedload-dominated mobile-bed processes. The generalization of the model to the treatment of suspended load is in progress.

2. BED PROFILE COMPUTATION

The basic one-dimensional equations that govern the flow over a mobile bed, in a wide rectangular channel, are: the continuity equation for water, the momentum equation for water and the continuity equation for sediment. Under the hypothesis that the Froude number, Fr, is less then 1, as in many practical cases, the orders-of-magnitude difference in celerity of bed perturbation waves and of water waves justify the uncoupling of the two processes [De Vries, 1973; Holly and Rahutl, 1990]. It is assumed the water-movement quasi-steady and uniform during the transient stages. The inertial terms are neglected and the system of equations is written in the following form [Termini et. al., 1997]:

- continuity equation for water:

$$\frac{\partial q}{\partial x} = 0 \tag{1}$$

- momentum equation for water:

$$gH\frac{\partial z}{\partial x} + gH\frac{n^2 q^2}{H^{10/3}} = 0 \tag{2}$$

- continuity equation for sediment:

$$\frac{\partial q_s}{\partial x} = -\frac{\partial z}{\partial t} \tag{3}$$

being z the bed elevation with respect a datum plane, t the time and x the longitudinal coordinate; H is the flow depth, n is the Manning roughness coefficient, q_s is the specific volumetric transport rate. In this condition, only bed-load is considered and it is expressed by using the Bagnold formula. The system 1-3 is solved by using an implicit finite difference scheme. The details of the model are reported in previous works [Termini et al., 1997; Termini and Bonvissuto, 1999(a); Termini and Bonvissuto, 1999(b)].

2.1 Hydraulic sorting

In order to solve the system 1-3, the considered river reach is divided into $N_{\Delta x}$ space intervals. For each interval Δx the initial sediments distribution is supposed to be known and the grains-size distribution curve is defined. The variation of the grains-size distributions, due to armouring process, at time step (t+Δt) is estimated by assuming a Gaussian distribution of the probability to stay on the bed of each grain of diameter D_i. It is calculated as follows [Termini and Bonvissuto, 1999(b)]:

$$P_{i,j} = \frac{1}{\sqrt{2\pi}} \int_{-\infty}^{X_{i,j}} \exp\left(\frac{X^2}{2}\right) dx$$

with
$$X_{i,j} = \frac{\left(\frac{\xi}{\eta}\right)_{i,j} - 1}{\sigma} \qquad i=1,\ldots, N_d \ ; \ j=1,\ldots, N_{\Delta x} \tag{4}$$

being N_d the number of diameters, σ the standard deviation of the distribution and X the variable of integration; η_i is the flow intensity, given by the ratio between the mobility number

$Y = \dfrac{\tau}{\gamma_s D_i}$ (being τ the average bottom shear stress and γ_s the specific weight of sediment)

and the critical mobility number, Y_{cr}, calculated knowing the diameter Di and the water density ρ [Yalin, 1992].The parameter ξ_1, called "hiding factor", allows to take into account the influence of the grains of the whole mixture on the transport of the grain of diameter Di. It is calculated in function of the ratio between the diameter Di and the median sediment diameter (D_{50}), both estimated in the space interval j.

Thus, the new fraction of the sediment with diameter Di with diameter Di remained on the bed in the space interval j, at time step(t+Δt), is calculated as:

$$\left(F_{i,j}\right)^{t+\Delta t} = \left(F_{i,j}\right)^{t} P_{i,j} \tag{5}$$

being $(F_{i,j})^t$ the fraction of grains with diameter D_i in the space interval j at time known t. The new grains-size distribution in the space interval j is estimated and new median sediment diameter is determined.

When long extents, with remarkable differences of sediments distributions, are examined, several space intervals Δx have to be considered in order to obtain reliable results.

2.2 *Selection of the numerical space and time intervals*

If the bed variations are analyzed, the numerical space and time intervals have to be opportunely selected. The bed perturbations have to travel, with a celerity C, within a time of Δt and for a distance of Δx. Thus, at the beginning of the simulation, it is imposed that the Courant number is equal to one [Park and Jain, 1987]:

$$C_{ou} = 1 = C\dfrac{\Delta t}{\Delta x} \tag{6}$$

It is:

$$\dfrac{\Delta x}{\Delta t} = \left[\dfrac{\dfrac{dq_s}{dv}}{1 - Fr^2} \dfrac{v}{H} \right] \tag{7}$$

where the term in square brackets represents the bed perturbations celerity [De Vries, 1973]; v is the average flow velocity. If the bed load is estimated by using the Bagnold formula, after simple passages [Termini and Bonvissuto, 1999(a)], the (7) can be written in the form:

$$\dfrac{\Delta x}{\Delta t} = \dfrac{\dfrac{b}{\gamma_s}\dfrac{\rho}{c}\left(\dfrac{3}{2}\dfrac{v^2}{c^2} - \tau_{cr} \right)}{1 - Fr^2} \dfrac{v}{H} \tag{8}$$

being c the initial friction factor, b is a coefficient depending on flow regime [Yalin, 1992; Termini and Bonvissuto, 1999(a)], τ_{cr} is the critical shear stress estimated for the initial median sediment diameter.

The equation (8) shows that the bed perturbations celerity and the choice of the numerical space and time steps strongly depend on initial friction factor and thus on the initial bed roughness and flow depth.

3. SENSITIVITY ANALISYS

For a given flow rate the equation (8) shows the importance of a reliable roughness prediction for morphological computations. The flow depth estimation depends on the knowledge of the roughness coefficient [Yen, 1992], that should reflect the influence of the bank and the bed materials and the dimension of the bed forms [Van Rjin, 1984]. But, the knowledge of the roughness coefficient has some degree of uncertainty. An error in its value will produce a significant error on flow depth estimation, affecting both the water wave propagation celerity and the bed perturbations celerity, and, consequently, the numerical discretization to solve the system of equations.

3.1. Influence on numerical discretization

In this work the influence of the Manning roughness coefficient on the numerical results is analyzed. A channel reach with length of 5700 m, initial bed slope of 0.14% is considered (see Figure 1). It is supposed that no bed form are on the bed at the beginning of the bed variation process and the grains size distribution of sediments on the bed is, for all the channel reach considered, with geometric mean diameter of 0.1 mm and geometric standard deviation of 4.8. A constant flow rate, equal to 22,048 m³/s, is supposed at the upstream section.

Figure 2 illustrates the variation of the initial bed perturbations celerity C, with respect the average flow velocity v, for different values of Manning roughness coefficient ranging from 0.02 $m^{-1/3}$s to 0.2 $m^{-1/3}$s. In Figure 2 the variation of the initial friction factor c with Manning roughness coefficient is also reported.

As it is shown in Figure 2, because of the increasing roughness coefficient, the initial flow depth increases and the ratio between the bed perturbations celerity and the average flow velocity decreases. Consequently, because of equation (8), the time and the space intervals change. For a given space step Δx, the variation of the dimensionless time step $\Delta t/Tq$, being T_q the time with constant upstream flow rate q, with Manning roughness coefficient is reported in Fig. 3.

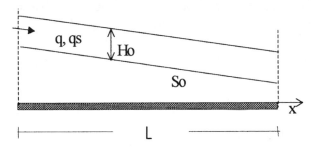

Fig. 1 Channel reach considered

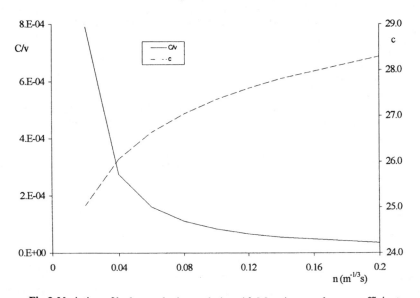

Fig.2:Variation of bed perturbations celerity with Manning roughness coefficient

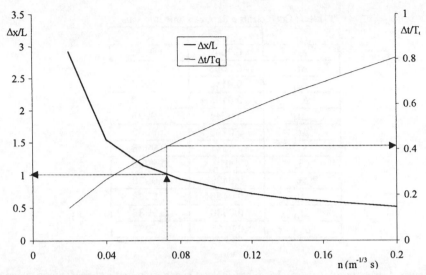

Fig. 3: Variation of the dimensionless space and time intervals with Manning roughness coefficient.

In the same figure the variation of the dimensionless space step Δx/L, for a given space step Δt, is also reported. Here L is the length of the channel reach considered.
As it is shown in Figure 3, for a given step Δt, because of (8), increasing Manning roughness coefficient the space step have to decrease; for a given step Δx, increasing Manning roughness coefficient, the time step have to increase. With small roughness coefficients, or small initial friction factors, the dimensionless space step could become bigger then 1, i.e. Δx>L. In this case it is imposed that Δx=L and the Courant number is less than one. On the other hand, the choice of the time step depends on the upstream boundary conditions. In fact, in order to analyze the bed deformation, the time step would be much smaller then the time in which the flow rate is constant (i.e. Tq>>Δt).

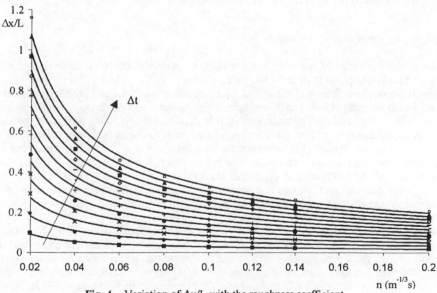

Fig. 4 – Variation of Δx/L with the roughness coefficient

Table.1: Coefficients *a* for given time intervals

	a	R²
Δt_1	0.0049	0.9938
Δt_2	0.0098	0.9938
Δt_3	0.0146	0.9938
Δt_4	0.0195	0.9938
Δt_5	0.0244	0.9938
Δt_6	0.0293	0.9938
Δt_7	0.0341	0.9938
Δt_8	0.039	0.9938
Δt_9	0.0439	0.9938
Δt_{10}	0.0488	0.9938
Δt_{11}	0.0536	0.9938
Δt_{12}	0.0585	0.9938

In Figure 4, the variations of the dimensionless space step with the Manning roughness coefficient, by assuming different values of Δt, are reported. The time step, starting from a value of 30 days, increases with steps of 30 days as indicated in Figure 4. For each value of Δt, the points on the plane ($\Delta x/L$ – n), can be interpolated by a curve with equation:

$$\frac{\Delta x}{L} = a\, n^{-0.7454} \tag{9}$$

The coefficient *a* varies with the time step Δt selected, as reported in Tab. 1. In Tab. 1 the coefficients R^2, are also reported.

The linear regression of the two variables *a* and Δt, gives the following equation:

$$a = \left(2 \times 10^{-9}\right)\Delta t + \left(2 \times 10^{-5}\right) \tag{10}$$

For a given value of Manning roughness coefficient n and selected the time step Δt (so that $Tq \gg \Delta t$), the equations (9) and (10) allow to determine the space step Δx that verify the equation (8).

3.2. Influence on bed profile estimation

The spatial discretization of the river reach considered strongly affect the numerical results, especially when large spatial extents are analyzed.

In order to show the influence of the spatial discretization on the final bed profile computation, the model has been applied to the same example of the previous section, by assuming a constant time step Δt. The final bed configuration estimated by Komura et al. (1967), observing the same channel reach considered, is used for the comparison with the numerical results. In order to describe the initial distribution of sediments on the bed, the channel is divided into three stretches of length L/3. The initial grains size distributions in every stretch are respectively with geometric mean diameter of 0.100 mm and geometric standard deviation of 4.8 (first stretch), with geometric mean diameter of 0.24 mm and geometric standard deviation of 2.9 (second stretch), with geometric mean diameter of 0.061 mm and geometric standard deviation of 3.6 (third stretch).

Three spatial discretizations of river reach, corresponding to different values of initial Manning roughness coefficient, are tested. For the first run three computational points are considered, for the second and the third runs respectively four and five computational points are considered. For each run the final bed configuration is estimated, by applying the model, and the percent error with respect the estimated one by Komura et al. [1967] (see also Park and Jain, 1987; Termini and Bonvissuto, 1999(b)) are determined at each computational point i, as:

314

$$E_{z_i} = 100 \times \left(1 - \frac{z_i}{z_{K,i}}\right) \tag{11}$$

being z_i the calculated final bed elevation at computational point i and $z_{K,i}$, the corresponding value estimated by Komura et al. (1967). In Tab. 2 the percent errors are reported.

Tab.2 – Percent error of the final bed profile

	$N_{\Delta x}$	$|E_{z1}|(\%)$	$|E_{z2}|(\%)$	$|E_{z3}|(\%)$	$|E_{z4}|(\%)$	$|E_{z5}|(\%)$	$\overline{E}\,(\%)$
run1	2	53.65	70.97	0.00			33.28
run2	3	28.39	39.08	62.91	0.00		32.59
run3	4	31.38	24.48	31.38	22.81	0.00	8.49

In the same table the average percent error is also reported. It is calculated as:

$$\overline{E} = \frac{1}{N_{\Delta x}+1} \sum_{i=1}^{N_{\Delta x}+1} E_{zi} \tag{12}$$

As it is shown the average error reduces increasing the number of the space steps.
In Tab. 3 the median sediment diameters estimated for each space interval Δx are reported. It seems clear that the evolution of the armouring process is different varying the space interval. This difference influences the final bed configuration computation.

Tab.3 – Median sediment diameters at the end of the runs

	$N_{\Delta x}$	$D_{50,\Delta x1}$ (mm)	$D_{50,\Delta x2}$ (mm)	$D_{50,\Delta x3}$ (mm)	$D_{50,\Delta x4}$ (mm)
run1	2	0.24	0.50		
run2	3	1.25	1.21	0.91	
run3	4	0.98	0.79	0.53	0.59

4. CONCLUSION

Forecasting of river bed variation is important for water resources. In order to obtain an accurate estimation of river bed variations the following aspects have to be considered: the hydraulic roughness and the variation of the grain-size distribution. An error in roughness coefficient value strongly affects the flow depth estimation and the bed perturbations celerity. This, in turn, produces an error in the numerical computations of the bed levels. In no-steady conditions the choice of the time step depends on the time with upstream constant flow rate. The spatial discretization has to be chosen in function of the time step selected, taking into account the Courant number limitation. For a given time step, the proposed relation between Manning roughness coefficient and the space step can be used.

REFERENCES

Cheng Nian-Sheng and Yee-Meng Chiew, "Pickup Probability for Sediment Entraintment", Journal of Hydraulic Engineering, vol. 124, n. 2, 1996.
Chin C.O., B.W. Melville and A.J. Raudkivi, "Streambed Armoring", Journal of Hydraulic Engineering, ASCE, Vol. 120 N. 8, 1994.

De Vries M., "River-bed variations – Aggradation and Degradation", Delft Hydraulic laboratory publication. N. 107, 1973.

Holly F.M. and J.L. Rahuel, "New Numerical/Physical Framework for Mobile-bed Modelling", Journal of Hydraulic Engineering, vol. 28, n.4, 1990.

Hsu S. M. and Forrest M. Holly, "Conceptual Bed-Load Transport Model and Verification for Sediment Mixtures", Journal of Hydraulic Engineering, vol. 118, n. 8, 1992.

Komura S. and D.B. Simons, "River-bed Degradation Below Dams", Proceeding of the American Civil Engineerings, HY2. paper 5335, 1967.

Park I. and S. C. Jain, "Numerical Simulation of Degradation of Alluvial Channel Beds", Journal of Hydraulic Engineering, vol. 113, n. 7, 1987.

Shen H. W. and Lu Jau-Yau, "Development and Prediction of Bed Armouring", J. Hydr. Engrg., ASCE, 109(4), 1983.

Termini D., M.S. Yalin, G. Bonvissuto, "A Numerical Method for the Computation of Bed Degradation Downstream a Dam". "Uprating and Refurbishing Hydropower Plants Conference" – Montreal (CANADA), 1/3 October 1997.

Termini D. and G. Bonvissuto, "La Degradatione del Fondo negli Alvei Naturali ed il Fattore di frizione". L'ACQUA, Vol. 2 1999 (a).

Termini D. and G. Bonvissuto, "Numerical Simulation of Bed Degradation and Bed Armouring", Proceedings of XXVIII IAHR Congress - Graz- Austria - 22-27 August, 1999(b).

Van Rijn L., "Sediment Transport, Part II: Suspended Load Transport", Journal of Hydraulic Engineering, vol. 110, n. 11, 1984.

Yalin M.S., *River Mechanics*, Pergamon Press. London, 1992.

Yen B.C., *Channel Flow Resistance: Centennial of Manning's Formula*, Yen B.C and University Of Illinois. USA 1992.

Stochastic Hydraulics 2000, Wang & Hu (eds) © 2000 Taylor & Francis, ISBN 90 5809 166 X

Probabilistic analysis of incipient motion of sediment particles

Sung-Uk Choi & Seungjoo Kwak
Department of Civil Engineering, Yonsei University, Seoul, Korea

ABSTRACT: Initiation of sediment particles by the flow can be classified by three different modes: rolling, sliding, and lifting. In this paper, threshold conditions of each mode obtained from force and momentum balances are investigated. Theoretical analysis of the conditions leads to curves similar to Shields diagram, resulting that the rolling and lifting thresholds yield the minimum and maximum values of dimensionless shear stresses, respectively. A probabilistic analysis of the same problem is carried out under the assumption that the velocity approaching the particle has a normal distribution. The results revealed that about 4% of the particles on the bed are lifted when the critical bed shear stress by Shields is applied.

Key words: Incipient motion of sediment, Sediment transport, Shields curve

1. INTRODUCTION

Physics involved in the initiation of sediment particles lying on the bed is featured by both deterministic and probabilistic characteristics. The deterministic nature comes from the view that the mean velocity or mean bed shear stress dominates the process. Whereas the probabilistic characteristic stems from the fact that turbulence is highly related with initiation of sediment particles.

Shields (1936) has been recognized as the first person who expressed the critical shear stress required for starting sediment particles in a dimensionless form. He obtained so called Shields diagram using similarity dynamics (Kennedy, 1995), which was also verified experimentally. Shields diagram has played a significant role as a milestone, and has been used most frequently in the study of sediment transport, channel degradation, and stable channel design. However, it is true that Shields diagram is subjective and not unique. The critical shear stress can be defined anywhere between rolling and lifting as indicated by Ling (1995).

Three different modes exist in the initiation of a sediment particle on the bed, i.e., rolling, sliding, and lifting. Among them, first two modes are closely relevant to bedload transport while the last does with both bedload and suspended load. That is, once the particle is picked up by the flow, turbulence level determines suspension of particles within the flow. If the turbulence level is high enough, then it will be a part of the suspended load. Otherwise the particle will settle down onto the bed.

In the present paper, three modes of incipient motions of sediment particles, namely, rolling, sliding, and lifting, are investigated by deterministic and probabilistic approaches. Threshold conditions of each mode are solved, and results are presented in a dimensionless form. Shields diagram is also revisited by taking the velocity approaching the particle as a distribution, which introduces the probabilistic view to the deterministic approach.

2. THEORETICAL ANALYSIS

Forces acting on a spherical particle lying on a bed are submerged weight, drag, lift, and force resisting motion of the particle. The submerged weight of the particle is written as

$$W_s = \rho R g \frac{\pi D^3}{6} \tag{1}$$

where ρ is the fluid density, R is the submerged specific gravity, g is the gravitational acceleration, and D is the particle diameter. The drag force F_d acting in the flow direction is given by

$$F_d = c_d A_p \frac{\rho u_b^2}{2} \tag{2}$$

where c_d is the drag coefficient, A_p is the projected area of the particle in the flow direction (= $\pi D^2 / 4$), and u_b is the instantaneous velocity approaching the particle on the bed. The non-uniformity of magnitude of fluid velocity acting on the particle results in the lift force. For simplicity, if we assume that velocity maximum and minimum occur at the top and at the bottom, respectively, then the lift force acts vertically upward. The lift force F_l can be expressed as

$$F_l = c_l A_p \frac{\rho u_b^2}{2} \tag{3}$$

Using the Coulomb's friction coefficient (μ), the resistance force is given in terms of normal force components. That is,

$$F_r = \mu(W_s - F_l) \tag{4}$$

Figure 1 shows four forces acting on the dangerously-placed particles (denoted by a and c in (a) and (b), respectively). For the sake of convenience in the analysis, two different arrangements of particles are considered. That is, Figure 1(a) illustrates a situation where particle a is rolling over particle b. In Figure 1(b), particle c is sliding on the surface of particle d, and thus only the friction between particles c and d is to be considered. Regarding lifting motion, either Figure 1(a) or (b) can explain the situation.

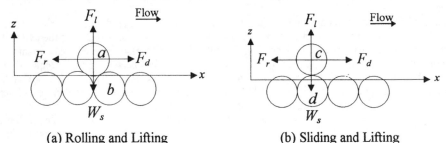

    (a) Rolling and Lifting   (b) Sliding and Lifting

Fig 1:. Schematic Sketch

For closely-packed three dimensional arrangement of spheres as seen in Figure 1(a), the following threshold condition for the particle a to roll over particle b can be obtained from the moment analysis (Coleman, 1967):

$$4 / \sqrt{2} F_d + F_l = W_s \tag{5}$$

The sliding motion of the particle c in Figure 1(b) occurs when the drag force is the same as the resisting force. Therefore, the threshold condition of particles' sliding can be written as

$$F_d = F_r \tag{6}$$

Another threshold condition for the particle to be picked up by the flow takes place when the lift force and the submerged weight of the particle are the same, i.e.,

$$F_l = W_s \tag{7}$$

In order to solve eqs.(5)-(7), the effective velocity acting on the particle ought to be estimated first. The viscous sublayer exists when the effective roughness height (k_s) is smaller than the thickness of the viscous sublayer which is usually taken as $10\nu / u_*$. The following relationship for velocity distribution holds in the viscous sublayer:

$$\frac{u}{u_*} = \frac{u_* z}{\nu} \tag{8}$$

where u is x-component velocity which is only a function of z herein. If we assume that the effective roughness height (k_s) due to grains is twice of their diameter, then eq.(8) holds for $Re_* < 5$. Whereas the following logarithmic law applies near the rough boundary:

$$\frac{u}{u_*} = \frac{1}{\kappa} ln\left(30\frac{z}{k_s}\right) \tag{9}$$

where κ is von Karman's constant. Note that no viscous sublayer exists if the particle diameter is greater than the half of boundary-layer thickness and eq.(9) is only valid for flows at $Re_* > 10.1$. For a smooth transition between two regions, the following formula provided by Reichardt (1951) is used for the effective normalized velocity f ($= u_{z=D/2} / u_*$) at the center height of the spherical particle:

$$f = \frac{1}{\kappa} ln\left(1 + \kappa z^+\right) - c\left(1 - e^{-z^+/11.6} - \frac{z^+}{11.6} e^{-0.33z^+}\right) \tag{10}$$

where $z^+ = u_* z / \nu$, $c = \kappa^{-1} ln(\kappa z_o^+)$, and $z_o^+ = u_* z_o / \nu$. Now, threshold conditions of particles' rolling, sliding, and lifting can be obtained by solving eqs.(5)-(7) with the help of normalized velocity by eqs.(10a) and (10b). This leads to the following conditions for rolling, sliding, and lifting respectively:

$$\tau_* = \frac{2}{3} \frac{1}{(\sqrt{2}c_d + 0.5c_l)} \frac{1}{f^2} \tag{11}$$

$$\tau_* = \frac{4}{3} \frac{\mu}{c_d + \mu c_l} \frac{1}{f^2} \tag{12}$$

$$\tau_* = \frac{4}{3} \frac{1}{c_l} \frac{1}{f^2} \tag{13}$$

In the above equations, τ_* ($= \tau_{bc} / \rho RgD$) is Shields parameter, which is the dimensionless critical bed shear stress.

In the present analysis, a relationship of c_d versus Re_* converted from the standard drag curve is used (Coleman, 1967; Ikeda, 1982). Here, Re_* is the friction Reynolds number defined by $u_* D / \nu$. Regarding the lift coefficient c_l, much less is known compared with the drag coefficient c_d. Many researchers proposed values ranging 0.1-0.4 for c_l through experimental measurements (Einstein and El-Samni, 1949; Chepil, 1958; Li et al., 1983; Patnaik et al., 1994). Chepil (1958) made extensive measurements of the drag and lift forces on hemispheres at friction Reynolds numbers from 24 to 3240. He found that the average ratio of the lift to drag force was

319

0.85 for $Re_* < 2500$, which is used in the present computation. The Coulomb's friction coefficient μ in eq.(4) is given by

$$\mu = \tan \phi \tag{14}$$

where ϕ is the angle of repose of particles. Wiberg and Smith (1985) stated that ϕ is not the mass angle of repose in the problem at hand, but should represent the difficulty of taking the particle out of the pocket where it is locating. Thus they used $\phi = 60$ deg. though the angle of repose of most sand particles is about 40 deg.

Eqs.(5)-(7) are solved to get threshold conditions for particles' incipient motions, and Shields parameter versus friction Reynolds number are drawn in Figure 2. Shields' experimental data as a curve are also presented for comparison's ake. The curves from threshold conditions for rolling, sliding, and lifting appear to be of similar shape. The figure indicates that the dimensionless bed shear stress for lifting is the highest, and that for rolling is the lowest in all flow ranges. The dimensionless bed shear stress for sliding is between lifting and rolling. Notice also that Shields' data lie between sliding and lifting thresholds in the range of $1 < Re_* < 10$ and they exactly coincide with the sliding threshold at high friction Reynolds number, which conforms to the result in Wiberg and Smith (1985).

3. PROBABILISTIC ANALYSIS

If we assume that the velocity approaching the particle (u_b) on the bed is normally distributed, then we have the probability density function of u_b such as (Cheng and Chiew, 1988; Nezu and Nakagawa, 1996)

$$f(u_b) = \frac{1}{\sqrt{2\pi}\sigma} exp\left[-\frac{(u_b - \bar{u}_b)}{2\sigma^2} \right] \tag{15}$$

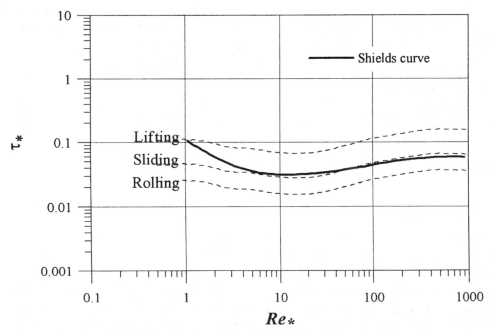

Fig 2:. Threshold Conditions for Rolling, Sliding, and Lifting

where $\bar{u}_b$ is the time-mean value of u_b and σ $(=\sqrt{(u_b-\bar{u}_b)^2}$) is the turbulence intensity. Nezu and Nakagawa (1996) proposed the following semi-empirical relationship for turbulence intensity in the close-to wall region:

$$\sigma / u_* = D_u \, exp(-z^+ / \mathrm{Re})\Gamma(z^+) + Cz^+\left[1-\Gamma(z^+)\right] \tag{16}$$

where $z^+ = u_* z / v$, $Re = u_* h / v$ (h = water depth herein), $\Gamma(z^+) = 1 - exp(-z^+ / B)$, and D_u, B, and C are empirical constants having values of 2.3, 10, and 0.3, respectively. In intermediate and outer regions, Nezu and Nakagawa (1996) obtained an expression for the turbulence intensity such as

$$\sigma / u_* = D_* \exp(-z / h) \tag{17}$$

where the empirical constant D_* has a value of 2.3 for intermediate region and 2.26 for outer region, respectively. Nezu and Nakagawa (1996) showed that turbulence intensity by eq.(17) is independent of Reynolds and Froude numbers from comparisons with measurement data. Kironoto and Graf (1994) obtained a constant value of $\sigma / u_* = 2$ in open-channel flow experiments with rough boundary condition. In the present computation, the relationship by eq.(16) is used for the turbulence intensity in the wall region, and a constant value of $\sigma / u_* = 2$ is used in the outer region.

In order for the particle to initiate a motion of rolling by the flow, the following inequality should hold from eq.(5): $4 / \sqrt{2}F_d + F_l > W_s$. That is, the probability of rolling is given by

$$P_r = P(4 / \sqrt{2}F_d + F_l > W_s) \tag{18}$$

where P_r ranges between 0 and 1. Rewriting eq.(18) with the help of eqs.(1)-(3) yields

$$P_r = P(u_b^2 > R) = P(u_b > \sqrt{R}) + P(u_b < -\sqrt{R}) \tag{19}$$

in which $R = \sqrt{2 / 3 \times RgD / (\sqrt{2}c_d + 0.5c_l)}$. By using eq.(15), eq.(19) can be rewritten as

$$P_r = 1 - \int_{-R}^{R} \frac{1}{\sqrt{2\pi}} \exp\left[\left(\frac{u_b - \bar{u}_b}{\sigma}\right)^2 / 2\right] d\left(\frac{u_b - \bar{u}_b}{\sigma}\right) \tag{20}$$

Integrating the above equation by using the error function enables one to estimate the rolling probability. Similarly, the sliding and the lifting probabilities are defined by, respectively,

$$P_s = P(u_b^2 > S) = P(u_b > \sqrt{S}) + P(u_b < -\sqrt{S}) \tag{21}$$

$$P_l = P(u_b^2 > L) = P(u_b > \sqrt{L}) + P(u_b < -\sqrt{L}) \tag{22}$$

which lead to the same expression as eq.(20) except that the parameter R should be replaced by $S = \sqrt{4 / 3 \times \mu RgD / (c_d + \mu c_l)}$ for sliding and $L = \sqrt{4RgD / (3c_l)}$ for lifting, respectively. It is worth while to indicate that each probability by eqs.(19), (21), and (22) can also be defined as the percentage of the number of particles in motion (rolling, sliding, or lifting) on a fixed area of bed surface (Einstein, 1950; Cheng and Chiew, 1998).

Figure 3(a)-(c) show probabilities of threshold conditions for rolling, sliding, and lifting, respectively. It is expected that the theoretical curves in Figure 2 correspond to curves of probability of 0.5 in Figure 3. Shields curve is also provided for comparisons in the figures. It is observed that Shields curve corresponds to a probability of 0.7, 0.45, and 0.04 for rolling, sliding, and lifting, respectively, at large values of Re_*. This means that sediment particles, if they are exposed to Shields critical stress, have about 70% chance of rolling, 45% chance of sliding, and 4% chance of lifting, respectively. If we assume that these processes are independent, it can be deduced that 59% of moving particles are transported by rolling, 38% by sliding, and only 3% by lifting at high Re_*. However, it should be indicated that the lifting contributes best to the amount of sediment transport because of traveling distance although the lifting probability is the lowest.

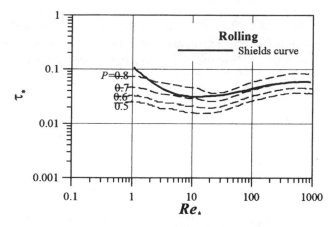

Fig 3(a):. Rolling Threshold for Different Probabilities

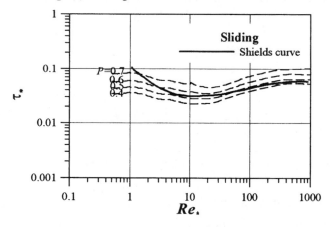

Fig 3(b):. Sliding Threshold for Different Probabilities

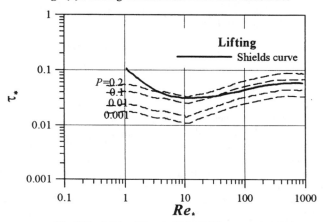

Fig 3(c):. Lifting Threshold for Different Probabilities

322

4. CONCLUSIONS

In this paper, three different modes in the initiation of sediment particles on the bed were investigated theoretically and probabilistically. Threshold conditions of each mode were obtained from force and momentum balances, and they were solved theoretically. Results were presented in a dimensionless form, i.e., critical bed shear stress versus friction Reynolds number. It was shown that rolling takes place at the lowest bed shear stress and lifting does at the highest one for a given flow condition. Shields curve was shown to lie between sliding and lifting curves at high friction Reynolds number. Probabilistic analysis was made possible by assuming that the velocity approaching the sediment particle is distributed normally. The analysis revealed that Shields data corresponds to greater than 70% probability of rolling, 45% of sliding, and 4% of lifting, respectively, at high friction Reynolds number.

REFERENCES

Cheng, N. -S. and Chiew, Y. -M. (1998). "Pickup probability for sediment entrainment." *Journal of Hydraulic Engineering*, ASCE, 124(2), 232-235.

Chepil, W. S. (1958). "Use of evenly spaced hemispheres to evaluate aerodynamic forces on soil surfaces." *Eos. Transactions of AGU*, 39(3), 397-404.

Coleman, N. L. (1967). "A theoretical and experimental study of drag and lift forces acting on a sphere resting on a hypothetical stream bed." *Proceeding, 12th Congress*, IAHR, Fort Collins, Colo., Vol. 3, 185-192.

Einstein, H. (1950). "The bed load function for sediment transportation in open channel flows." *Tech. Bull. 1026*, U.S. Department of Agriculture, Washington, D.C.

Einstein, H. A. and El-Samni, E. A. (1949). :Hydrodynamics forces on a rough wall." *Rev. of Modern Phys.*, 21, 520-524.

Ikeda, S. (1982). "Incipient motion of sand particles on side slope." *Journal of Hydraulic Division*, ASCE, 108(HY1), 95-113.

Kennedy, J. F. (1995). "The Albert Shields story." *Journal of Hydraulic Engineering*, ASCE, 121(11), 766-772.

Kironoto, B. A. and Graf, W. H. (1994). "Turbulence characteristics in rough uniform open-channel flow." *Proceeding of International Civil Engineers, Water, Maritime, and Energy*, London, England, 106(Dec.), 333-344.

Li, Z. Y., et al. (1983). "Laboratory investigation on drag and lift forces acting on bed spheres." *Proceeding, 2nd International Symposium of River Sedimentation*, Water and Power Press, Beijing, China, 330-340.

Nezu, I. and Nakagawa, H. (1996). *Turbulence in Open channel flows*. IAHR Monograph, A.A. Balkema, Rotterdam, The Netherlands.

Patnaik, P. C., Vittal, N., and Pande, P. K. (1994). "Lift coefficient of a stationary sphere in gradient flow." *Journal of Hydraulic Research*, Delft, The Netherlands, 32(3), 471-480.

Reichardt, H. (1951). "Vollstandige darstellung der tubulenten geschwindig-keitsverteilung in glatten leitungen." *Z. Angew. Math. Mech.*, Berlin, Germany, 31(7), 208-219 (in German).

Shields, A. (1936). "Application of similarity principles and turbulence research to bed-load movement." *Rep.*, W. P. Ott and J. C. van Uchelen, translators, California Inst. Of Technol., Pasadena, Calif.

Wiberg, P. L. and Smith, J. D. (1985). "A theoretical model for saltating grains in water." *Journal of Geophysical Research*, AGU, 90(c4), 7341-7354.

Stochastic Hydraulics 2000, Wang & Hu (eds) © 2000 Taylor & Francis, ISBN 90 5809 166 X

Shear stress and percentage of initiating particles on the bed

Zhang Xiaofeng
Department of River Engineering, Wuhan University of Hydraulic and Electric Engineering, Wuhan, China

Tamotsu Takahashi
Disaster Prevention Research Institute, Kyoto University, Japan

ABSTRACT: To overcome the defect of arbitrariness in judging the critical stress of sediment movement in experiments, quantitative method is needed to describe this gradual process. In this paper, an equation of the moving state probability considering the exchange time as suggested by Einstein is presented to describe the condition of initiation of particle motion. Relations between the pick up rate and the moving state probability of single particle, and the percentage of picked up particles from the bed per unit time and the shear stress are established. These results are not only helpful for understanding the phenomena but also applicable to measurement in experiments.

Key words: Shear stress, Bed sediment, Initiation probability of moving state of particles

1. INTRODUCTION

Because of turbulence in flow and randomness of particles' position on the bed, the process of initial motion of particles is stochastic in nature. The critical condition is not the one under which all the particles on the top layer begin to move suddenly. In fact, as observed, some move and some do not move. Kramer (1935) proposed a qualitative method using the following three intensities of motion as the critical condition: (1) week movement, (2) medium movement, (3) intense movement. Dou (1962), proposed to use the probabilities for particle incipient motions at values 0.0014, 0.0228, 0.1585 corresponding to Kramer's three critical conditions, respectively. Yalin (1972) suggested a parameter as the criterion of initial motion. Some researchers used limiting condition when the sediment discharge tends to zero by extrapolating the curve of observed rate of sediment transport versus shear stress, while Han (1984) used a small bed load transport rate as critical condition.

In this paper, a method is developed to establish the relationship between the shear stress and the percentage of particles initiating to move by describing the stochastic behavior of the initial motion. Sediment in natural rivers consists of mixtures of heterogeneous materials. Finer sediment tends to be cohesive. In this paper, however, as the first part of research on the critical condition, only non cohesive uniform sediment is concerned.

2. CONDITION FOR INITIATION OF PARTICLE MOTION

Consider a particle arbitrarily lying on the bed. The forces acting on this grain are hydrodynamic drag F_D, parallel to the bed, and lift F_L, normal to the bed which are the forces to move the grain, and submerged grain weight W, which resists to move. The equilibrium equation at the moment of initiation of motion gives

$$K_1 dF_D + K_2 dF_L = K_3 dW \tag{1}$$

in which d is the diameter of the grain, K_1d, K_2d, K_3d are the corresponding arms for the forces F_D, F_L and W, respectively.

At an instant, owing to the fact that the flow is turbulent, both F_D and F_L are random variables. Arms of forces K_1d, K_2d, and K_3d should also be treated as random variables because of arbitrary shape and position on the bed. But, for the sake of simplicity in mathematical treatment which helps to clarify the innate character of the problem, values of all these arms are replaced by their own mean values tentatively (Dou, 1962). Thus, random variables left in the equation (1) are F_D and F_L. Let the instantaneous driving moment be denoted by M,

$$M = K_1 d F_D + K_2 d F_L \tag{2}$$

In the previous research, many researchers think that the probability of incipient motion of a single grain equals the probability of $M \geq K_3 dW$. But removal of a particle needs an exchange time as suggested by Einstein (1950). During a period t, M might exceed $K_3 dW$ many times. But if the time interval of exceeding $K_3 dW$ is less than the exchange time Δt_*, the particle will only vibrate and yet will not be removed away from its position. So, only the particle that meets following two conditions simultaneously can be removed away: (1) M exceeds $K_3 dW$ and (2) time interval during which the moment exceeds $K_3 dW$ is longer than Δt_*. Denote Δt be the time interval of $M \geq K_3 dW$, according to the theory of probability, moving probability of a single particle p, can be given as

$$p = P[M \geq K_3 dW] \cdot P[\Delta t \geq \Delta t_* | M \geq K_3 dW] = p_1 \cdot p_2 \tag{3}$$

where $p_1 = P[M \geq K_3 dW]$, $p_2 = P[\Delta t \geq \Delta t_* | M \geq K_3 dW]$.

3. RELATION BETWEEN p_1 AND THE SHEAR STRESS

According to Zhang's (1995a, 1995b) research, F_D and F_L are normally distributed random variables. Assuming that F_D and F_L are independent variables, one obtains that M follows normal distribution. Following relations can be given:

$$\overline{M} = K_1 d \overline{F}_D + K_2 d \overline{F}_L \quad , \quad \sigma_M^2 = K_1^2 d^2 \sigma_1^2 + K_2^2 d^2 \sigma_2^2 \tag{4}$$

where F_D is the mean hydrodynamic drag force, F_L is the mean lift force, σ_1 is the standard deviation of drag force and σ_2 is the standard deviation of lift force. According to the experimental results (Fu 1992), relation between the standard deviations and the mean F_D and F_L are: $\sigma_1 = 1.11 F_D$, $\sigma_2 = 1.69 F_L$.

The mean drag force and the mean lift force can be expressed as follows (Fu 1992)

$$\overline{F}_D = C_D \alpha_1 d^2 \rho \frac{\overline{u}_b^2}{2} \quad , \quad \overline{F}_L = C_L \alpha_2 d^2 \rho \frac{\overline{u}_b^2}{2} \tag{5}$$

where C_D and C_L are the drag and the lift coefficients, respectively, α_1 and α_2 the coefficients, ρ the density of water and $\overline{u}_b$ is the mean characteristic velocity. Substitution of equation (5) into equation (4) yields

$$\overline{M} = (K_1 C_D \alpha_1 + K_2 C_L \alpha_2) d^3 \rho \frac{\overline{u}_b^2}{2} \quad , \quad \sigma_M = \sqrt{1.23 K_1^2 C_D^2 \alpha_1^2 + 2.86 K_2^2 C_L^2 \alpha_2^2} \, d^3 \rho \frac{\overline{u}_b^2}{2} \tag{6}$$

Dou (1962) gave $C_L = 0.1$, $C_D = 0.4$, $\alpha_1 = 2\pi/9$, $\alpha_2 = \pi/3$, $K_1 = 1/3$ and $K_2 = 1/2$. Substitution of those values into equation (6) gives

$$\overline{M} = 0.15 d^3 \rho \frac{\overline{u}_b^2}{2} \quad , \quad \sigma_M = 0.13 d^3 \rho \frac{\overline{u}_b^2}{2} \tag{7}$$

Accompanying with variation of the mean characteristic velocity $\overline{u}_b$, $\overline{M}$ and σ_M will change.

Note the resisting moment $M_C = K_3 dW = K_3 \frac{\pi}{6} (\gamma_s - \gamma) d^4$, which is independent of $\overline{u}_b$, probability p_1 is given by

$$p_1 = \int_{M_C}^{+\infty} \frac{1}{\sqrt{2\pi} \sigma_M} e^{-(x - \overline{M})^2 / 2\sigma_M^2} dx \tag{8}$$

Writing $t = (x - \overline{M})/\sigma_M$, $C = (M_c - \overline{M})/\sigma_M$, equation (8) becomes

$$p_1 = \int_C^{+\infty} \frac{1}{\sqrt{2\pi}} e^{-t^2/2} dt$$

(9)

Substituting equation (7) into C, one obtains

$$\overline{u}_b = \sqrt{\frac{2}{0.15 + 0.13C}} \sqrt{\frac{\pi}{6} K_3 \frac{\rho_s - \rho}{\rho} gd}$$

(10)

where ρ_s is the density of the grain and g is the acceleration due to gravity.

For the rough bed, vertical velocity distribution is given by (Keulegan 1938, Einstein 1950)

$$\frac{u}{u_*} = 8.5 + 5.75 \log\left(\frac{y}{K_s}\right)$$

(11)

where u is the velocity at distance y from the bed, u_* the shear velocity, K_s the size of the equivalent sand roughness. According to Dou (1962), $K_3 = \frac{1}{2}$ and evaluation of $\overline{u}_b$ is done at $y = K_s = d$. Then, relation between the shear stress τ and the probability p_1 on the bed can be obtained via the following formula

$$\frac{\tau}{(\rho_s - \rho)gd} = \frac{\rho u_*^2}{(\rho_s - \rho)gd} = \frac{\pi}{65.03 + 56.55C}$$

(12)

Namely, when the shear stress is known, equation (12) gives C, then by equation (9), probability p_1 can be calculated.

Regarding with the p_2, because proper method is not available, at present, to experimentally determine the exchange time Δt_*, an indirect method is introduced.

4. PICK UP RATE AND PROBABILITY OF MOTION OF SINGLE PARTICLE

According to the stochastic theory, movement process of a single particle can be described as Markov process with continuous time and intermittent states, and the waiting time at each state follows an exponent distribution (Yano et al 1968), i.e.

$$P[i; \tau] = e^{-\tau/T_i}$$

(13)

where i represents the state, T_i is the mean time interval in which a particular state i lasts during each step.

Let $i = 1$ and $i = 2$ represent the resting state and the moving state of the particle, respectively. Assuming that during the time interval $(t, t + \Delta t)$ state changing occurs at most once, equation (13) gives

$$P_{11}(\Delta t) = e^{-\Delta t/T_1} , \quad P_{22}(\Delta t) = e^{-\Delta t/T_2} , \quad P_{21}(\Delta t) = 1 - e^{-\Delta t/T_2} , \quad P_{12}(\Delta t) = 1 - e^{-\Delta t/T_1}$$

(14)

where $P_{ij}(\Delta t)$ is the probability of state change from the state i to the state j during the time interval $(t, t + \Delta t)$. Following equations can be obtained:

$$P_{12}(t + \Delta t) = P_{11}(t)P_{12}(\Delta t) + P_{12}(t)P_{22}(\Delta t)$$

(15)

$$P_{12}'(t) = \lim_{\Delta t \to 0} \frac{P_{12}(t + \Delta t) - P_{12}(t)}{\Delta t} = P_{11}(t)/T_1 - P_{12}(t)/T_2$$

(16)

Similarly, we obtain:

$$P_{21}'(t) = -P_{21}(t)/T_1 + P_{22}(t)/T_2 , \quad P_{11}'(t) = -P_{11}(t)/T_1 + P_{12}(t)/T_2 , \quad P_{22}'(t) = P_{21}(t)/T_1 - P_{22}(t)/T_2$$

(17)

According to the theory of Markov process, when $t \to \infty$, $P_{ij}(t)$ approaches stationary distribution, i.e.

$$\lim_{t \to \infty} P_{ij}(t) = P_j , \quad \lim_{t \to \infty} P_{ij}'(t) = 0$$

(18)

From equation (18), when $t \to \infty$, equations (16) and (17) become an identical equation:

$$P_1/T_1 - P_2/T_2 = 0$$

(19)

where P_1 is the probability of particle at the resting state, while P_2 is the probability at the moving state, and

$$P_1 + P_2 = 1 \tag{20}$$

From equations (19) and (20), P_1 and P_2 can be solved as following:

$$P_1 = \frac{T_1}{T_1 + T_2} \tag{21}$$

$$P_2 = \frac{T_2}{T_1 + T_2} = p \tag{22}$$

The mode of a single particle movement with time t along the river course x can be shown schematically as figure 1. Each step of movement includes two states: one is the resting and the other is the moving state. The mean time interval of each step is

$$T = T_1 + T_2 \tag{23}$$

in which T_1 is the mean resting time interval in one step, T_2 is the mean moving time interval in one step. Equation (22) means that p can be determined experimentally by measuring two parameters : (1) the mean moving time interval of each step, and (2) the mean total time interval of each step. From equation (22) and Nakagawa & Tsujimoto's (1975) results, relation between the pick up rate p_s and the moving probability p of a single particle is given by

$$p_s = \frac{p}{T_2} \tag{24}$$

5. RELATION BETWEEN p2 AND THE SHEAR STRESS

Substitution of equation (3) into (22) gives

$$p_2 = \frac{T_2}{Tp_1} \tag{25}$$

Concerning with T_2, many researchers made experimental studies on the saltation process using high-speed photography or video camera. Lee & Hsu (1994) and Hu (1995) used high-speed photography to investigate saltating particle motions, while Sun (1996) used video camera to search relation between T_2 and the shear stress. From the result obtained by Sun, following relation can be obtained:

$$T_2 \sqrt{\frac{\left(\rho_s/\rho - 1\right)g}{d}} = \frac{15.36}{1-p} \tau_*^{0.11} \left(\sqrt{\tau_*} - 0.072\right)^{0.12} \tag{26}$$

where $\tau_* = \dfrac{\tau}{(\rho_s - \rho)gd}$

The parameter T can be measured in the experiments on dispersion of labled or tracer particles. Einstein was the first to establish the model of probability distribution of particle dispersion.

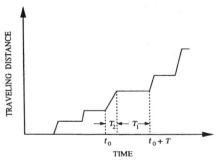

Fig.1: Process of Particle Movement

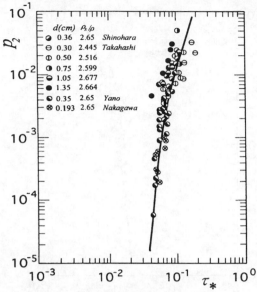

Fig. 2: Relation between p_2 and τ_*

Based on his result, several new models describing particle dispersion process were developed, (Han 1984, Hubbel & Sayre 1964). With experimental data (Shinohara & Tsubaki 1957, Takahashi 1966, Yano *et al.* 1968, Nakgawa & Tsujimoto 1975), and equations (9), (12), (25), and (26), relation between p_2 and τ_* is plotted in figure 2, which can be expressed as

$$p_2 = 2.016 e^{-\frac{0.35}{\tau_*}} \tau_*^{0.61} (\sqrt{\tau_*} - 0.072)^{0.12}$$

(27)

Substituting equations (9) and (27) into (3), relation between the moving probability of a single particle p and the shear stress τ_* is given by

$$p = 2.016 \tau_*^{0.61} (\sqrt{\tau_*} - 0.072)^{0.12} e^{-\frac{0.35}{\tau_*}} \int_{\frac{\pi}{56.55\tau_*} - 1.15}^{\infty} \frac{1}{\sqrt{2\pi}} e^{-t^2} dt$$

(28)

6. PERCENTAGE OF MOVING PARTICLES AND MOVING PROBABILITY OF SINGLE PARTICLE

Sediment on the bed consists of a huge amount of grains. In a certain area, number of sediment grains is countable. Supposing the total number in this area is n, possible numbers of sediment particles which will move are 0, 1, 2, ..., n. When the moving probability of a single particle is p, according to the probability theory, probability of just $k(0 \le k \le n)$ particles at moving state is

$$P\{X = k\} = P_n(k) = \binom{n}{k} p^k (1-p)^{n-k} , \qquad k = 0, 1, 2, ..., n$$

(29)

Expectation of random variable X in equation (29) is

$$E(X) = \sum_{k=0}^{100} k \binom{100}{k} p^k (1-p)^{100-k} = 100p$$

(30)

This means that average percentage of particles at moving state equals $100p$ and average percentage of particles removed from their position per unit time is $100p/ T_2$. These results can also be used to explain the physical meaning of equation (22). Considering a hundred particles on bed, at any moment, there are $100p$ particles at moving state. Average moving time interval of each particle is T_2. During the time interval T, all of the hundred particles have the chance to move once.

329

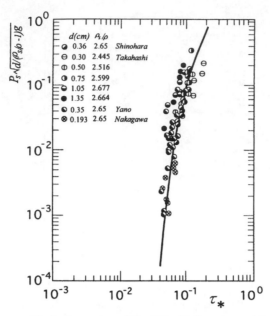

Fig.3: Comparison of Eq. (31) with Experimental Data

7. RELATION BETWEEN THE MEAN PERCENTAGE OF STARTING PARTICLES AND THE SHEAR STRESS

When the shear stress is known, probability p and percentage of particles at moving state P ($P = 100p\%$) can be calculated from equation (28). With equations (24), (26) and (28), percentage of particles picked up from the bed once per unit time is given by

$$P_x = 100p_x = 13.125(1 \quad p)\sqrt{\frac{(\rho_s / \rho \quad 1)g}{d}}\sqrt{\tau}\; e^{\frac{0.35}{\tau}} \int_{\left(\frac{\pi}{56.55\tau} \quad 1.15\right)}^{\infty} \frac{1}{\sqrt{2\pi}} e^{t^2/2}\, dt \tag{31}$$

That is: (1) at any moment, possibility of a particle at moving state is p, (2) during unit time, P_x percentage of particles move once a time. Calculated results and experimental data are shown in figure 3.

8. CONCLUSION

In this paper, following results were obtained:
(1) To remove the particle away from it's position, two conditions should be simultaneously met: one is that the instantaneous moment for motion acting on the particle exceeds the resisting moment; the other is that the time interval of exceeding is longer than an exchange time as suggested by Einstein. Thus, according to the theory of probability, moving probability of a single particle can be described by equation (3).
(2) According to the probability theory and the mechanical treatment and experimental data collected by other researchers, equation (28) that describes the relation between moving state probability of single particle and the shear stress was deduced.
(3) From Markov process, equation (24) that describes the relation between moving state probability and pick up rate was derived.
(4) Number of moving particles follows the binomial distribution. Relation of the moving probability with the average percentage of initiating particles can be described by equation (30).
(5) Equation (31), representing the relation between the percentage of particles picked up from the bed once per unit time and the shear stress, was established to describe the gradual process of sediment initiation.

REFERENCES

Dou, G. (1962). "Critical velocity of particles in bed". *Scientia Sinica*, 11(7), 999-1032 (in Russian).

Einstein, H. A. (1950). "The bed load function of sediment transportation in open channel flows." *U. S. Dept. Agri., Tech. Bull.* 1026.

Fu, R. (1992). "Drag and lift forces acting on uniform particles on bed and their stochastic properties." *Proc., National Conference on Sediment Transport Theory,* Beijing, China (in Chinese).

Han, Q. & He, M. (1984). Statistic Theory of Sediment Movement. Science Press, Beijing, 312-320 (in Chinese).

Hu, C. & Hui, Y. (1995). Open Channel Sediment Laden Flow Mechnics and Statistical Properties. Science Press, Beijing (in Chinese).

Hubbel, D.W. & Sayre, W.W. (1964). "Sand transport studies with radioactive tracers." *J. Hydr. Div.,* ASCE, 90(HY3), 39-68.

Keulegan, G. H. (1938). "Laws of turbulent flow in open channel." *J. Res.,* U. S. Nat. Bureau of Standards, 21, 701-741.

Kramer, H. (1935). "Sand mixtures and sand movement in fluvial models." *Trans.,* ASCE, 100, paper No. 1909, 798-878.

Lee, H. & Hsu, I. (1994). "Investigation of saltating particle motions." *J. Hydr. Engr.,* ASCE, 120(7), 831-845 .

Nakagawa, H. & Tsujimoto, T. (1975). "Study on mechanism of motion of individual sediment particles." *Proc.,* JSCE, (244), 71-80 (in Japanese).

Shinohara, M. & Tsubaki, T. (1957) "Transport mechanism of sediment on the river bed". *Annual of Applied Mechanics Institute*, Kyushu University, Japan, (10), 85-94 (in Japanese).

Sun, Z. (1996). "Stochastic theory of non uniform sediment transport." PhD thesis Wuhan University of Hydraulic and Electric Engineering, Wuhan, China (in Chinese).

Takahashi, M. (1966). "Experimental study on the gravel transportation (1st report)." *J. Japan Society Erosion Control Engineering,* 18(4), 5-14 (in Japanese).

Yalin, M. S. (1972) Mechanics of Sediment Transport, Pergamon Press, 74-110.

Yano, K., Tsuchiya, Y. & Michiue, M. (1968). "On the stochastic characterstics of transport mechanism of sand in a stream." *Annual of Disaster Prevention Research Institute,* Kyoto University, Japan, (11B), 61-73 (in Japanese).

Zhang, X. & Xie, B. (1995a). "Incipient velocity and incipient probability of particles." *J. Hydr. Engr.* Beijing, China, (10), 53-59 (in Chinese).

Zhang, X.(1995b). "Probability distribution of instantaneous flow drag and lift forces on particles." *J. Hydr. Engr.* Beijing, China, (supple), 155-158 (in Chinese).

Stochastic Hydraulics 2000, Wang & Hu (eds) © 2000 Taylor & Francis, ISBN 90 5809 166 X

Influences of the waterway project in the Yangtze River Estuary on sediment transport

Zhou Jifu, Li Jiachun & Liu Hedong
Institute of Mechanics, Chinese Academy of Sciences, Beijing, China

ABSTRACT: A waterway training project is being conducted in the Yangtze River Estuary to deepen and stabilize the navigation channel. In present paper, a 2D mathematical model is developed to study the dynamic processes of sediment transport around the project. Flow pattern, salinity distribution, sediment movement, bed deformation, and their interactions have been simulated simultaneously. In particular, the interaction of turbulence and sediment suspension is taken into consideration by introducing an entrainment function. And flocculation settling process in salty water is accounted for as well. The model is calibrated by using large amount of field data sets and then used to predict sediment transport and salinity distribution in the concerned area when the first phase project is completed.

Key words: Yangtze River Estuary, Waterway Project, Sediment Transport.

1. INTRODUCTION

The Yangtze River is the most important waterway of China. Its estuary is divided by the Chongming Island into the North Branch and the South Branch. The South Branch further bifurcates into the North and South Waterway. Again, the South Waterway bifurcates into the North Channel and the South Channel (see Fig. 1). Large ships enter into Shanghai Port via the North Channel. However, the minimum water depth in this channel is less than 7 m, because sediment deposit at the mouth and shape a mouth bar. The mouth bar naturally becomes a serious obstacle to modern navigation and shipping in the Shanghai Port, since the 3-rd or 4-th generation container ships need a waterway of 12.5-m at least. Consequently, a waterway deepening project has been proposed. The first phase of the project is to construct two levees on the base of Jiuduan and Hengshadong Shoals (see Fig. 1) along the sides of the North Channel in order to restrict water in the navigation waterway. At the same time, three groins are attached at each levee to facilitate sediment flushing for an 8.5-m-deep waterway maintenance. After completion of the whole project, a 12.5-m-deep waterway will be formed with routine dredge.

What about sedimentation in the navigation waterway? Is there any variation in the maximum turbidity? Whether or not salty water intrudes more upstream after completion of the project? These questions have drawn much attention from engineers and scientists. To resolve these problems, we have proposed a 2D mathematical model to simulate unsteady sediment and salinity transport in the area of concern. The model simultaneously simulates flow pattern, salinity distribution, sediment movement (including suspended load and bed load), bed deformation, and their interactions. In particular, the interaction of turbulence and sediment suspension is taken into consideration by introducing an entrainment function. And flocculation settling process in salty water is accounted for as well. A nested grid system with higher resolution in the area near the navigation channel is used to enhance the accuracy of computation.

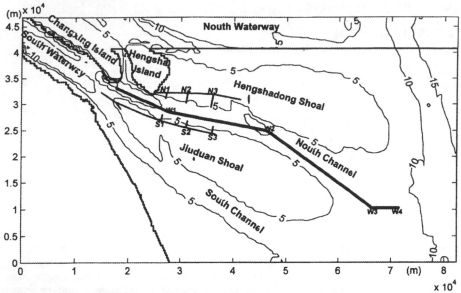

Fig. 1: General Bathymetry of the Yangtze River Estuary; Settings of the Hydro-Structures of the Project; Mesh Arrangement.

The simulated results demonstrate that: (1) with the marine structures erected, sediment particles would not be eroded from the shoals along the North Channel to the navigation waterway any more. Meanwhile, the groins shall locally increase current velocity in the waterway, which favors bed degradation. (2) The structures seem to exert little effect on the maximum salinity intrusion and the maximum turbidity in the navigation channel. And (3) both the annual deposition volume hindering navigation and the total annual deposition volume in the main navigation channel are no more than those under natural conditions. However, dredging is still necessary to maintain the designed navigation channel to the level of 8.5 m depth, but the cost seems to be affordable.

2. MATHEMATICAL MODEL

2.1 Governing Equations

The two dimensional flow is governed by following equations:

$$\frac{\partial \zeta}{\partial t} + \frac{\partial (Hu)}{\partial x} + \frac{\partial (Hv)}{\partial y} = 0 \tag{1}$$

$$\frac{\partial u}{\partial t} + u\frac{\partial u}{\partial x} + v\frac{\partial u}{\partial y} = fv - g\frac{\partial \zeta}{\partial x} - \frac{1}{\rho H}\tau_{bx} + \frac{1}{\rho H}\frac{\partial}{\partial x}(H\tau_{xx}) + \frac{1}{\rho H}\frac{\partial}{\partial y}(H\tau_{xy}) \tag{2}$$

$$\frac{\partial v}{\partial t} + u\frac{\partial v}{\partial x} + v\frac{\partial v}{\partial y} = -fu - g\frac{\partial \zeta}{\partial y} - \frac{1}{\rho H}\tau_{by} + \frac{1}{\rho H}\frac{\partial}{\partial x}(H\tau_{yx}) + \frac{1}{\rho H}\frac{\partial}{\partial y}(H\tau_{yy}) \tag{3}$$

where, ζ is the water level, H is water depth, u, v are velocity components, g is the gravitational acceleration, ρ is the density of water, $f = 2\Omega \sin \varphi$ is the Coriolis coefficient, Ω is the radian velocity of the Earth, φ is the attitude. τ_{bx}, τ_{by} are x-, y-components of bottom shear stress τ_b,

334

$$\tau_{bx} = \rho gu\sqrt{u^2 + v^2} / C^2$$

$$\tau_{by} = \rho gv\sqrt{u^2 + v^2} / C^2$$

$C = \dfrac{1}{n} H^{1/6}$ is the Chezy coefficient, n is Manning's roughness.

$$\tau_{xx} = 2\rho v_t \frac{\partial u}{\partial x}, \qquad\qquad \tau_{yy} = 2\rho v_t \frac{\partial v}{\partial y},$$

$$\tau_{xy} = \rho v_t \left(\frac{\partial u}{\partial y} + \frac{\partial v}{\partial x} \right), \qquad \tau_{yx} = \rho v_t \left(\frac{\partial v}{\partial x} + \frac{\partial u}{\partial y} \right)$$

are horizontal shear stresses. v_t is turbulent viscosity.

Two dimensional convection-diffusion equation for salinity is expressed as (Chian and Wan, 1986)

$$\frac{\partial (Hs)}{\partial t} + \frac{\partial (Hus)}{\partial x} + \frac{\partial (Hvs)}{\partial y} = \frac{\partial}{\partial x}\left(HD_{sx}\frac{\partial s}{\partial x} \right) + \frac{\partial}{\partial y}\left(HD_{sy}\frac{\partial s}{\partial y} \right) \qquad (4)$$

in which, s is salinity, D_{sx}, D_{sx} are horizontal diffusivity for salinity.

The 2-D suspended sediment equation takes the form

$$\frac{\partial (Hc)}{\partial t} + \frac{\partial (Huc)}{\partial x} + \frac{\partial (Hvc)}{\partial y} = \frac{\partial}{\partial x}\left(HD_x \frac{\partial c}{\partial x} \right) + \frac{\partial}{\partial y}\left(HD_y \frac{\partial c}{\partial y} \right) + S \qquad (5)$$

where, c is sediment concentration, D_x, D_y are horizontal sediment diffusivity, S is a source term. It is closely related to mechanism of the exchange of suspended load and bed load. It can be expressed as follows:

$$S = E_n \rho_s - \omega C_b \qquad (6)$$

Here, ρ_s is the density of sediment particles, ω is the settling velocity. In estuaries, sediment particles are generally so fine as to flocculate in salty water environment. The settling velocity of flocculated particles is generally very larger than that of an individual particle. Hence, it is important to take the flocculation effect on settling into consideration (Cao, 1997). E_n is called entrainment function, representing sediment flux in volume entrained up from unit bed area per unit time. A turbulent bursting-based entrainment function is derived by Cao (Tang, 1963) as follows:

$$E_n = PC_{0b}\sqrt{\frac{\rho_s - \rho}{\rho g}}d\left(\frac{\tau'}{\tau_c} - 1 \right)\frac{\tau'}{\rho_s - \rho} \qquad (7)$$

in which, C_{0b} is the volumetric concentration of bed material. P is a factor related to the averaged area of all bursts per unit bed area and the nondimensional bursting period. It varies from 370 to 480. d is the mean grain diameter. τ' is the skin friction of bed shear stress. τ_c is the critical incipient stress, which can be estimated by (ISDIW, 1997):

$$\tau_c = \frac{1}{77.5}\left[3.2(\rho_s - \rho)gd + \left(\frac{\gamma_b}{\gamma_{b0}} \right)^{10}\frac{k}{d} \right] \qquad (8)$$

where, $\gamma_{b0} = 1.6 \; g/cm^3$ is the bulk density of compact bed material, $k = 2.9 \times 10^{-4} \; g/cm$, γ_b is the bulk density of bed material.

C_b in equation (6) is the reference concentration. According to the exponential concentration profile, an expression for C_b can be derived as a function of the depth mean concentration:

335

$$C_b = K \left[\frac{\omega(H-a)}{\varepsilon_z} \left(1 - e^{-\frac{\omega}{\varepsilon_z}(H-a)} \right)^{-1} \right] c , \tag{9}$$

$K=1.08$, $\varepsilon_z = \frac{1}{6}\kappa u_* H$ is the vertical sediment diffusivity, $\kappa = 0.4$ is the von Karman's constant, a is reference height, taken as the mean diameter of bed grains.

The bed load transport rate is estimated by Dou's formula. Its x and y components read as

$$q_{bx} = \frac{K_0}{C_0^2} \frac{\rho_s \rho}{\rho_s - \rho} \left(\sqrt{u^2 + v^2} - u'_c \right) \frac{u(u^2 + v^2)}{\omega} \tag{10}$$

$$q_{by} = \frac{K_0}{C_0^2} \frac{\rho_s \rho}{\rho_s - \rho} \left(\sqrt{u^2 + v^2} - u'_c \right) \frac{v(u^2 + v^2)}{\omega} \tag{11}$$

in which,

$$u'_c = 0.264 \ln\left(11\frac{H}{K_s} \right) \sqrt{\frac{\rho_s - \rho}{\rho} gd_{50b} + 0.19\frac{\varepsilon_k + gH\delta}{d_{50b}}} \tag{12}$$

is the critical incipient velocity of bedload particles. d_{50b} is the mean diameter of bedload. $\varepsilon_k = 2.56cm^3/s^2$ is the coefficient of cohesive force, $\delta = 0.21 \times 10^{-4} cm$ is the thickness of pellicular water, K_s is bed roughness height.

$$K_s = \begin{cases} 0.5mm & d_{50b} \leq 0.5mm \\ d_{50b} & d_{50b} \geq 0.5mm \end{cases}$$

K_0 is a constant. $C_0 = 2.5 \ln\frac{11H}{K_s}$.

Due to sediment transport, sea bed will deform by scour or deposition. The bed elevation η_b should satisfy the following equation, derived from sediment conservation law:

$$\gamma_{s0} \frac{\partial \eta_b}{\partial t} + \frac{\partial q_{bx}}{\partial x} + \frac{\partial q_{by}}{\partial y} + S = 0 \tag{13}$$

where, γ_{s0} is dry specific weight of bed material.

2.2 Calibration of the Model

According to the feasibility study of the first phase project[4], we have assumed the following fundamental parameters for validations and predictions:

$\rho_s = 2650\,kg/m^3$; $d = 0.01mm$; $d_{50b} = 0.1mm$; $\gamma_{s0} = 1000\,kg/m^3$; $\rho = 1000\,kg/m^3$.

By comparing calculated results with measured data, we have $n = 0.015$ for flood duration and $n = 0.013$ for ebb duration. And the eddy viscosity is set to 150 m^2/s. Meanwhile, the sediment and salinity diffusivity are related to friction velocity:

$$D_x = D_{sx} = \beta u_{*x} H \qquad D_y = D_{sy} = \beta u_{*y} H$$

Here, $u_{*x} = \sqrt{\tau_{bx}/\rho}$, $u_{*x} = \sqrt{\tau_{by}/\rho}$ and $\beta = 200$.

The calculated domain is 82.3433 km long and 46.7945 km wide, and is meshed with space steps $\Delta x = 395.88m$, $\Delta y = 463.31m$ and refined ones $\Delta x' = 197.94m$, $\Delta y' = 185.32m$.

The time step is $\Delta t = 60s$ for flow calculation, $\Delta t = 120s$ for sediment and salinity transport and $\Delta t = 240s$ for bed deformation.

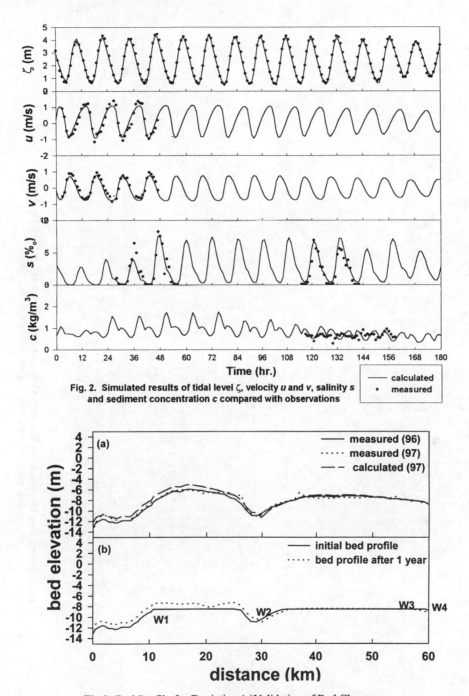

Fig. 2. Simulated results of tidal level ζ, velocity u and v, salinity s and sediment concentration c compared with observations

Fig.3: Bed Profile for Depicting (a)Validation of Bed Changes; (b) Prediction of Sedimentation of the Navigation Channed After the Projict

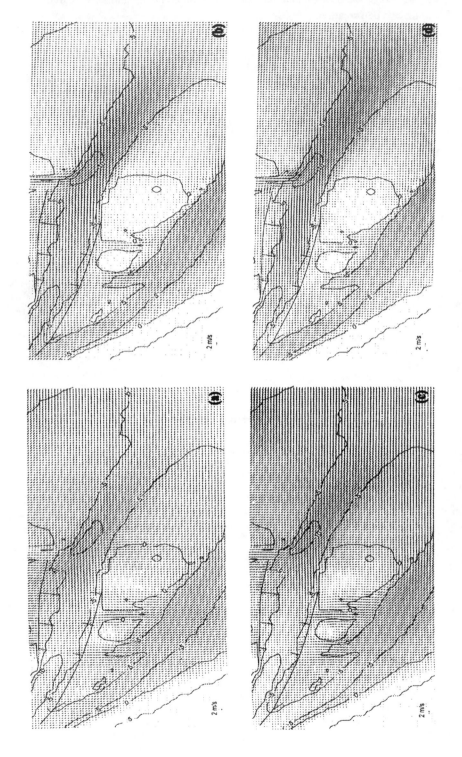

Fig. 4. The simulated flow patterns. (a) flood pattern before the project; (b) ebb pattern before the project; (c) flood pattern after the project; (d) ebb pattern after the project.

338

Fig. 2 shows comparisons of tidal level, velocity (two components), salinity and sediment concentration with measured data. The agreement of tide elevation and velocity is rather good both in phase and magnitude. Calculated salinity reflects the measured salinity variations well in phase with some deviation in magnitude, which is however acceptable for engineering purpose. The averaged magnitude of simulated sediment concentration coincides with the observation. Also, comparison of the calculated and measured profiles of the main navigation passage is shown in Fig. 3(a). It is seen that the calculated bed elevation is higher than the measured bed curve in the upstream reach from W2. However, since dredging is carried out annually to maintain 7m depth and the calculated sediment deposition volume above –7 m approximates to 11,000,000 m^3, very close to actual annual dredging volume 10,000,000 m^3 prior to the project.

3. RESULTS AND DISCUSSIONS

3.1 Effects of the project on flow pattern

The simulated flow patterns are shown in Fig. 4.

It is seen that, under natural conditions, over-shoal-flows exist above Jiuduan and Hengshadong Shoals, whose top elevations are about -5 m. Via the over-shoal-flows, the navigation channel exchanges water with the South Channel and the North Waterway. This water exchange does not favour navigation, since the going-out flow reduces sediment-carrying capability of currents in the navigation channel and the going-in flow carries bedload and the bedload very likely settles down in the navigation channel.

After the first phase of the project, an increase of current velocity in the navigation channel can be seen and a small portion of water flows along north side of north levee. The water exchanges are cut off due to existence of the levees whose top elevation is 2 m. Sediment loads are, of course, detained out of the navigation channel. When tidal level is high, over-levee-flows can be seen during flood period. Nevertheless, this overflow carries less sediment load into the navigation channel due to small sediment concentration in the upper water column. In addition, the sediment of upper part is generally very fine and can be easily carried away. Thus, the load carried by the over-levee-flow has less impact on the navigation channel.

3.2 Salty water intrusion

Salty water intrusion is closely related to tides and runoff discharge. Generally speaking, larger tidal range and smaller runoff discharge cause more upstream intrusion of sea water. Fig. 5 displays salinity contours of dry seasons and spring tides. Local effects of the project on salinity distribution can be seen mainly in the near area of the levees. For instance, the project reduces salinity near the east end of north levee to 15‰ from 20‰. However, it seems that the overall salinity distribution doesn't feel the existence of the marine structures. The most upstream location of 2‰ contour is still at the downstream end of the Changxing Island, unaltered by the project. Therefore, it can be concluded that the project shows little impact on the environment for agricultural and industrial water use.

3.3 Variations of maximum turbidity

Maximum turbidity is an important factor for the formation of mouth bar. Its variations is of significant concern. Our simulated results demonstrate that:

Under natural conditions, sediment concentration in the South Channel is reasonably higher than that in the North Channel, which agrees with the observations. In the vicinity between groins S2 and N2 exists an area of high sediment concentration with $c \approx 1.5$ kg/m^3, which is exactly the maximum turbidity. Its location is rather stable although it shifts up and down with tides. Sediment concentration above the shoals is generally higher with $c \approx 4.0$ kg/m^3. This is mainly attributed to the small depth and rather large velocity in these areas, causing sediment particles on the bed to be scoured into suspension.

When the project is completed, sediment concentration in the shallow shoal areas is below 2 kg/m³, much less than that under natural conditions. This is mainly because the over-shoal-flows are cut off, or its sediment concentration is very low even if over-levee-flows exist at high tidal level. As for the navigation channel, sediment concentration is slightly lower than that under natural conditions due to sediment detaining effect of the levees. But variations of the location and sediment concentration of the maximum turbidity are not striking.

3.4 The location and volume of sedimentation after the project

The influence of the project on bed deformation is depicted in Fig. 3(b). Under natural conditions, sediment deposition takes place in the upstream reach (between W1 and W2) of the navigation channel because the over-shoal-flows deliver sediment loads from the shoals to the channel. Hence, regular dredging is necessary to maintain demanded depth. In the downstream reach from W2, the channel bed is rather stable. With the levees erected, bed aggradation still happens between the levees. But two mild troughs are found between groins N2 and S2 and N3 and S3, because the groins narrow cross sections and hence increase local velocity. In the reach between W2 and W4, little bed changes can be seen before and after the project. This is due to the fact that the project doesn't influence flow patterns in this area very much.

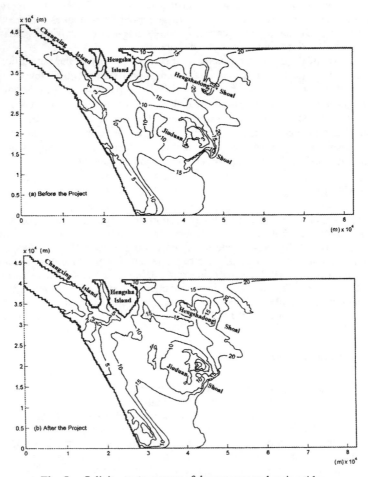

Fig. 5:   Salinity contour maps of dry seasons and spring tides

340

Therefore, dredging is still necessary to maintain the designed 8.5-m-deep waterway after completion of the first phase of the project. The annual dredging volume sums up to 10,000,000 m³ approximately. This dredging volume is close to that prior to the project. The expense is affordable.

4. CONCLUDING REMARKS

In this paper, a two dimensional mathematical model of salinity and sediment transport is established to simulate flow patterns, variations of maximum turbidity and bed deformation in the Yangtze River Estuary. The influences of the first phase waterway project are analyzed. The following conclusions can be drawn:

The hydro-structures of the project increase local velocity in the navigation channel and prevent sediment delivery from the shoals into the navigation channel. Hence, sedimentation in the navigation channel is abated.

The project has little effect on salty water intrusion.

Bed aggradation still happens in the navigation channel as it does under natural conditions. Although dredging is still necessary to maintain the designed water depth, the maintenance expense is within affordable range.

5. ACKNOWLEDGEMENT

The project is financially supported by the Institute of Coastal and Ocean Engineering, Hohai University. The authors are most grateful to professors N. Jin and Y. X. Yan for helpful discussions and collaboration. Professor Yan also provided us with useful data in the simulation.

REFERENCES

Cao Zhixian ,1997, Turbulent bursting-based sediment entrainment function. J. Hydr. Engrg., 123 (3) :pp 233-236.

ISDIN (Institute of Shanghai Design and Investigation of Waterway) ,1997, Feasibility Study of the Waterway Project in the Yangtze River Estuary.

Qian Ning, Wan Zhaohui, 1986, Mechanics of Sediment Transport(in Chinese). Science Press.

Tang Cunben, 1963, The law of sediment incipience. J. Hydr. Engrg.(in Chinese), 2:1-12 ·

Stochastic Hydraulics 2000, Wang & Hu (eds) © 2000 Taylor & Francis, ISBN 90 5809 166 X

Significance of mass transfer and morphology for solute discharge in streams

A.Gupta & V.Cvetkovic
Division of Water Resources Engineering, Royal Institute of Technology, Stockholm, Sweden

ABSTRACT: A Lagrangian methodology is used for evaluating solute discharge at specified control cross-sections of streams. The main advantage of using the Lagrangian approach is that recent theoretical developments have helped in clarifying the conceptual basis for decoupling physical transport from physical and chemical exchange processes. We account for advection, decay, diffusive process in the bed sediment with arbitrary sediment thickness and a first-order rate-limited kinetic sorption to the sediments in the storage zones. A general solution for solute discharge has been computed in the Laplace domain accounting for deterministic changes in morphology of streams. The results are illustrated for the fractional solute mass arrival (FMA) which depend nonlinearly on the downstream distance. The dimensionless kinetic storage parameter and bed parameter control the FMA. However, a potentially more significant effect on the amount of degraded solute mass is that of exchange with the sediment bed.

Key words: Streams; Solute transport; Diffusive exchange; Lagrangian formulation

1. INTRODUCTION

Environmentally hazardous chemicals of various kinds are released into rivers and streams since the beginning of industrial revolution. Solute migration in streams require consideration of retention mechanisms which involve different parts of the streams. Two main zones are recognized in streams; namely dead-end zones characterized by side pockets along the stream, eddy zones behind boulders etc., and bed sediment zone. Each zone is characterized by its individual residence times. In addition, morphology of streams also plays an important part in solute transport. In all natural streams, width, depth and velocity increases in the downstream direction as a function of discharge following simple power law functions (Leopold & Maddock, 1953).

In this paper, we investigate solute discharge (mass per unit time) in streams using a Lagrangian framework, where we consider three key mechanisms: advection (bulk movement along random flow paths), physico-chemical mass transfer processes (diffusion to the bed sediments with instantaneous sorption to the sediment bed and first-order rate-limited kinetic sorption in storage zones) and decay. Both advection and mass transfer processes are allowed to vary downstream due to changing morphological characteristics of the streams. A stochastic Lagrangian framework for reactive tracers has been developed and applied for quantifying transport in aquifers and fractured rock (e.g., Cvetkovic & Dagan, 1994; Cvetkovic *et al.*, 1999). Illustration example emphasizes the effect of physico-chemical mass transfer processes on fraction of mass arrival in the stream-sediment system.

2. LAGRANGIAN TRANSPORT FORMULATION

A reactive solute injected at a specified location in a stream is advected along random flow paths by the fluid flow and is also dispersed due to small scale velocity variations between adjacent streamlines. In addition, the solute is also subjected to various physico-chemical mass transfer processes both in the storage zones and in the bed sediment. The exchange of solute between the storage zones and the stream water is a much faster process than the exchange with the bed sediment. The current study aims to include the bed sediment processes as one-dimensional diffusive process, for reactive solutes displaying first-order decay, which are relatively more important for contaminants residing in the bed for a long time.

In the following, we neglect dispersion because the spreading of solutes due to kinetic processes is much more than the spreading due to dispersion alone. Thus, the governing one-dimensional mass balance Lagrangian equations are:

$$\frac{\partial C}{\partial t} + \frac{\partial C}{\partial T} = \left(\frac{\theta_b D_b \varsigma}{h^2}\right) \frac{\partial C_b}{\partial z}\bigg|_{z=0} + \alpha(C' - K_s C) - C\left(\frac{\partial \ln Q}{\partial T} + \lambda\right)$$

$$\frac{\partial C_b}{\partial t} = D_b \frac{\partial^2 C_b}{\partial z^2} + k(C_b' - K_b C_b) - \lambda C_b \tag{1}$$

$$\frac{\partial C'}{\partial t} = -\alpha(C' - K_s C) - \lambda C'$$

$$\frac{\partial C_b'}{\partial t} = -k(C_b' - K_b C_b) - \lambda C_b'$$

In (1), the first two equations are for total mobile concentration of solute in stream water, C [ML^{-3}] and bed sediment, C_b [ML^{-3}] respectively and the last two are for total immobile concentration of solute in dead-end zones, C'' [ML^{-3}] and bed sediment, C_b [ML^{-3}] respectively. The physico-chemical processes are specified by the source-sink terms on the RHS of (1). Further, Q [L^3T^{-1}] is the water discharge in stream, θ_b [-] is the porosity of the bed sediment, D_b [L^2T^{-1}] is the pore diffusivity in the bed, α [T^{-1}] is the rate coefficient for solute exchange in storage zones, K_s [-] and K_b [-] are dimensionless distribution coefficients for reversible sorption in storage zones and in bed sediment, k [T^{-1}] is the rate coefficient for solute exchange in bed sediment, w [L] is the width of the stream, h [L] is the depth of the stream ζ [L] is the bed sediment depth with $z = 0$ at the top of the sediment bed and λ [L] is the constant degradation rate of solute for stream water, storage zones and bed sediment. Further, t [T] is the time and T [T] is the advective travel time of solute from $x = 0$ to a position x (Figure 1).

For storage zones we consider first-order kinetic mass transfer and for bed sediment, mainly two models have been used to interpret experimental results: first-order kinetic exchange (e.g., Elliot & Brooks, 1997) and diffusive exchange (Wörman et al., 1998). For our purposes we assume diffusive flux which combines the effect of advection, dispersion and molecular diffusion into the sediment bed. Further, the diffusion model with variable bed thickness is similar to the first-order kinetic model for small bed thickness. The above equations are to be solved for specified boundary and initial conditions. The boundary condition in the bed sediment is $C_b = C_d = C/R$ at $z = 0$, and $dC_b/dz = 0$ at $z = \zeta(T)$; where C_d [ML^{-3}] is the dissolved solute concentration and R is called the retardation factor.

3. GENERAL SOLUTION

The basic solution of the transport problem can be obtained in the Laplace domain for instantaneous injection as (Gupta & Cvetkovic, 1999):

$$\frac{\hat{J}(s,T)}{\Delta M} = \hat{\gamma}(s,T) \tag{2}$$

where

$$\hat{\gamma}(s,T) = \exp\left[\int_0^T -\left(G_s + \psi_0 Z G_b^{1/2}\right)d\theta\right] \tag{3}$$

The solute mass flux, $J(t,T)=C(t,T)Q(T)$, with $Q(T)$ being the Lagrangian equivalent of water discharge at T and ΔM is the total injected solute mass. In (3) G_s and G_b accounts for the solute exchange with the storage zones and bed sediment respectively. Further, Z accounts for the effect of limited sediment zone. If the exchange with the storage zones is first-order rate-limited kinetic sorption, i.e., α is finite while the exchange with the bed sediment takes sufficiently long time as compared to the time scale of stream transport, equilibrium chemistry is assumed with $k \rightarrow \infty$. Furthermore, if the relative sediment depth available for stream-subsurface exchange is sufficiently large such that $\exp\left(2\varsigma\sqrt{G_b/D_b}\right) \rightarrow \infty$. Then,

$$G_s(s,T) = s + \lambda + \frac{\alpha K_s(s+\lambda)}{s+\alpha+\lambda}$$

$$G_b(s,T) = (s+\lambda)(R_b) \tag{4}$$

$$Z(s,T) \approx 1$$

$$_0 = \frac{\varsigma\theta_b\sqrt{D_b}}{Rh^2}$$

where $R_b = (1 + K_b)$, and is called the bed retardation factor. With these simplifications (3) reduces to

$$\ln\left[\hat{\gamma}(s,T)\right] = \int_0^T -\left[G_s + \psi_0[(s+\lambda)R_b]^{1/2}\right]d\theta \tag{5}$$

4. MORPHOLOGICAL RELATIONS

The morphological relations have been expressed as simple power-law functions along the length of the stream under the condition that different discharges of same frequency occur at cross-sections situated along the length of the stream. The dependent variables width, mean velocity and depth are related to discharge along the length of the stream as (Leopold & Maddock, 1953)

$$w = a\,Q^b \qquad\qquad h = c\,Q^f \qquad\qquad U = d\,Q^m \tag{6}$$

where U is mean velocity [LT^{-1}] and Q is the water discharge [L^{3}T^{-1}]. The parameters a, c and d are intercept coefficients for unity discharge and will vary among the streams. Further, b, f and m are exponent coefficients with average values as $b = 0.5$, $f = 0.4$ and $m = 0.1$ (e.g., Leopold & Maddock, 1953; Kolberg & Howard, 1995).

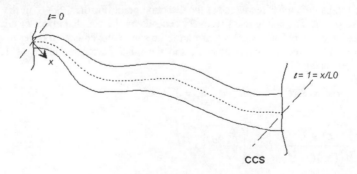

Fig. 1: Hypothetical flow path in a stream.

345

The relationship between suspended sediment load and discharge is available in the literature. The sediment at any one time may be a part of the suspended load and at other time part of bed load. However, it is the sediment size that determines the bed layer thickness and the bed load is a direct measure of the particle size it is capable of carrying. For illustration purposes, we assume that the relationship between bed sediment depth and discharge downstream is analogous to the relationship between suspended sediment load and discharge in a downstream direction; thus $\zeta = sQ^j$ and the constant s needs to be determined from field observations. Further, $j = 0.8$ ((Leopold & Maddock, 1953). However, it should be noted that the actual depth of the bed available for exchange depends on both geomorphologic and hydraulic factors (Elliot and Brooks, 1997). The length and the area of the stream are related by a power-law function as $x = nA^r$ where x is the length along the stream [L] and A is the cross-sectional area of the stream [L^2], and n and r are intercept and exponent coefficients respectively. Studies have shown the value of exponent r to be close to 0.6 (e.g., Gray, 1961).

From above relations and continuity equation $Q = AU$, the following are the new power-law relations are adopted for illustration purposes between morphology parameters and longitudinal length coordinate, x of the stream:

$$w(x) = a_0 x^{0.93} \quad h(x) = c_0 x^{0.74} \qquad U(x) = k_0 x^{0.19} \qquad \zeta(x) = s_0 x^{1.48} \tag{7}$$

where a_0 [L$^{0.07}$], c_0 [L$^{0.26}$], k_0 [L$^{0.81}$T^{-1}] and s_0 [L$^{-0.48}$] are the new constants and can be obtained from field measurements.

5. RESULTS

A reactive solute is injected as a pulse into a stream at $x = 0$ with varying hydraulic geometry (Figure 1). It is advected downstream, and is subject to physico-chemical mass transfer processes. We wish to characterize the solute discharge at a specified cross-section of the stream accounting for the effect of non-uniformities in stream morphology. An analytical solution of the transport problem can be obtained in the form of temporal moments. Temporal moments are useful for predictive purposes since they characterize solute discharge over the CCS at a given location and can also be used for reconstructing the entire breakthrough curve. Temporal moments of order "k" are defined as

$$m_k(T) = \int_0^\infty t^k J(t,T)dt = \Delta M \int_0^\infty t^k [\gamma(t,T)]dt = \Delta M(-1)^k \frac{\partial^k (\hat{\gamma})}{\partial s^k}\bigg|_{s=0} \tag{8}$$

Using (8), central temporal moments of any arbitrary order can be computed wherefrom solute discharge, J [MT^{-1}] can be evaluated, say in the form of a polynomial. Thus, the fractional factor of the recovered solute mass is defined as

$$\mu = \frac{m_0}{\Delta M} \tag{9}$$

where m_0 is the total solute mass recovered at the detection point. Further, $\mu = 1$ in the absence of decay, otherwise $\mu < 1$. The solute mass recovery depends on the coupled effect of advection and physico-chemical mass transfer processes that retard the solute movement. In order to evaluate μ, we first compute the advective travel time, T, and the other parameter in (5) as functions of stream morphology using (7)

$$T(x) = \int_0^x \frac{1}{U(x')}dx' = k_1 x^{0.81} \quad where \quad k_1 = 1.23 k_0^{-1}$$

$$\tag{10}$$

$$\int_0^T \frac{\varsigma}{h^2} d\theta = \int_0^x \frac{\varsigma(x')}{h^2(x')U(x')} dx' = \left(\frac{k_1 s_0}{c_0^2}\right) x^{0.81}$$

In (9) μ quantifies the discharged solute mass normalized with total injected solute mass, and is referred to as the fraction of mass arrival (FMA). Thus, the FMA for the combined effect of mass transfer and morphology in a dimensionless form is expressed as

346

$$\ln[\mu] = -K_1 \left[\lambda^* \left(1 + \frac{\alpha^* K_s}{\alpha^* + \lambda^*} \right) + Y\sqrt{\lambda^*} \right] l^{0.8} \tag{11}$$

The dimensionless parameters introduced above are defined as

$$\lambda^* = \lambda T_0; \; \alpha^* = \alpha T_0; \; l = x/L_0; \; K_1 = k_1 \frac{L_0^{0.81}}{T_0};$$

$$Y = \frac{s_0}{c_0^2} \frac{\theta_b \sqrt{D_b R_b}}{R} \sqrt{T_0} \tag{12}$$

where L_0 is the length of the stream up to the CCS and T_0 is the time for the solute to be advected to the CCS. The parameter Y relates the sediment depth, bed partitioning coefficient and diffusivity in the bed sediment. The two key dimensionless parameters in (11) are α^* and Y, referred to as the kinetic storage parameter and bed parameter respectively.

6. ILLUSTRATION EXAMPLE

In the following we illustrate the combined effect of physico-chemical mass transfer processes and morphology of streams on the FMA. In particular, we will quantify the relative effect of mass transfer in the storage zones and in the bed sediment as quatified by α^* and Y. The distribution coefficient in the storage zone is assumed as $K_s = 1.0$. Further, the intercept coefficient for mean velocity is fixed as $K_1 = 1.0$ which otherwise would be known from field measurements for the stream under consideration. The degradation rate λ can vary over a wide range depending upon the solute of interest. For illustration purposes we have assumed λ^* as 0.1 and 1.0. Clearly μ decreases as one moves downstream and also with increasing λ, as indicated in Figure 2a. The point of interest in Figure 2 is to compare the dependency of μ on α^* and Y.

Figure 2a shows the sensitivity of the FMA, μ along the stream length for varying kinetic parameter at $Y = 0.1$ and $\lambda^* = 0.1; 1.0$. It can be seen that stronger kinetic effects (i.e., lower kinetic rates) in the stream results in lower degradation of the solute, for fixed λ^* and Y. Since the exchange with the storage zones is kinetically controlled, there is a certain range of α^* which affects the solute transport beyond which the situation is in equilibrium. Furthermore, for equilibrium conditions, degradation rate is highest (i.e., lowest μ) for fixed λ and Y. It is apparent from Figure 2a, however, that the dependency of solute degradation on the mass transfer rate of solute in the storage zone is relatively small.

In Figure 2b, FMA is shown for Y varying from 0.1 to 25 at $\alpha^* = 0.1$ and $\lambda^* = 0.1$. A higher value of Y results in faster degradation of solute mass indicating that the bed parameter has a major role to play in retention/release of solute to the mainstream. However, in the absence of exchange with the bed sediment, the only parameter controlling the shape of the concentration breakthrough curve is the rate coefficient, α^* in the storage zones.

7. SUMMARY

A methodology for evaluating solute discharge for reactive solute transport in streams has been developed using a Lagrangian transport formulation. The main advantage of the approach is that the mass transfer processes can be decoupled form physical transport which simplifies the modelling considerably and enables us to obtain relatively simple semi-analytical expressions for solute transport.

The developed model relates advection, degradation and linear physico-chemical mass transfer based on the morphology of streams. The results have been presented for fractional mass arrival for a pulse injection, but can be extended to any arbitrary injection. The FMA is significant as an indicator of the relative amount of solute mass ultimately discharged across a particular CCS.

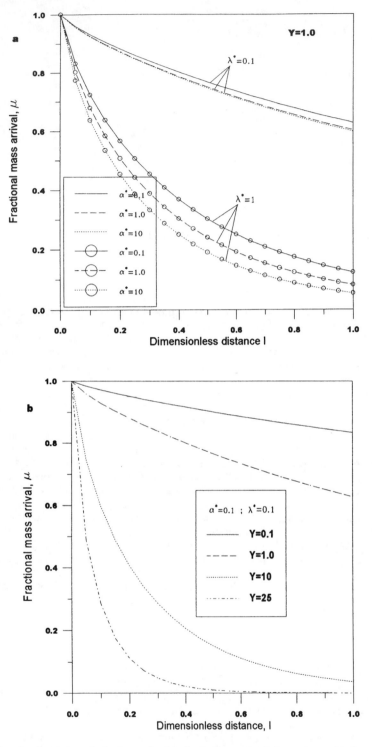

Fig.2: Fractional mass arrival along the longitudinal extent of the stream; (a) for varying kinetic parameter values at $\lambda^* = 0.1$ & 1.0 and $Y = 0.1$. (b) for varying bed parameter values at $\alpha^* = 0.1$ and $\lambda^* = 0.1$.

348

The results show that a higher value of either the kinetic parameter α^* in the storage zones or the bed parameter, Y results in faster reduction of μ. However, the results are more sensitive to uncertainties in the estimation of Y, clearly indicating that for larger values of Y, the relative effect α^* is minimized and that FMA will ultimately depend on the bed parameter. Thus, a pollutant discharged in a stream/river with even a small decay rate but a high diffusion and/or distribution in the bed may be completely degraded before arriving to a specified discharge section.

The methodology illustrated in this paper can be used as a first, simple quantitative assessment of contaminant plume migration in a particular stream for which hydraulic geometry relations can be estimated. Another important use of the results is for interpreting data from conducted tracer tests.

REFERENCES

Cvetkovic, V. & Dagan, G. (1994) Transport of kinetically sorbing solute by steady random velocity in heterogeneous porous formations. *J. Fluid Mech.* **265**, 189-215.

Cvetkovic, V., Dagan, G. & Cheng, H. (1999) Contaminant transport in aquifers with spatially variable hydraulic and sorption parameters. *Proc. R. Soc. Lond. Ser.* A **454**, 2173-2207.

Elliot, A. H. & Brooks, N. H. (1997) Transfer of nonsorbing solutes to a streambed with bed reforms: Theory. *Water Resour. Res.* **33**, 123-136.

Gray, D. M. (1961) Interrelationships of watershed characteristics. *J. Geophys. Res.* **66**, 1215-1223.

Gupta, A. & Cvetkovic, V. (1999) Temporal moment analysis of tracer discharge in streams: Combined effect of physico-chemical mass transfer and morphology. Submitted to *Wat. Resour. Res.*

Kolberg, F. J. & Howard, A. D. (1995) Active channel geometry and discharge relations in U. S. Peidmont and Midwestern streams: The variable exponent model revisited. *Water Resour. Res.* **31**, 2353-2365.

Leopold, L. B. & Maddock, T. Jr. (1953) The hydraulic geometry of stream channels and some physiographic implications. *U. S. Geol. Surv. Prof. paper* **252**.

Wörman, A., Forsman, J. & Johansson, H. (1998) Modelling retention of sorbing solutes in streams based on a tracer experiment using [51]Cr. *J. Environ. Engrg.* **124**, 122-130.

Stochastic Hydraulics 2000, Wang & Hu (eds) © 2000 Taylor & Francis, ISBN 90 5809 166 X

A computer model for the modified Einstein procedure and its application in two Iranian rivers

M. Shafai-Bajestan
Shahid-Chamran University, Ahwaz, Iran

M. Ostad-Askari
Dez Ab Consulting Engineering Company, Ahwaz, Iran

ABSTRACT : Modified Einstein procedure was developed by Colby and Hembree. They provided few nomographs to obtain the values of the integrals. Thus the procedure is complicated and time consuming. In this paper, a computer model was developed for faster and more accurate results. Then the bed, suspended and total load of two major Iranian rivers namely, Karoon and Karkh-e were comoputed by this model. The results shows that although the Modified Einstein procedure is based on shallow stream, the method with some modification gives reasonable results in large, deep sand bed steam as well.

Key words: Modified einstein, Sediments, bed - load, Karoon, Karkh-e

1. INTRODUCTION

Colby and Hembree(1955), Colby and Hubbel(1961) proposed a procedure to compute the total sediment transport rate, which includes wash load, for a given water discharge from measured depth – integrated suspended sediment samples and other flow and sediment characteristics. The Modified Einstein procedure is not a proper procedure for the design purpose, however it has been used to determine the ratio of bed load to the total load in an accurate range. Usually Many designers take this ratio in the order of 15 – 20 percent which is an arbitrarily value and it is not correct for all rivers. A brief review of this procedure is given here.

The first step is to calculate the suspended load discharge in the various size fractions per unit width in the sampled zone of the cross – section q'_{si} .

$$q'_{si} = \sum_i q'_{si} = c'_s \gamma \, Q'/B \tag{1}$$

in which c'_s is the mean measured sediment concentration, γ is the unit weight of water, Q' is the water discharge in the sampled zone and B is the river width. The relation between Q' and Q , the total discharge, is

$$\frac{Q'}{Q} = (1 - E') - 2.3 \frac{E' \log E'}{P_m - 1} \tag{2}$$

in which $E' = a'/d$ and a' is the distance from streambed to the sampler inlet tube, which depends on the type of the samplers, d is the flow depth, and P_m is given by the following equation:

$$P_m = 2.3 \log \frac{30.2 x d}{D_{65}} \tag{3}$$

The bed load for a given size fraction i_b per unit width, q_b , is given by:

$$i_b q_b = \frac{1}{2} \phi_* \, i_b \gamma_s \, \sqrt{g D_i^3} \, s \sqrt{G_s - 1} \tag{4}$$

in which γ_s is the sediment unit weight, G_s and D_i are the specific gravit and size of sediment particles, respecitvely. In this equation the term ϕ_* is the Einstein's intensity bed-load which has been divided by a factor of two to fit the observed river data more closely. The value of ϕ_* can be obtained from a monograph developed by Einstein(1950). The total load, Q_T , is determined from one of the following equations,

$$Q_{Ti} = B q'_{st} \frac{P_m J_1 + J_2}{P_m J'_1 + J'_2} \tag{5}$$

for the range of fine particle size and the relation

$$Q_{Ti} = i_b B q_b (P_m I_1 + I_2 + 1) \tag{6}$$

for the coarse particle size. The units of Q_{Ti} are dry weight per unit time. I_1, I_2, J_1, J_2 are obtained from the:

$$J_1 = \int_E^1 (\frac{1-y}{y})^{z'} dy \tag{7}$$

$$J_2 = \int_E^1 (\frac{1-y}{y})^{z'} Lny \, dy \tag{8}$$

$$I_1 = 0.216 \frac{E^{z'-1}}{(1-E)^{z'}} J_1 \tag{9}$$

$$I_2 = 0.216 \frac{E^{z'-1}}{(1-E)^{z'}} J_2 \tag{10}$$

in which $E = a/d$ and a is the bed layer thickness equal to $2D_i$. J'_1 and J'_2 have the same relation as J_1 and J_2 but E' is applied instead of E. The values of I_1 , I_2 , J_1 , J_2 is determined from nomographs provided by Einstein (1950) and Colby and Hembree(1955). The procedure is rather time consuming , and yet approximate because it is difficult to find exact values of the integrals from very complicated monographs. z' is the suspended – load exponent which is determined by trial and error to satisfy the following relation

$$\frac{q'_s}{i_b q_b} = \frac{I_1}{I_2} (P_m J'_1 + J'_2) \tag{11}$$

Colby and Hembree (1955) found that the value of z'_i can be evaluated using the following relation ,

352

$$\frac{z_i'}{z_1'} = \left(\frac{\omega_i}{\omega_1}\right)^{07} \tag{12}$$

in which z_i' is determined from $Eq\,(11)$ for the dominant grain size , ω_1 and ω_i are the fall velocity for the dominant grain size and grain size of D_i , respectively. For detail information related the modifled Einstein method readers may see, USBR (1955) , Simon and·Senturk(1992), and Yang(1996).

2. METERIAL AND METHODS

The main purpose of this study is to develope a mathematical model for the Modifled Einstein procedure. To do so first a method for the computation of integrals, Eqs(7) , (8) , (9) , and (10) should be applied. The numerical methods such as Simpson's rule is not applicable for these equation especially in deep sandbed rivers such as Karoon and Karkh –e rivers. This is due that in these rivers the value of $E = a/d = 2D_i/d$ is very small close to zero, therefore at the lower limit of the integral the value of $\frac{1-y}{y}$ becomes infinitive. To overcome this difficulties the I and $J's$ integrals can be written such,

$$\int_E^1 (\frac{1-y}{y})^{z'} = \int_E^\varepsilon (\frac{1-y}{y})^{z'}\,dy \;+\; \int_\varepsilon^1 (\frac{1-y}{y})^{z'}\,dy \tag{13}$$

in which ε has a value greater than E . The first integral of the right hand side of Eq .(13)

$$\int_E^\varepsilon (\frac{1-y}{y})^{z'}\,dy = \int_E^\varepsilon y^{-z'}(1-y)^{z'}\,dy \;=\; \int_E^\varepsilon y^{-z'}[1-zy+\frac{z(z-1)}{2}y^2 \;+...]\,dy \tag{14}$$

The right hand side of the above equations is the sum of a series of the integral which can be computed analytically, by eliminating the forth term and thereafter the J_1 and J_2 are difined as :

$$J_1 = F_1 + F_2 + F_3 + \int_E^1 (\frac{1-y}{y})\,dy \tag{15}$$

$$F_1 = \int_E^\varepsilon y^{-z'}\,dy = \begin{cases} \dfrac{1}{1-z'}(\varepsilon^{1-z'}-E^{1-z'}) & z' \neq 1 \\ Ln\varepsilon - LnE & z' \neq 1 \end{cases} \tag{16}$$

$$F_2 = \int_E^\varepsilon y^{1-z'}\,dy = \begin{cases} \dfrac{z'}{z'-2}(\varepsilon^{2-z'}-E^{2-z'}) & z' \neq 2 \\ -2(Ln\varepsilon - LnE) & z' = 2 \end{cases} \tag{17}$$

$$F_3 = \int_E^\varepsilon \frac{z'(z'-1)}{2}y^{2-z'}\,dy = \begin{cases} \dfrac{z'(z'-1)}{2(3-z')}(\varepsilon^{3-z'}-E^{3-z'}) & z' \neq 3 \\ -3(Ln\varepsilon - LnE) & z' = 3 \end{cases} \tag{18}$$

The J_2 can be evaluated with the same way, thus:

$$J_2 = G_1 + G_2 + G_3 + \int_\varepsilon^1 (\frac{1-y}{y})^{z'} Lny \; dy \tag{19}$$

$$G_1 = \begin{cases} \dfrac{E^{1-z'}}{1-z'}(LnE - \dfrac{1}{1-z'}) - \dfrac{E^{1-z'}}{1-z'}(Ln\varepsilon - LnE) & z' \neq 1 \\[3ex] \dfrac{1}{2}[(Ln\varepsilon)^2 - (LnE)^2] & z' = 1 \end{cases} \tag{20}$$

$$G_2 = \begin{cases} \dfrac{z'\varepsilon^{2-z'}}{z'-2}(Ln\varepsilon - \dfrac{1}{2-z'}) - \dfrac{z'E^{2-z'}}{z'-2}(Ln\varepsilon - \dfrac{1}{2-z'}) & z' \neq 2 \\[3ex] -[(Ln\varepsilon)^2 - (LnE)^2] & z' = 2 \end{cases} \tag{21}$$

$$G_3 = \begin{cases} \dfrac{z'(z'-1)e^{3-z'}}{2(3-z')}(Ln\varepsilon - \dfrac{1}{3-z'}) - \dfrac{z'(z'-1)E^{3-z'}}{2(3-z')}(LnE - \dfrac{1}{3-z'}) & z' \neq 3 \\[3ex] \dfrac{3}{2}[(Ln\varepsilon)^2 - (LnE)^2] & z' = 3 \end{cases} \tag{22}$$

using the above equations, the values of I_1, I_2, J_1' and J_2' can be computed.

In this model, the fall velocity of the sediment particle are computed from the Ruby's equation,

$$\omega_i = \frac{1}{D_i}[10.791D_i^3 + 36v^2]^{0.5} - 6\gamma \tag{23}$$

The kinematic viscosity v in m^2 / sec, is determined from:

$$v = 8.5194 \times 10^{-8} (LnT)^2 + 2.4 \times 10^{-8} Ln(T) + 1.7 \times 10^{-6} \tag{24}$$

In which T is the water temperature in centigrade.

The value of the ϕ_* is obtained from the following relation in which had been developed from the original Einstein's monograph,

$$\phi_* = E \times P[-1.071(Ln\psi)^2 + 0.569(Ln\psi) + 1.836] \qquad 0.9 < \psi \leq 27 \tag{25}$$

In which ψ is the intensity of shear stress and is equal to

$$\psi = \frac{(G_s - 1)D_{35}}{sd} \quad \text{or} \quad \psi = \frac{0.4(G_s - 1)D_i}{sd} \tag{26}$$

where s is the energy grade line, the large ψ value is used to find ϕ_*.

3. FIELD DATA

Data used in this study are as follow:
1) 80 data sets of measured suspended load during 1987 – 1995 in Ahwaz station, Karoon river.(Table.1)
2) 86 data sets of measured suspended load during 1987 – 1995 in Hamideh, station, Karkhe – e river.(Table.2)
3) 45 data sets of measured bed load from East Fork River published by Leopoled and Emmett(1976).(Table.3)

Table.1: Sediment transport computations for Karoon river

Computed QT (TON/DAY)	Qb (TON/DAY)	Measured Qs (TON/DAY)	Q (CMS)	Sample	Computed QT (TON/DAY)	Qb (TON/DAY)	Measured Qs (TON/DAY)	Q (CMS)	Sample
171892	3619	154942	3462	41	84238	880	76489	1233	1
68158	1864	59630	1821	42	5504	108	4936	414	2
98468	1786	89317	1988	43	4542	215	3729	419	3
259294	3007	235602	2760	44	5651	165	4961	495	4
87278	1217	79075	982	45	10154	370	8632	689	5
28559	665	25404	782	46	89000	1249	78955	1204	6
91893	821	85464	905	47	20195	351	17904	675	7
852101	579	847035	764	48	53610	1264	47171	1238	8
2943	150	2407	398	49	35247	631	31190	898	9
9373	150	8534	415	50	112684	654	106732	905	10
3798	64	3472	285	51	119450	511	114340	856	11
3547	43	3296	256	52	4060	140	3499	405	12
3640	174	2992	398	53	13168	326	11605	574	13
9591	113	8894	387	54	7228	121	6599	402	14
21487	228	20068	359	55	4952	119	4403	392	15
975	82	780	188	56	9321	207	8297	495	16
3615	217	2851	330	57	8398	161	7432	509	17
15574	138	14651	399	58	9706	288	8431	673	18
1917	41	1741	255	59	4967	140	4408	451	19
3880	107	3375	372	60	11182	238	9804	559	20
1861	81	1558	311	61	9029	481	7090	599	21
2434	91	2081	349	62	14298	540	2062	878	22
17114	212	15776	478	63	125665	2714	112889	3156	23
39122	893	33760	855	64	727144	2646	694142	2758	24
25610	504	22510	606	65	123753	2275	112469	2513	25
86602	915	79802	895	66	1773364	3436	1726420	3546	26
73375	1133	64823	1018	67	834536	3303	798203	3404	27
16833459	3541	1632009	2482	68	1258228	3507	1218676	3059	28
12587	181	11549	434	69	11681	214	10307	512	29
9813	74	9322	324	70	5416	35	5192	296	30
7390	74	6938	336	71	9145	128	8400	346	31
11054	77	10519	342	72	35871	854	31384	1009	32
17581	195	16386	434	73	63700	573	59221	774	33
28925	357	26279	551	74	2195	47	1962	277	34
53556	962	47410	938	75	1662	91	1347	346	35
28579	448	25593	621	76	2456	59	2185	294	36
34616	491	31319	645	77	1466	57	1258	280	37
67034	256	64931	488	78	3629	86	3201	390	38
19821	154	18804	395	79	2040	32	1872	258	39
17716	188	16533	482	80	105571	1222	94817	1621	40

4. RESULTS AND DISCUSSION

The bed load of East – Fork river first was computed by the computer model and compared to the measured data. The bed load computed using different ϕ_* values. The results show that for flow discharge less than $11\,m^3/\sec$, the ϕ_* should be multiple by a factor of 0.2 and for discharge of greater than $11\,m^3/\sec$, a factor of 0.5 gives more closely results to the measured values.

The total sediment transport load and bed load for the 80 sets of Karoon river data was computed by the computer model. The results shows that the ratio of bed load to the total load

Table.2: Sediment transport computations for Karkhe-e river

Computed		Measured		Sample	Computed		Measured		Sample
QT (TON/DAY)	Qb (TON/DAY)	Qs (TON/DAY)	Q (CMS)		QT (TON/DAY)	Qb (TON/DAY)	Qs (TON/DAY)	Q (CMS)	
7729	52	7008	75	44	376	2	351	50	1
197	5	160	31	45	274	0	261	41	2
672	8	586	42	46	837	2	795	59	3
277	0	258	28	47	28794	71	27342	174	4
363	1	331	26	48	86163	241	82279	443	5
110	1	96	24	49	144151	403	137267	432	6
109	0	100	21	50	17933	228	16043	308	7
5848	118	4914	139	51	20399	184	18583	278	8
21703	391	18158	253	52	61215	214	58075	410	9
71814	475	64496	272	53	85118	92	83063	465	10
190026	529	177561	369	54	101510	69	99610	382	11
822	37	650	83	55	587	27	453	52	12
1898	55	1564	77	56	227	10	178	42	13
49659	509	43649	255	57	200	5	171	41	14
624362	1418	587437	548	58	164	0	155	33	15
631	12	548	64	59	179	3	161	51	16
383	5	343	54	60	22629	205	20323	216	17
960	6	885	53	61	37605	258	34389	274	18
5270	9	4977	65	62	55208	548	49314	324	19
487	1	456	41	63	3748	78	3250	148	20
113	0	104	32	64	5889	54	5349	150	21
1068	0	1004	31	65	575	11	487	52	22
1146	2	1071	43	66	214	7	174	47	23
836	3	778	50	67	214	2	191	39	24
8679	69	7876	144	68	643	11	593	67	25
35763	113	33425	230	69	869	23	735	70	26
151948	437	144387	481	70	33530	278	30163	249	27
6126	66	5469	137	71	107056	485	99829	436	28
6700	30	6184	124	72	271152	766	255991	467	29
44545	141	41822	241	73	1741125	2308	1668184	1123	30
123262	61	115163	87	74	337640	845	321673	629	31
780	7	689	42	75	13732	86	12578	184	32
1720	6	1570	42	76	28039	254	24913	164	33
1527	5	1383	38	77	11070	358	8593	150	34
1518	3	1381	35	78	445	21	320	31	35
1445	5	1310	39	79	355	20	252	39	36
1771	8	1615	52	80	910	25	742	56	37
3199	33	2867	88	81	2753	137	2002	110	38
56739	599	51657	487	82	19366	256	16828	237	39
6564	223	5014	119	83	1149066	1878	1103743	1282	40
9298	243	7443	144	84	216198	813	203898	691	41
4089	241	2942	91	85	959787	2035	906583	815	42
1615	74	1229	62	86	5426	73	4709	125	43

varies with flow discharge and may vary from 2 percent to maximum of 11percent . This ratio for the 86 sets of Karkh – e river data was found to be in the range of 2 to 10 percent. During computaion of total load of Karoon and Karkh-e rivers, it was found that the value of z_t' can not be computed from Eq.12. The reason is that Eq.(12) is based on shallow gravel bed rivers, while the Karoon and Karkh-e rivers are deep sandbed rivers. In these rivers the dominant grain size is small and therefore ω_1 is very small so according to Eq.(12), the fall velocity for the bed load paricles become very large thus the total load is far away from the measured values. To overcome this difficultied, the z_t' values obtained with the same procedure as z_1' is computed using equation(11).

Q_b measured Q_b Computed	Q_b Computed		Q_b measured	V	$d.$	A	Q Cms	Sample
	$\phi_*=0.5$	$\phi_*=0.2$	(ton/day)	(m/sec)	(m)	(m∧2)		
0.42		6.3	2.6	0.61	0.28	4.03	2.4	1
0.39		18.0	7.1	0.76	0.48	6.99	5.3	2
3.43		17.8	61.2	0.78	0.51	7.42	5.8	3
0.71		8.4	6.0	0.83	0.58	8.45	7.0	4
0.40		10.0	4.0	0.84	0.59	8.64	7.2	5
0.63		12.3	7.8	0.87	0.65	9.47	8.2	6
3.17		31.5	99.9	0.90	0.70	10.20	9.1	7
2.57		40.4	103.9	0.92	0.74	10.80	9.9	8
2.32		44.2	102.6	0.92	0.74	10.90	10.0	9
0.70		19.9	13.9	0.92	0.75	10.90	10.1	10
0.82		14.9	12.2	0.92	0.75	10.90	10.1	11
2.48		49.5	122.7	0.94	0.78	11.40	10.7	12
1.13		21.8	24.5	0.94	0.78	11.50	10.8	13
0.22	76.4		16.8	0.96	0.81	11.80	11.3	14
0.36	37.6		13.5	0.99	0 88	12.90	12.8	15
0.71	70.5		50.1	1.00	0.90	13.10	13.1	16
0.51	85.5		44.0	1.03	0.96	14.00	14.4	17
0.72	48.4		34.9	1.06	1.02	15.00	15.8	18
0.25	130.0		32.9	1.06	1.03	15.00	15.9	19
0.38	62.5		24.0	1.06	1.04	15.20	16.1	20
0.19	132.2		24.8	1.06	1.04	15.20	16.1	21
0.20	151.4		30.4	1.07	1.06	15.50	16.6	22
0.26	112.0		29.7	1.09	1.11	16.30	17.8	23
0.11	256.2		27.6	1.19	1.17	17.10	19.2	24
2.82	139.9		393.9	1.13	1.21	17.70	20.0	25
0.36	112.0		40.1	1.16	1.27	18.60	21.5	26
1.13	196.3		221.9	1.15	1.27	18.60	21.5	27
0.80	131.5		105.8	1.16	1.28	18.70	21.7	28
0.63	158.9		100.5	1.17	1.30	19.00	22.2	29
0.40	204.9		81.7	1.18	1.33	19.40	22.9	30
1.54	177.0		273.2	1.18	1.34	19.50	23.1	31
0.11	227.7		26.1	1.20	1.38	20.20	24.3	32
0.62	171.1		105.8	1.20	1.38	20.20	24.3	33
1.55	169.6		262.1	1.21	1.40	20.25	24.8	34
0.07	216.6		16.4	1.22	1.43	20.90	25.5	35
1.15	189.8		219.1	1.22	1.44	21.00	25.6	36
0.68	172.3		116.9	1.22	1.44	21.10	25.7	37
0.17	228.4		38.6	1.23	1.47	21.50	26.5	38
0.14	254.1		36.2	1.25	1.52	22.30	27.9	39
0.85	177.0		149.8	1.27	1.57	22.90	29.0	40
0.09	204.9		21.8	1.27	1.59	23.30	29.7	41
0.86	332.0		284.3	1.28	1.60	23.40	29.8	42
0.76	198.2		149.8	1.28	1.62	23.60	30.3	43
0.32	312.4		99.3	1.31	1.68	24.60	32.2	44
1.15	320.0		367.5	1.41	2.01	29.50	41.5	45
0.87	Mean							avg.
0.85	σ							STD
0.97	r²							CV.

CONCULSION

A Computer model was developed in this study to evaluate the bed – load and total load of rivers. This model is fast and accurate. Although the original procedure has been developed from the shallow river data, this model was found to be applicable in deep sandbed rivers as well. The ratio of bed load to the total load of Karoon and Karkh - e was found to be vary with discharge.

REFERENCES

Colby, B.R. and C.H. Hembree (1955). " Computation of total sediment discharge, Niobrara River, Nebraska." USGS, water – supply paper No. 1357.

Colby, B.R. and D.W. Hubbell (1961). " Simplified methods for computation total sediment discharge with modified Einstein procedure. "USGS, water – supply paper No. 1593.

Einstein H.A. (1950). "The bed load function for sediment transportation in open channel flows." Technical Bulletin 1926, USDA, SCS.

USBR (1955). "Step method for computing total sediment load by the modified Eintein procedure. "USDI, Denver, Colorado.

Simon, D.B. and F. Senturk (1992). " Sediment transport technology." Book Craters Inc., Chelsea, Michigan, USA.

Yang, C.T. (1996), "Sediment transport: theory and practice." Mc Graw – Hill, New York, U.S.A.

Leopold, L.B. and W.W. Emmet, (1976). "Bed load measurments East Fork river, Wyoming, Dept. of Geology and Geophysics, University of California, Berkley, USA.

Stochastic Hydraulics 2000, Wang & Hu (eds) © 2000 Taylor & Francis, ISBN 90 5809 166 X

Use of radiometric tools and methods for studies of sediment transport in river mouths

B. Shteinman, Y. Kamenir, D. Wynne & A. Shatsov
Israel Oceanographic and Limnological Research, Yigal Allon Kinneret Limnological Laboratory, Tiberias, Israel

ABSTRACT: The natural radioactivity was applied to follow sediment movements in river mouths. Almost all studies of sediment dynamics are based on the use of radioactive sediments produced artificially through injection of radioactive isotopes and sorption of radioactive particles by river sediments. However, there may be legal obstructions in the use of such particles because of their possible health hazards. In addition, complex instruments are required to load these particles into the bed sediments. Use of the natural radioactive properties of sediments has none of the above problems. Since the integral gamma-radioactivity of the sea bottom sediments is comparatively low, it is possible to measure the higher radioactivity associated with the river alluvia, using multi-channel radiometers which continuously record the bottom sediment gamma-activity. A study carried out on sediment transport in the river mouth has shown that during periods of still weather distribution of the river sediment inside the mouth zone is produced by the velocity attenuation of the river jet flow. In general, the coarsest fractions fall to the bottom, near the river mouth bar. Further along the jet, the sedimentation rate and the particle mean size gradually diminish. A close correlation was noted between gamma-activity level and the percent concentration of the pellite fraction. A submarine radiometric survey of the seabed, made by recording gamma activity from a towed radiometer, lasts for some hours, enabling the relationship between sediment migration data and hydro-meteorological conditions to be formulated. In addition, not only can the area of river sediment accretion be determined by radiometric surveys, but the predominant fraction in the sediments can also be evaluated. Moreover, this study has shown that gamma radiation measurements could elucidate the effect of hydro-meteorological events. Measuring gamma radiation *in situ* would provide much more information. The use of natural radioactivity in studies of sediment transport in river mouths may be unique in providing information on sediment composition of large built-up forms, such as sand bars and spits, under different conditions; location of zones of maximal sediment accumulation; location of zones of erosion; change in littoral zone relief due to sediment transport and decomposition under varying conditions.

Key words: Sediment, Submarine radiometry, Gamma – activity, River mouth.

1. BACKGROUND

Methods based on the use of radioactivity currently find ever-growing application in hydrodynamics, hydrology, oceanology and many other fields of experimental studies. Radioactive isotopes have the widest use in studies of sediment dynamics, especially at the near-shore zone (Jaffry & Hours 1959; Martin et al. 1970).

Radioactive isotopes have turned out to be incomparable indicators providing means of investigation of shore dynamics, sand-wash in ports, the bed load and suspended load. They can provide information on the bottom sediment density and many parameters of their movement (Bunte & Ergenzinger 1989; Ferguson et al. 1996; Fernandes et al. 1991, Kamel & Johnson 1963; Owcharozik & Szpilowski, 1986).

Almost all studies of sediment dynamics are based on artificially produced radioactive sediments by injection of radioactive isotopes and sorption of radioactive particles by the river sediments. The basis of sediment translocation studies is very simple. Activated, natural sediment (or an artificial material with identical physical parameters (relative density, particle shape and size distribution) is loaded at the head of the selected part of the river or sea region. Then movement of the labeled particles is followed, using radiation detectors.

Several methods have been developed for production of radioactive tracers:

1) Irradiation of natural sediments in a nuclear reactor. An important advantage of this method is in the use of natural materials but it is limited by the reactor volume, which prevents the production of homogenous activity of all the material (Guidebook.1971).

2) Radioactive isotopes, such as Silver (Putnam & Smith 1956), Iridium and Barium (Smith & Parsons 1967), Chromium (Bueringer & Feldt 1964), Iron (Leontiev & Afanasiev 1959), Lead and Cesium (Fernandes et al. 1991; Rostan et al. 1997; Walling 1993; Wasson et al. 1987) are injected into the sediment surface.

Natural radioactive materials (e.g. uranium ore) are not suitable for application, due to drawbacks such as a very long half-life, instability of the structure, high toxicity, high cost, etc. From the above, it can be concluded that there are serious deficiencies in methods based on artificial isotopes. Moreover, there may be legal obstructions in the use of such particles because of their possible health hazards. In addition, artificial materials (e.g. glass) very seldom have hydrodynamic parameters equivalent to those of the natural sediment grains. Finally, complex and sophisticated instruments are necessary for loading of artificial radioactive materials into the bed sediments.

The use of natural radioactive properties of sediments eliminates the above problems. Many kinds of natural sediments contain a high concentration of radioactive isotopes (such as radium, strontium, rubidium, and potassium) which produce the natural radioactivity of such sediments. It had been originally considered that natural radiation was too weak to be measured accurately and natural radioactivity was considered only as a component of the background radiation (cosmic, water mass, bottom sediments), which should be determined at the beginning of each experiment. This article considers results of application of natural radioactive sediments.

2. MATERIALS (TOOLS) AND METHODS

Due to its low level, the natural radioactivity must be measured using very sensitive tools, with very high stability and low background noise. Such tools were developed at the Institute of Nuclear Geophysics and Nuclear Geochemistry, Ministry of Geology, USSR (Shatsov, 1969).

The equipment used here for measurement of the gamma-activity of bottom sediment is composed of the following components (Fig. 1): underwater container with the gamma-radiation detector (Fig. 1A); 2) control desk (Fig. 1B); 3) auto-recording unit; 4) power supply; 5) connecting cable (power supply, pulse recording).

The underwater radiometer is built as a strong, airtight, vibration resistant unit, fitted into a sealed cylinder, and is adapted for movement along the bottom, while being rapidly towed by ship. It is protected against spontaneous surfacing and breakaway from the screened carrying cable. Power is supplied by a satellite ship engine. Stable contact of the instrument with the bottom surface was obtained during all the measurements noted here. The recording unit was carried on board the vessel. Measurement profile coordinates were established with the help of buoy stations and shore signs. Detector stability was determined from background measurements and with a radioactive standard.

When radioactive elements are carried with the collodium-dispersed fraction, the silt particles and organic substrates take the sorbed potassium, uranium and thorium to the bottom. This process produces specific fine-grained structures characterized by a broad anomalous distribution of the radioactive matter.

This observation was used to look for river born sediments at the river mouth region (Shteinman 1994; Shteinman et al. 1992).

The integral gamma-radioactivity of the sea bottom sediments is comparatively low. Therefore, it is possible to measure the much higher radioactivity associated with the river alluvia, using multi-channel radiometers to continuously record gamma-activity of the bottom sediment (Fig. 2). The gamma-activity curve shows separate zones differentiated according to the type of bottom sediments.

Figure 1. Underwater container without the sleeve (A) and the on-board
recording unit (B) of the radiometrer

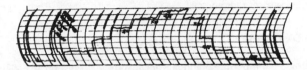

Figure 2. An example of the gamma-activity record
of the bottom sediments

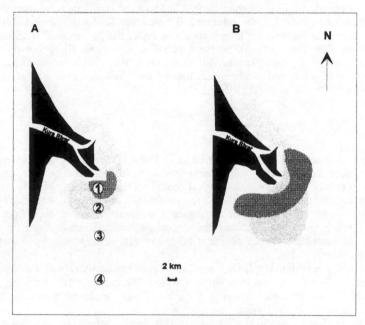

Figure 3. Charts of gamma - activity for the bed of the Kura River mouth,
obtained during a calm period (A) and after north-eastern wind, 18 m / s (B),
respectively:
1: maximum of gamma - radiation; 2: heightened intensity of radiation; 3: high intensity of radiation;
4: background radiation intensity

During periods of still weather, the river sediment distribution inside the mouth zone is produced by the velocity attenuation of the river jet flow (Shteinman & Kamenir 1998). In general, the coarsest fractions fall to the bottom near the river mouth bar. Further along the jet, the sedimentation rate and particle mean size gradually diminish. A close correlation was found between gamma-activity (γ) and the percent concentration of the pellite fraction (P):

$$\gamma = A \log P + B, \tag{1}$$

where A and B are empirical coefficients, constant for each fraction (Peisikov & Shatsov 1964).

To study sediment dynamics at the river mouth, methods, based on analysis of the gamma-radiation power spectra of natural radioactive elements were developed (Kostoglodov, 1979). This method is less informative than the integral gamma-activity analysis because of the much higher time consumption of gamma-spectrometry. Measurement of radioactive properties of sediment probes taken from separate points along the profile produces much higher stochastic then the continuous underwater radiometry.

3. RESULTS AND DISCUSSION

Gamma activity during a calm period and after winds, were obtained for the seabed of the Kura River mouth (Figures 3A and B). The results showed that diluted river silts with dimensions of less than 0.01 mm correspond to the area of maximum intensity, equal to 6.0 or more relative units. Sandy silt of 0.01-0.05 mm diameter corresponds to the area of lesser intensity (4.5-6.0 units). The area of moderate intensity (3.0-4.5 units) denotes the fine fractions of river origin. The regions of marine origin were characterized by low gamma-radiation intensity (< 3.0 units). Under conditions of relatively calm weather and large river discharge, gamma activity reflects the sedimentation character of the river mouth area under the influence of discharge currents, due to the river flow (Fig. 3A). Following a period of prevailing NE wind, river silts were displaced offshore, to a greater depth. During stormy periods, with high waves, muddy bed sedimentation takes place, and wind currents transfer the alluvium, causing displacement of the area of greatest gamma activity (Fig. 3B).

4. CONCLUSIONS

In comparison to other methods (i.e. sediment sampling), the radioactive method requires lesser effort and is more representative of long term conditions. The distribution of river alluvium at the discharge depends on hydrometeorological conditions that frequently change within several hours. It is clear that the data obtained at any specific time are not representative because they were not made under similar hydrometeorological conditions. A submarine radiometric survey of the seabed, made by towing a radiometer and recording gamma activity, lasts for some hours, allowing the relationship between sediment migration data and specific hydrometeorological conditions to be evaluated.

There is a close correlation between natural gamma radioactivity and the percent of silt content (analytically written as a logarithmic function) (Shatsov 1964). Therefore, radiometric surveys enable us to determine not only the area of river sediment accretion, but also the predominant fraction in the sediments.

A recent study (Shteinman 1994; Shteinman et al. 1992) carried out on sediment transport from the River Jordan into Lake Kinneret showed that probes taken before and after certain hydrometeorological events could elucidate the effect of those events. However, use of devices capable of measuring gamma radiation in situ would provide much more information. The use of natural radioactivity in studies of sediment transport in river mouths may be unique in providing information on: (1) sediment composition of large built-up forms, such as sand bars and spits, under different conditions; (2) location of zones of maximal sediment accumulation;(3) location of zones of erosion; (4) change in littoral zone relief due to sediment transport and decomposition under various conditions.

REFERENCES

Bueringer H. & Feldt W. 1964. Radiookologishe Studien als Erganzung zu den Sandwanderungsversuchen mit radioactiv markiert Sand (Cr-151). 155pp.

Bunte K. & Ergenzinger P. 1989. New tracer techniques for particles in gravel bed rivers. Bull. Soc. Geogr. Liege, Num. Spec. 25: 85-90.

Ferguson R., Hoey T., Watsen S. & Werrity A. 1996. Field evidence for rapid downstream running of river gravels through selective transport. Geology 24: 179-182.

Fernandes J.M., Badie C., Zhen Z. & Arnal I. 1991. Le ^{137}Cs traceur de la dinamique sedimentaire sur le prodelta du Rhone (Mediterranee nord-occidentale). In: Radionucl. Study Mar. Processes: Proc. Int. Symp. Norwich, 10-13 Sept. 1991. NY: 197-208.

Guidebook on Nuclear Techniques in Hydrology. 1971. International Atomic Energy Agency, Vienna, 256pp.

Jaffry P. & Hours R. 1959. Study of littoral transport by radioactive tracer methods. Cahiers Oceanogr. 11: 121-135.

Kamel A.M. & Johnson J.W. 1963. Tracing coastal sediment by naturally radioactive minerals. In: Proc. 5th Conf. Coastal Eng., Mexico City, 1962. Council Wave Res. Eng., Richmond, USA: 531-538.

Kostoglodov V.V. 1979. Gamma-spectrometric survey of marine bottom. Nauka, Moscow, 148pp. (in Russian).

Leontiev O.K. & Afanasiev V.N. 1959. The radioisotopes application for studies of movement of the sea sand sediments. Bull. Oceanogr. Commission AN USSR 3: 17-21 (in Russian).

Martin J.M., Meybeck M. & Heusel M. 1970. A study on natural radioactive tracers: an application to the Gironde Estuary. Sedimentology 14: 1-2.

Owcharozik A. & Szpilowski S. 1986. Application of radioisotope tracers for bed-load sediment transport studies in rivers and marine breaker zone. In: 3rd Work Meet. Radioisotop Appl. And Radiat. Process. Ind. Leipzig, 23-27 Sept. 1985. Proc. 2: 1259.

Peisikov Yu.V. & Shatsov A.N. 1964. Methods and tools for continuous radiometric submarine survey. In: Radiometry application for the survey of underwater oil stores. Neftegaz, Moscow: 7-47 (in Russian).

Putnam I.L. & Smith D.B. 1956. Radioactive tracer technique for sand and silt movements under water. Intern. J. Appl. Radioactive Isotopes 1: 129-137.

Rostan J.C., Juget J. & Brun A.M. 1997. Sedimentation rates measurements in former channels of the upper Rhone River using Chernobyl ^{137}Cs and ^{134}Cs as tracers. Sci. Total Environ. 193: 251-262.

Shatsov A.N. 1969. Marine Radiometry. Nedra, Moscow, 102pp. (in Russian).

Shteinman B.S. 1994. Sediment transport studies in River Mouths with the use of Natural Radioactivity. In: Proc. 7th Internatl. Biannual Conf. Physics of Estuaries and Coastal Seas. Woods Hole, USA: 1112-1119.

Shteinman B.S., Gutman A. & Gertner I. 1992. The Use of Natural Radioactivity in the Sediment Transport Study. In: Proc. 17th Conf. Nuclear Soc. Israel, Beer-Sheva Univ. Beer-Sheva, Israel: 122-126.

Shteinman B. & Kamenir Y. 1998. Decrease in suspended matter concentration with distance from the Jordan River mouth, Lake Kinneret: Hydrodynamical aspects. In: Advances in Hydro-Science and Engineering. Proc. 3rd International Conference on Hydro-Science and Engineering, Cottbus/Berlin, Germany (CD-ROM): 110-124.

Smith D.B. & Parsons T.V. 1967. Radioisotope techniques for determining silt movement from spoil grounds in the Firth of Forth. Isotopes in Hydrology. Symp. Proc. Vienna: 265-279.

Walling D.E. 1993. Use of cesium-137 as a tracer in the study of rates and patterns of floodplain sedimentation. IAHS Publ. 215: 319-328.

Wasson R.J., Clark R.L., Nanninga P.M. & Waters J. 1987. ^{210}Pb as a chronometer and tracer. Burinjuck Reservoir, Australia. Earth Surface Process and Landforms 12: 399-414.

Stochastic Hydraulics 2000, Wang & Hu (eds) © 2000 Taylor & Francis, ISBN 90 5809 166 X

Analysis on the increase of runoff and sediment yield due to reclamation in the Northwest of the Shanxi Province

Yong Liu & Dachuan Ran
Xifeng Water and Soil Conservation, Yellow River Committee, Shanxi, China

Shixiong Hu
International Research and Training Center on Erosion and Sedimentation, Beijing, China

ABSTRACT: The increase of flood and sediment yield due to the reclamation in the Northwest part of the Shanxi Province (Hekou—Longmen section, the upper reach) is analyzed in details in this paper. Based on the field investigation and data analysis, the evolution of the reclamation and its influence on soil erosion are studied. A new method to calculate the increase of flood and sediment yield due to reclamation is put forward. Analysis shows that the reclamation do not increase the flood significantly, but increase the sediment yield apparently at the rate of 6600 t/km^2 in the study area. The paper presents the sediment yield increase due to reclamation in 8 branches of the Yellow River Calculations show that the increase of sediment yield due to reclamation accelerated gradually from the 1960s to the 1990s. The recovery of the forest and grassland from the cultivated field and prohibition of deforestation is the precondition to rebuild the green Northwest China.

Key word: Reclamation, Increase in flood and sediment yield, The Northwest Shanxi Province.

1. INTRODUCTION

The human activities have two-blade function to the soil erosion—to control previous soil and water loss or accelerate the soil erosion. These two processes of positively prevention and passively execration of soil and water loss alternated in the whole human history. There are 5000 years long history of human civilization in the Yellow River Basin. With the progress of human society and development of productivity, excessive exploitation of natural resources results to the deterioration of ecological environment and serious soil erosion(Liang Sibao, 1990). In the northwest part of Shanxi Province, the rate of soil erosion increased by 0-100 times because of the intervention of human's activities.

The mode of new water and soil loss resulting from human activity is various, from the initial deforestation and exploitation of grass land to the reclamation on the steep slope, newly building the highway and railway, construction of kiln for bricks and house, mining and excessive grazing. The more and more kinds and larger and larger scale of the human's activities increase the sediment load in the river, which countervailed the reduction of sediment yield from the soil erosion control. Above all, the reclamation of the wasteland stands first in the list of human activities to increase sediment yield. In the previous studies of soil and water conservation in this area, the increase of sediment yield due to reclamation is paid less attention and weak ring of the whole study of the change of sediment and water resources in the Yellow River Basin. To analyze the increase effect of flood and sediment yield due to reclamation is very important to the calculation correctly the benefit of soil and water control measures, and very helpful to positively preventing the water and soil loss, rationally developing the water and land resources and protecting the environment.

The study area lies in the east to the Luliang Mountain, west to the main Yellow River, north

to the Hunhe River, south to the Sanchuan River valley. The area located in N37°30′ ~40°30
′ ,E110°30′ ~112°45′ ,including the Helinger and Qingshuihe counties in the Inner Mongolia,
the Youyu, Pianguan, Shenchi, Wuzhai, Kefeng, Hequ, Baode, Pinglu, Shuocheng, Xingxian,
Linxian, Lishi,Fangshan, Liulin,Zhongyang etc.15 counties in the Shanxi Province.

2. THE NATURAL AND SOCIAL CONDITIONS IN THE STUDY AREA

The reclamation of the wasteland is the consequent product of population explosion, and makes
the biggest contribution to the sediment increase. When the corn from the limit cultivated field
is not enough to meet the requirement of people, the only way in this area is to reclaim and
enlarge the cultivated field to raise the total production of corn.

As a dry, steep and poor mountain area, the population in the northwest part of the Shanxi
Province doubled from 1949, which means the increase of the cultivated field. From the
statistics of the population and cultivated field area in 1959,1985 and 1997, it can be seen that
the total area of cultivated field increased from 891.9 thousands hm^2 in 1959 to 1021.9
thousands hm^2 in 1985, decreased again to 896.97 thousands hm^2 in 1997. At the same time, the
populations of this area in 1959,1985,1997 are 1.68 million, 2.62 million, 3.06 million
respectively. The per capita area of the cultivated field is 0.53 hm^2 in 1959, decreased to 0.39
hm^2 in 1985,and 0.293 hm^2 in 1997. It means that one more person need to increase 0.487 hm^3
cultivated field during the 26 years from 1959 to 1985.All these increased farmland originate
from the reclamation of the waste land. In the study area, the reclamation activities not only
have continuity, but also have the suddenness of several large-scale acclamations. For example,
during 1959—1989, there are three times of large scale reclamation, which are: 1) Reclamation
to getting more grains for solving the starvation problems in 1960-1962; 2) Reclamation for
more corn production to respond for the slogans of "the corn production is the key link" and
"corn deposition preparing for the war and shortage of corn" in the 1966-1976; 3) Reclamation
on the steep slope in the early stage of 1980s, when the farmer have the right to cultivate their
contracted farmland, and they begin to reclaim the waste land without any restriction. The third
time is the largest one, and leads to no places to be reclaim in somewhere, for instance, the
cultivation rates in the Baode, Xingxian, Linxian counties are as high as over 45%, even 59.1%
in Liulin County. The cultivation rate is highly related with the population density. The higher
the population density is, the higher the cultivation rate is. The population density in the
northwest part of Shanxi Province is about 35.89—172.8 persons/km^2, the cultivation rate is
25.5—59.1%. Up to 1989, the population in the 8 branches of the Yellow River is 1.65 million,
among which includes 1.48 million farmers. The total cultivated land is 0.74 million hm^2, per
capita cultivated land 0.495 hm^2.

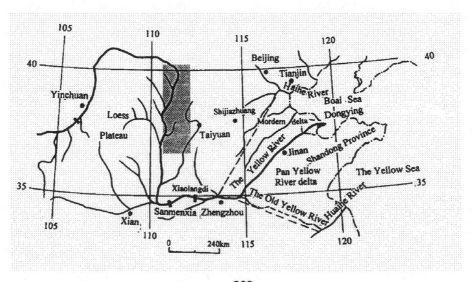

In the structure of the farmland, the slope land is high percentage, about 71.2%, and the rice field is 2.1%, and terrace field is 9.6%. Among the slope land, the proportion of over 15° slope land is higher than others. For instance of Bode County, over 15° slope land is 65.8% of all the slope land, and is 42.7% of the total cultivated field; The cultivation rate of 15°~25° slope land is 75.2%, that of over 25°~35° slope is 60%, that of over 35° is 8.7%。

Generally, the reclamation activity in the study area is from the hill slope to the gully slope and from the vicinity of the village to the faraway area. The reclamation rate is higher in the nearer area of the village. All where can be farmland have been cultivated. The farther farmland will be wasteland alternatively, but the management is very extensive cultivation. The investigation shows that in the area of high acclamation rate, there is cultivation through deforestation, for example, the silver chain forest is reclamation and planted the millet in the Qiaotou Town of Zhuajiachuan Basin of Baode County. The reclamation is all over the northwest of China and all is due to the increase of population. In the study area, The County with the most farmland from the reclamation is the Baode County, and the watershed is the Zhujiachuan Basin (Zuo Dakang, 1991)

3. THE IMPACT OF RECLAMATION TO THE SOIL EROSION

The research area is located in the Loess Plateau, and mainly composted of hills and gullies, which is broken and with gullies everywhere. The impacts of the reclamation activities includes ①reducing the vegetation coverage;②destroying the soil structure;③deteriorating the surface ecological environment.

3.1 Reclamation activities reduce the vegetation coverage

Without the human's intervening, the loess is covered by natural coverage, the water and soil loss is not serious. If the loess surface is cultivated, the natural vegetation will be destroyed completely. After the planting in the farmland, the vegetation coverage will be increased a little, but it is far from the natural vegetation coverage, even in the wasteland, the vegetation is impossible to recovery to the previous conditions.

Due to the decrease of the vegetation coverage, the reduction of the resisting capacity of raindrop's striking leads to rapid and serious soil erosion. At the same time, the low vegetation coverage makes the soil fertility decline, which will increase the erodibility of the soil.

3.2 Reclamation activities destroy the soil structure

Before the reclamation, the soil is not disturbed, the agglomerate structure can be formed, which has stable infiltration and high resisting force of raindrop's striking. After the reclamation, the soil surface is disturbed seriously, the infiltration rate is decreased, surface runoff is increased, therefore, erosion from overland flow become serious and serious. The steeper the slope land is, the more apparent the impact is. The ephemeral gully and rill erosion occurs due to the low resistance of the soil. For example, a rain storm of 47.8mm during 8 hours occurred in Xinxian County in Aug.31, 1994. 8 rills appeared on a slope land (slope:23°,area: 24 m^2) with millet. The maximum width of rill is 10 cm; the depth of rill is 12 cm. The erosion module is as much as 9720t/km^2. At the same time, there is no rill occurred in the neighboring slope land without reclamation. Another example in this area is a piece of slope land (slope:28°, coverage rate:5-10%) with millet, the soil erosion is very serious because of the steep slope and large catchment basin. During the same rain storm in 31st of August 1994, the lower part of the slope became a small trough as wide as 2-3 meters. The entire soil layer in the trough was destroyed, and two gullies with depth of 32 cm were cut into the mother rock, the maximum depth of the gully is as much as 47 cm. This kind of phenomena can be seen everywhere in the loess hillslope after rain storm. All these illustrate that the resistance of the soil layer reduces after reclamation. After years' wash from the overland flow, the numerous gullies and rills appear in the loess plateau and the surface broken into pieces, which result in more serious soil erosion.

367

3.3 Reclamation activities deteriorate the ecological environment

The data of plot indicate that the soil erosion in the slope farmland has a very close relation with the slope gradient. The steeper the slope is, the bigger the soil erosion module is. Table 1 is the data of plot of different slope in the Wangjiagou Valley, Lishi City from 1957—1966. It shows that there is no relations between the runoff and slope gradient, but the erosion increases with the increasing slope gradient, especially, the increasing range is bigger when the slope is more than 25°.

Table 1: the water and soil loss of slope farmland in different slope gradient

Slope	10°	15°	20°	25°	30°
Water loss(mm)	30.45	42.175	34.975	34.05	31.375
Soil erosion(t/km²)	2636	3440	3926	3957	7188

The slope gradient of the reclamation land varied with the scale of reclamation in the northwest Shanxi Province. The tendencies of reclamation is from the lower slope to steeper one, and from the near to the far area. After 1980s, most reclamation farmland concentrated on the steep waste slope above 25°. The investigation in the study area shows that the steepest farmland is above 35°, even 42°. The local farmers make best use of every bit of the space, even in the steep discarded earth slope after the highway construction. The steep slope (38--42°) and loose soil often lead to sever over erosion from overland flow, and also induce to the gravidity erosion, such as landslide and debris flow. Planting and reaping on the slope above 35° always make the soil particles roll down, especially under the action of strong wind.

The soil erosion also causes the serious loss of soil nutrient and soil degradation. 17 soil samples were got from the plots of 10-40° from 1989 to 1990, which locate in the slow slope land (10°) of hillslope top, and steep slope (10-35°) of upper part of hillslope, the valley slope and valley flat bottom. The plots are covered by corn, millet, potato and soybean. The result of the nutrient test shows that the losses of organic matter (OM), nitrogen (N) and phosphor (P) are 3.2%, 8.2% and 19.8% of the fertilization in the farmland, respectively. Moreover, the OM and N in the loss sediment are higher than that of the surface soil layer (0-10cm). It also illustrates that there is no erosion and nutrient loss in the terraced land. However, the soil erosion module of the slope land is 67.5 tons/hm² in annual average, which means the losses of 89kg ammonium nitrate, 226kg calcium phosphate per hectare, and the losses of 690kg stall fertilizer.

The reclamation on the steep slope results in the serious soil erosion and bare rock in the thin soil layer area. It is very difficult to recover the vegetation duo to no soil. For example, the mother rock is bare in the Nuanquan area of Inner Mongolia and earth–rock mountains of the northwest Shanxi Province, even no grass grow in these areas.

In the desert grassland and typical grass plain, the dessertification occurs because of reclamation in large scale, the desert area developed southward fast. The drought in the spring and summer occur more and more frequently in this area because of the low vegetation coverage. In conclusion, the reclamation caused the environmental deterioration and affects the human's life. More attention should be paid to the environmental protection and prohibition the reclamation of the steep slopes so as to the better recycling.(Liang Sibao, 1990).

4. CALCULATION OF THE INCREASING RUNOFF AND SEDIMENT YIELD DUE TO RECLAMATION

4.1 The calculation of reclamation area

1. According to the data of farmland area, calculating the slope farmland above 25°, which is equal to the area of reclamation .
2. Dispatch the reclamation area in different period by the population.

4.2 The index of the increasing runoff and sediment yield due to reclamation

The contrast is made between the reclamation slope and natural waste slope first. The statistic data of the runoff and sediment yield on the reclamation slope and natural waste slope can be used to calculate the index of the increasing runoff and sediment yield due to reclamation.

The data of the plot in the Suide soil and water conservation experimental station from 1958 to 1963 is analyzed in this paper. The data included 20 plot/year. Table 2 shows the average values of runoff are similar between the reclamation slope and natural waste slope, which means that the reclamation does not increase the runoff. However, the sediment yield of the reclamation slope is 6570 t/km^2 more than that of the natural waste slope. The contrast in the Wangjiagou watershed also indicate that he sediment yield of the reclamation slope is 6686 t/km^2 more than that of the natural waste slope. The sediment increases at the same level. Therefore, 6600 t/km^2 will be used as the index of the increasing runoff and sediment yield due to reclamation in the study area.

Table 2: Contrast of the runoff (R) and sediment yield (S) between the reclamation slope (RS) and natural waste slope (NWS) in Suide area of the Shanxi Province

Year	R from RS (m^3/km^2)	R from NWS (m^3/km^2)	R_{RS}-R_{NWS} (m^3/km^2)	S from RS (t/km^2)	S from NWS (t/km^2)	S_{RS}-S_{NWS} (t/km^2)
1958	47950	64700	-16750	20930	16520	4410
1959	95890	74580	21310	58940	25980	32960
1960	5330	6580	-1250	2880	2010	870
1961	49040	41290	7750	29040	16760	12280
1962	11020	13860	-2840	6490	10830	-4340
1963	32600	42220	-9620	4410	11180	-6770
Average	40160	40580	-420	20450	13880	6570

4.3 The calculation of the reclamation area and indexes

The sediment increase of 8 branches of the Yellow River can be calculated by the reclamation area and indexes (see table 3).

Table 3: The sediment increase of 8 branches of the Yellow River due to the reclamation

River	Reclamation area (km^2)				Annual sediment increase (10^{4}t)			
	60s	70s	80s	90s	60s	70s	80s	90s
Hunhe	13.67	16.65	19.55	22.15	9.02	11.0	12.9	14.62
Pianguan	8.12	9.86	11.61	13.11	5.36	6.51	7.66	8.65
Xianchuan	6.78	8.22	9.86	11.26	4.47	5.43	6.51	7.43
Zhujiachuan	14.77	17.92	21.14	24.14	9.75	11.83	13.95	15.93
Zhangyi	8.08	9.82	11.55	13.08	5.33	6.48	7.62	8.63
Weifen	15.64	19.02	22.38	25.54	10.32	12.55	14.77	16.86
Qiushui	16.55	20.09	23.61	26.81	10.92	13.26	15.58	17.69
Sanchuan	17.65	21.64	25.23	28.63	11.65	14.28	16.65	18.89
Total	101.26	123.22	144.93	164.72	66.82	81.34	95.64	108.7

Table 3 shows that the sediment increase due to the reclamation in the northwest Shanxi Province during 1990-1996 ascends gradually. The sediment increase due to the reclamation in 1990s is 1.087 million tons, 12% more than that of 1980s, 25% more than that of 1970s, and 38.5% more than that of 1960s. The sediment increase value due to reclamation in the four branch rivers (Zhujiachuan, Weifen, Qiushui, and Sanchuan) reached 150 thousand tons in 1990s. Therefore, in order to have green Northwest China and protect the environment in the development in the west China, it is necessary to adjust the agricultural structure, return the

369

reclamation slope to the grassland and forest, and prohibit the destroying of grass and deforestation.

ACKNOWLEDGEMENT

Financial support is from Ministry of Science and Technology of China under grant No. G1999043604, and the Foundation supporting study on the Water and Sediment Variation in the Yellow River.

REFERENCES

Liang Sibao, 1990,The water and soil loss and conservation the Loess Plateau in Qing and Ming Dynasty, China Water and soil conservation, No.6;
Zuo Dakang (eds), 1991, The collections of the environment evolution and laws of water-sediment movement in the Yellow River Basin, No.1, The Geology Press.

Water resources management

Stochastic Hydraulics 2000, Wang & Hu (eds) © 2000 Taylor & Francis, ISBN 90 5809 166 X

'Armriver' system of Armenia water-resources systems planning

Kamo Aghababyan
Institute of Water Problems and Hydraulic Engineering, Yerevan, Armenia

ABSTRACT: The "Armriver" common system for solution the water-resources issues has been developed for the Republic of Armenia. There have been considered as separate circles 18 main river basins of the Republic. For each circle there have been considered three stage water resources management hierarchy where concrete water resources problems are solved.

Key words: Mathematical modelling, Water resources management, Reservoir system, Reservoir release rule, Irrigation, Water supply,

1. INTRODUCTION

Armenia is the smallest of the former Soviet republics (Fig.1). Armenia covers a total area of about 29,800 km². It is a landlocked country, located on the extreme South of the Caucasus between 39° and 41° North and between 44° and 46° East. Armenia enjoys a variety of climatic conditions depending upon altitude. The Ararat valley is characterised by dry hot summers and cold dry winters, with annual precipitation not exceeding 300 mm. Temperatures decrease with altitude and are on annual average 10^0C in Yerevan and 4^0C at Lake Sevan. Correspondingly, growing periods vary between 250 days in the lower valleys and 170 days around Lake Sevan. Monthly precipitation is highest during the period April to June and lowest during July to September. Crop water deficit during May to August ranges between 200 to 700 mm, which cannot be supplied from soil moisture alone, and irrigation is therefore necessary for crop growth. Armenia has limited rainfall, surface water and groundwater. Totally average rainfall is 620 mm/year over the country. The natural quality of surface water ranges between good to low. Mineralisation of river water is generally between 0.6 to 1.2 g/l and pH values range between 7.2 and 8.3. Man-made pollution is limited to some specific reaches of rivers and originates from industry, particularly the mining industry, and inlet of sewage to the rivers (Hovhannesyan, 1995). Groundwater quality is generally good except in some parts of the Ararat valley close to the Araks river and north of city Artashat, where the level of salinity is problematic.

2. ANALYSIS AND RESULTS

Due to its peculiarities and complex structure water economy is classified to the class of complex systems, and its management needs sophisticated methods. For Armenian conditions there has been created a scheme of investigation problems in four hierarchic levels (fig.2):

1. Planning of water economy in global scale (zero level);
2. Planning of water resources systems (first level);
3. Planning of water resources subsystems (second level);
4. Planning of water resources elements (third level).

Fig.1: Map of the Armenian rivers.

Solving of planning problems for each level is connected with great difficulties depending on the complexities of water resource systems, a row of parameters and great volume of information being indefinite. Because of these reasons highly cited problems' solution is to be considered with the help of mathematical modelling.

In spite of the fact that mathematical modelling of water resources management problems aren't widely applied in the Republic of Armenia, there have been carried out definite investigations and created some software (Chilingaryan, 1973; Aghababyan, Shnaydman, 1990; Aghababyan, 1995).

The task of water resources systems (WRS) management can be formulated in the following way: to provide all the water consumers and users with water according to their consuming regimes, with the quantity and the quality in accordance with their design probabilities. On this purpose some tasks are set: to develop mathematical models for the main river basins of the Republic of Armenia, to carry out analyses of water resources situations with the help of these models, to elaborate management rules, and to control the system.

374

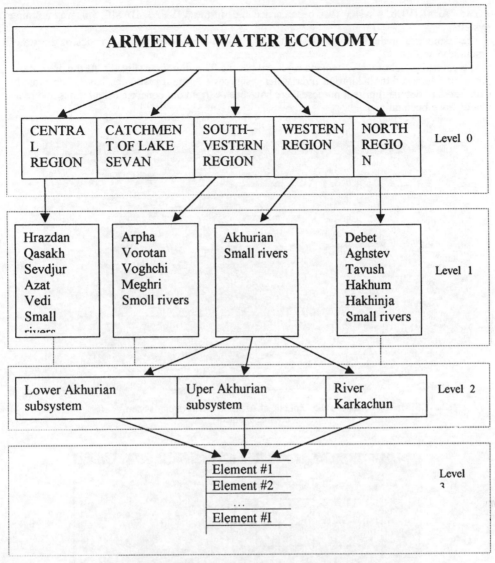

Fig.2: Hiearchical schem of Water Resource Systems of Armenia.

That was considered at the first level of the hierarchy, for which there has been developed the definition of parameters and an interconnected system of simulation and optimisation (linear programming) models(Loucks D....,1981, Pryazhinskaya V.G., 1988). For levels with already known parameters there has been developed a management simulation model.

The "Armriver" system will be instrumental for carrying out (Chilingaryan L.A., Aghababyan K.A., 1995,1996) :
- different hydrologic calculations for the rivers of Armenia;
- water-resources calculations (taking into consideration the water quantitative, regime and qualitative aspects);
- statistical processing of the data (correlation and regression analyses);
- Reservoir release rule design (for separate operating and cascade reservoirs);
- Flood routing;
- Mathematical modeling of the flow hydrological rows by the Monte-Carlo method. For the distribution within a year the so-called method of "fragments" has been used.

The "ArmRIVER" system has been carried out in the VISUAL BASIC 5.0 programming system.

The issues are methodically substantiated taking into consideration peculiarities of the water resourses use.

The control panel of the system management has the folowing view (Fig. 3). At the present phase problems of the Akhurian river water resources use main planning have been mostlly elaborated in details. For that purpose there have been carried out concrete calculations and real results have been gainen.

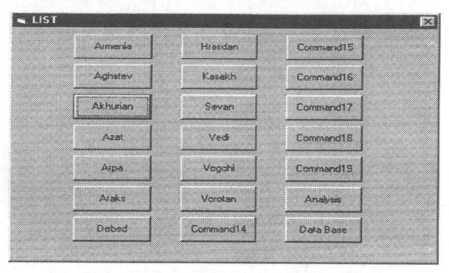

Fig.3: Main problems of the Republic of Armenia "ArmRiver" system water resources management.

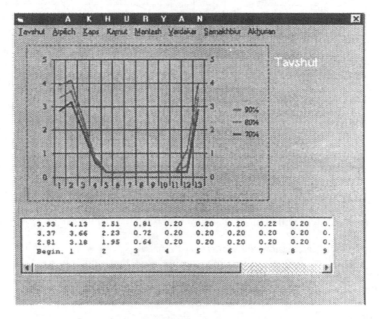

Fig.4: The Tavshut reservoir release rules

Fig. 4. Shows the Tavshut reservoir release rule based on storage volumes in the basin of the Akhurian river.

REFERENCES

Aghababyan K.A., Shnaidman V.M., 1990: Selection of irrigation reservoir parameters using interrelated optimization and simulation models.-Journal "Water Resources". Moscow. 168-177. (In Russian)

Aghababyan K.A., 1995: The simulation model of complex use of water resources on river basin in conditions of the Republic of Armenia.- XXVI-th IAHR Congress. London. Vol. 4, 284-289.

Chilingarian L.A., 1973: About the linear optimisation model of seasonal runoff control on reservoir cascades. "The problems of hydroenergetics and water economy". Alma-Ata. Issue 13, Nauka, 135-145. *(In Russian)*

Chilingaryan L.A., Aghababyan K.A., 1995: The methodological prerequisits of the research of complex utilization of water resources of Armenia. Part 2. Water resources tasks' hierarchy and prerequisits of their methodological solution with the help of mathematical modelling. Mag. Herald of agricultural sciences. -Yerevan. N4, p.147-154. (In Armenian. Summary in English, In Russian).

Chilingaryan L.A., Aghababyan K.A., 1996: Methodological prerequisits of research and complex use of water resources of Republic of Armenia. Part 1. Common and comparative characteristics of the water economy, water resources and their use Republic of Armenia. Mag. Agroscience.- Yerevan. N3-4.p.p.117-126. (In Armenian. Summary in English, In Russian).

Hovhannesyan K.K., 1995: Irrigation water quality of the Republic of Armenia.-Regional Conference on Water Resources Management. Isfahan.

Loucks D., Stedinger J., Hait D., 1981: Water Resources Systems Planning and Analysis.-Prentice-Hall, New Jersey.

Pryazhinskaya V.G., (Editor), 1988: Mathematical modelling on managment of water resources.-Nauka, Moscow. *(In Russian)*

Reznikovskiy A., (Editor), 1989: The hydrological foundations of hydroenergetics.- Energy, Moscow, 262p. *(In Russian)*

Stochastic Hydraulics 2000, Wang & Hu (eds) © 2000 Taylor & Francis, ISBN 90 5809 166 X

Water resources management in seasonal wetlands of North-East Bangladesh

Dhali Abdul Qaium
Water Resources Planning Organization (WARPO), Bangladesh

ABSTRACT: Bangladesh, the lowest riparian of the Ganges-Brahmaputra- Meghna basin, drains approximately 1.7 million square kilometer area with average annual runoff of approximately 175 billion cubic meter. Average Land elevation of the country is about 12.50 meter above mean sea level and it faces two extreme conditions : overabundence of surface runoff during wet season and scarcity during dry season. Overabundence of water during wet season creats flood situation on atmost one third of the country in a normal year while in extreme condition may rise upto two third (example 1988 & 1998 flood in Bangladesh). Flood situation creates seasonal wetland in low lying areas which during wet season turns into water bodies sustaining wetland acqutic flora and funa while during dry season the same is converted into agricultural land. Over population, as elsewhere in the Globe, and consequently, over exploration of natural resources, of this unique geological feature. The natural and traditional water management system of these seasonal wetland has been changed due to human interference which should be managed in a sustainable way for pressure these unique geological feature.

Key Words: Ganges- Brahmaputra - Meghna Basin, Flash flood, Haor project, Flood control and Drainage project, Seasonal wetland

1. INTRODUCTION

Bangladesh is the lowest riparian of the Ganges-Brahmaputra-Meghna (GBM) basin (Figure-1). Territorial area of the country stands to 147570 square kilometere (sq.km.) lying between latitude $20°34'$ to $26°38'$ (North) and $88°01'$ to $92°41'$(East); the area is approximately 8% of 1.70 million sq.km, the catchment of GBM basin.(Yogacharya,1999; BSS, 1999; Faisal et al.,1999)

Bangladesh possesses flat and plain land terrain with very little undulation in north-east, east and south-east corner. Land elevation varies from 90 meter (in north west corner) to mean sea level (in the south and north-east depression); 50% of the land lies below 12.5 meter while 30% remains below 5 meter (Master Plan Organization, 1986)

Table 1 : Seasonal and regional variation of rainfall in Bangladesh (in mm)

Region	Annual	Wet Season (June –October)	Dry Season (November –May)
NW	1966	1811	156
NE	2961	2593	368
SW	1975	1792	183
SE	2921	2661	260

Table 2: Major River Flows (in m³/s)			
Flow Type	**Ganges** (Harding Bridge)	**Brahmaputra** (Bahadurabad)	**Meghna** (Bhairab Bazzar)
Mean Low Flow	790	19,557	1,332
Mean Annual Flow	8,544	37,550	6,748
Mean Peak Flow	51,625	65,491	14,047
Highest Peak Flow	76,000	98,600	19,800

Note : Data of Post Farraka Era

Table 1 presents the seasonal and regional variation of rainfall and Table 2 shows the discharges of three major rivers in the country (Seminar preceedings , 1988; Brammer, 1993)

Annual rainfall within the country can generate only 123.50 cubic kilometer (Cu.km) of runoff but total annual drainage volume of the GBM basin that passes through the river system of Bangladesh stands to 1481 cu.km (Ali, 1999). This volume , if accumulated on the entire country (assuming the land surface as a horrizontal plain) would constitute a water column of 10 meter height.

The above picture may indicates the availability of fresh water for the entire country throughout the year but temporal and spatial variation of rainfall and cross-border flow is a major impadiment to the development of Bangladesh.

Eighty percent of the local rainfall and transboundary flow occurs within the period of June to October but major concentration appears in June, July & August (Seminar preceedings , 1988)

Obviously, flood is an annual event in Bangladesh and from time immortal, people are coping with the phenomena. About 25% land mass of the country is inundated every year under normal condition; locally the situation is termed as "Barsha". In this process, livelihood, economic activities, cropping pattern and settlements are well adjusted and adopted.(Brammer, 1993) But in abnormal flood which is commonly known as "Bonna", about 60% of the country goes under water (Brammer, 1993) Disruption of communication, loss of agricultural crops, damage to homestate, non- availability of safe drinking water and flood is associated with this type of natural calamity. Catastropic flood as happened in 1954, 1955, 1970, 1974, 1987, 1988 and 1998 causes immese suffering to the people with loss of lives, damages of all infrastructures and substantially the hampering the progress of the whole nation. For example, about 30 million (out of 111.4 million) people and 60% of the country was severely affected by flood for long three month (July-September) (BCAS,198)

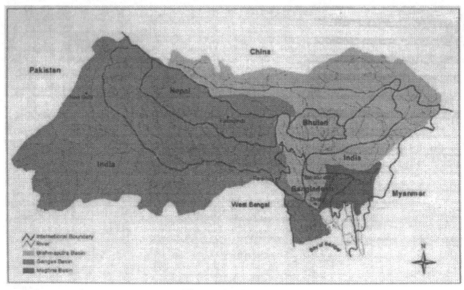

Fig.1: The GBM Basin

380

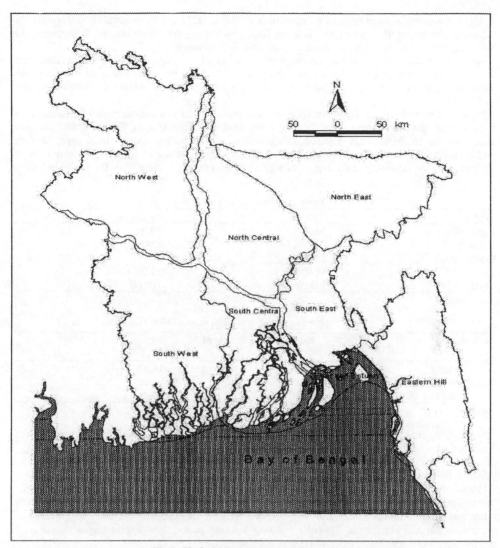

Fig.2: Hydrological Regions of Bangladesh

2. SEASONAL WETLAND

Land, that goes under water during "Barsha", can be termed as seasonal wetland. The striking feature of these wetland is seasonal inundation (depending on land elevation the area may be inundated for few days to as long as 9 months). These wetlands (during inundation period) create an aquatic environment favourable for open water fishery, easy navigation, varity of flora and funa but agricultural production is hampered depending to rising and recession of water.

Bangladesh is divided into six hydrological region based on river network (figure-2).

Seasonal wetlands are located in all the regions but almost 60% area of North East region and 13% area of South-West region is formed by this type of land.

3. NORTH EAST REGION

The Region comprises of an area of 24200 sq.km. Approximately 50% of the land lies below 5meter while 25% exibits elevation of less than 3 meter (above mean sea level).

The entire area is criss-crossed by numerous rivers; notable are the Surma, the Ushiyara, the Khowai, the Manu, the Sutang, the Juri, the Jonai, the Piyain, the Jadukata, the Someswari, the Knaysa, the Dauki, the Banlia, the Dhanu, the Old Brahmaputra.

Almost all the rivers (except the old Brahmaputra) of the region possess flashy character (sudden rise of water level within a very short say 29 hours even in the month of February. All these rivers meet together at the tail end of the region (south – east corner) and takes the name "Meghna".

Hijal and koroch are the dominant trees in these wetland that can withstand full inundation for long period; the trees form habitat for fish, water fowl and other species. Economics of the area as elsewhere in Bangladesh a based on agriculture and fisheries; Out of total area, 66% is cultivable and 39% is single cropped (Boro rice). Dependency of agricultural products & activities on the economy of the region as well as of the country is clear from the tables below:

Table 3: Sectoral share of GDP

Sector	National	Regional
Agriculture	36.00	33.00
Industry	8.73	8.00
Construction	5.67	6.00
Services	49.60	53.00
Total	100.00	100.00

Table 4 : Rice Production

Indicator	National	Regional	National (%)
Area (sq.km.)	147570	24200	16.40
Rice Area (million-ha)	10.44	1.36	13.00
Rice Production (million- ton)	17.85	3.30	18.48

Single rice crop "Boro" production and its vulnerability due to flash floor is crucial for livelihood of 17.66 million people of the region.

The wetland in the North-East region is locally known as Haor; a large bowl-shaped depression encircled by natural levees of rivers.

Haors are unique geological feature; the entire landmass remains dry for about six months (December- May) while for the rest of the year, it goes under water except homestates. During monsoon season, homestates look like islands. At that time, boats are the only means of communication in Haor area. Agriculture during dry season and finishing during monsoon time are the major economic activities of Haor people.

Agricultural activities in Haor area is also limited to single rice production in one calender year. Rice variety locally known as "Boro" is transplanted in December-January and harvested in April-May. This rice variety is vulnerable to flash flood.

4. HAOR PROJECT :

To ensure safe harvest of "Boro" rice, human intervention in the form of :
- raising (upto a certain height to resist entry of river water due to flash flood into the Haor) and strengthening natural levees of river encircling the Haor.

and
- constructing water control structure at intersection of natural levees and drainage channel of the Haor

is a century-old practice termed as "Haor Project".

Management of water resources in Haor area is limited only to "Haor Projects." Gradual inundation of landmass during pre-monsoon period, submergence of the entire area during monsoon, recession of water level during post-monsoon period and flash flood during March-April-May in some wet-year are natural phenomenons beyond human control.

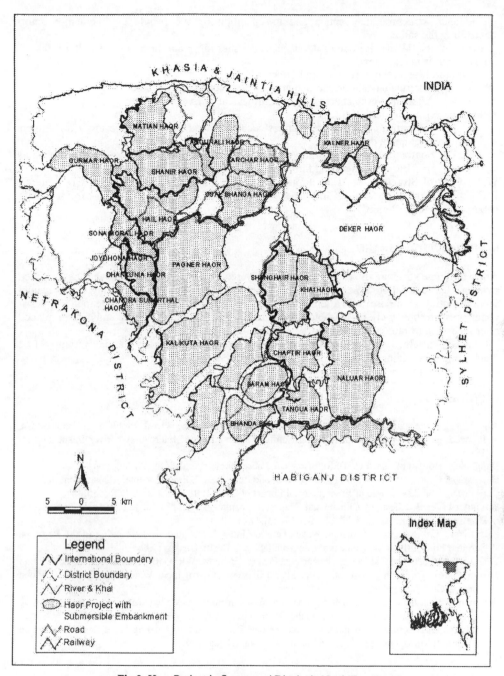

Fig.3: Haor Project in Sunamgonj District in North-East Region

5. EFFECT OF HAOR PROJECT

Structural intervention (levees and structures) has significant positive impact on agriculture but minor negative impact is noticed in fisheries and navigation.

In fact, water management in Haor area is a highly complex phenomena. Though natural events like flash flood, drought and other climatological factors have definite role on the phenomena, structural intervention and its operation by the people are the major actors controlling the same.

Some of the Water Resources Management issues of Haor area as a result of human intervention is listed below:

- Structural Intervention (Levees and Structures)
  - Obstruct navigation and fish migration except peak monsoon.
  - Affect non-project areas due to confinement of river flow.
- Structural Intervention (Structures)
  - Fishermen and high land farmers except delayed drainage while low land farmers want quick drainage.
  - Boatman and low land farmers desire inundation (after Boro harvesting) as soon as possible while high land farmers expect detailed inundation.

Water Control Structures of "Haor Projects" are the traditional ones built in FCD Projects; R.C.C. structure with vertical lift gates. This type of structure is not in-harmony with submersible "Haor Project"

6. MITIGATION MEASURE

A flexible type structure in "Haor Project" may overcome the issues listed above. The proposed structure will consist of : a permanent R.C.C. bed in conformity with the existing canal/channel section (acting as a base) and a temporary water tight super-structure strong enough to resist water pressure but easily removeable as soon as "Boro" harvesting is over and again replaced after recession of monsoon water.

Proposed structure will be eco-friendly; no adverse effect on fisheries and navigation will be there while minimizing the other conflicting management issues.

REFERENCE

Ali R.M.E.. International River Basin Management to solve the Flood Problem of Bangladesh. Proceeding of the Regional Seminar on Conflict Management of International River Basin. Dhaka, 1999.

Bangladesh Bureau of Statistics (BSS), Statistical Pocket book, Bangladesh, Dhaka, June 99.

Bangladesh Centre for Advanced Studies (BCAS) and Resources & Environmental Management Services (REMS). The Assessment of Environmental Impact of flood 1998 on Dhaka City Report.

Brammer H. et.el. Effects of Climate and Sea-Level Changes on the Natural Resources of Bangladesh. Briefing Document No. 3. BUP, CEARS, CRU, Dhaka, 1993.

Faisal I.M. et.al. Managing Common Waters between India & Bangladesh. Proceeding of the Regional Seminar on Conflict Management of International River Basin. Dhaka, 1999.

FPCO. North East Regional Water Management Plan (FAP-6). MoWR, GoB. Dhaka, September, 1993

Master Plan Organization. National Water Plan. Ministry of Irrigation, Water development & Flood Control, GoB. December, 1986.

Seminar Proceedings. International Seminar on Water Resources Management and Development in Bangladesh with particular reference to the Ganges river, MoWR, GoB. Dhaka, March, 1988.

Yogacharya K.S. Conflict Management in International River Basin. Proceeding of the Regional Seminar on Conflict Management of International River Basin. Dhaka, 1999.

Stochastic Hydraulics 2000, Wang & Hu (eds) © 2000 Taylor & Francis, ISBN 90 5809 166 X

Stochastic modelling of fluctuations of evapotranspiration and water supply deficit for corn and wheat fields

Ye. M. Gusev & O. Ye. Busarova
Institute of Water Problems, Russian Academy of Sciences, Moscow, Russia

ABSTRACT: The temporal series of evapotranspiration from corn and wheat fields under natural conditions and under the condition of sufficient water supply were obtained for about 30 sites situated in the step and forest-step zones of the European part of the Former Soviet Union. In so doing, the values of evapotranspiration were calculated using a simplified physically based model which describes the processes of heat- and water transfer within a soil - vegetation - atmosphere system. The analysis of monthly values of evapotranspiration and water supply deficit for corn has shown that the problem of stochastic modelling interannual dynamics of these quantities is connected with creating a technique for modelling series of correlated positive and zeroth random quantities. Here, such technique, based on modelling integral probability of fluctuations of evapotranspiration and water supply deficit, is proposed. Model's parameters were obtained and mapped for the above mentioned region.

Key words: Evapotranspiration, Water supply deficit, Stochastic modelling.

1. INTRODUCTION

Solution of the problem of rational use of inland water resources requires improving the quality of stochastic modelling not only of the input component of water resources defined by the river runoff volume, but also of the consumption components of water use. For agriculture and especially for irrigated farming, the accuracy of forecasting water consumption for irrigation depends on the accuracy of predicting evapotranspiration from irrigated agricultural field. Such characteristics of agricultural water management as water supply deficit, irrigation water consumption, and other connect closely with evapotranspiration from agricultural ecosystems. In particular, the physiological water supply deficit for agroecosystems can be determined as the difference between evapotranspiration from irrigated agricultural field and that from a field under natural conditions (without irrigation).

As a rule, the estimates of water balance components for a certain area (especially long-term ones) can only be probabilistic. Evapotranspiration is also known to significantly vary with time. Thus, in stochastic hydrology, in addition to the problem of developing stochastic models for river runoff, the problem of stochastically describing evapotranspiration as a water balance component also arises. To solve this problem it is necessary (i) to create stochastic model that simulates fluctuations of annual and monthly evapotranspiration from agricultural fields and (ii) to find values of appropriate model parameters.

2. TECHNIQUE FOR OBTAINING LONG-TERM SERIES OF EVAPOTRANSPIRATION AND WATER SUPPLY DEFICIT

Construction of stochastic models describing geophysical and hydrological processes is usually based on the data of temporal series of experimentally determined (observed) characteristics with different degrees of discreteness. Classical methods for probabilistic description of dynamics of surface and groundwater resources use the data of temporal series of directly measured runoff volumes (annual, monthly, etc.). However, application of this approach for studying the regularities of evapotranspiration variations is impossible because the absence of measured long-term series of evapotranspitration from different agricultural ecosystems. Therefore, it is necessary to develop a technique for theoretically calculating such series of evapotranspiration.

To solve this problem, we must use a physically based model for water transfer in a "soil - vegetation - atmosphere" system, which require relatively small set of parameters. In this study, the evapotranspiration calculations are based on A.I.Budagovskii's model [1] with same modifications [4]. According to this model, the evaportanspiration rate for the warm period of a year is calculated as follows:

$$E_C = \phi_0 E_{T0}[1 - \psi(L)] + M E_{S0}\psi(L) \quad , \tag{1}$$

where

$$M = 1 - (1 - \gamma V)\exp[-P/(E_{S0}\psi(L))] \quad , \quad \phi_0 = V/V_{cr} \quad , \quad V_{cr} = \alpha + \beta E_{T0} \quad ;$$

E_C is the evapotranspiration; E_{T0} is the potential transpiration; E_{S0} is the potential evaporation from a soil; ψ is the function of the leaf area index L [1]; P is the precipitation; V is the available water stores in 1-m soil layer (total water stores minus water stores corresponding to wilting point); V_{cr} is the critical available water stores in 1-m soil layer; α, β, and γ are empirical parameters [1].

The values of E_{S0} can be evaluated by means of approximate approach based following relationship $E_{S0} = f(d)$ [7], where d is the deficit of air humidity. After that the values of E_{T0} can be simplify determined on the basis of close relationship between E_{S0} and E_{T0} [1]:

$$E_{T0} = \eta E_{S0} \quad , \tag{2}$$

where η is the empirical coefficient varying within a narrow range; in present study, we use its average value, 1.15.

The soil water stores is evaluated in accordance with (1) and the water balance equation for 1-m soil layer

$$dV/dt = P - E_C + Q - Y \quad , \tag{3}$$

where t is the time, Y is the surface runoff and Q is the rate of water exchange between the 1-m soil layer and underlying layers. In so doing the measured value of soil water stores after snow-cover loss V_{in} is used as initial value. It should be noted that considered model has been used for the areas of the steppe and the forest-steppe zones of the European part of the Former Soviet Union. In these zones summer surface runoff Y, as a rule, can be neglected and water exchange between the upper 1-meter soil layer and underlying layers Q is also relatively small. The latter can be explained by the two reasons: (i) large depth to water table and (ii) infiltrated water usually does not penetrate deeper than 1-m soil layer. Therefore, Q and Y were accepted to be equal to zero.

For the cold period of a year, we assume that

$$E_C \approx E_{S0} \tag{4}$$

The calculation of V and E_C is conducted with a ten-day time step, after that monthly values of E_C are calculated.

Thus, principal factors determining temporal variations in the evapotranspiration are P, d, and V_{in}. The series of their monthly values were taken for a number (about 30) of sites situated in the steppe and forest-steppe zones of the European territory of Russia. The data of agromeorological stations were used. It should be noted that available series of measured soil water stores are rather short and poor. That is why they were updated by stochastically modeled values. Input data on P, d, and V_{in} allow us to simulate the long-time series (about 50 years) of monthly values of evapotranspiration for selected sites and for two crops: wheat and corn, which are the most representative crops for area under consideration. The values of function ψ for these crops were taken from [3].

The dynamics of the monthly values of evapotranspiration was calculated for two cases: (i) E_C^{ir}, for the condition of sufficient moistening (in this case, we assumed that the field is irrigated at the moment when the water stores decrease to V_{cr}); and (ii) E_C^{nat}, for natural conditions (without irrigation). The water supply deficit DE was estimated as follows:

$$DE = E_C^{ir} - E_C^{nat} \tag{5}$$

3. RESULTS OF SIMULATING LONG - TERM SERIES OF ANNUAL AND MONTHLY VALUES OF EVAPOTRANSPIRATION AND WATER SUPPLY DEFICIT

The physically based simulation of the long - term series of annual and monthly values of evapotranspiration and water supply deficit allow us to derive the following conclusion. The coefficients of correlation between of annual values both for E_C^{nat} and for DE are not large ($\pm$ 0.05÷0.15), and their order corresponds to relative error of their determination. That is why the statistical modeling annual values of E_C^{nat} and DE can be performed on the basis of simulated series of independent random variables, which are distributed in accordance with the relevant function of the unconditional cumulative distribution U (or cumulative frequency $P=1-U$) of E_C^{nat} - and DE-values. As a rule, these functions can be approximated by the functions of lognormal or gamma distributions (Figure 1).

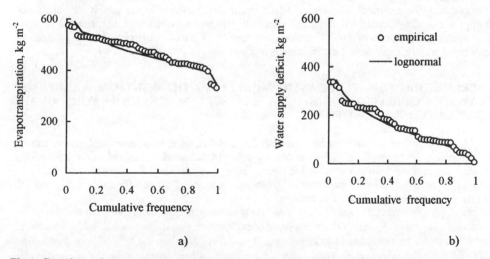

a)   b)

Fig.1: Cumulative frequencies (empirical and approximated by lognormal function) of annual values of evapotranspiration for natural conditions (a) and water supply deficit (b) for corn-sown areas for the Kamennaya Step' Station.

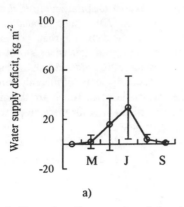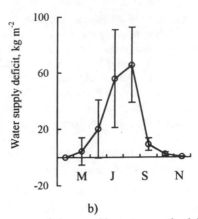

a)   b)

Fig.2: Examples of temporal variation in the average of the monthly water supply deficit and its standard deviations for corn-sown areas for the stations: Glukhov (a), and Kamennaya Step' (b).

The coefficients of variation of annual values of E_C^{nat} are also not high (0.07÷0.16), whereas the coefficients of variation of annual values of DE are significantly higher (0.4÷1.1). Hence, the temporal variability of water supply deficit cannot be neglected in the probabilistic forecasting of its annual values.

The analysis of monthly values of evapotranspiration and water supply deficit for wheat and corn has shown (see, example, in Figure 2) that their variability is larger than that for annual values, and their values for subsequent months are correlated significantly. Thus, the coefficients for variation of monthly values both of E_C^{nat} and of DE for mentioned the crops are as high as 0.1÷0.3 and 0.6÷1.3, respectively. The coefficients of correlation between monthly values for both E_C^{nat} and DE for subsequent months range from 0.6 to 0.8. The obtained results show that it is necessary to develop a technique for stochastic modelling the series of essentially nonstationary, correlated variables for simulating monthly values of both evapotranspiration and water supply deficit for considered crops.

In addition, the empirical unconditional function of distribution of water supply deficit cannot be approximated by classical functions of probability (lognormal or gamma distribution) because of available zero values of DE. Therefore, the specific form of approximation of probability function of water supply deficit distribution is needed.

4. TECHNIQUE FOR STOCHASTIC MODELLING OF MONTHLY VALUES OF EVAPOTRANSPIRATION AND WATER SUPPLY DEFICIT FOR GRAIN AGRICULTURAL ECOSYSTEMS

Stochastic modelling annual dynamics of both E_C and DE, as it was mentioned above, can be performed on the basis of simulating stationary series of independent random variables because of weak correlation between their subsequent values. Contrariwise, for monthly values, in so doing, it is necessary to take into account the nonstationarity of series of values of E_C and DE and their correlation for subsequent months.

One of possible ways for solving this problem within correlation theory is to model the series of monthly values of E_C and DE on the basis of information about their joint distribution. Here, we will consider an example of modelling monthly values of DE with the help of joint distribution $f(DE_i, DE_{i+1})$, where i is the number of a month. When constructing the algorithm for simulating the random variable DE_i, it is convenient to use its cumulative distribution function U_i (or cumulative frequency P_i) which can be obtained by the following change of variables

$$U_i(DE_i) = 1 - P_i(DE_i) = \int_0^{DE_i} p_i(DE_i')dDE_i' \quad , \tag{6}$$

where $p_i(DE_i')$ is the unconditional (marginal) distribution density of monthly values of water supply deficit for i-th month. This change of variables leads to the task of constructing a series of variables U_i, such that each of them is uniformly distributed within [0, 1] and correlated with other. The values of DE_i can be easily obtained from simulated values of U_i in accordance with Equation 6.

One of the ways to construct the series of uniformly distributed and correlated random variables $u=U_i$ and $v=U_{i+1}$ is based on a bilinear expansion of their joint distribution density $f(u,v)$ in the Legendre polynomials [6]:

$$f(u,v) = \frac{1}{\pi}\int_0^\pi (1-\lambda)dz\{\{1-2\lambda[(2u-1)(2v-1) + \sqrt{(u-u^2)(v-v^2)}4cos(z)] + \lambda^2\}^{3/2}\}^{-1} \quad , \tag{7}$$

where λ is the coefficient of correlation between the random variables u and v.

The cumulative conditional distribution function of v at given u, which is necessary to simulate the series of monthly values of DE, can be obtained from the well-known formula

$$G(v|u) = \int_0^v f(u,v')dv' / \int_0^1 f(u,v')dv' \tag{8}$$

If u and v are random variables uniformly distributed in the interval [0, 1], then G represents a variable of the same type. This allow one to simulate a series of monthly values of DE with specified characteristics $P(DE)$ for each month and λ for the subsequent months [5].

The outlined procedure for simulation of monthly values of DE requires, according to the described algorithm, the information about the marginal functions $P(DE)$, as well as the coefficient of correlation between cumulative frequencies of water supply deficit for subsequent months. The present study has shown that the coefficients of correlation between cumulative frequencies of water supply deficit for subsequent months are close to λ of values of DE. Since most realizations of DE have zero values, classical approximation of the cumulative frequency functions $P(DE)$ (normal, lognormal, gamma distribution, etc.) are unsuitable in this case. In this study, we used the following function to approximate the relationship between DE and its cumulative frequency function P:

$$DE = 0 \quad \text{for } 1 \le P \le P_0 \quad , \tag{9}$$

$$P = P_0 \frac{1 + (a/b)^4}{1 + [(a+DE)/b]^4} \quad \text{for } P > P \ge 0 \quad , \tag{10}$$

where P_0, a, and b are empirical parameters. The examples of cumulative frequency function for the water supply deficit is shown in Figure 3.

Constructed long-time series of water supply deficit allow us to determine the parameters P_0, a, b, and λ, to systematize and represent them in the form of map-schemes for difference months for the above mentioned region [2]. The examples of such the map-schemes are given in Fig.4. Thus, the algorithm, described by Equations (6) - (10), and information about parameters P_0, a, b, and λ represent the base for stochastic modelling seasonal dynamics of water supply deficit for agricultural ecosystems.

The series of monthly values of E_C or DE simulated by the suggested algorithm, provide an approach for estimating the statistical characteristics of the dynamics of evapotranspiration and water supply deficit for many problems associated with the forecasting, management, and optimization of the utilization of water resources in agricultural ecosystems.

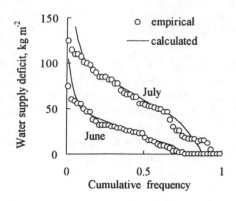

Fig.3: Empirical and calculated by Equations (9), and (10) cumulative frequency functions of monthly values of water supply deficit of corn-sown areas for the Kamennaya Step' Station.

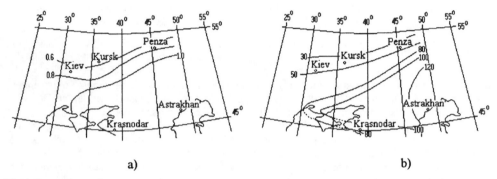

a)   b)

Fig.4: Parameters of cumulative frequency function of water supply deficit: P_0 (a) and b (in mm/month) (b) for July.

ACKNOWLEDGEMENTS

The present work was supported by the Russian Foundation for Basic Researches

REFERENCES

Budagovsky, A.I. 1989. Principles of the method of calculating the duty of water and irrigation regimes. *Water Resources*, 16 (1), 27-35.

Busarova, O.Ye. and Gusev, Ye.M. 1998. Application of stochastic modeling to forecast water consumtion deficit. *Water Resources*, 25 (3), 254-258.

Dzhogan, L.Ya. and Lozinskaya, E.A. 1993. A method for averaging leaf relative area in mesoscale assessment of heat and moisture exchange between the underlying surface and the atmosphere. *Water Resources*, 20 (6), 612-619.

Gusev, Ye.M., Busarova, O.Ye. and Nasonova O.N. 1996. On the problem of constructing stochastic models of variation in the evapotranspiration rate from the land surface. *Water Resources*, 23 (1), 1-7.

Gusev, Ye.M., Busarova, O.Ye. and Nasonova O.N. 1997. On the possibility of statistical modelling of water-consumption deficit in agroecosystems. *Water Resources*, 24 (2), 128-133.

Sarmanov, I.O. 1975. New forms of correlation for hydrologic application. *Water Resources*, 2 (2), 183-195.

Zubenok, L.I. 1976. Evaporation from continents (In Russian). Gidrometeoizdat, Leningrad. 264 pp.

Stochastic Hydraulics 2000, Wang & Hu (eds) © 2000 Taylor & Francis, ISBN 90 5809 166 X

Water supply management for Hsinchu area of Taiwan

Wen-Cheng Huang & Tung-Hsin Chang
Department of Harbor and River Engineering, National Taiwan Ocean University, Keelung, Taiwan, China

ABSTRACT: The aim of this research is to examine the deficit problem of the water-supply system of Hsinchu area, location of Taiwan's "Silicon Valley". Simulation results indicate that Hsinchu is an area without sufficient water to meet the current demand, and water obtained from alternative (Yungheshan reservoir) will be necessary. Further, the expansion plans for industrial use should be curtailed to refrain from economic damages, if the proposed Baoshan II reservoir is not completed by 2006.

Key words: Deficit; Simulation; Reservoir

1. INTRODUCTION

Taiwan is abundant in rainfall with about 2,510 mm/yr. However, water resources distribution is very uneven, of which 78% comes from the wet season (from May through October) and only 22% is available during the dry season (from November to April). Droughts and floods frequently occur every year.

Hsinchu Science-based Industrial Park (HSIP), Taiwan's "Silicon Valley" in producing integrated circuits, computers and peripherals, telecommunication, and optoelectronics etc., was established in 1980. At present, there are 245 companies gathered in this science park with combined annual sales of more than US$13 billion. The current demand for municipal and industrial uses is about 270,000 m^3/day, and the estimated demands in Hsinchu area after 2006 may rise above 480,000 m^3/day.

The main purpose of this paper is to examine the deficit problem of the water-supply system of Hsinchu area. Within the work, general evaluation tools are emphasized to enable the analysts to make inferences about the impact of water shortage on a water-supply system.

2. GENERAL EVALUATION TECHNIQUES

2.1 Duration Curve

A flow duration curve presents the percent of time that flow is equal to or greater than indicated values. Since division from weirs lacks for storage, therefore, the time unit of flow duration curve should be the day so as to show minimum flows.

2.2 Drought Frequency Curve

The intensity duration frequency (IDF) used in drought frequency curves of low flow is useful to the water supply design. The quantity of water needed to meet designated recurrence interval can be determined from the frequency of minimum flows for different lengths (Linsley et al. 1992).

2.3 Exceedence Probability Curve

Instead of daily flow, a 10-day-long flow is adopted here as the basis in analyzing the reliability of water demand in each period. The exceedence probability of X exceeding or equaling a specific value $x_{(m)}$ in period t is

$$\Pr[X \geq x_{(m)}] = \frac{m}{k+1} \tag{1}$$

where

X = streamflow at any specific site, a random variable in period t
$x_{(m)}$ = a possible value of X, the mth largest observation
k = the total number of observations

2.4 Truncation Level Curve

Run theory based on the truncation level (Yevjevich 1967; Chang and Kleopa 1991) can be employed to evaluate the severity of shortage (run-sum) and its duration of water deficit (run-length) at any specific withdrawl site. The rate of the run-sum to the run-length indicates the average deficit of water supply, if shortage occurs. It may present the shortage tolerances of water supply for municipal and industrial uses.

2.5 Shortage Index

A general shortage index defined by U.S. Army in 1963 is

$$SI = \frac{100}{n} \sum_{i=1}^{n} \left(\frac{S}{D} \right)^2 \tag{2}$$

where

SI = shortage index
S = annual shortage (annual demand-annual supply)
D = annual demand
n = number of years considered

Obviously, the shortage index is weighted down with the shortage ratio. With larger annual shortage ratio, more significant effect of shortage index will be. Normally, SI=1 is commonly used herein in water resources simulation analysis.

3. CASE STUDY

Restrictions have been placed on use of groundwater to prevent land subsidence. At present, Taiwan Water Supply Corporation (TWSC) diverts surface water from the Touchien Chi Basin to the Hsinchu city (Fig.1). Within the sources of supply, since the existing Baoshan reservoir is small with an active storage of 5.35 MCM only, most of the daily water demand (as high as 200,000 m^3/day) is diverted directly from the stream (nearby Lung'en weir completed in 1999) to treatment facilities. However, adequate river flows in Hsinchu area are not available during the dry season. Additional large reservoirs herein thus play a very important role in retaining excess surface water flow from the wet season for use during the dry season. Baoshan II reservoir with 31.34 MCM capacity is currently planned to build in order to meet the water requirement. So far, if streamflow in the Touchien Chi cannot meet the demand of Hsinchu city, Yungheshan reservoir (27.39 MCM capacity) situated at the Chunggang Chi Basin, neighbor basin of Touchien Chi Basin, could provide potable water at most 120,000 m^3/day.

3.1 Hydrologic Analysis

According to the Water Law of Taiwan, water users must request water rights from the national water authorities. After having subtracted downstream registered water rights requested for irrigation and environmental conservation from the daily flow data (1980-1997), the flow duration curve of Lung'en weir shows that around 32% of time over a year the demand cannot be met, particularly during the dry season. Also, the curve indicates 3% of time flow is waterless. By viewing the low-flow frequency curves during periods of various lengths (Fig.2), the overall slopes become quite steep because of the flow subtraction by water rights. For the required water 200,000 m³/day, obviously, flow would be insufficient to satisfy the demand every year. That is, a storage over the water demand to avert shortage occurrence during the drought period would be necessary.

Based on the run-theory analysis corresponding to various truncation levels (50,000 ~ 300,000 m³/day herein), the more the demand is, the greater the run-sum and run-length will be, as shown in the truncation level curves (Fig.3). In addition, by using of the exceedence probability curves of 10-day-long streamflow nearby the Lung'en weir, Fig.4 provides the reliabilities corresponding to various water demands (50,000 ~ 300,000 m³/day). It appears the reliabilities to pump the flow at Lung'en weir are as high as 90% during the wet season. Nevertheless, the reliabilities drop promptly below 50% from November to January (within the dry season), if the demand water is above 150,000 m³/day.

3.2 Simulation Analysis

Work related to reservoir operation has been studied by many authors (Loucks and Falkson 1970; Tai and Goulter 1987; Huang et al. 1991; Huang 1996; Mujumdar and Ramesh 1997; Huang and Yang 1999). In actual practice, reservoir operation in Taiwan is generally examined by using simulation models, because of its simplicity and flexibility.

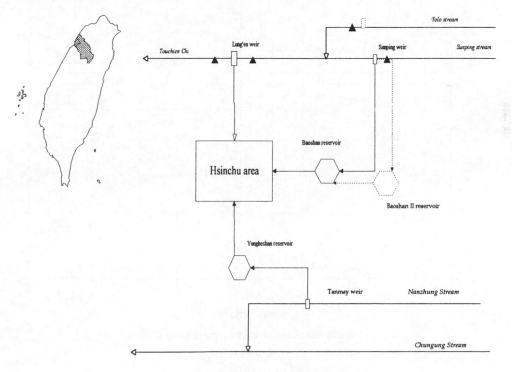

Fig.1: Hsinchu area water supply system

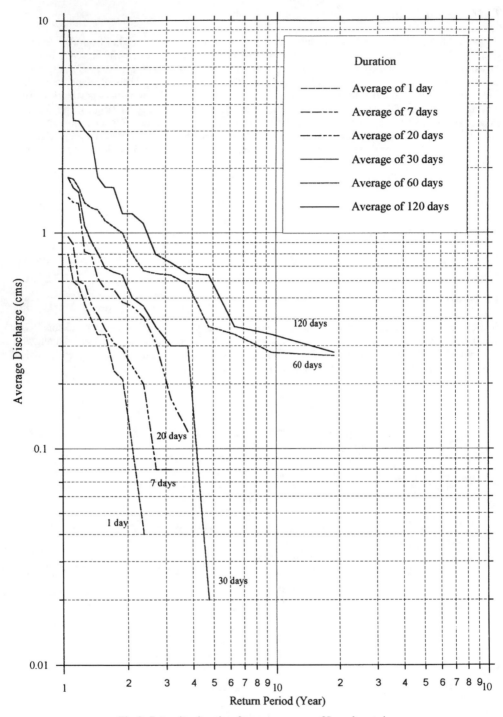

Fig.2: Intensity-duration-frequency curves of Lung'en weir

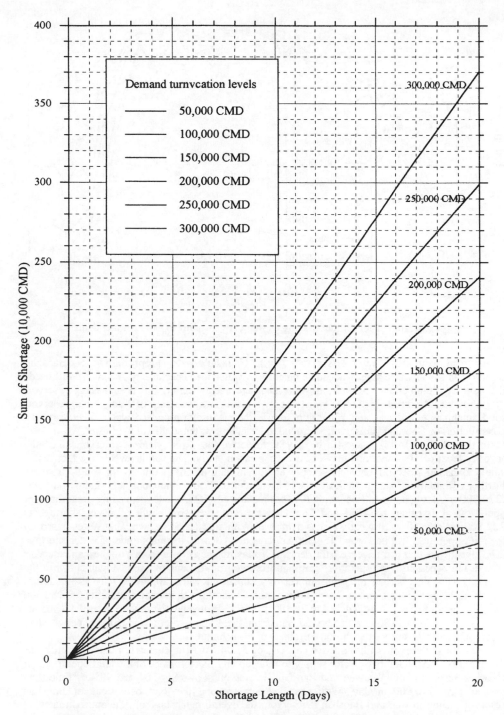

Fig.3: The relationship between sum of shortage and shortage length of Lung'en weir

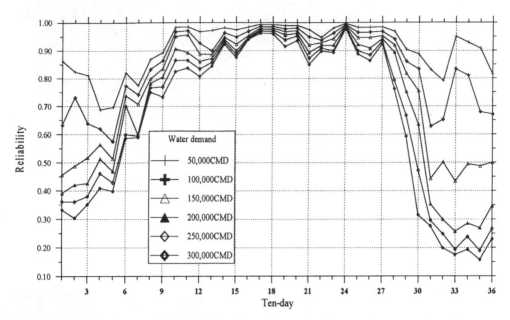

Fig.4: Streamflow reliabilities of Lung'en weir corresponding to various water demands

As mentioned above, reservoirs play a key to Hsinchu's water supply. However, Baoshan reservoir is too small to supply adequate quantity of water for Hsinchu city. Today, it still needs the help of Yungheshan reservoir, if water shortage event occurs. But Yungheshan reservoir originally was designed for another water-supply system. The assistance of Yungheshan reservoir for Hsinchu is temporary, and 120,000 m³/day is the maximum promised water. Therefore, the construction of Baoshan II reservoir is crucial to Hsinchu's water supply in the future.

In this study, SI=1 is suggested by TWSC to derive the "firm" yield of reservoirs, i.e., the severity level of 10% shortage may occur every year in study. The result reveals the fact that a single Baoshan reservoir only can supply 85,200 m³/day as SI=1, due to its small capacity. Further, simulation results indicate the current demand for municipal and industrial uses (270,000 m³/day) can be fulfilled with 70% of time and SI=7.27, if only depending upon the joint operation of Baoshan reservoir and Lung'en weir. The average amount of water shortage is 227,800 m³/day, if insufficient water is available. With specific SI=1, available water based on the joint operation will be around 154,300 m³/day only. Clearly, the magnitude of water shortage is so significant that it cannot meet the demand. Today, the assistance from Yungheshan reservoir would be necessary. In operating Baoshan reservoir and Lung'en weir, together with the operations of Yungheshan reservoir subject to a constraint with 120,000 m³/day as maximum supply to Hsinchu, the average deficit ratio of annual shortage to annual demand is 5.72%, and the shortage quantity can be reduced to 74,100 m³/day. Meanwhile, the time to meet the demands increases from 70% to 85% with SI=0.57, satisfying the designated criterion (SI=1). Yungheshan reservoir evidently softens the shortage problem of Hsinchu area.

Baoshan II reservoir with an active capacity of 31.34 MCM is currently planned to increase the quantity and reliability of water supply to Hsinchu city. Performance of the joint operations of Baoshan reservoir, Lung'en weir and Baoshan II reservoir shows SI=0.04 and 99% of time flow greater than 270,000 m³/day. With specified SI=1 among the joint operations of Baoshan reservoir, Lung'en weir and Baoshan II reservoir, the overall percentage of time approximates 92 percent that flow exceeds 387,900 m³/day. Together with the maximum quantity provided by Yungheshan reservoir (120,000 m³/day), the total amount could meet the estimated demands (480,000 m³/day) of Hsinchu area after 2006. However, the expansion plans of HSIP should be curtailed to refrain from economic damages, if the proposed reservoir is not completed by 2006.

396

4. CONCLUDING REMARKS

Water is the key to Hsinchu Science-based Industrial Park (HSIP), Taiwan's "Silicon Valley". Simulation results show that the current demand for municipal and industrial uses (270,000 m^3/day) can be fulfilled with 70% of time and shortage index SI=7.27, if only depending on the joint operations of Baoshan reservoir and Lung'en weir. Together with the Yungheshan reservoir, the time to meet the demand increases to 85% with an acceptable value of SI=0.57. Nevertheless, the expansion schedule of HSIP should be revised, if the planned Baoshan II reservoir cannot be done by 2006 to meet the water requirement above 480,000 m^3/day in the future.

ACKNOWLEDGMENTS

This research was funded by the Taiwan Water Supply Corporation.

REFERENCE

Chang, T.J. and X.A. Kleopa (1991), "A proposed method for drought monitoring.", Water Resources Bulletin, 27(2), pp.275-281.

Loucks, D.P. and L.M. Falkson (1970), "A comparison of some dynamic, linear and policy iteration methods for reservoir operation." Water Resources Bulletin, 6(3), pp.384-400.

Huang, W.C. and F.T. Yang (1999), "A handy decision support system for reservoir operation in Taiwan." American Water Resources Association, 32(6), pp.1221-1232.

Huang, W.C. (1996), "Decision support system for reservoir operation." Water Resources Bulletin, 32(6), pp.1221-1232.

Huang, W.C., R. Harboe, and J.J. Bogardi (1991), "Testing stochastic dynamic programming models conditioned on observed or forecasted inflows." Journal of Water Resources Planning and Management, ASCE, 117(1), pp.28-36.

Linsley, R.K., J.B. Franzini, D.L. Freyberg and G. Tchobanoglous (1992), Water-Resources Engineering, McGraw-Hill, Inc., Singapore.

Mujumdar, P.P. and T.S.V. Ramesh (1997), "Real-time reservoir operation for irrigation." Water Resources Research, 33(5), pp.1157-1164.

Tai F.K. and I.C. Goulter (1987), "A stochastic dynamic programming based approach to the operation of a multi-reservoir system." Water Resources Bulletin, 23(3), pp.371-377.

U.S. Army Corps of Engineers (1963), "Estimating long-term storage requirements and firm yield of rivers." Technical Bulletin Number 9, Sacramento District.

Yevjevich, V. (1967), "An objective approach to definitions and investigations of continental hydrologic droughts." Hydrology papers, No. 23, Colorado State University, Colorado.

Stochastic Hydraulics 2000, Wang & Hu (eds) © 2000 Taylor & Francis, ISBN 90 5809 166 X

Safety and risk of the water supply in the Moscow region

Vladimir I.Klepov
Water Problems Institute, Russian Academy of Sciences, Moscow, Russia

ABSTRACT: The water resource system (WRS) for urban residence and industrial water supply is a complicated scientific, technological and economic problem. Both special and multipurpose reservoirs have been created for the purpose of water supply. Many large cities such as Beijing, Madrid, London, Melbourne, Sydney, San Paolo, Tokyo, Bombay, and Moscow use reservoirs as the main water resources WRS is required of high guarantee rate of water yield. The guarantee rate of water supply to Moscow is about 95-97% in a number of consecutive years.

Key words: Water supply, Safety, Moscow, Reservoirs

1. INTRODUCTION

Since the moment of the construction, the system of water supply of the Moscow Region has been designed to operate using mainly surface flow. As a result, a network of canals, pipelines, large reservoirs, pumping stations has been constructed, which has changed raidally the natural hydrological regime in the basin. At the same time, intensive water withdrawal with numerous water intake facilities has lagerly disturbed the river flow regime. The Moscow water supply system, which has long been developed intensively, is now on the verge of its ecological exhaustion. Further operation and development of this system necessitates searching for ways to dicrease ecological danger.

2. REGIONAL WATER RESOURCES AND THEIR USE

Surface water resources of the Moscow Region, used for its water supply, include the Volga river flow from its source to the Ivankovo Reservoir, the flow of the Volga's right-hand tributary the Vasusa River, the flow of Moscow River and its tributaries the Istra, Rusa and Oserna to the Rublevo Reservoir. The total catchment area is about 56.000 km3. The total amount of water is 340 m3/S or 10.7 km3 per annum on the average. The alimentation of these rivers is mainly with meltwater and summerfall precipitation. The role of groundwater alimentation is not important. The flow regime is characterised by high spring floods and summer flow period. During the high flow period (March - May) 65-70% of the annual flow passes. In the Rublevo Reservoir site, near the hydropower station, closing the Moskvoretsk part of the Moscow water supply system (located on the Moskva River) the maximum mean annual water discharge of 75.3 m3/s was registered in 1933-1934, the minimum one (20.8 m3/s) in 1921-1922, the average longterm discharge of 544 m3/s was registered in April 1948-1949, the minimum (monthly) one 6.6 m3/s in August-September of 1938-1939.

At present, the water resource system (WRS) of the Moscow Region includes 8 reservoirs in neighbouring river basin, as well as river reaches, canals, pumping stations and other hydraulic facilities. The main parameters of the reservoirs, regulating the river flow, are listed in Table 1. The flow regulation ensured the increase in the reliability of water supply of the Region. For exemple, for a 95% reliability (according to the number of uninterrupted years) in the Ivankovo site of the Volga River the regulated flow is three times larger, than the natural one, in the Rublevo site of the Moskva River- 4.0 times, in the Zubtsov site on the Vasusa River - 4.0 times larger, than the natural one. The total water yield of the system grew 3.4 times.

Table 1: The main parameters of the reservoirs

Name of the reservoir	Level mark		Reservoir capcity (mil.m^3)		Water surface area at normal pool level, km^2	Year of construction
	normalpool level (m)	dead volume level (m)	total	useful		
Verkhnevolshsk	206.5	303.0	562	487	181	1830
Ivankovo	124.0	119.5	1120	813	327	1937
Istra	170.0	159.0	183	171.5	33.6	1935
Moshaisk	183.0	170.0	235	221.4	31.0	1964
Rusa	182.5	169.0	219.8	215.7	32.7	1968
Oserna	180.5	169.0	143.8	140.0	23.1	1968
Vasusa	180.25	170.5	540	430	106	1978
Yausa	215.0	212.0	290	130	51	1978

3. INCREASED DEMANDS TO THE RELIABILITY OF WATER SUPPLY

The estimation of the available water resources in the Region shows that the further development of its water supply system can be implemented either by interbasin water transfer, or by using more ground water masses in the process of water supply, thus increasing the part of the ground water in the existing water supply balance. One way or another, these measures can have negative enviromental consequences, and that is why, they require serious scientific studies.

At the same time, lately, resource-saving technologies have been introduced all over the world. This refers, particularly, to water-saving technologies. In solving water supply problems, this is connected with considirable reduction in specific water consumption. Undoubted, a similar process will also occur in the Moscow Region.

However, reduction in industrial water consumption will inevitably lead to increased demands to the reliability of uninterrupted water supply, because the less the amount of water spent per unit product, the larger the economical damages, will be caused by water deficit. This also can be referred to social and economical consequences of interruption in the water supply, since the more economically the water is used for municipal or enviromental purposes, the larger the social or ecological damage from the violation of normal water supply.

The reliability of water yield can be increased by two means: (1) increasing the regulatory volume of the reservoir; (2) involving additional water sources. The latter can cause new ecological problems. Our investigation has shown, that the risk of interruption in water supply can be decreased by applying a new scientific approach to the combined use of surface and groundwater flow.

4. THE RESULTS OF THE RESEARCH

4.1. System simulation model

To study the guarantee rate of water yield of Moscow region under various management, a flow mathematical model was used to analyse the systems with many reservoirs. The model is able to analyse the amount and safety of water yield in the Upper Volga basin, conservative water releases in the Vuoksa basin, etc. Some parts of WRS are viewed as elements of a graph (network) - nodes and arcs. A node is a point of junction or branching inside the network, i.e.

the analog of the reservoir, junction of channels and river beds. Reservoirs are viewed as nodes capable of keeping water within certain boundaries. Water movement is executed along the arcs (connections) - analogs of river bed, channels, etc. with a limited capacity

4.2. Safety characteristics of the Moscow region

The study is aimed at identifying the availability of the guaranteed water yield deficit for certain hydrological conditions. The guaranteed water yield deficit is the difference between available water resources and demanded water yield during any period of time considered (month, year, etc.). The guaranteed water yield deficit can probably take place under hydrological conditions of water deficiency. The water yield deficit therewith will be increased as the volume of water yield grow.

To identify the guaranteed water yield deficit, the realization comparison of the river flow for the period from 1914 till 1995 has been performed with a month division within a year for the low water period (June - February) and with ten-day intervals for the flood period (March - June) and comparison of guaranteed water yield values given variantly to each of the three reservoir subsystems. Low levels of the guaranteed water yield are those, when there is no deficit even under extremely low water conditions. The upper level considers the water yield which exceeds the initial one by 30 - 40 %.

There is an simulation calculation for each value of the water yield which allows to identify specific years and months within the initial long-term period during which the demanded claim on water cannot be satisfied completely ore partially with available water resources in the system itself and in its confluents.

The following hydrological indices are received as a result of calculation of water yield management variants considered:
- Volume characteristics of water yield deficit per each period of time (month, year),
- Temporal characteristics of water yield deficit expressed in a number of month and years when water resources deficiency is observed

Safety characteristics of Moscow Region WRS water yield are giveen in Table 2. The results of the research show that dependable water yield supply. in a number of uninterrupted years (usually used in practice) does not reflect the peculiarities of flow regulation and does not ensure sufficient safety of water supply. That is why it is necessary to consider other indices of available water supply: duration of uninterrupted periods, volume of the water consumed, the level ("depth") of guaranteed water yield decrease under low water conditions. Only comprehensive analysis of the water yield value with the help of different criteria of water supply can assist in identifying the efficiency of WRS functioning and its water economic safety.

Table 2: Water yield reliability of Moscow region WRS

The Upper Volga subsystem				The Moskvoretskaya subsystem				The Vazuzskaya subsystem			
Water yield m^3/s	Water supply, %			Water yield m^3/s	Water supply, %			Water yield m^3/s	Water supply, %		
	number of uninterrupted years	duration	volume		number of uninterrupted years	duration	volume		number of uninterrupted years	duration	volume
60	98,5	100	100	27	98,5	100	100	15	98,5	100	100
64	97,0	99,7	99,9	28	97,0	99,5	99,8	16	97,0	9,9	99,7
66	97,0	99,4	99,8	29	97,0	99,1	99,6	17	97,0	98,5	99,5
70	97,0	99,0	99,7	30	97,0	98,5	99,4	18	95,5	98,1	99,3
74	97,0	98,4	99,6	31	95,5	98,1	99,3	19	95,5	97,7	99,1
76	95,5	97,6	99,5	32	94,0	97,6	99,1	20	94,0	97,3	98,9
78	95,5	96,5	99,3	33	91,0	97,0	98,9	21	89,6	96,6	98,6
80	94,0	95,2	99,0	34	89,6	96,2	98,8	22	86,6	94,8	97,8

4.3. Calculations of the combined use of surface and ground water flow

For the task under consideration, the most interesting are low-flow years and their groups, when there is deficit of surface waters. In order to reveal such years, the comparison has been carried out of the water inflow to the reservoir, changing from month to month and from year to year, on the one hand, and the firm water yield, assigned in variants, on the other hand. As a result of the calculations, water management indices were obtained, characterizing possible water deficit - the volume of non-supplied water in each of the interrupted time intervals (month, year), concrete interrupted years in the investigated hydrological series and the number of the deficient months in each of these years were also determined. Table 3 presents one of the fragments of water deficit characteristics depending on the increase in the firm water yield of a surface flow WRS.

Table 3: Relation of the water yield, reliability and deficit of water yield

Water yield,	Reliability,	Deficit of water yield in the years:				
m³/s	(%)	1921	1922	1939	1940	1964
71	97	12/7	12/3	0	0	0
73	95	16/7	16/3	16/6	7/1	0
79	95	20/7	20/3	20/6	9/1	0
83	94	22/7	22/3	22/6	10/1	22/6

It follows from the Table, that in order to guarantee firm water yield for the Moscow water supply, amounting to 83 m3/s at a 97% reliability, it is necessary to add additional water from groundwater sources during interrupted years: in 1921/22 - 10 m^3/s during 10 months, during other interrupted periods - the volumes, corresponding to the values in the last line of Table 3.

The results, obtained under the above regime of ground water withdrawal, should be compared with the results, obtained under traditional, permanent water withdrawal. The calculations have shown, that in accordance with the estimated operational reserves, under permanent water withdrawal, the capacity of the Dubna area is 8.1 m3/s, and in this case the resources of this area will be completely exhausted in 50 years.

5. CONCLUSION

By now natural water resources used for Moscow water supply region have been practically exhausted. River resources time transfer and flow regulation with the help of reservoirs made it possible to increase water supply of Moscow. The index analysis of natural and regulated flow for the main closing sites of reservoirs shows the following. The value of safe water yield under regulated conditions in comparison with the natural ones has been increased and ranges to 3.0 in the Ivankovo site, the Volga basin, 4.0 in the Rublievo site, the Moscow river, 4.8 in the Zibtsov site, the Vasusa river for safety of 95%. The total water yield of the whole system has been increased by 3.4 times.

Use and management of the water resources in the Upper Volga basin is aimed at the increased reliability of Moscow water supply. This problem has some peculiarities derived from a number of factors, such as: the highest density of population, concentration of industrial enterprises and developed infrastructure in the comparatively small area on the one hand and deficiency of local water resources (surface and ground) on the other hand. Ecological conditions both in the city itself and in the places of water resources formation have a great impact on Moscow water supply.

Stochastic Hydraulics 2000, Wang & Hu (eds) © 2000 Taylor & Francis, ISBN 90 5809 166 X

Model assessment of a novel system of reservoir sedimentation control

J.H.Loveless & A.M.Siyam
University of Bristol, UK

ABSTRACT: The paper describes some results of a physical and mathematical model study of a novel system for the control of reservoir sedimentation. The proposed system consists of a submerged intake and weir connected to the main dam by closed conduits as shown in Figure 1. Experimental tests were conducted on a physical model of the proposed system and the results were compared to SSIIM a numerical model which has already shown promise in modelling reservoir sedimentation. SSIIM was found to be successful in modelling the velocity field and the sedimentation process for a diffused inlet jet, but for an attached jet, which is more likely in practice, the observed length of the circulation zone in the physical model was shorter. This mismatch can be corrected by accepting lower convergence criteria. Though SSIIM slightly underestimated the outflow concentration through the pipe and overpredicted that passing over the main dam, the results looks encouraging.

Key words: Reservoir sedimentation control, Physical model, Numerical model, Experiments

1.INTRODUCTION

Once a reservoir is created by building a dam on a river, the gradual filling up of the reservoir with sediment is inevitable unless measures are taken to control it. Alarming rates of storage depletion have been reported world-wide and especially in drought prone areas. (Bechteler 1996). As suitable sites for new reservoirs become rarer and the cost of replacement becomes high in addition to environmental, social and political concerns, more research effort on the problem of reservoir sedimentation will be needed. The key objectives have always been focused on understanding the processes of reservoir sedimentation and devising control measures and operational strategies that help to sustain the useful life of reservoirs.

It is known that the processes of reservoir sedimentation start with sediment erosion by different means and transportation through the river system as bed load or suspended load. As the flow enters the reservoir, it decelerates and its sediment transport capacity diminishes. The coarser material deposits first forming deltas at the head of the reservoir and the finest material settles last forming the bottom sediment which spreads throughout the reservoir. The deposition pattern depends mainly on: the inflow hydrograph; sediment inflow; sediment characteristics; reservoir configuration; regional geography and reservoir operation.

Briefly current methods for controlling reservoir sedimentation can be summarised as: reducing the influx of sediment by construction of upstream check dams and watershed management; minimising sediment deposition through reduction of trap efficiency by drawdown sluicing; flushing; density current venting and removal of deposited sediment by dredging (Fan 1985). However, in many cases, the success of these measures is either limited in efficiency or cost prohibitive (Mahmood 1987).

Recently, the use of a pipeline to transport the incoming or newly deposited sediment in the reservoir has been more commonly advocated (Eftekharzadeh & Laursen 1990, Singh & Durgunoglu 1991, Hotchkiss & Huang 1994). In this paper a similar system to that suggested by Singh & Durgunoglu (1991) is proposed. It features a submerged intake dam (check dam) and a pipeline system which is meant to serve the purpose of blocking the advancement of the delta into the reservoir and transporting the incoming or flushed sediment to the downstream side of the reservoir.

Laboratory tests were conducted to explore the performance of such a system when the reservoir is undergoing sedimentation. The programme of research was conducted at the University of Bristol, Department of Civil Engineering by the authors between October 1996 and November 1999. Performance of the system under other operational conditions is described in Siyam (2000). A three-dimensional mathematical model called SSIIM was used to compare directly with the laboratory experiments. This is seen as a useful duplication in order to validate both the numerical and physical models which could be used independently for assessing the performance of prototype sedimentation control systems.

2. THE NUMERICAL MODEL

The SSIIM program which is an abbreviation for Sediment Simulation In Intakes with Multiblock option, is a 3-Dimensional Computational Fluid Dynamics (CFD) program developed by Olsen (1997, 1999) in 1993. The SSIIM model has many versions and is available free of charge. In this study the last two versions of the model, SSIIM14 and SSIIM15b were used. The program has been designed to solve water and sediment flow in rivers and reservoirs and presents graphical plots of the velocity and sediment concentration fields with changes in channel geometry due to the morphological processes.

The SSIIM program solves the Navier-Stokes equations with the k-ε model on a 3-D non-orthogonal structured grid. It uses the Control Volume Method (CVM) for the discretisation of the equations with either a power-law or Second Order Upwind (SOU) scheme. The SIMPLE (Semi-Implicit Method for Pressure-Linked Equations) algorithm that was developed by Patanakar and Spalding (1972) is used by SSIIM for the pressure coupling. When the velocity field is obtained the convection-diffusion equation for the sediment in 3-D is solved for different sediment sizes. Next the morphological processes are predicted based on an imposed boundary condition at the bed.

The program has two main input files, but also offers a wide range of other input files and detailed output result files. The program has the capability to present graphical plots of intermediate calculations as well as comparing the finally computed result with the measured data.

3. THE PHYSICAL MODEL

All the experiments were conducted in a rectangular flume 13.75 m long, 0.75 m wide, and 0.9 m deep. The inlet part of the flume, which simulates a short river reach, has a narrower width of 0.30 m and length of 1.1 m. A schematic layout of the flume is shown in Figure 2. Figure 2a presents the plan view and Figure 2b shows the side view of the flume. A submerged intake and weir 0.16 m in height was constructed across the flume at a distance of 3.0 m from the dam towards the downstream end of flume and it is shown in general view of the flume in Figure 3. The part of the flume between the submerged weir and the short river reach had a false sloping bottom to simulate a fan. The average slope was 2.4%. The remaining part of the flume between the submerged weir and the dam was horizontal. Three perspex pipes, diameter 45 mm, fitted with bellmouth entries were also laid on the bottom of the flume and passed through the dam to facilitate bypassing of water and sediment from the front of the submerged weir. Perspex pipes were chosen to enable visual observation of the mode of flow in the pipes during operation. Each pipe had a separate rectangular control gate just on the downstream side of the submerged dam. Also downstream of the main dam all the pipes were joined to a larger 75 mm manifold for the purpose of measuring the combined outflow when any or all of the pipes were in operation.

All the flow passing through the reservoir was returned to a sump tank and then re-circulated to the flume by means of two 500 gpm capacity pumps. The water was delivered first to an over headtank before it entered the flume. The flow rate was adjustable by two control valves fitted to the delivery pipe at locations before and after the overhead tank. Baffles were used at the inlet to distribute the inflow evenly across the flume. Special fine and light weight plastic material was used to simulate the suspended sediment and flow velocities and sediment concentrations at some selected sections of the flume were measured. A rolling carriage was used to hold the measuring equipment. These were, the velocity profile measuring probe (ADV), the concentration measuring probe (turbidity meter) and a digital point gauge for both water level and bed profile measurements.

The main dam was hinged so that it could be adjusted to achieve the desired reservoir water level. Also a 10 cm wide low level sluice gate and 40 mm diameter sluice pipe were installed in the dam. The sluice gate was adjusted manually while the sluice pipe had a valve control.

4. COMPARISON OF THE RESULTS FROM THE TWO MODELS

In total five sedimentation runs and two clear water tests as were carried out as shown in Table 1. Runs 1,2, 8 and 11 were carried out with an additional jet diffuser configuration at the inlet to the flume. Run 3 was conducted with an unhindered attached wall jet and recirculation zone configuration. The test procedures for all tests were basically the same except the inflow water and sediment discharges, the outflow through pipes, the initial bed level profiles and reservoir water levels could all be varied from one test to another. All the sediment concentration measurements were carried out using the turbidity meter. Samples of inflow and outflow discharge were usually collected and their sediment concentrations were later measured using the turbidity meter. At the end of Run 1 samples of the deposited sediment were collected from different sections along the flume and their grain size distributions were determined. In all tests the bed profiles were measured before and after the test using an ADV probe. Sufficient time was allowed at the end of the test for the sediment to be deposited from suspension. Apart from limited local bed movement at the upstream portion of the flume, the entire flume in the sedimentation tests was in a deposition mode.

Table 1 Summary of the clear water and sedimentation experiment

Run No.	Q_{in} (l/s)	Q_p (l/s)	Q_d (l/s)	depth H(m)	Operation type	inlet configuration
CL1	11.52	3.84	7.68	.31	clear water	undiffused jet
CL2	11.52	3.84	7.68	.31	clear water	diffused jet
1	11.86	3.9	7.96	0.31	sedimentation	diffused jet
2	6.46	1.98	4.48	0.35	sedimentation	diffused jet
3	11.52	3.8	7.72	0.314	sedimentation	undiffused jet
8	6.35	3.56	2.79	0.206	sedimentation	diffused jet
11	9.55	3.29	6.26	.365	sedimentation	diffused jet
Q_{in} - inflow (l/s); Q_p - flow through diversion pipes (l/s); Q_d - flow passing over the submerged weir (l/s); H - reservoir depth at the main dam						

As expected, in run 2, where the depth was greater than in Run 1, there was a strong sediment deposition tendency. The highest rate occurred right in front of the diffuser (see Figure 4). The short clear space in front of the diffuser was created by the small water jets. The computed bed profile fits quite well with the measured. The outflow sediment concentration through the pipe, C_p, and over the main dam, C_d seem to be slightly under and over estimated (see Figure 5). The main reason may be the water jets created by the diffuser at the upstream end which produce extra turbulence in the flow. This in turn will produce extra shear stresses at the bed that prevent sediment from settling or moving as bed load for some distance down the flume. Though the diffuser configuration can be reproduced in SSIIM, it requires a much finer grid than the one used here with a longer time of computation for both water flow field and sediment calculations. Another source of error was that the outflow sediment through pipe was taken as the computed concentration at the centre of the bed cell. This in fact is about 2 to 3 cm above the channel bed at the upstream face of the submerged dam and near to the pipe entrance. Considering the fact that, from the Rouse profile, sediment concentration gets higher close to bed, it can be expected that the actual inflow sediment concentrations through the pipe are on the higher side of the computed values. Conversely the actual outflow over the dam was expected to be lower than the computed values. This is due to the fact that the actual depth of flow over the spillway is smaller than the height of the upper most cell in the SSIIM model.

Run 8 represents an advanced stage of reservoir sedimentation where the delta deposits have reached the submerged dam. Figure 6 shows that though the top set region of the delta had risen during the test the delta front region had not moved forward. This may be attributed to the presence of the submerged dam with its bypassing system. The SSIIM model was able to show this effect. During run 8 and run 11 and attempt was made to measure longitudinal concentration profiles using the turbidty meter rather than the conventional method of measuring points along a vertical. The measured results were compared with the computed using the SSIIM model. Figure 7, indicates an exponential decay of sediment concentration along the length of the flume.

The apparent fluctuation in the measured profiles and its discrepancy with the computed can easily be explained by the fact that the turbidty meter, which is stationed on the carriage, was manually dragged with the carriage along the flume and on average a single reading was logged every 3 to 4 cm apart. The effect of changing location of the probe tip might have some effect on the result if not done at a certain rate. By contrast in single point measurements an average of more than 50 readings were taken. However, the trend of exponential decay is evident in both the physical and numerical tests.

5. DISCUSSION & CONCLUSIONS

The overall performance of the submerged dam and the bypass pipes for the five sedimentation tests is summarised in Table 2. The table shows the computed and the measured overall trap efficiency during the each run. This is the percentage of the total inflow sediment deposited in the flume. The trap efficiency without the pipe was calculated assuming that the outflow concentration would be that measured or computed over the main dam. The reduction in trap efficiency due to the bypass system is the difference between these two trap efficiencies.

In all, the computed figures of the trap efficiencies compared very well with the measured ones. However, both the computed and the measured reduction in trap efficiency were low. This was due to the fact that most of the sediment was deposited in the upstream region ahead of the submerged dam. Consequently the sediment concentrations that reached the submerged dam were very low and tended to be uniform in the vertical direction. Also it must be understood that the desilting facility is meant to protect the downstream region of the reservoir. i.e. the region between the submerged intake and weir and the main dam.

Table 2 Summary of measured and computed trap efficiencies of the five sedimentation runs

description	Run 1		Run 2		Run 3	
	measured	SSIIM	measured	SSIIM	measured	SSIIM
inflow sediment (kg)	62.77	62.32	58.55	61.00	76.77	82.38
outflow through pipe (kg)	5.15	4.71	3.16	2.63	12.06	13.21
outflow over spillway	4.72	4.27	3.29	4.02	11.21	14.39
over all trap efficiency	84.27	85.58	88.99	89.11	69.69	66.50
trap efficiency without pipe	90.01	90.88	92.66	91.40	80.58	76.77
reduction in trap efficiency	5.74	5.30	3.67	2.29	10.89	10.27
run time	39.00	43.50	84.00	84.00	88.00	88.00
%age of bypass flow	32.88	32.88	30.65	30.65	32.99	32.99
description	Run 8		Run 11			
	measured	SSIIM	measured	SSIIM		
inflow sediment (kg)	45.03	44.44	20.33	22.00		
outflow through pipe (kg)	5.48	4.92	1.55	1.23		
outflow over spillway	1.05	1.52	1.28	1.37		
over all trap efficiency	85.50	85.51	86.06	88.19		
trap efficiency without pipe	96.36	94.66	91.51	91.62		
reduction in trap efficiency	10.86	9.15	5.44	3.42		
run time	32.00	32.00	20.50	20.00		
%age of bypass flow	56.06	56.06	34.45	34.45		

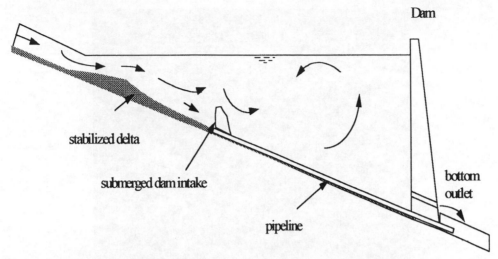

Fig. 1 Layout of the Submerged dam intake closed conduit system.

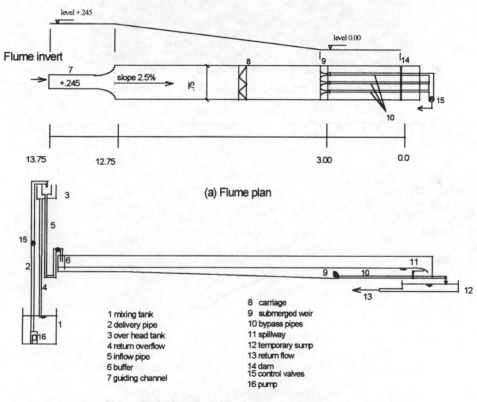

(a) Flume plan

1 mixing tank
2 delivery pipe
3 over head tank
4 return overflow
5 inflow pipe
6 buffer
7 guiding channel

8 carriage
9 submerged weir
10 bypass pipes
11 spillway
12 temporary sump
13 return flow
14 dam
15 control valves
16 pump

(b) Longitudinal section of the flume

Fig. 2. A schematic layout of the laboratory flume

Fig.3. General view of the laboratory flume looking downstream

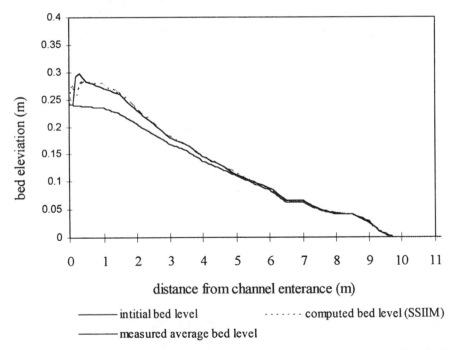

Fig. 4. Comparison of measured and computed longitudinal bed profile along the centre of the flume for Run 2

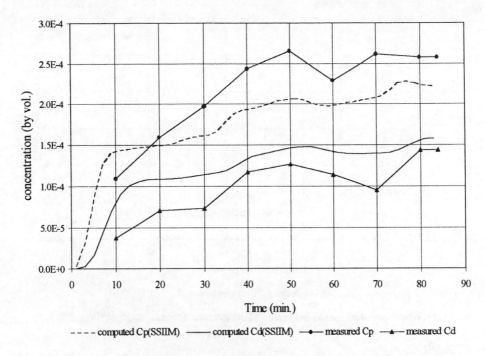

---- computed Cp(SSIIM) ——— computed Cd(SSIIM) —•— measured Cp —▲— measured Cd

Fig. 5. Comparison of measured and computed outflow sediment concentration for Run 2

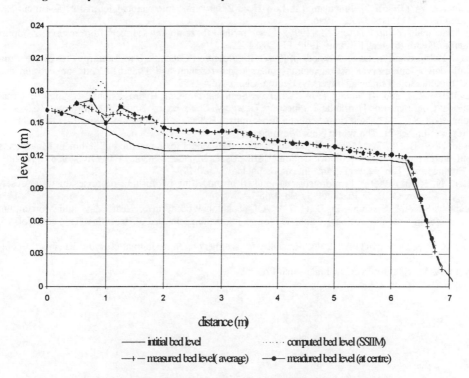

——— intitial bed level ·········· computed bed level (SSIIM)

—+— measured bed level(average) —•— meadured bed level (at centre)

Fig. 6. Comparison of measured and computed longitudinal profile
of bed change for Run 8 along the centre of the flume

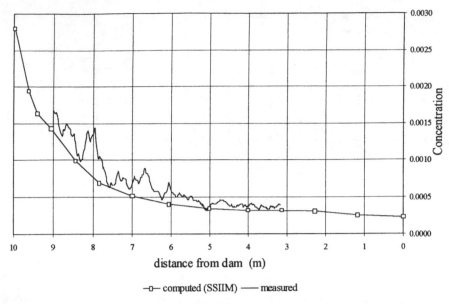

distance from dam (m)

—□— computed (SSIIM) ——— measured

Fig. 7. Comparison of measured and computed longitudinal sediment concentration: Run 8

REFERENCES

Bechteler, W. (1996), "Sedimentation in Low Head Reservoirs," International Journal of Sediment Research, Vol. 11, No. 3, Dec. 1996.

Eftekharzadeh, S. and Laursen M. (1990), " New method for removing sediment from reservoirs," Journal of Hydro Review, February 1990, PP. 80-84

Fan, J 1985 Lecutre notes on Methods of Preserving Reservoir Capacity,,from Lecture notes of the training course on reservoir sedimentation" International research and Training Centre on Erosion and Sedimentation, Tsinghua University, Beijing, China, Nov. 1985.

Hotchkiss, R.H. and Xi Haung, 1994, "Reservoir Sediment removal: Hydrosuction Dredging," International Conference on Hydraulic Engineering, George V. Cotroneo, ed. And Ralph R. Rumer, ed., 1994.

Mohamood, K;, 1987, "Reservoir Sedimentation: impact, Extent, and Mitigation," World Bank Technical report Number 71, The World bank, Washington, D.C.

Olsen, N. R. B. (1997), "A Three-Dimensional Numerical Model for Simulation of Sediment Movements in Water Intakes with Multiblock Option," SSIIM User's Manual, Version 21.5a, Division of Hydraulic Engineering, The Norwegian Institute of Technology, April 1997.

Olsen, N. R .B, (1999), "Two-dimensional numerical modelling of flushing processes in water reservoirs," Journ. Of Hydr. Res. Vol. 37 No. 1 pp 3-13

Patankar, S. V. & Spalding, D. B. (1972) "A Calculation Procedure for Heat, Mass and Momenttum Transfer in three-Dimensional Parabolic Flows," International Journal of Heat Mass Transfer, Vol. 15, p. 1787

Singh K. P. & Durgunoglu A. 1991, " Remedies for sediment buildup, Journal of Hydro Review, December 1991, PP. 90-97.

Siyam, A. M. 2000Reservoir Sedimentation Control

Risk analysis of hydraulic structures

Stochastic Hydraulics 2000, Wang & Hu (eds) © 2000 Taylor & Francis, ISBN 90 5809 166 X

Invited lecture: Risk assessment and minimization in steep catchments

Helmut Scheuerlein
University of Innsbruck, Austria

ABSTRACT: Mountainous catchments are exposed to harsh natural conditions and characterized by highly specialized ecosystems. This paper presents a risk assessment and minimization in steep catchments. Risks are from natural events like torrential flood, debris flow, landslide and avalanches. Human activities also induce risks. Structural measures and non-structural measures are employed to mitigate the risks.

Key words: Risk analysis, Mountainous catchments, Probabilistic approach, Hazard zone mapping, Structural measures planning

1. INTRODUCTION

Mountainous regions are zones of specific natural conditions concerning topography, geology and climate. Besides, they are significantly influenced by anthropogenetic impact like settling activities, landuse, economical interests and cultural habits. Natural ecosystems are strongly specialized and vulnerable as they have to survive in a rather harsh environment with limited supply of nutrients. Human pressure upon the ecosystems endangers their habitats and basis of life.

Living conditions are dangerous for man and his property, too. As a consequence, man tries to minimize hazards for himself by introducing appropriate protection measures. Protection measures are not only expensive and take considerable time to reach their desired efficiency, they may also interfere with existing ecosystems. Hence, careful coordination of anthropocentric demands and ecological constraints deserve adequate consideration.

2. NATURAL CONDITIONS OF MOUNTAIN AREAS

The soil conditions in mountain watersheds are mostly the result of the youngest geological phase in the region. In the Alps, f.i. the quaternary age left huge depositions of loose material in most of the valleys when the glaciers melted.

The erosivity of the soils is also highly influenced by the existing vegetation. Erosivity is high at barren land and low at complete vegetation cover. The role of forest depends on its mixing with shrubs and other lower vegetation.

The hydraulic characteristics of mountain streams are significantly determined by two different hydrological events. Advective rainfalls are responsible for flood events of several days duration in the main streams of the catchment. Convective rainfalls may create flash floods of short duration but high intensity in small mountain streams as well as landslides and debris flow events. A third factor which is important for the hydraulic features of mountain streams must be seen in the snowfall. Depending on the altitude a large percentage of the total precipitation falls as snow. During winter time (in glacial cachments also in summer) the snowcover stores the precipitation. In spring and summer, snowmelt can contribute considerably to floods in mountain streams, particularly when it is superimposed with convective rainfall events. Besides, during winter and spring time, the possibility of hazardous avalanches is a substantial risk factor in high mountain areas.

3. ANTHROPOLOGICAL IMPACT

In the past, mountain areas have been occupied by man rather slowly and with careful observation of the given natural conditions and by considering the potential risks one had to encounter by settling in such areas. People knew about the objective danger of nature and settled only at rather safe places. They could not risk to make mistakes as this would have ruined their existence. There was no insurance and no refunding through desaster aid programmes at that time. In the Alps the economic recession after the Napolean wars together with the permanent population growth in the middle of the 19. century resulted in increasing pressure towards exploitation of mountain valleys as settling areas and for food production. As a consequence, also rather risky locations were used for housing, road construction and other infrastructural measures.

As long as the mountain valleys were not too densely populated and settling was restricted to rather safe places, landuse was restricted to pastures for cattle and limited exploitation of forests for construction material for housing. During this time the ecological balance of use and regeneration of natural resources was more or less in equilibrium.

The development of the alpine valleys in the last four decades must be seen in close connection with the development of tourism. Construction of cable cars, preparation of ski-tracks and the provision of accomodation for tourists with continuously increasing comfort were measures which had considerable impact on the alpine environment. Probably the most severe effect, however, were the infrastructural measures which were necessary to develop tourism towards an economically feasible industry with efficient highways, parking areas, water supply, waste water and solid waste disposal and so on. Together with the flooding of previously low developed mountain valleys by mass tourism the call for appropriate protection of the new infrastructure became more and more powerful.

4. HYDRAULIC RISK POTENTIAL IN STEEP CATCHMENTS

4.1 Risk induced by natural events

The most important natural events with extraordinary high risk potential for man and his infrastructure are:
- torrential floods
- debris flow
- landslides
- avalanches.

All of them are somehow related to water. Floods and debris flow are events occurring in water courses, landslides and avalanches may happen anywhere presumed the slope is steep enough to provide conditions for gliding. Landslides and avalanches are rather local events lasting a few seconds only. Torrential floods and debris flow are events of extended range and duration. In the following, attention shall preferably be focussed on torrential floods and debris flow. Torrential watersheds can be divided into three typical and completely different reaches (Fig. 1):
- upper reach   or   source area
- middle reach   or   gorge
- lower reach   or   debris cone

As far as risk for man and his infrastructure is concerned the most essential zone is the lower reach of the torrent. This is the place where man has his settlements, his pastures, his roads and all his other facilities. Consequently, these objects are often harmed by debris flow or channel shifting due to progressive sediment deposition.

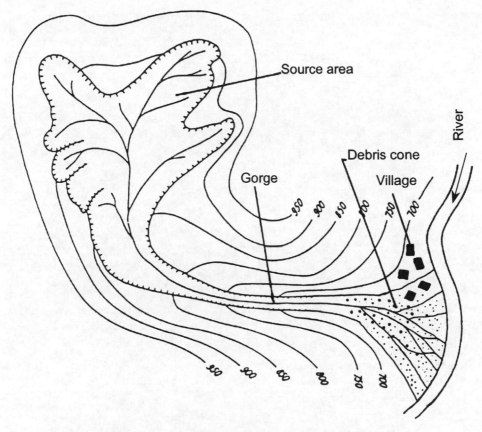

Fig.1: Schematic feature of a typical torrential watershed

4.2 Risk induced by human activities

The generation of extraordinary natural events which may result in desastrous effects on man and his infrastructure is not directly influenced by human activities nor is it within the range of man's command. But human activities may have considerable influence on the boundary conditions of natural events at the earth surface as f.i.

- deforestation
- landuse change
- reduction or elimination of flood retention areas
- reduction or elimination of sediment retention areas.

Human activities of this kind are particularly dangerous when they take place in the upper part of mountainous catchments as they then affect the most vulnerable part of the whole system. Evaluation of the risk potential associated with human activities is difficult as neither the kind nor the extension nor the side-effects of human impact is reliably predictable. All that could be done is to hypothetically anticipate various szenarios of limited actions and to try to make an estimate about the consequences.

5. MITIGATION MEASURES TO MINIMIZE RISK

Minimization of damages can be achieved by means of two strategies: structural and non-structural measures.

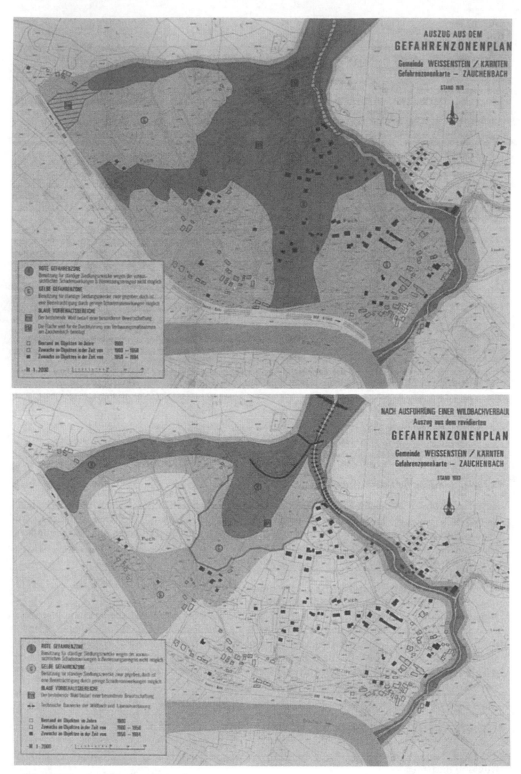

Fig. 2: Hazard zone map a) before structural mitigation measures b) after structural mitigation measures

5.1. Structural measures

Torrent control by means of structures comprises sustainable measures to protect man and his infrastructure against damage or loss. The measures differ considerably according to their location in the torrential system.

Main task in the **upper reach** of a torrential watershed is the prevention of erosion. The aim is to create a closed vegetation cover. Forest protection or rehabilitation measures may be helpful when erosion has not reached critical conditions yet. If, however, erosion has already reached an advanced stage, stronger measures like check dams and hill–side slope protection measures have to be applied.

Main task in the **middle reach** is the stabilization of the stream channel. The rather steep slope provides favorable conditions for lateral and vertical erosion which has to be controllled by means of appropriate measures.

This can be achieved by constructing a chain of consecutive drop structure with the consequence that channel slope, shear stress and erosive power are reduced. The drop structures are not determined to function as storage basins. Their sole task is reduction of the channel slope. The slope reduction can be achieved by few high drops or many low drops.

Main task in the **lower reach** is the retention of sediment, particularly in the case of debris flow events. This means that a debris retention basin of sufficient size has to be provided which is equipped with an automatically functioning control structure to release water and hold back sediment and wooden parts. As far as the construction of the basin is concerned, favorable natural topographical features should be used wherever possible. In the worst case an artificially shaped basin with surrounding dams or walls must be constructed. Downstream of the retention basin care has to be taken that the adjacent conveying channel to the receiving river is capable to discharge the design flood without overtopping of the banks.

5.2 Non-structural measures – hazard zone mapping

Non-structural measures comprise hazard zone mapping as the main component. Hazard zone mapping is in Austria instrumental since 30 years. The reason why it was developped must be seen in the economic upswing in the sixties combined with booming tourism. Both resulted in considerable extension of settling activities. The danger of torrential floods and avalanches was ignored; instead protective measures against natural hazards were expected. The exploding costs for torrent and avalanche control made it necessary to evaluate and compare risk potential and costs for protection. The first hazard zone map was established in 1968.

A hazard zone map consists of two components: a cartographic part and a report. The cartographic part consists of a normal topographical map 1:50000, 1:25000 or 1:20000 covering the area to be evaluated plus the relevant catchment areas. It additionally comprises the real hazard zone map 1:5000 where all endangered zones (on the basis of a flood of 150 years of reoccurrence) are indicated according to the following criteria:

The **Red Zone** indicates those areas which are endangered through torrents or avalanches in such a way that their permanent use for housing or roads cannot be tolerated.

The **Yellow Zone** indicates all the other areas within the range of torrents or avalanches whose permanent use for housing or roads might be affected somehow at rare natural events (f.i. 150 years flood).

Besides these hazard zones some other areas of special importance are indicated in the map:

The **Blue Zone** indicates areas which are needed for special technical, forestry-related, and biological measures or which require special landuse techniques in order to guarantee a certain protective function.

The **Brown Zone** indicates areas with potential tendency towards landslides or stone fall.

The **Violet Zone** indicates areas which are essential for the conservation of the soil or a certain topographical formation which is worthwhile to be protected.

An example of hazard zone mapping is given in Fig. 2.

The report accompanying the cartographic part contains
- the description of the maps
- the justification of the evaluation
- the interpretation of the consequences for the people living in the area.

In order to guarantee unbiased treatment and equal conditions countrywide, the hazard zone map is evaluated and checked by a special commission consisting of representants of
- the Austrian government
- the Federal State
- the mayor of the affected community
- and the head of the torrent and avalanche control authority.

The hazard zone map is also presented to the public with the possibility to articulate objections which in turn have to be discussed and decided upon by the commission. Meanwhile hazard zone mapping covers the whole Austrian territory having mountainous watersheds. It should be noted that in Austria new projects of torrent or avalanche control measures are only accepted and financed when the respective hazard zone map has been used as the basis of the planning.

6. RISK MANAGEMENT IN MOUNTAINOUS WATERSHEDS

6.1 Risk assessment

Flood events in mountainous environment are highly dynamic and characterized by a considerable potential of morphological changes, course shifting, erosion and deposition processes. These natural phenomena are basically neutral. They are evaluated and classified as favorable or unfavorable by man according to his subjective judgement. In the understanding of man, flood events are apt to affect or even damage his life and his property; hence they must be considered as risk factor. Evaluation of the risk potential can be carried out on the basis of hazard zone mapping. If hazard zone maps are not available they have to be worked out particularly for this purpose.

Classical hazard zone maps in Austria are set up for flood events of 150 years of reoccurrence. This has been decided by law. In principle any other rare event could be chosen, too. It might even make sence to create hazard zone maps for more than one event in order to be more flexible as far as restrictive rules for the use of these zones are concerned. In this case, the meaning of the zones defined for a specific event and the restrictive rules associated with their handling have to be formulated newly, according to the probability of occurrence of the chosen event.

So far identification of hazard zones in Austria is carried out to a large extent by engineering judgement through experienced experts on the basis of flood routing calculations. This, however, is a somehow unsatisfying approach as it rather considers linear causality of one cause (f.i. 150 years flood) and its effect (f.i. the appertaining water levels). The influence of other parameter which may also play a role in the system are either neglected or are incorporated by means of engineering judgement. This honorable but admittedly somewhat unscientific approach could be given a more logical background if a probability concept combined with systems analysis would be applied. In systems analysis a "failure" is defined as a condition at which the system looses its functionality. In a torrential system endangering or even destruction of objects worthwhile to be preserved may be such a condition. In the sense of stochastics "failure" is an event which corresponds with a certain probability P of occurrence. According to PLATE, 1993, the techniques of systems analysis and probabilistic concepts may very well be applied to hydraulic engineering problems. Formulation of complex causality chains and the possibility to link them together to "failure trees", "event trees", or "decision trees" may be used to calculate the range of the probability of failure by considering all influences involved. Fig. 3 shows an example of a failure tree describing the risk that an object situated in a torrential watershed may be destroyed. The main causes for a disastrous top event are systematically arranged and their interactions within the emerging network are pursued up to the final event.

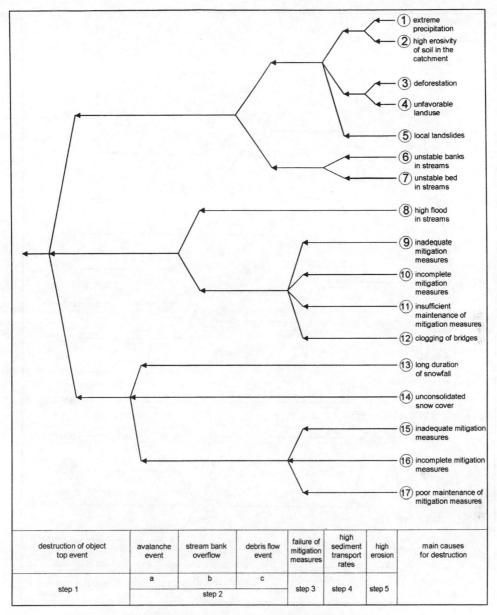

destruction of object top event	avalanche event	stream bank overflow	debris flow event	failure of mitigation measures	high sediment transport rates	high erosion	main causes for destruction
	a	b	c				
step 1		step 2		step 3	step 4	step 5	

Causes:
1. extreme precipitation
2. high erosivity of soil in the catchment
3. deforestation
4. unfavorable landuse
5. local landslides
6. unstable banks in streams
7. unstable bed in streams
8. high flood in streams
9. inadequate mitigation measures
10. incomplete mitigation measures
11. insufficient maintenance of mitigation measures
12. clogging of bridges
13. long duration of snowfall
14. unconsolidated snow cover
15. inadequate mitigation measures
16. incomplete mitigation measures
17. poor maintenance of mitigation measures

Fig. 3: Failure tree concerning a disastrous top event

In Fig. 4 the complex system of causes and effects is linked by means of logic AND/OR switches. By means of this decision tree the probability of a disastrous failure event can be calculated if the entrance probabilities of the various causes are known. In Fig. 5 such a calculation has been carried out for the three main disastrous top events "debris flow", "stream bank overflow" and "avalanche". A final logic coupling of these three events is neither necessary nor helpful as either one of the three events may occur independently from the other two. With the example it can be shown that the resulting probability for the top event is determined by the highest entrance probability which is passed on through OR switches. This provides the possibility of efficient risk reduction by appropriate measures to minimize the respective entrance probability.

419

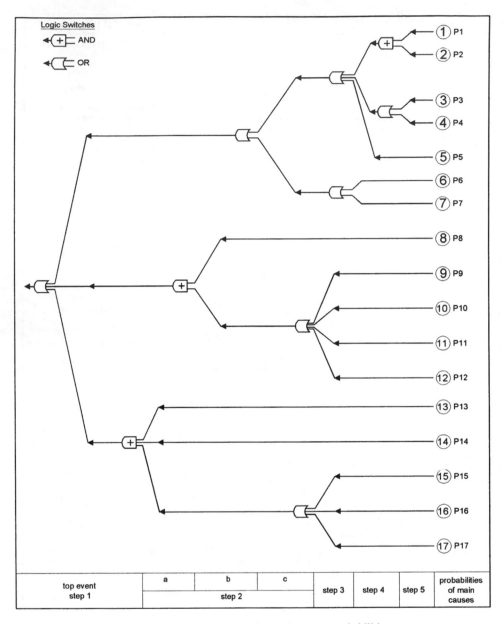

Fig. 4: Decision tree with logic switches and entrance probabilities

6.2 Risk minimization

The risk potential can be reduced by the implementation of appropriate structural mitigation measures. However, structural mitigation measures need time and money. In addition, structural measures interfere with natural processes and possibly might have undesired ecological side-effects. The feasibility of protective measures has to be examined by considering the following criteria:

– possible risk reduction
– required measures
– costs
– ecological side-effects

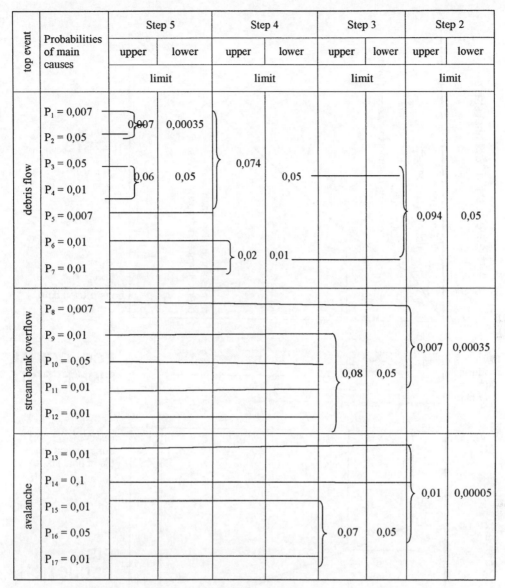

top event	Probabilities of main causes	Step 5		Step 4		Step 3		Step 2	
		upper	lower	upper	lower	upper	lower	upper	lower
		limit		limit		limit		limit	
debris flow	$P_1 = 0{,}007$	0,007	0,00035	0,074	0,05			0,094	0,05
	$P_2 = 0{,}05$								
	$P_3 = 0{,}05$	0,06	0,05						
	$P_4 = 0{,}01$								
	$P_5 = 0{,}007$								
	$P_6 = 0{,}01$			0,02	0,01				
	$P_7 = 0{,}01$								
stream bank overflow	$P_8 = 0{,}007$					0,08	0,05	0,007	0,00035
	$P_9 = 0{,}01$								
	$P_{10} = 0{,}05$								
	$P_{11} = 0{,}01$								
	$P_{12} = 0{,}01$								
avalanche	$P_{13} = 0{,}01$					0,07	0,05	0,01	0,00005
	$P_{14} = 0{,}1$								
	$P_{15} = 0{,}01$								
	$P_{16} = 0{,}05$								
	$P_{17} = 0{,}01$								

Fig.5: Probability of various hazardous top events insteep catchment (ref. to Fig. 3 and 4)

Fig. 6 shows how the complex interrelations of the various factors of influence can be displayed qualitatively in a simplified way (SCHEUERLEIN et al, 1996). The example chosen for the display is a typical torrent with high sediment potential in the upper and middle reach and an alluvial fan with adjacent human infrastructure in the lower reach (f.i. as considered in Fig. 3 to 5). The general concept concerning mitigation mea-sures comprises checkdams in the upper reach, drop structures in the middle reach, and a sediment retention basin with subsequent conveying channel to the receiving river in the lower reach. Final stage of the measures is the capability to cope with a flood of 150 years of reoccurrence (HQ $_{150}$).

The chosen display of costs and construction time as relative values compared to total costs and time illustrates also recommendations concerning the magnitude and the stage of implementation of the measures.

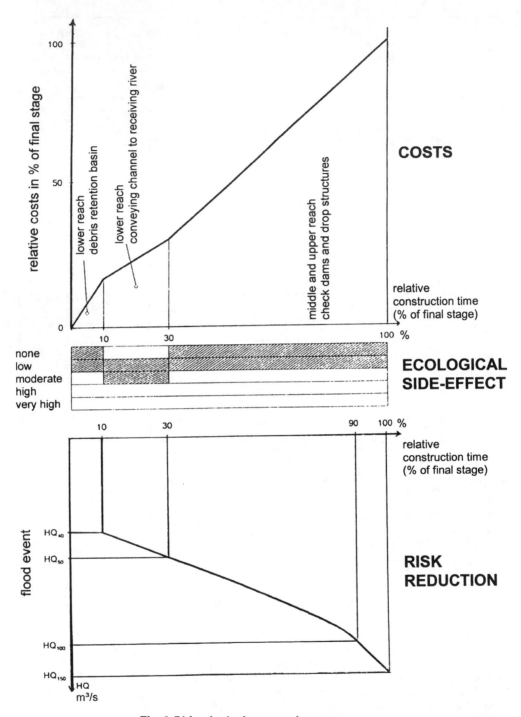

Fig. 6: Risk reduction by structural measures

As far as ecological side-effects are concerned an exact verification is hardly possible. The chosen verbal evaluation is a substitute for so far not available other criteria. A possibility of a more analytic way of consideration might be an approach on the basis of the fuzzy logic methods.

Application of diagrams according to Fig. 6 may serve as a helpful tool for decision-making concerning the selection of mitigation measures. As can be taken from the lower diagram it may also be used to show to what extent the progress of mitigation measures is apt to improve the safety against failure (f.i. 30 % of final stage corresponds with HQ_{50}, 90 % with HQ_{100}, etc). This may also be a helpful information for cost and benefit analysis (LINSLEY, et al, 1997).

7. CONCLUSIONS

Mountainous catchments are exposed to harsh natural conditions and characterized by highly specialized ecosystems. Although basically not really favorable for human beings, man has settled there and adjusted himself to this environment or has found ways to render his existence possible.

He has introduced structural measures to protect himself against disastrous events like floods, debris flow and avalanches. Additionally in some mountain regions he has established non-structural measures like hazard zone maps which are used as a helpful tool to avoid risks.

With respect to risk assessment the methods of systems analysis and probabilistic concepts can be used to consider and verify complex stochastic processes.

REFERENCES

Bundesministerium für Land- und Forstwirtschaft, 1984; 100 Jahre Wildbachverbauung in Österreich, Eigenverlag, Wien.

LINSLEY, R., FRANZINI, J., FREYBERG, D. and TCHOBANOGLOUS, G.; Water Resources Engineering, 4th edition, McGraw Hill International Editions, Civil Engineering Series, Singapore 1997.

PLATE, E., Statistik und angewandte Wahrscheinlichkeitslehre für Bauingenieure, Ernst & Sohn, Berlin, 1993.

SCHEUERLEIN, H. und HANAUSEK, E., 1996; Risiko Management im alpinen Wasserbau – Die Optimierung von Schutzmaßnahmen mit Hilfe von Gefahrenzonenplänen in Tirol, Internationales Symp. INTERPRAEVENT 1996, Germisch-Partenkirchen, Proceed. Band 3, S. 19-28.

Stochastic Hydraulics 2000, Wang & Hu (eds) © 2000 Taylor & Francis, ISBN 90 5809 166 X

Application of the Parker-Klingeman model for bedload transport in small mountain streams

Wen C.Wang
Multech Engineering Consultants, San Jose, Calif., USA

David R.Dawdy
San Francisco, Calif., USA

Peggy Basdekas
US Forest Service, Klamath Falls, Oreg., USA

ABSTRACT: The Parker-Klingeman bedload transport model is an engineering tool, based on dimensional analysis, for gravel transport in streams. Its use for bedload prediction is shown for small mountain streams in Oregon. The fitting of the two parameters of the model is described. Bedload size distribution for any discharge and for the average annual load can be predicted.

Key words: Bedload, Parker-Klingeman model, Gravel bed streams

1. INTRODUCTION

The U.S. Forest Service has been collecting streamflow and sediment discharge data in the Winema National Forest since 1993. Bedload data have been collected by wading with a Helley-Smith sampler. For six stations, intensive bedload data have been collected, with 10-20 measurements made over a snowmelt season. Surface and subsurface bed material data have been collected at each site, and the data analyzed using the Parker-Klingeman (PK) model (Parker and Klingeman, 1982). The statistics for the intensive gaging sites are shown in Table 1.

Table 1: Statistics for the Gaging Sites

Stream Name	Drainage Area Square km	Pavement		Subpavement	
		D50 mm	D84 mm	D50 mm	D84 mm
Annie Creek	72.0	16.2	37.7	9.6	28.3
Cherry Creek	41.2	68.4	171.1	23.3	154.0
Fivemile Creek	91.7	71.1	146.3	10.3	53.2
North Fork Sprague River	93.3	153.8	217.5	17.8	67.9
Paradise Creek	64.8	39.2	57.9	17.8	49.7
Spencer Creek	93.8	87.6	192.5	19.5	86.8
South Fork Sprague River	160.9	86.0	183.5	30.2	139.2

The PK procedure includes the effect of a "hiding factor". The term "hiding factor" is used to describe the fact that when there is a mixture of particles on the stream bed, the larger particles hide the smaller particles. Thus, the smaller particles are harder to move than would be predicted by the usual equations based on bed shear which assume a uniform bed material, such as that of Meyer-Peter and Mueller. Similarly, because the larger particles project into the flow more than they would if there were uniformly large particles on the bed, larger particles are moved more easily than otherwise predicted. The result is a more uniform movement of particles of all sizes, which is termed "almost equal mobility".

The PK procedure includes a physically-based semi-empirical equation with two calibration parameters. Those two parameters are: first, a reference critical shear value, TRS50, the shear at which the median diameter of bed material moves, and second, an exponent which relates the shear value required to move any other size present in the bed material to TRS50. The prediction of the size distribution of the bedload is based on the distribution of a parent distribution. The parent distribution may be for the pavement material on the bed or the sub-pavement material under the pavement.

Parker and Klingeman's equation 21 is as follows:

$$\frac{TRS(I)}{TRS50} = (\frac{DG(I)}{DMREF})^{-PEXP}$$

(1)

where TRS(I) is critical shear for size of material DG(I), and TRS50 is the critical shear for the reference size of material, DMREF, taken as the D50 for either the pavement or subpavement material. The exponent, PEXP, and the reference shear stress, TRS50, in their equations 22 and 27 (TRS50=0.0876 for subpavement and =0.035 for pavement material) are related. They also are related to the Wr*, a dimensionless bedload, for which Parker and Klingeman choose 0.002 (p. 1412). The value of 0.002 is a "small but measureable bedload movement" used to determine the reference shear stress, TRS50. In a previous paper it was shown that the PK parameters must be calibrated (Dawdy, Wang, and Basdekas, 1998), using Coloerado data. This conclusion is borne out by examples from Oregon in this paper. If the size distribution of the bedload is determined by a proper choice of exponent, the volume transported can be fixed by a proper choice of TRS50. This means that with a good set of data the Parker and Klingeman empirical approach can be calibrated for a wide set of conditions.

Determination of a proper "calibrated" reference shear stress depends upon the determination of energy slope. Thus, slope must be known in order for TRS50 to be used to predict the bed load without calibration. Error in determination of the energy slope and the reference size, D50, have a similar effect on prediction. If D50 is increased, a change in reference shear, TRS50, must be made to predict with equal accuracy. Thus, both the size distribution of the parent material and energy slope must be accurately determined in order to use the Parker and Klingeman method without calibration. However, if parameters are calibrated to data, the calibrated parameter values will compensate for any errors in measurement of slope and D50, and the resulting parameters may be used to predict bedload movement for that site.

A prediction using the Meyer-Peter and Mueller formula has a PK exponent of zero. This results in the prediction of too small bedload sizes and too great a potential sediment load when used to predict gravel movement. The exponent in Parker and Klingeman's Equation 21 must be different from 1.0 (the Parker and Klingeman paper uses 0.982 with the sub-pavement distribution used for prediction). That exponent determines how the size distribution of the bedload is related to that of the parent material. The size distributions of the two potential parent distributions (the pavement and the sub-pavement) are different. Therefore, the exponent should be different for the prediction equations based on the two bed material distributions. The pavement exponent should be smaller than that of the sub-pavement because the bedload size distribution is more different from that for the pavement than that for the sub-pavement. The farther the exponent differs from one, the more the size of the bedload differs from the parent distribution, either pavement or sub-pavement.

2. APPLICATION

For each of the intensive sites two parameters of the PK model were calibrated. The calibration procedure minimized the squared deviations of the logarithms of the measured and predicted bedloads for each size fraction and for the total load. However, there was a constraint that the predicted bedload for all measurements must equal the measured load. An iterative procedure was used to meet these two conditions.

An inspection of the resulting scatter diagrams of the residuals of the data indicated that there was a trend, with a similar trend for each set of data. Therefore, the data were stratified by discharge for each site, and the optimal parameters determined for each discharge group. The PK exponent was then set to the average for all groups at a site, and the value of TRS50 determined for each group of measurements. Figure 1 shows a typical example of the variation of TRS50 with discharge. The numbers next to the points are the number of measurements used to define the point on the figure. There were 15 bedload measurements made for Annie Creek and 13 for the Sprague. In every case analyzed so far, the value of TRS50 increases with discharge. This apparent change results from the assumptions of the PK model, and indicate an adjustment in the solution is needed. Meanwhile, calibration of the variation in the characteristic shear is required. Figures 2 and 3 show the comparison of predicted bedload discharge with measured bedload for Annie Creek and for the South Fork of the Sprague River in southern Oregon. With calibration, the PK model fits both the size distribution and the total bedload discharge. Note that the PK exponent varies for the two calibrated conditions, 0.863 for Annie Creek and 0.954 for the Sprague, compared to 0.982 for the Oak Creek data used in the original PK paper. The armoring of the bed to ensure that the bedload supply equals the bedload discharge causes the relation of the parent distribution to the bedload distribution to change, which changes the exponent.   Thus, the PK model is a good model for prediction of bedload discharge for gravel-bed rivers, but it requires bedload measurements for calibration.

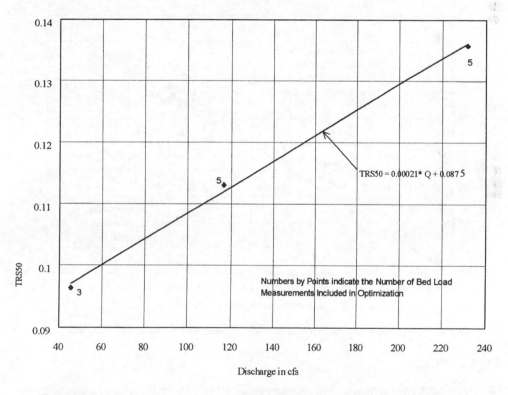

Fig.1:   Relation of Parker-Klingeman Characteristic Shear to Discharge for South Fork Sprague River, Oregon

427

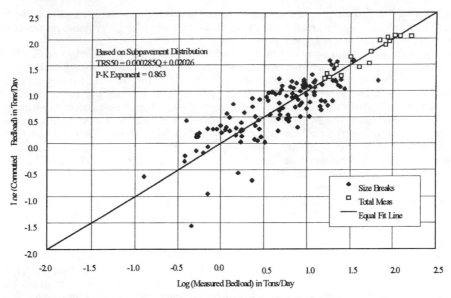

Fig.2: Comparison of Predicted with Measured Bedload for Annie Creek, Oregon

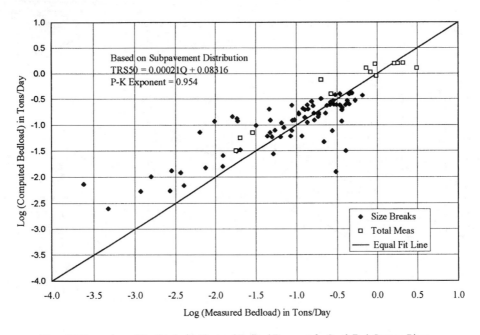

Figure 3 Comparison of Predicted with Measured Bedload Transport for South Fork Sprague River

REFERENCES

Dawdy, David R., Wang, Wen C., and Basdekas, Peggy, (1998) "Calibration of the Parker-Klingeman model with variable shear", Proceedings, Seventh International Symposium on River Sedimentation, Hong Kong, China, A. A. Balkema, Rotterdam, December 1998.

Parker, Gary, and Klingeman, Peter C. (1982) "On why Gravel Bed Streams are Paved", Water Resources Research (18)5, 1409-1423.

Stochastic Hydraulics 2000, Wang & Hu (eds) © 2000 Taylor & Francis, ISBN 90 5809 166 X

On evaluating the effectiveness of the soil water management in the forest-steppe and steppe zones

Nadezhda A. Shumova
Water Problems Institute, Russian Academy of Sciences, Moscow, Russia

ABSTRACT: The Paper pesented is the generalize estimates of the effectiveness of the human activity in an effort to reduce the intensity and recurrence of droughts. The "dry agriculture" technique such as retention of the meltwater in the fields, autumn ploughing, fallow and surface soil mulching with crop residues is under investigation. The evapotranspiration model with standard meteorological data and leaf area index is used to obtain the agricultural field water supply under natural water supply conditions and "dry agriculture" technique. The agricultural field water supply was analyzed for large scale spatial trends over many years for the spring wheat fields represented by 45 agrometeorological stations in the forest-steppe and steppe zones of the Former Soviet Union. For 6 agrometeorological stations the agricultural field water supply for individual years within a period over 20 years was examined. A comparison between the agricultural field water supply under natural water supply condition and "dry agriculture" technique has been carried out. The calculation showed a rather low natural agricultural field water supply in the forest-steppe and steppe zones. Retention of the meltwater in the fields, autumn ploughing and fallow tends to increase the agricultural field water supply by 10 to 20%. The surface mulching tends to increase the agricultural field water supply by about 1.7 times.

Key words: Agriculture, Agricultural field water supply, Drought, Water resources management, "Dry agriculture" technique

1. INTRODUCTION

Standard practice to evaluate the effectiveness of the "dry agriculture" technique is to compare the yields from experimental and control plots. When generalizing the results of experiments beyond the limits of the conditions under which they were conducted, considerable difficulties occur. The foundation of the present approach is the calculation of the quantitative consumption of water by agroecosystems during their growth and development. The really regularities of the formation of the soil-hydrology conditions with their variability from year to year are considered too.

2. METHODS

According to Budagovsky & Shumova (1976) the agricultural field water supply parameter is determined

$$WSP = ET/ETO \tag{1}$$

were: WSP - water supply parameter,
ET - actual transpiration [mm day^{-1}],
ETO - potential transpiration (plants are well provided with water) [mm day^{-1}].

The water supply parameter *WSP* shows to what extent soil water provides the development of vegetation cover. If the parameter *WSP* is equal to 1, it indicates, that plants are well provided with water. If the parameter *WSP* is below 1, it indicates that the drought took place.

Difference between potential and actual transpiration shows water deficit in absolute values

$$dET = ETO - ET \tag{2}$$

were: dET - transpiration deficit [mm day^{-1}].

The use of the empirical relation between yield and the water supply parameter, which for arid regions have a practically acceptable accuracy, is possible. The approximate relationship between the water supply parameter *WSP* and relative yield is:

$$Y/YO = (WSP - a)/(1 - a) \tag{3}$$

where: Y/YO - relative yield,
  Y - actual yield,
  YO - optimal yield (plants are well provided with water),
a - coefficient, for cereal grains its value changes from 0.2 to 0.35 depending on the agrotechnical standard.

The relationship (3) indicate that yield declines to a greater extent than transpiration when drought occurs. With relationship (3) we can estimate the yield for different values of the agricultural field water supply.

The evapotranspiration model has been used for obtaining actual and potential transpiration. Standard observational data of agrometeorological stations (such as air temperature, air humidity deficit, net radiation, wind speed, precipitation and initial soil water storage) and leaf area index are used for calculations. The leaf area index is determined by the simplest methods (Shumova, 1994) from the data of the agrometeorological stations concerning the average height of plants and their number per unit of the sown area or from the generalized curves of the leaf area index.

Average of many years values of *ET* and *ETO* for 45 agrometeorological stations of the forest-steppe and steppe zones (and partially outside their ranges) of the European territory of the Former Soviet Union for the spring wheat crops under natural water supply conditions and under "dry agriculture" technique have been obtained (Fig. 1). Among these, 6 agrometeorological stations reflected all variety of natural conditions of the forest-steppe and steppe zones. *ET* and *ETO* values were determined for individual years within a period of over 20 years. These data have been used for obtaining the water supply parameter *WSP* under natural conditions and "dry agriculture" technique.

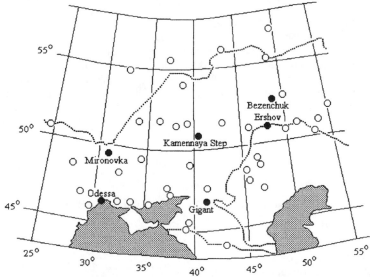

Fig 1: Locations of the agrometeorological stations (ₒ - normal data, ● - years of time). The dotted lines indicate the boundaries of the forest-steppe and steppe zones.

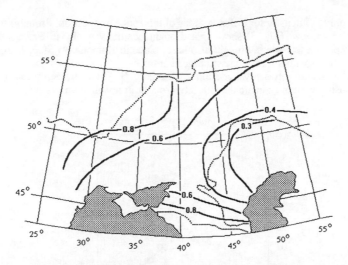

Fig.2: Water supply parameter under natural conditions *WSP-natural*.

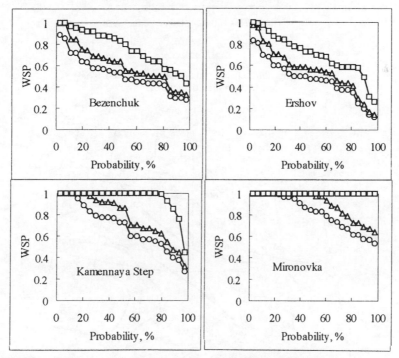

Fig.3: Probability curves of the water supply parameter under natural conditions and "dry agricultural" technique (o- *WSP-natural,* -△ *WSP-runoff,* □ - *WSP-mulching*).

3. RESULTS AND DISCUSSION

3.1 Natural water supply

Under natural water supply conditions average water supply parameter *WSP-natural* varies from 0.8 to 0.3 (Fig. 2). These values indicate a rather low natural water supply of crops in the forest-steppe and steppe zones. The droughts are uniformly distributed in the forest-steppe and steppe

431

zones and differ only in intensity for different areas of the zones in question. Probability curves of the parameter *WSP-natural* (Fig. 3) show, that drought occurs annually in Bezenchuk, Ershov and Odessa. Except in a few years, drought also takes place in Kamennaya Step, Mironovka and Gigant. Occasionally droughts have a catastrophic character.

Fig. 4 gives relationship between water supply parameter *WSP* and transpiration deficit *dET*. Using this relationship we can estimate the intensity drought in absolute values too.

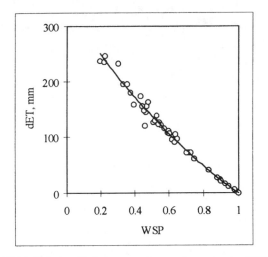

Fig.4: Relationship between water supply parameter *WSP* and transpiration deficit *dET*.

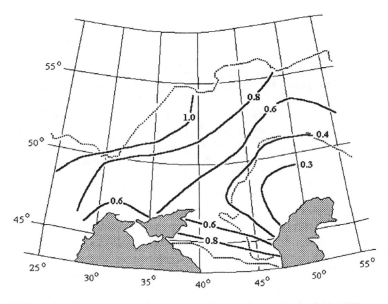

Fig.5: Water supply parameter under retention of the meltwater in the fields *WSP-runoff*.

3.2 Retention of the meltwater in the fields

Retention of the meltwater in fields tends to increased spring available soil water storage, and it turns to increase the available soil water storage over the growing season. The spring runoff data

432

by Komarov (1959) were used to calculate *WSP-runoff*. It must be kept in mind that average annual spring runoff on the northern boundary of the steppe zone is only 60 mm according to Komarov (1959), falling to 20-30 mm for its southern boundary. In evaluating the effectiveness of the retention of meltwater in the fields we reasoned that in actual practice one can use no more than 80% of these values. Drought boundary (*WSP* = 1.0) came into view (Fig. 5). Probability curves of the parameter *WSP-runoff* (Fig. 3) show that recurrence of droughts is reduced. Comparisons of *WSP-runoff* and *WSP-natural* show that agricultural field water supply increased 1.18 times (Fig. 6a) in the case of retention of meltwater in the agricultural fields.

3.3 Autumn Ploughing

Autumn ploughing leads to more full up take of the meltwater by the soil and also increases the spring available soil water storage. Koronkevich (1973) reported the spring runoff losses because of autumn ploughing, (63% in the forest-steppe zone and 89% in the steppe zone). Using these estimates one can calculate the agricultural field water supply under autumn ploughing. The agricultural field water supply under autumn ploughing increased 1.14 times (Fig. 6b).

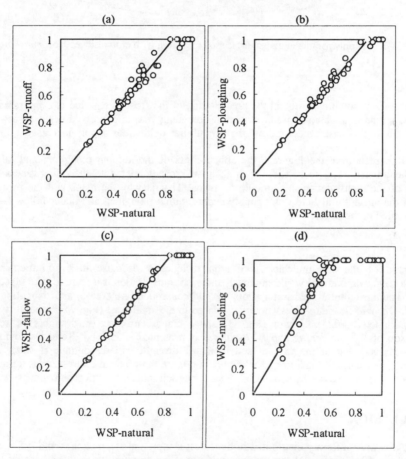

Fig.6:Comparisons of the average quantities of the parameter *WSP-natural* and (a) *WSP-runoff*, (b) *WSP-ploughing*, (c) *WSP-fallow*, (d) *WSP-mulching*.

433

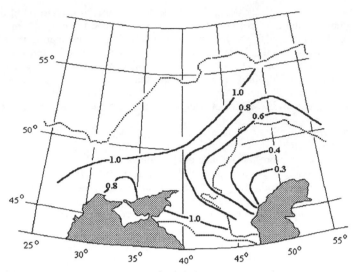

Fig.7: Water supply parameter under surface soil mulching with crop residues *WSP-mulching.*

3.4 Fallow

In the case of fallow, practically all the precipitation of the frost-free period is evaporated. This means the spring available soil water storage in foregoing year is conserved in the spring of the current year. In this case (under fallow) the agricultural field water supply increased 1.19 times (Fig. 6c).

As will be seen from the Figure 6 the effectiveness of the autumn ploughing and fallow to reduce the intensity and recurrence of droughts is closely related to the effectiveness of the retention of the meltwater in the agricultural fields. That is why the maps and the probability curves of the agricultural field water supply under autumn ploughing and under fallow have not been presented in the paper.

3.5 Mulching

In the case of the soil mulching with crop residues a decrease in evaporation by soil (nonproductive evaporation) would obviously increase transpiration if it were to take place under drought. The agricultural field water supply would consequently increase, approximating to the optimum. Drought boundary is essentially displaced in the south-east (Fig. 7). Probability curves of the parameter *WSP-mulching* (Fig. 3) show that recurrence of droughts is reduced. Comparison of *WSP-mulching* and *WSP-natural* is presented in Fig. 6d. Referring to Fig. 6 it will be observed that in the case of soil mulching drought will discontinue if water supply parameter under natural conditions *WSP-natural* no more than 0.6. In the rest cases agricultural field water supply increases by about 1.67 times under soil mulching with crop residues.

4. CONCLUSION

The droughts are uniformly distributed in the forest-steppe and steppe zones and differ only in intensity for different areas of the zones in question. The agrohydrological methods of the soil water management reduce the intensity and recurrence of droughts. The agricultural field water supply increased (i) 1.18 times in the case of retention of meltwater in the agricultural fields, (ii) 1.14 times under autumn ploughing, (iii) 1.19 times under fallow, (iv) 1.67 times under surface mulching with crop residues.

REFERENCES

Budagovsky, A.I. & Shumova, N.A. (1976) "Methods for Analysis of Evapotranspiration and Estimates of Efficiency of Its Regulation", Vodnye Resursy, No 6, pp. 83-98 (in Russian).

Komarov, V.D. (1959) Spring Runoff of Flatland Rivers of the European USSR, Conditions of Its Formation and Forecasting Methods, Gidrometeoizdat, Leningrad (in Russian).

Koronkevich, N.I. (1973) Water Balance Remaking, Nauka, Moscow (in Russian).

Shumova, N.A. (1994) "Leaf Area Index in Estimating Evapotranspiration from Spring Wheat Crops", Vodnye Resursy, Vol. 21, No 6, pp. 697-703 (in Russian).

Stochastic Hydraulics 2000, Wang & Hu (eds) © 2000 Taylor & Francis, ISBN 90 5809 166 X

Risk analysis of reservoir level during rehabilitation

Shiang-Jen Wu & Jinn-Chuang Yang
Department of Civil Engineering, Chiao-Tung University, Hsin-Chu, Taiwan, China

Yeou-Koung Tung
Department of Civil Engineering, Hong Kong University of Science and Technology, Kowloon, China

ABSTRACT: After five decades of service since 1953, the storage capacity of Ah-gong-dein reservoir in Taiwan has been reduced by about 71% due to accumulated sedimentation. Its ability to serve flood control function has been greatly compromised. The Taiwan Department of Water Resources currently is undertaking measures to vitalize its flood control function over an eight-year period. The paper addresses probabilistic behavior of reservoir surface level during the rehabilitation period. In particular, randomness of monthly rainfall amounts is considered in developing function relationship between monthly probability of water surface elevation exceeding certain target level and various factors. Information such as these is useful to examine the implications of proposed rehabilitation schemes on overtopping risk and to monitor the required revitalization progress and reservoir storage condition.

Key words: Storage capacity, Flood control function, Rehabilitation, Revitalization, Sedimentation, Risk analysis, Reservoir level

1. INTRODUCTION

Ah-gong-dein reservoir is situated in Yan-Chao County, Kaohsiung. It is a concrete dam, with crest elevation at 42m, located at the confluence of Wang-Lai River and Chou-Shui River. The upstream catchment area is 31.87 km². Reservoir volume below 26m is the dead storage. The reservoir has a morning-glory emergency spillway with its crest elevation at 34.5m having a maximum discharge of 90cms. The outlet tunnel to withdraw water for irrigation and water supply is at 26m elevation and has a maximum capacity of 5cms.

Ah-gong-dein reservoir has flood control as its primary function, supplemented by irrigation and water supply. Ever since its commission in 1953, the reservoir has subject to rapid sedimentation build-up. In 1991, a field survey showed that the original design storage capacity of 2.05Mm³ has reduced to 0.59Mm³, a reduction of 71 the original capacity. Over the year, the sediment build-up in reservoir has reached a level that threatens the dam safety and has severely hampered the flood control and other functions of the reservoir.

In 1993, a preliminary study was conduct on an eight-year reservoir rehabilitation project, which include both operational and engineering means. Operationally, the reservoir will store water only during the dry season (October-March) whereas, during the high flow season, the reservoir will be empty to allow direct passage of incoming sediment laden water. To achieve this, the emergency spillway is to be lowered from its original 34.5m to 27m. Additional engineering means also include dredging reservoir sediment and the construction of a new inter-basin overflow spillway at 37m elevation to divert excess water. After the completion of the rehabilitation work, the study estimated that the reservoir could protect downstream from floods up to 10,000-year and restore the functions for irrigation and water supply.

Since the rehabilitation will take over a period of eight years, the main objective of this study is to have a preliminary assessment of the probabilistic characteristics of monthly water surface level as affected by hydrologic factors and reservoir properties. The latter particularly are affected by the progress of various rehabilitation schemes mentioned above. Information derived from the study could provide useful insight for the flood protection performance during the period of rehabilitation work

2. PROBAILISTIC ANALYSIS OF RESERVOIR WATER LEVEL

Evaluation of reservoir hydrologic responses involves analysis of rainfall-runoff process in upstream watershed of the reservoir, storage-stage-outflow relationship, and reservoir routing (see Fig. 1). The probabilistic analysis of reservoir water level is the determination of the probability that the maximum water stage in reservoir exceeds some specified target levels. In overtopping risk analysis, the target water level is the dam crest elevation.

As shown in Fig. 1, it is clear that the maximum reservoir water level is dependent on many meteorological, watershed, and reservoir characteristics, which are inherently random and/or subject to uncertainty. As a result, the maximum water level in the reservoir under a hydrologic load is uncertain following certain probabilistic laws.

Mathematically, the probability that the maximum reservoir water level (H) would exceed a specified critical stage (h_c) can be expressed as

$$p_c = Pr\,[H \geq h_c\,] \tag{1}$$

in which H is a function of several random factors such as storm depth, duration, and it temporal pattern, rainfall-runoff mechanism, initial reservoir stage, and reservoir storage-outflow rating relationship.

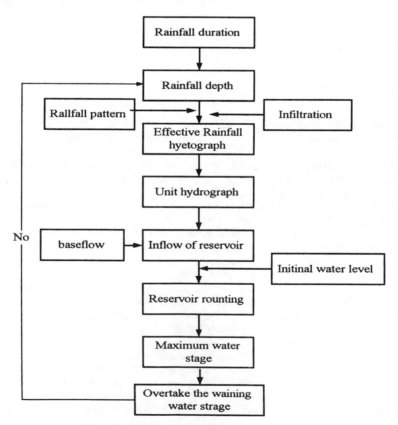

Fig.1: Flowchart of risk analysis of reservoir water level

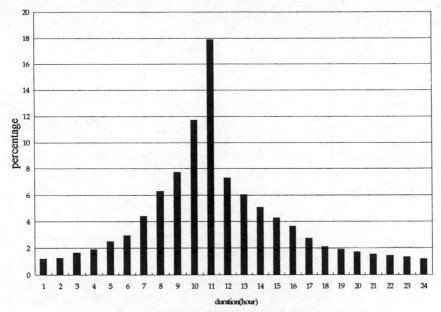

Fig.2: Rainfall hyetograph used

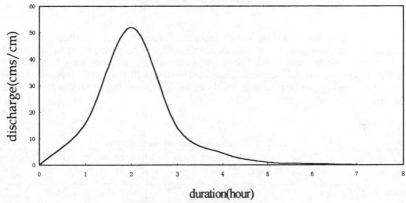

duration(hour)

Fig.3: 1-hr unit hydrograph used

3. CASE STUDY

Of particular interested in this probabilistic study is to evaluate the effect of reservoir rehabilitation project on the safety aspect of dam. Among the various components affecting the maximum water level in Ah-gong-dein reservoir, only the inherent randomness of monthly maximum rainfalls of different durations is considered. Uncertainties associated with storm duration, temporal pattern, unit hydrograph, reservoir stage-discharge rating curve, and initial water level are ignored (see Figs. 2-4). Table 1 lists the conditions under which the monthly probability of reservoir water level exceeding various specified target stages are computed.

Based on the existing (1998) storage condition, the monthly probability of maximum water level exceeding the target stage of 40m (2m below the dam crest) resulting from a 24h storm are shown in Fig. 5. As can be seen that the top three months having the highest exceedance probability are July-September, which is in the heat of typhoon season. Therefore, in the consequent analysis, focuses of reservoir water level are placed on these three months.

439

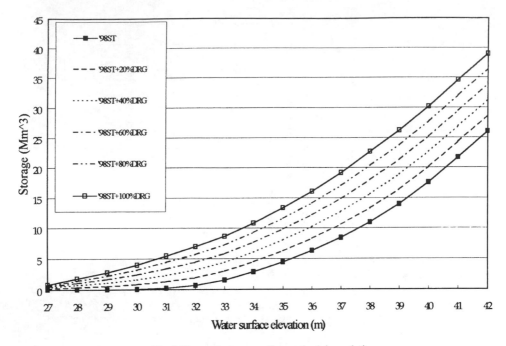

Fig.4: Stage discharge sediment dredging relation

Because the investigation considers only inherent randomness of monthly maximum rainfall amounts of different durations, the threshold rainfall depths corresponding to various durations and initial water levels beyond which the target water level is exceeded are computed (see Fig. 6). From the threshold rainfall amount under various conditions, the corresponding monthly probability of exceeding the target water level can be calculated based on the proper distribution describing the monthly maximum rainfall.

Table 1:.Conditions used in risk analysis

	Before completion of inter-basin transfer spillway	After completion of inter-basin transfer spillway
Initial water level	27-40m	27-37m
Target water level	40-42m	37-40m
Rainfall duration	24h-48h-72h	
Rainfall pattern	Shown in Fig. 2	
Reservoir storage	Existing (1998) storage + (0-100%) total dredging	
Baseflow	1cms	
Infiltration index	2.2 mm/h	

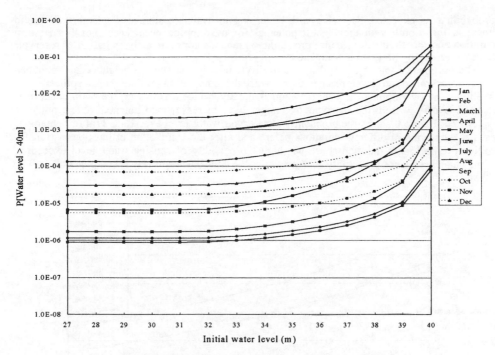

Fig.5: Monthly [H>40m] by a 24h storm under'98 storage condition

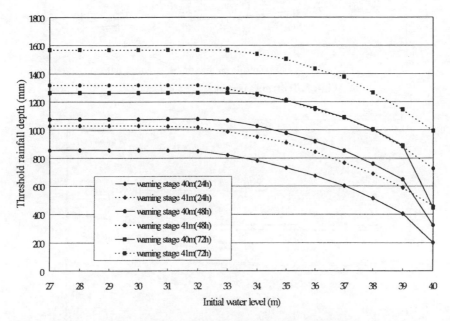

Fig.6: Threshold rainfall depths of various durations exceeding different
Target reservoir stages under `98 storage condition

Figure 7 illustrates the variation of probability that the maximum water level exceeds h_c=40m in July-September with respect to the percentage of completion of sediment dredging under initial water stage of 27m and three storm durations. The target-level-exceedance

441

probability, as expected, decreases with increasing in available storage. September (by solid lines) is the month with the highest potential for overtopping occurrence. For fixed storm duration and month, the exceedance probability is reduced approximately by half after reservoir sediment dredging program is completed.

The purpose of inter-basin spillway is to divert flow to the river in neighboring watershed when reservoir water level reaches 37m. With the completion of inter-basin spillway, the likelihood of overtopping the dam (at 42m) is greatly reduced. However, the main concern is that the overflow from the reservoir that might have increase flood potential to the adjacent watershed. Figure 8 shows the probability of reservoir water level exceeding 37m in September caused by storms of different durations before and after the completion of inter-basin transfer spillway. The longer duration storm has higher probability of causing water level exceeding 37m. This exceedance probability decreases after the inter-basin spillway is completed

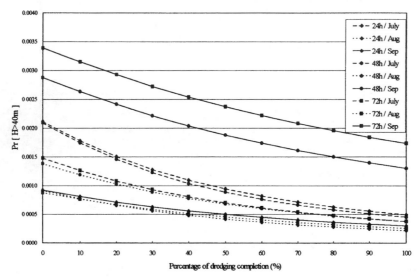

Fig7: Pr[H>41m]vs.percentage of completing

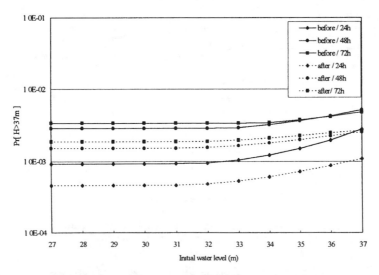

Fig.8: Pr[H>37m] in september before and after the completion
Of inter-basin transfer spillway under `98 storage

442

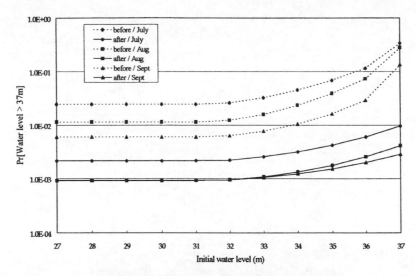

Fig.9: Probability of overflowing inter-basin transfer spillay by 24h stom before and after the completion of dredging

Figure 9 shows a comparison of the probability of occurrence of overflowing inter-basin spillway by a 24h storm in the July-September corresponding to the existing storage condition and that after the completion of sediment dredging. Clearly, it is seen that, after the completion of dredging, the risk of overflowing inter-basin spillway is reduced approximately by 10 times.

4. CONCLUSIONS

The study presents a methodological framework to assess the monthly probabilistic characteristics of reservoir water level corresponding to different rehabilitation schemes under various hydrologic conditions. The quantification of reservoir level exceeding a certain water level provides useful insights for Department of Water Resources in Taiwan to assess, plan, and monitor progress of rehabilitation project.

However, one should be aware that the study only presents a simplistic approach by considering only the randomness of maximum monthly rainfalls. Other uncertainties also exist and may have important implications and impacts on the risk evaluation of reservoir performance. The inclusion of other factors presents many technical challenges that would have to be further developed.

ACKNOWLEDGMENTS

The study is sponsored by Provincial Department of Water Conservancy Bureau, Taiwan.

REFERENCES

Water Resources Dept., MOEA, (1985), Ah-Gong-Dein Reservoir Improvement Planning Report (in Chinese)

Jong, Y.L. (1988), "Study on Spillway Safety due to Uncertainties of Rainfall its Pattern," Master Thesis, National Chaio Tung University, Hsinchu, Taiwan

Yen, B.C. and Tung, Y.K. (1993), Reliability and Uncertainty Analysis in Hydraulic Design, ASCE, New York

Telford, T. (1996), Floods and Reservoir Safety, The Institution of Civil Engineers, London, UK

US Army Corps of Engineers, (1996), "Flood-Runoff Analysis," Technical Engineering and Design Guides, no 19.

Stochastic Hydraulics 2000, Wang & Hu (eds) © 2000 Taylor & Francis, ISBN 90 5809 166 X

Modelling of extreme precipitation events in risk analysis of water systems

F. Kuipers & M. Kok
Department of Civil Engineering and Management, University of Twente, Enschede, Netherlands

C.J.M.Vermeulen
HKV Consultants, Lelystad, Netherlands

ABSTRACT: In this paper we are interested in a methodology for the determination of failure probabilities of water systems. To estimate the probabilities of exceedance for extreme water levels due to heavy rainfall a probabilistic method is used. In this method, the time varying factors that cause an extreme event are identified. By combining these stochastic variables, events with a known probability are generated that include all possible extreme events.

For polders, the identified stochastic variables are the initial soil wetness, the availability of structures, the rainfall depth and rainfall intensities. The crucial question is the modelling of rainfall events. Here, a data driven approach is followed. A nine-day duration is chosen, so that events can be considered mutually independent. The rainfall depth and the intensity pattern are identified as stochastic variables. The main feature of intensity pattern is the degree of concentration, which determines the maximum n-day depth fractions of the event.

Key words: Precipitation modelling, Rainfall, Polders, probabilistic method, Stochastic variables

1. INTRODUCTION

The Netherlands is situated on the delta of three of Europe's main rivers: the Rhine, the Meuse and the Scheldt. As a result of this, the country has been able to develop into an important, densely populated nation. Large parts (approximately 50 %) of the Netherlands are below mean sea level and water levels that may occur on the rivers Rhine and Meuse. Therefore, pumping stations are needed.

Heavy rainfall can, due to drainage or pump capacity constraints, cause increased surface and ground water levels, which can cause a lot of damage. In the autumn of 1998, these events caused damage of about \$ $400*10^6$, mainly in the agricultural sector. The government has compensated these damages under the Dutch Compensation Law for Damage Caused by Disasters and Serious Accidents (WTS). In 1998 also, emergency measures were applied such as flooding polders in order to prevent greater flood damage in other regions.

In this paper we are interested in a methodology for the determination of failure probabilities of water systems. To estimate the probabilities of exceedance for extreme high water levels due to rainfall a probabilistic method is used. The method is applied to surface water levels in polders.

We first describe the water system and the relation between rainfall and extreme water levels. After an overview of several design methods, the probabilistic method is discussed. Subsequently, the modelling of extreme rainfall events is detailed, followed by a report on an application of the method. Finally, we present the conclusions.

2. DESCRIPTION OF THE WATER SYSTEM

In this paper we consider a special type of water system: polders. Polders are used for several functions, like housing, industry, nature and agriculture. To optimally fulfil all these functions, a target surface water level is maintained in the polders. Around this target level some deviations have to be tolerated. To prevent the soil from getting too wet (because the farmers cannot then access the land by heavy machinery) often drains or trenches are used to lead ground water to the surface water.

In Fig.1 a cross-section of a typical polder-reservoir system is projected. The figure shows that polders are often surrounded by a reservoir that can direct the water towards the river or sea. Because the reservoir's water level exceeds the surface water level in the polders, dikes and pumps are needed.

2.1 Relation between rainfall and high surface water levels

Because the surface water covers only a small part of the polder area (a typical value is 5%), most of the rain will fall on the ground. The soil wetness at the start of the rainfall event influences the storage capacity. In a dry situation, a relatively large amount of rainfall is stored in the ground and only a small part will drain to the surface water. In a wet situation, when storage capacity is reached, even a small rainfall amount can lead to a relatively large surface water level rise.

In Fig 2 the most important water flows are shown for a polder containing drains. The rainfall-runoff characteristics of the polder will govern the rate of transportation of the rainfall to the surface water. For instance in well-drained sandy areas, rainfall will be transported faster than in poorly drained clay soils. The pumping stations pump the water to the reservoir. When the pumping capacity is lower than the inflow to the surface water, the surface water level will rise.

For two reasons, the pumping stations do not always function at full capacity. Firstly, the pumps can be (partly) out of use due to damage or maintenance. Secondly, when the reservoir's water level is considered too high, the pumping may be stopped, because relatively high water levels in the reservoir can be a huge danger for the reservoir's embankments and surrounding areas.

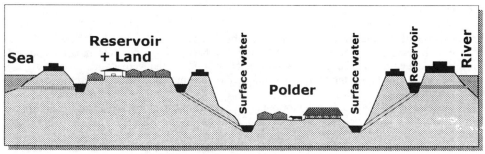

Fig.1 Schematisation of a polder-reservoir system

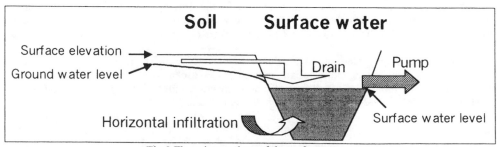

Fig 2 Flows into and out of the surface water

446

3.DESIGN METHODS

There are several methods to design a water system for extreme wet conditions. Firstly, a deterministic approach can be followed, using rules of thumb. Secondly, it can be decided to work with design rainfall events. Using a model these events can be used to design for instance the drainage levels and the pumping stations. A third method is to base the design on a calculation of probabilities of exceedance for water levels. We are interested here in exceedance frequencies from approximately 1/5 to 1/100 or 1/1000 year[-1]. In Fig 3 a possible result of such a calculation is shown.

A commonly used approach to calculate probabilities of exceedance for water levels is to simulate long term meteorological conditions (for instance over thirty years) with a model of the water system of interest. This approach is especially suitable for calculating the probabilities of exceedance for water levels that occur at relatively high frequencies, like once in one to five years. For more extreme water levels, coincidence in the probability of occurrence of extreme events in the simulated period starts to play an important role because it is a highly non-linear system. Hence, arbitrary choices need to be made for extrapolation.

In the probabilistic method presented in this paper more attention is given to the events with probabilities of exceedance from about once in five years and downwards. The stochastic characteristics of the main factors leading to extreme events are used to generate extreme events with a known probability of occurrence that are simulated with a model. The generated events cover the most important situations that can lead to extreme surface water levels. Because extreme events are simulated no choices need to be made for extrapolation in calculating the probabilities of exceedance for water levels.

4.A PROBABILISTIC METHOD

The probability distribution for surface water levels above the target level due to rainfall in polders is derived in four steps:

1. Find all events leading to a surface water level higher than the target level;
2. Determine the maximum surface water level for each event;
3. Determine the probability of each event;
4. The probability of exceedance of water level h_1 is found by summation of the probabilities for all events that have maximum water level higher than h_1.

Step 1: Extreme events

In figure 3 an example of an extreme event is shown. The event is preceded by a period in which the surface water level is 'under control'. Some deviations around the target level have to be tolerated. The event begins at the moment that the average surface water level exceeds the target level with a certain margin (here 0.10 m) and ends when the average surface water level is 'under control' again for a sufficiently long time.

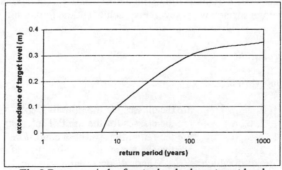

Fig 3 Return periods of water levels above target level

447

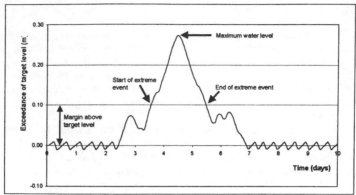

Fig 4. Illustration of an exceedance event: the course of the surface water level.

An event is described as a combination of the stochastic variables that represent the main time varying factors determining the event's maximum surface water level. These stochastic variables are discretised, and a discretion step (class) for each variable determines an event.

For polder areas the identified factors are:

1. the rainfall depth and rainfall intensity pattern.
2. the initial soil wetness;
3. and the availability of structures (pumps);

It is however possible to include other factors, like wind or tides (if the polder pumps directly to a water system under influence of tide).

As will be discussed below precipitation events are modelled by a fixed duration of nine days, so that rainfall events can be considered independent, and the stochastic variables of rainfall depth and intensity pattern.

The initial soil wetness indicates the situation at the start of the precipitation event. Depending on the polder system, more or fewer classes can be used to describe the different conditions. One possible classification is dry, average and wet. The probabilities of occurrence for the classes are obtained from historical data or model simulation.

The availability of structures indicates whether the pumps function, partially function or do not function. Pumps may, at times, not function because of damage or maintenance or because the reservoir's water level is considered too high. The classification for the availability of structures depends on the type of structure. For a pumping station with two pumps, a possible classification is 0%, 50% and 100 %. Probabilities are obtained from maintenance data of the water boards.

Step 2: The event's maximum water level

The event is simulated with a rainfall-runoff model of (part of the) polder of interest, that calculates the area-average surface water level. For each event the maximum surface water level is derived.

Step 3: The event's probability

We define the following variables:

D_i	$i = 1, 2, ..., m$	are the classes for the rainfall Depth
I_j	$j = 1, 2, ..., n$	are the classes for the rainfall Intensity pattern
S_k	$k = 1, 2, ..., o$	are the initial Soil wetness conditions
A_l	$l = 1, 2, ..., p$	are the classes for the Availability of structures
V_{ijkl}	is the event in which D_i, I_j, S_k and A_l occur simultaneously.	
h_{ijkl}	is the maximum water level in event V_{ijkl}	

448

Define:

$\Pr\{D_i\}$ as the probability that a nine-day period contains a depth in class D_i;

$\Pr\{I_j \mid D_i\}$ as the probability that a nine-day period has intensity pattern as defined in class I_j;

$\Pr\{S_k\}$ as the probability that the soil wetness at the start of a nine-day precipitation period is in class S_k;

$\Pr\{A_l\}$ as the probability that during a nine-day period structures function as defined in class A_l.

The intensity pattern is defined as conditional upon the rainfall depth while the Rainfall, the Availability of Structures and the Initial Soil Wetness are assumed to be mutually independent and independent of the rainfall. Therefore, the event's probability is calculated as:

$$\Pr\{V_{ijkl} \text{ in nine days}\} = \Pr\{D_i\} * \Pr\{I_j \mid D_i\} * \Pr\{S_k\} * \Pr\{A_l\}$$

Because nine-day periods can be considered mutually independent, and the year contains 365/9 nine-day periods, the event's year probabilities can be calculated using the formula:

$$\Pr\{V_{ijkl} \text{ in a year}\} = 1 - \left(1 - \Pr\{V_{ijkl} \text{ in nine days}\}\right)^{\frac{365}{9}}$$

Step 4: The probabilities of exceedance

The probability of exceedance of water level h_j is found by summation of the probabilities for all events that have maximum water level higher than h_j. The formula is given below:

$$\Pr\{h > h_j\} = \sum_{ijkl} \Pr\{V_{ijkl} \mid h_{ijkl} > h_j\}$$

5. MODELLING EXTREME PRECIPITATION EVENTS

The generated precipitation events are applied as uniform events in the area of the water system of interest to calculate the event's maximum surface water level (if the area is small we do not have to take into account the spatial variations [Buishand, 1980]). The modelling therefore should contain the rainfall event's characteristics that determine this level. As discussed above, a fixed nine-day duration is used and the depth and intensity pattern are considered as stochastic variables. These three factors are discussed below. For the analysis hourly rainfall data are used from KNMI-station de Bilt, in the period 1907-1998.

5.1 The event's duration

The variation in rainfall event's duration is important for the response of the surface water level. For instance, two days of 30 mm will have a stronger impact than six days of 10 mm. This variation can be modelled in two ways.

One approach is to consider the rainfall duration as a stochastic variable. In this approach, probabilities for the discretised durations and (conditional) probabilities for the rainfall depth and intensity pattern have to be found. An important feature of the durations however is that they are dependent. An extreme two-day event can for instance be part of an extreme eight-day event. As a consequence of different runoff characteristics, for a relatively fast responding polder the two-day event can cause an extreme event, while for a relatively slow responding polder the eight-day event causes an extreme event. This means that both precipitation events should be included in the rainfall modelling, since we do not know a priori which period is most important.

An alternative approach is to use a fixed duration and vary depth and intensity pattern. In this approach, one stochastic variable is dropped out while still all rainfall events with duration shorter than the chosen fixed duration can be modelled. Also, in combination with the initial soil wetness conditions, longer events can be taken into account.

449

The choice for a nine-day duration is based on several factors. Firstly, the duration should be long enough to be able to contain most events. Analysing rainfall data, it appeared that nine days was the longest period for which the maximum depth exceeded the pump capacity of 14 mm/d. Secondly, as discussed before, nine-day periods can be considered mutually independent, which is necessary for transferring nine-day probabilities to yearly probabilities. The conclusion that nine-day precipitation events can be considered mutually independent is based on a calculation of the autocorrelation of daily precipitation depths [Kuipers, 1999]. Furthermore, the nine-day duration corresponds to KNMI observations of duration of passing fronts and combinations of fronts. The nine-day duration also corresponds to observed high water periods.

5.2 The precipitation depth

This stochastic variable represents the total nine-day precipitation depth. The depths are divided into classes and for each class a representative value is chosen that will be used as input for the model. Depths below a certain value are discarded because they would not cause extreme events. The probability distribution is found by a fitting a Gumbel distribution on the yearly maxima [HKV, 1999].

5.3 The intensity pattern

The intensity pattern indicates how the event's depth is distributed with time. The intensity pattern is analysed considering entropy and concentration. Based on the latter, rainfall events are generated.

Entropy is a measure used in information theory to indicate the amount of information in the form of peaks in a series of observations, where relatively high entropy indicates a relatively high amount of information. This measure can for instance be used to scan radio signals from space. When a nine-day precipitation period is seen as a sequence of nine daily precipitation depth observations, entropy can be calculated as: [Kullback, 1997]

$$E = \sum_{i-1}^{9} \frac{d_i}{d} \ln\left(\frac{d_i}{d}\right) \qquad \text{where:} \quad d_i = \text{rainfall depth on day i (mm)}$$

$$d = \sum_{i-1}^{9} d_i = \text{total nine-day rainfall depth (mm)}$$

From the formula it follows that entropy reaches the maximum value 0 when the total depth is contained in one day. The minimum is reached when every day the same depth occurs and equals -2.20.

For 138 nine-day precipitation periods with a total depth above 50 mm, the entropy is calculated, where difference is made for classes of total nine-day depth. The left plot of Fig.5 shows frequencies of exceedance for entropy for rainfall event with depths between 50 and 70 mm. The figure shows that for most events, entropy is between -1,5 and -2,0 and seldom exceeds a value of -1.

Entropy seems to be a valuable variable to describe rainfall events. In the model we want to have the same entropy as in the data. However, because it does not account for the relative position of the daily depths, entropy alone is not a sufficient measure.

As discussed above it is also important that the generated rainfall events represent short (one to two days) as well as relatively long (five to eight days) extreme events. A rainfall characteristic that includes the maximum n-day depth fractions (for n= 1,2, ..., 8) is the degree of concentration. A relatively concentrated event contains a higher fraction of the total event's depth in a shorter time.

In the right plot of Fig.5 frequencies of exceedance are shown for maximum n-day depth-fractions for nine-day rainfall periods with total depth between 50 and 70 mm. From this graph, intensity patterns are generated with different degrees of concentration. For this purpose, the frequencies of exceedance are divided into classes. The probability of occurrence for each class is calculated by subtracting the lower boundary value from the upper boundary value and for each class a representative value is chosen. For instance a 50% concentrated event contains the maximum n-day fractions corresponding to the frequency of exceedance of 0.5, which are

450

listed in Fig.6. Using the alternating block method [Chow, 1998] the daily intensities are found, as shown in Fig.6.

6.APPLICATION OF THE METHOD

The probabilistic method as described in this paper is applied in a study of the water system of the Northeast Polder [HKV, 2000]. This polder in the IJssel-lake, with a total area of $4.8*10^8$ m^2 was disclaimed from the IJssel-lake in the first half of the 20th century. The analysis revealed that this water system is relatively sensitive for concentrated rainfall events, while relatively evenly distributed events cause little problems.

The Low Department is a polder in the Northeast polder, with an area of $3.2*10^8$ m^2 and target level NAP-5.70 m.. Tollebeek is a relatively low-lying sub-polder in the Low Department, with an area of 2.7 10^7 m^2 and target level NAP-6.20 m.. In autumn 1998, Tollebeek was confronted with extreme high water levels. Isolating Tollebeek from the Lower Department by means of structures is considered as an option to avoid extreme water levels in the future.

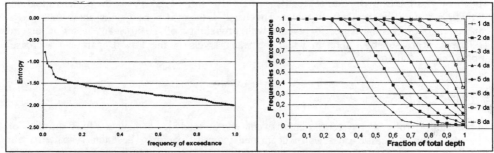

Fig.5: Characteristics of nine-day rainfall events with depth between 50 and 70 mm: frequencies of xceedance for entropy (left) and frequencies of exceedance for maximum n-day depth-actions (right).

n	maximum n-day depth fraction
1	0.42
2	0.56
3	0.64
4	0.73
5	0.82
6	0.9
7	0.96
8	1
9	1

Fig.6: Maximum n-day depth-fractions and daily intensities for a 50%-concentrated rainfall event of 50 mm.

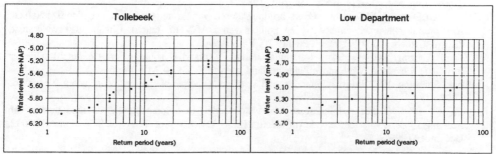

Fig.7: Return periods of water levels in Tollebeek and the Low department in present situation.

451

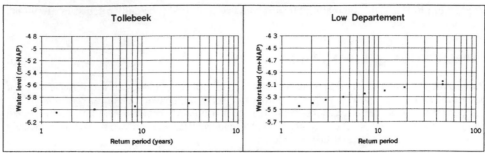

Fig.8: Return periods of water levels in Tollebeek and the Low Department for the situation in which Tollebeek is isolated.

Fig.7 and Fig.8 show the calculated return periods for average surface water levels in Tollebeek and the Low Department for steps of $2.5*10^{-2}$ m in. Fig.7 shows the present situation and Fig.8 shows the situation in which Tollebeek is isolated. The graphs show that, as a consequence of this measure, for events with probability of 1/100 year[-1] water levels in Tollebeek decrease about 0.65 m while the water levels in the Low Department increase approximately 0.05 m.

7. CONCLUSION

In risk analysis of polder systems, rainfall events can be modelled as having a fixed duration of nine days and stochastic variables of depth and intensity pattern. The intensity patterns are based on the concepts of entropy and concentration. Together with other stochastic variables and a model of the water system, the system's failure probability can be calculated.

The main advantage of the method is that it includes all possible extreme events, and not the events that have randomly occurred in the past. Because of this, no choices need to be made for extrapolation of water levels to relatively low probabilities. In addition, the relative contributions of the stochastic variables can be calculated.

The probabilistic approach can be used for all kinds of water systems and gives opportunities to compare the risks of the water system with the accepted safety standards. Finally, it's a rational approach, hence it allows inclusion of more detailed knowledge of the different factors. For example, additional data on failure of structures or changes in rainfall characteristics because of climate change can be incorporated in the analysis.

REFERENCES

Buishand, T.A. and C.A. Velds, 1980. Neerslag en Verdamping (in dutch). Koninklijk Nederlands Meteorologisch Instituut, De Bilt. 1980.

Chow, Ven Te.,1988. Applied hydrology. McGraw-Hill series in water resources and environmental engineering. McGraw-Hill Book Company. 1988.

HKV, SC-DLO and KNMI, 1999. Overschrijdingskansen van Waterstanden in het Noordzeekanaal, Amsterdam-Rijnkanaal en boezem en polders van HAGV (in dutch). Clients: Rijkswaterstaat, directie Utrecht and DWR, report PR 119.10. 1997-1999.

HKV and Alterra, 2000. Evaluatie waterhuishouding Flevoland – hoofdrapport (in dutch). Clients: Provincie Flevoland, Waterschap Zuiderzeeland, report PR 255.30. March 2000.

Kullback, Solomon, 1997. Information theory and Statistics. Dover Publications, Inc, Mineola, New York. 1997.

Kuipers, F., 1999. Overschrijdingskansen van waterstanden in poldergebieden, de modellering van neerslag in de stochastenmethode (in dutch). Thesis University of Twente, HKV report RO 032. December 1999.

Stochastic Hydraulics 2000, Wang & Hu (eds) © 2000 Taylor & Francis, ISBN 90 5809 166 X

An empirical wave forecasting model based on the neural network

Takao Ota, Akira Kimura & Yoshinori Hagi
Faculty of Engineering, Tottori University, Japan

ABSTRACT : A multiple regression model based on the neural network is applied to short-term wave forecasting. Three-layered neural network and back-propagation learning rule are used to develop the forecasting model. A series of learning data, which is necessary to construct the network, consist of difference of atmospheric pressure (input) and significant wave height (output). The forecasting models are made with respect to 4 points along the coast of the Japan Sea. The results show that the problem of the usual regression models is solved fairly.

Key words: Wave forecasting , Neural network , Multiple regression model

1.INTRODUCTION

Short-term wave forecasting is important to secure the port and harbor works. There are two kinds of forecasting method, one is the numerical model using the energy balance equation and the other is the empirical method based on the meteorological data. The empirical method which is based on the multiple regressive equation or the multivariate auto-regressive equation has been proposed as a simple wave forecasting model. However, this method has drawback in that it requires comparatively many data. Furthermore, the relation between the inputs and the outputs is restricted to the linear function. When the natural phenomena including the sea waves are treated, it seems that a nonlinear function is suitable for the expression of the relation. In this study, the neural network, which can express the relation between inputs and outputs without assuming a function explicitly, is applied to the wave forecasting.

2.NEURAL NETWORK

The forecasting model is described basically by the following equation.

$$y(t) = f\{x_1(t-u), x_2(t-u), \cdots, x_n(t-u)\} \tag{1}$$

We consider to express the function f with the neural network in this study. The neural network is a mathematical model emulating the information processing of human brain. A three-layered network (an input layer, a hidden layer and an output layer), shown in Fig. 1, is used in this study. The units of the hidden layer and the output layer contain the threshold value θ_j, γ and the nonlinear transfer function f(x). The coefficients of weight (V_{ik}, W_k) are set on the paths between the units. The output of this network is given as (Nakano et al. 1989; Mase et al. 1995)

$$\bar{y} = f(S) \quad , \quad S = \sum_{k=1}^{m} w_k H_k - \gamma \ ,$$

$$H_k = f(U_k) \quad , \quad U_k = \sum_{i=1}^{n} v_{ik} x_i - \theta_k \tag{2}$$

453

where S is input to the output layer, H_k is output from k-th unit of the hidden layer, U_k is input to k-th unit of the hidden layer and x_i is input to i-th unit of the input layer. The sigmoid function is used as the transfer function. It is expressed by

$$f(X) = \frac{1}{1+\exp(-2X/u_0)} \tag{3}$$

in which u_0 is a constant. The neural network is completed by determining θ_j, γ, W_k and V_{ik}.

The process of determining these values is called 'learning', and they are modified so as to minimize the error between given data (learning data) and the output of the network. The evaluation function on the learning process is described by

$$E_t = \sum_{p=1}^{N} E_p = \frac{1}{2} \sum_{p=1}^{N} (y^p - \bar{y}^p)^2 \tag{4}$$

where y^p is the output data in p-th learning data and $\bar{y}^p$ is the output of the network corresponding to p-th input data. To minimize E_t, the steepest decent procedure is adopted. The differentials of E_t regarding γ, w_k, θ_k and v_{ik} are given as follows:

$$\frac{\partial E_t}{\partial \gamma} = -\frac{2}{u_0} \sum_{p=1}^{N} (\bar{y}^p - y^p)\bar{y}^p(1-\bar{y}^p) \tag{5}$$

$$\frac{\partial E_t}{\partial w_k} = \frac{2}{u_0} \sum_{p=1}^{N} (\bar{y}^p - y^p)\bar{y}^p(1-\bar{y}^p)H_k^p \tag{6}$$

$$\frac{\partial E_t}{\partial \theta_k} = -\left(\frac{2}{u_0}\right)^2 \sum_{p=1}^{N} (\bar{y}^p - y^p)\bar{y}^p(1-\bar{y}^p) \times w_k H_k^p(1-H_k^p) \tag{7}$$

$$\frac{\partial E_t}{\partial v_{ik}} = \left(\frac{2}{u_0}\right)^2 \sum_{p=1}^{N} (\bar{y}^p - y^p)\bar{y}^p(1-\bar{y}^p) \times w_k H_k^p(1-H_k^p)x_i^p \tag{8}$$

The parameters are changed in the learning process by

$$\delta w_k = -\alpha \frac{\partial E_t}{\partial w_k} \quad , \quad \delta \gamma = -\alpha \frac{\partial E_t}{\partial \gamma}, \delta v_{ik} = -\alpha \frac{\partial E_t}{\partial v_{ik}} \quad , \quad \delta \theta_k = -\alpha \frac{\partial E_t}{\partial \theta_k} \tag{9}$$

where α is the learning rate which is determined by the linear search. The learning procedure is called back-propagation.

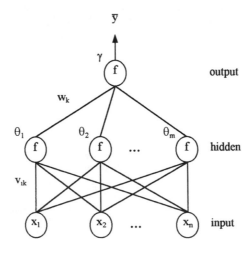

Fig.1: Structure of neural network

454

Wakkanai

Matsumae   Hakodate

Atsumi

Wajima

Kashima   Kyogamisaki

Izuhara

Fukuejima   Tottori

Fuku   Hamada

wave obs. ●

weather st. △

Fig.2: Observation points

3.LEARNING DATA

The data of atmospheric pressure at several ten points is used in the usual empirical wave forecasting model (Kobune et al. 1990). In this study, the object of the prediction (output) is the significant wave height at several points along the coast of the Japan Sea. The severe sea state which may inflict damage to the coastal structures is mainly caused by the monsoon in winter in the Japan Sea. Because the monsoon is generated by the typical distribution of atmospheric pressure in winter, we consider that the significant wave height can be predicted on basis of difference of the atmospheric pressure. We use the data of wave and atmospheric pressure which is issued by Japan Meteorological Agency. Fig.2 illustrates the picked-out points where the wave data or the meteorological data are obtained. The combinations of the points to calculate difference of the atmospheric pressure are given as follows:

· Wakkanai - Hakodate, Wajima, Izuhara, Fukue
· Hakodate - Wajima, Izuhara, Fukue
· Wajima - Hamada, Izuhara, Fukue
· Tottori - Izuhara, Fukue
· Hamada - Izuhara, Fukue

We calculate the cross-correlation coefficients between the differences of atmospheric pressure about above 14 combinations and the significant wave height. Then we obtain the lag time when the value of the cross-correlation coefficient is maximum. Table 1 shows the results of Atsumi as an example. In this table, the positive lag time means that the change of the difference of atmospheric pressure precedes that of the significant wave height. We consider that the output of the forecasting model is the significant wave height in 6 hours and 12 hours. Thus we select the combinations from the above, which have high cross-correlation to the significant wave height and the lag time over 6 or 12 hours. Table 2 and 3 represent the details of learning data for 6-hour and 12-hour prediction. Because there are no combinations whose lag time is over 12 hours for Kashima and over 6 hours for Fukuejima, we can not prepare the learning data.

Table.1: Cross-correlation (Atsumi)

	Oct.'91-Mar.'92		Oct.'93-Mar.'94		Oct.'96-Mar.'97	
	R	τ (hrs)	R	τ (hrs)	R	τ (hrs)
Hakodate - Wakkanai	0.24	12	0.36	18	0.22	16
Wajima - Wakkanai	0.60	1	0.73	2	0.61	2
Izuhara - Wakkanai	0.72	8	0.80	8	0.76	8
Fukue - Wakkanai	0.70	10	0.79	9	0.75	9
Wajima - Hakodate	0.75	4	0.85	4	0.78	4
Izuhara - Hakodate	0.79	10	0.87	10	0.85	10
Fukue - Hakodate	0.77	12	0.85	11	0.84	11
Hamada - Wajima	0.61	17	0.75	16	0.70	15
Izuhara - Wajima	0.65	19	0.75	17	0.71	16
Fukue - Wajima	0.65	21	0.74	19	0.71	18
Izuhara - Tottori	0.60	21	0.69	19	0.66	17
Fukue - Tottori	0.63	24	0.68	21	0.67	20
Izuhara - Hamada	0.58	25	0.60	21	0.57	19
Fukue - Hamada	0.62	28	0.60	24	0.62	24

Table.2: Learning data (for 6-hour prediction)

output (wave data)	term	input (difference of atmospheric pressure)
Matsumae	Jan. '96	Fukue - Wakkanai, Fukue - Hakodate, Izuhara - Hakodate
Atsumi	Feb. '94	Fukue - Wakkanai, Fukue - Hakodate, Izuhara - Wakkanai, Izuhara - Hakodate
Kyogamisaki	Feb. '94	Fukue - Hakodate, Izuhara - Wajima, Hamada - Wajima
Kashima	Jan. '95	Fukue - Wajima, Fukue - Tottori, Izuhara - Tottori, Izuhara - Hamada

Table.3: Learning data (for 12-hour prediction)

output (wave data)	term	input (difference of atmospheric pressure)
Matsumae	Jan. '96	Fukue - Wajima, Izuhara - Wajima Hamada - Wajima
Atsumi	Feb. '94	Fukue - Wajima, Izuhara - Wajima, Izuhara - Tottori, Hamada - Wajima
Kyogamisaki	Feb. '94	Fukue - Wajima, Fukue - Tottori, Izuhara - Tottori

4.RESULTS OF LEARNING AND FORECASTING

A multi regression model for the wave forecasting is constructed by using the neural network and the learning data. The number of learning data (N) is 744 (January) or 672 (February). It is necessary to give the number of unit in the hidden layer (m) and the learning repetition (N_l), however, there is nothing to do but determine by the rule of trial and error. The influence of them on the learning and forecasting is investigated about the data of Atsumi as a example, in which

456

the cross-correlation between the difference of atmospheric pressure and the significant wave height is higher than the others. The learning and forecasting are performed about the cases of m=4,8,12 and N_l=1000,3000,5000,7000,10000. Because the output of the neural network is restricted to [0,1], the output data of the learning data is normalized by its maximum. Fig.3(a) shows the learning error (value of the evaluation function, eq.(4)) and (b) illustrates the fit rate of the forecasting. The fit rate is given by a criterion (Goto et al. 1993);

$$|H_p - H_o| \leq 0.3\,(m) \qquad [H_o \leq 1.0\,(m)]$$
$$|H_p - H_o|/H_o \leq 0.3 \qquad [H_o > 1.0\,(m)] \tag{10}$$

where H_p is the predicted significant wave height (output of network) and H_o is the observed one.

Because the fit rate is maximum at N_l =3000 in all cases, it is considered the optimum number of learning repetition. The number of unit in hidden layer (m) is not determined from the above results, however, we adopt twice as many as the input data following Mase et al. (1995). Fig. 4 shows the results of learning and forecasting at Atsumi and Fig.5 shows those of Kyogamisaki. The solid line is the observed data and the dotted line is the output of neural network. The time lag between the observed and predicted, which is the serious defect of the usual regression models, is solved fairly in the above results of forecasting. The number of unit in hidden layer (m) and the fit rate of the forecasting are displayed in table 4. The forecasting model in this study is considered accurate enough for practical application in the case of 6-hour forecasting.

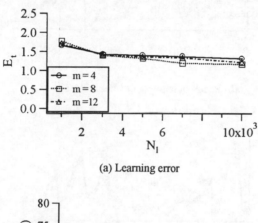

(a) Learning error

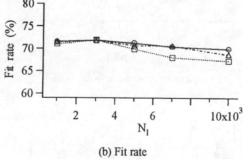

(b) Fit rate

Fig.3: Influence of m and N_l

5.CONCLUSIONS

A multiple regression model based on the neural network is applied to the short-term wave forecasting. Tree-layered neural network and back-propagation learning rule are used to construct the forecasting model. The object (output) is the significant wave height at several points along the coast of the Japan Sea in winter. The differences of atmospheric pressure are

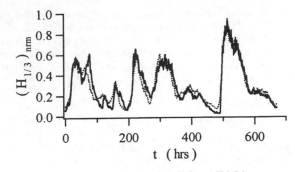

(a) Result of learning (6-hour, '94.2)

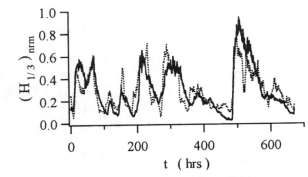

(b) Result of learning (12-hour, '94.2)

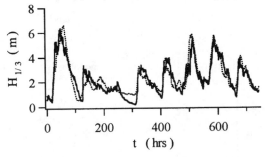

(c) Result of forecasting (6-hour, '97.1)

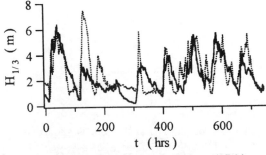

(d) Result of forecasting (12-hour, '97.1)

Fig.4: Results of learning and forecasting (Atsumi)

458

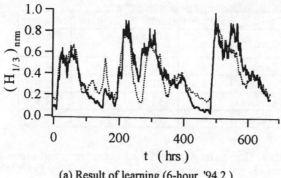

(a) Result of learning (6-hour, '94.2)

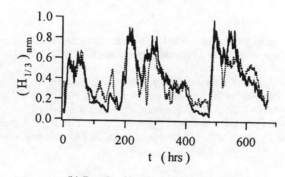

(b) Result of learning (12-hour, '94.2)

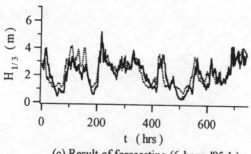

(c) Result of forecasting (6-hour, '95.1)

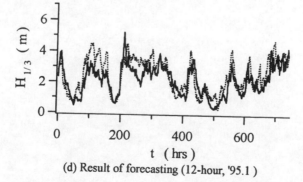

(d) Result of forecasting (12-hour, '95.1)

Fig.5: Results of learning and forecasting (Kyogamisaki)

459

Table.4: Fit rate of forecasting

point	6-hour forecasting		12-hour forecasting	
	m	Fit rate (%)	m	Fit rate (%)
Matsumae	6	59.5	6	34.9
Atsumi	8	71.9	8	43.0
Kyogamisaki	6	65.5	6	54.8
Kashima	8	72.4		

adopted as the input data. The results show that the forecasting model solves fairly the defect of the usual regression models and has the availability in the short-term forecasting.

REFERENCES

Goto, C., H. Shibaki, T. Aono and T. Katayama (1993): Multiple regression models described in physical parameters for the purpose of wave forecasting, Proceedings of JSCE, No.473/II-24, pp.45-53. (in Japanese)

Kobune, K., N. Hashimoto and Y. Kameyama (1990): On reliability of wave forecasting by empirical wave forecasting models, Tech. Note of the Port and Harbour Res. Inst., No.673, pp.1-42. (in Japanese)

Mase, H., M. Sakamoto and T. Sakai (1995): Neural network for stability analysis of rubble-mound breakwaters, J. Wtrwy., Port, Coast.and Oc. Eng., ASCE, 121(6), pp.294-299.Nakano, K., K. Inuma and W. Kiritani (1989): Neurocomputer, Gijyutu-Hyoronsha Co. Ltd. (in Japanese)

Stochastic hydrology

Stochastic Hydraulics 2000, Wang & Hu (eds) © 2000 Taylor & Francis, ISBN 90 5809 166 X

Invited lecture: Integrating hydrologic uncertainties for overtopping risk of dams

Y.-K.Tung
Department of Civil Engineering, Hong Kong University of Science and Technology, Kowloon, China

J.C.Yang, C.H.Chang & S.J.Wu
Department of Civil Engineering, Chiao-Tung University, Hsin-Chu, Taiwan, China

ABSTRACT: In dam safety evaluation and design, overtopping is one of the main concerns, especially for earthfill dams. The likelihood of overtopping incidence largely depends on the spillway capacity, flood control operation rules, and flood-producing hydrologic factors. Hydrologic factors that affect inflow to a reservoir include rainfall depth, duration, frequency, storm pattern, and rainfall-runoff transferring mechanism. In reality, these hydrologic factors are subject to uncertainty due to inherent randomness of hydrologic processes involved as well as sampling errors. This paper presents a methodological framework to integrate various uncertainties in hydrologic aspects of determining reservoir inflow hydrograph. Specifically, uncertainties in rainfall depth-frequency, storm pattern and unit hydrograph are considered to quantify the uncertainty features of reservoir inflow hydrograph which, in turn, are used to assess the overtopping risk of a dam. The methodology has been applied it to evaluate overtopping risk of several dams in Taiwan.

Key words: Dam safety, Hydrologic uncertainty, Overtopping risk

1. INTRODUCTION

Reservoirs created by dams have been playing key role in many water resource projects around the world. Their existences often serve multiple purposes and multiple objectives of water resource development including but not limited to flood control, water supplies, power generation, recreation uses, and etc.. Due to the enormity in size of many structures and the serious consequences of their failures, the issue of dam safety has always been discussed and reviewed in engineering and public circles.

Like any hydraulic structures, dams are placed in natural environments subject to multitude of geophysical forces that are random by nature. The safety of dams over their expected service life may be affected by hydrologic, hydraulic, structural, and construction and installation aspects. There are several modes that dam failure could occur (Cheng, 1993). Broadly speaking, they can be categorized into overtopping, piping/leakage, sliding, spillway damage, and others. A comprehensive assessment of the safety of a dam would require integrated efforts of specialists from various disciplines. The scope of this paper focuses on the assessment of overtopping probability of a dam considering uncertainties of various hydrological factors. Note that the occurrence of overtopping of a dam does not necessarily imply its structural failure leading to the breach of the dam. Uncertainties in hydrological aspects include inherent randomness of various hydrologic processes, model and parameter uncertainties, as well as data inconsistency, non-homogeneity, and incompleteness (Yen et al., 1986; Mays & Tung, 1992).

Dam overtopping incidences largely depend on the spillway capacity, flood control operation rules, and flood-producing hydrologic factors such as rainfall depth, duration, frequency, storm pattern, and rainfall-runoff transferring mechanism. This paper presents a methodological framework that integrates uncertainties of various hydrologic factors in determining reservoir inflow hydrograph for overtopping risk assessment.

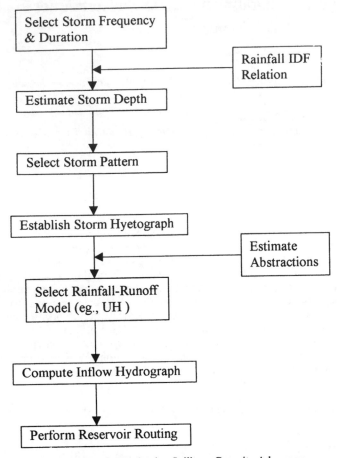

Fig.1:. Procedure for Evaluating Spillway Capacity Adequacy

2. EVALUATION OF SPILLWAY CAPACITY ADEQUACY

Figure 1 shows a typical procedure in engineering practice for designing the spillway capacity and evaluating its adequacy. The procedure involves specifications of representative rainstorm depth-duration-frequency relationship, temporal distribution of storm pattern, rainfall-runoff mechanism, and reservoir storage-discharge characteristics as well as its operations. Due the presence of uncertainties in rainfall depth-frequency, storm pattern and unit hydrograph, the derived reservoir inflow hydrographs are also subject to uncertainty. In this paper, uncertainty features of reservoir inflow hydrograph are quantified which, along with the reservoir storage-outflow characteristics, are used to assess the overtopping risk of a dam.

3. UNCERTAINTY ANALYSIS OF RAINFALL IDF RELATION

In a lump-system approach, basin-wide rainfall IDF relationship for the drainage area upstream of the reservoir is used in determining storm depth of a particular duration and frequency. Depending on the availability of rainfall data within or around the study basin, at-site and/or regional frequency analyses would have to be performed (Stedinger et al., 1993).

For the point (or at-site) analysis, point rainfall IDF relationships are established for gauged locations using annual maximum rainfalls of various durations via proper hydrologic frequency analysis. In this study, the following IDF model is adopted

$$P_{t,T} = (G + H \times \log(T)) \frac{A}{(B+t)^C} \tag{1}$$

in which $P_{t,T}$ = rainfall quantity (rainfall depth, intensity, or intensity ratio) of duration t and return period T; $\theta=(G, H, A, B, C)$ are model parameters. Due to limited amount of point rainstorm data, uncertainty exists in the derived point IDF relations and, hence, model parameters are subject to uncertainty. In this case, procedure such as bootstrap resampling technique (Efron & Tibshirani, 1993) can be applied to quantify the uncertainty features, such as mean, standard deviation, and correlation of the model parameters.

For the regional analysis, point IDF model parameters are used to delineate homogenous regions within which regional equations for IDF model parameters as related to basin physiographical/meteorological characteristics are developed using proper regression analysis. To facilitate uncertainty analysis, statistical features associated with the derived regional equations must be retained.

The point and regional IDF model parameters can be combined at a gauged location by

$$\theta_j = w_{j,s} \times \theta_{j,s} + w_{j,r} \times \theta_{j,r}, \text{ for location j} \tag{2}$$

in which θ_j = weighted IDF model parameters at gauged location j; $w_{j,s}$ and $w_{j,r}$ are, respectively, the weighting factors for the at-site parameters, $\theta_{j,s}$, and regional parameters, $\theta_{j,r}$. The weights can be determined based on the relative uncertainty associated with the at-site and regional parameters. Based on Eq.(2), the basin-wide model parameters, θ_B, can be estimated as

$$\theta_B = \sum_j a_j \times \theta_j \tag{3}$$

where a_j = weight at gauge location j which can be determined by the contributing area (e.g., Thissen polygon) or others. Knowing the uncertainty features of at-site and regional parameters, the uncertainty features of basin-wide parameters can be quantified.

4. UNCERTAINTY ANALYSIS OF STORM PATTERNS

Storm pattern is needed to distribute rainfall depth of duration t and return period T calculated from an adopted IDF model. Ideally, typical storm patterns in the study area should be identified and applied. This can be done by analyzing historical storm events in the area. To avoid the scaling effect of storm duration and depth on inherent storm pattern, non-dimensionalized rainfall mass curves can be derived from historical storms as

$$0 \le \tau' = \tau/t \le 1 \; ; \qquad 0 \le d'_\tau = \Sigma_t \, \delta_t/d \le 1 \tag{4}$$

in which τ' and d'_τ are dimensionless cumulative time and rainfall amount at a particular time instant τ within the duration t of a storm; $\Sigma_t \delta_t$ is the cumulative rainfall amount up to time t; and d is the total rainfall depth for the storm event. Based on the dimensionless rainfall mass curves, representative storm patterns are identified by statistical cluster analysis.

According to the representative storm patterns classified, factors affecting the occurrence of storm patterns are investigated. The uncertainty features of storm pattern can be defined by statistical properties of dimensionless rainfall mass ordinates for each representative storm pattern.

5. UNCERTAINTY ANALYSIS OF RAINFALL-RUNOFF MODELS

Rainfall-runoff models provide mechanisms that transform rainfall of a storm event to a possible surface runoff from the study area either in the form of peak discharge or a complete hydrograph. In dam overtopping risk evaluation, models that allow producing runoff hydrograph from rainfall hyetograph are used. Hydrologic rainfall-runoff models of this type are many, ranging from simple unit hydrograph (UH) to sophisticated watershed simulation models. Without overly complicating the task, this paper considers the conventional UH model.

465

Uncertainty associated with the derived UH of a watershed is attributed to various errors from using limited numbers of storm samples, UH model and its parameters, processing rainfall and runoff data, and measurements of rainfall and streamflow. Therefore, UH derived can vary from one storm to another. Furthermore, derived UH ordinates oscillate and unstable. In this study, the UH representative of the drainage basin upstream of the reservoir is derived by multiple-storm analysis (Zhao et al., 1994) from available rainfall and runoff data using ridge least square procedure as

$$\mathbf{u} = \left(\sum_{r=1}^{R} \mathbf{P}_r^t \, \mathbf{P}_r + k \, \mathbf{I} \right)^{-1} \left(\sum_{r=1}^{R} \mathbf{P}_r^t \, \mathbf{q}_r \right) \tag{5}$$

in which $\mathbf{u}$ = vector of UH ordinates of a selected duration derived from R storm events; $\mathbf{P}_r$ = matrix of effective rainfall hyetograph ordinates of the r-th storm event; $\mathbf{q}_r$ = column vector of direct runoff hydrograph ordinates of the r-th runoff event; $\mathbf{I}$ = an identity matrix; k = ridge constant; and superscript 't' is the transpose of a matrix or vector. In Eq.(5), the optimal ridge constant can be determined by minimizing the mean-squared errors of the UH vector, $\mathbf{u}$, or direct runoff hydrograph vectors, $\mathbf{q}_r$ (Zhao et al., 1994)

Although multiple-storm analysis yields a single UH, its uncertainty is not removed due to the presence of uncertainties mentioned above. To quantifying uncertainty associated with the multi-event UH by Eq.(5), storm re-sampling scheme developed by Zhao et al. (1997) can be utilized. By storm re-sampling scheme, R storm events selected for UH derivation are treated as random events based on which random sampling with replacement is used to generate R bootstrapped storm events for determining the corresponding multi-event UH. The process is repeated a number of times from which the statistical properties of multi-event UH ordinates are summarized.

6. EVALUATION OF DAM OVERTOPPING RISK

In summary, random variables affecting overtopping risk of a dam in this study include rainfall IDF model parameters, ordinates of rainfall mass curve and of UH. The problem is multivariate by nature in that random variables involved are non-normal and are correlated. Furthermore, the process requires reservoir routing which makes analytical-oriented reliability analysis approaches, such as first-order analysis, less applicable. Hence, Monte-Carlo simulation procedure as shown in Fig. 2 is applied in this study. Note that the overtopping risk obtained at the end of flowchart is a conditional probability for a given storm duration and return period. The annual overtopping risk for particular storm duration under consideration can be calculated by

$$p_f(t) = \int_0^1 p_f(t, F) \, dF \tag{6}$$

in which $p_f(t)$ = annual overtopping probability associated with t-hr storm; and $p_f(t,F)$ is the conditional overtopping probability under the t-hr, T-yr storm with F being the corresponding cumulative distribution function value, i.e., $F=1-(1/T)$. To consider further for the unconditional annual overtopping risk, one can integrate $p_f(t)$ over the probability density function that describes the random occurrence of storm duration at the study area.

The technical challenge encountered in this study is to the generation of constrained multivariate non-normal random variables. To generate unconstrained multivariate non-normal random variables, the procedure developed by Chang et al. (1994) are applied by which the marginal PDFs of involved random variables and their correlation coefficients are preserved. The procedure is practical due to the difficulty of obtaining the plausible joint PDF for correlated non-normal random variables. For the constrained case, probabilistically realizable storm patterns and UHs must satisfy the following constraint

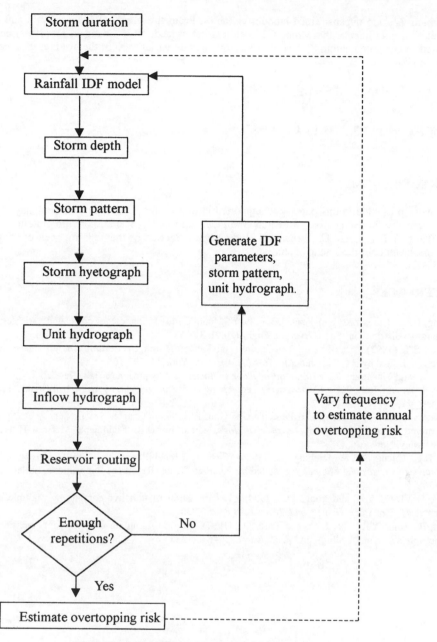

Fig.2: Flowchart of Dam Overtopping Risk Evaluation

$$\sum_t a_t X_t = c \tag{7}$$

in which a_t is the known weighing factor for random variable X_t and c is a known constant. By utilizing log-likelihood ratio transformation of the original random variables, i.e.,

$$Z_t = \ln\left(\frac{a_t X_t}{a_{t*} X_{t*}}\right), \text{ for } t \neq t* \tag{8}$$

in which Z_t's are unconstrained random variables. From the marginal PDFs of Z_t's and their correlation coefficients, the Monte-Carlo simulation procedure of Chang et al. (1994) can be applied to generate random Z_t for $t \neq t^*$. Then, the random variable X_t of the original scale can be obtained by

$$x_t = a_t \times \exp(z_t) \Big/ 1 + \exp\left(\sum_{t \neq t^*} z_t\right), \text{ for } t \neq t^* \tag{9a}$$

$$x_t = a_t \Big/ 1 + \exp\left(\sum_{t \neq t^*} z_t\right), \text{ for } t = t^* \tag{9b}$$

ACKNOWLEDGMENT

The work presented in this paper was supported in part by the Taiwan Power Company (TPC) and in part by the Hong Kong Research Grant Council (RGC). The authors are grateful to Mr. J.D. Yang, Y.T. Lai, and C.T. Chang of Hydrographic Section of the TPC for their continuous support throughout the course of study.

REFERENCES

Chang, C.H., Tung, Y.K., and Yang, J.C. (1994). "Monte Carlo simulation for correlated variables with marginal distributions," *J. of Hydraul. Engr.*, 120(2): 313-331.

Cheng, S.T. (1993). "Statistics on dam failures" In: *Reliability and Uncertainty Analysis in Hydraulic Design*. Edited by B.C. Yen and Y.K. Tung, ASCE, New York, NY. 97-106.

Efron, B. and Tibshirani, R.J. (1993). *Introduction to Bootstrap*, Chapmann & Hall, New York.

Mays, L.W. and Tung, Y.K., *Hydrosystems Engineering and Management*, McGraw-Hill Book Co., New York, NY, 1992.

Stedinger, J.R., Vogel, R.M., and Foufoula-Georgiou, E. (1993) "Chapter 18: Frequency Analysis of Extreme Events." In: *Handbook of Hydrology*, edited by D.R. Maidment, McGraw-Hill Book Company, New York.

Yen, B.C., Cheng, S.T. & Melching, C.S. (1986) "First-order reliability analysis." In: *Stochastic and Risk Analysis in Hydraulic Engineering*, edited by B.C. Yen, Water Resources Publications, Littleton, CO. 1-36,.

Zhao, B., Tung, Y.K. and Yang, J.C. (1994). "Determination of unit hydrographs by multiple storm analysis," *J. of Stoch. Hydrol. and Hydraul.*, 8(4): 269-280.

Zhao, B., Tung, Y.K., Yeh, K.C., and Yang, J.C. (1997). "Storm resampling for uncertainty analysis of a multiple-storm unit hydrograph," *J. of Hydrol.*, 194: 366-384.

Stochastic Hydraulics 2000, Wang & Hu (eds) © 2000 Taylor & Francis, ISBN 90 5809 166 X

Stochastic vs. deterministic approaches of modelling hydrological time series

A.W.Jayawardena
Department of Civil Engineering, University of Hong Kong, China

ABSTRACT:In this paper, an attempt is made to compare the performance of the traditional stochastic approach of analysis and prediction of hydrological time series with that of a deterministic approach. The latter approach is non-linear and uses the dynamical systems theory. Analysis is carried out in the re-constructed phase space and prediction is done in the local neighbourhood. There is strong evidence to suggest that better predictions can be made by the new approach.

Key words: Stochastic modelling, Dynamical systems, Chaos, Noise reduction

1. INTRODUCTION

Traditionally, linear auto-regressive moving average (ARMA) modelling approaches have been widely used in the past for analysis and prediction of hydrological time series. They are well established but the underlying assumptions such as linearity and randomness of the processes may not always be true. Non-linear stochastic approaches such as threshold auto-regressive models have been proposed, but the complexity of the models as well as the lack of insight into the processes limit their application.

An alternative approach is to view the time series as a realization of a non-linear deterministic system and to model and predict using the dynamical systems approach. In this type of analysis, the data is treated as one component of the observable. Analysis is done in the re-constructed phase space, and a great deal of information, which can be useful for modelling and prediction, could be extracted from the invariant measures computed in this space. Prediction is done using an evolutionary type of relationship valid usually (but not necessarily) in the local neighbourhood.

The developments in the dynamical systems approach, which lead to the theory of chaos, have come mainly from mathematics and physics. Their application to hydrology is relatively new, and there are many yet unresolved issues. For example, the effect of noise in the phase space re-construction and the computation of invariant measures, the choice of the optimal time delay and the embedding dimension, and the choice of the optimal number of neighbourhoods for prediction are some of the topics of current research interest. In this paper, an attempt is made to highlight some of these and other related issues in the dynamical systems approach of modelling and to compare the results of prediction made by the traditional stochastic approach and the dynamical systems approach.

2. TOCHASTIC APPROACH

2.1 Decomposition of time series

In the stochastic approach, a time series in general is considered to be composed of both stochastic and deterministic components. The modelling procedure involves decomposing the series into its constituent components, modelling each component individually, and synthesizing back into the original form using the models for each component. In general, the deterministic component consists of trends and/or periodic parts. The stochastic component consists of a dependent structure and a random independent part. The different components can be synthesized as follows:

Time series = Trend + Periodic part + Dependent stochastic part + Independent (residual) random part (1)

Trend, if present, can be modelled by a polynomial of which the straight line is the simplest, and the periodic part can be modelled by a Fourier series. Certain tests are however necessary to determine whether an apparent trend is significant or not. Similarly, a cumulative periodogram test can be carried out to determine the number of significant harmonics needed to adequately describe a periodicity. Both these types of tests are well documented in the literature. Removal of the deterministic component from the original series is a pre-requisite to use linear modelling approaches, which are defined for stationary time series.

2.2 Linear ARMA modelling

A linear stochastic ARMA (p,q) model for a stationary time series can be expressed as (Box et al., 1994):

$$\phi_p(B)z_t = \theta_q(B)a_t \qquad (2)$$

where $\phi_p(B)$ and $\theta_q(B)$ and are polynomials of degree p and q in B; p and q represent the order of the model, and B is the backshift operator which has the property, $B(z_t) = z_{t-1}$, z_t is the observation at time t, and a_t is the random shock (white noise).

Once a model has been identified for a stationary time series using tools such as the auto-correlation function and the partial auto-correlation function, certain diagnostic checks should be carried out to ensure that the remaining residual series is random and that the conditions of stationarity and invertibility are satisfied. As there could be more than one model satisfying the above conditions, it is then necessary to look for the most parsimonious one.

2.3 Parsimony criteria

A widely used method for the determination of the order of a model is the Akaike Information Criterion (*AIC*), given by (Akaike, 1974)

$$AIC(p,q) = N\,ln(\sigma_a^2) + 2(p+q) \qquad (3)$$

where σ_a^2 is the maximum likelihood estimate of the variance of the residual series and N is the length of the data set. The model that gives the minimum value of *AIC* is the parsimonious one. Other criteria include Mallows C_p (Mallows, 1973), Bayesian Information Criterion, BIC (Schwarz, 1978) and the Generalized Degrees of Freedom, GDF (Ye, 1998).

2.4 Prediction

Lead-time predictions are normally obtained by using the Box-Jenkins type of predictor given by

$$\hat{z}_t(l) = \phi_1[z_{t+l-1}] + ... + \phi_p[z_{t+l-p}] - \theta_1[a_{t+l-1}] - .... - \theta_q[a_{t+l-q}] + a_{t+l} \qquad (4)$$

where ϕ_i and θ_i are the coefficients of the model. More details on the stochastic modelling and prediction can be found in the author's earlier work (Jayawardena and Lai, 1989).

3. DETERMINSITIC APPROACH

3.1 Dynamical systems and chaos

An alternative strategy to the stochastic approach of modelling is to view the time series as a realization of a non-linear deterministic system. Such dynamical systems possess extraordinarily rich spectra that are indistinguishable from those of random processes (May, 1976). Recent studies reveal that certain types of time series which appear to be originating from stochastic processes can be modelled better by assuming them to be driven by deterministic processes (Farmer and Sidorowich, 1987; Rodriguez-Iturbe et al., 1989; Casdagli, 1989; Abarbanel et al., 1990; Sugihara and May, 1990; Smith, 1992; Jayawardena and Lai, 1994; Sivakumar et al., 1999; Jayawardena and Gurung, 2000; amongst others). By treating the system that generates the time series as a dynamical one, it is possible to uncover the dynamics of the system, and make more realistic predictions in the short term. Such systems have stable properties, which are predictable with certainty at times but may become 'chaotic' under certain initial conditions. The question then is how to identify whether a series is 'chaotic' or not.

Chaotic systems are necessarily deterministic but not all deterministic systems are chaotic. The first step in treating the time series as chaotic is to diagnose the system; i.e. to determine whether the series is driven by a low dimensional deterministic system, or, by a stochastic system which can have an infinite number of dimensions. This can be done by computing several invariant measures such as the fractal dimension, the correlation dimension, the Lyapunov exponent, and the Kolmogorov entropy. These are all computed in the re-constructed phase space.

3.2 Phase space re-construction

A dynamical system can be described by a phase space diagram whose trajectories describe its evolution from some initial state, which is assumed to be known. Fig. 1 shows the phase space diagram for a data set used in this study. If the trajectories converge to a single sub-space regardless of the initial conditions, then, it is called an attractor. An attractor can be multi-dimensional, and lies in an m-dimensional phase space but has dimension less than m. The dynamics of a time series $x_1, x_2, \ldots, x_n$ are fully captured or embedded in the m-dimensional phase space ($m \geq d$, where d is the dimension of the attractor) defined by the vector

$$Y_t = \{x_t, x_{t-\tau}, x_{t-2\tau} \ldots\ldots\ldots\ldots, x_{t-(m-1)\tau}\} \tag{5}$$

where τ is the delay time. According to the embedding theorem of Takens (1981), a d-dimensional attractor can be embedded into a $(2d+1)$-dimensional phase space to evaluate the characteristics of the dynamical system.

3.3 Invariant measures

There are several invariant measures associated with dynamical systems. Table 1 summarizes the typical values of the three commonly used invariant measures, namely, the correlation dimension, the Kolmogorov entropy and the Lyapunov exponent. The correlation dimension is a variant of the fractal dimension which gives some idea about the complexity of the system and the degrees of freedom required to model it, the Kolmogorov entropy gives a measure of the rate of loss of information, and the Lyapunov exponent gives a measure of the average rate of exponential divergence of initially nearby orbits in the phase space.

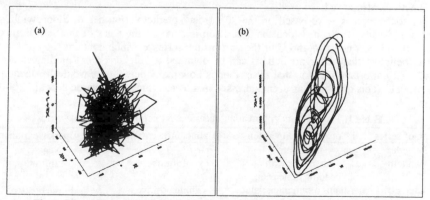

Fig 1: Phase space diagram for S Index data (a) Raw data (b) Noise reduced data

471

These measures of invariants can be computed in the phase space by a number of methods. A widely used method for the computation of the correlation dimension is that proposed by Grassberger and Procaccia (1983a) in which a correlation integral which varies approximately linearly with distance in the log-log scale was defined. The slope of the linear part of this variation is the correlation exponent and its saturation value is the correlation dimension. Wolf et al. (1985) have proposed a method for the estimation of the Lyapunov exponent. Grassberger and Procaccia (1983b) proposed a method for the computation of the second order Renyii entropy K_2 which has characteristics similar to that of Kolmogorov entropy but much easier to compute.

3.4 Effect of noise and noise reduction

The above methods of computing the invariant measures work well for data that are noise free. However, most real data are often contaminated with noise and it is a difficult task to separate the data into noise and signal. The method of re-construction of the phase space would also be unworkable in the presence of noise (Broomhead and King, 1986). Most techniques of computing invariant measures and predictions fail if the data contain as little as 2% noise (Schreiber, 1993a).

The computation of the noise indicators requires knowledge of the clean signal. For theoretical data sets, this is not a problem. For noise corrupted practical data sets however, the clean signal is unknown. Methods of determining noise levels have been proposed by Cawley and Hsu (1992), Schouten et al. (1994) and Schreiber (1993a), among others. Non-linear noise reduction techniques have been proposed by Schreiber and Grassberger (1991), Grassberger et al. (1993), and Schreiber (1993b). In particular, the application of some of these techniques to hydrological time series can be found in the studies carried out by Porporato and Ridolfi (1997), Kawamura et al. (1998), Sivakumar et al. (1999) and Jayawardena and Gurung (2000).

3.5 Prediction

Prediction of a deterministic system can be made using the evolutionary equation of the form

$$\hat{y}_{i+1} = f(y_i) \qquad (6)$$

where $\hat{y}_{i+1}$ is the predicted value, f is the predicting (mapping) function, and y_i is the embedded vector. The prediction process therefore involves an accurate estimation of the mapping function f, which transforms the present and past values to the future value. In a chaotic system, the predictive power is lost very quickly because of sensitivity to initial conditions.

The mapping function can be estimated using local models in which the function approximation at each time step is done from data sets of the local neighbourhood only, in a piecewise manner, or, global models in which the function approximation is done for the whole domain. Local models include linear or polynomial function approximations in the local neighbourhoods whereas global models are generally of the polynomial type although radial basis functions also have been used.

Earlier local models were based on the "0th order" predictor (Farmer & Sidorowich, 1987; Sugihara and May, 1990) in which the prediction is done on the basis of the behaviour of the series in the closest neighbourhood of the vector time series x_t which contains the current value x_t. It is believed that better predictions can be obtained if "higher order" predictors are used instead. An important question that arises then is how many neighbours would be needed for a better model? This question has been addressed in a recent study (Jayawardena et al., 2000).

Table 1: Typical values of invariant measures for certain types of time series

Type of series	Correlation dimension	Kolmogorov entropy	Lyapunov exponent
Periodic	Constant	0	Negative
Random	Increases with embedding dimension, m	Infinite	Infinite
Chaotic	Attains a saturation value after certain m	Finite positive	Finite positive

Table 2: Results of ARMA modelling (the numbers inside the brackets indicate the order of differencing and the period of harmonic analysis)

Time series	Transformation	Model order	AIC	Correlation coefficient	NMSE ($\times 10^{-3}$)
Mekong at Nong Khai	Differencing(2)	AR(14)	50785	0.99	1.26
Mekong at Pakse	Periodicity(365) + Differencing(1)	MA(15)	54631	0.917	1.05
S Index	-	AR(14)	10074	0.67	342

4. APPLICATION AND RESULTS

4.1 Data sets

The two approaches of prediction were applied to daily discharges of Mekong River at Nong Khai (17.87°N and 102.72°E, Basin area, 302,000 km², GRDC # 2969090) in Thailand, and of the same river at Pakse (15.12°N and 108.80°E, Basin area, 545,000 km², GRDC # 2469260) in Lao for the period April 1980 to December 1991, and the monthly mean sea surface temperature (SST) anomaly over the region bounded approximately by 6°N-6°S and 180°-90°W for the period January 1872 to December 1986, which has been defined as S Index by Wright (1984) and used to identify climatic anomalies attributed to El-nino and Southern Oscillation. The first three data sets were obtained from the Global Runoff Data Centre (GRDC) in Germany and the last one from a table compiled by Wright (1989). A few missing records of the data sets were replaced by the long-term averages. Both raw data as well as noise reduced data (Jayawardena and Gurung, 2000) were used in this study.

4.2 Prediction by stochastic approach

The annual periodicities in the daily data sets were removed by differencing and harmonic analysis. The series were then tested for stationarity using the auto-correlation function. Model identification, parameter estimation and diagnostic checks were done using tools such as the auto-correlation function, partial auto-correlation function, the *AIC*, and the modified Portmanteau statistic Q (Ljung and Box, 1978). A summary of the results of stochastic modelling is given in Table 2 and illustrated in Fig. 2. Lead-time predictions for 25 time steps were made using the Box-Jenkins type predictor for each series. The correlation coefficient is between the noise reduced observed series and the predicted one whereas the NMSE is defined as follows (in Tables 2 and 3):

$$NMSE = \frac{1}{N\sigma^2} \sum_{i=1}^{N} (\hat{x}_i - x_i)^2 \tag{7}$$

where $\hat{x}_i$ and x_i are the predicted and observed values, and σ is the standard deviation.

4.3 Prediction by dynamical systems approach

The prediction process of a time series by the dynamical systems approach requires knowledge of three parameters; the time delay τ, the embedding dimension, d_e, and the number of nearest neighbours, N_B. Despite numerous suggestions, there is no rigorous method of determining an optimal value for τ. In this study, a value of unity was assumed for all the data sets, and the embedding dimension was chosen to satisfy the theorem of Takens (1981). An estimate of the optimal value of the embedding dimension can also be obtained by the False Nearest Neighbour (FNN) method proposed by Abarbanel (1995). The third parameter N_B, can take values greater than or equal to unity. A value of $N_B > 1$, is expected to give a better prediction, but the question then is how to determine the optimal number of neighbours. In the results presented in this study, the number of neighbours were varied to yield the maximum correlation coefficient between the observed and the predicted series. The results are shown in Table 3 and Figs. 3. In a separate study (Jayawardena et al., 2000) it has been shown that the choice of the optimum number of neighbours using a criterion based on the Generalized Degrees of Freedom (GDF) proposed by Ye (1998) yield better predictions.

For all the data sets used in this study, the dynamical systems approach of prediction outperformed the stochastic approach. Even with the raw data, the correlation coefficients are higher in the dynamical systems approach than in the stochastic approach. With noise reduced data, the performance is improved and with some data the NMSE is one order smaller for the dynamical systems approach.

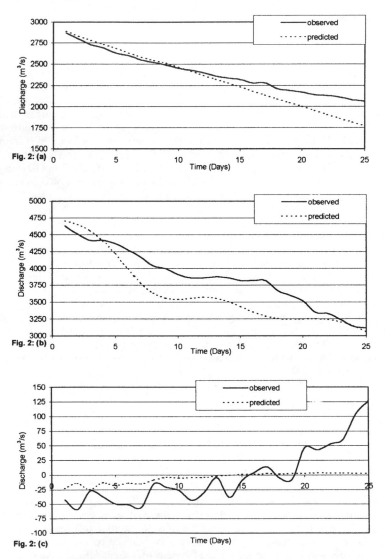

Fig 2: (a) Observed and predicted daily discharges in Mekong River at Nong Khai
(Prediction by ARMA modelling, Day 1 corresponds to Decemeber 7, 1991)
(b) Observed and predicted daily discharges in Mekong River at Pakse
(Prediction by ARMA modelling, Day 1 corresponds to Decemeber 7, 1991)
(c) Observed and predicted monthly S Index (Prediction by ARMA modelling,
Month 1 corresponds to Decemeber, 1984)

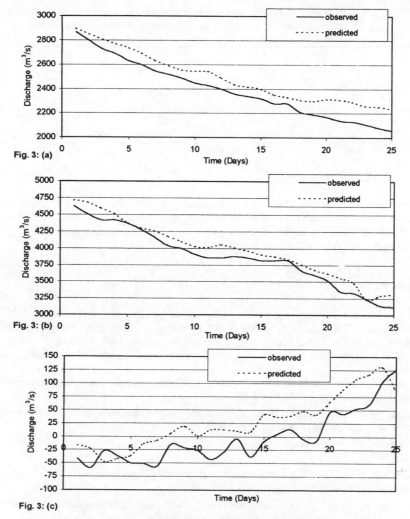

Fig 3: (a) Observed and predicted daily discharges in Mekong River at
Nong Khai(Prediction by 0th order predictor using raw data, Day 1
corresponds to Decemeber 7, 1991)

(b) Observed and predicted daily discharges in Mekong River at Pakse
(Prediction by 0th order predictor using raw data, Day 1 corresponds to
Decemeber 7, 1991)

(c) Observed and predicted monthly S Index (Prediction by 0th order
predictor using raw data, Month 1 corresponds to Decemeber, 1984)

5. CONCLUSION

The comparison of the two approaches of prediction of hydrological time series point to the
conclusion that the dynamical systems approach is a feasible alternative to the traditional
stochastic approach. In fact better short-term predictions can be obtained by the dynamical
systems method for all the data used in this study. There is strong evidence to suggest that the
seemingly stochastic series may be the outcome of low dimensional deterministic systems. Care
should however be taken in confirming such data as driven by deterministic processes because
low dimensional attractor and finite positive Lyapunov exponents have also been observed in
auto-correlated noise. Further work is needed to resolve many such issues and to consolidate
the results obtained.

Table 3: Results of prediction by the deterministic approach (Raw refers to raw data; I, II , and III refer to 3 different methods of noise reduction of which details can be found in the paper by Jayawardena and Gurung (2000))

Time series	Correlation coefficient				NMSE ($\times 10^{-5}$)			
	Raw	I	II	III	Raw	I	II	III
Mekong at Nong Khai	0.991	0.992	0.998	0.994	88	29	35	44
Mekong at Pakse	0.991	0.994	1.000	0.995	17	10	48	42
S Index	0.881	0.972	0.966	0.988	26380	7300	25050	21300

ACKNOWLEDGEMENTS

The computational work contained in this paper was carried out by the author's graduate student, A. B. Gurung. His contribution to the research is highly appreciated. The project was sponsored by the Hong Kong Research Grants Council Grant No 524/96E and their financial support is gratefully acknowledged.

REFERENCES

Abarbanel, H.D.I. (1995): Analysis of observed chaotic data, 2nd Ed., Springer, New York.

Abarbanel, H. D. I., R. Brown, and J. B. Kadtke (1990). Prediction in chaotic nonlinear systems: Methods for time series with broad band Fourier spectra. Physical Review, A 41 (4), 1782-1807.

Akaike, H. (1974). A new look at the statistical model identification. IEEE Trans. Automatic Control, 19, 716-723.

Box, G. E. P., Jenkins, G.M., and Reinsel, G.C. (1994): Time series analysis: Forecasting and control. 3rd Edition, Prentice Hall, Englewood Cliffs, NJ.

Broomhead, D. S. and King, G. P. (1986): Extracting qualitative dynamics from experimental data, Physica D, 20, 217-236. Reprinted in Ott et al. (1994).

Casdagli, M. (1989): Nonlinear prediction of chaotic time series, Physica, D, 35, 335-356. Reprinted in Ott et al. (1994)

Cawley, R. and Hsu, G. H. (1992): Local geometric projection method for noise reduction in chaotic maps and flows, Physical Review, A, 46(6), 3057-3082.

Farmer, J. D.& Sidorowich, J. J. (1987): Predicting chaotic time series. Physical Review Letters, 59, 845-848.

Grassberger, P. & Procaccia, I.(1983a): Measuring the strangeness of strange attractors. Physica, D, 9, 189-208.

Grassberger, P. & Procaccia, I.(1983b): Estimation of Kolmogorov entropy from chaotic signals, Physical Review, A, 28(4), 2591-2593.

Grassberger, P., Hegger, R., Kantz, H., Schaffrath, C. and Schreiber, T (1993): On noise reduction methods for chaotic data, Chaos 3, 127-141. Reprinted in Ott et al. (1994)

Jayawardena, A.W. and Lai Feizhou (1989). Time series analysis of water quality data in Pearl River, China, Journal of Environmental Engineering, ASCE, 115(3), 590-607.

Jayawardena, A. W. and Lai, F. Z. (1994): Analysis and prediction of Chaos in Rainfall and Streamflow Time Series, Journal of Hydrology, 153(1-4), 23-52.

Jayawardena, A. W. and Gurung, A. B. (2000) : Noise reduction and prediction of hydrometeorological time series: dynamical systems approach vs. stochastic approach, Journal of Hydrology, 228 (3-4), 242-264.

Jayawardena, A. W., Li, W. K. and Xu Penchang (2000): Model selection for prediction of hydrological time series (In preparation)

Kawamura, A. McKerchar, A. I., Spigel, R. H. and Jinno, K. (1998): Chaotic characteristics of the Southern Oscillation Index time series, Journal of Hydrology, 204, 168-181.

Ljung, G.M. and Box, G.E.P. (1978): On a measure of lack of fit in time series models, Biometrica, 65(2), 297-303.

Mallows, C. L. (1973): Some comments on C_p, Technics, 15, 661-675.

May, R. M. (1976): Simple mathematical models with very complicated dynamics, Nature, 261, 459-467.

Ott, E., Sauer, T. and Yorke, J. A. (1994): Coping with chaos: analysis of chaotic data and the exploitation of chaotic systems, John Wiley and Sons, Inc.

Porporato, A. and Ridolfi, L. (1997): Non linear analysis of river flow time sequences, Water Resources Research, 33(6), 1353-1367.

Rodriguez-Iturbe, I, Power, B.F.D., Sharifi, M.B. and Georgakakos, K. P. (1989): Chaos in rainfall, Water Resources Research, 25(7), 1667-1675.

Schouten, J.C., Takens, F. & van den Bleek, C. M. (1994) Estimation of the dimension of a noisy attractor. Physical Review, E, 50, 1851-1861.

Schreiber, T. (1993a): Determination of the noise level of chaotic time series, Physical Review, E, 48(1), R13-R16.

Schreiber, T. (1993b): Extremely simple noise reduction method, Physical Review, E, 47(4), 2401-2404.

Schreiber, T. and Grassberger, P. (1991): A simple noise-reduction method for real data, Physical Letters, A, 160, 411-418.

Schwarz, G. (1978): Estimating the dimension of a model, The annals of statistics, 6, 461-464.

Sivakumar, B., S. Y. Liong, C. Y. Liaw, and K. K. Phoon (1999). Singapore rainfall behavior: chaotic? Journal of Hydrologic Engineering, ASCE 4 (1), 38-48.

Smith, L.A.(1992): Identification and prediction of low dimensional dynamics, Physica, D, 58, 50-76.

Sugihara, J. and R. M. May (1990). Nonlinear forecasting as a way of distinguishing chaos from measurement error in time series. Nature, 344 (19), 734-741. Reprinted in Ott et al. (1994).

Takens, F. (1981): Detecting strange attractors in turbulence, In: Rand, D. A., Young, L.S. (Eds). Dynamical systems and turbulence, Lecture notes in mathematics, Proceedings of a symposium held at the University of Warwick, 1979/80, vol 898, Springer, New York, 366-381.

Wolf, A., Swift, J. B., Swinney, H. L. and Vastano, J. A. (1985): Determining Lyapunov exponent from a time series, Physica D, 16, 285-317.

Wright, P.B. (1984): relationship between the indices of the southern oscillation, Monthly Weather Review, 112, 1913-1919.

Wright, P.B. (1989): Homogenized long-period southern oscillation indices, International Journal of Climatology, 9, 33-54.

Ye, Jianming, (1998): On measuring and correcting the effects of data mining and model selection, Journal of the American Statistical Association, 93(441), March 1998, 120-131.

Stochastic Hydraulics 2000, Wang & Hu (eds) © 2000 Taylor & Francis, ISBN 90 5809 166 X

Stochastic modeling of rainfall-runoff with green-ampt infiltration model

Gary C.C.Chan
Department of Civil Engineering, Hong Kong University of Science and Technology, China

Yeou-Koung Tung
Technology Clear Water Bay, Kowloon, Hong Kong, China

ABSTRACT:Infiltration plays an important role on surface and subsurface flow during rainstorm events. Over the years, many models have been proposed to quantify infiltration. From previous studies, significant variations of soil-water retention parameters for infiltration models are present. This implies that the quantification of infiltration could potentially subject to significant amount of uncertainty. This paper considers random nature of soil-water retention parameters in the Green-Ampt equation to quantify uncertainty features of infiltration and effective rainfall hyetograph. Based on the uncertainty analysis, important soil-water retention parameters are identified.

Key words: Infiltration, Rainfall-runoff modeling, Stochastic analysis, Sensitivity analysis

1. INTRODUCTION

Modeling of subsurface flow and infiltration is very important for many engineering problems. Infiltration causes subsurface runoff to enter different parts of the environment, which undergo different processes. The infiltration also has the function of eliminating pollutant, reducing floods, replenishing groundwater supplies and maintaining aquatic habitats. Among many infiltration models of varying degrees of sophistication, the Green-Ampt infiltration model is frequently applied in hydrologic study for quantifying infiltration during storm events. Mathematical expressions of the Green-Ampt infiltration model are

$$f_t = K\left(1 + \frac{\eta \psi_f}{F_t}\right) \tag{1}$$

$$F_t - \eta \psi_f \ln\left(1 + \frac{F_t}{\eta \psi_f}\right) = K t \tag{2}$$

in which f_t and F_t are, respectively, instantaneous potential infiltration rate and cumulative infiltration at time t; η is available porosity (the differences between total porosity, ϕ, and initial volumetric water content, θ_i); ψ_f is suction head or capillary pressure head at the wetting front; and K is unsaturated hydraulic conductivity.

2. SOIL PROPERTIES FOR GREEN-AMPT INFILTRATION MODEL

Application of the Green-Ampt infiltration model requires knowing capillary pressure head at the wetting front (ψ_f), available porosity (η) and hydraulic conductivity (K). They can be computed by some basic soil-water retention parameters, such as initial volumetric water content (θ_i), total porosity (ϕ), saturated hydraulic conductivity (K_s), residual volumetric water content (θ_r), by the following equations (Brakensiek, 1977; Rawls et al., 1983).

$$\eta = \varphi - \theta_i - \theta_r \tag{3}$$

$$\psi_f = \left(\frac{2+3\lambda}{2+2\lambda}\right)\frac{\psi_b}{2} \tag{4}$$

$$K = \frac{K_s}{2} \tag{5}$$

in which λ is the pore size distribution parameter and ψ_b is the bubbling pressure or air-entry value and both parameters are related to the van Genuchten soil-water characteristic model parameters α and N as

$$\psi_b = \frac{1}{\alpha} \tag{6}$$

$$\lambda = N - 1 \tag{7}$$

Evaluations on the basic soil-water retention parameters (K_s, θ_r, α and N) have been reported by Rawls et al. (1983) and Carsel and Parrish (1988). They exhibit significant amount of variability due to inherent heterogeneity of soils in the field. Some data on soil-water retention parameter variability have been published (Nielsen et al., 1973; Warrick et al., 1977; Brakensiek et al. 1981). From the joint probability analysis of Carsel and Parrish (1988), the basic soil-water retention parameters (K_s, θ_r, α and N) were found to be correlated having non-normal marginal distributions. The distributional properties of individual soil-water retention parameters and their correlations are available for various soil types in USDA soil textural triangle.

3. STOCHASTIC MODELING PROCEDURES

To performance stochastic analysis of rainfall-infiltration-runoff process, Monte Carlo simulation (MCS) is used (Ang and Tang, 1984). It is suitable for the uncertainty analysis of flow problems in which the statistical features and the corresponding distribution or the joint probability distribution of the model parameters are known.

In this study, the basic random soil-water retention parameters considered are K_s, θ_r, α, N and ϕ. From the available information in Carsel and Parrish (1988), the first four parameters in a particular soil type are treated as correlated, non-normal random variables. The total porosity, ϕ, is assumed to be independent of all remaining four soil-water retention parameters due to lack of reliable information either empirically or physically.

By the MCS, probabilistically realizable random soil-water retention parameters for a selected soil type are generated according to their probabilistic characteristics. The generated soil-water retention parameters are used in the Green-Ampt equation to calculate pertinent infiltration quantities and the corresponding rainfall excess from a given rainfall hyetograph. The process is repeated a large number of time from which statistical features of model outputs will be summarized and analyzed.

4. NUMERICAL ILLUSTRATION

Table 1 shows the distributional information of five basic soil-water retention parameters for sandy clay and loamy sand. Furthermore, their correlations in the normal-transformed space are shown in Table 2. In this study, a 3-hr deterministic rainfall hyetograph of symmetric shape, peaked at 90min, was used. In MCS, 1000 repetitions were made in generating random soil-water retention parameters for different soil types from which the corresponding model outputs such as potential infiltration rates, cumulative infiltration, and cumulative rainfall excess at some selected times are extracted and analyzed.

480

Table.1: Statistical Information for Soil-Water Parameters (Rawls et al., 1983; and Carsel and Parrish, 1988)

	φ	K_s	$θ_r$	α	N
Sandy clay	0.380[*] (0.050[**]) NOR[#]	0.120 (0.280) LN	0.100 (0.013) SB	0.027 (0.017) LN	1.230 (0.100) LN
Loamy sand	0.410 (0.090) NOR	14.590 (11.360) SB	0.057 (0.015) SB	0.124 (0.043) NOR	2.280 (0.270) SB

      *   Mean values of the parameter in original space
      ** Standard deviation of the parameter in original base
  #   Distribution type: Normal (NOR); Log-Normal (LN); Log-Ratio (SB)

Table.2: Correlation coefficients of soil-water parameters in normal-transformed space   (Carsel and Parrish, 1988)

(a)   Sandy clay

	φ	K_s	$θ_r$	α
K_s	0			
$θ_r$	0	0.939		
α	0	0.957	0.937	
N	0	0.972	0.928	0.932

(b)   Loamy sand

	φ	K_s	$θ_r$	α
K_s	0			
$θ_r$	0	-0.359		
α	0	0.986	-0.301	
N	0	0.730	-0.590	0.354

As shown in Figs. 1 and 2, it is expected that the mean potential infiltration rate, f_t, and cumulative infiltration, F_t, of loamy sand are significantly greater than that of sandy clay. This is primarily due to high clay content in sandy clay as compared with that of in loamy sand. Because of the lower expected infiltration for sandy clay, the corresponding expected cumulative rainfall excess (RE_t) is considerably higher.

Fig.3 shows the time variation of coefficient of variation (COV) of f_t, F_t and RE_t. For loamy sand, COV of RE_t before t=90min is significantly higher than those after t=90min primarily due to very low expected rainfall excess for t<90min as shown in Fig. 2. Because of high expected rainfall excess values throughout the entire storm as compared with those of loamy sand, their corresponding values of COV for sandy clay are relatively low. Similar argument can be used to explain the COV behavior of F_t associated with the two soils because $Var(F_t)=Var(RE_t)$. However, it is interesting to see that, between the two soil types considered, the values of COV for potential infiltration rate are relatively close.

5. SENSITIVITY ANALYSIS

Based on the simulated infiltration model outputs and the corresponding soil-water retention parameters in the MCS for different soil textures, the issue of model output sensitivity with respect to change in model parameters can be addressed. In this study, a global sensitivity analysis was conducted by which mathematical expressions relating the model outputs (i.e. f_t, F_t and RE_t) to model parameters (i.e. φ, K_s, $θ_r$, α and N) are established by multiple regression technique at three selected times, namely t=45, 90 and 135min. To accurately assess sensitivity levels of model outputs, it is necessary to establish a representative relationship.

During the process of developing representative regression equations, the 1000 sets of soil-water retention parameters were standardized to avoid scale effect of different model parameters.

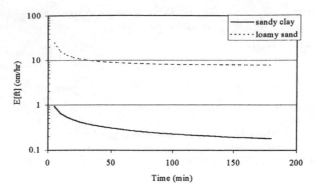

Fig.1: Mean Potential Infiltration Rate(f_t)

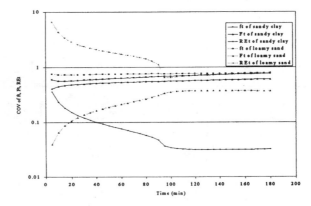

Fig.2: Mean Cumulative Infiltration (F_t) and Rainfall Excess (RE_t)

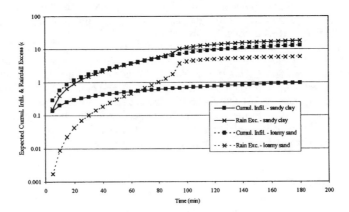

Fig.3: Coefficient of Variation of f_t, F_t and RE_t

In this study, the overall general mathematical expression is expressed as

$$Y = \beta_0 + \beta_1 \varphi' + \beta_2 Ks' + \beta_3 \theta_r' + \beta_4 \alpha' + \beta_5 N' + \beta_6 \varphi'^2 + \beta_7 Ks'^2 + \beta_8 \theta_r'^2 + \beta_9 \alpha'^2 + \beta_{10} N'^2$$
$$+ \beta_{11} \ln(N'^2) + \beta_{12} \ln(Ks'^2) \ln(N'^2) \qquad (8)$$

in which $x' = (x - x)/s_x$, a standardized variable with sample mean and standard deviation equal to x and s_x, respectively. The three quantities for the dependent variables (Y) in the regression equations are f_t, F_t and RE_t. The resulting values of regression coefficients for sandy clay and loamy sand, along with standard error (S_e) and adjacent R^2, are listed in Table 3 for different Green-Ampt infiltration model outputs at various selected times. In the great majority of the cases considered, the resulting regression equations are adequate enough to describe the general behavior of the model because of high R^2 values.. Although the equation form as shown in Eq. (8) may not be best for all soil types and model outputs as seen from varying R^2 values. However, it probably represents the best yet simple form for describing the relationship between the model outputs and inputs for conducting global sensitivity study.

The regression coefficients β_1-β_5 represent sensitivity of model outputs due to one unit standard deviation change of the corresponding soil-water retention parameter. The sign of regression coefficients shows the direction of change. In all cases considered, the first-order sensitivity coefficients, except K_s for F_{90} and F_{135}, are significantly higher than the second-order sensitivity coefficients for both soil types. Among the five soil-water retention parameters, the Green-Ampt model outputs for sandy clay is the least sensitive to residual moisture content, θ_r, while the most sensitive to saturated hydraulic conductivity, K_s. For both soil types, the regression coefficients have exactly the opposite signs between the cumulative infiltration and rainfall excess as one might expect.

For loamy sand, the picture is a bit more complicated than the sandy clay because the relative sensitivity of Green-Ampt model outputs to soil-water retention parameters varies with model output type and the time. For potential infiltration rate, f_t, it is the least affected by the variation of residual water content, θ_r. However, its importance supercedes that of total porosity for the cumulative infiltration, F_t.

One could also examine the trend of regression coefficients with respect to time to have some insight about the behavior of the model. From the first-order sensitivity coefficients, the sensitivity of potential infiltration rate with respect to change in all five soil-water retention parameters decreases in time whereas the cumulative infiltration has increasing trend in time, but in a diminishing rate.

Because all independent variables are in the standardized scale, one can make direct cross comparison of Tables 3(a) and 3(b) to examine the relative importance of soil-water retention parameters between different soil types. In all cases considered, the Green-Ampt model outputs are much sensitive to the five soil-water retention parameters in loamy sand than in sandy clay.

(a) Sandy Clay — **Table.3:** Sensitivity coefficients from regression analysis

	β_0	ϕ'	K_s'	θ_r'	α'	N'	ϕ'^2	$K_s'^2$	$\theta_r'^2$	α'^2	N'^2	$\ln(N'^2)$	β_{12}	S_e	R^2
f_{45}	0.348	0.060	0.222	0.022	-0.049	0.054	-0.006	-0.005	0.007	-0.005	-0.017	0.009	0.003	0.034	97.1
f_{90}	0.257	0.042	0.180	0.016	-0.034	0.038	-0.004	-0.003	0.005	-0.003	-0.012	0.006	0.002	0.023	97.9
f_{135}	0.217	0.034	0.162	0.013	-0.028	0.031	-0.003	-0.003	0.004	-0.002	-0.010	0.005	0.001	0.019	98.4
F_{45}	0.487	0.087	0.258	0.033	-0.069	0.081	-0.009	-0.008	0.011	-0.007	-0.251	0.013	0.004	0.049	95.6
F_{90}	0.707	0.124	0.406	0.047	-0.099	0.115	-0.012	-0.011	0.016	-0.009	-0.035	0.019	0.005	0.069	96.4
F_{135}	0.883	0.152	0.534	0.057	-0.122	0.140	-0.015	-0.014	0.019	-0.011	-0.043	0.023	0.006	0.085	96.8
RE_{45}	2.456	-0.087	-0.258	-0.033	0.069	-0.081	0.009	0.008	-0.011	0.007	0.025	-0.013	-0.004	0.049	95.6
RE_{90}	7.569	-0.124	-0.406	-0.047	0.099	-0.115	0.012	0.011	-0.016	0.009	0.035	-0.019	-0.005	0.069	96.4
RE_{135}	14.95	-0.152	-0.534	-0.057	0.122	-0.140	0.015	0.136	-0.019	0.011	0.043	-0.023	-0.006	0.085	96.8

(b) Loamy Sand

	β_0	ϕ'	K_s'	θ_r'	α'	N'	ϕ'^2	$K_s'^2$	$\theta_r'^2$	α'^2	N'^2	$\ln(N'^2)$	β_{12}	S_e	R^2
f_{45}	9.354	0.896	7.125	0.096	-0.312	0.176	-0.017	-0.268	-0.021	0.110	0.036	0.016	0.003	0.641	99.1
f_{90}	8.154	0.352	6.274	0.036	-0.130	0.070	-0.010	-0.071	-0.007	0.048	0.010	0.004	0.001	0.217	99.9
f_{135}	7.862	0.222	6.076	0.033	-0.074	0.063	-0.003	-0.043	-0.003	0.040	0.004	0.003	0.001	0.139	99.9
F_{45}	2.784	0.088	-0.809	0.187	0.864	0.630	-0.020	-0.064	0.001	0.036	-0.042	-0.011	-0.003	0.187	84.8
F_{90}	7.115	0.217	-0.368	0.429	1.941	1.422	-0.027	-0.755	0.005	0.216	-0.068	0.017	0.014	0.360	96.7
F_{135}	11.69	0.284	1.051	0.515	2.687	1.879	-0.048	-1.443	-0.002	0.361	-0.068	0.047	0.028	0.711	96.8
RE_{45}	0.158	-0.088	0.809	-0.187	-0.864	-0.630	0.020	0.064	-0.001	-0.036	0.042	0.011	0.003	0.187	84.8
RE_{90}	1.162	-0.217	0.368	-0.429	-1.941	-1.422	0.027	0.755	-0.005	-0.216	0.068	-0.017	-0.014	0.360	96.7
RE_{135}	4.141	-0.284	-1.051	-0.515	-2.687	-1.879	0.048	1.443	0.002	-0.361	0.068	-0.047	-0.028	0.711	96.8

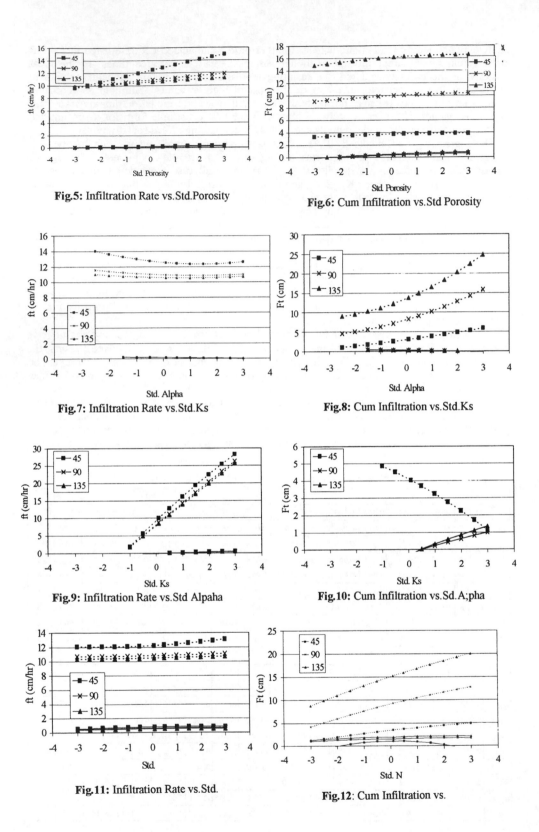

Fig.5: Infiltration Rate vs.Std.Porosity

Fig.6: Cum Infiltration vs.Std Porosity

Fig.7: Infiltration Rate vs.Std.Ks

Fig.8: Cum Infiltration vs.Std.Ks

Fig.9: Infiltration Rate vs.Std Alpaha

Fig.10: Cum Infiltration vs.Sd.A;pha

Fig.11: Infiltration Rate vs.Std.

Fig.12: Cum Infiltration vs.

With a much simpler, yet relatively accurate, regression equation as Eq.(8), one can easily evaluate the sensitivity of the Green-Ampt model outputs with respect various model parameters. Because of some interactive terms are involved in Eq.(8), the picture of sensitivity may not be as clear as a simple linear and quadratic equation. For visualization, Some graphical displays of f_t and F_t with varying standardized soil-water parameters for both sandy clay (solid lines) and loamy sand (dash lines) by the regression equations are shown in Figs. 5 to 12.

6. CONCLUSIONS

Stochastic modeling of the Green-Ampt infiltration model using Monte Carlo simulation was conducted. The procedure allows quantification of uncertainty features of infiltration and rainfall excess that would be used in the design and analysis of various hydraulic structures. Information such as this is essential to have important insight about the safety of the structures. Furthermore, stochastic analysis provides information needed to examine the sensitivity of model outputs with respect to change in model inputs/parameters subject to uncertainty. The paper demonstrates the framework using the Green-Ampt infiltration equation and shows how to identify important soil-water model.

REFERENCES

Ang, A.H., and Tang, W.H. (1984), Probability Concepts in Engineering Planning and Design, Volume II – Decision, Risk, and Reliability, John Wiley & Sons, NY.

Brakensiek, D.L. (1977), "Estimating the effective capillary pressure in the Green and Ampt infiltration equation," Water Resour. Res., 13(3), 680-682.

Brakensiek, D.L., Engleman, R.L., and Rawls, W.J. (1981), "Variation within texture classes of soil water parameters," Trans. of Am. Soc. Ag. Eng., 24(2), 335-339.

Carsel, F.C., and Parrish, S.P. (1988), "Developing joint probability distributions of soil water retention characteristics," Water Resour. Res., 24(5), 755-769.

Chow, V.T., Maidment, D.R., and Mays, L.W. (1988), Applied Hydrology, McGraw-Hill Book Company, Singapore.

Fredlund, D.G., and Rahardjo, H. (1993), Soil Mechanics for Unsaturated Soils, John Wiley & Sons, NY.

Nielsen, D.R., Biggar, J.W., and Erh, K.T. (1973), "Spatial variability of field measured soil-water properties," Hilgardia, 42(2), 215-259.

Rawls, W.J., Brakensiek, D.L., and Miller, N. (1983), "Green-Ampt infiltration parameters form soils data," J. of Hydraul. Eng., 109(1), 62-70.

van Genuchten, M.T. (1980), "A closed-form equation for predicting the hydraulic conductivity of unsaturated soils," Soil Sci. Soc. J. 44, 892-898.

Warrick, A.W., Mullen, G.J. and Nielsen, D.R. (1977), "Scaling field-measured soil hydraulic properties using a similar media concept," Water Resour. Res., 13(2), 355-362.

Stochastic Hydraulics 2000, Wang & Hu (eds) © 2000 Taylor & Francis, ISBN 90 5809 166 X

Stochastic response of runoff model to mutually dependent rainfall input

Gaku Tanaka & Mutsuhiro Fujita
Graduate School of Engineering, Hokkaido University, Sapporo, Japan

Masaru Kaido
Obayashi Corporation, Tokyo, Japan

ABSTRACT: Using three runoff models, we present differential equations whose solutions provide the first four moments of discharge under the impact of mutually dependent rainfall inputs. This paper shows that the regression coefficient, ρ, plays an important role in the stochastic response property of discharge. These equations enable estimation of the probability density function of discharge.

Key words: Stochastic property, Runoff model, AR(1) process, Rainfall.

1. INTRODUCTION

The planning of flood-control projects requires a design flood. To estimate the design flood, we must provide the probability density function of discharge. However, it is very difficult to estimate the probability density function of discharge even if the stochastic property of rainfall is known. Fujita et al.[1] used the storage function runoff model and proposed a method of estimating the probability density function of discharge by calculating the first four moments of discharge respectively on the condition of mutually independent rainfall inputs. However, rainfall that causes floods is a non-stationary and mutually dependent random variable. Therefore, the results derived by Fujita et al (1994) do not provide sufficient solutions.

Using several runoff models, such as the Kinematic wave model and two types of storage function runoff model obtained from lumping the Kinematic wave model, we derived differential equations that gave the first four moments of discharge under the impact of non-stationary and mutually dependent rainfall inputs. These differential equations enable estimation of the probability density function of discharge. In this paper, we assume that rainfall inputs belong to the first order auto-regressive process, AR(1). This assumption is verified by the statistical analysis of observed rainfall data. The results obtained in this paper show that the regression coefficient, ρ, plays an important role in the stochastic response property of discharge. It is possible to estimate the design flood.

2. STOCHASTIC PROPERTY OF RAINFALL

Figure 1 shows the schematic illustration of the calculation process for runoff. That is, the calculated discharge is an output that passes through two low pass filters, a discrete system and a runoff system. If rainfall input is a random variable, discharge output is also described by a random process. Observed rainfall, $r_{d,i}$, as a discrete time series is related to rainfall, r, by the following equation.

$$r_{d,i} = \frac{1}{\Delta t} \int_{(i-1)\Delta t}^{i\Delta t} r(\tau) d\tau \tag{1}$$

$$r_d(t) = \sum_i r_{d,i} \{U(t - (i-1)\Delta t) - U(t - i\Delta t)\} \tag{2}$$

$r_{d,i}$: observed rainfall; U : step function;

Δt : time interval of $r_{d,i}$; i : integer

$r_{d,i}$ $(i = 1,2,3,\cdots)$ arises from accumulating a variable r over a period of time, Δt . If r is the random variable, $r_d(t)$ is a random step function that has a time interval, Δt . r and $r_{d,i}$ are described by

$$r(t) = \bar{r}(t) + \tilde{r}(t) , \quad E\{\tilde{r}(t)\} = 0 \tag{3}$$

$$r_{d,i} = \bar{r}_{d,i} + \tilde{r}_{d,i} , \quad E\{\tilde{r}_{d,i}\} = 0 \tag{4}$$

In which the bar sign and the tilde sign show the mean and the deviation from this mean. From eqs. (1), (3) and (4), the following equations are obtained.

$$\bar{r}_{d,i} = \frac{1}{\Delta t} \int_{(i-1)\Delta t}^{i\Delta t} \bar{r}(\tau)d\tau \tag{5}$$

$$\tilde{r}_{d,i} = \frac{1}{\Delta t} \int_{(i-1)\Delta t}^{i\Delta t} \tilde{r}(\tau)d\tau \tag{6}$$

Figure 2 shows an example of $r_{d,i}$ at Jozankei Dam basin. The solid line in this figure indicates the third order moving average. Figure 3 shows the auto-correlation function of $r_{d,i}$ under the assumption that the third order moving average in Figure 2 is the mean of $r_{d,i}$. From Figure 3, the auto-correlation function of $r_{d,i}$ is expressed by

$$E\{\tilde{r}_{d,i}\tilde{r}_{d,j}\} = \sigma_{r_d}^{\;2} \rho^{|i-j|} \tag{7}$$

ρ : regression coefficient,

$\sigma_{r_d}^{\;2}$: variance of $r_{d,i}$

Eq. (7) suggests that the time series of $\tilde{r}_{d,i}$ is described by the first order auto-regressive process, AR(1).

$$\tilde{r}_{d,i} = \rho\tilde{r}_{d,i-1} + N_i \tag{8}$$

N_i : noise component

It is easy to calculate the higher order cumulant function of $r_{d,i}$ from eq. (8). Here, special attention should be paid to the difference between the stochastic properties of $r_{d,i}$ and r . Tanaka et al (1997) proposed a method for estimating the higher order cumulant function of r . If we focus on the second order cumulant function of r , then $E\{\tilde{r}(\tau_1)\tilde{r}(\tau_2)\}$ satisfies eq. (6).

$$E\{\tilde{r}_{d,i}\tilde{r}_{d,j}\}$$
$$= \frac{1}{\Delta t^2} \int_{(i-1)\Delta t}^{i\Delta t} \int_{(j-1)\Delta t}^{j\Delta t} E\{\tilde{r}(\tau_1)\tilde{r}(\tau_2)\}d\tau_2 d\tau_1 \tag{9}$$

$E\{\tilde{r}(\tau_1)\tilde{r}(\tau_2)\}$ may be estimated by solving this integral equation (9). The left side of eq. (9) is known and is derived easily from eq. (7). In this study, we adopt the following equations presented by Tanaka et al (1999a) as higher order cumulant function of r .

$$E\{\tilde{r}(\tau_1)\tilde{r}(\tau_2)\}$$
$$= \frac{\Delta t}{c} \sum_{n=0}^{\infty} \sigma_{r_d}^{\;2} A\delta(\tau_1 - \tau_2 - n\Delta t)e^{-\gamma(\tau_1-\tau_2)} \tag{10}$$

$$E\{\tilde{r}(\tau_1)\tilde{r}(\tau_2)\tilde{r}(\tau_3)\}$$
$$= \frac{\Delta t^2}{c^2} \sum_{m=0}^{\infty}\sum_{n=0}^{m} \mu_{r_d 3} A\delta(\tau_1 - \tau_2 - n\Delta t) \tag{11}$$
$$\times \delta(\tau_2 - \tau_3 - (m-n)\Delta t)e^{-\gamma(\tau_1+\tau_2-2\tau_3)}$$

$$E\{\tilde{r}(\tau_1)\tilde{r}(\tau_2)\tilde{r}(\tau_3)\tilde{r}(\tau_4)\}$$

$$= \frac{\Delta t^3}{c^3} \sum_{l=0}^{\infty} \sum_{m=0}^{l} \sum_{n=0}^{m} B\delta(\tau_1 - \tau_2 - n\Delta t)$$
$$\times \delta(\tau_2 - \tau_3 - (m-n)\Delta t)$$
$$\times \delta(\tau_3 - \tau_4 - (l-m)\Delta t) \qquad (12)$$
$$\times \left\{ (\mu_{r_d 4} - 3\sigma_{r_d}^{\ 4}) e^{-\gamma(\tau_1 + \tau_2 + \tau_3 - 3\tau_4)} \right.$$
$$+ 2\sigma_{r_d}^{\ 4} e^{-\gamma(\tau_1 + \tau_2 - \tau_3 - \tau_4)}$$
$$\left. + \sigma_{r_d}^{\ 4} e^{-\gamma(\tau_1 - \tau_2 + \tau_3 - \tau_4)} \right\}$$

in which,

$$\gamma = -\frac{\log|\rho|}{\Delta t}, \quad A = \begin{cases} 1 & (\rho > 0) \\ (-1)^n & (\rho < 0) \end{cases},$$
$$B = \begin{cases} 1 & (\rho > 0) \\ (-1)^{l+m+n} & (\rho < 0) \end{cases}, \quad \tau_1 \geq \tau_2 \geq \tau_3 \geq \tau_4$$

c : constant with a time dimension;

$\mu_{r_d 3}$, $\mu_{r_d 4}$: third and fourth order moments of $r_{d.i}$

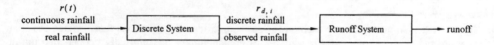

Fig.1: Schematic illustration of calculation process for runoff

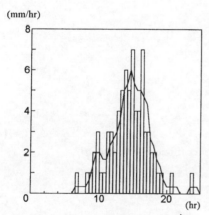

Fig. 2: Observed rainfall at Jozankei Dam basin in 1994

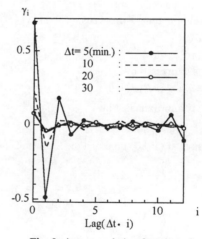

Fig. 3: Auto-correlation function of observed rainfall

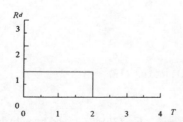

Fig. 4: Rectangular average rainfall

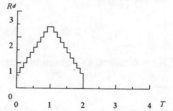

Fig. 5: Traiangular average rainfall

489

3. STOCHASTIC RESPONSE OF RUNOFF MODEL

The Kinematic wave model is described by

$$\frac{\partial h}{\partial t} + \frac{\partial q}{\partial x} = r \tag{13}$$

$$h = \varepsilon q^p, \quad 0 \le x \le l \tag{14}$$

$$s(t) = \int_0^l h(x,t)dx \tag{15}$$

h : water depth; q : discharge per unit width; r : rainfall (rainfall intensity); l : slope length; s : storage per unit width; ε, p : constants

In addition to the above runoff model, two storage function runoff models derived from lumping eqs. (13) and (14) are considered.

$$\frac{ds_h}{dt} + q_h = r \tag{16}$$

$$s_h = k_1 q_h^{p_1} \tag{17}$$

$$s_h = k_1 q_h^{p_1} + k_2 \frac{dq_h^{p_2}}{dt} \tag{18}$$

s_h : storage; q_h : depth of runoff;

k_1, k_2, p_1, p_2 : constants

In the next section, we show a method for deriving differential equations that provide the first four moments of discharge under the impact of non-stationary and mutually dependent rainfall inputs.

3.1 Stochastic response of Kinematic wave model

By eliminating h from eqs. (13) and (14),

$$\frac{\partial(\varepsilon q^p)}{\partial t} + \frac{\partial q}{\partial x} = r \tag{19}$$

q is expressed by

$$q = \bar{q} + \tilde{q}, \quad E\{\tilde{q}\} = 0 \tag{20}$$

q^p is approximated by

$$q^p = \alpha_1 \bar{q} + \beta_1 \tilde{q} \tag{21}$$

Bras et al (1980) proposed the following equations for coefficients, α_1 and β_1.

$$\alpha_1 = \bar{q}^{p-1} \left\{ 1 + \frac{1}{2} p(p-1) \frac{E(\tilde{q}^2)}{\bar{q}^2} \right.$$

$$\left. + \frac{1}{6} p(p-1)(p-2) \frac{E(\tilde{q}^3)}{\bar{q}^3} \cdots \right\} \tag{22}$$

$$\beta_1 = \frac{\bar{q}^{p+1}}{E(\tilde{q}^2)} \left\{ \frac{E(\tilde{q}^2)}{\bar{q}^2} + \frac{1}{2} p(p-1) \frac{E(\tilde{q}^3)}{\bar{q}^3} \right.$$

$$\left. + \frac{1}{6} p(p-1)(p-2) \frac{E(\tilde{q}^4)}{\bar{q}^4} \cdots \right\} \tag{23}$$

Eq. (24) is obtained by substituting eqs. (3), (20) and (21) into eq. (19).

$$\frac{\partial \{\varepsilon(\alpha_1 \bar{q} + \beta_1 \tilde{q})\}}{\partial t} + \frac{\partial(\bar{q} + \tilde{q})}{\partial x} = \bar{r} + \tilde{r} \tag{24}$$

By taking the expectation of eq. (24), eq. (25) is obtained. Eq. (26) is derived by subtracting

490

eq. (25) from eq. (24).

$$\frac{\partial(\varepsilon\alpha_1\overline{q})}{\partial t} + \frac{\partial\overline{q}}{\partial x} = \overline{r} \tag{25}$$

$$\frac{\partial(\varepsilon\beta_1\widetilde{q})}{\partial t} + \frac{\partial\widetilde{q}}{\partial x} = \widetilde{r} \tag{26}$$

Eq. (26) can be decomposed as the following simultaneous differential equations.

$$\frac{d\widetilde{q}}{dt} + \frac{1}{\beta_1}\frac{\partial\beta_1}{\partial t}\widetilde{q} = \frac{\widetilde{r}}{\varepsilon\beta_1} \tag{27}$$

$$\frac{dx}{dt} = \frac{1}{\varepsilon\beta_1}, \quad x \geq x_0, \quad t \geq t_0 \tag{28}$$

(x_0, t_0): origin of a characteristic curve

The solution of eq. (27) is

$$\widetilde{q} = e^{-\int^t D_1(\tau)d\tau} \int_{t_0}^t \frac{\widetilde{r}(\tau_1)}{\varepsilon\beta_1} e^{\int^{\tau_1} D_1(\tau_2)d\tau_2} d\tau_1 \tag{29}$$

$$D_1(t) = \frac{1}{\beta_1}\frac{\partial\beta_1}{\partial t} \tag{30}$$

After raising eq. (29) to the second, third and fourth power and taking the expectation of each of them, it is possible to estimate the second, third and fourth order moments of q. If we focus on the second order moment of q, σ_q^2, we can derive

$$\sigma_q^2 = e^{-2\int^t D_1(\tau)d\tau} \int_{t_0}^t \int_{t_0}^t \frac{E\{\widetilde{r}(\tau_1)\widetilde{r}(\tau_2)\}}{\varepsilon^2\beta_1(\tau_1)\beta_1(\tau_2)}$$

$$\times e^{\int^{\tau_1} D_1(\tau_3)d\tau_3 + \int^{\tau_2} D_1(\tau_4)d\tau_4} d\tau_2 d\tau_1 \tag{31}$$

The following equations are obtained from eqs. (10) and (31).

$$\frac{d\sigma_q^2}{dt} + \frac{2}{\beta_1}\frac{\partial\beta_1}{\partial t}\sigma_q^2 = \frac{\Delta t}{(\varepsilon\beta_1)^2}(\sigma_{r_d}^2 + 2Z_{1,1}) \tag{32}$$

$$\frac{dU_{1,1}}{dt} + \frac{1}{\beta_1}\frac{\partial\beta_1}{\partial t}U_{1,1} = 0, \quad U_{1,1}(0) = 1 \tag{33}$$

where, $t_1 = t - t_0$, $U_{1,2} = \beta_1 U_{1,1}$

$$Z_{1,1} = \begin{cases} 0, & [t_1/\Delta t] < 1 \\ \sigma_{r_d}^2 \sum_{i=1}^{[t_1/\Delta t]} \frac{\rho^i U_{1,2}(t)}{U_{1,2}(t-i\Delta t)}, & [t_1/\Delta t] \geq 1 \end{cases}$$

[*] : Gauss Notation

It is possible to estimate σ_q^2 by solving eqs. (28), (32) and (33). For further details of third and fourth order moments of q, μ_{q3} and μ_{q4}, the reader should refer to Tanaka et al (1999a)

3.2 Stochastic response of storage function runoff model, eqs. (16) and (17)

We consider q_h and s_h to be random variables.

$$q_h = \overline{q}_h + \widetilde{q}_h, \quad E\{\widetilde{q}_h\} = 0 \tag{34}$$

$$s_h = \overline{s}_h + \widetilde{s}_h, \quad E\{\widetilde{s}_h\} = 0 \tag{35}$$

By eliminating q_h from eqs. (16) and (17),

$$\frac{ds_h}{dt} + D_2 s_h^{m_1} = r \tag{36}$$

Where, $D_2 = \left(\dfrac{1}{k_1}\right)^{m_1}$, $m_1 = \dfrac{1}{p_1}$

The same approximation is used for $s_h^{m_1}$.

$$s_h^{m_1} = \alpha_2 \bar{s}_h + \beta_2 \tilde{s}_h \tag{37}$$

Coefficients, α_2 and β_2, are obtained by substituting m_1, $\bar{s}_h$ and $\tilde{s}_h$ for p, $\bar{q}$ and $\tilde{q}$ in eqs. (22) and (23), respectively. Using eqs. (3), (35) and (37), eq. (36) is rewritten as follows:

$$\frac{d(\bar{s}_h + \tilde{s}_h)}{dt} + D_2(\alpha_2 \bar{s}_h + \beta_2 \tilde{s}_h) = \bar{r} + \tilde{r} \tag{38}$$

By the same way as in the previous section, eqs. (39) and (40) are obtained.

$$\frac{d\bar{s}_h}{dt} + D_2\alpha_2 \bar{s}_h = \bar{r} \tag{39}$$

$$\frac{d\tilde{s}_h}{dt} + D_2\beta_2 \tilde{s}_h = \tilde{r} \tag{40}$$

The solution of eq. (40) is

$$\tilde{s}_h = e^{-\int_0^t D_2\beta_2 d\tau} \int_0^t \tilde{r}(\tau_1) e^{\int_0^{\tau_1} D_2\beta_2 d\tau_2} d\tau_1 \tag{41}$$

On the other hand, the following equations are obtained from eqs. (17), (34) and (37).

$$\bar{q}_h + \tilde{q}_h = D_2(\alpha_2 \bar{s}_h + \beta_2 \tilde{s}_h) \tag{42}$$
$$\bar{q}_h = D_2\alpha_2 \bar{s}_h \tag{43}$$
$$\tilde{q}_h = D_2\beta_2 \tilde{s}_h \tag{44}$$

After raising eq. (41) to the second, third and fourth power and taking the expectation of each of them, it is possible to estimate the second, third and fourth order moments of s_h. Furthermore, the stochastic properties between q_h and s_h are obtained by using eqs. (43) and (44). Let us focus on the second order moment of q_h, $\sigma_{q_h}^2$. The following equations can be derived.

$$\sigma_{s_h}^2 = e^{-2\int_0^t D_2\beta_2 d\tau} \int_0^t \int_0^t E\{\tilde{r}(\tau_1)\tilde{r}(\tau_2)\}$$
$$\times e^{\int_0^{\tau_1} D_2\beta_2 d\tau_3 + \int_0^{\tau_2} D_2\beta_2 d\tau_4} d\tau_2 d\tau_1 \tag{45}$$

$$\sigma_{q_h}^2 = (D_2\beta_2)^2 \sigma_{s_h}^2 \tag{46}$$

Using eq. (10) as $E\{\tilde{r}(\tau_1)\tilde{r}(\tau_2)\}$, the following equations are obtained.

$$\frac{d\sigma_{s_h}^2}{dt} + 2\left(\frac{1}{k_1}\right)^{m_1} \beta_2 \sigma_{s_h}^2 = \Delta t(\sigma_{r_d}^2 + 2Z_{2,1}) \tag{47}$$

$$\frac{dU_{2,1}}{dt} + \left(\frac{1}{k_1}\right)^{m_1} \beta_2 U_{2,1} = 0, \quad U_{2,1}(0) = 1 \tag{48}$$

in which,

$$Z_{2,1} = \begin{cases} 0, & [t/\Delta t] < 1 \\ \sigma_{r_d}^2 \displaystyle\sum_{i=1}^{[t/\Delta t]} \frac{\rho^i U_{2,1}(t)}{U_{2,1}(t-i\Delta t)}, & [t/\Delta t] \geq 1 \end{cases}$$

It is possible to estimate $\sigma_{q_h}^2$ by solving eqs. (46) to (48). For further details of third and fourth order moments of q_h, $\mu_{q_h 3}$ and $\mu_{q_h 4}$, the reader should refer to Tanaka et al (1999a)

3.3 Stochastic response of storage function runoff model, eqs. (16) and (18)

Eq. (49) is derived by eliminating s_h from eq. (16) and (18).

$$k_2 \frac{d^2 q_h^{P_2}}{dt^2} + k_1 \frac{dq_h^{P_1}}{dt} + q_h = r \tag{49}$$

$q_h^{P_1}$ and $q_h^{P_2}$ are considered to be approximated by the following equations.

$$q_h^{P_1} = \alpha_3 \bar{q}_h + \beta_3 \tilde{q}_h \quad (50) \quad q_h^{P_2} = \alpha_4 \bar{q}_h + \beta_4 \tilde{q}_h \tag{51}$$

Coefficients, α_3, β_3, α_4 and β_4, in the above equations are obtained by substituting p_1, $\bar{q}_h$, $\tilde{q}_h$ and p_2, $\bar{q}_h$, $\tilde{q}_h$ for p, $\bar{q}$ and $\tilde{q}$ in eqs. (22) and (23), respectively. The following equations are derived in the same way.

$$k_2 \frac{d^2 \alpha_4 \bar{q}_h}{dt^2} + k_1 \frac{d\alpha_3 \bar{q}_h}{dt} + \bar{q}_h = \bar{r} \tag{52}$$

$$k_2 \frac{d^2 \beta_4 \tilde{q}_h}{dt^2} + k_1 \frac{d\beta_3 \tilde{q}_h}{dt} + \tilde{q}_h = \tilde{r} \tag{53}$$

Eqs. (52) and (53) can be simplified as follows:

$$\frac{d^2 \bar{V}}{dt^2} + f_1(t) \frac{d\bar{V}}{dt} + g_1(t)\bar{V} = \bar{r} \tag{54}$$

$$\frac{d^2 \tilde{V}}{dt^2} + f_2(t) \frac{d\tilde{V}}{dt} + g_2(t)\tilde{V} = \tilde{r} \tag{55}$$

$$\bar{V} = k_2 \alpha_4 \bar{q}_h \quad (56) \quad \tilde{V} = k_2 \beta_4 \tilde{q}_h \tag{57}$$

Where,

$$f_1(t) = \frac{k_1 \alpha_3}{k_2 \alpha_4}, \quad g_1(t) = \frac{1}{k_2} \left\{ k_1 \frac{d}{dt} \left(\frac{\alpha_3}{\alpha_4} \right) + \frac{1}{\alpha_4} \right\}$$

$$f_2(t) = \frac{k_1 \beta_3}{k_2 \beta_4}, \quad g_2(t) = \frac{1}{k_2} \left\{ k_1 \frac{d}{dt} \left(\frac{\beta_3}{\beta_4} \right) + \frac{1}{\beta_4} \right\}$$

Eq. (55) is transformed into simultaneous differential equations by introducing complex coefficients, $f_3(t)$ and $g_3(t)$, to solve eq. (55) for $\tilde{V}$.

$$\frac{d\tilde{V}_1}{dt} + f_3(t)\tilde{V}_1 = \tilde{r} \quad (58) \quad \frac{d\tilde{V}}{dt} + g_3(t)\tilde{V} = \tilde{V}_1 \tag{59}$$

Where,

$$f_3(t) = F(t) + jG(t), \qquad g_3(t) = H(t) + jI(t)$$

j : imaginary unit

By comparing eq. (55) with eqs. (58) and (59), the following equations are obtained.

$$\frac{dH}{dt} + (f_2 - H)H = g_2 - I^2 \tag{60}$$

$$\frac{dI}{dt} + (f_2 - 2H)I = 0 \tag{61}$$

$$F = f_2 - H \quad (62) \quad G = -I \tag{63}$$

Eqs. (58) and (59) are solved as follows:

$$R_e \{\tilde{V}\} = C_1 W_1 + S_1 W_2 \tag{64}$$

$$\frac{dW_1}{dt} + HW_1 = W_5 W_3 + W_6 W_4 \tag{65}$$

$$\frac{dW_2}{dt} + HW_2 = W_6 W_3 - W_5 W_4 \tag{66}$$

$$\frac{dW_3}{dt} + FW_3 = C_1\tilde{r} \tag{67}$$

$$\frac{dW_4}{dt} + FW_4 = S_1\tilde{r} \tag{68}$$

$$\frac{dC_1}{dt} = -IS_1, \quad C_1(0) = 1 \tag{69}$$

$$\frac{dS_1}{dt} = IC_1, \quad S_1(0) = 0 \tag{70}$$

in which, $W_5 = C_1^2 - S_1^2$, $W_6 = 2C_1S_1$, and the symbol "R_e" represents the real part. If we focus on the second order moment of q_h, $\sigma_{q_h}^2$, the following equations can be derived after raising eqs. (57) and (64) to the second power and taking the expectation of each of them.

$$\sigma_V^2 = (k_2\beta_4)^2\sigma_{q_h}^2 \tag{71}$$

$$\sigma_V^2 = E\{(C_1W_1 + S_1W_2)^2\} \tag{72}$$

We must obtain $E\{W_1^n W_2^m\}$ (n, m: integer, $n + m = 2$) in order to estimate $\sigma_{q_h}^2$ through eqs. (71) and (72). By using eqs. (10) and (65) to (68), the differential equations for $E\{W_1^n W_2^m\}$ are obtained as follows:

$$\frac{dU_{3,i}}{dt} + 2FU_{3,i}$$
$$= \Delta t(\sigma_{r_d}^2 C_1^{3-i} S_1^{i-1} + U_f^2 Y_i) \qquad , \quad i = 1 \sim 3 \tag{73}$$

$$\begin{cases} \dfrac{dU_{3,i+3}}{dt} + (F+H)U_{3,i+3} \\ \quad = W_5 U_{3,i} + W_6 U_{3,i+1} + \Delta t U_h Y_{i+3} \\ \dfrac{dU_{3,i+5}}{dt} + (F+H)U_{3,i+5} \\ \quad = W_6 U_{3,i} - W_5 U_{3,i+1} + \Delta t U_h Y_{i+5} \end{cases} \tag{74}$$

in which, $i = 1, 2$

$$\frac{dE\{W_1^2\}}{dt} + 2HE\{W_1^2\}$$
$$= 2W_5 U_{3,4} + 2W_6 U_{3,5} \tag{75}$$

$$\frac{dE\{W_1W_2\}}{dt} + 2HE\{W_1W_2\}$$
$$= W_5(U_{3,6} - U_{3,5}) + W_6(U_{3,7} + U_{3,4}) \tag{76}$$

$$\frac{dE\{W_2^2\}}{dt} + 2HE\{W_2^2\}$$
$$= 2W_6 U_{3,6} - 2W_5 U_{3,7} \tag{77}$$

$$\frac{dU_f}{dt} + FU_f = 0, \quad U_f(0) = 1 \tag{78}$$

$$\frac{dU_h}{dt} + HU_h = 0, \quad U_h(0) = 1 \tag{79}$$

$$\frac{dU_{w,i}}{dt} = W_{i+4}\frac{U_f}{U_h}, \quad i = 1, 2 \tag{80}$$

where, $Y_1 = 2Z_{3,1}(1,1)$,

$$Y_2 = Z_{3,1}(1,2) + Z_{3,1}(2,1), \quad Y_3 = 2Z_{3,1}(2,2)$$

$$\begin{cases} Y_{i+3} = C_1^{2-i} S_1^{i-1}\left(Z_{3,2}(1,1) + Z_{3,2}(2,2)\right) \\ Y_{i+5} = C_1^{2-i} S_1^{i-1}\left(Z_{3,2}(2,1) - Z_{3,2}(1,2)\right) \end{cases}, \quad i = 1,2$$

$$Z_{3,1}(n_1, n_2)$$

$$= \begin{cases} 0, & [t/\Delta t] < 1 \\ \sigma_{r_d}^2 \displaystyle\sum_{i=1}^{[t/\Delta t]} \rho^i u_{n_1}(t) u_{n_2}(t - i\Delta t), & [t/\Delta t] \geq 1 \end{cases},$$

$$Z_{3,2}(n_1, n_2)$$

$$= \begin{cases} 0, & [t/\Delta t] < 1 \\ \sigma_{r_d}^2 \displaystyle\sum_{i=1}^{[t/\Delta t]} \rho^i X_{n_1}(t,i) u_{n_2}(t - i\Delta t), & [t/\Delta t] \geq 1 \end{cases}$$

$$u_1 = \frac{C_1}{U_f}, \quad u_2 = \frac{S_1}{U_f},$$

$$X_i(t,j) = U_{w,i}(t) - U_{w,i}(t - j\Delta t), \quad i = 1,2$$

For further details of third and fourth order moments of q_h, $\mu_{q_h 3}$ and $\mu_{q_h 4}$, the reader should refer to Tanaka et al (1999b)

4. COMPARISONS OF THE STOCHASTIC PROPERTIES OF DISCHARGE

In the previous section, we presented the differential equations whose solutions provide the stochastic properties of discharge. The validity of these differential equations had been already checked in the works by Tanaka et al (1999a,1999b) This section shows the comparisons of the stochastic properties of discharge. The calculation is simplified by introducing the following quantities.

$h = h_* H$, $\quad q = q_* Q$, $\quad r = r_* R$,

$t = t_* T$, $\quad x = x_* X$

Where, $h_* = \varepsilon(\bar{r} l)^p$, $\quad q_* = \bar{r} l$, $\quad r_* = \bar{r}$,

$\quad t_* = \varepsilon \bar{r}^{p-1} l^p$, $\quad x_* = l$

As a result, eqs. (13) to (18) are converted into nondimensional equations.

$$\frac{\partial H}{\partial T} + \frac{\partial Q}{\partial X} = R \tag{81}$$

$$H = Q^p, \quad 0 \leq X \leq 1 \tag{82}$$

$$S = \int_0^1 H(X,Y)dX \tag{83}$$

$$\frac{dS}{dT} + Q = R \tag{84}$$

$$S = K_1 Q^{p_1} \tag{85}$$

$$S = K_1 Q^{p_1} + K_2 \frac{dQ^{p_2}}{dT} \tag{86}$$

The dimensional coefficients, k_1 and k_2, are expressed by

$k_1 = K_1 \varepsilon l^{p_1}$, $\quad k_2 = K_2 \varepsilon^2 l^{2p_1} \bar{r}^{2p_1 - p_2 - 1}$

Here, we consider a nondimensional rainfall, R, instead of the dimensional rainfall, r. $R_{d,i}$ is related to R by

$$R_{d,i} = \frac{1}{\Delta T} \int_{(i-1)\Delta T}^{i\Delta T} R(\tau)d\tau \tag{87}$$

The mean of $R_{d,t}$, $\overline{R}_{d,t}$, is expressed by

$$\overline{R}_{d,t} = \frac{1}{\Delta T} \int_{(t-1)\Delta T}^{t\Delta T} \overline{R}(\tau)d\tau \tag{88}$$

Furthermore, a nondimensional random variable, $\widetilde{R}_{d,t}$, is expressed by

$$\widetilde{R}_{d,t} = \rho \widetilde{R}_{d,t-1} + N_t \tag{89}$$

N_t: uncorrelated random variable

The second, third and fourth order moments of nondimensional observed rainfall, $R_{d,t}$, are

$$\sigma_{R_d}^{\ 2} = \frac{\sigma_N^{\ 2}}{1-\rho^2} \quad (90) \qquad \mu_{R_d3} = \frac{\mu_{N3}}{1-\rho^3} \tag{91}$$

$$\mu_{R_d4} = \frac{6\rho^2 \sigma_N^{\ 2} \sigma_{R_d}^{\ 2} + \mu_{N4}}{1-\rho^4} \tag{92}$$

The probability density function of N_t in eq. (89) is expressed by

$$f(N) = \begin{cases} \lambda e^{-\lambda\left(N+\frac{1}{\lambda}\right)} & -\frac{1}{\lambda} \leq N \\ 0 & elsewhere \end{cases} \tag{93}$$

$$E(N) = 0, \quad \sigma_N^{\ 2} = \frac{1}{\lambda^2}, \quad \mu_{N3} = \frac{2}{\lambda^3}, \quad \mu_{N4} = \frac{9}{\lambda^4}$$

λ: const.

The parameters are set for the computation as follows (Tanaka etal,1999b)

$$\Delta T = 0.1, \quad \sigma_{R_d}^{\ 2} = \begin{cases} 0.25 & 0 \leq T \leq 2 \\ 0 & elsewhere \end{cases}$$

$$p = p_1 = 0.6, \quad K_1 = \frac{1}{1+p_1},$$

i) Rectangular average rainfall input (Figure 4)

$$\overline{R} = \begin{cases} 1 & 0 \leq T \leq 2 \\ 0 & elsewhere \end{cases}, \quad p_2 = p_1^{1.5}, \quad K_2 = 0.1p_1^{-0.2}$$

ii) Triangular average rainfall input (Figure 5)

$$\overline{R} = \begin{cases} 0.5 + 2T & 0 \leq T \leq 1 \\ 4.5 - 2T & 1 \leq T \leq 2, \quad p_2 = 0.4509, \\ 0 & elsewhere \end{cases}$$

$$K_2 = 0.09608$$

Figures 6 to 9 show the comparisons of the first two moments of discharge from three runoff models. The variance of discharge, $\sigma_Q^{\ 2}$, increases with increases in the value of regression coefficient, ρ (Figures 7 and 9). These figures suggest that ρ plays an important role in the stochastic response property of discharge.

5. DISCUSSION AND CONCLUSIONS

Using three runoff models, the Kinematic wave model and two types of storage function runoff model, we presented the differential equations whose solutions provide the first four moments of discharge under the condition that rainfall is a mutually dependent random variable. These equations are derived on the rough assumption that observed rainfall, $r_{d,t}$, is described by the

AR(1) process.   This study showed that the regression coefficient, ρ, plays an important role in the stochastic response property of discharge.   We don't mention the probability density function of rainfall input.   Hashino (1986) proposed that the transition probability of a reduced variate of hourly rainfall is approximated by Freund's bivariate exponential probability distribution.

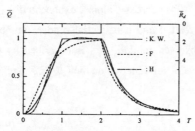

Fig. 6: Comparison of the mean of discharge (rectangular rainfall input)

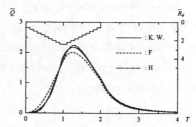

Fig. 8: Comparison of the mean of discharge (triangular rainfall input)

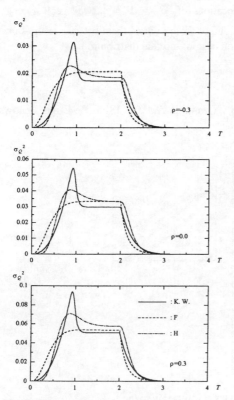

Fig. 7: Comparison of the variance of discharge (rectangular rainfall input)

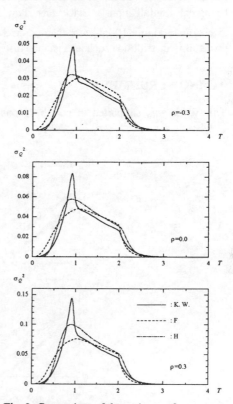

Fig. 9: Comparison of the variance of discharge (triangular rainfall input)

Its probability density function is expressed by
$$g(r_{d,i-1}, r_{d,i})$$
$$= \begin{cases} \alpha\beta e^{-\beta r_{d,i-1} - (2\alpha-\beta)r_{d,i}} & 0 < r_{d,i} < r_{d,i-1} \\ \alpha\beta e^{-\beta r_{d,i} - (2\alpha-\beta)r_{d,i-1}} & 0 < r_{d,i-1} < r_{d,i} \end{cases} \qquad (94)$$
α, β : positive coefficient

$$\bar{r}_d = \frac{\alpha + \beta}{2\alpha\beta} \quad (95) \quad \sigma_{r_d}^2 = \frac{3\alpha^2 + \beta^2}{4\alpha^2\beta^2} \tag{96}$$

$$\rho_* = \frac{1 - k^2}{1 + 3k^2}, \quad k \equiv \frac{\alpha}{\beta} \tag{97}$$

The auto-correlation function of $r_{d,i}$ described by eq. (94) decreases almost exponentially. If $\rho_* = 0$ in eq. (97), it is evident that $r_{d,i}$ is the uncorrelated and exponential random number. These results coincide with our approximation that observed rainfall, $r_{d,i}$ is described by the AR(1) process. It is possible to estimate the probability density function on the (β_x, β_y) plane. β_x and β_y are expressed by

$$(\beta_x, \beta_y) = \left(\frac{\mu_{Q3}^2}{\sigma_Q^6}, \frac{\mu_{Q4}}{\sigma_Q^4} \right) \tag{98}$$

Figure 10 shows examples of estimation of the probability density function of discharge for triangular rainfall input. Black dots show the locations of β_x and β_y at the peak average discharge. Figure 10 indicates that the discharge output from each of them belongs to the same probability distribution such as a gamma distribution or a log-normal distribution.

ACKNOWLEDGMENT

This work was supported in part by Monbusho's Grants-in-Aid for JSPS Research Fellows (1999).

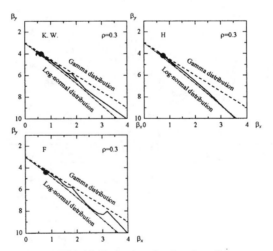

Fig. 10: Examples of estimation of the
probability distribution (triangular rainfall input)

REFERENCES

Bras, R. L. and Georgakakos, K. P.; Real time non-linear filtering techniques in streamflow forecasting: A statistical linearization approach, *Third International Symposium on Stochastic Hydraulics*, Proceedings, pp. 95-105, 1980.

Fujita M. , N. Shinohara, T. Nakao and M. Kudo; Stochastic response of storage function model for flood runoff, *Stochastic and Statistical Methods in Hydrology and Environmental Engineering*, Vol. 2, pp. 241-254, 1994.

Hashino M.: Stochastic formulation of design storm pattern and proposal of conditional probability-rainfall intensity formula, Proceedings of JSCE, Vol. 369/II-5, pp. 139-146, 1986. (in Japanese)

Hoshi K. and I. Yamaoka: A relationship between Kinematic wave and storage routing models, Proceedings of the 26th JCH, pp. 273-278, 1982. (in Japanese)

Tanaka G., M. Fujita, M. Kudo, K. Uchijima and K. Hasegawa: The impact of mutually dependent rainfall input to non-linear runoff system, *International Conference on Large Scale Water Resources Development in Developing Countries: New Dimensions of Prospects and Problems*, Proceedings, pp. MM246-MM254, 1997.

Tanaka G., M. Fujita and S. Kumagai: Estimating of the probability density function of discharge for flood runoff, *2nd International Conference on Water Resources*, Handbook and Proceedings, Vol. 2, pp. 1160-1166, 1999a.

Tanaka G., M. Fujita, M. Kaido and S. Kumagai: Mutually dependent rainfall input and its stochastic response, *International Congress on Modelling and Simulation, Modelling the Dynamics of Natural, Agricultural, Tourism and Socio-economics Systems*, Proceedings, Vol. 1, pp. 111-116, 1999b.

Stochastic Hydraulics 2000, Wang & Hu (eds) © 2000 Taylor & Francis, ISBN 90 5809 166 X

Investigating inherent probabilistic characteristics of rainfall IDF models

Yeou-Koung Tung & Baohong Lu
Department of Civil Engineering, Hong Kong University of Science and Technology, Kowloon, China

ABSTRACT: Rainfall intensity-duration-frequency (IDF) relationships are often used by hydrologic engineers for planning, design, and performance analysis of hydraulic structures. Various empirical forms of IDF relationships have been developed based on at-site annual maximum rainfall data of different durations. This paper examines the probabilistic characteristics of several commonly applied rainfall IDF models in hydrologic engineering. In particular, their associated probability distribution functions are derived. This leads to an important finding that, for each IDF model under consideration, there exist bounds on the model parameters that have to be satisfied in order to ensure model's compliance of probabilistic law. A numerical example is given to demonstrate the important implication of the inherent constraint on the fitted IDF model.

Key words: Probabilistic characteristics, Rainfall, IDF model

1. INTRODUCTION

Rainfall intensity-duration-frequency (IDF) relationships are widely used in water resources engineering for planning, designing and operating of water resources projects. At a given location, such relationship is often expressed in the form of a series of rainfall intensity-duration curves of different return periods. Over a geographical region, rainfall IDF relationships are represented by a series of rainfall intensity (or depth) isohyetal maps (e.g., Frederick et al., 1973).

To avoid extracting rainfall information from maps or figures, rainfall IDF models for estimating rainfall intensity (or depth) of different durations and frequencies have been proposed in many locations and regions (Raudkivi, 1979; Gert et al. 1987). Procedures for establishing at-site IDF model from a regional IDF isohyetal maps have been developed (Chen, 1983; Froehlich, 1995a,b).

In general, IDF model can be expressed as

$$y_{t,T} = h(t,T|\theta) \tag{1}$$

in which $y_{t,T}$ is rainfall quantity corresponding to a specific return period (T) and storm duration (t); $h(t,T|\theta)$ is a general expression of an IDF model; and θ is a vector of model parameters. Three commonly used rainfall IDF models are listed in the 2nd row of Table 1. All three IDF models are empirical formulas without much theoretical basis. They describe the general behavior that the average rainfall intensity reduces with an increase in duration.

Due to the presence of return period term in an IDF model, it is probabilistic. This paper derives the cumulative distribution function (CDF) and probability density function (PDF) corresponding to each individual rainfall IDF model. Information such as these are essential for a better understanding of its probabilistic characteristics which, in turns, can be incorporated in the development of effective and consistent rainfall IDF models.

Table 1:. CDF, PDF and lower bound corresponding to three rainfall IDF models

Model Type	Model-1	Model-2	Model-3
Model Form	$y_{t,T} = [g + h \times \log(T)]\dfrac{a}{(t+c)^d}$	$y_{t,T} = \dfrac{a \times T^b}{(t+c)^d}$	$y_{t,T} = \dfrac{a \times T^b}{(t^d + c)}$
CDF, $F(y_t)$	$F(y_t) = 1 - \dfrac{1}{10^{\frac{1}{h}\left[\frac{y_t(t+c)^d}{a} - g\right]}}$	$F(y_t) = 1 - \left[\dfrac{(t+c)^d}{a} y_t\right]^{-\frac{1}{b}}$	$F(y_t) = 1 - \left[\dfrac{(t^d + c)}{a} y_t\right]^{-\frac{1}{b}}$
PDF, $f(y_t)$	$f(y_t) = \dfrac{\ln(10) \times (t+c)^d}{h \times a \times 10^{\frac{1}{h}\left[\frac{y_t(t+c)^d}{a} - g\right]}}$	$f(y_t) = \dfrac{\left[\dfrac{(t+c)^d}{a} y_t\right]^{-\frac{1}{b}}}{b \times y_t}$	$f(y_t) = \dfrac{\left[\dfrac{(t^d+c)}{a} y_t\right]^{-\frac{1}{b}}}{b \times y_t}$
Lower Bounds $y_{t,min}$	$y_{t,min} = \dfrac{a \times g}{(t+c)^d}$	$y_{t,min} = \dfrac{a}{(t+c)^d}$	$y_{t,min} = \dfrac{a}{(t^d + c)}$

2. UNCONDITIONAL PDF AND CDF

Assume that the observed annual maximum rainfalls of duration t are statistically independent and identically distributed. From Eq. (1), the return period T for the rainfall quantity of duration t can be expressed as

$$T = h^{-1}[y_{t,T} \mid \theta] \tag{2}$$

From the relation between the return period and the exceedance probability, the following relationship holds

$$T = \frac{1}{P\{Y_t \geq y_{t,T}\}} = \frac{1}{1 - P\{Y_t \leq y_{t,T}\}} = \frac{1}{1 - F(y_{t,T})} \tag{3}$$

in which $F(y_{t,T}) = P\{Y_t \leq y_{t,T}\}$ is the CDF of random rainfall quantity Y_t of duration t. Then, the CDF for the general form of rainfall IDF model, after dropping subscript 'T', can be obtained as

$$F(y_t) = P\{Y_t \leq y_t\} = 1 - \frac{1}{h^{-1}[y_t \mid \theta]} \tag{4}$$

Hence, the corresponding PDF of the t-hr rainfall quantity, $f(y_t)$, can be obtained as

$$f(y_t) = \frac{d}{dy_t}[F(y_t)] = \frac{1}{[h^{-1}(y_t \mid \theta)]^2} \times \frac{d}{dy_t}[h^{-1}(y_t \mid \theta)] \tag{5}$$

The unconditional CDFs and PDFs associated with the three commonly used IDF models are shown, respectively, in the 3rd and 4th rows of Table 1.

From the CDFs associated with each IDF model, it is easily seen that $F(y_t=\infty)=1$. On the other hand, there exists a finite valued lower bound, $y_{t,min}$, which can be derived by solving $F(y_{t,min})=0$. The lower bound associated with each of the three IDF models are shown in the last row of Table 1. The presence of lower bound requires that the minimum observed rainfall quantity for an individual duration must be greater than or equal to the theoretical lower bound, $y_{t,min}$.

Otherwise, it will result in a negative CDF value for part of rainfall observations that violates the basic axiom of non-negativity of probability. Since the lower bound, $y_{t,mn}$, is a function of IDF model parameters, this reveals a fact that an inherent constraint exists to a rainfall IDF model. The parameter estimation of empirical rainfall IDF model in practice must consider this intrinsic lower bound constraint to ensure the feasibility of observed data are within the range under consideration.

3. NUMERICAL ILLUSTRATION

For illustration, rainfall intensity-based IDF Model-2 (see Table 1) is applied to fit annual maximum rainfall data of fifteen durations (15-min, 30-min, 1-hr, 2-hr, 4-hr, 6-hr, 8-hr, 12-hr, 18-hr, 24-hr, 2-day, 3-day, 4-day, 5-day, 7-day) at the Hong Kong Observatory (HKO) over the period of 1884-1939, and 1947-1990 (Lam and Leung, 1994). In the records, short-duration rainfall data (i.e., 15-min, 30-min) are available only after 1947. Data recording was interrupted in 1940-1946 due to World War II during which time Hong Kong was occupied by Japan.

The conventional approach estimates the parameters of an IDF model by first conducting rainfall frequency analysis for different durations. Then, rainfall intensity for different pre-selected return periods and durations are extracted to fit the IDF model. In this paper, nine return periods (5-, 10-, 100-, 200-, 500-, 1000-, 2000-, 5000-, and 10000-year) are considered. The least-squared criteria used in model parameter estimation is

$$\text{Minimizing} \quad D(\theta) = \sum_T \sum_t \left[y_{t,T} - \frac{a \times T^b}{(t+c)^d} \right]^2 \tag{6a}$$

$$\text{Subject to} \quad y_{t,min|T_L}(\theta) \le p_{t,min|T_L}, \text{ for all durations} \tag{6b}$$

in which $\theta=(a,b,c,d)$; $p_{t,mn|TL}$ and $y_{t,mn|TL}(\theta)$ are, respectively, observed and theoretical lower bounds for duration t corresponding to a specified lower design standard T_L-year.

Due to nonlinear relationships of the objective function and constraint, direct nonlinear search schemes such as simulated annealing and conjugate gradient algorithm are used for estimating the parameters. Numerical implementation of the conjugate gradient method to estimate rainfall IDF model parameters revealed that the algorithm can easily converge to a local optimum, which may be quite far away from the global one. This is partly because the objective function is not well-behaved in the parameter space. The other contributing factor is that the relative magnitude of the objective function gradient associated with the model parameters. To improve the solution optimality, a global optimization scheme called the simulated annealing was also implemented. The simulated annealing scheme is a random search scheme that is designed to avoid being trapped in a local optimum. Numerical experience for solving the problem in hand indicate that, when the simulated annealing and the conjugate gradient methods are used in tandem, the optimal model parameter values can be obtained in a rather effective manner.

With the LS criterion as stated in Eq.(6), the optimal model parameters by using simulated annealing algorithm, in conjunction with conjugate gradient scheme with and without considering the inherent lower bound constraint are shown in Table 2.

An examination of Table 2 reveals that the model performance, in terms of RMSE, without considering the lower bound constraint is better than that with the constraint. The CDF curves of rainfall IDF Model-2 for 30- and 60-min rainfall intensities based on the constrained and unconstrained optimal model parameters are plotted in Fig. 1.

Table 2: Optimal parameters of the conventional approach for Model-2

	a	b	c	d	$\|\nabla D(\theta)\|$	$D(\theta)$	Method
Uncon-strained	1922.5	0.1090	77.2	0.6711	26794.5	12.359	S.A.
	1922.5	0.1079	77.2	0.6688	277.4	12.349	C.G.
Constrained	2428.5	0.2326	77.4	0.9175	273503.8	29.754	S.A.
	2428.5	0.2327	77.4	0.9173	271138.3	29.739	C.G.

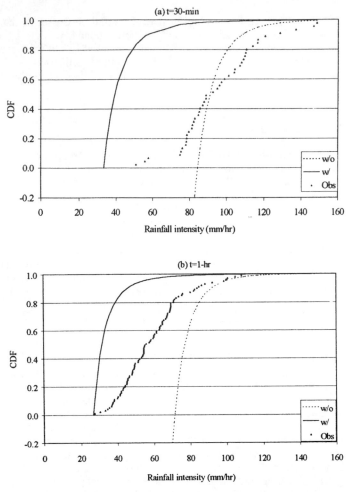

Fig.1: CDFs of intensity-based Model-2 under T_L=1-year

It can be seen From Fig.1 that the theoretical lower-bound values corresponding to the optimal unconstrained IDF parameters are larger than many of the observed rainfall intensities. Consequently, for observed rainfall intensities whose values are smaller than the theoretical low bound (indicated by vertical chain-dash line) would have negative values associated with the CDF. This implies that the unconstrained optimal IDF model parameters are infeasible in the context of violating the inherent constraint in IDF Model-2.

4. SUMMARY

The paper presents a study investigating probabilistic characteristics of rainfall IDF models frequently used in water resources engineering. The probability density function and cumulative distribution function of three commonly used rainfall IDF models are derived. Aside from the PDF and CDF associated with a rainfall IDF model, this study also points out the existence of lower bound constraints that are inherently attached to the IDF model, which are frequently ignored. As shown in the paper, the presence of these lower bound constraints has important theoretical and practical implications on the model parameter estimation and the subsequent applications in the design and analysis of water resource projects. Furthermore, knowledge about the PDF associated with an IDF model allows breaking away from using the conventional parameter estimation procedure that does not directly utilize the observed rainfall data.

ACKNOWLEDGMENTS

The study is sponsored by Hong Kong Research Grants Council, Project no. HKUST 6035/97E, on "Analysis of Rainstorm Characteristics in Hong Kong."

REFERENCES

Chen, C.L. (1983) 'Rainfall intensity-duration-frequency formulas', J. of Hydraul. Engrg., ASCE, 109(12), 1603-1621.

Frederick, R. H., Myers, V. A., and Auciello, E. P. (1973) 'Five- to 60-minute precipitation frequency for the Eastern and Central United States'. US National Weather Services. Washington D.C..

Froehlich, D. C. (1995a) 'Long-duration-rainfall intensity equations', J. of Irrig. & Drain. Engrg, ASCE, 121(3), 248-252.

Froehlich, D. C. (1995b) 'Intermediate-duration-rainfall intensity equations', J. of Hydraul. Engrg., ASCE, 121(10), 751-756.

Aron, G., Dunn, C. N., Wall, D. J, White, E.L (1987) 'Regional rainfall intensity-duration-frequency curves for Pennsylvania', Water Resources Bulletin, 23(3), 479-486.

Lam, C.C. and Leung, Y.K. (1994) "Extreme rainfall statistics and design rainstorm profiles at selected locations in Hong Kong." Technical Note no.86, Hong Kong Observatory.

Raudkivi, A. J. (1979) Hydrology: An Advanced Introduction to Hydrological Processes and Modeling. Pergamon, Oxford, UK.

Stochastic Hydraulics 2000, Wang & Hu (eds) © 2000 Taylor & Francis, ISBN 90 5809 166 X

Study on a stochastic approach to the rainfall runoff process

Hiroshi Hayakawa, Kunihide Uchijima & Bing Qian Wei
Department of Civil Engineering, Kitami Institute of Technology, Japan

Mutsuhiro Fujita
Graduate School of Engineering, Hokkaido University, Sapporo, Japan

ABSTRACT: The rainfall runoff process is mainly dominated by weather conditions such as the topography of a drainage basin and the heterogeneity of basin conditions. Assuming that the most general representation of topography and rainfall is a random field, a region of space and time within which the value of a variable is defined by a probability distribution, it is suitable for discussing the hydrologic response to adopt a stochastic approach instead of a deterministic approach. As the rainfall runoff process is generally described by a differential equation, we are introducing a random differential equation into the runoff process. This paper focuses on the stochastic response of a runoff process and derives statistical properties of discharge variation theoretically.

Key words: Runoff process, Topography, Stochastic approach, Discharge variation

1. INTRODUCTION

A drainage basin is composed of a channel network and its associated subbasins, where a subbasin consists of one channel and its associated hillslopes (Fig. 1). Therefore, it is natural that the distributed runoff models are built according to the channel network geomorphology. For a small drainage basin, a subbasin response is more important than a channel network response. As a basin increases in size, the hydrological response of a drainage basin becomes increasingly dominated by the channel network response. The main problems in the application of distributed models are the choices of an appropriate subbasin size.

As subbasins' area and its' channel length are considered in random variables (Shreve, 1966), they are suitable for discussing the hydrologic response of a drainage basin to adopt a stochastic approach instead of a deterministic approach. As the rainfall-runoff process is generally described by a differential equation, we are introducing a random differential equation into rainfall runoff models. We focus on the stochastic response of the runoff process and device the mean value and variance of discharge theoretically under conditions that define a subbasin area is defined by a probability distribution (Hayakawa *et al.*, 1996). This paper denotes an improved method that provides an opportunity to achieve the third and fourth moments of discharge variation. From these results, it is possible to decide the probability distribution of the discharge variation.

2. STOCHASTIC RESPONSE OF RUNOFF PROCESS

As shown in Fig.1, a drainage basin can be divided into smaller subbasins based on channel network geomorphology. The discharge at the outlet, $Q_n(t)$, is simply represented as Eq. (1),

$$Q_n(t) = \sum_{i=1}^{n} q_i(t - \tau_i) \tag{1}$$

where $q_i(t)$ is the discharge of the i-th subbasin with the subbasin area A_i and its channel (link) length L_i, τ_i is the travel time from the i-th subbasin, and n is the number of subbasins. The travel time τ_i of the discharge from the subbsains stands for the lag time of the discharge $q_i(t)$ arriving at the outlet. Equation (1) represents the basic distributed runoff model.

As the subbasin area and its channel length are considered in random variables (Shreve, 1966), we have derived theoretically the mean value and variance of discharge at the outlet of the drainage basin under conditions that define the subbasin area and channel length by a probability distribution (Hayakawa *et al.*, 1996). For estimating the probability distribution of the discharge variation by a method of moments, it is necessary to obtain the third and fourth moments of discharge variation. This paper reports an improved method for evaluating the third and fourth moments of discharge variation.

To investigate the rainfall runoff process, the storage routing model proposed by Hoshi *et al.* (1982) has often been adopted in Japan. This model is described by Eqs. (2) and (3),

$$\frac{dS}{dt} + q = r \tag{2}$$

$$S = K_1 q^{P_1} + K_2 \frac{dq^{P_2}}{dt} \tag{3}$$

in which S(mm), q(mm/hr) and r(mm/hr) denote storage, discharge and rainfall intensity, respectively. The storage coefficients K_1, K_2 and the storage exponents P_1, P_2 are expressed as Eqs. (4)-(6),

$$K_1 = 2.823\left(N/\sqrt{s}\right)^{0.6} A^{0.24} \tag{4}$$

$$K_2 = 0.2835 K_1^2 \bar{r}^{-0.2648} \tag{5}$$

$$P_1 = 0.6, \quad P_2 = 0.4648 \tag{6}$$

where N, s, $\bar{r}$, and A are the equivalent roughness coefficient, the hillslope angle, mean rainfall intensity, and the subbasin area, respectively. Equations (2) and (3) can be rewritten as a nonlinear differetial equation as shown in Eq. (7).

$$\frac{d^2 K_2 q^{P_2}}{dt^2} + \frac{dK_1 q^{P_1}}{dt} + q = r \tag{7}$$

As subbasin area A is considerd in a random variable, the storage coefficients K_1, K_2 are also considered in random variables. As Eq. (7) is, therefore, described as a random differential equation, the discharge $q(t)$ then becomes a random variable even if the rainfall $r(t)$ is in a steady state.

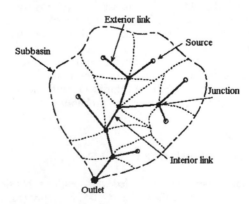

Fig. 1: Schematic drainage basin with channelnetwork of magnitude 6

508

Now, consider that the storage coefficients K_1, K_2 and the discharge $q(t)$ consist of a mean value and the deviation from its mean as shown in Eqs. (8)-(10),

$$K_1 = \overline{K}_1 + \tilde{K}_1, \qquad E[\tilde{K}_1] = 0 \tag{8}$$

$$K_2 = \overline{K}_2 + \tilde{K}_2, \qquad E[\tilde{K}_2] = 0 \tag{9}$$

$$q = \overline{q} + \tilde{q}, \qquad E[\tilde{q}] = 0 \tag{10}$$

where $E[x]$ operator denotes the expected value of a random variable x. Statistic properties of the storage coefficients K_1, K_2 are expressed in Eqs. (11)-(13), respectively.

$$E[\tilde{K}_1^2] = \sigma_{K_1}^2, \qquad E[\tilde{K}_2^2] = \sigma_{K_2}^2 \tag{11}$$

$$E[\tilde{K}_1^3] = \mu_{3K_1}, \qquad E[\tilde{K}_2^3] = \mu_{3K_2} \tag{12}$$

$$E[\tilde{K}_1^4] = \mu_{4K_1}, \qquad E[\tilde{K}_2^4] = \mu_{4K_2} \tag{13}$$

If the storage exponents P_1, $P_2 \neq 1$, it is difficult to solve Eq. (7) theoretecally because of the nonlinear terms q^{P_1}, q^{P_2} in this equation. So, we assume that these terms are approximately expressed by a linear equation as shown in Eqs. (14), (15).

$$(\overline{q} + \tilde{q})^{P_1} = \alpha_1 \overline{q} + \beta_1 \tilde{q} \tag{14}$$

$$(\overline{q} + \tilde{q})^{P_2} = \alpha_2 \overline{q} + \beta_2 \tilde{q} \tag{15}$$

Bras *et al.* (1980) proposed Eqs. (16) and (17) in regard to the random variable with an exponent, q^{P_i} (i=1,2).

$$\alpha_i = \overline{q}^{P_i-1} \left\{ 1 + \frac{1}{2} P_i (P_i - 1) \frac{E[\tilde{q}^2]}{\overline{q}^2} + \frac{1}{6} P_i (P_i - 1)(P_i - 2) \frac{E[\tilde{q}^3]}{\overline{q}^3} + \cdots \right\} \tag{16}$$

$$\beta_i = \frac{\overline{q}^{P_i+1}}{E[\tilde{q}^2]} \left\{ P_i \frac{E[\tilde{q}^2]}{\overline{q}^2} + \frac{1}{2} P_i (P_i - 1) \frac{E[\tilde{q}^3]}{\overline{q}^3} + \frac{1}{6} P_i (P_i - 1)(P_i - 2) \frac{E[\tilde{q}^4]}{\overline{q}^4} + \cdots \right\} \tag{17}$$

By substituting Eqs. (8)-(10), (14)-(17) into Eq. (7), we can rewrite as follow:

$$\frac{d^2}{dt^2} \left\{ (\overline{K}_2 + \tilde{K}_2)(\overline{Q} + \tilde{Q}) \right\} + \frac{d}{dt} (\overline{K}_1 + \tilde{K}_1)(\frac{\alpha_1}{\alpha_2} \overline{Q} + \frac{\beta_1}{\beta_2} \tilde{Q}) + \frac{\overline{Q}}{\alpha_2} + \frac{\tilde{Q}}{\beta_2} = r \tag{18}$$

where $\overline{Q}, \tilde{Q}$ are replaced for simplification following discussion that will be presented later.

$$\overline{Q} = \alpha_2 \overline{q}, \quad \tilde{Q} = \beta_2 \tilde{q} \tag{19}$$

Taking the expectation of Eq. (18) yields Eq. (20).

$$\frac{d^2}{dt^2} \left\{ \overline{K}_2 \overline{Q} + E[\tilde{K}_2 \tilde{Q}] \right\} + \frac{d}{dt} \overline{K}_1 \frac{\alpha_1}{\alpha_2} \overline{Q} + \frac{\beta_1}{\beta_2} E[\tilde{K}_1 \tilde{Q}] + \frac{\overline{Q}}{\alpha_2} = r \tag{20}$$

Subtracting Eq. (20) from Eq. (18) yields the deviation of the discharge.

$$\frac{d^2}{dt^2} \left\{ \overline{K}_2 \tilde{Q} + \tilde{K}_2 \overline{Q} + \tilde{K}_2 \tilde{Q} - E[\tilde{K}_2 \tilde{Q}] \right\} + \frac{d}{dt} \left\{ \overline{K}_1 \frac{\beta_1}{\beta_2} \tilde{Q} + \tilde{K}_1 \frac{\alpha_1}{\alpha_2} \overline{Q} + \tilde{K}_1 \frac{\beta_1}{\beta_2} \tilde{Q} - \frac{\beta_1}{\beta_2} E[\tilde{K}_1 \tilde{Q}] \right\} + \frac{\tilde{Q}}{\beta_2} = 0 \tag{21}$$

Equations (20) and (21) are also rewritten as follows:

$$\frac{d^2 \overline{Q}}{dt^2} + f_1(t) \frac{d\overline{Q}}{dt} + g_1(t) \overline{Q} = h_1(t) \tag{22}$$

$$\frac{d^2 \tilde{Q}}{dt^2} + f_2(t) \frac{d\tilde{Q}}{dt} + g_2(t) \tilde{Q} = h_2(t) \tag{23}$$

where,

$$f_1(t) = \frac{\overline{K}_1}{\overline{K}_2} \frac{\alpha_1}{\alpha_2} \tag{24}$$

509

$$g_1(t) = \frac{1}{\overline{K}_2} \left\{ \overline{K}_1 \frac{d}{dt}\left(\frac{\alpha_1}{\alpha_2}\right) + \frac{1}{\alpha_2} \right\} \tag{25}$$

$$h_1(t) = \frac{1}{\overline{K}_2} \left[r - \frac{d^2}{dt^2} E[\tilde{K}_2 \tilde{Q}] - \frac{d}{dt}\left\{ \frac{\beta_1}{\beta_2} E[\tilde{K}_1 \tilde{Q}] \right\} \right] \tag{26}$$

$$f_2(t) = \frac{\overline{K}_1}{\overline{K}_2} \frac{\beta_1}{\beta_2} \tag{27}$$

$$g_2(t) = \frac{1}{\overline{K}_2} \left\{ \overline{K}_1 \frac{d}{dt}\left(\frac{\beta_1}{\beta_2}\right) + \frac{1}{\beta_2} \right\} \tag{28}$$

$$h_2(t) = \frac{1}{\overline{K}_2} \left[\frac{d^2}{dt^2}\left\{ E[\tilde{K}_2 \tilde{Q}] - \tilde{K}_2 \overline{Q} - \tilde{K}_2 \tilde{Q} \right\} + \frac{d}{dt}\left\{ \frac{\beta_1}{\beta_2} E[\tilde{K}_1 \tilde{Q}] - \tilde{K}_1 \frac{\alpha_1}{\alpha_2}\overline{Q} - \tilde{K}_1 \frac{\beta_1}{\beta_2}\tilde{Q} \right\} \right] \tag{29}$$

Equations (22) and (23) are, therefore, represented in the basic differential equation for the mean value and the deviation of discharge, respectively.

Hayakawa *et al.* (1996) proposed the method for calculating the mean value and variance of discharge. The mean value of discharge is solved by Eqs. (20) or (22), where $E[\tilde{K}_1\tilde{Q}]$, $E[\tilde{K}_2\tilde{Q}]$ are obtained by multiplying the both sides of Eqs. (21) or (23) by $\tilde{K}_1$ or $\tilde{K}_2$ and taking the expectation.

2.1 Variance of discharge

Equation (21) or (23) is the basic differetial equation for the deviation of discharge and practically used to solve simultaneous differential equations such as Eqs. (30) and (31).

$$\frac{d\tilde{Q}}{dt} = \tilde{Q}_1 \tag{30}$$

$$\frac{d\tilde{Q}_1}{dt} + f_2(t)\tilde{Q}_1 + g_2(t)\tilde{Q} = h_2(t) \tag{31}$$

Hayakawa *et al.* (1996) replace Eq. (21) as a simultaneous differential equation of Eqs. (30)' and (31)',

$$\frac{d\tilde{Q}}{dt} + g_2(t)\tilde{Q} = \tilde{Q}_1 \tag{30'}$$

$$\frac{d\tilde{Q}_1}{dt} + f_2(t)\tilde{Q}_1 = h_2(t) \tag{31'}$$

and after solving $\tilde{Q}$ and $\tilde{Q}_1$ theoretically, multiplying by itself and taking the expectation, the variance $\sigma_Q^2(t)$ $(= E[\tilde{Q}^2(t)])$ of the discharge is derived. Because this method must be calculated with complex numbers, we introduced the same method used for calculating the mean value into the variance in spite of the fact that it solves simultaneous differential equations of many unknowns.

Multiplying Eqs. (30) and (31) by $\tilde{Q}$ and $\tilde{Q}_1$ and taking the expectation gives the following differential equations:

$$\frac{dE[\tilde{Q}^2]}{dt} = 2E[\tilde{Q}_1\tilde{Q}] \tag{32}$$

$$\frac{dE[\tilde{Q}_1\tilde{Q}]}{dt} + f_2(t)E[\tilde{Q}_1\tilde{Q}] = E[h_2(t)\tilde{Q}] - g_2(t)E[\tilde{Q}^2] + E[\tilde{Q}_1^2] \tag{33}$$

$$\frac{dE[\tilde{Q}_1^2]}{dt} + 2f_2(t)E[\tilde{Q}_1^2] = 2E[h_2(t)\tilde{Q}_1] - 2g_2(t)E[\tilde{Q}_1\tilde{Q}] \tag{34}$$

The variance $\sigma_Q^2(t)$ $(= E[\tilde{Q}^2(t)])$ is obtained by solving as simultaneous differential equation of

510

Eqs. (32)-(34), in which the higher order terms $E[K_1^i\tilde{Q}](=X_i)$, $E[K_2^i\tilde{Q}](=Y_i)$, $E[K_1^i K_2^j \tilde{Q}](=Z_{ij})$, and $E[K_1^i K_2^j](=W_{ij})$ are obtained when the mean value of discharge is solved by neglecting a higher order term than the fourth order term as follows:

$$\frac{d^2 X}{dt^2} + f_2(t)\frac{dX}{dt} + g_2(t)X = \frac{1}{\overline{K}_2}\left[\frac{d^2}{dt^2}\{B_1 Y_1 - B_2\overline{Q} - B_3\}\right.$$

$$\left. + \frac{d}{dt}\left\{B_1\frac{\beta_1}{\beta_2}X_1 - B_4\frac{\alpha_1}{\alpha_2}\overline{Q} - B_5\frac{\beta_1}{\beta_2}\right\}\right] \tag{35}$$

where,

$$\begin{aligned}
\mathbf{X} &= [X_1, Y_1, X_2, Y_2, Z_{11}, X_3, Y_3, Z_{21}, Z_{12}, X_4, Y_4, Z_{31}, Z_{22}, Z_{13}]^T \\
\mathbf{B}_1 &= [0, 0, \alpha_{K_1}^2, \alpha_{K_2}^2, W_{11}, \mu_{2K_1}, \mu_{2K_2}, W_{21}, W_{12}, \mu_{4K_1}, \mu_{4K_2}, W_{31}, W_{22}, W_{13}]^T \\
\mathbf{B}_2 &= [W_{11}, \alpha_{K_2}^2, W_{21}, \mu_{2K_2}, W_{12}, W_{31}, \mu_{4K_2}, W_{22}, W_{13}, 0, 0, 0, 0, 0]^T \\
\mathbf{B}_3 &= [Z_{11}, Y_2, Z_{21}, Y_3, Z_{12}, Z_{31}, Y_4, Z_{22}, Z_{13}, 0, 0, 0, 0, 0]^T \\
\mathbf{B}_4 &= [\alpha_{K_2}^2, W_{11}, \mu_{2K_1}, W_{12}, W_{21}, \mu_{4K_1}, W_{13}, W_{31}, W_{22}, 0, 0, 0, 0, 0]^T \\
\mathbf{B}_5 &= [X_2, Z_{11}, X_3, Z_{12}, Z_{21}, X_4, Z_{13}, Z_{31}, Z_{22}, 0, 0, 0, 0, 0]^T
\end{aligned} \tag{36}$$

in which $[\ ]^T$ denotes a transposed matrix. For evaluating the terms $E[h_2(t)\tilde{Q}]$, $E[h_2(t)\tilde{Q}_1]$ of $h_2(t)$ involved, some consideration is needed. Because $h_2(t)$ involves the differetial terms $d^2\tilde{Q}/dt^2$, $d\tilde{Q}/dt$ of a random variable $\tilde{Q}$, it is impossible to adopt the method multiplying Eq. (29) by $\tilde{Q}$ and $\tilde{Q}_1$, and calculating the expectation, so that it is often desirable to reduce the order of $h_2(t)$ by substituting Eqs. (30) and (31) into $h_2(t)$ as follows:

$$h_2(t) = \frac{1}{\overline{K}_2}\left[\frac{d^2 Y_1}{dt^2} - \tilde{K}_2\frac{d^2\overline{Q}}{dt^2} + f_2(t)\tilde{K}_2\tilde{Q}_1 + g_2(t)\tilde{K}_2\tilde{Q} + \frac{d}{dt}\frac{\beta_1}{\beta_2}X_1 - \tilde{K}_1\frac{d}{dt}\frac{\alpha_1}{\alpha_2}\overline{Q}\right.$$

$$\left. - \frac{\beta_1}{\beta_2}\tilde{K}_1\tilde{Q}_1 - \tilde{K}_1\tilde{Q}\frac{d}{dt}\frac{\beta_1}{\beta_2}\right] - \frac{\tilde{K}_2 h_2(t)}{\overline{K}_2^2} \tag{37}$$

$$h_2(t) = \frac{1}{\overline{K}_2}\left[A_1(t) - A_2(t)\tilde{K}_2 + A_3(t)\tilde{K}_2\tilde{Q}_1 + A_4(t)\tilde{K}_2\tilde{Q} + A_5(t) - A_6(t)\tilde{K}_1\right.$$

$$\left. - A_7(t)\tilde{K}_1\tilde{Q}_1 - A_8(t)\tilde{K}_1\tilde{Q}\right] - \frac{\tilde{K}_2 h_2(t)}{\overline{K}_2^2} \tag{37'}$$

where the coefficients $A_i(t)$ ($i=1,8$) are distinct values. By multiplying Eq.(37) by $\tilde{Q}$ and $\tilde{Q}_1$, and taking the expectation, the terms $E[h_2(t)\tilde{Q}]$, $E[h_2(t)\tilde{Q}_1]$ are obtained as shown in Eq. (38) with respect to $E[h_2(t)\tilde{Q}]$,

$$E[h_2(t)\tilde{Q}] = \frac{1}{\overline{K}_2}\left[-A_2 E[\tilde{K}_2\tilde{Q}] + A_3 E[\tilde{K}_2\tilde{Q}_1\tilde{Q}] - A_4 E[\tilde{K}_2\tilde{Q}^2] - A_6 E[\tilde{K}_1\tilde{Q}] - A_7 E[\tilde{K}_1\tilde{Q}_1\tilde{Q}] + A_8 E[\tilde{K}_1\tilde{Q}^2]\right]$$

$$- \frac{1}{\overline{K}_2^2}\left[A_1 E[\tilde{K}_2\tilde{Q}] - A_2 E[\tilde{K}_2^2\tilde{Q}] + A_3 E[\tilde{K}_2^2\tilde{Q}_1\tilde{Q}] - A_4 E[\tilde{K}_2^2\tilde{Q}^2] + A_5 E[\tilde{K}_2\tilde{Q}] - A_6 E[\tilde{K}_1\tilde{K}_2\tilde{Q}]\right.$$

$$\left. - A_7 E[\tilde{K}_1\tilde{K}_2\tilde{Q}_1\tilde{Q}] + A_8 E[\tilde{K}_1\tilde{K}_2\tilde{Q}^2]\right] + \frac{1}{\overline{K}_2^3}\left[A_1 E[\tilde{K}_2^2\tilde{Q}] - A_2 E[\tilde{K}_2^3\tilde{Q}] + A_3 E[\tilde{K}_2^3\tilde{Q}_1\tilde{Q}]\right.$$

511

$$-A_4 E[\tilde{K}_2^3 \tilde{Q}^2] + A_5 E[\tilde{K}_2^2 \tilde{Q}] - A_6 E[\tilde{K}_1 \tilde{K}_2^2 \tilde{Q}] - A_7 E[\tilde{K}_1 \tilde{K}_2^2 \tilde{Q}_1 \tilde{Q}] + A_8 E[\tilde{K}_1 \tilde{K}_2^2 \tilde{Q}^2]\Big]$$

$$-\frac{1}{\bar{K}_2^4}\Big[A_1 E[\tilde{K}_2^3 \tilde{Q}] - A_2 E[\tilde{K}_2^4 \tilde{Q}] + A_5 E[\tilde{K}_2^3 \tilde{Q}] - A_6 E[\tilde{K}_1 \tilde{K}_2^3 \tilde{Q}]\Big] + \frac{1}{\bar{K}_2^5}(A_1 + A_5)E[\tilde{K}_2^4 \tilde{Q}] \quad (38)$$

in which the new terms $E[\tilde{K}_1 \tilde{Q}^2]$, $E[\tilde{K}_1 \tilde{Q}_1^2]$, etc., are calculated reflexively as shown in Eqs. (39)-(41).

$$\frac{dE[\tilde{\mathbf{K}}\tilde{Q}^2]}{dt} = 2E[\tilde{\mathbf{K}}\tilde{Q}_1 \tilde{Q}] \quad (39)$$

$$\frac{dE[\tilde{\mathbf{K}}\tilde{Q}_1 \tilde{Q}]}{dt} + f_2(t)E[\tilde{\mathbf{K}}\tilde{Q}_1 \tilde{Q}] = E[\tilde{\mathbf{K}}h_2(t)\tilde{Q}] - g_2(t)E[\tilde{\mathbf{K}}\tilde{Q}^2] + E[\tilde{\mathbf{K}}\tilde{Q}_1^2] \quad (40)$$

$$\frac{dE[\tilde{\mathbf{K}}\tilde{Q}_1^2]}{dt} + 2f_2(t)E[\tilde{\mathbf{K}}\tilde{Q}_1^2] = 2E[\tilde{\mathbf{K}}h_2(t)\tilde{Q}_1] - 2g_2(t)E[\tilde{\mathbf{K}}\tilde{Q}_1 \tilde{Q}] \quad (41)$$

$$\tilde{\mathbf{K}} = \Big[\tilde{K}_1, \tilde{K}_2, \tilde{K}_1^2, \tilde{K}_1 \tilde{K}_2, \tilde{K}_2^2, \tilde{K}_1^3, \tilde{K}_1^2 \tilde{K}_2, \tilde{K}_1 \tilde{K}_2^2, \tilde{K}_1^3\Big]^T \quad (42)$$

Furthermore, $E[\tilde{\mathbf{K}}h_2(t)\tilde{Q}]$, $E[\tilde{\mathbf{K}}h_2(t)\tilde{Q}_1]$ in Eqs. (40) and (41) are similar to $E[h_2(t)\tilde{Q}]$ and $E[h_2(t)\tilde{Q}_1]$, and Eq. (43) indicates the result of $E[\tilde{K}_1 h_2(t)\tilde{Q}]$.

$$E[\tilde{K}_1 h_2(t)\tilde{Q}] = \frac{1}{\bar{K}_2}\Big[A_1 E[\tilde{K}_1 \tilde{Q}] - A_2 E[\tilde{K}_1 \tilde{K}_2 \tilde{Q}] + A_3 E[\tilde{K}_1 \tilde{K}_2 \tilde{Q}_1 \tilde{Q}] - A_4 E[\tilde{K}_1 \tilde{K}_2 \tilde{Q}^2] + A_5 E[\tilde{K}_1 \tilde{Q}]$$

$$- A_6 E[\tilde{K}_1^2 \tilde{Q}] - A_7 E[\tilde{K}_1^2 \tilde{Q}_1 \tilde{Q}] + A_8 E[\tilde{K}_1^2 \tilde{Q}^2]\Big] - \frac{1}{\bar{K}_2^2}\Big[A_1 E[\tilde{K}_1 \tilde{K}_2 \tilde{Q}] - A_2 E[\tilde{K}_1 \tilde{K}_2^2 \tilde{Q}]$$

$$+ A_3 E[\tilde{K}_1 \tilde{K}_2^2 \tilde{Q}_1 \tilde{Q}] - A_4 E[\tilde{K}_1 \tilde{K}_2^2 \tilde{Q}^2] + A_5 E[\tilde{K}_1 \tilde{K}_2 \tilde{Q}] - A_6 E[\tilde{K}_1^2 \tilde{K}_2 \tilde{Q}]$$

$$- A_7 E[\tilde{K}_1^2 \tilde{K}_2 \tilde{Q}_1 \tilde{Q}] + A_8 E[\tilde{K}_1^2 \tilde{K}_2 \tilde{Q}^2]\Big] + \frac{1}{\bar{K}_2^3}\Big[A_1 E[\tilde{K}_1 \tilde{K}_2^2 \tilde{Q}] - A_2 E[\tilde{K}_1 \tilde{K}_2^3 \tilde{Q}]$$

$$+ A_5 E[\tilde{K}_1 \tilde{K}_2^2 \tilde{Q}] - A_6 E[\tilde{K}_1^2 \tilde{K}_2^2 \tilde{Q}]\Big] - \frac{1}{\bar{K}_2^4}(A_1 + A_5)E[\tilde{K}_1 \tilde{K}_2^3 \tilde{Q}] \quad (43)$$

Finally, the variance $\sigma_Q^2(t)(= E[\tilde{Q}^2(t)])$ is obtained by solving simultaneous differitial equations (32)-(35) and (40)-(42) of many unknowns, and of course, the variance $\sigma_q^2(t)$ from subbasins is transformed in Eq. (44) from Eq. (19).

$$\sigma_q^2(t) = \sigma_Q^2(t) / \beta_2(t)^2 \quad (44)$$

2.2 Third and fourth moments of discharge

The third moment $\mu_{3Q}(t) (= E[\tilde{Q}^3(t)])$ of discharge is obtained by multiplying Eqs. (30) and (31) by $\tilde{Q}$ and $\tilde{Q}_1$ and taking the expectation as follows:

$$\frac{dE[\tilde{Q}^3]}{dt} = 3E[\tilde{Q}_1 \tilde{Q}^2] \quad (45)$$

$$\frac{dE[\tilde{Q}_1 \tilde{Q}^2]}{dt} + f_2(t)E[\tilde{Q}_1 \tilde{Q}^2] = E[h_2(t)\tilde{Q}^2] - g_2(t)E[\tilde{Q}^3] + 2E[\tilde{Q}_1^2 \tilde{Q}] \quad (46)$$

512

$$\frac{dE[\tilde{Q}_1^2\tilde{Q}]}{dt} + f_2(t)E[\tilde{Q}_1^2\tilde{Q}] = 2E[h_2(t)\tilde{Q}_1\tilde{Q}] - 2g_2(t)E[\tilde{Q}_1\tilde{Q}^2] + 2E[\tilde{Q}_1^3] \tag{47}$$

$$\frac{dE[\tilde{Q}_1^3]}{dt} + 2f_2(t)E[\tilde{Q}_1^3] = 3E[h_2(t)\tilde{Q}_1^2] - 3g_2(t)E[\tilde{Q}_1^2\tilde{Q}] \tag{48}$$

in which $E[h_2(t)\tilde{Q}^2]$, $E[h_2(t)\tilde{Q}_1\tilde{Q}]$, and $E[h_2(t)\tilde{Q}_1^2]$ are calculated by multiplying Eq. (29) by $\tilde{Q}^2$, $\tilde{Q}_1\tilde{Q}$ and $\tilde{Q}_1^2$ and taking the expectation in the same manner as the variance. The fourth moment $\mu_{4Q}(t)\,(= E[\tilde{Q}^4(t)])$ is similar to above method.

3. DISCUSSION AND CONCLUSION

For verifying this theoretical method, we performed a comparison of the theoretical results with the simulation results. As the probability distribution of subbasin area A is considered as a Gamma distribution (Hayakawa *et al.*, 1996), the statistical properties of the storage coefficient K_1, K_2 are determined from Eqs. (4), (5) when given the statistical properties of A.

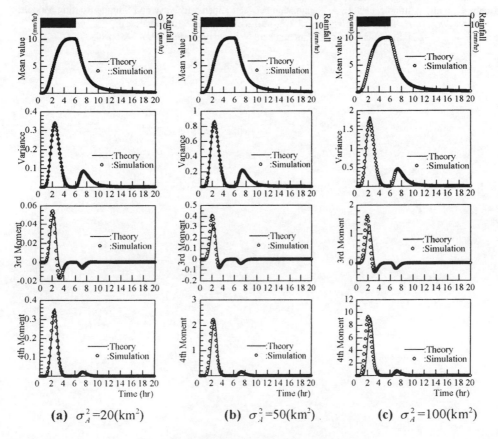

(a) $\sigma_A^2 = 20(\text{km}^2)$ **(b)** $\sigma_A^2 = 50(\text{km}^2)$ **(c)** $\sigma_A^2 = 100(\text{km}^2)$

Fig.2: Stochastic response of subbasin discharge
(Comparison of theoretical results to simulation)

513

Consider a drainage basin area A_t=200(km^2) with the number of subbasin n=11, that is, the mean $\overline{A}$ =18.18(km^2). If the variance of the subbasin is σ_A^2=20, 50, 100 (km^4), then the mean, variance, and third and fourth moments of discharge from the subbasins are illustrated as shown in Fig.2, respectively, under conditions of rectangular rainfall with the mean intensity $\overline{r}$=10(mm/hr) and the basin factor of $N/\sqrt{s}$=1.0. The results show that the theoretical results agree with the simulation results. When the variance of the subbasin decreases, the agreement of the higher moment increases as shown in the case of σ_A^2=20 (km^4). Therefore, this proposed method for estimating stochastic response of a runoff process is a very valid method.

REFFERENCES

Bras, R.L. and K.P. Geogakakos (1980), Real time nonlinear filtering techniques in streamflow forecasting – A statistical linearization approach -, *Third IAHR Int. Symp. on Stochastic Hydraulics, Tokyo*, pp.95-105.

Hayakawa, H., M. Fujita and K. Uchijima (1996), Stochastic Response of Distributed Runoff Models based on Characteristics of A Drainage Basin Topography, *Journal of Hydraulic, Coastal and Environmental Engineering, JSCE*, No. 545/II-36, pp.11-22 (in Japanese).

Hoshi, K. and I. Yamaoka (1982), A Relationship between Kinematic Wave and Storage Routing Models, *Proceedings of Hydraulic Engineering, JSCE*, Vol. 26, pp.273-278 (in Japanese).

Shreve, R.L.(1966), Statistical law of stream numbers, *Journal of Geology*, Vol.74, pp.17-37.

Stochastic Hydraulics 2000, Wang & Hu (eds) © 2000 Taylor & Francis, ISBN 90 5809 166 X

A study on the reproduction of embayment made of vegetation

T. Abe, A. Numata & Z. L. Zhu
Department of Civil Engineering, Tohoku Institute of Technology, Sendai, Japan

ABSTRACT: The flow visualization using the particle tracking velocimetry and the numerical analysis using a turbulence model were conducted with the flow field area of a river course that had a vegetated-zone with an embayment. From the point of view of creating and preserving a hospitable environment for aquatic life, we studied a method of estimating the ratio of the flow velocity inside a vegetated-zone with an embayment to the flow velocity outside the zone. The flow velocity was controlled to allow investigation of the structure and the density of the vegetation, in which we made use the transparency of the vegetation.

Key words: PTV-measurement, Turbulence model, Open-channel flow, Embayment, Vegetation

1. INTRODUCTION

The natural river course has rapids, pools, and embayments, all of which offer a suitable environment as a habitat for aquatic life. Water plants also grow forming stands, thus functioning as working ecosystem. Previous policies for river reconstruction have focused on taking countermeasures against emergencies such as floods or droughts. However, today in Japan, there is a demand for an overall policy whose aim is to reproduce the river environment as a multipurpose system thus functioning as a habitat for plant and animal species, as well as a recreational area.

Vigorous experimental and numerical studies are being conduced so that previous studies have presented the fundamental idea concerning how to deal with vegetation when analyzing the flows that occur within the vegetated channel. They have also indicated how the vegetation influences these flows. Successive studies have been just conducted on the improvement of the river system and how the vegetation is utilizes within these channels. For example, vegetation is expected to function as a spur dike thus reducing the current velocity around the area. However, studies on the effects of the vegetation's spur dike have only just begun (Fukuoka *et al.* 1996). Tu *et al.* (1994) also conducted experiments on the fluctuation in the velocity inside and outside an embayment using a non-transparent wall. However, the flowing condition around a transparent vegetated embayment differs from that generated around a wall- type embayment (Zhu *et al.* 1998).

This study suggests that an improved river system requires a vegetative river channel with a diverse natural environment that has a vegetated zone, a vegetated embayment and a spur dike made of vegetation. The purpose of this study is to create a habitat consisting of vegetation, which can be used to collect information necessary for maintaining its function in order to establish an evaluating system of the habitat. Then, hydraulic experiments were constructed using horizontal shear flows generated in the vegetated channel, based on the flow visualization using the technique of PTV, and numerical analysis using a modified k-ε model was also conducted. Ikeda's method (1991), which dealt with vegetation in the numerical analysis system, was followed. Results from the study have allowed for the presentation of a method of using the characteristics of vegetation to estimate and control the velocity ratio of the flows inside and outside the embayment with vegetation, as well as the significant basic technological information on the momentum transport.

2. OUTLINE OF THE STUDY

2.1 The outline of the hydraulic experiment

The characteristics of the flow field in the vegetated channel: A schematic view of an imitative flow field in a channel having a vegetated embayment and the experiment conditions are shown in Fig. 1 and Table 1, respectively. The materials of the bottoms of most rivers are not uniform; besides the downstream has a compound cross-section channel which consists of a main channel and a flood plain. To construct an imitative compound cross-section of a channel, a rough board containing varying sizes of sand was placed on the riverbed, where a gradient of 1/1000 was established in the riverbed. Furthermore, a porous material having a void of 95% was used, as an "imitative vegetation", on the imitative flood plain of the left bank. This is called a "vegetated channel". The sites with $b_2/b_1=0$, $b_2/b_1=1$, $0< b_2/b_1<1$ represent a vegetated zone, a vegetated spur dike, and a vegetated embayment, respectively. The flow field that was chosen for this study has an underwater vegetation, $H/K>1$. However, it is a sub-critical flow containing predominantly horizontal shear flows and a low water level. Experimentation involved changing the ratio of the width of the main channel with respect to the width of the river, b/B. This ratio is equivalent to the ratio of the vegetated zone to the area of the whole channel. In the case that the aspect ratio is 0.4 b/B, horizontal vortexes occur irregularly, and a value of 0.8 b/B suggests that horizontal vortexes occur regularly. A value of 0.6 b/B shows a transitional tendency (Fukuoka *et al.* 1989).

The technique of flow visualization and image processing: The Particle Image Velocimetry (PIV) flow visualization technique is a combination of a flow visualization technique and a technique of image processing. Upon adopting this technique for evaluating of the flow, we need to select a proper method of analyzing data and visualizing the flow in order to establish a system of evaluation, according to the regime of the flow field and the required accuracy. This study used a low density PIV analysis because it is suitable for analyzing the low density of particles. As to the analysis methods for image processing, this study used the four successive times tracking method, which corresponds to the particle tracking velocimetry (PTV) flow visualization technique (Tsuda *et al.*, 1992; Abe *et al.*, 1997). The measurement with the PTV technique was conducted at horizontal cross sections of 0.5 y/d and 1 y/d depth. It is important to note that the measurements of velocity using the fiber-optic laser Doppler anemometer (LDA) technique were also carried out. The results of the PTV analysis on the mean velocity in the main stream are shown in Fig. 2. There is very little difference, regarding the main streams in the upstream and downstream sides of the embayment where local flows are generated, between the results of the PTV analysis and those of the LDA technique. It can be inferred that the result of the PTV analysis corresponds to that obtained from site measurement.

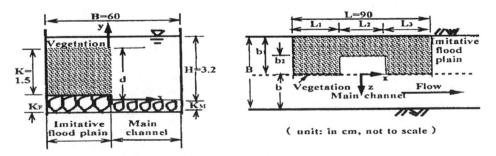

Fig. 1: The schematic view of imitative compound open-channel with a concave vegetated-zone

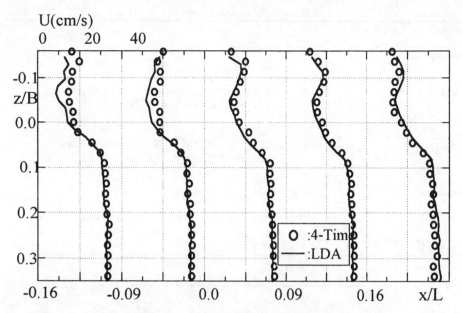

Fig.2: The comparison between the PTV and the LDA measurements: the case C2 (b/B=0.4)

2.2 Outline of the numerical analysis

Turbulence model: As demonstrated by Obi *et al.* (1992), the k-εt urbulence model has a problem in reproducing the turbulence in the flow field where strong acceleration occurs, because of overestimation of the turbulent energy.   However, as demonstrated by previous studies (Rodi, W., 1984; Nezu *et al.*, 1987), it is possible to accurately identify each constant included in the model. We inspected the propriety of our system by measuring reattachment distances of the standard k-εm odel in the test of a backward facing step flow without a free surface. Upon modification of the model for the open-channel flow with low Reynolds number, it was adopted to the flows in the vegetated embayment, to develop a numerical analysis system that controls the flow velocity both inside and outside the embayment. The basic equation governing a steady two-dimensional turbulent flow in this study may be expressed, in Cartesian tensor notation, as follows:

$$\partial(U_i\Phi - \Gamma(\partial \ /\partial x_i))/\partial x_i = S \tag{1}$$

where $\Phi(U, V, k, \varepsilon)$: hydraulic quantity, Γ: turbulent diffusion coefficient, S: the generation-destruction term of Φ. The diffusion coefficient and constant term of the transportation equation are shown in Table 2. In the equation of continuity, "$\Phi=1$" is adopted.

Table.1: The experimental conditions

Case	C2	C4	C6
Q (cm³/s) x10³	3.76	3.70	4.84
H (cm)	3.3	3.2	3.3
H/K	2.2	2.1	2.2
B/H	18.2	18.4	18.2
b/B	0.4	0.6	0.8
b_2/L_2	1/3	1/3	1/3
Re x10³	5.8	5.7	7.5
Fr	0.33	0.34	0.43
U*(cm/s)	1.79	1.77	1.79

Re=UH/ν, Fr=U/(gH)$^{1/2}$, U*=(gHI)$^{1/2}$, U=Q/BH

517

Equations	Φ	Γ	S
continuity equation	1	0	0
U - equation	U	$\upsilon + \upsilon_t$	$- gI + \partial[(\upsilon + \upsilon_t)\,(\partial U/\partial x)]\,/\partial x + \partial[(\upsilon + \upsilon_t)\,(\partial W/\partial x)]\,/\partial z$ $+ (f\,/H)\,U\,(U^2+W^2)^{1/2} + (K/H)\,F_x$
W - equation	W	$\upsilon + \upsilon_t$	$\partial[(\upsilon + \upsilon_t)\,(\partial U/\partial z)]\,/\partial x\ \partial[(\upsilon + \upsilon_t)\,(\partial W/\partial z)]\,/\partial z$ $+ (f\,/H)\,W\,(U^2+W^2)^{1/2} + (K/H)F_z$
k - equation	k	$\upsilon_t\,/\sigma_k$	$G + C_{fk}\,(F_x\,U + F_z\,W) - \varepsilon$
ε - equation	ε	$\upsilon_t\,/\sigma_\varepsilon$	$(\varepsilon\,/k)\,[C_{1\varepsilon}\,\{G + C_{f\varepsilon}\,(F_x\,U + F_z\,W)\} - C_{2\varepsilon}\,\varepsilon]$

$F_x = [C_d a\,U\,(U^2+W^2)^{1/2}]\,/2;\quad F_z = [C_d a\,W\,(U^2+W^2)^{1/2}]\,/2$

Table 3:. The conditions for calculations

Case		C2	C4	C6
b/B		0.4	0.6	0.8
b_2/L_2			1/3	
H(cm)		3.3	3.2	3.3
$R_e \times 10^3$		5.8	5.7	7.4
$F_M \times 10^{-3}$			2.6	
$C_d a$			1.4	
grid size	Δx		2.17	
(cm)	Δz		1.0	

Practice of numerical analysis: Since the object of analysis was the flow field having predominant horizontal shear flow, two-dimensional calculation was conducted. After the basic equation was integrated by the control-volume, discrete with the staggered grids was conducted. And then the pressure field and the flow velocity field were determined by a semi-implicit method for pressure linked equation method. The area covered by the calculation, was four times as long as the length of the vegetation zone. As the boundary conditions, the experimental values of the upstream side of the vegetated zone were adopted. The fluctuation on the surface of the water was not counted in the calculation. In order to calculate the values, the resistance coefficient "C_d" and the sheltered area of water flow "a" have to be determined. However, since the vegetation model in this study consisted of a porous material, it was impossible to determine the resistance coefficient and the sheltered area of water flow of the vegetation according to the definition shown by Ikeda *et al.* (1991). Then we regarded the $C_d a$ values as a parameter indicating the density of the vegetation, and set the $C_d a$ so that calculated values corresponded to the experimental values. The conditions for calculation were determined in accordance with the experimental conditions shown in Table 3. The boundary conditions were set so that the velocity gradient would be zero in both direction of "x" and "z" at the edges of the upstream and downstream, and "no-slip" was the condition for the sidewall. The area for calculation was 390 cm and 60 cm in the stream wise (x) and span wise (z) directions, respectively. The size of the calculation grid was 2.17cm x 1.0 cm (number of the grids was 180 x 60). Calculation was conducted in two stages: firstly, one for laminar flows and secondly for turbulent flows. As the initial conditions on the velocity field, the velocity distribution was uniform in the stream wise direction, and in the span wise direction was set at zero. The coefficient of eddy viscosity was given as $v = U_0\,H\,/\,Re$, where "Re" represents the Reynolds number. The bottom friction coefficient of the main channel was determined from the results obtained from the velocity distribution in the experiment. The bottom friction coefficient of the flood plain is usually three or four times as large as that of the main channel. So we set it as three times as large.

__Reproduction of the mean velocity field:__ The results of the calculation of the mean current velocity U, as well as the experimental results, were shown in Fig. 3. According to the figure, both results coincide with each other quite well. This means that our numerical model is suitable for grasping the real mean flow field that occurs around the vegetated embayment. The relationship between the calculated values and the experimental ones, regarding the uniform velocity area inside the embayment and the front mainstream area, proves the reliability of evaluation the mean flow velocity both inside and outside the embayment through calculations.

However, regarding the evaluation of the turbulent area at the mouth (z=0) of the embayment, there were some differences regarding structures such as the gradient of the velocity and the width of the turbulent mixing area. Moreover, the differences detected were regarding the downstream edge in the embayment. Furthermore, although the experimental values of the velocity distribution in the embayment are supplemented with an infection point, it is not reproduced by the calculated values. The evaluation by calculation can't be perfect for reproducing the elaborate structure of the flows. The difference between the calculated values and the experimental values of the mean current velocity in the span wise direction is often great in the areas where the influence of the irregularity of the flows is greater. The area around the boundary of the vegetation in the upstream side is influenced by the outgoing flows of the vegetated zone. The area around the upstream side in the embayment is influenced by the vortexes shedding from the boundary of the vegetation, in addition to the flows coming from the vegetated zone. As a result, the effects of the local flows generated under such circumstances become stronger compared to the span wise mean current velocity. However, our numerical analysis system does not sufficiently explain such circumstances.

3. RESULTS AND DISCUSSION

3.1 The mechanism of momentum transport occurring around the vegetated embayment

Figure 4 shows the values of non-dimensional Reynolds stress at the upstream side, $x/L= -0.11$ near the embayment, for each case (C2, C4, C6). The "y /d" in the figure means the height from the riverbed. "0.5 y/d" indicates the half height of the vegetation, and the y/d of 1.0 is equivalent to the top of the vegetation. In the case C2, as shown in the figure, the Reynolds stress values outside the embayment in the upstream side of the embayment are almost zero both at the height of 0.5 y/d and 1.0 y/d. On the contrary, inside the embayment the Reynolds stress is four times as strong as the friction velocity. At the height of 0.5 y/d, the value is positive, which means the momentum transport is made from the outside to the inside of the embayment. At the height of 1.0 y/d the value is negative, which means the momentum transport is from the inside to the outside.

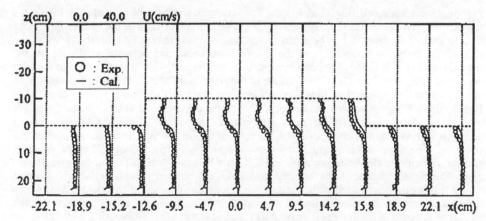

Fig. 3: The comparison between PTV-measurements and calculated results by a
modified k-εmodel in the mean flow field (Case C2: b/B=0.4)

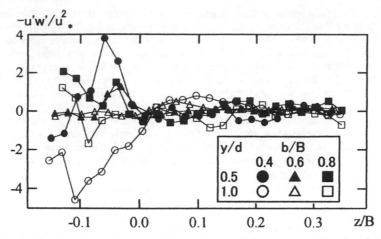

Fig. 4: The distributions of non-dimensional Reynolds stress at the upstream side (x/L= -0.11)

In conclusion the momentum transport from the inside of the embayment to the main stream is predominant at the top of the vegetation; at the medium height, horizontal vortexes shedding from the edge of the vegetated zone cause a momentum transport from the main stream to the inside of the embayment.

3.2 Estimation and control of the velocity ratio inside and outside the vegetated embayment

We calculated the mean velocity inside the embayment, U_I, and the mean velocity outside the embayment, U_O; that is, the main stream, and got the ratio of U_I/U_O. And then we investigated the propriety of these values as data for the estimation and control system for the flow velocity. For this purpose, we selected the following items as parameters: (1) Parameters regarding the shape of the vegetated zone: the aspect ratio, b/B, which indicates the ratio of the area of the vegetated zone to the whole area of the channel; the aspect ratio, b_2/L_2, which indicates the shape of the embayment. (2) Parameters regarding the resistance of the flow: the ratio of the bottom friction coefficient f_F/f_M, the vegetation's density or the form drag coefficient C_da, and the ratio of the height of the vegetation to the depth of the water K/H. The velocity ratio, U_I/U_O, was calculated with uniform flow in the center of the vegetated embayment: x/L=0, unless there is a proviso.

The case of a uniform density of vegetation

(1)The aspect ratio b/B: Fig. 5 shows the fluctuation of the ratio of the velocity inside the embayment to that outside the embayment, U_I/U_O, in relation to the ratio of the width of the main stream to that of the river, b/B. The mean current velocity was calculated with the flows in the center of the embayment, x/L=0, and the experimental value was obtained at the top of the vegetation, y/d=1. The shape of the embayment, b_2/L_2, is constant. The "b/B=0" refers to the case in which vegetation covers the whole surface of the river, and "b/B=1" refers to the case of an imitative compound channel lacking vegetation. According to the figure, in the range from 0.4b/B to 0.8 b/B, the calculated values coincide with the experimental values. When the b/B is less than 0.4, the flow velocity increases. Conversely, as the ratio of the vegetation becomes lesser; that is, b/B becomes larger than 0.4, the ratio of the current velocity does not change but approaches a constant. Accordingly, if the vegetated zone is narrower than half of the channel, the mean velocity inside the embayment keeps a little less than half of that in the mainstream region. This constant value of the velocity ratio U_I/U_O depends on the difference in roughness between the flood plain and the main channel in the imitative compound channel, b/B=1, which has no vegetation. In this situation the influence of the vegetation on the velocity ratio is smaller than that of the bottom friction coefficient. As a result, it can be said that by using vegetation, we can control the flow velocity in the embayment. A similar change in the velocity ratio was detected in the case of the embayment having a different span wise depth and whose mouth had a different length (b_2/L_2=1/4).

520

(2) Bottom friction coefficient, f_F/f_M: The flow velocities around the embayment depend on the bottom friction resistance against the riverbed and the vegetation resistance. We investigated the velocity ratio by changing the friction coefficient ratio with the values of C_da as a parameter that was set so that the calculation conditions were the same as that for the experiment of the case C2. The relations between these calculated values and the experimental results are shown in Fig. 6. The experimental values were obtained from the case C2, C4, and C6 at the upstream edge, center, and downstream edge of the embayment. The friction coefficient ratio was calculated with $f_F/f_M = (U_M/U_F)^2$. According to the figure, as the friction coefficient ratio becomes larger, the velocity ratio decreases thus making a straight declining line in the graph. Since the value of the vegetation resistance C_da (1.0~2.0) corresponds with that gained under Ikeda's theory (1991); C_da =1.4, it can be said that the method of calculation in our model is a reliable one for estimating the flow velocity around the embayment.

(3) The ratio of the height of the vegetation to the depth of the water K/H: Fig. 7 shows the fluctuation of the mean flow velocity around the vegetated embayment depending on the ratio of the height of the vegetated zone to the depth of the water. "K/H=1" refers to the case in which the water depth is equal to the height of the vegetation. According to the calculated results, it is shown that in the flow where the vegetation is submerged in water, as the density of the vegetation becomes larger, the decreasing rate of the mean velocity inside the embayment, depending on the increase of the height of the vegetated zone, becomes large. The decreasing rate of the velocity is larger when the values of K/H are smaller. In the case of a certain height of the vegetation, the shallower the water becomes, the greater the vegetation's influence on the velocity.

The case of a non-uniform density of vegetation
We made local changes in the vegetation's resistance to flows by varying the density of vegetation, thus trying to control the current velocity inside the embayment. We calculated the values with a flow field having a vegetated zone containing a non-uniform density. The vegetated zone was divided into three blocks: the upstream part, center part and downstream part. The resistance coefficients of these blocks are C_{d0}, C_{d1} and C_{d2}. The conditions for the calculation were the same as before, except for the vegetation's resistance coefficient.

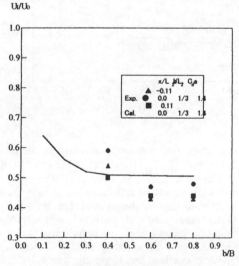

Fig. 5: U_I/U_O vs. b/B (f_F/f_M=3)

521

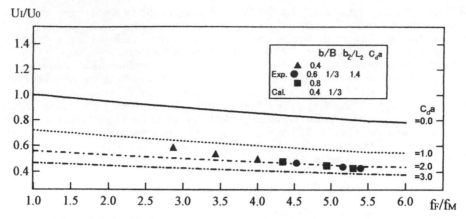

Fig. 6: Relationship the velocity ratio U_I/U_O and the bottom friction coefficient f_F/f_M

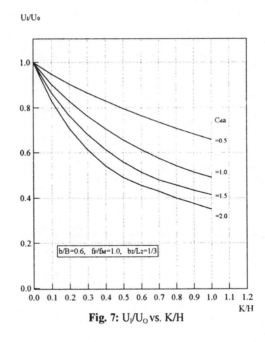

Fig. 7: U_I/U_O vs. K/H

Fig. 8 shows the fluctuation of the flow velocity depending on the change in the resistance coefficient, $C_{d0}a$, in the upstream part. "C_{d1}" of the center of the vegetation is set as the same as "C_{d2}" of the downstream part. The value of f_F/f_M are shown as parameters on the figure. "$C_{d0}a=0$" refers to the flow fields without the vegetated embayment, and then "$f_F/f_M=1$" and "$f_F/f_M\neq1$" correspond to a section with uniform roughness and the imitative compound channel, respectively. "$C_{d0}=1$" refers to the embayment with uniform vegetation According to the figure, even if the resistance coefficient of the upstream part is set lower than those of the center and downstream part, the effect of the vegetation to reduce the flow velocity is great. Moreover, as the resistance coefficient of the upstream becomes larger, the flow velocity inside the embayment decreases rapidly making a logarithmic line in the graph. However, the value of the velocity ratio, after decreasing to a certain value, remains the same. So there is a limitation to controlling the velocity by completely changing the resistance coefficient in the upstream edge .

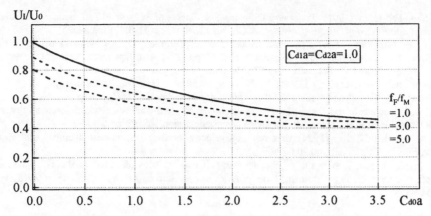

Fig. 8: Relationship the velocity ratio U_I/U_O and the vegetation's density $C_{do}a$

4. CONCLUSIONS

The results of this study are summed up as follows.

Hydraulic characteristics of the river course with a vegetated embayment:
(1) Since the vegetated embayment is transparent, flows can be expected to occur within the embayment. Unlike an embayment with low-level transparency, the larger the area occupied by the vegetated zone within the whole channel, the faster the current velocity in the embayment.
(2) Remarkable, the momentum transport mechanism which occurs around the embayment is that while the momentum transport from the inside of the embayment to the main stream is predominant at the height of the top of the vegetation, at half of the depth in the upstream side of the embayment, the momentum transport from the main stream to the inside of the embayment occurs due to the vortexes with coherent structure shedding from the edge of the vegetation.

Estimation and control of the mean-flow field inside the vegetated embayment:
(1) The velocity in the embayment decreases as the ratio of the vegetated zone to the channel decreases.
(2) The velocity inside the embayment can be reduced by making the density of the vegetation thicker, to increase the form drag. Moreover, increasing the difference in the friction coefficients between the flood plain and the main channel is effective for reducing the velocity inside the embayment.
(3) The velocity inside the embayment can be controlled by locally changing the vegetation's density

ACKNOWLEDGMENT

The authors wish to express their gratitude to Messrs. A. Aihara for his help with the experiments and data processing.

REFERENCES

Abe T., A. Aihara and T. Hiraya (1997): Turbulence measurement using the PTV-technique in imitative compound open-channel flows with vegetated zone, *The 1st Pacific Symp. on Flow Visualization and Image Processing*, Vol. 1, pp. 89-98.
Fukuoka S., and K. Fujita (1989): Prediction of flow resistance in compound channels and its application to design of river courses, Proc. of JSCE, No.411, II-12, pp.63-72.
Fukuoka S., A. Watanabe and M. Ohhashi (1996): A study on utilization of vegetations as a spur dike, Annual Meeting of JSCE, II-113, pp.226-227 (in Japanese).

Ikeda, S. and N. Izumi (1991): Transverse Diffusion Coefficients of Suspended Sediment, *J. of Hydraulic, Coastal and Environmental Engineering, JSCE,* No.434,II-16, pp.47-55(in Japanese).

Nezu, I., H. Nakagawa (1987): Numerical Calculation of Turbulent Open-Channel Flows by Using a Modified k- ε Turbulence Model, *J. of JSCE,* No.387, pp.125-134(in Japanese).

Obi, S., M. Peric and G. Scheuerer (1992): Numerical Study on the Turbulent Flow around a Two-Dimensional Square-Sectioned Obstacle, *JSME,* B, No.58-555, pp.81-86 (in Japanese).

Rodi, W. (1984): Turbulence models and their application in hydraulics, *IAHR,* 104p.

Tsuda, N., T. Kobayashi and T. Saga (1992): Development of a real-time velocity measurement system for high Reynolds fluid flow using digital image processing design, *ASME*: pp.9-14.

Tu H., N. Tamai and K. Kan (1994): Unsteady-flow velocity variations in and near an embayment, *Proc. of Hydr. Eng., JSCE,* Vol. 38, pp.703-708.

Zhu Z. L., A. Aihara, T. Abe and A. Numata (1998): Characteristics of turbulent shear flows near the vegetated zone with an embayment, *Proc. of 11th Cong. of Asia and Pacific Division of IAHR,* Vol. 2, pp.547-556.

Stochastic Hydraulics 2000, Wang & Hu (eds) © 2000 Taylor & Francis, ISBN 90 5809 166 X

Stochastic modeling of unsaturated flow in bounded domains

Marco Ferrante & Bruno Brunone
Department of Civil and Environmental Engineering, University of Perugia, Italy

T.-C.Jim Yeh
Department of Hydrology and Water Resources, University of Arizona, USA

ABSTRACT: In this paper a numerical model is used to study uncertainty propagation of water flow through unsaturated soils. This model is based on a first-order Taylor series expansion of the discretized Richards' equation and considers soil hydrologic properties as stochastic processes in space. To examine the influence of boundary conditions and heterogeneity on the pressure head variance, profiles of the head variance during one-dimensional vertical infiltration cases are generated. Dependence of pressure head variance on the flow conditions (drying or wetting) is examined for different soil models. Effects of deterministic boundary condition on pressure head variance profiles are also analyzed.

Key words : Stochastic model, Taylor series, Unsaturated flow, Soil properties.

1. THE NUMERICAL MODEL

The one-dimensional (1-D), *h*-based, vertical Richards' equation is generally used to predict the flow in unsaturated soils:

$$\frac{\partial}{\partial z}\left(K(h)\frac{\partial(h-z)}{\partial z} \right) = C(h)\frac{\partial h}{\partial t} \tag{1}$$

where z is the positive downward vertical coordinate, K is the hydraulic conductivity, h is the soil-water pressure head, $C = \partial\theta / \partial h$ is the specific moisture capacity, and θ is the moisture content. h is negative for unsaturated flow.

Because (1) is a non linear parabolic partial differential equation, analytic solution is possible only for some special cases. Therefore, numerical methods are typically used to integrate the Richards' equation. Using a fully implicit finite element scheme, eq. (1) can be expressed in a matrix form:

$$\mathbf{P}(\mathbf{h}_i,\mathbf{p})\mathbf{h}_i = \mathbf{Q}(\mathbf{h}_i,\mathbf{p})\mathbf{h}_{i-1} + \mathbf{f}(\mathbf{h}_i,\mathbf{p},\mathbf{u}) \tag{2}$$

In (2), $\mathbf{h}_i$ is the vector of h values at the time t_i at the n nodes, with the subscript i denoting the time level; $\mathbf{p}$ is the vector of the m parameters used to define the soil hydraulic properties; $\mathbf{u}$ is the boundary condition vector; $\mathbf{P}$ is the matrix associated with unsaturated hydraulic conductivity values and moisture capacity terms evaluated at $\mathbf{h}_i$; $\mathbf{Q}$ is the matrix associated with the moisture capacity term evaluated at $\mathbf{h}_i$; the vector $\mathbf{f}$ is related to the boundary conditions and the gravity term.

As shown in a previous paper (*Ferrante and Yeh*, 1999), expanding (2) in Taylor series around the mean up to the first order and taking the expected value, the following approximate mean equation is obtained:

$$P(\langle \mathbf{h}_i \rangle, \langle \mathbf{p} \rangle) \langle \mathbf{h}_i \rangle = Q(\langle \mathbf{h}_i \rangle, \langle \mathbf{p} \rangle) \mathbf{h}_{i-1} + \mathbf{f}(\langle \mathbf{h}_i \rangle, \langle \mathbf{p} \rangle, \langle \mathbf{u} \rangle) \tag{3}$$

where $<>$ denotes the expected value of variables and parameters. Eq. (3) is the mean equation of the first-order analysis and it coincides with that of the deterministic approach, when the mean values for the parameters and the variables are used.

Subtracting eq. (3) from the Taylor expansion of eq. (2) up to the first order results in a perturbation equation; it can be used to estimate the dependence of the head covariance matrix $\mathbf{R}_{hh}$ on the parameter and boundary condition covariance matrices, $\mathbf{R}_{pp}$ and $\mathbf{R}_{uu}$, respectively. Terms in the $\mathbf{R}_{pp}$ covariance matrix depend on the parameters that are used to define the soil hydraulic properties and on their covariance and cross-covariance functions. In the one-dimensional case, the boundary condition covariance matrix, $\mathbf{R}_{uu}$, is a two-by-two matrix with the variance of the boundary condition on the diagonal and zero on the off-diagonal terms.

Assuming that the soil parameters and the boundary conditions are uncorrelated, the perturbation equation can be written as (*Ferrante and Yeh*; 1999):

$$\mathbf{R}_{hh} = \left\langle \mathbf{h}'_i \mathbf{h}'^T_i \right\rangle = \mathbf{B}_i \, \mathbf{R}_{pp} \, \mathbf{B}^T_i + \mathbf{C}_i \, \mathbf{R}_{uu} \, \mathbf{C}^T_i \tag{4}$$

$\mathbf{B}_i$ and $\mathbf{C}_i$ matrices depends on matrices $\mathbf{P}$, $\mathbf{Q}$, and $\mathbf{f}$ and on their derivatives with respect to $\mathbf{h}_{i-1}$ and $\mathbf{h}_i$ as shown in the above referenced paper.

2. SOIL MODELS

The integration of (1) requires the functions $K(h)$ and $C(h)$ to be defined, the latter being derived from the $\theta(h)$ function:

$$C(h) = \frac{d\theta}{dh} \tag{5}$$

The two constitutive functions, $K(h)$ and $\theta(h)$, define the soil model. The parameters of these functions depend on the unsaturated soil hydraulic characteristics and can be determined from experimental data.

Many different models have been proposed and are used. Because of its simplicity, in most stochastic models the Gardner [1958] soil model is used, that describes the constitutive relationship between K and h with an exponential function:

$$K(h) = K_s \, e^{\alpha h} \tag{6}$$

K_s is the saturated hydraulic conductivity and α is a parameter related to the soil pore distribution. Russo [1988] derived a parametric expression for the other soil constitutive relationship, $\theta(h)$, consistent with the Mualem's [1976] theory:

$$\theta(h) = \theta_r + (\theta_s - \theta_r) \left[e^{\alpha h/2} (1 - \alpha h/2) \right]^{2/(m+2)} \tag{7}$$

where θ_s e θ_r are the saturated and residual water contents, respectively, and m is a parameter that accounts for the dependence of the tortuosity, usually taken to be known.

Mantoglou and Gelhar [1987a,b] in their stochastic analysis of transient flow in unsaturated soils used a linear relationship between θ and h:

$$\theta = \theta_s + ch \tag{8}$$

corresponding a constant value, c, for the specific soil moisture capacity, $C(h)$.

An exponential relationship can also be used, similar to that proposed by Russo but not consistent with the Mualem's theory,

$$\theta(h) = \theta_r + (\theta_s - \theta_r) e^{\alpha h} \tag{9}$$

that yields useful simplifications in the Richards' equation integration [*Srivastava and Yeh*; 1991].

The soil model proposed by Brooks and Corey [1964] is based on the empirical relationships:

$$\theta(h) = \begin{cases} (\theta_s - \theta_r)(-\alpha h)^{-\beta} & , |h| < 1/\alpha \\ (\theta_s - \theta_r) & , |h| \geq 1/\alpha \end{cases} \tag{10}$$

and

$$K(h) = \begin{cases} K_s(-\alpha h)^{-(2+3\beta)} & ,|h| < 1/\alpha \\ K_s & ,|h| \geq 1/\alpha \end{cases} \tag{11}$$

where α is the inverse of the air-entry pressure head value.

Another commonly used soil model has been proposed by van Genuchten (1980), based on the relationships:

$$\theta(h) = \left\{ 1 / \left[1 - (\alpha|h|) \right]^\beta \right\}^{1-(1/\beta)} \tag{12}$$

and:

$$K(h) = K_s \frac{\left\{ 1 + (\alpha|h|)^{n-1} \left[1 - (\alpha|h|)^n \right]^{-1+(1/\beta)} \right\}^2}{\left[1 - (\alpha|h|)^\beta \right]^{m(1-(1/\beta))}} \tag{13}$$

The Brooks and Corey (BC) and the van Genuchten (vG) soil models have more parameters than those based on the Gardner equations and they usually perform better in fitting the experimental data on a wider range of h values. Some formulae have been proposed to convert parameters from one of these two soil models to the other one (*Ma et al.*; 1999) for a comparison between methods). The Gardner-Russo (GR) model as well as the Gardner-Exponential (GE) and the Gardner-Linear (GL) models, because of their simpler mathematical expression, are used to obtain analytical solutions for the deterministic and stochastic analysis of flow in unsaturated soils.

In the present work results for the five different soil models are shown.

3. MATERIALS AND METHODS

Numerical tests were performed with regard to drying and wetting processes for different boundary conditions and soil models.

The total length of the 1-D domain, 500 cm, is divided into 250 elements with a length of 2 cm each. Soil parameter $f = log(K_s/<K_s>)$ is assumed to be second-order stationary Gaussian processes. Mean parameter values for the five soils are given in Table I.

Table 1: Deterministic parameter values for the different soil models.

	K_s (cm/min)	α (cm^{-1})	β	θ_s (cm^3/cm^3)	θ_r (cm^3/cm^3)
Gardner	$7.42034 \ 10^{-5}$	$2.38612 \ 10^{-2}$			
Russo (GR)				0.1946	0.0946
Exponential (GE)				0.4196	0.0946
Linear (GL)				0.4196	0.0946
van Genuchten (vG)	$2.78255 \ 10^{-2}$	$4.95288 \ 10^{-2}$	1.8109	0.4196	0.1342
Brooks and Corey (BC)	$2.78255 \ 10^{-2}$	$7.93026 \ 10^{-2}$	0.6384	0.4717	0.1342

Some of the data in Table I are identical to those of the Bet Dagan soil (*Russo and Bouton*; 1992); other parameter values were estimated in order to achieve similar constitutive functions for the different soils. A best fitting algorithm was used to help in achieving this goal. It is based on the minimization of the area enclosed between two curves, bounding the parameter values to be in an acceptable range. The $K(h)$ and $C(h)$ functions are shown in figures 1 and 2 for the five different soils.

At the soil surface a constant flux or a constant head boundary condition is used, and at the bottom boundary condition a constant pressure head equal to zero, representing the water table, is specified.

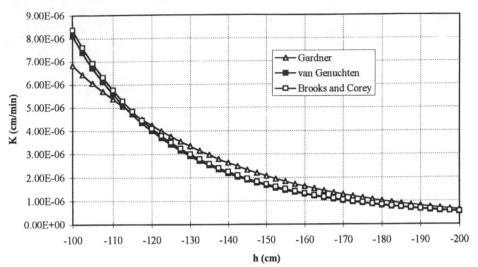

Fig.1: Hydraulic conductivity functions, K, for the different soil models, in the pressure head range considered in the simulations.

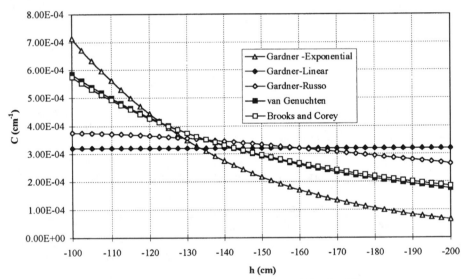

Fig.2: Specific moisture capacity functions, C, for the different soil models, in the pressure head range considered in the simulations.

In order to compare front propagation for wetting and drying cases, only two flux values are chosen for the constant flux boundary condition, q_a=7.469 10^{-6} and q_b=5.792 10^{-7}cm/min. The wetting front is originated changing at t=0 from q_b to q_a and, viceversa, the drying front is originated changing from q_a to q_b. For the head surface boundary conditions values, those corresponding to the steady-state unit gradient distribution for q_a and q_b are chosen. In this manner a variation of the mean pressure head ranging approximately from −200 to −100 cm is considered for all five soils and all cases.

The steady-state consistent with the prescribed initial boundary conditions is assumed for the initial conditions. Then, the surface boundary condition is suddenly changed, for t=0, from the initial value to the new constant one, either for the constant flux or for the constant head boundary condition. This originates a wetting or drying front propagating into the soil until a final steady-state condition is obtained.

4. RESULTS AND DISCUSSION

Figure 3a shows head profiles during a drying scenario, caused by a change of the flux boundary condition at the surface. As it can be seen, differences in front propagation velocity and diffusivity are evident. Initial and final steady-state profiles coincides for the three soil models GE, GL, and GR, because they share the same Gardner constitutive function and the $C(h)$ function does not affect the steady-state.

Initial mean pressure head profiles for vG and BC soils are very similar to each other. Differences between these profiles and initial steady-state profiles for the other soil models can be explained with differences in K values for different constitutive functions (see $K(-100)$ and $K(-200)$ values for Gardner, Brooks and Corey, and van Genuchten models on figure 1). On the other hand, although vG and BC constitutive functions are very similar, still differences in the front propagation velocity are evident.

On figure 3b pressure head variances due to the uncertainty only on f value are shown for the pressure head profiles in figure 3a, for GE, GR, and GL models. Variance of f is assumed to be equal to 1 for all soil models. For a better comparison, the normalized variance of h, $\sigma_h^2 / \sigma_{h,max}^2$, is plotted, where $\sigma_{h,max}^2$ is the maximum of σ_h^2 during front propagation in each soil. $\sigma_{h,max}^2$ values are: 1.045×10^{-6} cm^2 (GE), 1.835×10^3 cm^2 (GR), 1.513×10^{-7} cm^2 (GL), 9.584×10^2 cm^2 (vG), and 2.125×10^3 cm^2 (BC).

Profile for t=0 corresponds to the stochastic initial steady-state condition, originated by the deterministic boundary conditions in the soil with stochastic hydraulic characterization [*Yeh*; 1989].

The initial steady-state pressure head variance value is in a very good agreement with the analytical solution by Yeh [1985]. Initial and final steady-state pressure head variance profiles are exactly the same for each of the three soil models GE, GL, and GR, independently of mean pressure head value. This finding is consistent with that of Yeh (1989), for Gardner soil model in steady-state conditions, when only f is a stochastic variable. Similar to the mean pressure head, the σ_h^2 steady-state profile for the three soils is exactly the same.

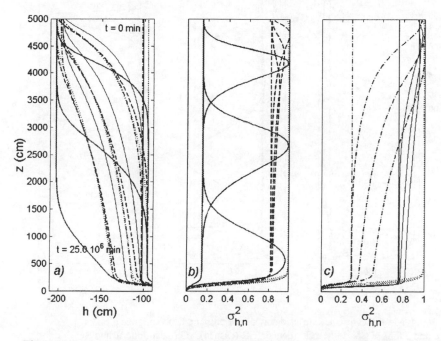

Fig.3: Mean pressure haed (*a*) and normalized pressure head variance profiles (*b* and *c*), for GE (solid line), GL (dotted), GR (dashed), vG (dash-dotted), and BC (thin) soil models, for a drying scenario and a constant flux top boundary condition.

529

The pressure head variance peaks during transient are close to the locations of the maximum head gradient and move with them. Their values usually increase with time. The lower diffusion in front propagation for GE, comparing to GR and GL soils, yields a narrower zone interested by transient conditions and hence a narrower pressure head variance "wave". In figure 3c normalized pressure head variance profiles are shown for GL, vG and BC soil models. The general behavior is similar to that described in figure 3b, but for initial and final steady-state pressure head variance profiles that do not coincide for Bc and vG soils. This finding does not contradict the Yeh [1985] finding that applies to Gardner soil model.

In figure 4a, mean head profiles at different times are shown for the wetting scenario, when the flux boundary condition at the surface is suddenly changed. Comparing this figure to figure 3a, differences in front propagation velocity and diffusivity are more evident between the soil models. The GE model in this case shows a lower diffusivity than all other models; it can be shown that for this soil, the diffusivity value does not depend of wetting or drying conditions [*Srivastava and Yeh*; 1991]. Initial and final steady-state profiles coincides for the three soil models GE, GL, and GR, as shown for the wetting case.

The lower diffusivity of the wetting process narrows the extension of the zone where transient takes place. As a result, pressure head variance "waves" shown in figure 4b and 4c are narrower than that for wetting case. Pressure head variance peaks are still located close to the highest mean pressure head gradient. For the profiles shown in figures 4a and b $\sigma^2_{h,max}$ values are: 6.706 x 10^{-7} cm^2 (GE), 1.339 x 10^4 cm^2 (GR), 1.731 x 10^{-7} cm^2 (GL), 1.304 x 10^3 cm^2 (vG), and 1.367 x 10^4 cm^2 (BC). A drying front propagation is shown in figure 5a, originated by a sudden change in the constant pressure head condition at the top boundary. Mean pressure head profiles are very similar to those in figure 4a, when a constant flux boundary condition is used.

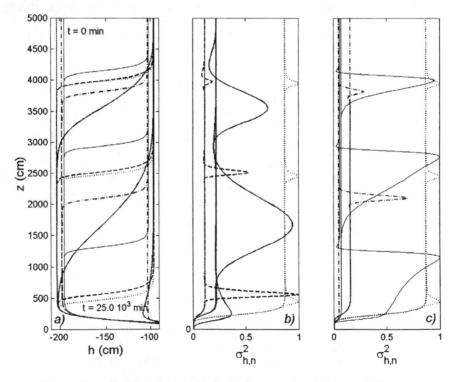

Fig.4: Mean pressure head (*a*) and normalized pressure head variance profiles (*b* and *c* for GE (solid line), GL (dotted), GR (dashed), vG (dash-dotted), and BC (thin) soil models, for a wetting scenario and a constant flux top boundary condition.

Notwithstanding, differences are in pressure head variances shown in figure 5b and 5c, with $\sigma^2_{h,max}$ values of 1.036×10^4 cm^2 (GE), 3.009×10^3 cm^2 (GR), 3.009×10^{-3} cm^2 (GL), 1.587×10^2 cm^2 (vG), and 3.581×10^3 cm^2 (BC). Because of the deterministic boundary condition at the surface, corresponding to a stochastic boundary condition of constant zero pressure head variance, differences are mostly located close to the surface. Not considering the area close to the boundaries, pressure head variance in steady-state condition is not equal to zero; furthermore peak values are greater than those associated with the constant head boundary condition. In this case, uncertainty in pressure head may be explained considering that the flux corresponding to the prescribed pressure head at the surface is also dependent on the soil hydraulic properties, i.e. f. A stochastic nature of f and the deterministic boundary condition result in a random flux under steady-state conditions and hence in a pressure head variance far from the boundaries. Transient conditions and mean head gradient yield an increase in pressure head variance.

5. CONCLUSIONS

In this paper results from a numerical model are presented, referring to an unsaturated front propagation in a bounded domain. Mean pressure head and pressure head variance profiles are shown for different soil models, boundary conditions, and wetting or drying cases. Only f parameter is considered to be second-order stationary Gaussian processes in space.

Results seem to enforce the conclusion that there is a strong proportionality of pressure head variance and pressure head gradient during transients, when f is a stochastic variable. The constant flux boundary condition seems not to modify the pressure head variance distribution, if compared to that for infinite domain; on the contrary, a constant head boundary condition may strongly affect the variance profiles, in transient and in steady-state conditions, not only close to the boundary. Using different soil models results in different pressure head profile variance, although general behaviors are similar.

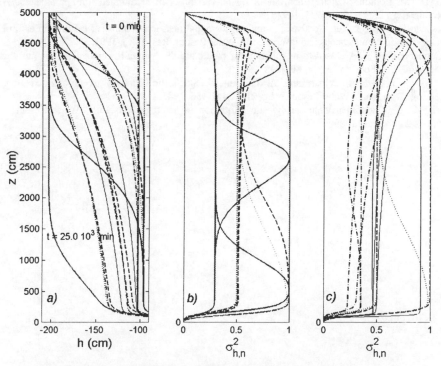

Fig.5: Mean pressure head (a) and normalized pressure head variance profiles (b and c), for GE (solid line), GL (dotted), GR (dashed), vG (dash-dotted), and BC (thin) soil models, for a drying scenario and a constant head top boundary condition.

531

ACKNOWLEDGMENTS

This research has been funded by University of Perugia, project "Criteri ottimali per la valutazione e l'utilizzazione delle risorse idriche", and by National Research Council of Italy, strategic project "Criteri delle disponibilità di acqua da utilizzare a scopo potabile".

REFERENCES

Brooks, R.H., and A.T. Corey, Hydraulic properties of porous media, Hydrol. Pap. 3, Colo. State Univ., Fort Collins, 1964.

Ferrante, M. and T.-C. J. Yeh, Head and flux variability in heterogeneous unsaturated soils under transient flow conditions, Water Resour. Res., 35(5), 1471-1479, 1999.

Gardner, W.R., Some steady-state solutions of unsaturated moisture flow equations with application to evaporation from a water table, Soil. Sci., 85, 228-232, 1958.

Ma, Q., J.E. Hook, L.R. Ahuja, Influence of three-parameter conversion methods between van Genuchten and Brooks-Corey functions on soil hydraulic properties and water-balance predictions, Water Resour. Res., 35(8), 2571-2578, 1999.

Mantoglou, A., and L. W. Gelhar, Capillary tension head variance, mean soil moisture content, and effective specific soil moisture capacity of transient unsaturated flow in stratified soils, Water Resour. Res., 23(1), 47-56, 1987 a.

Mantoglou, A., and L. W. Gelhar, Effective hydraulic conductivities for transient unsaturated flow in stratified soils, Water Resour. Res., 23(1), 57-67, 1987 b.

Mualem, Y., A new model for predicting the hydraulic conductivity of unsaturated porous media, Water Resour. Res., 12, 513-522, 1976.

Russo, D., Determining soil hydraulic properties by parameter estimation: On the selection of a model for the hydraulic properties, Water Resour. Res., 24, 453-459, 1988.

Russo, D., M. Bouton, Statistical analysis of spatial variability in unsaturated flow parameters, Water Resour. Res., 28(7), 1911-1925, 1992.

Sryvastava, R., T.-C. J. Yeh, Analytical solutions for one-dimensional, transient infiltration toward the water table in homogeneous and layered soils, Water Resour. Res, 27(5), 1991.

van Genuchten, M.T., A closed-form equation for predicting the hydraulic conductivity of unsaturated soils, Soil. Sci. Soc. Am. J., 44, 892-898, 1980.

Yeh, T. -C. J., L. W. Gelhar, and A. L. Gutjahr, Stochastic Analysis of unsaturated flow, 1, Heterogeneous soils, Water Resour. Res., 21(4), 447-456, 1985.

Yeh, T. -C. J., One-dimensional steady-state infiltration in heterogeneous soils, Water Resour. Res., 25(10), 2149-2158, 1989.

Stochastic Hydraulics 2000, Wang & Hu (eds) © 2000 Taylor & Francis, ISBN 90 5809 166 X

Time series analysis of monthly rainfall, mean air temperature and carbon dioxide in Japan

Masahiko Hasebe & Takashi Daidou
Department of Civil Engineering, University of Utsunomiya, Yoto, Japan

Takanori Kumekawa
Utsunomiya Technical High School, Kyoumachi, Japan

Sumiko Nejou
JR East Consultants Company Limited, Yoyogi, Tokyo, Japan

ABSTRACT: In this paper, the time series analysis of monthly rainfall, mean air temperature and monthly carbon dioxide in Japan are performed. The forecasting of future values of their time series will be estimated and finally the influence of greenhouse effect from relationship between rainfall or air temperature and carbon dioxide is also investigated. Consequently, from the cross correlation of relationship between air temperature and carbon dioxide, it is guessed that carbon dioxide is influenced by seasonal change and that the pattern of cross correlation is distinguished between summer season and winter season.

Key words: Rainfall series, Air temperature, Carbon Dioxide seasonal change

1. INTRODUCTION

Carbon dioxide is increasing little by little according to rapid development of productive activity of agriculture and industry since the Industrial Revolution in the latter half of the eighteenth. It is assumed that the increase of carbon dioxide (CO_2) is mainly caused by productive human activity such as consumption of fossil fuel and by change of land use with forest destruction and so on.

Carbon dioxide is much amount of gas volume but also the highest effect of green house effect for long life in atmosphere. It is estimated that the concentration of CO_2 gas contributes 60% over against rising trend of air temperature on the earth by green house effect. The second report of IPCC(1996) is said that mean concentration of carbon dioxide all over the world in 1994 goes up from 280 ppmv (before The Industrial Revolution) to 358 ppmv. The most important problem to grasp a change of carbon dioxide concentration in atmosphere is on taking a suitable countermeasure and prediction of changing future climate. In this study, firstly the hydrological characteristics are grasped through time series analysis of monthly rainfall, mean air temperature and carbon dioxide. Next, it is investigated the relationship between air temperature, mean rainfall and CO_2 in Utsunomiya. And finally, from these the relation, it is investigated that the deportment of carbon dioxide influences on mean air temperature and rainfall.

2. ANALYSIS METHOD

Generally, the time series of monthly precipitation, mean air temperature and carbon dioxide (CO_2) in the air are divided into three components, that is, trend component, periodic one and stochastic one.

Trend component is mainly analyzed by least square method. Existence of trend of this component is checked by a statistical analysis.

Periodic component is analyzed by harmonic analysis. So many problems in harmonic analysis involve functions, such as meteorological or economic quantities, whose period is either a day or a year, it is customary to assume that data are available at intervals of a period.

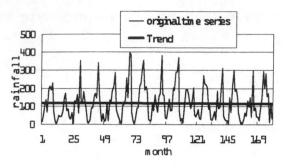

Fig.1: Mean monthly rainfall

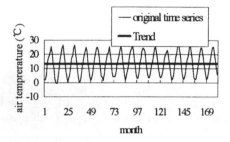

Fig.2: Mean monthly air temperature

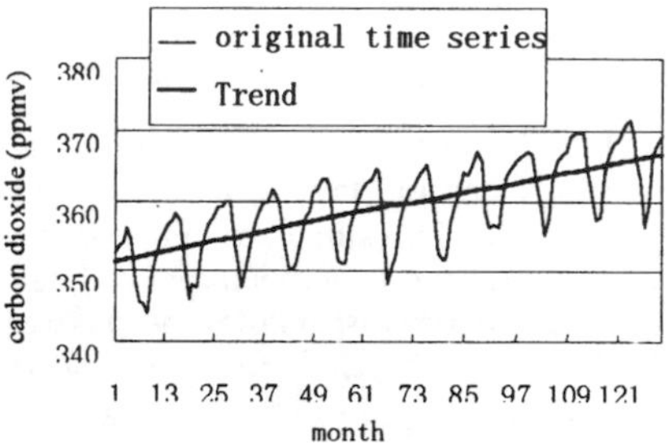

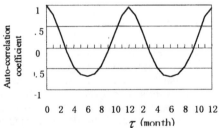

Fig.3: Monthly carbon dioxide

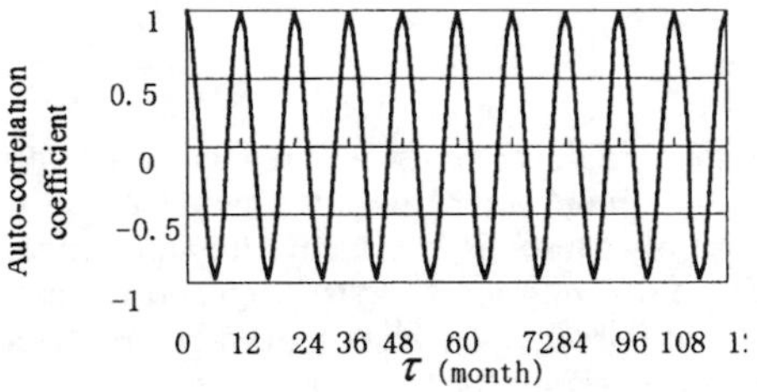

Fig.4: Auto correlation of original time series of
air temperature

Fig.5: Auto correlation of original time series of
carbon dioxide

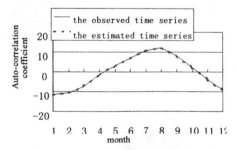

Fig.6: Monthly air temperature

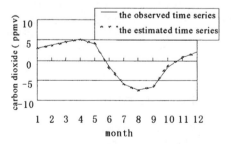

Fig.7: Monthly carbon dioxide

Stochastic component is that a statistical phenomenon that evolves in time according to probabilistic laws is called a stochastic process. We shall often refer to it simply as a process, omitting the word stochastic. The time series to be analyzed may then be thought of as one particular realization, produced by the underlying probability mechanism, of the system under study.

In this paper, the time series analysis of monthly rainfall and mean air temperatures from 1891 to 1997 year at Utsunomiya City and monthly mean carbon dioxide from 1987 to 1998 in Japan, are performed. The forecasting of future values of their time series will be estimated by analyzing them and finally the influence of greenhouse effect from relationship with rainfall or air temperature and carbon dioxide is also investigated.

3. TIME SERIES ANALYSIS.

3.1 Trend components

The time series above mentioned (y_t) is divided into three components, that is,
trend component (T_t), periodic component (P_t) and stochastic component (ε_t) given by equ.(1).

$$y_t = T_t + P_t + \varepsilon_t \tag{1}$$

Raw original data of mean rainfall, mean air temperature and carbon dioxide are shown in Fig.1,2 and 3.
Regression curves ($y_t = b \| x_t + a$) of these time series are estimated by method of least squares and as shown in figures. For checking existence of trend, regression coefficients are checked as following.

Regression coefficient of population of coefficient b calculated by samples N on the assumption that probability function is t-distribution is considered as β. And then check of existence of trend is as follows.

$$t = (b - \beta)/S_b \tag{2}$$

where S_b is as follows.

$$s_b^2 = \frac{\sum_{i=1}^{N} \left\{ (y_i - \overline{y})^2 - b^2 (x_i - \overline{x})^2 \right\}}{(N-2)\sum_{i=1}^{N}(x_i - \overline{x})^2} \tag{3}$$

where N are numbers of samples, x and y are mean values of input and output data.

Consequently, trends of time series of monthly precipitation and mean carbon oxide exist to be a rise, but trend of time series of both monthly mean air temperature not exists.

3.2 Periodic components

Period components of these time series (in case of monthly rainfall and CO_2, the residual time series subtracted trend component from raw time series) are determined by harmonic analysis. Harmonic analysis is fitting by Fourier series judging from auto-correlation function.

From results calculated by auto-correlation as shown in Fig.4 and 5, a period of year (12 months) in three time series exists. The results of these time series fitted by Fourier series are shown in Fig.6 and 7.

The reason why periodic component of CO_2 exists is assumed as seasonal change of carbon dioxide. Namely, this seasonal change may be almost due to biology of land in the north hemisphere. But, in the south hemisphere, as there is almost sea, it is thought smaller than that of the north hemisphere.

Consequently, this seasonal change is contributed to the activity on a life of land.

3.3 Stochastic components

For reference, auto-correlation of the stochastic time series subtracted trend and periodic component from raw time series is shown in Fig.8 in case of CO_2.
Consequently, stochastic component is assumed to be white noise.

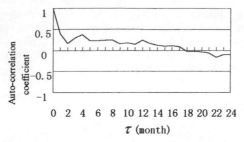

Fig.8: Auto correlation of stochastic component (CO_2)

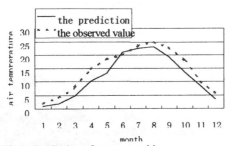

Fig.9: Prediction of mean monthly
air temperature

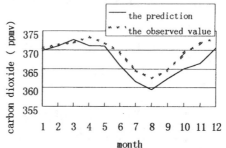

Fig.10: Prediction of monthly carbon dioxide

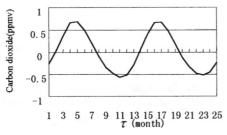

Fig.11: Relationship between rainfall
and carbon dioxide

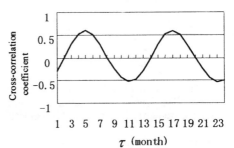

Fig.12: Relationship between rainfall
and carbon dioxide

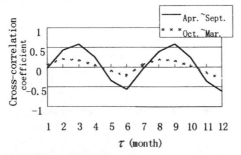

Fig.13: Cross correlation of summer period and winter period

3.4 Forecasting of mean air temperature and carbon dioxide.

Lastly, the forecasting of mean air temperature and carbon dioxide are shown in Fig.9 and 10.

536

4. THE CROSS CORRELATION

The cross correlation of the relationship between carbon dioxide concentration and rainfall except trend component is shown in Fig.11. As coefficients of cross correlation are high, the physical mean is maybe vague. The cross correlation of the relationship between carbon dioxide and air temperature is shown in Fig.12. The cross correlation coefficient is high as same case of Fig.11. It is assumed that there is seasonal influence in the between carbon dioxide concentration and air temperature.

Next, for investigating physical mechanism, cross correlation of separated time series into two seasons (summer period and winter one) is shown in Fig.13.

It is clear that these figures are different tendency. Cross correlation in summer is high but it in winter is low.

As cross relation coefficients are different in these figures, it is thought that carbon dioxide is due to air temperature. Land use of the north hemisphere is different with that one of the south hemisphere. And a biological activity of both hemispheres is different.

5. MAIN RESULTS AND CONCLUSIONS

Main results and conclusions obtained from this analysis are as follows.
  (a) Trend component of time series of monthly rainfall and carbon dioxide exist to be a rise, but trend component of the time series of mean temperature does not exist.
  (b) There are seasonal component (unit of year) in both monthly rainfall, mean air temperature and carbon dioxide as periodic component.
  (c) Stochastic components will be considered to be white noise on the whole and this component is to be fitted AR model.
  (d) Judging from this time series analysis, the estimated value of air temperature and CO_2 are almost in agreement with these observed values. And if this time series is assumed to be non stationary, the forecasting precision of the estimated value is better than stationary time series.
  (e) From the cross correlation of the relation between carbon dioxide and air temperature, it is guessed that carbon dioxide is influenced by seasonal change and that the pattern of cross correlation is distinguished between summer season and winter one. It is thought that their contribution is due to an activity of geology on land.

REFERENCES

Box G.P., and G.M.Jenkins 1976, Time series analysis; forecasting and control, HOLDEN-DAY.
Hasebe.M., 1977, Forecasting and analysis of non-stationary hydrologic time series by Box & Jenkins, Proc., of JSCE, No. 261,59-66 (in Japanese).
Hasebe.M.,1981, Runoff analysis by stochastic process theory, Technical Report No.28 (Tokyo Institute of Technology), 1-141

Stochastic Hydraulics 2000, Wang & Hu (eds) © 2000 Taylor & Francis, ISBN 90 5809 166 X

Stochastic long-memory modelling of seasonal hydrologic time series: A case study

Salvatore Grimaldi

Dipartimento di Idraulica, Trasporti e Strade, University of Rome 'La Sapienza', Italy

ABSTRACT: The extension of stochastic modelling to daily hydrological series assures a bigger sample for study, but entails certain methodological problems not present in series with higher aggregations (monthly, annual). The case study proposed (daily mean discharge time series of the Tiber River) attempts to examine these problems. In fact in this paper a careful deseasonalization analysis is developed, the presence of the so-called Hurst-Effect has suggested the application of the fractional long memory models and the deseasonalization of the variance applied on the series highlighted the necessity of considering the seasonal fractional models.

The comparison of the results of the analysis with the analogous ones obtained by applying traditional short-memory models highlights the suitability of the long-memory approach for the case study considered here.

Key words: Stochastic modelling, Long-modelling, Deseasonalization, Hydrological series, Long-memocy modelling.

1.INTRODUCTION

The estimation of the flood frequency of the Tiber River plays a key role in determining the flood risk for the city of Rome. The evaluation of the expected peak flow with assigned frequency is made necessary by the incommensurable economic and historical value of some parts of the city adjoining the river. Daily river flows of the Tiber were recorded at the Ripetta cross-river section, in the centre of Rome, from 1921 to 1983. Since the sample size available (64 years) is not large enough to reliably estimate the peak flows for very large return periods, we have applied a stochastic model for simulating synthetic river flows observations.

The seasonal component obtained with classical estimation method usually presents, for daily hydrological series, a shape similar to that shown in figure 1. The high variability present, caused by the daily data and by the paucity of the observed years (relative to the data variability), isn't acceptable both physically and stochastically; in fact if one accepts a seasonal component with such variability, one assumes as deterministic what is certainly a stochastic component. To avoid this problem one can refer to STL method (see paragraph 2) which appears more suitable for this case study.

The R/S statistic method, shown in figure 2, seems to show the probably presence of the so-called Hurst-Effect in the series, which would indicate a possible long-term persistence. In the paragraphs 2 and 4 we describe the techniques used and present the results obtained in order to identify this particular correlation structure.

The ascertained presence of this characteristic led to consider the fractional ARMA models and the methodology used to develop this modelling (see paragraph 3).

The results relative to the case study proposed are described in the paragraph 4.

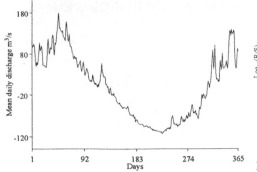
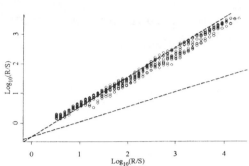

Fig.1: The seasonal component of the daily river flow of the Tiber River.

Fig.2: R/S statistic deseasonalized mean daily Tiber River flow. The two solid lines represent the slopes corresponding to H=0.5 and H=1.

2. THE PRELIMINARY ANALYSIS

As well known from literature, the preliminary analyses have the aim to transform the series to satisfy the hypothesis of stationarity and normality. For the nonstationarity in variance and for the normality one can use the Box and Cox transformations:

$$Y_t = (X_t^\lambda - 1)/\lambda; \text{ for } \lambda = 0, Y_t = \log(X_t) \tag{1}$$

For the hydrological series the nonstationarity in mean usually is represented by seasonal variability of the observed values. This characteristic may be removed from the series by certain deseasonalization techniques. Among these the possible applications of the STL method seem to be the most efficient for the case study proposed here.

The basic STL procedure (Seasonal-Trend Decomposition Procedure Based on Loess, Cleveland et al. 1990) is developed by certain steps, repeating iteratively (inner loops):

Step 1:Detrending. The second cycle onwards (having calculated the trend with first cycle) the original series Y_t is removed from the trend T^k k=inner loop number.

Step 2:Cycle-subseries Smoothing. The resulting series is subdivided in to the m subseries (m=period) to which the loess technique is applied.

Step3:Low-Pass Filtering of Smoothed Cycle-Subseries. Having recomposed the series C^k with transformed subseries the following are applied to it: twice a filter of moving average of interval m, a filter of moving average of interval 3 and finally a loess smoothing. The output L^k represents the temporary series trend.

Step4:Detrending of Smoothed Cycle Subseries. The seasonal component is temporarily defined by the difference $S^{k+1} = C^{k+1} - L^{k+1}$

Step5: Deseasonalizing. The original series is deseasonalised $Y - S^{k+1}$

Step 6: Trend Smoothing The loess smoothing is adequately applied to the deseasonalised series to define the trend component T^{k+1}.

The basic STL just described presents a high variability, therefore a post-smoothing with a variable interval of the seasonal component may be done applying the loess technique.

A further development of the STL is represented by the procedure that gives robustness to it (Robust STL). One can operationally execute certain "outer loops" (from 1 to 10): having completed the inner loops one can obtained the residual series $R = Y - T - S$ to which a filter is applied to give it robustness. This filter is characterised by the weight function:

$$\rho_t = B(|R_t|/h) \quad B(u) = \begin{matrix} (1-u^2)^2 & \text{for } 0 \le u \le 1 \\ 0 & \text{for } u > 1 \end{matrix} \quad h = 6 \times median(|R_t|) \tag{2}$$

which permits the exclusion of the outliers present in the series. After such subtraction the inner loops are redone to repeat the outer loops iteratively.

540

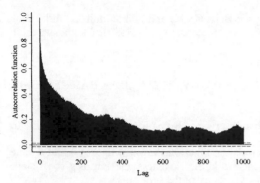

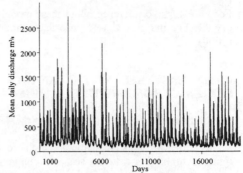

Fig.3: Autocorrelation function of the deseasonalizated river Tiber River mean daily discharge series.

Fig.4: Mean discharge time series of the Tiber River observed at Roma Ripetta gauge (1930-1983)

Once the transformations required to have a stationary series have been made, if a slowly decreasing autocorrelation function is present (see figure 3), an identification analysis to ascertain the possible presence of the Hurst-Effect is needed. This analysis consists in using some heuristic methods that permit the detection long memory and provide a rough estimate of the H exponent value. In the case study proposed the R/S method, the aggregated variance method, the differencing variance method, the absolute values method, the Higuci method, the residuals of regression method, the periodogram method have been applied; for a description of the heuristic procedures applied here, see Taqqu et.al (1995).

3.THE STOCHASTIC MODEL USED

In order to study a series with the Hurst-Effect, an important extension of the classical ARIMA models, named FARIMA (fractionally ARIMA) models, was introduced (Granger and Joyeux (1980), Hosking (1981)). Allowing the differencing order of the model to be fractional, this model type displays long range dependence, enabling it to reproduce the Hurst phenomenon. The model structure is the same as the classical short memory models:

$$\phi(B)(1-B)^d X_t = \theta(B)\alpha_t \qquad (3)$$

where B is the backward shift operator $(BX_t=X_{t-1})$, $\phi(B)$ is the autoregressive polinomial, $\theta(B)$ is the moving average polinomial, α_t is the noise term and d is the differencing order.

But in the case of the FARIMA model in (1) the index d can be fractional and not only integer. In fact the d value is related to the Hurst parameter H by the relationship d=H-0.5. This model is thus capable of reproducing the autocorrelation structure of natural processes displaying both short and long-term persistence.

In certain series one can find a weak seasonal stochastic component which can not be remove by the classical methods (Salas et al, 1982). For such cases an extension of the seasonal short memory models and of the long memory models was introduced (Montanari et. al. 2000). The general form of these models is:

$$\phi(B)\Phi(B^s)(1-B)^d(1-B^s)^{d_s} X_t = \theta(B)\Theta(B^s)\alpha_t \qquad (4)$$

where $\Phi(B^s)$ and $\Theta(B^s)$ are respectively the seasonal autoregressive and moving average polinomials and d_s is the seasonal differencing order.

Also in this case the only difference with the seasonal short memory models is the seasonal differencing orders (d_s) that can be fractional.

541

For an extensive reference about these models see Beran(1994) and Samorodnitsky and Taqqu (1994). For the procedure to use these models see Montanari et al. (1997), Montanari et al. (2000).

4.THE CASE STUDY: THE ANALYSIS OF DAILY MEAN DISCHARGE TIME SERIES AT ROMA RIPETTA GAUGE

The daily mean discharge time series of the Tiber observed at Roma Ripetta gauge from 1930 to 1983 (19723 data) is considered.

This series presents the shape and the autocorrelation function shown in figures 4 and 5. The figure 5 highlights a strong seasonal component, while the skweness and kurtosis coefficient values (respectively: 3.64, 24.11) show the non-gaussianity of the series.

The transformations (1) have been applied with λ equal to -0.75, this value has been chosen using a skweness test (Cromwell J.B. et al., 1994).

As shown in figure 1 the mean seasonal component has high variability and therefore has been defined applying the STL technique with ten "outer loops" and with a smoothing of 120 lags (see figure 6).

The variance seasonal component, which has the same variability problems of the mean seasonal component, has been defined (see figure 7) applying the sum of harmonics regression (Kottegoda N.T., 1980).

Figure 3 shows the autocorrelation function of the transformed and deseasonalized series.

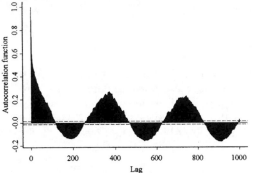
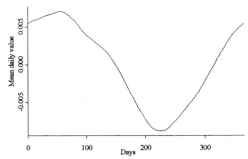

Fig.5: Autocorrelation function of the Tiber River observed at Roma Ripetta gauge (1930-1983)

Fig.6: Mean seasonal component of the transformed series.

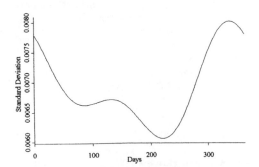
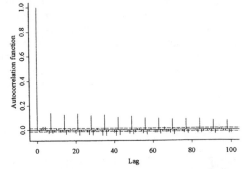

Fig.7: Variance seasonal component of the transformed series.

Fig.8: Residual autocorrelation function of the ARMA(1,0) model.

542

Table 1: H parameter values obtained with the heuristic estimation methods

R/S	Var. Agg.	Var. Diff.	Val. Ass.	Res. Regr.	Higuci	Periodog	Mean Value
0.945	0.861	1.065	0.864	0.991	0.861	1.069	0.951

Table 2: Parameter and index values of the seasonal long memory models developed for the deseasonalizated river Tiber mean daily discharge series

FARIMA$(1,d,1)\times(1,d_s,1)_7$	H=0.913	(0.877, 0.945)	$\sigma^2 = 0.153696$	
	$\phi_1 = 0.502$	(0.45 , 0.553)	Q = 187.7 (lag=100)	Q=29.7 (lag=20)
	$\theta_1 = 0.105$	(0.071 , 0.139)	$\chi^2_{(95\%)} = 117.6$	$\chi^2_{(95\%)} = 23.7$
	$\Phi_1 = 0.279$	(0.231 , 0.326)	BIC = -16002.7	
	$H_s = 0.829$	(0.795, 0.863)	Explained variance 78%	
FARIMA$(1,d,0)\times(1,d_s,1)_7$	H=0.973	(0.946, 0.999)	$\sigma^2 = 0.1538684$	
	$\phi_1 = 0.347$	(0.316 , 0.379)	Q = 205.3 (lag 100)	Q=35.1 (lag=20)
	$\Phi_1 = 0.278$	(0.234 , 0.324)	$\chi^2_{(95\%)} = 118.7$	$\chi^2_{(95\%)} = 25$
	$\Theta_1 = 0.535$	(0.477 , 0.594)	BIC = -15997.4	
	$H_s = 0.831$	(0.796, 0.865)	Explained variance 78%	
FARIMA$(2,d,0)\times(1,d_s,1)_7$	H=0.925	(0.893, 0.959)	$\sigma^2 = 0.1537253$	
	$\phi_1 = 0.385$	(0.349 , 0.421)	Q =192 (lag=100)	Q=30.6 (lag=20)
	$\phi_2 = 0.036$	(0.021 , 0.051)	$\chi^2_{(95\%)} = 117.6$	$\chi^2_{(95\%)} = 23.7$
	$\Phi_1 = 0.279$	(0.232, 0.326)	BIC = -16001	
	$\Theta_1 = 0.529$	(0.47 , 0.589)	Explained variance 78%	
	$H_s = 0.829$	(0.795, 0.864)		
FARIMA$(2,d,1)\times(1,d_s,1)_7$	H=0.621	(0.534, 0.708)	$\sigma^2 = 0.1533805$	
	$\phi_1 = 1.532$	(1.415, 1.65)	Q =142.5 (lag=100)	Q=21.5 (lag=20)
	$\phi_2 = -0.558$	(-0.66 , -0.46)	$\chi^2_{(95\%)} = 116.5$	$\chi^2_{(95\%)} = 22.4$
	$\theta_1 = 0.845$	(0.8 , 0.891)	BIC = -16016	
	$\Phi_1 = 0.26$	(0.216, 0.305)	Explained variance 78%	
	$\Theta_1 = 0.544$	(0.487 , 0.601)		
	$H_s = 0.856$	(0.821, 0.892)		

As written in the previous paragraph the shape of this function suggests using the identification methods to detect the presence of long memory. Table 1 shows the value of the H parameter obtained applying the seven estimation methods listed in paragraph 2. One can see that all methods point out the presence of long memory persistence.

Applying the methodology described in Montanari et al. (1997) and in Montanari et al. (2000), the series has been filtered with a fractional difference of $d=H_{mean}-0.5=0.45$ and then an ARMA(p,q) model has been built. The residuals of some estimated models shows a weak seven order periodicity as one can see in figure 8. The relative periodicity component calculated with the previous method is not meaningful, therefore the application of a seasonal fractional model became necessary. Table 2 shows the results of the most plausible configurations; the parameters' values and the relative confidence interval, the residual variance estimation, the Portmonteau Q statistic value, the BIC index value (Piccolo, 1990) and the explained variance of the filter have been calculated.

Therefore the best model seems to be the FARIMA $(2,d,1)\times(1,d_s,1)_7$:

$$(1-1.53B+0.56B^2)(1-0.26B^7)(1-B^7)^{0.36}(1-B)^{0.62}X_t=(1-0.84B)(1-0.54B^7)a_t \qquad (3)$$

The whiteness of the residuals can be confirmed by the cumulative periodogram shown in figure 9.

From the point of view of the application, the goodness of the model built may be judged only by verifying that the simulated series are characterised by a probability t density and by an ACF similar to those of the observed series. This simulation is made by an equi-probable sampling of the residuals and by the infinity moving average representation of he model (3) (Montanari A, el al., 1997; Hipel K.W. McLeod A.I., 1994).

543

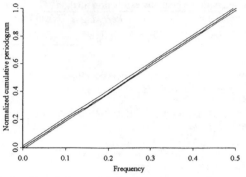

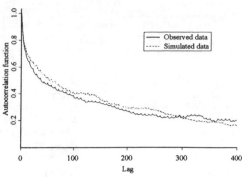

Fig.9: Cumulative periodogram test on the residual of the FARIMA $(2,d,1)\times(1,d_s,1)_7$ model. The two straight lines are the 95% confidence bands.

Fig.10: Comparison between the autocorrelation function of observed data and simulated data with the FARIMA $(2,d,1)\times(1,d_s,1)_7$

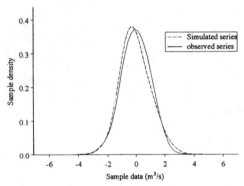

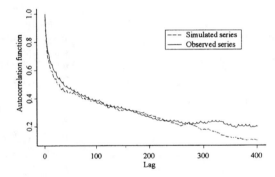

Fig.11: Comparison between the sample density of observed data and simulated data with the FARIMA $(2,d,1)\times(1,d_s,1)_7$

Fig.12: Comparison between the autocorrelation function of observed data and simulated data with the ARMA (15,7)

Figures 10 and 11 show comparison between the density function and the ACF relative to the real and simulated series, highlighting the suitability of the model chosen.

5. CONCLUSIONS

The obtained results with the case study developed in this work permit certain conclusions and considerations.

The necessity of a particular deseasonalization in mean and in variance for the series with low aggregation stands confirmed.

Only with a fractional model can one obtained a simple modelling. In order to obtain the same comparison between the autocorrelation functions shown in figure 9 using the short memory model one has to develop a very onerous identification procedure. For this case study the best short memory model seems to be a ARMA(15,7) with 8 parameters and this model can provide the comparison between the autocorrelation functions shown in figure 12.

It has been shown that the deseasonalization in variance applied to the series highlights a weak periodicity. This particular but common characteristic for the daily discharge time series can be modelled only with a seasonal fractional differencing.

However, further problems must be solved before a perfect modelling of daily hydrological series can be obtained. These are usually non-gaussian series. The application of Box and Cox

544

transformations partially permit us to overcome this difficulty, in fact when one applies the inverse transformation one can have a strong alteration of the series.

In order to avoid these distortions we can study more carefully the transformations or we can use the non-gaussian models. These are the subjects of ongoing works.

ACKNOWLEDGEMENTS.

The author would like to thank Alberto Montanari for his help in series modelling.

REFERENCES

Beran J. Statistics for long-memory processes, Chapman and Hall, New York, 1994

Cleveland B.R., W.S. Cleveland, J.E. McRae, and I. Terpenning, STL: A seasonal trend decomposition procedure based on loess, J. Off. Stat., 6, 3-73,1990

Cromwell J.B, W.C. Labys, M. Terraza. Univariate Tests for Time Series Models, Sage Universiy Paper series on Quantitative Applications inthe social sciences, 1994

Granger, C.W.J., and R. Joyeux, An introduction to long range time series and fractional differencing, J. Time Ser. Anal.,1,15-30, 1980.

Hipel K.W. A.I. McLeod, Time Series Modelling of Water Resources and Enviramental Systems, Elsevier Science, 1994

Hosking, J.R.M., Fractional differencing, Biometrika, 68,165-176, 1981.

Kottegoda, N.T., Stochastic Water Resources Technology, Macmillan, New York,1980

Montanari A., Rosso R., Taqqu M.S., Fractionally differenced ARIMA models applied to hydrologic time series: identification, estimation and simulation, Water Resour. Res., vol. 33, 1035-1044, 1997

Montanari A., Rosso R., Taqqu M.S., A seasonal fractional ARIMA model applied to the Nile river monthly flows at Aswan, to appear on Water Resour. Res., 2000

Piccolo D., Introduzione all'analisi delle serie storiche, La Nuova Italia Scientifica, 1990

Salas, J.D., C.D. Boes, and R.A. Smith, Estimation of ARMA models with seasonal parameters, Water Resour. Res., 18, 1006-1010, 1982

Samorodnitsky, G.,and M.S. Taqqu, Stable Non-Gaussian Random Processes: Stochastic Models With Infinite Variance,Chapman and Hall, New York, 1994

Taqqu M.S., Teverovsky V., Wilinger W., Estimators for long-range dependence: an empirical study, Fractals, Vol. 3, No.4 (1995) 785-788.

Stochastic Hydraulics 2000, Wang & Hu (eds) © 2000 Taylor & Francis, ISBN 90 5809 166 X

Statistical evaluation of the empirical equations that estimate hydraulic parameters for flow through rockfill

Seyed Mahmood Hosseini

Civil Engineering Department, Ferdowsi University, Mashhad, Iran

ABSTRACT: In this paper, the empirical equations that estimate the hydraulic parameters for non-linear flow through rockfill are evaluated using a series of independent data collected in the laboratory. In this regard, three different uniform soils ranging in size from 8.7 to 25.7 mm have been selected and three random samples drawn from each material. The physical characteristics such as size distribution, porosity and shape factor have been measured for the 9 samples. Permeameter tests have been conducted on the samples to create a set of reliable *hydraulic gradient vs. bulk velocity* data. Statistical measures are used to compare the permeameter data with those predicted by the empirical equations. The study shows that McCorquodale *et al.* and Stephenson equations, which some subjective parameters related to the surface characters of the material, have been incorporated in their structures, can give good results.

Key words: Rockfill, Non-linear flow, Non-Darcy flow

1. INTRODUCTION

Use of rock as a construction material for hydraulic structures such as rockfill dams, gabions, and breakwaters is a common practice across the world. A common characteristic of rockfill structures is that flow through the porous media often deviates from Darcy's law and a non-linear or non-Darcy flow equation is applicable.

Estimation of hydraulic parameters under non-linear flow conditions is often a problem for rockfill structures. This is due to the fact that these lumped parameters are macroscopic manifestation of physical properties of the media such as size and size distribution of the materials, porosity, orientation, and shape and roughness of the grains. The average hydraulic effects of these characteristics are hard to quantify and therefore uncertainty will be an inherent ingredient of the estimated parameters.

A commonly used method to obtain the hydraulic parameters is the use of empirical relations based on physical properties of the media. Although the research in this area has been extensive, there is no general agreement on one specific equation (Hansen *et. al*, 1995). In this study six main equations in the literature are examined using the independent permeameter data collected in the laboratory. The equations considered in this study are Ergun equation, McCorquodale *et al.* equation, Stephenson equation, Adel equation, Wilkins equation and Martins equation.

2. NON-LINEAR FLOW EQUATIONS

The first theories developed to account for non-linear effects in porous media are models more or less intuitive and empirical in nature. The first equation to account for non-linear effects was proposed by Forchheimer (Bear, 1972) who suggested the following one-dimensional form:

$$i = aV + bV^2 \tag{1}$$

where i is hydraulic gradient, V is bulk velocity and a and b are constants. Another commonly used non-linear equation is the Missbach equation (Scheidegger, 1974):

$$i = lV^\lambda \tag{2}$$

where l and λ are constants which depend on media and fluid properties. λ is variable between 1 and 2 and changes from case to case. A different approach is considered to be the concept of introducing a friction factor for the porous media that can be obtained from a friction factor-Reynolds number diagram similar to the Moody diagram for pipe flow. This approach results in an equation similar to Eq. (1).

Parallel to research on the theoretical and physical explanation of non-linear flow (Cvetkovic, 1986; Hassanizadeh and Gray, 1987; Irmay, 1958), extensive research has also been done to relate the coefficients in these equations to fluid and porous media properties. Hansen *et al.* (1995) give a good review of different non-linear equations with specified parameters. The equations below are the most widely used equations in the literature.

Ergun equation (Ergun, 1952):

$$i = \frac{150(1-n)^2 v}{n^3 g d^2} V + \frac{1.75(1-n)}{n^3 g d} V^2 \tag{3}$$

McCorquodale *et al.* equation (McCorquodale *et al.*, 1978):

For low Reynolds numbers, i.e. $R_p = \dfrac{Vm}{vn} \leq 500$ or $R_w = \dfrac{Vmn^{1/2}}{v} \leq 125$

$$i = \frac{4.6v}{gnm^2} V + \frac{0.79}{gn^{1/2}m} V^2 \tag{4a}$$

For high Reynolds number, i.e. $R_p > 500$ or $R_w > 125$

$$i = \frac{70v}{gnm'^2} V + \frac{0.27\left(1 + \dfrac{f_\varepsilon}{f_o}\right)}{gn^{1/2}m'} V^2 \tag{4b}$$

Stephenson equation (Stephenson, 1979):

$$i = \frac{800v}{gnd^2} V + \frac{k}{n^2 g d} V^2 \tag{5}$$

Adel equation (developed in 1987) as reported by Bakker and Meijers (1988):

$$i = \frac{160v(1-n)^2}{gn^3 d_{15}^2} V + \frac{2.2}{gn^2 d_{15}} V^2 \tag{6}$$

In Eqs. (3) to (6), i is hydraulic gradient, V is bulk velocity, n is porosity, v is kinematic viscosity, g is gravitational acceleration, d is harmonic mean particle size, R_p is pore Reynolds number, R_w is Ward Reynolds number, k is the friction factor in the turbulent region of flow (= 1 for smooth polished marble, = 2 for semi-rounded stones and = 4 for angular rocks) and d_{15} is the grain size which 15% of the particles, by weight, are smaller than. Also:

f_ε = Darcy friction factor for rock and permeameter, but the wall effect removed from the data,

f_o = Darcy-Weisbach friction factor for a hydraulically smooth surface functioning at the same Reynolds number (obtained from a Moody diagram for pipe flow), $f_\varepsilon / f_o \approx 1.5$ for crushed rock (Hansen *et al.*, 1995). According to an example given in the reference (McCorquodale *et al.*, 1978) values of 1.15 and 1.75 are expected for river gravel and crushed rock, respectively,

m' = effective pore hydraulic radius,

$$m' = \frac{Total\ effective\ volume\ of\ voids}{Total\ effective\ surface\ area} = \frac{n}{6\int_0^1 (1-n)\dfrac{\alpha df}{D(f)} + \dfrac{c_w}{R_x}} \tag{7}$$

m = pore hydraulic radius (the same as above with wall effect removed from the equation).
where:
f = accumulative frequency of the granular material, $D(f)$ = grain size finer than f by weight, R_x = hydraulic radius of permeameter, α = Ratio of the surface area of the particle to the surface area of a sphere of the same volume. This parameter can be estimated from Fig. 8 of the reference (McCorquodale *et al.*, 1978) and c_w = empirical coefficient to account for wall effect (= -0.5)

All of the above equations follow the Forchheimer-type constitutive relationship. However, the two equations below have also been used which have a power-law structure. The equations are arranged to give the hydraulic gradient.

$$V_v = Wm^{0.5}i^{0.54}$$
(8)

Wilkins equation (developed in 1952) as reported by Garga *et al.* (1989):

$$i = \frac{1}{m^{0.93}}\left(\frac{V}{Wn}\right)^{1.85}$$
(9)

where i is hydraulic gradient, V is bulk velocity, m is mean hydraulic radius, W is a constant (= 5.243 in SI system of units), V_v is velocity in the voids or interstitial velocity (V/n) and n is porosity. The key parameter in Wilkins equation is the mean hydraulic radius which is defined as:

$$m = \frac{ed}{6r_e}$$
(10)

where e is void ratio, d is particle diameter which can be calculated as harmonic mean according to hydraulic radius theory (Vukovic and Soro, 1992) and r_e is the relative surface area efficiency (Garga *et al.*, 1989), a coefficient that accounts for the deviation from a smooth spherical shape (= 1 for sphere, $\approx$ 1.6 for crushed limestone, up to 2 for crushed rock).

Martins equations (Martins, 1990):

$$V_v = \frac{K_m}{C_u^{\gamma}}\sqrt{2gedi}$$
(11)

$$i = \frac{C_u^{2\gamma}}{2n^2K_m^2ged}V^2 \qquad Re = \frac{4V_vm}{v} > 300$$
for
(12)

where i is hydraulic gradient, V is bulk velocity, $C_u = d_{60}/d_{10}$ is coefficient of uniformity, γ is a constant (=0.26), n is porosity, K_m is a constant (= 0.56 for angular materials and = 0.75 for rounded materials), g is gravitational acceleration, e is void ratio, V_v is velocity in the voids (V/n), d is mean particle diameter, v is kinematic viscosity and $m = ed/c'$ is mean hydraulic radius where c' is a constant (= 6.3 for rounded particles and = 8.5 for angular particles). For transition zone, Reynolds number between 10 and 300, Martins proposes an empirical table that relates $\eta = V_v quadratic/V_v transition$ to parameter $ed\sqrt[3]{i}$.

3. EXPERIMENTAL WORK

Three types of materials selected in this study were different in size ranging from 8.7 mm to 25.7 mm. Materials were obtained from a sand and gravel quarry. Although each material had been mechanically sorted in the quarry, it was washed and completely mixed in the laboratory to produce a media as homogeneous as possible. Three samples were randomly drawn from each material. Each sample was large enough to fill the permeameter (about 25 kg). For each sample the size distribution, particle density, porosity and shape factor were determined. The porosity considered for each of the three samples was the in-situ porosity measured in the permeameter. To estimate the shape factor, three major axes were measured for the particles using a digital caliper, and the average axes lengths calculated for each sample. Shape factors (SF) were estimated using the relationship $SF = c^*/\sqrt{a^*b^*}$ where a^* is length in longest direction and b^*, c^* are lengths measured in mutually perpendicular medium and short directions, respectively.

For each material, the three random samples were tested in the permeameter. The permeameter was made of a vertical PVC pipe 1.0 m long and 152 mm inside diameter containing the media and head losses were determined using two piezometer taps 717 mm apart. Discharge was measured using a 38 mm diameter orifice in the supply line. At least 17 discharges were tested for each sample and totally 248 *hydraulic gradient-bulk velocity* data were collected for all samples.

4. EXPERIMENTAL RESULTS AND SUMMARY

Table 1 summarises the properties of the three samples randomly drawn from each of the three materials.

Table 1:. Material Properties

Material		d_{50} (mm)	d_{10} (mm)	Coef. of Uniformity (-)	Coef. of Concavity (-)	Particle Density (g/cm^3)	Porosity (-)	Shape Factor (-)
	(1)	8.7	5.7	1.63	1.06	2.76	0.477	0.49
Small	(2)	8.5	5.6	1.61	1.03	2.76	0.483	0.49
	(3)	8.5	5.6	1.61	1.06	2.74	0.489	0.42
	(1)	21:2	15.3	1.44	1.14	2.75	0.456	0.52
Medium	(2)	21.0	15.0	1.46	1.13	2.69	0.458	0.54
	(3)	21.1	15.1	1.46	1.13	2.70	0.459	0.54
	(1)	27.4	20.7	1.38	1.04	2.62	0.443	0.47
Large	(2)	25.6	20.0	1.36	0.96	2.63	0.443	0.56
	(3)	27.6	20.5	1.41	1.05	2.54	0.443	0.48

Most parameters related to the equations discussed in Section 3 are reported in Table 1. However, some of these equations have parameters that must be selected by engineering judgement. These parameters may act as the main source of uncertainty. The following is the list of these parameters together with rational behind selecting them:

a) f_ε / f_o in McCorquodale *et al.* equation = 1.4 for medium and large material and 1.6 for small material based on shape and surface characters.

b) α in McCorquodale *et al.* equation = 1.4 for medium and large material and 2.0 for small material based on shape and surface characters of the materials and using Fig. 8 in the reference (McCorquodale *et al.*, 1978)

c) k in Stephenson equation = 2.75 for medium and large material and = 3.5 for small material based on the shape and surface characters of the materials.

d) r_e in Wilkins equation = 1.4 for medium and large material and = 2.0 for small material based on the shape and surface characters of the materials and recommendations made by Wilkins.

e) K_m in Martins equation = 0.68 for medium and large material and = 0.61 for small material based on the shape of the materials and recommendations made by Martins.

All other parameters were calculated either from size distribution curve or by direct measurement the way described previously.

5. DATA ANALYSIS

In this section, the physical properties associated with each empirical equation, reported in Section 3.3.1, are applied to each equation to find the hydraulic gradients for all velocity values corresponding to permeameter tests. The simulated hydraulic gradients resulting from different equations are then compared to the corresponding observed values to see which equation has the best performance.

Fig. 1 shows typical *i(simulated) vs. i(observed)* curve for Stephenson equation. Full line in the graphs show the best-fit line (prediction line) which is obtained by applying the least square method to the data. In an ideal situation the slope of such curves should be one (45° line shown by dashed line) while the intercept is zero. The curve is for nine samples resulting from three different materials.

In addition to this method of analysis, the mean absolute relative error was also calculated for all equations. Table 2 shows the summary of the results.

Table 2:. Examination of Different Non-linear Loss Equations

Equation	Slope of Prediction Line	Mean Absolute Relative Error (in %)	Prediction
Ergun equation	0.635	32	Underestimate
McCorquodale et al. equation	0.978	11	Underestimate
Stephenson equation	1.035	9	Overestimate
Adel equation	0.948	12	Underestimate
Wilkins equation	1.218	28	Overestimate
Martins equation	0.530	48	Underestimate

6. CONCLUSIONS

By looking at the results reported in Table 2, following conclusions can be made:

1. Ergun and Martins equations underestimate the hydraulic gradient based on the data analysis conducted in this study. This conclusion is in agreement with the study conducted by Hansen *et al.* (1995).

2. McCorquodale *et al.* and Stephenson equations, which some subjective parameters related to the surface characters of the material, have been incorporated in their structures, can give good results. A reasonable estimation of these subjective parameters can be made by following the recommendations made by the developers of the equations. McCorquodale *et al.* equation is more computationally intensive and needs adjustment of *a* and *b* based on Reynolds number. This equation is recommended for experimental work in the laboratory because it accounts for wall effect and deals with low Reynolds number flow using a separate equation developed for this range.

3. Adel equation considers a small size as the characteristic size of the domain and does not incorporate any effect of shape in its structure. However, its result is quite comparable with those of McCorquodale *et al.* and Stephenson equations. Equations such as this are also quite useful especially in the presence of the size distribution and the absence of any information about the surface characters of the materials.

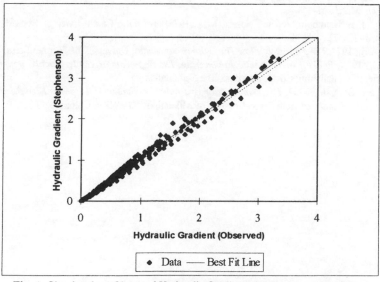

Fig. 1: Simulated vs. Observed Hydraulic Gradients for Stephenson equation

Although some agreement is found between the conclusions made in this study with those made by Hansen *et al.* (1995), these studies do not encompass all situations and therefore should not be considered absolute as absolute knowledge is not attainable for open systems. Therefore, the modeler should be aware of the possible reasons for this discrepancy and the consequences of using a specific equation. Different model structures, different error values in experimental work and different methodologies for statistical analysis of data (selection of dependent and independent variables) may explain the reasons for such a discrepancy in the results obtained from different equations.

ACKNOWLEDGEMENTS

Financial support for the data analysis of this research was provided by a research grant from Ferdowsi University, Mashhad, Iran. This support is gratefully acknowledged. The author would also like to acknowledge Dr. Douglas Mark Joy from School of Engineering, University of Guelph, Canada for his technical and financial support during collecting the experimental data.

REFERENCES

Bakker, K.J., and Meijers P., 1988, "Stability Against Sliding of Flexible Revetments", in *Modelling Soil-Water-Structure Interactions*, Kolkman *et al.* (ed.), Balkema, Roterdam.

Bear, J., 1972, *Dynamics of Fluids in Porous Media*, American Elsevier Publishing Company, NY.

Cvetkovic, V.D., 1986, A Continuum Approach to High Velocity Flow in Porous Medium. *Transp. Porous Media*, 1, 63-97.

Garga, V.K., Hansen, D., and Townsend, D.R., 1989, Considerations in the Design of Flow Through Rockfill Drains, *14th Mine Recl. Symp.*, Min. Assoc. of B.C., Cranbrook, British Columbia.

Hansen, D., Garga, V.K., and Townsend, D.R., 1995, Selection and Application of One-dimensional Non-Darcy Flow Equation for Two-dimensional Flow Through Rockfill Embankments, *Can. Geotech. J.*, 32, 223- 232.

Hassanizadeh, S.M., and Gray, W.G., 1987, High Velocity Flow in Porous Media, *Transp. Porous Media*, 2, 521-531.

Irmay, S., 1958, On the Theoretical Derivation of Darcy and Forchheimer Formulas, *Trans. of Am. Geophys. Union*, 39(4), 702-707.

Martins, R., 1990, Turbulent Seepage Flow Through Rockfill Structures, *Water Power and Dam Construction*, March, 41-45.

McCorquodale, J.A., Hannoura, A., and Nasser, M.S., 1978, Hydraulic Conductivity of Rockfill, *J. Hydr. Res.*, 16(2), 123-137.

Scheidegger, A.E., 1974, *The Physics of Flow Through Porous Media*, University of Toronto Press, Toronto.

Stephenson, D., 1979, *Rockfill in Hydraulic Engineering, Developments in Geotechnical Engineering 27*, Elsevier Scientific Publishing Co., Amsterdam, The Netherlands.

Vukovic, M., and Soro, A., 1992, *Determination of Hydraulic Conductivity of Porous Media from Grain-SizeComposition*, Translated from Serbo-Croation by Milandinov, D., WRP, Colorado, U.S.A.

Stochastic Hydraulics 2000, Wang & Hu (eds) © 2000 Taylor & Francis, ISBN 90 5809 166 X

Parameter reduction of decomposed MAR(1) procedure

Chavalit Chaleeraktrakoon & Raviwan Artlert
Department of Civil Engineering, Thammasat University, Pathumthani, Thailand

ABSTRACT: Generation of multisite seasonal flows that can reproduce the flow properties at various aggregate time scales has often encountered the problem of parameter estimation. The number of parameters may easily exceed available flow data. This paper proposes a new scheme for reducing the parameters of a stochastic procedure (i.e., singular value decomposition combined with multivariate autoregressive model of order one). The proposed approach uses the criterion of percentage cumulative variance to select the significantly standardized principal components, and fits the autoregressive model to the significant ones. It was applied to four sequences of monthly flows in the Chao Phraya river basin. Results have demonstrated that the proposed scheme has less parameters than the existing one. Further, it can preserve the basic statistics, and some drought related properties of historic monthly and annual flows adequately.

KEY WORDS: Seasonal flow generation, Singular value decomposition (SVD), Multivariate AR(1) [MAR(1)]

1.INTRODUCTION

Seasonal flows are basic information for planning, design and operation of various water resource systems. They may be used for estimating reservoir storage capacities, optimal operating policies and long-term deficit risks in water supply and hydroelectric generation. To obtain reliable estimates of the system characteristics, simulation studies of water resources systems against many samples of generated flows at multiple sites are usually suggested in practice. Often, most simulation studies require generated flows that reproduce the means, standard deviations, temporal and spatial correlation coefficients and skewness coefficients of historic flows at various aggregation time scales. The requirements may cause problems when modeling historic seasonal flows at a number of sites simultaneously. The number of model parameters may easily exceed available flow data.

Several condense disaggregation models are widely used to generate multisite seasonal flows (e.g.; Lane, 1982; Grygier and Stedinger, 1988; Santos and Salas, 1992). Recently, Chaleeraktrakoon (1999) has proposed an alternative stochastic procedure for modeling such flows. The procedure transforms the matrix of normalized seasonal flows into its standardized principal components (SPC) with singular value decomposition (SVD), and fits the first order of multivariate autoregressive model [MAR(1)] to the SPC matrix. When dealing with seasonalflows at the large number of locations, the SVD/MAR(1) procedure has reduced the number of parameters by modeling the MAR(1) to each sub-matrix of successive SPC.

A new scheme for reducing the SVD/MAR(1) parameters is proposed in this study. The proposed method uses the criterion of cumulative variance to separate the SPC matrix into statistically significant SPC and remaining noises, and model the MAR(1) to the significant one.

Table.1: Number of parameters and degrees of freedom of the multisite SVD/MAR(1) procedures (Chao Praya data)

Model	Number of parameters	Degrees of freedom
Simultaneous-decomposition	242	25
Stage-decomposition	1058	13
Chaleeraktrakoon(1999)	914	24

Table.2: Comparison of month to month correlation coefficients of generated and historical flows at station Y.1C.

Month	1	2	3	4	5	6	7	8	9	10	11	12
q	0.456	0.163	0.484	0.113	0.439	0.490	0.562	0.474	0.594	0.747	0.808	0.719
Simultaneous decomposition												
$\bar{q}$	0.255	0.347	0.368	0.230	0.441	0.429	0.569	0.570	0.659	0.740	0.843	0.748
s	0.184	0.178	0.162	0.199	0.162	0.154	0.116	0.139	0.082	0.069	0.068	0.097
$\bar{q} + s$	0.439	0.526	0.530	0.429	0.603	0.583	0.685	0.709	0.741	0.809	0.912	0.845
$\bar{q} - s$	0.071	0.169	0.206	0.031	0.279	0.275	0.453	0.431	0.577	0.671	0.775	0.652
Stage decomposition												
$\bar{q}$	0.268	0.340	0.375	0.260	0.408	0.396	0.531	0.524	0.622	0.715	0.826	0.726
s	0.167	0.165	0.160	0.195	0.182	0.157	0.135	0.138	0.093	0.069	0.058	0.095
$\bar{q} + s$	0.435	0.505	0.534	0.455	0.590	0.552	0.666	0.661	0.715	0.784	0.884	0.821
$\bar{q} - s$	0.101	0.176	0.215	0.065	0.227	0.239	0.395	0.386	0.528	0.645	0.768	0.632
Chaleeraktrakoon (1999)												
$\bar{q}$	0.311	0.353	0.387	0.256	0.457	0.421	0.551	0.569	0.645	0.731	0.848	0.741
s	0.164	0.180	0.177	0.200	0.191	0.140	0.125	0.126	0.083	0.058	0.061	0.104
$\bar{q} + s$	0.475	0.533	0.565	0.457	0.648	0.561	0.676	0.694	0.729	0.788	0.909	0.845
$\bar{q} - s$	0.148	0.172	0.210	0.056	0.266	0.280	0.426	0.443	0.562	0.673	0.787	0.637

The existing and proposed approaches are compared using four monthly flow sequences in the Chao Phraya river basin. Results have shown that the number of parameters for the proposed scheme is smaller than that for the existing one. Further, it can preserve the basic statistics and drought properties of historic flows at monthly and annual time scales adequately.

2. PROPOSED MULTISITE SVD/MAR(1) PROCEDURE

Let $Q = [Q_{v\tau}] =$ an $p \times n\omega$ matrix of seasonal flow data at multiple sites where $p =$ the available record period in years, $n =$ the number of sites, and $\omega =$ the number of seasons. The proposed SVD/MAR(1) procedure begins with transforming the seasonal flow matrix Q into the normalized flow matrix X flows as (Loucks, et al., 1981):

$$X_{v\tau} = \ln[Q_{v\tau} - a_\tau] \tag{1}$$

where $a_\tau =$ the lower bound parameter of a lognormal distribution during each site and season , and $\tau = 1, ..., n\omega$. The parameter a_τ is estimated by quantile lower-bound estimation method (Stedinger, 1980):

$$a_\tau = \frac{[X_\tau(1)][X_\tau(p)] \quad [X_\tau(m)]^2}{X_\tau(1) + X_\tau(p) \quad 2X_\tau(m)} \tag{2}$$

554

Table 3: Comparison of monthly cross correlation coefficients of generated and historical flows at station P.12 and Y.1C.

Month	1	2	3	4	5	6	7	8	9	10	11	12
q	0.132	0.491	0.560	0.684	0.757	0.563	0.533	0.819	0.569	0.546	0.322	-0.003
Simultaneous decomposition												
$\bar{q}$	0.167	0.350	0.509	0.673	0.683	0.561	0.532	0.756	0.606	0.509	0.355	0.034
s	0.185	0.168	0.136	0.126	0.122	0.136	0.130	0.079	0.123	0.153	0.184	0.181
$\bar{q}+s$	0.352	0.517	0.645	0.799	0.805	0.696	0.662	0.835	0.729	0.662	0.539	0.215
$\bar{q}-s$	-0.019	0.182	0.373	0.547	0.562	0.425	0.401	0.677	0.482	0.356	0.171	-0.148
Stage decomposition												
$\bar{q}$	0.048	0.202	0.455	0.502	0.472	0.446	0.380	0.703	0.504	0.303	0.281	0.096
s	0.176	0.147	0.135	0.159	0.142	0.152	0.130	0.102	0.133	0.205	0.175	0.222
$\bar{q}+s$	0.224	_0.349_	0.589	_0.660_	_0.614_	0.598	_0.511_	_0.805_	0.637	_0.508_	0.457	0.318
$\bar{q}-s$	-0.128	0.055	0.320	0.343	0.330	0.294	0.250	0.600	0.370	0.098	0.106	-0.126
Chaleeraktrakoon (1999)												
$\bar{q}$	0.172	0.352	0.542	0.637	0.665	0.547	0.549	0.759	0.607	0.523	0.347	0.038
s	0.190	0.161	0.131	0.137	0.109	0.121	0.118	0.073	0.106	0.142	0.163	0.160
$\bar{q}+s$	0.363	0.513	0.674	0.773	0.774	0.668	0.667	0.832	0.713	0.665	0.510	0.198
$\bar{q}-s$	-0.018	0.192	0.411	0.500	0.556	0.426	0.431	0.685	0.501	0.380	0.185	-0.121

Table 4: Comparison of annual auto-correlation coefficients of generated and historical flows.

Station	P.12	W.3A	Y.1C	SK
q	0.174	0.281	-0.097	-0.027
Simultaneous decomposition				
$\bar{q}$	0.227	0.169	-0.016	-0.019
s	0.155	0.168	0.168	0.171
$\bar{q}+s$	0.382	0.337	0.152	0.152
$\bar{q}-s$	0.072	0.000	-0.184	-0.191
Stage decomposition				
$\bar{q}$	0.115	0.127	-0.093	-0.028
s	0.154	0.178	0.164	0.160
$\bar{q}+s$	0.269	0.304	0.071	0.132
$\bar{q}-s$	-0.040	-0.051	-0.257	-0.188
Chaleeraktrakoon (1999)				
$\bar{q}$	0.144	0.161	-0.052	-0.062
s	0.170	0.164	0.169	0.138
$\bar{q}+s$	0.314	0.325	0.117	0.076
$\bar{q}-s$	-0.026	-0.003	-0.221	-0.200

Table 5: Comparison of annual cross correlation coefficients of generated and historical flows.

Station	P.12&W.3A	P.12&Y.1C	P.12&SK	W.3A&Y.1C	W.3A&SK	Y.1C&SK
q	0.860	0.685	0.675	0.789	0.666	0.793
Simultaneous decomposition						
$\bar{q}$	0.790	0.609	0.644	0.768	0.613	0.721
s	0.080	0.126	0.108	0.073	0.118	0.089
$\bar{q}+s$	0.868	0.735	0.752	0.841	0.731	0.810
$\bar{q}-s$	0.710	0.483	0.536	0.695	0.495	0.632
Stage decomposition						
$\bar{q}$	0.773	0.588	0.581	0.737	0.584	0.683
s	0.071	0.110	0.120	0.084	0.128	0.103
$\bar{q}+s$	<u>0.843</u>	0.697	0.701	0.820	0.712	<u>0.786</u>
$\bar{q}-s$	0.702	0.478	0.461	0.653	0.456	0.580
Chaleeraktrakoon (1999)						
$\bar{q}$	0.773	0.588	0.641	0.763	0.618	0.721
s	0.069	0.104	0.098	0.077	0.123	0.096
$\bar{q}+s$	0.843	0.692	0.739	0.840	0.741	0.817
$\bar{q}-s$	0.704	0.483	0.543	0.686	0.495	0.625

Table 6: Comparison of monthly drought properties of generated and historical flows at station P.12.

Properties	Mean D	Max. D	Mean V	Max. V
q	7.27	10.00	2328	3520
Simultaneous decomposition				
$\bar{q}$	7.22	9.80	2309	3522
s	0.40	0.43	145	186
$\bar{q}+s$	7.62	10.23	2454	3708
$\bar{q}-s$	6.82	9.37	2163	3336
Stage decomposition				
$\bar{q}$	7.23	9.80	2302	3510
s	0.40	0.43	145	189
$\bar{q}+s$	7.63	10.23	2448	3699
$\bar{q}-s$	6.84	9.37	2157	3321
Chaleeraktrakoon (1999)				
$\bar{q}$	7.29	9.78	2342	3494
s	0.43	0.42	157	185
$\bar{q}+s$	7.71	10.20	2498	3680
$\bar{q}-s$	6.86	9.36	2185	3309

in which $X_\tau(1)$ = the maximum flow of X_τ, $X_\tau(p)$ = the minimum flow and $X_\tau(m)$ = the median. The next step is to standardize the transformed flow matrix X into an $p \times n\omega$ matrix $\tilde{X}$. In the following, two different decompositions of $\tilde{X}$ using SVD are briefly described.

2.1 Simultaneous Decomposition

Consider the $p \times n\omega$ matrix $\tilde{X}$. If the number of sites n is small, this matrix will be nonsingular ($r \leq n\omega$, r = the rank of $\tilde{X}$). Hence, it will be decomposed into the SPC matrix with SVD as (Cavadias, 1986):

$$Y = \tilde{X}\Gamma\left[\Delta^{1/2}\right]^{-1} \tag{3}$$

where Y = the $p \times r$ matrix of SPC with orthogonal columns $[(1/p)Y^{T}Y = I_{p}$, in which I_{p} = the identity matrix of order $p]$, Γ = the $n\omega \times r$ matrix of eigenvectors with orthonormal columns and rows ($\Gamma^{T}\Gamma = I_{p}$ and $\Gamma\Gamma^{T} = I_{n\omega}$), and Δ = the $r \times r$ diagonal matrix of nonnegative eigenvalues. Note that the diagonal element Δ_{jj} of Δ represents for the variance of the j'th component Y_{j}, and is arranged in descending order ($\Delta_{11} > ... > \Delta_{jj} > ... > \Delta_{rr}$). The total variance of Y ($\sum_{j} \Delta_{jj}$) is equal to that of $\tilde{X}$ ($n\omega$). The eigenvalue and associated eigenvector matrices are now estimated by

$$R\Gamma = \Gamma\Delta \tag{4}$$

in which R = the $n\omega \times n\omega$ correlation matrix of X $[R = (1/p)\tilde{X}^{T}\tilde{X}]$.

On the other hand, if the large number of sites is considered, the matrix $\tilde{X}$ will be singular ($r \leq p$). In this case, the eigenvalue matrix Δ is computed as

$$S\Pi = \Pi\Delta \tag{5}$$

where Π = the $p \times r$ eigenvector matrix and S = the $p \times p$ product matrix $[S = (1/p)\tilde{X}\tilde{X}^{T}]$. Note that it is advantageous to compute Δ based on S because S has smaller dimensions than R. The following step is to calculate the eigenvalue matrix Γ by

$$\Gamma = (1/p)^{1/2}\tilde{X}^{T}\Pi(\Delta^{1/2})^{-1} \tag{6}$$

Then, the SPC matrix Y is calculated as

$$Y = (p)^{1/2}\Pi \tag{7}$$

The resulting SPC matrix is then divided into the statistically significant SPC sub-matrix $Y1$ of size $p \times k$ (k = the number of statistically significant components), and the noise sub-matrix $Y2$ of dimensions $p \times (r - k)$ using the criterion of percentage cumulative variance g_{k}. This criterion is commonly used in factor analysis for reducing the number of variables. The cumulative variance criterion g_{k} can be written as (Joliffe, 1986):

$$g_{k} = \frac{100}{n\omega}\sum_{j=1}^{k}\Delta_{jj} \tag{8}$$

Its value is selected and used as the cut-off level between the significant and noise SPC. In general, the criterion g_{k} ranges from 70-90%. If we use the small value of g_{k} ($\cong 70\%$), only few SPC (Y_{j}, for $j = 1, ..., k$) will be statistically significant. In other words, the large number of variables can be reduced. However, some important statistical characteristics of variables may be loss. On the other hand, if the large g_{k} ($\cong 90\%$) is used, only the small number of variables can be removed. In this study, the cumulative variance criterion of 85% is select in order to reproduce the over-the-year correlation properties of $\tilde{X}$ adequately.

The next step is to fit an MAR(1) model to the sub-matrix $Y1$ as

$$Y1_{v} = Y1_{v-1}\Phi + U_{v} = Y1_{v-1}\Phi + Z_{v}H \tag{9}$$

where $Y1_{v} = 1 \times k$ row vector of $Y1$, $U_{v} = 1 \times k$ row vector of residuals, $Z_{v} = 1 \times m$ row vector of an SPC matrix Z (Z = the SPC matrix of U, m = the rank of U, $m \leq k$), $\Phi = k \times k$ parameter matrix, and $H = m \times m$ parameter matrix. The parameter matrices are estimated by

$$\Phi = M^{-1}M_{1} = M_{1} \tag{10}$$

and

$$H = \Psi^{1/2} \Omega^T \Sigma^{1/2}$$ (11)

in which M = the $k \times k$ lag zero correlation matrix of $Y1$, M_1 = the $k \times k$ lag one correlation matrix of $Y1$ [$M_1 = (1/p) Y1_{v-1}^T Y1_v$], Σ = the $k \times k$ diagonal matrix of variances of U, Ψ = the $m \times m$ diagonal matrix of eigenvalues, and Ω = the $k \times m$ matrix of eigenvectors of L (L = the $k \times k$ correlation matrix of U).

The next computational step is to generate the SPC matrix Z of U following $N(0,1)$ [$N(0,1)$ = the distribution of standard normal). Then, calculate backward to obtain the generated seasonal flow matrix $\hat{Q}$. This procedure has the number of parameters of $2m^2$, excluding the eigenvalue and eigenvector matrices Δ and Γ. If we consider the size of the multisite SVD/MAR(1) procedure based on the concept of regression, the procedure has the $r - k$ degrees of freedom left for computing H (Grygier and Stedinger, 1988).

2.2 Stage decomposition.

The procedure separates the $p \times n\omega$ standardized matrix $\widetilde{X}$ of multisite transformed flows into the flow sub-matrix $\widetilde{X}'$ for each site ($i = 1, ..., n$). The following step is to transform $\widetilde{X}^i$ to its SPC Y^i with either equations (3)-(4) or (5)-(7). The SPC matrix of each site Y^i is now divided into the significant components $Y1'$ and associated noises $Y2'$ based on the criterion described above. The next step is to concatinate the matrix $Y1'$ for all sites (i.e., $Y1 = Y1^1: Y1^2: ... : Y1':$ ... : $Y1^n$). The significant SPC matrix $Y1$ of multisite flow data is modeled using the MAR(1) approach [equations (9)-(11)].

The generated seasonal-flow matrix $\hat{Q}$ is obtained by calculating backward from the synthetic SPC matrix $\hat{Z}$. The stage decomposition procedure has $2\sum_i m_i^2$ ($i = 1, ..., n$) parameters. Its degrees of freedom left for estimating H is $r - \sum_i m_i$.

3. APPLICATION OF THE SVD/MAR(1) PROCEDURES

The proposed multisite SVD/MAR(1) procedure (e.g., simultaneous and stage decomposition), and the existing one (Chaleeraktrakoon, 1999) were applied to 37 years (1955-1991) of four monthly-flow series at station P.12 (Ping River), W.3A (Wang River), Y.1C (Yom River) and SK (Nan River), referred to hereafter as Chao Phraya data. Their parameters were estimated. Table 1 shows the number of parameters and degrees of freedom of the three procedures for the considered data set. It indicates that the stage-decomposition procedure is the most expensive. For the proposed simultaneous-decomposition and existing procedure, although their degrees of freedom are approximately the same; the proposed one, however, has the smaller number of parameters.

One hundred sets of synthetic monthly flows with the same size as that of the historic samples used were generated and compared with appropriate historic flows. The comparisons would be assessed on the reproduction of basic and drought properties of monthly and annual flows. If the considered property q was within the assessed interval $\bar{q} \pm s$ ($\bar{q}$ and s are the mean and standard deviation of the correspondingly generated one $\hat{q}$), it was inferred that the observed statistic q was preserved adequately. Results of the comparisons are as follows.

Generally, the proposed and existing procedures preserve monthly mean, standard deviation and skewness coefficient well. Table 2 compares the generated and historic month-to-month correlation coefficients (Yom River). It is evident that the three models can preserve the correlations adequately. Table 3 shows the the generated and historic monthly cross-correlation coefficients between monthly flow at station P.12 and Y.1C. It is seen that the stage-decomposition procedure cannot preserve several cross-correlation. While, the others can reproduce the properties. For the annual basic statistics, all competitive models yield good results (see Tables 4-5). Table 6-7 presents the historic and generated drought statistics (e.g., average and maximum drought magnitudes, and average and longest drought duration) of the monthly and annual flows. It demonstrates that all procedures reproduce adequately the drought related characteristics of historic flows at monthly and annual time scales.

Table 7: Comparison of annual drought properties of generated and historical flows at station W.3A.

Properties	Mean D	Max. D	Mean V	Max. V
q	2.22	6.00	1045	3322
Simultaneous decomposition				
$\bar{q}$	2.97	6.90	1264	3326
s	1.12	2.89	557	1452
$\bar{q} + s$	4.09	9.79	1821	4778
$\bar{q} - s$	1.85	4.01	707	1874
Stage decomposition				
$\bar{q}$	2.93	7.59	1191	3385
s	1.33	3.47	621	1707
$\bar{q} + s$	4.27	11.06	1812	5092
$\bar{q} - s$	1.60	4.12	570	1678
Chaleeraktrakoon (1999)				
$\bar{q}$	3.29	7.90	1407	3718
s	1.39	3.45	738	1779
$\bar{q} + s$	4.69	11.35	2146	5498
$\bar{q} - s$	1.90	4.45	669	1939

4.SUMMARY AND CONCLUSIONS

This paper proposes a new scheme for reducing the parameters of decomposed MAR(1) models for generating seasonal flows at multiple site. The parameter reduction is mainly based on the criterion of percentage cumulative variance, the technique commonly used in factor analysis. The proposed and existing procedures were applied to four series of monthly flows of the Chao Praya data. The concurrent period of records is 37 years (1955-1991). Results have indicated that the stage-decomposition procedure has largest numbers of parameter and does not reproduce the monthly cross-correlation coefficients of observed flows. The existing and the simultaneous-decomposition procedures can preserve the monthly and annual basics statistics and drought properties adequately. However, the proposed simultaneous-decomposition procedure is advantageous since it has smaller number of parameters.

REFERENCES

Cavadias, G.S. (1986), "A multivariate seasonal model for streamflow simulation." In: *La recherche en hydrologie au Quebec*, edited by V-T-V. Nguyen, and Y. Faucher, Les Presses de l'Universite du Quebec, 35, 58-77.

Chaleeraktrakoon, C. (1999). "Stochastic Procedure for Generating Seasonal Flows." *Journal of Hydrologic Engineering*, ASCE, 4(4), 337-343.

Grygier, J.C., and Stedinger, J.R. (1988). "Condensed disaggregation procedures and conservation correlation for stochastic hydrology." *Water Resource Research*, 24(10), 1574-1584.

Joliffe, I.T. (1986). *Principal Component Analysis*, Springer – Verlag, New York Inc., N.Y.

Lane, W.L. (1982). "Corrected parameter estimates for disaggregation schemes." in: *Statistical Analysis of Rainfall and Runoff*, edited by V.P. Singh, Water Resources Publications, Littleton, C.O., 505-530.

Lucks, D.P., Stedinger, J.R., and Haith, D.A. (1981). *Water resource system planning and analysis*, Prentice-Hall, Englewood Cliffs, N.J.

Matalas, N.C. (1967). "Mathematical assessment of synthetic hydrology." *Water Resource Research*, 3(4), 937-945.

Santos, E.G., and Salas, J.D. (1992). "Stepwise disaggregation scheme for synthetic hydrology." *Journal of Hydraulic Engineering,* ASCE, 118(5), 765-784.

Stedinger, J.R. (1980). "Fitting lognormal distributions to hydrologic data." *Water Resources Research,* 16(3), 481-490.

Stochastic Hydraulics 2000, Wang & Hu (eds) © 2000 Taylor & Francis, ISBN 90 5809 166 X

Physical basis of stage-discharge ratings

David R. Dawdy
San Francisco, Calif., USA

Walt Lucas
US Forest Service, Klamath Falls, Oreg., USA

Wen C. Wang
Multech Engineering Consultants, San Jose, Calif., USA

ABSTRACT: The Manning's equation can be used to determine the shape of the stage discharge rating based on physical principles. The analysis constrains the parameters of the rating. For within-bank flows with channel control, the rating is a parabola, and plots as a straight line on logarithmic paper. The physical factors, which determine the parameters of the rating, are described.

Key word: Stage-discharge rating, Relative roughness, Width-depth ratio, Channel shape.

1. INTRODUCTION

Stage-discharge ratings are needed to convert continuous records of stage to discharge. The rating is based on the discharge measurements made in the field, but the resulting rating should be based on physical principles rather than an empirical fit to the measurements. In particular, for stages for which channel control holds and the stage is controlled by friction in the channel rather than a controlling riffle or structure downstream, Manning's equation should hold:

$$Q = A \, R^{2/3} \, S^{1/2} / n \tag{1}$$

Where   A = area in square meters
          B = hydraulic radius in meters
          S = energy slope of the stream
          n = Manning's resistance coefficient

If Mannings' equation holds, the shape of the stage-discharge rating is constrained, and can be estimated before any measurements are made. If flow depth can be adequately represented by hydraulic radius, the above equation becomes:

$$Q = (W \, S^{1/2} / n) \, R^{5/3} = P \, R^{5/3} \tag{2}$$

where W = channel width in meters

$$P = W \, S^{1/2} / n$$

Eq. 2 can be alternatively expressed as the standard rating curve equation. The rating curve plots as a straight line on logarithmic paper with the hydraulic radius being represented by flow depth, which is the gage height, GH, minus a gage height of effective zero flow, e. The equation is then:

$$Q = P \, (GH-e)^{N} \tag{3}$$

The first estimate for e can be the lowest point in the measurement cross section, but preferably should be estimated at the control cross section. This approach is covered by Rantz et. el. (1982).

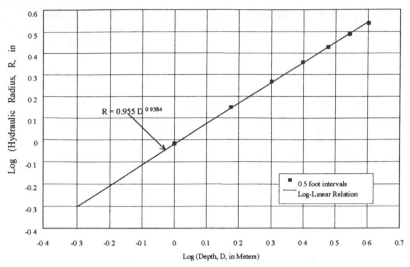

Fig.1: Relation of Hydraulic Radius to Depth for a Rectangular Channel

For demonstration purpose, assume a rectangular channel 50 meters wide with a gravel bed with D84 of the bed material of 20 mm. Using the Limerinos equation (Limerinos, 1970) to compute Manning's n:

$$n = 0.0761 \, R^{1/6} / (1.06 + 2.03 \, \text{Log} \, (R / D84)) \tag{4}$$

It can be computed that n varies from 0.0174 for R = 0.5 m to 0.0167 for R = 4 m. If the slope is 0.0144 and n is taking as 0.02, P is equal to 300 (= 50 x $0.0144^{1/2}$ / 0.02). The relation of hydraulic radius to flow depth, D, is plotted in Figure 1, and a least square fit for R from 0.5 m to 4 m yields R = 0.955 $D^{0.9384}$, so that N should be 1.57 (= 0.9384 x 5/3). Based on determinations from Manning's equation, a rating equation may be established to fit the computed discharges as:

$$Q = 306 \, (GH - 0.03)^{1.6} \tag{5}$$

The value of 0.03 is the gage height of effective zero flow, or e, which best converts the data into a straight line-plot on logarithmic paper. "e" is very seldom the actual point of zero flow on the downstream control for the gaging station. There is usually an amount of depth which must be subtracted from the gage height to convert gage height to effective head. It is this adjustment for effective zero flow which causes the fitted rating to deviate slightly from the previously estimated value.

The example shown is simplistic, because we assumed the Manning's equation applied to determine the discharges, then showed that we could use the Manning's equation to determine the parameters for the stage-discharge rating. However, the concepts presented apply to real data on real streams. The formulation of the Manning's equation can be used to estimate the stage-discharge rating before collecting data, and can be used to test the reasonableness of any rating equation resulting from the discharge measurements.

2. RELATED PHYSICAL FACTORS

Three factors which most affect the shape of the stage-discharge rating are the shape of the channel, the width-depth ratio and the relative roughness and its change with stage. Most natural channels are not rectangular, so that the area increase more rapidly than for a rectangular one. This causes most ratings to have an exponent greater than the theoretical 5/3 for converting head to discharge, and normal value for open channel flow is about 1.8 rather than 5/3. However, if

562

the stream is rectangular in cross section or if is quite narrow in general, so as to have a small width-depth ratio, says less than 10, the exponent will tend to be reduced below 1.8. Resistance to flow (Manning's n) is a function of relative roughness. If the ratio of the depth to the D84 of the surface bed material is small at bankfull discharge, says less than 5 or so, the effect of changes in relative roughness can be large, causing the exponent of the rating curve to be greater than 1.8. This is because the resistance becomes less with increasing stage, thus, the velocity increases as a result of changes in depth and, thus, resistance.

An example of a station where a small width-depth ratio affects the exponent in the rating curve is shown in Figure 2. Coyote Creek is in the Upper Klamath Basin, Orgeon, U.S.A.. Its width is 1.5 m and its depth 0.75 m at bankfull discharge of 0.28 cms. The exponent for the within-bankfull flow has a value of 1.6 because the width-depth ratio is so low. The D84 for the surface bed material is 8.4 mm, so that the variation of relative roughness does not affect the exponent. Note that the effect of the width-depth ratio on this small stream is the same as that for our hypothetical much larger stream above which had a similar width-depth ratio at bankfull stage.

An example of a station where a change in relative roughness affects the exponent is seen in the rating curve shown in Figure 3 for the South Fork of the Sprague River, which is also in the Upper Klamath Fall Basin. The D84 of the surface bed material is 152 mm, and the ratio of R/D84 is 3.07 with a hydraulic radius from a cross section at an elevation somewhat below bankfull and it increases to 3.92 at somewhat above bankfull discharge of 9 cms. This yields a change from 0.049 to 0.044 for Manning's n and a change in the exponent of the rating curve to 2.2 prior to making discharge measurements. The exponent is calculated as 1.8 + Log ((ratio of hydraulic radius$^{1.8}$) / (ratio of Manning's n calculated from Limerinos equation)). In addition, the P constant multiplier in equation 3 can be estimated to be 11.4 m (at the highest width in the cross-section or at 1 m depth if available) x 1.516 ($S^{1/2}/n$) = 17.3, where n is estimated to be 0.037 at a hydraulic radius of 1 m and S is 0.06 as determined from survey. This physically based equation is thus

$$Q = 17.3 \ (D)^{2.19} \tag{6}$$

as compared to the following graphically fitted equation:

$$Q = 20.10 \ (GH - 0.275)^{2.20} \tag{7}$$

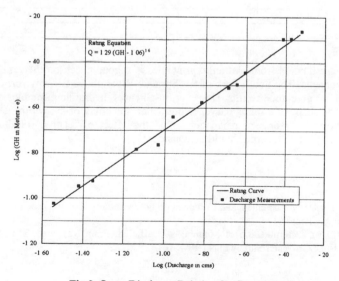

Fig.2: Stage-Discharge Relation for Coyote Creek,

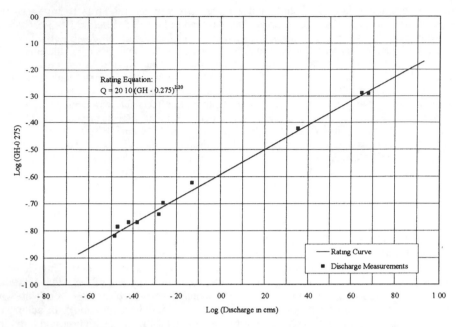

Fig.3: Stage-Discharge Relationship for South Fork of the Sprague

A comparison of the discharges for the in-bank flows for discharges from 2.83 to 5.66 cms are:

Q from Measured Rating (Eq. 7) Cms	Q from Estimated Rating (Eq. 6) cms	Error %
1.93	2.15	+12
3.00	3.28	+9
4.30	4.70	+9
5.92	6.37	+8

Thus, the rating estimated from hydraulic principles is a good check on the rating based on discharge measurements. The South Fork of the Sprague River above Brownsworth Creek was chosen because it did give good results. However, in all cases, hydraulic principles can be used to check the general shape of a rating curve. In particular, the effect of the width-depth ratio on the relation of hydraulic radius to mean depth and the effect of relative roughness on the variation of Manning's n with depth should be considered.

REFERENCES

Rantz, S.E., et. al., (1982), Measurement and Computation of Streamflow: Computation of Discharge, v.2: USGS Water Supply Paper 2175.

Limerinos, J.T., (1970), Determination of Manning's Coefficient From Measured Bed Roughness in Natural Channels, USGS Water Supply Paper 1898-B.

Stochastic Hydraulics 2000, Wang & Hu (eds) © 2000 Taylor & Francis, ISBN 90 5809 166 X

Chaos characteristics and its application to prediction of runoff time series

Xiekang Wang, Duo Fang & Shuyu Cao
National Laboratory of Hydraulics, Sichuan University, Chengdu, China

ABSTRACT: In the recent years, many researchers in different fields often apply the stochastic and nonlinear theories to study data time series. The stochastic and nonlinear methods of analyzing hydrological data such as discharge, water stage, sediment yield are different from the traditional deterministic methods. In this study, the Fractal Dimension, Lyapunov and Kolmogorov index etc. for the daily runoff time series have been analyzed, the preliminary results appear that the daily runoff time series has chaotic characteristics. Further, predictions are made by applying the local approximation method to daily runoff time series, from the comparison between the predictions and observations of daily runoff data, the daily runoff time series can be convincingly modeled by the time delay embedding approach of the chaotic theory.

Key words: Chaotic characteristics, Stochastic, Sediment yield, Fractal dimension index, Lyapunov index, Kolmogorov index, Time delay embedding approach, Nonlinear time series.

1 INTRODUCTION

Analysis of time series from dynamical systems is one of the most important issues in engineering practice. The common methods such as AR, MA, ARMA can be used to analyze and predict the linear time series data, however, these methods can not be applied for the nonlinear time series. In recent years, The chaos characteristics of non-linear dynamic systems have been recognized, there is a new technique for time series analysis because in many instances the time series can be viewed as a dynamic system with a low-dimensional attractor that can be re-constructed using the ' time delay embedding method '. The dynamic systems theory approach has been applied to analyze solar radio pulsation data (Kurths and Herzel, 1987) and runoff (Wang Wenjun et al,1994) which are focused on the calculation of the geometric and dynamic constant of an underlying attractor from a time series of observations, such as the fractal dimension and the Lyapunov exponents. The calculated constant can be used to diagnose whether the time series is chaotic or not.

Non-linear prediction of chaotic time series has been explored by Farmer and Sidorowich(1987). In this study, the correlation dimensions, the largest Lyapunov exponent and Kolmogorov entropy are used to find the chaotic nature of the daily runoff series at Linshan Gauging Station of Yanking county in Sichuan province . The time delay embedding method is used to reconstruct an attractor from a time series and make predictions.

2 INDICATORS OF CHAOTIC BEHAVIOR

The basic parameters of chaotic behavior are the fractal dimension of the attractor, positive Lyapunov exponents and positive finite Kolmogorov entropy. The definition and calculating method of these parameters are as follows:

2.1 Reconstruction of the phase-space

A method of reconstructing a phase-space from a single time series has been proposed by Packard et al.(1980). The dynamics of a time series $\{x_1, x_2, \cdots, x_n\}$ are fully captured or 'embedded' in the m-dimensional phase-space(m>d, where d is the dimension of the attractor) defined by

$$Y_t = \{x_t, x_{t-\tau}, x_{t-2\tau}, \cdots, x_{t-(m-1)\tau}\}$$

(1)

where τ is the delay time. According to the embedding theory, a d-dimensional attractor can be embedded into a (2d+1) dimensional phase-space to evaluate the characteristics of the dynamic system.

The delay time τ needs to be appropriately chosen. If τ is too small, then the $x_t, x_{t-\tau}, \cdots$ are not independent, resulting in a loss of information and characteristics on the attractor structure. If τ is too large, i.e. much larger than the information decay time, then there is no dynamic correlation between the state vectors, thus causing a loss of information on the original system. The choice of τ is usually made with the help of the auto-correlation function of the time series $\{x_1, x_2, \cdots, x_n\}$. Another method is to take the time lag that first generates a zero auto-correlation if the auto-correlation function crosses the zero line (Mpitsos et al.,1987).

2.2 Correlation dimension

Grassberger and Procaccia (1983a) defined the correlation function C(r) as

$$C(r) = \frac{1}{N_{ref}} \sum_{j=1}^{N_{ref}} \frac{1}{N} \sum_{i=1}^{N} H(r - \|Y_i - Y_j\|), i \neq j$$

(2)

where H is the Heaviside step function, with $H(u) = 1$ for $u > 0$, and $H(u) = 0$ for $u \leq 0$; N is the number of points in (Y_t); N_{ref} is the number of reference points taken from (Y_t); r is the radius of the sphere centered on Y_i or Y_j. The norm $\|Y_i - Y_j\|$ may be any of the three usual norms, the maximum norm, the diamond norm, or the standard Euclidean norm (the maximum norm is just the maximum absolute difference between the elements of Y_i and Y_j, whereas the diamond norm is the sum of all the absolute differences). The Euclidean norm is adopted in this study as it is the usual way to calculate the distance between two points.

If an attractor for the system exists, then, for positive r, it can be shown that

$$C(r) \cong \alpha r^\upsilon$$

(3)

where υ is called the correlation exponent and α is a constant. The correlation exponent υ may be estimated by the slope of a straight line in the plot of $\ln[C(r)]$ versus $\ln(r)$. For random processes, υ varies linearly with increasing m, without reaching a saturation value, whereas for deterministic processes, the value of υ levels off after a certain m. The saturation value d is defined as the correlation dimension of the attractor or a time series.

The nearest integer above the saturation value d provides the minimum number of embedding dimensions of the phase-space necessary to model the dynamics of the attractor.

2.3 Lyapunov exponent

Another measure of the chaotic nature of a time series is the Lyapunov exponent that gives the average exponential rate of divergence or convergence of nearby orbits in phase-space. Some authors (Rodriguez-Iturbe et al., 1989) have reported that the existence of a positive Lyapunov exponent implies the presence of chaotic dynamics . Very often, a system having at least one positive Lyapunov exponent is considered to be chaotic. Numerical algorithms for the computation of the Lyapunov exponents from a time series have been given by wolf et al.(1985) and Eckmann et al.(1986). In this study, the algorithm and the computer program given by wolf et al.(1985) which gives the non-negative exponents are adopted. The largest Lyapunov exponent λ_1 is defined as

$$\lambda_1 = \frac{1}{N\Delta t} \sum_{j=1}^{M} \log_2 \frac{L'(t_j)}{L(t_{j-1})}$$

(4)

where Δt is the time interval between two successive observations, M is the number of replacement steps, N is the total number of points in the sequence (Y_t), $L(t_{j-1})$ is the Euclidean distance between the point $\{x(t_{j-1}), x(t_{j-1-\tau}), x(t_{j-1-2\tau}), \cdots, x[t_{t-1-(m-1)\tau}]\}$ and its nearest neighbor, and $L'(t_j)$ is the evolved length of $L(t_{j-1})$ at a time t_j.

When $\lambda_1 > 0$, it means that the time series has at least one positive Lyapunov exponent and the series is chaotic. When $\lambda_1 \leq 0$, the time series is a regular motion process.

2.4 Kolmogorov entropy

The Kolmogorov entropy of a time series gives a lower bound to the sum of the positive Lyapunov exponents. An estimate for the Kolmogorov entropy K is K_2, which may be obtained from the set of correlation functions $C_m(r)$ (Grassberger and Procaccia,1983b). It is computed as the limit as $m \to \infty$ of the distance (in log-log coordinates) between successive correlation curves $C_m(r)$ and $C_{m+1}(r)$, i.e.

$$K_2(m) \cong \lim_{r \to 0}(\frac{1}{\Delta t}\{\log[C_m(r)] - \log[C_{m+1}(r)]\})$$ (5)

and

$$K_2(m) \cong \lim_{r \to \infty}[K_2(m)]$$ (6)

The K_2 entropy and Komogorov entropy are thought to have the same qualitative behavior, i.e. to be zero for a regular system, to be positive and finite for chaotic systems and to be infinite for stochastic processes.

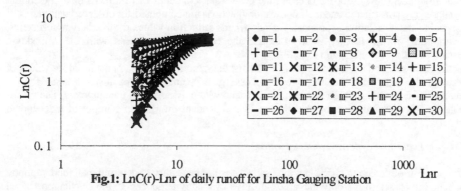

Fig.1: LnC(r)-Lnr of daily runoff for Linsha Gauging Station

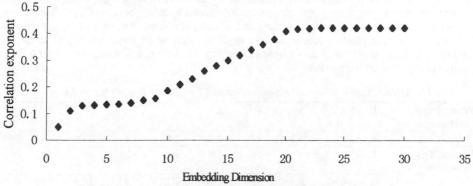

Fig.2: Variation of correlation exponent with embedding dimension of daily runoff data for Linshan Gauging Station

567

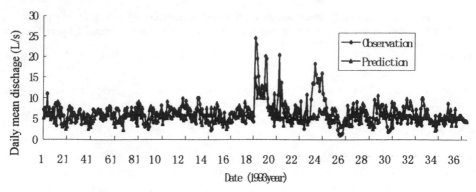

Fig.3: The Comparsion between predictions and observations of daily runoff

3 THE CHAOTIC ANALYSIS OF DAILY RUNOFF TIME SERIES

3.1. Data used

The various methods of diagnosing chaotic time series were applied to analyze daily runoff in Yanking county, Sichuan province. The daily runoff data consist of a 12 year record (1982 to 1993, with the number of data points n=4180) at Linshan gauging station.

3.2 Dimension calculation

From the analysis of the autocorrelation functions for the observed daily runoff data series, the value of τ is three lags (days). For the chosen time delay values, a series of m-dimensional embedding phase spaces are constructed and their correlation functions are computed for observed data series. It was seen that the correlation exponents ν tend towards saturation values for the daily runoff series. The variations of the LnC(r) versus Lnr and the correlation exponents with the embedding dimension are shown in Fig.1 and Fig.2.

For the daily runoff data, it can be seen (Fig.2) that the saturation values are achieved for values of m between 20 and 30 in the linear region. These give estimates of the lower bounds for the dimensions of embedding phase space sufficient to model the dynamics of the attractors. The results indicate that the dynamic system of daily runoff is dependent on a larger number of independent variables.

3.3 Largest Lyapunov exponent and κ_2 entropy

For the purpose of detecting chaos in a time series, it is not necessary to determine all the Lyapunov exponents, but to determine only the largest positive Lyapunov exponent. In the calculations based on the algorithm of Wolf et al.(1985): (a) reconstructing the dynamics in a finite dimensional space; (b) obtaining the tangent maps to this reconstructed dynamics by a least-squares fit; (c) deducing the Lyapunov exponents from the tangent maps. In this study, The mean of a series of λ_1 generated in different dimensional phase-spaces from m=1 to 20 is taken as an estimated largest Lyapunov exponent for daily runoff time series (Table 1). The original daily runoff series was found to have positive largest Lyapunov exponents.

Table.1: Estimated largest Lyapunov exponent ($\times 10$-2bits/s) for daily runoff data

Embedding Dimension	1	2	3	4	5	6	7	8	9	10
λ_1	8.5	7.6	5.3	4.1	3.2	2.8	1.4	0.82	0.74	0.66
Embedding Dimension	11	12	13	14	15	16	17	18	19	20
λ_1	0.5	0.42	0.34	0.25	0.23	0.19	0.15	0.12	0.09	0.08
Mean of Estimated largest Lyapunov exponent : λ_1 =1.87										

Table.2: The estimated K_2 entropy ($\times 10\text{-}2$) for daily runoff at Linsha Gauging Station

Site	Embedding Dimension									Mean
	21	22	23	24	25	26	27	28	29	
Linshan	5.5	5.6	4.8	4.0	5.1	4.6	4.7	4.8	4.5	4.7

For a series of embedding dimensions, Eq.(5) is used to evaluate the quantity $K_2(m)$ for the original daily runoff series. All K_2 values were found to be positive and finite. Table 2 lists the last nine K_2 values computed using the correlation function in the last ten embedding phase-spaces.

4 PREDICTION OF DAILY RUNOFF

For chaotic data, Packard et al (1980), Farmer and Sidorowich (1987) have suggested a forecasting technique. The first step is to create a state vector $X(t)$; the next step is to assume a functional relationship between the current state $X(t)$ and the future state $X(t+T)$ can be written as follows:

$$X(t+T) = f_t(X(t))$$

(7)

The problem then is to find a predictor f_t of f_t which for chaotic series will necessarily be non-linear. There are several possible approaches for determining f_t. The least-squares fitting of a mth order polynomial is one such approach. However, such global polynomial fitting leads to intractable difficulties when the number of parameters is large, and will be the same in case of large m, they also have the problem of errors magnifying with longer prediction times.

A promising alternative is the "local approximation" (Farmer and Sidorowich, 1987; Casdagli, 1989) approach which uses only nearby states to make predictions.

The basic idea in the local approximation method is to break up the domain into local neighborhoods and fit parameters in each neighborhood separately. To predict $X(t+T)$ based on $X(t)$ (an m-dimensional vector) and past history, k nearest neighbors of $X(t)$ are found on the basis of the minimum values of $\|X(t) - X(t')\|$, with $t' < t$. If only one such neighbor is considered, the prediction of $X(t+T)$ would be $X(t'+T)$. Since k neighbors are chosen, the prediction of $X(t+T)$ could be taken as a weight average of the k values of $X(t'+T)$. For convenience, the prediction of $X(t+T)$ is considered as a scalar. The value of k is determined by trial and error.

In this study, the following variation of the local approximation method is attempted. Namely, averaging several neighbors in the form

$$\overline{x(t,T)} = \frac{1}{k}\sum_{t=1}^{k} X(t'+T)$$

(8)

The number of neighbor k is determined by experimentation.

In the approximations used for non-linear prediction, Eq. (8) gave the satisfactory predictions in this study. Trial values of the number of selected relevant neighboring points k=10,15,20,25,30,40,50,60,70,80, etc., were used. The optimal value of k=60 was adopted for prediction calculation. For k>60, the improvement made was very marginal.

The predictions were made using the original observed daily runoff data at Linshan gauging station. Firstly, the early year data (from 1982 to 1992) are used to identify the pattern of behavior and determine the relevant parameters; secondly, the last year (1993) daily runoff data are used for comparison between the observations and predictions, see to Fig.3.

5 CONCLUSION

The analysis of daily runoff time series indicates the presence of chaos characteristics according to indicators of chaotic behavior such as the fractal dimension of the attractor, positive Lyapunov exponents and positive finite Kolmogorov entropy. There is convincing that daily runoff time series could be modeled by the time delay embedding method of chaotic theory. It should, however, be pointed out that study on the chaotic property of daily runoff time series and the method for prediction of daily runoff time series is still in the preliminary stages in this study and need further investigation.

ACKNOWLEDGEMENTS

This work is supported financially by the National Science Foundation of China (Grant No. 59890200 & 49831010).

REFERENCES

Casdagli, M., 1989. Nonlinear prediction of chaotic time series. Physica D, 35:335-356.

Eckmann, J.P., Kamphorst, S.O., Ruelle, D. et al, 1986. Lyapunov exponents from time series. Phys. Rev. A, 34(6):4971-4979.

Farmer,J.D. and Sidorowich, J.J.,1987. Predicting chaotic time series. Phys. Rev. Lett., 59(8):845-848.

Grassberger,P., and Procaccia, I., 1983a. Measuring the strangeness of strange attractors. Physica D,9:189-208.

Grassberger,P., and Procaccia, I., 1983b. Estimation of the Kolmogorov entropy from a chaotic signal. Phys. Rev. A, 28:2591-2593.

Kurths,J. and Herzel,H.,1987. An attractor in a solar time series. Physica D,25:165-172.

Mpitsos,G.J., Creech,H.C., Cohan, C.S. et al,1987. Variability and chaos: neurointegrative principles in self-organization of motor patterns. In:Hao Bai-Lin(Editor), Directions in Chaos, Vol.1. World Scientific,pp.162-190.

Packard,N.H., Crutchfield,J.P., Farmer,J.D. et al,1980. Geometry from a time series. Phys. Rev.Lett.,45:712-716.

Rodriguez-Iturbe, I., De Power,B.F., Sharifi,M.B. et al,1989. Chaos in rainfall. Water Resour. Res., 25(7):1667-1675.

Wang Wenjun, Ye Min and Chen Xianwei,1994. Quantiative analysis of chaotic characters for the Yangtze river flow time series. Advances In Water Science,5(2):87-94.(In Chinese).

Wolf A. et al (1985). Determining Lyapunov exponents from a time series. Physica D,16:285-290.

Stochastic Hydraulics 2000, Wang & Hu (eds) © 2000 Taylor & Francis, ISBN 90 5809 166 X

An object-oriented hydrologic simulator for the miomote river basin

Y.C. Wang
Newjec Incorporated, Tokyo, Japan

O. Minemura
Miomote River Development Office, Niigata Prefecture Government, Japan

ABSTRACT: Water resources planning and management involve both technical and nontechnical people, which suggests a need for highly intuitive, easily used river basin simulation models. In this paper, a PC-based hydrologic simulator is developed for the Miomote river basin. This simulator allows the user to manipulate windows, menus, dialog boxes, and other graphic objects and responds interactively to the user's input actions. The users' experience indicates that the simulator is useful in developing river basin policy and management plans, providing hydrologic information for water resources project design, and explaining the complex hydraulic and hydrologic behavior of river systems.

Key words: Water Resources Management , River Basin Simulation , Rainfall-Runoff Model

1. INTRODUCTION

Water resources planning and management have long been considered to be only the job of water resources engineers and scientists. In recent years, however, the development of river basin policy and management plans requires taking into account not only the technical feasibility and economic effectiveness of water resources projects but also the impacts of the projects on the natural environment, ecosystem, and societal requirements. Most water resources projects are planned for and financed by a governmental unit or public utility. Many such projects become controversial political issues and are debated by concerned individuals and groups (e.g., environmentalists, social scientists, economists, and politicians) whose understanding of the basic aspects of the problem is limited. This situation suggests a need for highly intuitive, easily used river basin simulation models that would allow these individuals and groups to develop a conceptual and intuitive understanding for the hydrologic and hydraulic behavior of complex river basin systems. The simulation approach to understanding river basin behavior necessitates a physically realistic simulator that can be easily used by both technical and nontechnical people. A useful simulator has several potential purposes; it may be used as a tool for investigation, for policy making, or for engineering design (McCuen 1989).

In this paper, a PC-based, interactive hydrologic simulator is developed for the Miomote river basin to meet this need. This simulator enables both the technical and nontechnical people to easily examine how flood frequency, different flood hydrographs, the parameters of a rainfall-runoff model, and flood regulation method affect policy making in water resources development.

The Miomote river is located in the Niigata prefecture of northeastern Japan, with a catchment area of $664.3 km^2$ (Figure1). The river basin is surrounded by a chain of mountains, making it one of the regions with the most heavy rainfall and snowfall in Japan. The comprehensive development of the river basin started in 1945 and the Miomote dam, with flood control as its main purpose, was constructed in 1953. Unfortunately, in 1967, a heavy storm caused a flood much greater than the design flood of the downstream levees, resulting great damage. Thus, the

Niigata prefecture government decided to construct dams and improve channels so as to protect the Miomote river basin from a 100-yr flood. Now, the Okumiomote dam is under construction and will be completed this year. The channel improvement of the Miomote river is being carried out, and the planning of the Takane dam located at the tributary of the Miomote rive is in progress.

2. RAINFALL FREQUENCY ANALYSIS

In the Miomote river basin, flow records are short and the number of river gaging stations is considerably insufficient, thus making it impossible to directly estimate design flood through the statistical analysis of streamflow data. In this case, three alternative methods are usually used to estimate flood frequency (Linsley 1979). The first method is the use of empirical formulas. The second possibility is to estimate the probable flood frequency at the ungaged point from data nearby gaging stations. The third alternative is to utilize an analysis of rainfall frequency and compute the flood flows from the rainfall by use of a rainfall-runoff model. The third method is used in this paper, assuming that the estimated runoff has the same frequency as the rainfall.

Frequency analysis of rainfall is complicated by the fact that one may be concerned with rainfall of various durations and over various areas. In this paper, considered is the rainfall of two-day duration and over the whole catchment area because the duration of most storms in Japan is within 2 days and the design flood at the outlet of the Miomote basin is used as the design standard for dams and levees. The average two-day rainfall on the basin is computed by use of the daily rainfall data in six rain gaging stations (Figure1). The Thiessen area of each rain gaging station is shown in Table 1.

Table 1. Thiessen Area of Each Rain Gaging Station

Rain Gaging Station	Murakami	Takane	Miomote	Washigasuyama	Okumiomote	Saruta
Thiessen Area (km²)	59.2	183.5	67.8	73.2	174.1	106.5

A maximum two-day rainfall in each year of 35 years (1942-1976) is used for rainfall frequency analysis. The result of rainfall frequency analysis indicates that the maximum two-day rainfall is lognormally distributed. The two-day rainfalls with various return periods are shown in Table 2.

Table 2. Two-Day Rainfalls with Various Return Periods

Return Period	100-yr	80-yr	50-yr	30-yr	20-yr	10-yr	5-yr	2-yr
Two-Day Rainfall(mm)	380	365	335	305	280	240	197	133

In order to take into consideration the spatial distribution of rainfall, the Miomote basin is divided into 14 sub-basins. On the other hand, 7 representative hyetographs (Table 3) are selected for each sub-basin to reflect the temporal variation of rainfall.

Table 3. Seven Representative Floods

Date of	July 12	Aug. 28	July 30	Aug. 7	July 8	July 8	June 25
Flood Occurence	1964	1967	1969	1969	1972	1974	1978

Let Rg be the maximum two-day rainfall of a representative hyetograph, and Rs be the two-day rainfall with a N-yr return period. If Rg is not equal to Rs, the representative hyetograph is modified by multiplying its hourly rainfall by a coefficient k ($k = Rs / Rg$). Finally, the modified representative hyetographs of each sub-basin are used for computing flood flows through the rainfall-runoff model to be stated in the subsequent section.

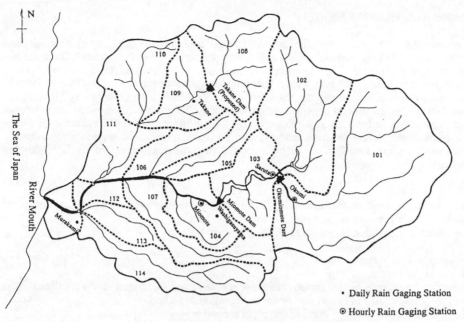

Fig. 1: Miomote River Basin and its Subdivision

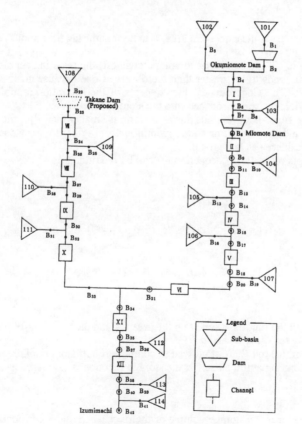

Fig.2: Miomote River Network Model

3. RAINFALL-RUNOFF MODEL

The unit hydrograph method is widely used for streamflow estimate because of its simplicity and applicability. The basic principle of the unit hydrograph is such that the catchment behaves like a linear system whose impulse response function is a unique, time-invariant, linear transformation. In fact, however, a strict linear relationship between rainfall and runoff does not exist. As a result, a lot of nonlinear rainfall-runoff models have been proposed by hydrologists (Singh 1995). In the present paper, the so-called storage function method, a nonlinear storage model used widely in Japan as a hydrologic design tool, is adopted for runoff estimate.

The basis of the storage function method is the assumption that the storage in a catchment is the power function of outflow or runoff from the catchment as expressed in the following equation.

$$S(t) = k \cdot q^p(t) \tag{1}$$

where k and p are constant coefficients, $S(t)$ in millimeter is the storage per unit area at time t, and $q(t)$ in millimeter per hour is the runoff per unit area at time t.

The continuity equation is expressed as

$$\frac{dS(t)}{dt} = f \cdot r(t) - q(t) \tag{2}$$

where $r(t)$ is the average rainfall on the catchment at time t, and f is termed inflow coefficient that is similar to the concept of runoff coefficient in the rational method.

The catchment outflow $Q(t)$ in cubic meter per second is

$$Q(t) = \frac{1}{3.6} f \cdot A \cdot q(t - T_l) + Q_b \tag{3}$$

where A in square kilometer is the catchment area, T_l in hour is the lag time and Q_b in cubic meter per second is the base flow.

When equations (1) to (3) are employed to compute runoff, the resulting runoff on the rising limb of the hydrograph tends to be greater than the observed one because no loss of rainfall is considered in equation (2). Consequently, the concepts of the saturation rainfall Rsa and the first-stage runoff coefficient f_1 are introduced into the storage function method.

When the sum of the first t-hour rainfalls during a storm is smaller than Rsa, the runoff occurs only on the area f_1A, and the rainfall on the remaining area $(1- f_1)A$ does not contribute to the runoff and, thus, is considered as lost rainfall.

Based on the concepts of Rsa and f_1, equation (3) can be rewritten as

$$Q_1(t) = \frac{1}{3.6} f_1 \cdot A \cdot q_1(t) \tag{4}$$

$$Q_2(t) = \frac{1}{3.6}(1 - f_1) \cdot A \cdot q_2(t) \quad (\text{If } \sum_{t=1}^{t} r(i) \leq R_{sa}, \ q_2(t) = 0) \tag{5}$$

$$Q_0(t) = Q_1(t) + Q_2(t) + Q_b \tag{6}$$

$$Q(t) = Q_0(t - T_l) \tag{7}$$

where $q_1(t)$ and $q_2(t)$ are the runoffs at time t on the area f_1A and the area $(1-f_1)A$ respectively and equations (1), (2) apply to both $q_1(t)$ and $q_2(t)$.

Equations (1), (2) and (4) to (7) are employed to compute runoff from rainfall.

For the storage routing in natural channels, the storage function is expressed as

$$S_c(t) = k_c \cdot Q_c^{p_c}(t) \tag{8}$$

where $S_c(t)$ in cubic meter is the storage volume of the river reach under consideration at time t, $Q_c(t)$ in cubic meter per second is outflow from the reach at time t, and k_c, p_c are coefficients.

The continuity equation of the reach is

$$\frac{dS_c(t)}{dt} = I_c(t) - Q_c(t) \tag{9}$$

where $I_c(t)$ in cubic meter per second is inflow into the reach at time t.

Considering the lag effect of the reach, the outflow from the reach can be expressed as

$$Q_c'(t) = Q_c(t - T_{cl}) \tag{10}$$

where T_{cl} in hour is the lag time of the reach.

Equations (8) to (10) are applied to storage routing in channels.

In principle, the parameter k, p, T_l, f_l, Rsa for each sub-basin and k_c, p_c, T_{cl} for each river reach should be calibrated by use of observed rainfall and flood data. In the Miomote basin, however, there are few river gaging stations, thus making it difficult to calibrate all the parameters. As a result, these parameters are estimated using empirical formulas that have been proved to be effective on other river basins. The estimated values of these parameters are listed in Table 4.

Table 4. Estimated Model Parameters

Sub-Basin	A (km²)	Rsa (mm)	f_l	p	k	T_l (hour)	River Reach	p_c	k_c	T_{cl} (hour)
101	174.5	120.0	0.5	0.333	38.0	0.50	I	0.6	15.00	0.070
102	89.0	120.0	0.5	0.333	35.0	0.21	II	0.6	7.00	0.030
103	42.2	120.0	0.5	0.333	18.0	0.00	III	0.6	7.00	0.030
104	34.9	200.0	0.5	0.333	21.0	0.00	IV	0.6	2.50	0.000
105	18.4	200.0	0.5	0.333	31.0	0.00	V	0.6	4.50	0.020
106	20.0	200.0	0.5	0.333	31.0	0.00	VI	0.6	3.50	0.020
107	36.5	200.0	0.5	0.333	48.0	0.39	VII	0.6	9.50	0.040
108	60.5	120.0	0.5	0.333	26.0	0.00	VIII	0.6	9.50	0.040
109	41.3	200.0	0.5	0.333	24.0	0.00	IX	0.6	5.75	0.025
110	30.7	200.0	0.5	0.333	31.0	0.00	X	0.6	5.75	0.025
111	37.2	200.0	0.5	0.333	23.0	0.00	XI	0.6	5.75	0.020
112	12.6	200.0	0.5	0.333	34.0	0.00	XII	0.6	5.75	0.020
113	13.9	200.0	0.5	0.333	46.0	0.00				
114	52.6	200.0	0.5	0.333	45.0	0.50				

The validity of the estimated parameters was verified by simulating 5 floods at the Miomote dam site and a part of simulation results are shown in Figure 3.

4. OBJECT-ORIENTED SIMULATOR DEVELOPMENT

The first step for development of an object-oriented simulator is to analyze the users' requirements in order to extract objects. The users' main requirements are summarized as follows.

1. The Okumiomote dam is one of the most important components in the Miomote river basin, because both the Niigata prefecture government (decision maker) and the riparian citizens wish to know intuitively why the dam is necessary, how the dam size is determined, and what benefits can be expected by construction of the dam.

2. The statistical concepts of design flood such as return period, representative floods are difficult to understand even for technical people. The simulator should be capable of allowing the users to examine how different design floods affect the project size.

3. The Miomote dam has been playing an important role in flood control and hydropower utilization since its construction. The construction of the Okumiomote dam, however, would

greatly alter inflow into Miomote reservoir. The simulator must be designed in such a way that the users are able to easily observe the change of inflow into the Miomote reservoir and check whether a new operation method is required for the Miomote reservoir.

4. The planning of the Takane dam is in progress. Thus, the simulator should enable the users to compare the difference between the flood flows with and without the Takane dam.

5. The rainfall-runoff model is the basis of the simulator. The users are interested in the effect of model parameters on hydrograph and relationship between outflows from the upstream dams and the peak flows in the downstream channels.

After defining the users' requirements or objects, the next step is to develop an object-oriented software system or simulator. Since the simulator is used by both technical and nontechnical people engaged in planning water resources development, it should be visually based, conceptually simple, and sufficiently true to the real world to offer a good understanding of how a river basin functions. In other words, the simulator needs to be graphically interactive and have the appearance of a computer game, yet be based on an accurate network model of the river.

To meet this need, the object-oriented simulator for the Miomote river basin is developed using object-oriented programming techniques. The simulator includes two main system components: numerical computation subsystem and graphic user interface (GUI). The numerical computation subsystem consists of a package of C++ programs using classes to describe rainfall-runoff relation in any sub-basin, storage routing in any river reach or reservoir and the processing of input and output data. The numerical computation is complex and not accessible to nontechnical people. Thus, the GUI is designed according to the users' requirements. The resulting simulator (Figure 4) has both the appearance and the function of standard Windows software. It allows the user to manipulate windows, menus, dialog boxes, and other graphic objects and responds interactively to the user's input actions

5.APPLICATION TO MIOMOTE BASIN

1. Effect of rainfall-runoff model parameters on hydrograph

The rainfall-runoff model is one of the most important components in developing river basin policy and management plans. It is difficult, however, explain to those who have little experience in hydrology how these parameters affect the design flood hydrograph and, further, the size of related projects. The simulator is particularly well suited for developing an intuitive understanding of physical meaning of these parameters. For example, the estimated value of parameter k for sub-basin 101 is 38 and the resulting peak flow from the sub-basin is $2,347m^3/s$. By setting k=100, one can observe that the peak flow becomes $757m^3/s$, but the hydrograph becomes much flatter than that in the case of k=38. This implies that k is a parameter reflecting the effect of catchment storage. Similarly, one can make an experimentation with any other parameters.

2. Determination of flood control capacity of Okumiomote dam

The Okumiomote dam is a multipurpose dam with flood control as its main purpose. It is expected that the Okumiomote dam will enable the 100-yr flood to safely flow through point B42 (Figure 2), the standard site for flood control plan in the river basin. By means of the simulator, one can readily determine the peak discharges of 100-yr flood at point B42, given the flood control capacity of the Okumiomote dam. Conversely, if the expected peak discharge of 100-yr flood at point B42 is specified, the flood control capacity of the dam can be determined. The flow capacity at point B42 after channel improvement is $3,400m^3/s$. The simulation result by use of the simulator indicates that a flood control capacity of $54,000,000m^3$ is required for the Okumiomote dam in order to satisfy the condition that the peak flow of 100-yr flood at point B42 does not exceed $3,400m^3/s$.

576

3. Flood Regulation Method of Miomote Reservoir

The Miomote reservoir has a gated sluiceway and is operated by adjusting the height of the gate to release a discharge not exceeding 700m³/s. However, the decision makers wish to fix the height of the gate in order to avoid the operation risk caused by subjective judgement. The simulator can be used to examine how the adjustable and fixed gates influence the effect of flood regulation and determine, if the fixed gate is preferable, the height of the gate.

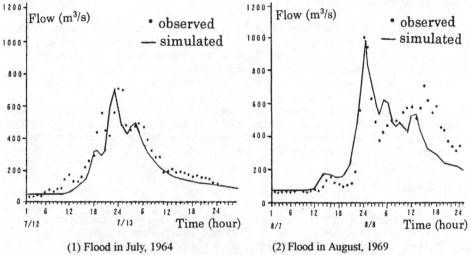

(1) Flood in July, 1964   (2) Flood in August, 1969

Fig. 3:. Runoff Simulation Using Estimated Model Parameters

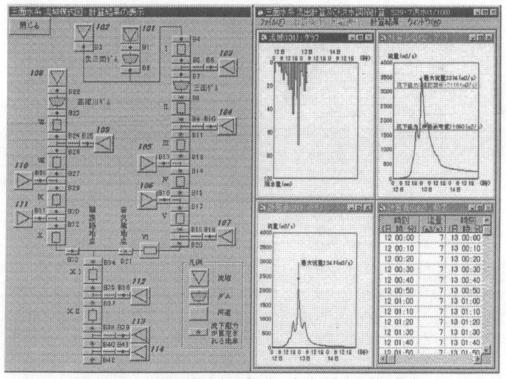

Fig. 4:. Windows-Based Hydrologic Simulator for the Miomote River Basin

4. Channel Improvement

In addition to dam construction, channel improvement is required for protecting the river basin from 100-yr flood. The simulator allows the users to observe the hydrograph and the flow capacity at any point. By comparing peak flows and flow discharges, the users can know where channel improvement is necessary.

6.CONCLUSIONS

The hydrologic simulator for the Miomote river basin is a computer-based river basin simulation system accessible to a broad range of people exploring water resources planning and management problems. The system sends the user's instructions to and receives information automatically from the numerical computation subsystem, thus allowing the user to only interact with the GUI. The users' experience in the use of the simulator indicates that the simulator is very useful in developing river basin policy and management plans, providing hydrologic information for water resources project design, and explaining the complex hydraulic and hydrologic behavior of river systems. Future research in this area includes the integration of GIS interface, which would lead to modeling of rainfall-runoff with real-time, spatially distributed rainfall and runoff information.

REFERENCES

Linsley, R.K., and Franzini, J.B. (1979). Water-resources Engineering. McGraw-Hill Book Company.
McCuen, R.H. (1989). Hydrologic analysis and design. Prentice-Hall, Englewood Cliffs, N.J.
Singh, V.P. (1995). Computer model of watershed hydrology. Water Resources Publications.

Stochastic Hydraulics 2000, Wang & Hu (eds) © 2000 Taylor & Francis, ISBN 90 5809 166 X

Pattern classification of hyetographs and its application into flood control planning

Toshiharu Kojiri & Kunio Tomosugi
DPRI, Kyoto University, Gokasho, Japan

Takayasu Shiode
Mitsui Construction Company, Tokyo, Japan

ABSTRACT: Pattern classification methodology is applied into hydrological time sequences such as hyetograph or hydrograph to analyze their feature patterns. Classification objectives are proposed and optimal number of classified patterns is decided through AIC technology. Then the fluctuation of peak discharge against the different hyetograph under with same total volume is considered. Occurrence probability of hydrological event is evaluated from the viewpoint of feature pattern without the traditional probability approach. Moreover flood control planning matching the designed occurrence probability is discussed.

Keyword: Pattern classification, Occurrence probability, Flood control planning

1. INTRODUCTION

The present flood control system is planned with peak discharge at the downstream reference point under the condition of designed return period. However the embankment break might be caused with not only high flood peak but also long duration period of high water level. We have to consider the feature pattern, in other words, the temporal and spatial distribution of flood hydrograph. The pattern recognition concept is useful for time-dependent and multi-point phenomena in vector treatment. Unny et al. (1981) proposed the application of classification methodology into long-term hydrographs with traditional approach. Kojiri et al. (1992) used the pattern classification to evaluate and predict the drought severity. As there are several important circumstances such as (i) the objective functions suited to classification purpose and (ii) the optimum number of classified patterns. First, we will extract the suitable objective functions by applying pattern classification methodology of ISODATA. The flood hydrographs must be shifted according to the feature patterns to identify the initial time for calculation. Second, the relationship between hyetograph and hydrograph is evaluated through run-off analysis for flood control. Lastly, the flood control planning matching the designed occurrence probability is proposed.

2. CLASSIFICATION PROCEDURES AND OBJECTIVE FUNCTIONS

To classify the flood hydrograph with hourly unit different from yearly one, the objective function should be defined. Usually the summation type of Euclidian distance between sample data and representative one, the so-called cluster center, is the standard classification function (Tou and Gonzalez, 1974). In the case of flood event, the following functions are obtained to find the differences on intensity, fluctuation and other attributes. The first one is the root square of residuals for whole periods as follows;

$$OF1(i,m) = \min\_ D\sqrt{\sum_t \{X(t-D,i)-CL(t,m)\}^2} \qquad (1)$$

where, $X(t,i)$ is the data value of sample i at time t and $CL(t,m)$ is the calculated value of classified representative pattern m (cluster center) at time t, D is the shifting distance of sample data against the cluster center because the initial time for calculation in flood hyetograph or hydrograph is not fixed among all data. It might be changed according to the changing cluster center. The fuzzy set is introduced to discriminate the attributes of considered hyetographs in time and space. The differences on feature and peak are represented as follows;

$$K(i,m) = \min\_ D \max\_ t |X(t-D,i)-CL(t,m)| \qquad (2)$$

$$P(i,m) = |\max\_ t\{X(t,i)\} - \max\_ t\{CL(t,m)\}| \qquad (3)$$

$$RK(i,m) = 1 - K(i,m)/IK \qquad (4)$$

$$RP(i,m) = 1 - P(i,m)/IP \qquad (5)$$

where, $RK(.)$ and $RP(.)$ are the fuzzy membership functions with gradient IK and IP, respectively. The integrated objective function is formulated as follows;

$$OF2(i,m) = \min\{RK(i,m), RP(i,m)\} \qquad (6)$$

The following function is defined to evaluate the total volume;

$$S(i,m) = \sum_t X(t,i) - \sum_t CL(t,m) \qquad (7)$$

$$RS(i,m) = 1 - S(i,m)/IS \qquad (8)$$

where, $RS(.)$ the fuzzy membership functions with gradient IS and $S(i,m)$ is the residual of sample data I against cluster center m. The last objective function at reference point is formulated depending on the engaged objectives as follows;

$$OF3(i,m) = \min\{RK(i,m), RP(i,m), RS(i,m)\} \qquad (9)$$

The spatial evaluation is also formulated with the same fuzzy set approaches. The modified cluster center is obtained through the averaging method.

$$CL(t,m) = \frac{\sum X(t-Dopy(i,m),i)}{ICT(m)} \qquad (10)$$

where, $Dopy(i,m)$ is the optimum shifting distance in time of sample data i against cluster center m and $ICT(m)$ is the number of sample data in cluster m.

We applied the proposed classification procedures into the real river basin in Japan. The considered area consists of the elliptical shape of 163 km² and the summit is 413 m elevation located near Kyoto. Twenty-seven samples are classified through five objective functions such as i) root-square of eq.(1), ii) standardized root-square (the residual in eq.(1) is divided by the value of cluster center), iii) feature pattern of eq.(2), iv) fuzzy feature and peak of eq.(6), and v) fuzzy feature, peak and total volume of eq.(9). For spatial analysis, thirty-seven data will be added. The maximum iteration number and the initial combination of clusters are set as 50 and 50, respectively. The parameters of division, consolidation and operation are 1.0-7.0, 15.0-35.0 and 0.00001, and the feasible combination among those parameters are applied in computer to obtained the optimum number of cluster centers on AIC (Kojiri et al., 1992). Figure 1 shows the AIC sequence against cluster number in the case of fuzzy evaluation on feature pattern, peak and total volume. Though theoretically the classified number should be less than one forth of total data number, the optimum one is decided as eight. Other cases also show the same results. Figure 2 shows the outlines of classified sample data in each pattern. Pattern 8 has only one sample because the feature pattern is quite strange against other centers because of long duration periods and small total volume. Pattern 4 has special feature of three peaks. Pattern 3 looks normal feature of one peak and pattern 5 is bigger than pattern 3 with same feature. Classified characteristics through other objective functions are summarized in Table 1.

To consider the spatial distribution of precipitation and to compare averaging, minimizing and maximizing approaches, same functions are applied adding another thirty-seven data.

Table 1: Characteristics of objective functions

Objective function	Characteristics of classified results
i) Square-root	Fluctuation of sequence is not evaluated
ii) Normalized square-root	Total volume is underestimated
iii) Feature	Multi-peak is not evaluated
iv) Fuzzy feature and peak	Total volume is not considered
v) Fuzzy feature, peak and total volume	Many patterns are classified

Table 2: Spatial characteristics of objective functions.

Objective function	Characteristics of classified results
i) Square-root with averaging	As fluctuation of sequence is ignored, spatial distribution is not evaluated
ii) Square-root with maximization	Less total volume is identified
iii) Fuzzy feature, peak and total volume with averaging	Though spatial difference is discriminated, multi-peak is not evaluated
iv) Fuzzy feature and peak with minimization	Centralized rainfall is identified

Calculation conditions are set as same as the previous one. Concretely, the approach of root-square takes two methods such as spatial averaging and spatial maximization. On the other hand, the fuzzy feature-peak-total volume takes same two methods, too. The analyzed characteristics are listed in Table 2. The fuzzy membership functions with same linear types are applied after several trials for identification. Spatial distribution, especially centralized rainfall and multi-peak, must be discriminated.

As the classification for temporal and spatial distribution of hyetograph describes same results, we decided the objective function of fuzzy feature-peak-total volume as the optimum one among the considered functions.

As run-off process in mountain area yields non-linear effect on discharge, hydrographs are classified by identifying the peak discharge with the following objective function;

$$OF4(i,m) = \left| \max\_t\{CL(t,m)\} - \max\_t\{t,m)\} \right|$$ (11)

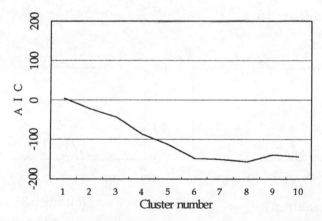

Fig.1: AIC sequence for cluster number in the case of fuzzy, peak and total volume.

581

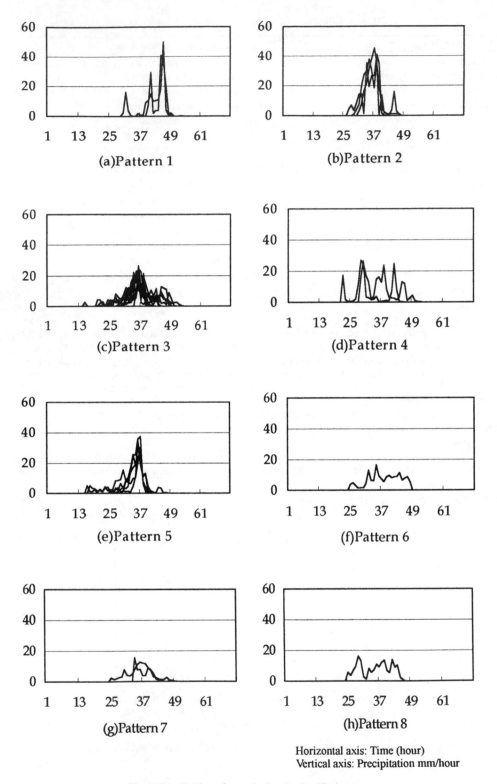

Horizontal axis: Time (hour)
Vertical axis: Precipitation mm/hour

Fig.2: Distribution of sample data in classified patterns.

The classification data and conditions were same as in the case of hyetographs. As the AIC sequence did not show the minimum point, the optimum number was decided as visually rational limit of five. While many hydrographs consist of one peak after smoothing through run-off process, the difference of peak discharge was discriminated into several levels such as bigger than 300 m³/sec and less than 50 m³/sec. Moreover, the long duration or multi-peak was not found.

Table 3 shows the relationship between precipitation and discharge patterns through the fitness concept where the considered event is judged by its occurrence possibility comparing with other events (Kojiri et al., 1999). As pattern 3 and 5 had many sample data (see Fig. 2), the hydrograph patterns were scattered into three patterns. Pattern 1, 2, 6, 7, and 8 of hyetographs were converted into the coincident patterns because of few sample data. Hyetograph pattern 2 was done into hydrograph pattern 1 because of normal and centralized features. In other words, from the viewpoint of rainfall intensity and duration periods, hyetograph pattern 1, 2, 6 & 8 and 7 described their characteristics as combinations of (strong, short), (strong, long), (week, long) and (week, short). According to the temporal distribution with those characteristics, hydrographs were classified.

Table 3: Relationship between precipitation and discharge patterns.

		Pattern	number	of	discharge		
		1	2	3	4	5	Sum
	1					O	2
	2	O					3
Pattern	3		O	O	O		11
number	4		O				2
Of	5	O	O			O	5
precipitation	6				O		1
	7			O			2
	8				O		1
	Sum	4	8	3	8	4	27

3. ESTIMATION OF FLOOD RISK

The designed rainfall for flood control planning is decided with required return period meaning the occurrence probability that is calculated through total volume of historical hyetographs. However, the flood risk is dominated by both distributions of flood peak and duration because flood inundation is caused by overflow or embankment break. By considering the present embankment circumstances constructed with concrete in Japan, the peak discharge is the most important factor for planning. So, the occurrence probability is equivalent with the occurrence feasibility of feature pattern. Assuming that the considered hyetograph belongs to the classified pattern with specific attributes and the objective function is obeyed with log-normal distribution, the occurrence probability is formulated with multiplication of the occurrence probability as cluster and the occurrence probability of hyetograph in it's cluster as follows.

$$P(X) = \sum_{m}^{N} PC(m)PD(X/m) \tag{12}$$

$$PC(m) = ITC(m)/ND \tag{13}$$

$$PD(X/m) = \frac{1}{\sqrt{2\pi}\zeta(m)} \int_0^{OF((X,m))} \frac{1}{u} \exp\left[-\frac{1}{2}\left\{ \frac{\ln(u)-\lambda(m)}{\zeta(m)} \right\}^2 \right] du \tag{14}$$

where, ND is the total number of sample data, $\lambda(m)$ is average and $\lambda(m)$ is the standard deviation in cluster m.

Figure 3 shows the historical flood hyetographs with same total volume of about 143mm. The occurrence probabilities through the traditional approaches are 0.54 and 0.5, respectively. Sample 1 is the steady rainfall with less than 30mm/hour. However sample 2 got the centralized rainfall of about 50mm/hour. The peak discharges are 484m³/sec and 698m³/sec, respectively. The occurrence probabilities through proposed one are 0.20 and 0.06.

4. APPLICATION INTO FLOOD CONTROL PLANNING

Traditionally, flood control facilities are planned to prevent the designed hydrograph, which is decided to cover the necessary return period (the occurrence probability). The existing maximum flood (hydrograph) is extended according to this value as the most risky flood case. As mentioned in the previous chapter, hyetographs with same total volume yield different types of hydrographs depending on feature patterns. Herein, two approaches are proposed. One is to extend the cluster center as representative feature to the designed probability. As nine patterns obtained in spatial analysis are handled as representative hyetographs, peak discharges and feature patterns of hydrographs are calculated through run-off model. The other is to assume that hyetograph distributes around the classified cluster center and to simulate the modified hyetograph with random number obeying normal or lognormal distribution. A number of simulations might be taken if simulated hyetographs does not satisfy the necessary occurrence probability. Finally, the maximum peak discharge is extracted as the designed high flood.

Table 4 shows the generated peak discharge through two approaches against nine cluster centers for spatial distribution data. Run-off analysis was carried out with the kinematic wave method in the divided sub-catchment. Overflow and embankment break were not happened at any points. All historical and simulated data were defined as yearly maximum flood. As pattern 7 is totally small, the extension rate was too big to be found among ten thousand simulations with random number. Then, the discharge of 1133.85 m³/sec is estimated as designed flood peak.

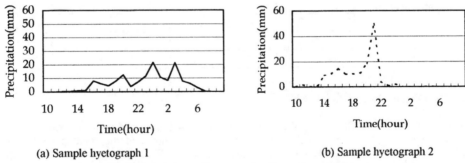

| (a) Sample hyetograph 1 | (b) Sample hyetograph 2 |

Fig.3: Comparison of hyetograph with same total volume.

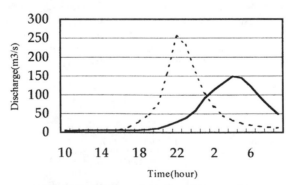

Fig.4: Comparison of simulated hydrographs.

584

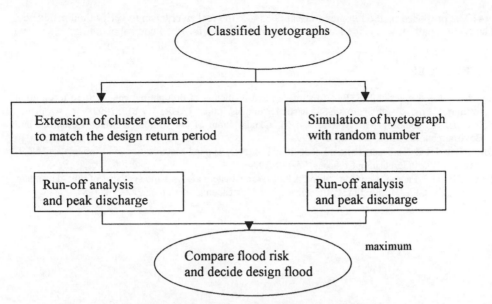

Fig.5: Decision procedures of designed hydrograph.

Though the occurrence probability of discharge of the existing maximum flood whose peak discharge was 1404.0 m³/sec was evaluated as 0.004, the traditional approach yields 1717.0 m³/sec as designed flood. The traditional extension one against single flood feature might generate the rare feature pattern because it focuses on only total volume. Both of peak discharge and duration periods must be considered in future because of main reason of flood risk.

Table 4: Simulated results of hydrograph with designed occurrence probability.

(a) Extension approach

Pattern	Extension rate	Peak discharge (m3/sec)
1	1.15	1050.26
2	1.81	520.02
3	1.14	232.14
4	1.20	555.04
5	1.49	393.15
6	1.19	421.24
7	3.00	850.65
8	1.22	272.82
9	1.32	396.02

(b) Simulation approach

Pattern	Peak Discharge (m³/sec)
1	1133.85
2	397.18
3	254.52
4	637.59
5	589.00
6	440.66
7	-
8	297.81
9	456.30

5. CONCLUSIONS

We discussed the procedures of pattern classification for hydrological events and its application into flood control planning. To sum up, the following results were obtained.

1) The allowable functions were evaluated and the optimal one was decided through its performance and AIC for classification.

2) The occurrence probabilities of hyetographs with same total amount were compared from flood control viewpoint. Risky hyetograph for flood inundation was described with relationship between hyetograph and hydrograph.

3) The significant hydrograph under the consideration of feature pattern and peak discharge was extracted.

4) The proposed methodologies were applied into the real river basin to verify their efficiency. The feature pattern was judged as one of important factors for flood control planning.

REFERENCES

Kojiri, T., T.E. Unny and U.S. Panu: Estimation and prediction of drought by using pattern recognition technique, Proc. 6th Int. Symp. on Stochastic Hydraulics, Taipei, IAHR, 711-817, 1992.

Kojiri, T., K. Tomosugi, K. Sakurai and M. Okuda, Proc. of Int. Conf. on Water Resources and Environment Research, 1186-2001, 1999

Tou, J.T. and R.C. Gonzalez: Pattern Recognition Principle, Applied Mathematics and Computation, No.7, Addison Wisely Publishing Company, 75-109, 1974

Unny, T.E., U.S. Panu, C.D. Macinees and A.K.K. Wong: Pattern analysis and synthesis of time-dependent hydrologic data. Advanced in Hydro-science, Vol.12, Academic Press Inc., 222-244, 1981

Stochastic Hydraulics 2000, Wang & Hu (eds) © 2000 Taylor & Francis, ISBN 90 5809 166 X

Applying stochastic hydrology theory to study the sample errors of electric indexes of hydroelectric station system

Li Ailing
Beijing General Hydrology Service, China

ABSTRACT: An observed series is in fact a sample of the population of the hydrological sequence. Errors exist if the series is used to represent the population. Therefore, there are sample errors in the values of electric indexes for long-term operation of hydropower systems calculated according to the observed series. In order to assess the reliability of the long-term optimal operation of multiple reservoir systems, It is necessary and to estimate the range of the sample errors. This paper, applying the theory and method of stochastic hydrology, studies the sample errors in electric indexes of stochastic optimal operation of multiple reservoir systems. As a case study, the stochastic nature of the five hydroelectric reservoir systems on the upper reaches of the Yellow River in China is analyzed. Firstly, a stochastic optimal operation model is set up, then, an annual and monthly stochastic runoff model are developed, and finally, the sample errors are calculated and analyzed according to the observed and synthetic series.

Key words: Sample errors, Optimal operation model, Reservoir, Runof

1. INTRODUCTION

Generally, the operating indexes of hydroelectric reservoirs are calculated according the observed series. On the one hand, the length of observed series is too short sometimes for the analysis of these indexes, on the other hand, observed series is in fact a sample of hydrological population, the values of these indexes calculated according to the observed series are sample values. Therefore sample errors exists when it is used as values of the population. The purpose of this paper is to analyses the sample errors of the stochastic optimal operation of multiple hydroelectric reservoir systems according to the theory and method of stochastic hydrology. Finally, an example of multiple reservoir systems in series on the upper reaches of the Yellow River in China is given.

Five hydroelectric reservoirs on the upper reaches of the Yellow River in China, named Long Yangxia, Liu Jiaxia, Yan Guoxia, Ba Panxia and Qing Tongxia with installed capacity of 1.280, 1.160, 0.352, 0.160 and 0.272 million KW respectively, serve mainly for electricity production also for irrigation, flood control and icicle control etc. In which Long Yangxia reservoir (LYR) is a multi-annual one, Liu Jiaxia reservoir (LJR) an annual one, Yan Guoxia reservoir (YGR), Ba Panxia reservoir (BPR) and Qing Tongxia reservoir (QTR) are the daily ones respectively. There are three main inflows in the 5-reservoir systems, and they are respectively inflow into LYR, Q_{1t}, interval inflows between LYR and LJR, IB_{1t}, and between YGR and BPR IB_{2t}. (Figure 1.). The stochastic nature of the reaervoir systems are analyzed in this paper.

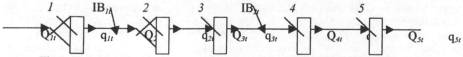

Fig.1: Hydroelectric reservoirs in series on the upper reaches of the Yellow River

2. STOCHASTIC OPTIMAL OPERATION MODEL

2.1 Problem formulation

Because YGR, BPR and QTR are the daily reservoir, and their operation exhibit little influence on the electricity of the 5-reservoir systems in long period time scale, such a stochastic optimal operation can be simplified as that of the 2-reservoir systems of LYR and LJR. Let t ($t = 0, 1, 2, ..., T$) denote the dispersed time index and T the total number of time steps in a year. The time interval $[t, t+1]$ can be, for instance, a month (in this case, $T=12$). Maximization of expected annual electricity is selected as the objective function of long-term optimal operation of the 5-reservoir systems:

$$F_t(V_{1t}, V_{2t}) = \max_{q_{1t}, q_{2t}} \sum P(Q_{1t}, Q_{2t})[N_t(Q_{1t}, Q_{2t}, V_{1t}, V_{2t}, q_{1t}, q_{2t}) + F_{t+1}(V_{1t+1}, V_{2t+1})] \tag{1}$$

$$V_{min it} \leq V_{it} \leq V_{max it} \qquad (i=1,2) \tag{2}$$

$$q_{min it} \leq q_{it} \leq q_{max it} \qquad (i=1,2) \tag{3}$$

$$N_{min it} \leq N_{it} \leq N_{max it} \qquad (i=1,2,...,5) \tag{4}$$

$$N_{min t} \leq N_t = \sum_{i=1}^{5} N_{it} \leq N_{max t} \tag{5}$$

$$N_{it} = \eta_i q_{it} h_{it} \tag{6}$$

$$V_{it} + Q_{it} = V_{i t+1} + q_{it} \qquad (i=1,2) \tag{7}$$

$$Q_{2t} = q_{1t} + IB_{1t} - Y_{1t} \tag{8}$$

$$Q_{3t} = q_{2t} = q_{3t} \tag{9}$$

$$Q_{4t} = q_{3t} + IB_{2t} - Y_{2t} = q_{4t} \tag{10}$$

$$Q_{5t} = q_{4t} - Q_{3t} = q_{5t} \tag{11}$$

$$t = \begin{cases} \sum_{i=1}^{5} N_{it} & (\sum_{i=1}^{5} N_{it} \geq N_p) \\ \sum_{i=1}^{5} N_{it} - \alpha(N_p - \sum_{i=1}^{5} N_{it})^{\beta} - M & (\sum_{i=1}^{5} N_{it} < N_p) \end{cases} \tag{12}$$

where $F_t(V_{1t}, V_{2t})$ is dynamic programming optimal value function with storage V_{1t} of LYR and storage V_{2t} of LJR from the beginning of time step t to the end of the terminal time step T; $F_{t+1}(V_{1 t+1}, V_{2 t+1})$ is expected annual electric energy with storage V_{1t+1} of LYR and storage V_{2t+1} of LJR from the end of time step t to the end of the terminal time step T; $N_t(Q_{1t}, Q_{2t}, V_{1t}, V_{2t}, q_{1t}, q_{2t})$ is current electric energy with storage V_{1t}, inflow Q_{1t}, release q_{1t} of LYR and storage V_{2t}, inflow Q_{2t}, release q_{2t} of LJR at time step t; $P(Q_{1t}, Q_{2t})$ is joint distribution function for Q_{1t} and Q_{2t}; V_{it}, $V_{min it}$, $V_{max it}$ are storage, allowable lower and upper limits on storage of the ith reservoir at time step t, respectively; q_{it}, $q_{min it}$, $q_{max it}$, are release, allowable lower and upper limits on release of the ith reservoir at time step t, respectively; N_{it}, $N_{min it}$, $N_{max it}$ are power production, allowable lower and upper limits on power production of the ith reservoir at time step t, respectively; N_t, $N_{min t}$, $N_{max t}$ are power production, allowable lower and upper limits on power production of the 5-reservoir systems at time step t, respectively; IB_{1t}, IB_{2t} are interval inflows between LYR and LJR and between YGR and BPR at time step t, respectively; Y_{1t}, Y_{2t}, Y_{3t} are flow requirements for irrigation between LYR and LJR , between YGR and BPR, and between BPR and QTR at time step t, respectively; Q_{1t}, Q_{2t}, Q_{3t}, Q_{4t}, Q_{5t} are inflows of the ith reservoir at time step t, respectively; q_{1t}, q_{2t}, q_{3t}, q_{4t}, q_{5t} are releases of the ith reservoir at time step t, respectively; α, β, M are punitive coefficient; η_i is power coefficient of the ith reservoir; h_{it} is head of water of the ith reservoir at time step t, N_p is firm power for the 5-reservoir systems.

2.2 Recursive solution procedure

Equation (1) is a stochastic dynamic programming optimal problem and can be written as the following equation:

$$F_t(V_{1t},V_{2t}) = \max_{q_{1t},q_{2t}} \sum P(Q_{1t})P(Q_{2t}/Q_{1t})[N_t(Q_{1t},Q_{2t},V_{1t},V_{2t},q_{1t}q_{2t}) + F_{t+1}(V_{1t+1},V_{2t+1})] \tag{13}$$

Let:

$$N(V_{1t},V_{2t}) = \sum P(Q_{1t})P(Q_{2t}/Q_{1t})N_t(Q_{1t},Q_{2t},V_{1t},V_{2t},q_{1t},q_{2t}) \tag{14}$$

substitution of equation 14 into equation 13, then, equation 13 can be written as:

$$F_t(V_{1t},V_{2t}) = \max_{q_{1t},q_{2t}}[N(V_{1t},V_{2t}) + \sum P(Q_{1t})P(Q_{2t}/Q_{1t})F_{t+1}(V_{1t},V_{2t})] \tag{15}$$

therefore, the recursive function can be set up as the following:

$$F_t(V_{1t},V_{2t}) = \begin{cases} \max_{q_{1T},q_{2T}}[N(V_{1T},V_{2T}) + \sum P(Q_{1T})P(Q_{2T}/Q_{1T})F_{T+1}(V_{1T+1},V_{2T+1})] & t = T \\ \max_{q_{1t},q_{2t}}[N(V_{1t},V_{2t}) + \sum P(Q_{1t})P(Q_{2t}/Q_{1t})F_{t+1}(V_{1t+1},V_{2t+1})] & t \neq T \end{cases} \tag{16}$$

where $F_{T+1}(V_{1T+1},V_{2T+1})$ is dynamic programming optimal value function at the end of terminal time step T, and can be supposed a initialized value at the beginning of iteration.

Backward recursive calculation through the stages gives optimal releases q_{it} for all stages $t = T \ldots 1$. These releases represent a now nominal policy to initiate the next iteration of the algorithm. Storage values corresponding to this nominal policy are computed in a forward run through the stages using equation 7 by making use of known storage at the beginning of stage 1, computed releases, and expected inflows. The backward and forward runs of the algorithm are repeated until convergence is achieved.

3. ANNUAL AND MONTHLY STOCHASTIC RUNOFF GENERATION MODEL

There are three inflows in the 5-reservoir systems, and they are inflow into LYR, interval inflows between LYR and LJR and between YGR and BPR respectively. In this paper, a three-variable AR(1) model is selected to simulate the annual inflows, and a disaggregation model is used to disaggregate the annual inflows into monthly inflows.

3.1 The stochastic model of annual runoff

The three-variable AR(1) model can be described by the equation:

$$Z_t = AZ_{t-1} + B\varepsilon_t \tag{17}$$

where A is a coefficient matrix of 3×3 dimensions describing auto-correlation and cross-correlation for lag 1 among three variables; B is also a coefficient matrix of 3×3 dimensions describing auto-correlation and cross-correlation for lag 0 among three variables; ε_t is a matrix of 3×1 dimensions of the independent random variable with zero-mean and one-deviation; Z_t and Z_{t-1} are a matrix of 3×1 dimensions of annual runoff at time t and $t-1$ respectively.

Equation (17) is timed on the left side by $Z_{t-1}{}^T$, and run a expectation operation:

$$E(Z_t Z_{t-1}{}^T) = AE(Z_{t-1} Z_{t-1}{}^T) + BE(\varepsilon_t Z_{t-1}{}^T) \tag{18}$$

where E is the expectation operator.

Equation (18) can be written as the following because of ε_t a independent random variable:

$$A = R(1) R(0)^{-1} \tag{19}$$

where, $R(0)$ and $R(1)$ are the correlation coefficient matrixes of 3×3 dimensions for lag 0 and 1 of annual runoff.

Equation (17) is timed on the left side by $Z_t{}^T$, and run a expectation operation:

$$E(Z_tZ_t^T)=AE(Z_{t-1}Z_t^T) + BE(\varepsilon_tZ_t^T) \tag{20}$$

considering equation (17) gives:

$$E(B\varepsilon_tZ_t^T)=E[B\varepsilon_t(AZ_{t-1}+B\varepsilon_t)^T] \tag{21}$$

Equation (21) can be written as the following because of ε_t a independent random variable:

$$E(B\varepsilon_tZ_t^T)=E(BB^T) \tag{22}$$

substituting the value of $E(B\varepsilon_tZ_t^T)$ from equation (22) into equation (20) gives:

$$BB^T=R(0)-AR(1)^T \tag{23}$$

suppose B a below triangle matrix and define:

$$D=R(0)-AR(1)^T =(d_{ij})_{3\times3} \tag{24}$$

then, coefficient matrix B can calculated as the following:

$$b_{ij} = \begin{cases} d_{ji}/b_{jj} & (j=1,i=1,2,3) \\ [d_{ij}-\sum_{k=1}^{j-1}(b_{jk})^2]^{1/2} & (j=2,3,i=j) \\ [d_{ij}-\sum_{k=1}^{j-1}b_{jk}b_{ik}]/b_{jj} & (j=2,3,i=j+1,\cdots,3) \end{cases} \tag{25}$$

according to equation (19) and (25), the coefficient matrix A and B can be calculated.

Equation (17) has supposed ε_t the standard normal random distribution function, in fact, most hydrological factors are distributed non-normally, therefore, it is necessary to correct the series of ε_t into the series of ξ_t with a expected distribution of the observed series. The value of Cs of the ξ_t series can be calculated by the following equation:

$$Cs_\xi=\frac{\sum_{i=1}^{n}\varepsilon_{it}^3}{(n-3)} \tag{26}$$

where n is the length of observed series. The random variable ξ_t can be obtained from the standard normal random distribution variable ε_t [$\sim N(0,1)$] through the W-H transform:

$$\xi_t = \frac{2}{Cs_\xi}(1+\frac{Cs_\xi\varepsilon_t}{6}-\frac{Cs_\xi^2}{36})^3 -\frac{2}{Cs_\xi} \tag{27}$$

According to the 32-year observed datum of annual runoff in the 5-reservoir systems, the coefficient matrixes A, B and Cs_ξ are calculated, the stochastic model of annual runoff is set up, and 50 groups of synthetic series at equal length to the observed series are simulated.

3.2 The disaggregate model of monthly runoff

The monthly runoff can be simulated by the typical disaggregate model through disaggregating the annual runoff into monthly runoff. Firstly, the statistic value of $Z_t=\sum_{i=1}^{3} Z_{it}$ of synthetic series can be obtained; then, a typical year, whose annual runoff value is the nearest to the above statistic value Z_t, is selected from observed series; finally, the monthly runoffs for three inflows are simulated according the discharge distribution of the typical year.

4. ANALYSIS FOR SAMPLE ERRORS OF ELECTRIC INDEXES

According to the operation plan of the 5-reservoir systems in series on the upper reaches of the Yellow River can be obtained through to the stochastic optimal operation model (1), 50 groups of synthetic series and observed series are calculated according to the operation plan. The results of sample errors of electric indexes for the long-term operation are given in Table 1

Table.1: The results of sample errors of electric indexes for the long-term operation

Reservoir Indexes		LYR	LJR	YGR	BPR	QTR	Total
Average Annual Electricity	Observed	5.316	6.097	2.19	1.13	1.216	15.952
	Synthetic	5.380	6.203	2.220	1.142	1.235	16.180
	Errors	0.330	0.238	0.065	0.026	0.026	0.667
Guarantee Percent for Firm power	Observed	95	Guarantee Percent for Irrigation	Observed	90	Firm power (Kw)	1,300,000
	Synthetic	96		Synthetic	93		
	Errors	3		Errors	5		

Note: The unit of average annual electricity is Billion KW • h, and errors is standard deviation here.

Table 3 shows that the total electricity energy for long-term operation is 16.180 Billion KW •h, the sample errors is 0.667 Billion KW •h, and the relative errors is 4%. The electricity of LYR, LJR, YGR, BPR and QTR is 5.380、6.203、2.220、1.142 and 1.235 billion KW • h respectively, the relative errors is 6%、4%、3%、2% and 2% respectively. The relative errors of LYR is the largest, that of LJR is the second largest, and that of other reservoir is relatively small. The reason is that LYR and LJR , with large regulation ability as multi-annual reservoir and annual reservoir respectively, will choose the flexible operation way in order to obtain the largest electricity benefit for the hydroelectric reservoir systems under various probable inflows, therefore, the electricity energy of LYR and LJR will have a remark difference for different synthetic series.

Table 3 shows also that the results of electric indexes for long-term operation for synthetic series and observed series are nearly, therefore, the results are reasonable and can be accepted.

5. CONCLUSIONS

The values of electric indexes for long-term operation of hydropower systems calculated according to the observed series exist the sample errors. This paper, applying the theory and method of stochastic hydrology, contributes to the study on the sample errors in electric indexes of stochastic optimal operation of multiple reservoir systems. A real example of the 5-reservoir systems on the upper reaches of the Yellow River in China is analyzed and calculated. it is obvious that the different operation plan will have the different errors, and at the same, It can seen that individual reservoir of multiple reservoir systems has different errors owing to the difference in regulation and character. It is believed that the sample errors in electric indexes obtained by other operation plan or in other hydrological indexes can also be studied through such a method mentioned in this paper.

REFERENCES

Braga, B. P. F., W. W-G. Yeh, L. Becker, M. T. E. Barros (1991) Stochastic optimization of multiple reservoir system operation. *Journal of Water Resources Planning and Management*, 117(4): 471-481

Ding Jing, Deng Yuren (1988) Stochastic Hydrology. *Science and Technology University of Cheng Du Press, Cheng Du*

F. A. EL-Awar, J. W. Labadie, T. B. M. J. Ouarda (1998) Stochastic differential dynamic programming for multi-reservoir system control. *Stochastic Hydrology and Hydraulics*, 12: 247-266

Katz. R. W. (1981) Estimating the order of a Markov Chain. *Technometrics*, 23: 243-249

Salas, J. D. et al (1980) Applied modeling of hydrologic time series. *Water Resources Publications. Littleton, Colorado*

J. Mohapl (1998) A stochastic model of water-table elevation. *Stochastic Hydrology and Hydraulics*, 12 223-245

Kelman J, Stedinger J, Cooper L, Hsu E. Yuan S (1990) Sampling stochastic dynamic prpgramming applied to reservoir operation. *Water Resources Research*, 26(3): 447-454

Turgeon A (1980) Optimal Operation of Multireservoir Power Systems with Stochastic Inflows. *Water Resource Research*, 16(2): 275-283

Stochastic Hydraulics 2000, Wang & Hu (eds) © 2000 Taylor & Francis, ISBN 90 5809 166 X

Preliminary measurement of model tunnel flow

P.J.Li
Hydropower Laboratory, Norwegian University of Science and Technology (NTNU), Trondheim, Norway

B.Svingen
Water Resource, Civil and Environmental Engineering, SINTEF, Trondheim, Norway

ABSTRACT: Preliminary measurement of stationary flow and simple oscillation flow in a special build model tunnel with smooth wall has been made. LDV and ADV instrument have been applied in the measurement to get the 2D and 3D velocity respectively. The mean velocity profiles of the tunnel flow and corresponding RMS values with different flow rate were presented. The profile from ADV and LDV basically agree with each other and conform to the theoretical profile. The local pressure and differential pressure along the tunnel have been measured at the same time with velocity measurement.

Key words: Stationary flow, Oscillation flow, Velocity profile, Tunnel flow

1. INTRODUCTION

Many hydropower plants in Norway were built under ground. Underground tunnel system collects water from reservoirs and transfer it into underground powerhouse. The flow characteristic in the tunnel is significant to the operation of the power plant, especially when oscillation occurred in the flow.

Acoustic Doppler Velocimeter (ADV) and Laser Doppler Velocimeter (LDV) have been used in flow measurement for some years. ADV was mostly used in open channel flow measurement. Voulgaris, G. et al (1998) compared application of ADV and LDV in turbulence measurement of flow in a 17-m flume. The limitation of ADV is mainly on measuring boundary layer and its sampling volume. Song, T. et al (1997) measured vertical velocity distribution in a open channel with ADV. The agreement between the theoretical distribution of vertical velocity and measured data with ADV was reasonable good. Sukhodolov, Alexander et al (1999) used ADV to get the turbulence structure in an ice-covered sand bed river. This is similar to the case of a closed tunnel we would work on. The mean velocities measured generally conform to the distributions derived from a two-dimensional flow model at least in the central part of the flow. A miniature ADV was used in the measurement of 3D velocity by Weber-Shirk, Monroe et al (1995). This is also an application of ADV for open channel flow measurement. Owen, F. Kevin et al (1991) carried out a measurement of fluorescent dye concentration in a 16 by 24-inch water tunnel with LDV. The measurements show that the LDV can provide new insight into the structure and shear layer flows. Wijetunge, Janaka J. et al (1998) investigated the effects of sediment transport on the fluid velocities and turbulence in oscillatory flow with LDV. The measurements were made over flat beds in an oscillatory flow water tunnel. Dick J.E. et al (1991) measured the velocities in oscillatory flow with LDV. The velocity profile of oscillatory flow was analyzed afterwards. All of these measurements confirmed that ADV and LDV could be used in the flow measurement for various purposes.

The purpose of this measurement is to prepare for further measurement of oscillatory flow in model tunnel and head loss estimation. The measurement was carried out in Norwegian Hydrology Technology Laboratory (NHL), Trondheim. The measurement was part of a project in hydropower and was sponsored by Norwegian Research Council (NFK).

2. SETUP OF RIG

The measuring system includes model tunnel, water supply system and data acquisition system. There are three different rig setups totally. Two of them have been tested preliminarily. The difference between these two setups is mainly on the water supply system and the velocity measurement method, which is a part of data acquisition system. Different configurations are used for different flow style measurement.

2.1 Model tunnel

The model tunnel is composed of 8 sections and made of plastic glass. The first section and the last section are in trapezoidal and acted as simple expansion of inlet and contraction of outlet respectively. The other 6 sections in between are in rectangular. The cross section of rectangular sections is 230 mm * 320 mm. The length of each section is 3 meters.

2.2 Water supply system

There are two configurations of water supply system. The water comes from top tank in NHL in the first case, in which the static head of flow keeps constant, but only stationary flow is possible to produce. In the second configuration, the water tank and connection pipe build up a closed circuit with the model tunnel on site. Pumps were installed in the tank to produce different flow styles for the model tunnel, such as oscillatory flow, stationary flow, etc.

2.3 Data acquisition system

Data acquisition system consists of sensors, data acquisition unit, control PC and control software.

♦ Sensors: including velocity sensors, pressure cells, differential pressure sensors and a flow meter. Velocity sensors are ADV (Acoustic Doppler Velocimeter) from Nortek AS and 2D LDV (Laser Doppler Velocimeter) from Dantec Measurement System. Differential pressure cells are from Rosemount Controls and Fuji Electric. MagMaster Electromagnetic Flow meter from ABB Kent-Taylor was used to get the flow rate at the inlet tube. Pressure cells are from IMT Industrie-MessTechnik GMBH.

♦ Data acquisition unit: HP Data Acquisition / Switch Unit, Data Acquisition Board from National Instrument, and connection board from Flow Design Bureau (FDB)

♦ Control PC: control PC for LDV sub-system, control PC for ADV unit, control PC for other data collection

♦ Control software: BSA Flow V1.4 from Dantec Measurement System, ADVlab V2.7 from Nortek, and LabView V5.0 from National Instrument.

3. VELOCITY PROFILE MEASURED

Velocity profiles of stationary flow were measured with ADV and LDV. Velocity profiles of oscillation flow in different frequency were measured with LDV. The results are shown as follows.

3.1 Velocity Profiles of stationary flow by ADV and LDV

Velocity profile of stationary flow in the model tunnel with smooth wall has been measured with LDV and ADV simultaneously. The LDV gave out the 2D-velocity profile and the ADV gave out the 3D-velocity profile.

Figures 1--2 (next page) showed the ADV measured velocity V_x and V_y with flow rate changing from Q = 5.0 l/s to 87.6 l/s. The RMS of velocity V_x and V_y were shown in figures 3—4 (next page). Velocity V_z and its RMS are not shown here. First, V_z was not measured by LDV later and V_z was not compared between the results of ADV and LDV. Second, V_z was very small compare to V_x and very similar to V_y.

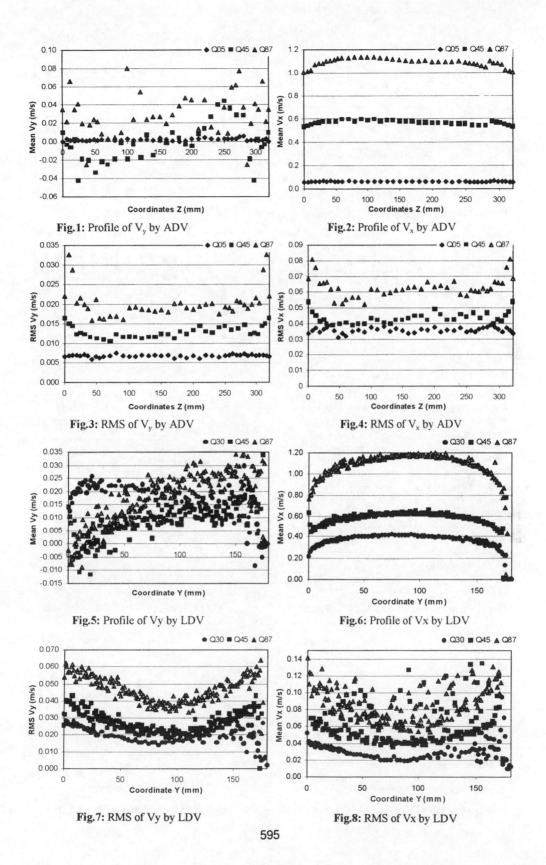

Fig.1: Profile of V_y by ADV

Fig.2: Profile of V_x by ADV

Fig.3: RMS of V_y by ADV

Fig.4: RMS of V_x by ADV

Fig.5: Profile of Vy by LDV

Fig.6: Profile of Vx by LDV

Fig.7: RMS of Vy by LDV

Fig.8: RMS of Vx by LDV

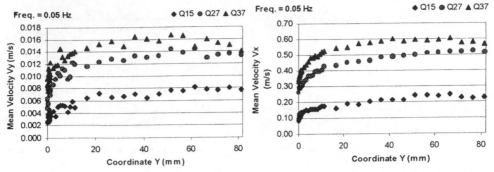

Fig.9: Profile of Vy by LDV   f = 0.05Hz

Fig.10: Profile of Vx by LDV f = 0.05Hz

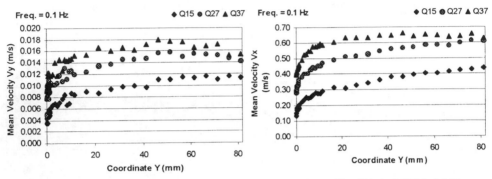

Fig.11: Profile of Vy by LDV f = 0.1Hz

Fig.12: Profile of Vx by LDV f =0.1 Hz

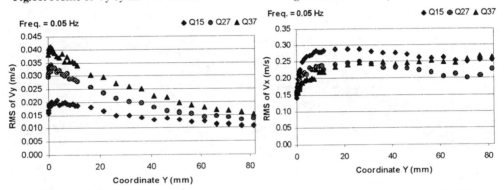

Fig.13: RMS of Vy by LDV f = 0.05Hz

Fig.14: RMS of Vx by LDV f =0.05 Hz

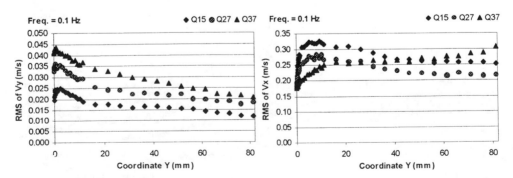

Fig.15: RMS of Vy by LDV f = 0.1Hz

Fig.16: RMS of Vx by LDV f =0.1 Hz

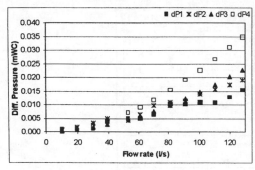
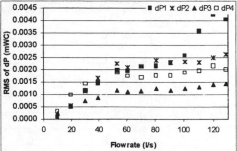

Fig.17: Mean dP of stationary flow   **Fig.18:** RMS of dP of stationary flow

Figures 5-6 (previous page) showed the LDV measuring results with flow rate changing from $Q = 30$ l/s to 87 l/s. The RMS of velocity V_x and V_y by LDV were shown in figures 7—8 (previous page).

3.2 Velocity profiles of oscillation flow by LDV

The flow was activated and oscillated in the model tunnel with smooth wall by two accurately controlled pumps installed in a closed loop with model tunnel. The 2D-velocity profiles of oscillation flow were measured with LDV. Velocity profiles in x and y directions with oscillating frequency of 0.05 Hz were shown in figures 9—10. Velocity profiles in x and y directions with oscillating frequency of 0.1 Hz were shown in figures 11—12.

The RMS of velocity vectors also has been given out from LDV analysis software. RMS of V_y and V_x at frequency of 0.05 Hz was shown in figures 13-14 (previous page). RMS of V_y and V_x at frequency of 0.01 Hz was shown in figures 15-16.

Due to the characteristics of oscillatory flow, measuring is very time consuming with LDV. To save measuring time, we only measured half of the tunnel cross section in this case. The profiles shown in figures 9—16 were half of the whole profiles of the rectangular tunnel. We supposed that the other half profiles would be symmetry to the measured half. The cross section of model tunnel is right rectangular, which is symmetry across the middle section from geometry point of view. The flow should be symmetry from hydraulic point of view. From the measured profile for whole cross section shown in figure 5—8, we could confirm this hypothesis to certain extent.

All of the velocity profiles showed above are mean velocity. For the oscillatory flow, the velocity of flow was changing in sine wave always. The form of wave was not presented here.

3.3 Head loss measured along the model tunnel

Differential pressures and local pressures along the model tunnel were measured at the same time while we were measuring the velocity. We had four dP sensors along the tunnel. The distance between the two inlet points of these dP cells was 3 meters when we measured the stationary flow. As we measured the oscillatory flow, the distance of dP cell inlet points were changed into different value for four censors: dP1—9 meters; dP2—6 meters; dP3—6 meters; dP4—3 meters. Six pressure sensors were installed along the tunnel with distance in between of 3 meters.

Figure 17 shows the differential pressure measured along the model tunnel with stationary flow in it. Figure 18 is the RMS curve of these differential pressures. Figure 19 gives out the measured mean dP of oscillation flow.

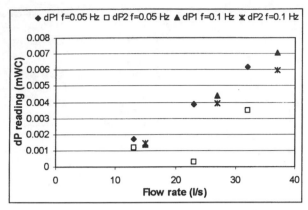

Fig.19: Mean of dP readings at f = 0.05 Hz and f = 0.1Hz

4. SUMMARY OF THE PRELIMINARY MEASUREMENT

Generally, the velocity profiles from ADV and LDV conformed to the theory analysis results, they are similar to each other basically. Obviously, the results are much better than the results from ADV, especially near to the boundary layer.

Due to the limitation of the construction of ADV, it is difficult to move it from boundary to boundary of the model. It causes problems. One of the problems is that it is impossible to get the velocity profile next to the boundary, where the boundary layer located and it is very important to the velocity profile of the turbulent flow. The other problem is that we could not move the sensor to the top wall of the tunnel as to the bottom wall of the tunnel. We could only suppose that the profile near the top wall is the same as the profile near the bottom wall. From the view of symmetry, this should be true.

The ADV disturbed the flow to some extent. The sensor is 5 cm away from the sampling volume, the disturbance was not very significant normally by sensor itself. But when insert the sensor into the flow, a slot had to be opened on the top wall of the tunnel, it was across the tunnel width and it is a little bit lower than the normal surface, this disturbed the flow a lot. It was shown in the profiles from ADV, the mirrored top boundary was a lit bit away from the profile of neighbored area, result in the profile was not symmetry between the upper and lower part.

Normal distribution function of samples was checked randomly (not shown here). The results from ADV are well distributed in the whole profile. The results from LDV are different. At the boundary point (on the wall), the distribution was not good enough, the distribution in between is very good. The reason for this result is:

1) It is difficult to determine the point on the wall in this model by naked eyes. The wall is made of glass and the laser beam is very strong.
2) The reflection of the wall to the laser beam is very strong when the sampling point is near to the wall. The noise from reflection made the signal unstable.

The sensor of LDV was moved across the tunnel automatically and controlled accurately. The ADV sensor was moved manually and position was measured by common scale. This would cause different errors in the measurement.

The sampling volume of ADV is much bigger than that of LDV. It is difficult and meaningless to measure the velocity with ADV by very small step, such as 1 mm as that in the case of LDV.

The velocity profile by ADV is across the tunnel in vertical direction, and the velocity profile by LDV is across the tunnel in horizontal direction. They are normal to each other. This caused some difference.

When the flow rate is becoming bigger, the turbulent near the boundary is very strong, that ADV sensor vibrated in the flow. This would cause a lot of measuring error.

598

The alignment of ADV sensor to the flow direction is fatal to the measurement. But there is not an effect way to align the sensor to the flow exactly at this moment. The alignment was only observed by the naked eyes. It is not accurate.

The RMS value of velocity in the central part was smaller than the value from the side part flow. This is due to the velocity in the central part is bigger and more stable than the velocity near the boundary.

5. CONCLUSION

The velocity profiles measured conform to that of theoretical profile generally. The results from LDV are much better than the results from ADV in the case of closed tunnel flows measurement. Differential pressures measured along the tunnel increase with the flow rate both in stationary flow and oscillation flow. Further work should focus on the analysis of head loss in the tunnel.

REFERENCE

Dick, J.E.; Sleath, J.F.A. 1991, Velocities and concentrations in oscillatory flow over beds of sediment, in *Journal of Fluid Mechanics* v 233 Dec 1991 p 165-196

Owen, F. Kevin; Orngard, Gary M.; Neuhart, Dan H., 1991, A laser fluorescence anemometer system for the Langley 16- by 24-inch water, *International Congress on Instrumentation in Aerospace Simulation Facilities,* Dec 1991 Sponsored by: IEEE Aerospace & Electronic Systems Soc Publ by IEEE p 403-412

Parthasarathy, R.N.; Muste, M., Velocity measurements in asymmetric turbulent channel flows, *Journal of Hydraulic Engineering* v 120 n 9 Sept 1994 ASCE p 1000-1020

Ribberink, Jan S.; Al-Salem, Abdullah A. , 1995, Sheet flow and suspension of sand in oscillatory boundary layers , *Coastal Engineering* 25 3-4 Jul 1995 Elsevier Science B.V. p 205-225

Song, T.; Chiew, Y.M., 1997, Vertical velocity distribution in steady non-uniform and unsteady open-channel flow, *Journal of Hydrodynamics* v 9 n 3 1997 China Ocean Press p 49-64

Sukhodolov, Alexander; Thiele, Michael; Bungartz, Heinz; Engelhardt, Christof, 1999, Turbulence structure in an ice-covered, sand-bed river, *Water Resources Research* v 35 n 3 Mar 1999 American Geophysical Union p 889-894

Voulgaris, G.; Trowbridge, J.H., 1998, Evaluation of the acoustic Doppler velocimeter (ADV) for turbulence measurements, *Journal of Atmospheric and Oceanic Technology* 15 1 pt 2 Feb 1998 p 272-289

Weber-Shirk, Monroe; Brunk, Brett; Jensen-Lavan, Anna; Jirka, Gerhard; Lion, Leonard W., 1995, Simulating turbulence in natural , *International Water Resources Engineering Conference - Proceedings* 2 Aug 14-18 1995 1995 Sponsored by: ASCE ASCE p 1620-1625

Wijetunge, Janaka J.; Sleath, John F.A., Effects of sediment transport on bed friction and turbulence Journal of Waterway, Port, *Coastal and Ocean Engineering* v 124 n 4 July-Aug 1998 ASCE p 172-178

Flood estimation and control

Stochastic Hydraulics 2000, Wang & Hu (eds) © 2000 Taylor & Francis, ISBN 90 5809 166 X

Invited lecture: Risk zonation and loss accumulation analysis for floods

Wolfgang Kron
Münchener Rückversicherungs-Gesellschaft, München, Germany

ABSTRACT: Worldwide, flood is the number one cause of losses from natural events. The insurance of the flood risk has attracted more and more attention in recent years, not only in the fields of science and politics but also within the insurance industry itself. Even if the list of great natural disasters is headed by storms and despite the relatively low density of insurance against flooding the large number of local flood events constitutes a major factor for the insurance industry. Flooding comes in very different ways, ranging from storm surges to flash floods and from sometimes country-wide river floods to glacial lake outburst and dam break floods. These different types are important when insurance concepts are regarded, for which adverse selection plays an important role.

Various concepts for the difficult field of flood insurance have been investigated and discussed in different countries. In Germany, the insurance industry, in a combined effort, has launched a project to identify zones of similar risk resulting from river floods for the whole country. The identification of possible clients and the calculation of premiums are two uses of the developed model as well as the assessment of probable maximum losses. Only river floods are considered in the model, flash floods and storm surges are excluded; flood control measures are not taken into account either.

Key words: Risk zonning, Flood loss, Insurance on flood risk, River flood

1. INTRODUCTION

In most parts of the world, flooding is the leading cause of losses from natural phenomena and is responsible for a greater number of damaging events than any other type of natural hazard. Roughly half of all losses due to nature's forces can be attributed to flooding. Flood damage has been extremely severe in recent decades and it is evident that both the frequency and intensity of floods are increasing. In the past ten years losses amounting to more than US$ 250bn have had to be born by societies all over the world to compensate for the consequences of floods. There are countries, such as China, in which flooding is a frequent, at least annual event, and others, such as Saudi Arabia, where inundation is rare but its impact sometimes no less severe. No populated area in the world is safe from being flooded. However, the range of vulnerability to the flood hazard is very wide, in fact wider than for most other hazards. Some societies (communities, states, regions) have learnt to live with floods. They are prepared. Others are sometimes completely taken by surprise when a river stage (or the sea) rises to a level neighbouring residents have never experienced before in their lives.

Why natural catastrophes are becoming more and more frequent and severe, although protection and preparedness measures have been improving. Besides public and individual measures, insurance is an important factor in reducing the exposure of individuals, enterprises and even whole societies to natural hazards. Proper insurance can considerably mitigate the effects of extreme events on them and avoid their being ruined.

2. NATURAL DISASTER STATISTICS

Reinsurance companies, due to their worldwide activities, are among the best sources for natural disaster statistics (Kron, 2000). Their analyses focus on three aspects: the number of people affected (fatalities, injured, homeless), the overall economic damage to the country hit, and the losses covered by the insurance industry.

Natural disasters with thousands of deaths almost always hit poor countries and are caused by earthquakes (Table 1). The one aspect (poverty) is related to the higher vulnerability in less developed countries (poorer quality of structures, more people), the other (earthquakes) to the sudden onset of such events, which strike without warning. In the past (more than 50 years ago), floods were responsible for a huge number of deaths. With the exception of storm surges this is not so anymore today. The table of the deadliest disasters during the past 30 years contains only three great water-related disasters, the 1970 and 1991 Bangladesh storm surges, and the recent flood and debris-flow event in Venezuela. For no other type of natural disaster have early warning methods become more operational, more reliable and hence more effective than for meteorological and hydrological disasters. A 1994 Bangladesh storm surge that ran up to a height comparable to the one in 1991 cost the lives of only 200 people. This reduction in the number of victims has mainly been a consequence of improved early warning methods based on better storm forecast models together with the availability of elevated shelters that allowed people on low land to flee the flood waters (Kron et al., 2000). Nowadays, geogenic disasters (earthquakes, volcanic eruptions, landslides) pose the deadliest threat. In addition to their extremely sudden onset, the prediction of major geological events is difficult or even impossible, and in most cases there is no time left for warning. In contrast to this, hydrological events almost always build up relatively slowly. Usually, even the few minutes an approaching flash flood leaves for people to flee may be enough for many to save their lives. However, the Venezuela floods showed that this is not always the case. The extreme risk to which people exposed themselves by settling on and below highly unstable slopes combined with unusual rainfall were the reasons for the shocking event that killed more than 30,000 people (some estimates go even as high as 50,000) just before Christmas 1999.

The statistics for losses display a different picture: the record economic losses (Table 2) occur mostly in rich countries. While two earthquakes still lead the table, floods, which usually affect much larger areas than earthquakes and occur much more frequently, have at least the same importance. Especially in China they cause almost every year billions of dollars of losses for the economy and severe distress in the nation. Not only the great disasters display such a tendency, but also the accumulated annual amount of losses from the many small and medium-sized events. On average, floods cause as much damage as all other destructive natural events together. Additionally, one should bear in mind that the financial means societies all over the world spend on flood control (sea dikes, levees, reservoirs, etc.) is a multiple of the costs they devote to protection against other impacts from nature. For the insurance industry storms are clearly the most critical loss events (Table 3), occurring exclusively in rich countries, although earthquakes – e.g. a major event in California, which may cost the insurance companies several billion dollars – represent the greatest loss potentials.

The tables reveal that all but two of the economic and insured losses occurred in the last third of the regarded 30-year period clearly indicating an increase in these events. An analysis of all great natural disasters in the past half century (Munich Re, 1999) shows that the losses generated by natural disasters have been exploding since the sixties (Fig. 1). Great natural disasters are those in which the affected areas are clearly unable to help themselves and require interregional or international aid. This is normally the case when there are thousands of fatalities, hundreds of thousands of people made homeless, or substantial economic losses (depending on the economic circumstances in the affected country). Only 27 such catastrophes were counted in the sixties, but this number rose to 63 in the eighties and 87 in the nineties. The increase took place more or less in two steps, which becomes quite clear if the averages of 15-year periods are regarded. The period 1955-1969 produced 2.5 great natural disasters per year, 1970-1984 about 4.3, and the past fifteen years 8.7 (Fig. 1, upper part). The graphs for economic and insured losses (Fig. 1, middle part) show a continuous and constantly accelerating upward trend. The total losses from great natural disasters accumulated to almost US$ 609bn in the years from 1990 to 1999, which is – when inflation is taken into account – nearly nine times as much as in the sixties (US$ 71bn). Even more dramatic is the increase in the insured losses: US$ 109bn (last ten years) versus about US$ 7bn (sixties) yields a factor of over 16. The main causes for this development are: the increasing concentration of people and values in areas that are exposed to unfavourable natural conditions, the increasing vulnerability of structures and goods, the – often unjustified – trust in protection systems, and the changes in environmental conditions including climate change (Munich Re, 1999; Berz, 1999). The disproportional increase in insured losses may be attributed primarily to an increasing insurance density.

Table.1: The ten deadliest natural disasters of the past 30 years (not including droughts)

Rank	Year	Event	Country	Fatalities
1	1970	Storm surge	Bangladesh	300 000
2	1976	Earthquake (Tangshan)	China	290 000
3	1991	Storm surge	Bangladesh	140 000
4	1970	Earthquake, landslide, tsunami	Peru	67 000
5	1990	Earthquake	Iran	40 000
6	1999	Floods, debris flows	Venezuela	>30 000
7	1988	Earthquake	Armenia	25 000
8	1985	Volcanic eruption, lahar	Colombia	23 000
9	1976	Earthquake	Guatemala	22 000
10	1999	Earthquake (Izmit)	Turkey	>20 000

Table.2: The ten costliest natural disasters of the past 30 years
(original values, not adjusted for inflation)

Rank	Year	Event	Country/Region	Economic losses US$ bn
1	1995	Earthquake (Kobe)	Japan	100
2	1994	Earthquake (Northridge)	USA	44
3	1998	Floods	China	30
4	1992	Hurricane Andrew	USA	27
5	1996	Floods	China	24
6	1993	Flood (Mississippi)	USA	16
7	1990	Winter storms	Europe	15
	1991	Floods	China	15
	1995	Floods	North Korea	15
	1999	Floods, debris flows	Venezuela	15

Table.3: The ten costliest natural disasters of the past 30 years for the insurance industry
(original values, not adjusted for inflation)

Rank	Year	Event	Country/Region	Insured losses US$ bn
1	1992	Hurricane Andrew	USA	17.0
2	1994	Earthquake (Northridge)	USA	15.3
3	1990	Winter storms	Europe	9.8
4	1991	Typhoon Mireille	Japan	5.2
5	1989	Hurricane Hugo	Caribbean, USA	4.5
6	1999	Winter storm Lothar	Europe	4.0
7	1998	Hurricane Georges	Caribbean, USA	3.4
8	1987	Winter storm	Western Europe	3.0
9	1995	Earthquake (Kobe)	Japan	3.0
10	1995	Hurricane Opal	USA	2.1

Great Natural Disasters 1950 - 1999

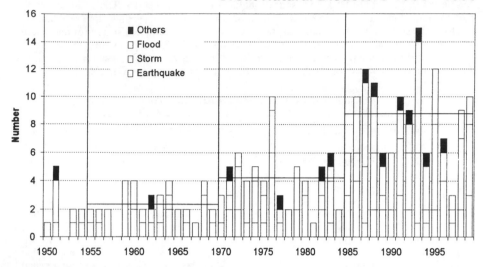

Economic and insured losses with trends

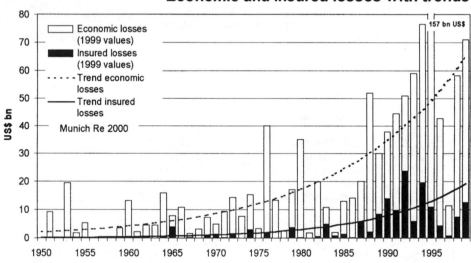

Great Natural Disasters 1950 - 1999

Decade comparison

	Decade 1950-1959	Decade 1960-1969	Decade 1970-1979	Decade 1980-1989	Decade 1990-1999	Factor 90s : 60s
Number	20	27	47	63	87	3.2
Economic losses	39.6	71.1	127.8	198.6	608.5	8.6
Insured losses	0	6.8	11.7	24.7	109.3	16.1

Losses in US$ billion - 1999 values

NatCat *SERVICE*

Munich Re, REF/Geo - January 2000

Fig. 1: Development of great natural disasters during the past half century

3. THE ROLE OF REINSURERS

Billion-dollar catastrophes cannot be born by a local insurance market without major damage to the insurance industry itself. Even in strong markets such as the United States, great events leave their traces. Hurricane Andrew wiped out 13 primary insurance companies in the American Southeast in 1992. The burden from claims exceeded by far the capacity of these companies, and they went bankrupt. To prevent such things from happening and to protect themselves from bankruptcy, insurance companies must assess the maximum probable losses they may be confronted with and prepare for them. One – often the main – aspect of preparation is to seek reinsurance. Reinsurance is nothing but insurance for insurance companies.

While most insurers concentrate their business on a particular country or region (e.g. the United States, Europe) reinsurance companies do business worldwide. How effectively this idea of transferring local losses via the reinsurance sector to a worldwide system works is shown by the example of another hurricane, Gilbert, which hit the Caribbean in 1988. Jamaica in particular suffered great losses; its economy was hit by losses amounting to about US$ 1bn, of which 70% was insured. These US$ 700m would have destroyed the Jamaican insurance industry completely. It survived because nearly 99% or US$ 690m was reinsured and was therefore paid by the world's reinsurance industry. For the local companies a mere 10-million-dollar obligation remained.

A reinsurance rate of more than 95 percent is typical for developing countries. In developed countries, reinsurance rates range between 50 and 90%, depending on the strength of the primary insurance companies in the region. Since reinsurance costs money, large primary companies tend to keep a larger portion of the risk themselves. Two examples: the series of winter storms in Europe in 1990 cost insurers US$ 9.8bn, of which the reinsurers paid 6.4 billion (65%); of Hurricane Andrew's 17-billion-dollar insured losses bill 50% was paid by the reinsurance sector.

4. TYPES OF FLOOD

In insurance contracts, flooding is defined as a temporary covering of land by water as a result of surface waters escaping from their normal confines or as a result of heavy precipitation. When it comes to insurance cover it is very important to distinguish between the different causes for flooding. There are three main types of flood and a number of special cases (Munich Re, 1997). The main types are: storm surges, river floods, and flash floods; special cases include tsunami, waterlogging, backwater (e.g. caused by a landslide that blocks a water course), dam break floods, glacial lake outbursts, groundwater rise, debris flow events and others.

Storm surges can occur along the coasts of seas and big lakes. They bear the highest loss potential of water-related natural events, both for lives (cf. Table 1) and for property. Improved coastal defence works have prevented huge losses in developed regions during the recent past, but the loss potential of storm surges remains very high.

River floods are the result of intense and/or persistent rain for several days or even weeks over large areas sometimes combined with snowmelt. The ground becomes fully saturated and the soil's capacity to store water is exceeded. It behaves as if it were sealed and the precipitation runs off directly into creeks and rivers. The same effect is produced by frozen ground, which also prevents the water from infiltrating the soil. River floods build up gradually, though sometimes within a short time. The area affected can be very large in the case of flat valleys with wide flood plains. In narrow valleys the inundated area is restricted to a small strip along the river, but water depths are great and flow velocities tend to become high, with the result that mechanical forces and sediment transport play a major role as a cause of damage. Although inundation due to river floods starts from a water course and is somewhat confined to its valley, the areas affected can be far greater than those hit by storm surges.

Flash floods sometimes mark the beginning of a river flood, but mostly they are local events relatively independent of each other and scattered in time and space. They are produced by intense rainfall over a small area. The ground is not usually saturated, but the infiltration rate is much lower than the rainfall rate. Typically, flash floods have an extremely sudden onset with a surge rushing down a valley that may not even have a creek at its bottom. A flood wave can propagate very quickly to locations even some tens of kilometres away, where the rainstorm is not even noticeable. From this fact comes the – probably true – saying that "in a desert more people drown than die of thirst". Forecasting flash floods is almost impossible, with lead times for warnings in the order of seconds to minutes. Although flash floods occur in a relatively small area and last only a few hours (sometimes minutes), they have an incredible potential for destruction.

5. INSURANCE PROBLEMS ASSOCIATED WITH FLOODS

The basic problem in flood insurance is the difference in the demand for cover from potential clients who are exposed to flooding and the offer made by the insurance sector (Kron, 1999a). Those who wish to buy flood insurance do not get it because their exposure to flood is too high. Those whom the insurance companies are willing to give cover are not interested in insurance because they live far away from a river or at a higher elevation.

Most people have a certain – and they think good – perception of the flood hazard they are exposed to. The ones who have already experienced flooding on their property are aware of the threat, others – even if they live close to a river – ignore the danger or just do not believe that they can be affected at all. Often their perspective is wrong, though. About half of all losses from floods occur far away from major rivers and outside major events that hit large areas and whole river systems. Instead, these loss events are restricted to a small, local area, they can be extremely intense and they happen very often, though not at the same site. Practically no place in the world is safe from this kind of flooding. Even buildings on the slopes high above the valley floor may be damaged by excessive rainfall that does not infiltrate into the ground but runs off on the surface and right into the houses. If this is made known to the majority of the people, the conditions for effective flood insurance are good.

The phenomenon that only owners of frequently flooded property – mostly along rivers – seek flood insurance is called *adverse selection* or *antiselection*. For this group of people two basic conditions of insurance are violated:
1. There is no spatial spread of the insured risks; the community of insureds is a relatively small group of people who all have a high flood risk.
2. Temporal compensation of losses is not possible because flooding occurs too often, so that loss events cannot be regarded as unexpected events. Unexpectedness, however, is a necessary condition for any insurance cover.

These two factors lead to premiums being so high that insurance does not make sense anymore for the client. Hence, there is no insurance solution that can possibly make insurance companies settle all the losses that may be incurred. Instead, a certain amount has to be borne by the insured before the insurance becomes effective, i.e. deductibles must be introduced. Such a structure has advantages for both the insurer and the insured. The insurer does not have to settle masses of small losses and saves – besides loss compensation money – a lot of administrative costs. The client may only become insurable at all if he pays a share of the losses.

The relatively uniform probability of flash floods in time and space makes this type of event insurable without any problem. The community of insureds is large, the frequency of people being hit by extreme events is low. Therefore, the premiums can be kept low, too. The aspects of the two types of inland flooding with respect to insurance are summarized in Table 4. In the case of the storm surge hazard the antiselection effect is even more severe; therefore storm surges are, in general, not insurable.

6. FLOOD ZONATION AND ACCUMULATION CONTROL

Premiums for flood insurance must reflect the individual exposure. It would be unfair and inexplicable to clients if each member of an insured community had to pay the same premium not taking into account the individual risk his property is exposed to. In mass business – i.e. for private homes and small businesses plus their contents – the effort required to assess the exposure of a certain building must be seen in the context of the annual premium income for one such object, which is in the range of perhaps US$ 50–100. Therefore, an individual assessment of the risk and the calculation of an individual premium for these objects are impossible, so that the premium must be fixed on the basis of a flat-rate assumption. For this, zones with a similar flood risk must be identified and/or defined, within which the premiums are constant.

Table 4:. Comparison of river flooding and flash floods

River flooding	Flash floods
• Only a relatively small proportion of the cover for buildings and contents in any given insurance market is exposed. • The areas affected are always the same. • Only people in these areas seek insurance (adverse selection). • Flooding on a specific river at almost regular intervals cannot be regarded as an unforeseeable event. • If an insurance company wished to sell individual policies on a voluntary basis, the insurance premiums would have to be so high that policyholders would normally find them prohibitive. • Flash floods are caused by local storms and can occur almost anywhere.	• Consumer demand for insurance protection could already be quite large or could be developed on a broad front. • Adequate premiums can be calculated with a relatively high degree of reliability. • The necessary geographical spread of risks is given. • Flood damage caused by flash floods is insurable.

Flood zonation is a tedious and difficult task. The German insurance industry has started to establish a rating system that defines the exposure of all areas of the country to river floods according to three exposure classes:

I   high exposure   Areas on flood plains that are affected by floods with recurrence intervals of up to 10 years; objects in these areas are in general not insurable, but under certain conditions they may become so.

II   moderate exposure   Areas that are affected by floods in the recurrence interval range of 10 to 50 years. Objects in these areas are basically insurable.

III   small exposure   Areas that are affected less that once per 50 years on average; objects there are insurable.

Zonation does not consider the flash flood hazard. The risk from this type of flooding is assumed to be uniform all over Germany, because the spatially varying extreme rainfall intensities is thought to be more or less compensated by the required design assumptions for storm water systems and river works. Natural water courses have also usually adapted a regime that reflects the local hydrologic situation. As a consequence, in regions with higher rainfall intensities the discharge capacities of the drainage systems and channels are also higher.

One major task of insurance and reinsurance companies is accumulation control. The assessment of the probable maximum loss (PML) and a business strategy that accounts for this loss is most important for the survival of a company in the case of a very extreme event. The company must decide on the reserves it needs and its reinsurance requirements. PML calculations are based on scenarios that assume a major event hitting a large area or an area with a high concentration of values. It is not obvious beforehand which scenario will determine the worst case for a given company as the expected losses depend on the company's portfolio, and particularly on the spatial distribution of its liabilities. For each company a different scenario may determine the PML.

PML models have been available for many years as a means of calculating maximum losses from earthquakes and windstorms. For the analysis of floods, such tools were not available until recently. Flood events are much more influenced by small-scale and local aspects, which include soil conditions and topography, the exact location (elevation) and the effectiveness of flood control measures. Therefore, such models require considerably more detail and sophistication.

7. FLOOD PML ASSESSMENT FOR GERMANY

A newly developed model makes it possible for the first time to carry out accumulation analyses of flood events occurring in Germany. Eight different accumulation scenarios were chosen, on the basis of which liability data or even "as if" portfolios with projections for the future can be analysed. The aim is to determine what liabilities are affected and to estimate the probable losses for fictitious 10-year to 200-year flood scenarios.

The flood PML model considers only river floods. Flash floods after torrential rain are not included on account of the fact that they occur locally and therefore play a subordinate role in accumulation considerations. Floods caused by storm surges are not considered either since they are not insurable at present because of their gigantic loss potential.

The model considers areas that are inundated during extreme floods. With the aid of a geographic information system (GIS), they are superimposed on settlement areas and the liabilities allocated to them. The model has five components:
- flow statistics
- river network
- terrain model
- land-use (settlement areas)
- post code areas

Estimation of the accumulation losses to be expected is carried out in five steps. First the discharges for given frequencies are determined using hydrologic stochastic regionalization (hydrology). With the components river network and digital terrain model (DTM) the water levels and the corresponding flooded areas can then be determined (hydraulics). As only property damage is of interest, flooded settlement areas are identified (spatial analysis). The expected losses are estimated from the number of objects affected and loss averages (portfolio analysis). This is done on the basis of post code areas because portfolio data are aggregated in this form. The final step consists of summing up the loss values for all post codes in the regarded flood accumulation zone to obtain the probable maximum loss (accumulation analysis). When choosing appropriate methods, it should be kept in mind that both computing time and availability of data are limiting factors. Therefore, simplifying assumptions concerning the underlying physics have to be made. In the following the five steps are described in a little more detail.

7.1 Hydrology: discharges

The PML scenarios are based on the assumption that the T-year flood discharge occurs simultaneously at all locations in every water course of the considered river network. The assumption of simultaneity is justified although such a scenario is not possible in reality. The probability that this will happen is close to $1/T$ only for small catchments. The larger the area and the corresponding river network, the smaller the probability that a T-year flood will occur everywhere at the same time. A 100-year flood peak in each of two rivers, for instance, will practically always generate a flood peak with a much lower frequency than 100 years downstream of their confluence. However, if the simultaneity condition is somewhat relaxed, it is theoretically possible that a 100-year peak will pass any location of a river basin during a single flood event, e.g. during a period of several days. Flood scenarios that comprise areas of several thousands of square kilometres cannot be associated with a probability of one in 100 years; their occurrence probability is much smaller.

The so called "ArcDeutschland" river network (1:500,000; total length of the rivers included: 35,110 km), which is digitally available for the whole area of Germany was taken as base information. Flow records are available only at certain points of the river system where there are gauging stations. To construct the required scenarios, extreme discharges have to be known for every cross-section, however. A regionalization procedure had to be developed, with the aid of which the 10-, 20-, 50-, 100-, 200-year discharges could be estimated for any given location.

First frequency analysis was performed for selected gauges by fitting the five parameter Wakeby and Pearson III distributions to the discharge series of 322 gauges of three German regions (248 Bavaria, 29 Rhineland-Palatinate, 45 Lower Saxony) (Kleeberg et al., 1998).

Parameters were estimated using L-moments. The model chosen consists of two steps: the first part connects the drainage area A_E at a given river cross-section to the cumulative length L_c of all water courses upstream of the regarded point. L_c can be determined with the GIS using the digital river network. In a second step the actual regression for the discharge quantiles is executed, in which the independent variable is A_E (derived in the first part of the regression from L_c). The simple model uses linear regression with T-year discharge $Q_T = 0$ for $A_E = 0$. The samples for determining the coefficients were weighted by their size. The obtained coefficient of determination for $A_E = f(L_c)$ was $r^2 = 0,9566$ at a significance level of 99.9% . Including other variables such as mean annual rainfall and parameters describing topographical features did not improve the goodness of fit and were therefore discarded. With this regression relationship quantiles may be obtained for any given point along the river network corresponding to the return periods of $T = 10, 20, 50, 100$, and 200 years.

7.2 Hydraulics: flooded areas

On the basis of the discharge values thus derived the corresponding flood stages and the flooded areas can be calculated. Channel cross-sections and local slopes were extracted from a digital elevation model of Germany (DHM-M745, 1:50,000, horizontal resolution 30 m). Profiles were taken at about every 100 m orthogonally to the flow direction; in this way more than half a million cross-sections had to be considered. Existing flood control measures (e.g. dikes) were not taken into account; they do not guarantee prevention of flooding of a certain area and they also might be bypassed by water bursting the banks upstream.

The hydraulic calculations were done with a simple, one-dimensional, stationary hydraulic model based on the Manning-Strickler relationship. Manning's n was assumed to be constant for all rivers with values of $n = 0.04$ for the 10-year scenario and $n = 0.067$ for the scenarios in the range of 20 to 200 years. Flooded areas along the river were obtained by interpolating between the flow-widths at the profile locations considered. Once this step had been completed, the flooded areas were known for the different flood scenarios all over Germany. The total area of the 100-year flood zones amounted to 16,646 km^2, which is 4.7% of Germany's total area of 356,974 km^2.

7.3 Spatial analysis: flooded settlement areas

The insured values can be assumed to be located within the boundaries of settlements. By integrating land-use information in the GIS and superimposing it on the flooded areas, the settlement areas affected by flooding can be identified. Insurance data are usually aggregated on the basis of administrative areas, typically post code areas. In Germany the five-digit post code areas are used, which are roughly 10,000 in number. Although this information is very rough when dealing with floods, it is the only data base available and had therefore to be used. So far no distinction has been made between different types of settlement areas, although the information – in the form of GIS layers – is available. A refined breakdown into residential, commercial and industrial areas for the accumulation analysis of different classes of insurance is planned for the future. The result of the spatial analysis is a percentage of settlement area flooded within each post code zone and for each scenario.

7.4 Portfolio analysis: affected insurance contracts

The distribution of liabilities in the portfolio to be analysed is supplied by the insurer in the form of aggregated figures for each of the five-digit post code areas. The exact location of the insured objects is not known. Therefore one has to assume that the liabilities, i.e. the total sum insured within a post code zone, is distributed uniformly over the settlement area of this zone. For a single post code this assumption would definitely contain too much uncertainty. If however, as is the case in accumulation analysis, large regions are regarded with many post codes, the assumption of uniformly distributed liabilities is reasonable on average, particularly for mass business. In industrial business where relatively few objects with high concentrations of values at certain spots are regarded, this assumption may not be valid anymore. For a post code zone i the expected loss for a given flood scenario is:

$$L_i = \frac{S_{f,i}}{S_i} \cdot r \cdot s \cdot (SI)_i \tag{1}$$

where

L_i	total expected losses
$S_{f,i}$	flooded settlement area
S_i	settlement area
r	loss frequency
s	average loss (in percent of sum insured)
$(SI)_i$	total sum insured (liabilities)

All terms in the equation except the sum insured are subject to the respective scenario. The term "loss frequency" accounts for the fact that not each house located within the flooded area will suffer damage. Some objects are on a – maybe artificially – elevated position that is not shown in the digital elevation model, others may successfully apply individual flood protection measures and thus avoid damage. In the term "average loss", which is given as a percentage of the total value of a building (or its contents), the results of extensive loss analyses and the experience on loss susceptibility in the different classes of insurance gathered over the years are incorporated.

7.5 Accumulation analysis: probable maximum loss

The last step in the analysis reveals the probable accumulation losses in different loss accumulation zones (LAZ's) for the portfolio under consideration. The accumulated losses are found by simply adding the losses expected in each post code area.

However, it is extremely unlikely that a flood event will hit all or most of Germany at one and the same time. Extreme events are usually limited to specific regions, e.g. individual river basins. Consequently, loss accumulation zones have to be defined. It seemed reasonable to choose eight such zones (Fig. 2). Five of them correspond to Germany's large river basins (Rhine, Danube, Oder, Elbe, Weser-Ems). Three further zones (South, Central, North) were defined as being zones that comprise parts of more than one basin. The central zone, for instance, embraces – besides the northern sub-catchments of the Danube in Bavaria – the catchment areas of the Main and Neckar, the areas on the left bank of the Rhine north of Karlsruhe, and the Middle and Lower Rhine Valley. This zone corresponds approximately to the area mainly affected in the 1993 Christmas flood.

The accumulation analysis is carried out separately for each of these loss accumulation zones. Fictitious events based on the modelled discharges serve as scenarios. This means, for example, that a 100-year discharge is assumed for the 100-year scenario along all rivers in the loss accumulation zone being analysed. This produces a probable accumulation loss for every scenario. The resulting values are plotted on a PML graph (Fig. 3) that shows the relative losses for each scenario and forms the central part of the analysis. The critical region for the portfolio examined here is LAZ 4 (Elbe). It would have to be covered against a probable maximum loss of about DM 2.5m.

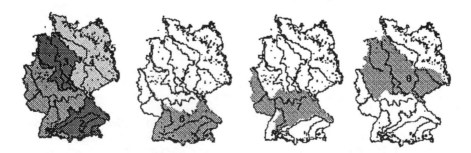

Fig. 2: Loss accumulation zones (LAZ's) for Germany
LAZ: 1 Rhine, 2 Danube, 3 Weser-Ems, 4 Elbe, 5 Odra, 6 South, 7 Central, 8 North

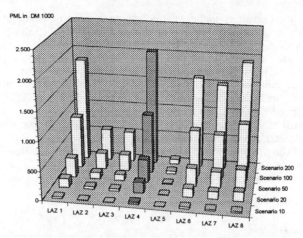

Fig. 3: Probable maximum losses (PML) for eight loss accumulation zones (cf. Fig. 2) and five flood scenarios corresponding to 10 to 200-year discharges

8. LIMITS OF THE ANALYSIS

The accuracy of the analysis is certainly not sufficient for local consideration of the flood risk. It must be clearly seen that the aim of the calculations is accumulation control and therefore treatment of large areas. The various components (regionalization procedure, DTM, river network, hydraulic model, allocation of liabilities, loss averages, etc.) are each subject to quite a high degree of uncertainty. The spatial resolution and the accuracy of the information in the individual components of the flood model are geared specifically to the question of accumulation. Small-scale observations are not intended, let alone risk assessment for individual objects. Flood prevention measures, which are not considered in the model, may play a decisive role. This is in particular true in the case of scenarios with short return periods (10, 20, and possibly 50 years). On the other hand, dikes may also fail. Finally, as mentioned already, the applied discharge return periods cannot be compared directly with loss return periods and are therefore of restricted use in the process of premium calculation.

9. CONCLUSIONS

Flooding has become an important topic for the insurance industry and its significance will continue to grow in the future. The increasing demand for insurance cover and the pressure for proper insurance concepts from all sides is forcing the insurance industry to develop solutions for flood cover. Various countries have already established insurance schemes for this type of hazard, some in the form of insurance pools, others on an individual basis. The types of contract range from obligatory to completely voluntary coverage, and from all-risk policies to flood-only policies. There are advantages and disadvantages in all these concepts and none can be declared the best. It is certainly advisable, however, to offer multi-hazard packages, thus combining the flood risk with other risks such as earthquake, landslide, windstorm, hail, subsidence, snow-load, etc. to avoid adverse selection.

In Germany, the insurance industry is currently in the process of promoting flood insurance and has started to tackle the problem by establishing flood risk zones. This is being done in a concerted action not only by the whole community of German insurers (and some reinsurers) but also in close cooperation with public water resources authorities. Despite the fierce competition in the market the intention is clearly to come up with a unique zonation system valid for all companies that will even help the state in its efforts to enforce land-use planning that is compatible with the flood hazard.

Parallel to the primary insurers that need risk zoning for the purposes of acquisition and designing a premium structure, reinsurers – as part of their service to the primary insurance companies and in the interest of their own business – need risk zoning to calculate the expected losses that the insurance industry might face as the result of an extreme event threatening a company's existence. It was with this in mind that Munich Re in cooperation with two universities developed the world's first flood loss accumulation model for an entire country . The model has been operational for Germany since 1999. A similar model is currently being developed for the United Kingdom. Initial steps have also been taken in generating a model for China. This model will be much less comprehensive than the models for the other two countries, however. It will concentrate on certain regions, such as the provinces in the middle and lower Yangtze River area. The production of this flood model for the People's Republic will heavily depend on the future development of the insurance market in the country, which, however, is expected to flourish within a few years.

REFERENCES

Berz, G. (1999): *Flood disasters: lessons from the past – worries for the future*. Invited lecture at XXVIII IAHR Congress, Graz, Aug. 22-27, 1999 (unpublished)

Kleeberg, H.-B., Willems, W., Stricker, K. (1998): *Geoinfosystem "Überschwemmung Deutschland" (Geoinformation system "Flood Germany")*. Report to Munich Reinsurance company, 16 pp (in German)

Kron, W. (1999a): *Insurance aspects of river floods*. In:. Proceedings of the European Expert Meeting on the Oder-Flood 1997 – RIBAMOD concerted action. A. Bronstert et al. (Eds.), European Communities, 135-150

Kron, W. (1999b): *Reasons for the increase in natural catastrophes: The development of exposed areas*. In: topics 2000: Natural catastrophes – the current position. Munich Reinsurance Company, 82-94

Kron, W. (2000): *Natural Disasters: Lessons from the past – concerns for the future*. In: Proc. of the CALAR Conference Living with Natural Hazards, Vienna, Jan. 17-19, 2000.

Kron, W., Berz, G., Smolka, A. (2000): *Benefits of early warning from the viewpoint of the Insurance industry*. Proceedings of the World Conference on Early Warning, Potsdam, Sep. 7-11, 1999 (in print)

Munich Re (1997): *Flooding and insurance*, Munich Reinsurance Company, 79 pp.

Munich Re (1999): *topics 2000: Natural catastrophes – the current position*. Munich Reinsurance Company, 126 pp

Stochastic Hydraulics 2000, Wang & Hu (eds) © 2000 Taylor & Francis, ISBN 90 5809 166 X

Upper Salzach stream care plan, a compromise between improve flood protection and attain near-natural ecological conditions

Helmut Mader & Konrad Bogner
Institute for Water Management, Hydrology and Hydraulic Engineering, University for Natural Resources, Vienna, Austria

ABSTRACT: The ecologically oriented *Upper Salzach Stream Care Plan* represents an optimal response to several use demands. The potential morphological status of the river Salzach is described by the selected methodical process of a coordinated and structured investigation of historical data and comparison waters analyses. Out of historical vouchers the potential stream course morphology is derived. Out of analyses from reference distances within type-specific comparison waters the potential bed morphology is shown. Out of the results of the deficit analysis for the present morphological conditions of Salzach river, the target status for further measures is established as a section of the interdisciplinary definition of the guiding view.

Key words:Flood protection, River restoration, Guiding view, Bed load, Deficit analysis,

1. INTRODUCTION

The ecologically oriented *Upper Salzach Stream Care Plan* represents an optimal response to several use demands.The main goal of the project, which took into account the potential effects of conventional river engineering measures, was to analyze the need of riparian property to find out a compromise between the improvement of flood protection and the desire to attain near-natural ecological conditions.

A coordinated study of historical data and comparative stream analyses has been carried out, describing the potential and actual condition of the upper Salzach river morphology. The sediment input from the tributaries is analyzed by a stochastic long-term simulation. Both, the annual amounts of sediment transport and the bed loads from peak events of the tributaries have been determined with the help of digital terrain models, statistical regionalisation, and peak value analyses, supported by field reconnaissance for the upper Salzach basin.

As a result from the present paper, the potential morphology of the stream course and the bed morphology and the sediment input from the tributaries lay the foundation for the definition of the guiding view for upcoming river restoration measures, s. a. the enlarging of the channel profile, the increasing of the variability and the heterogeneous near natural bed design.

Via numerical simulations of the sediment flow within the Salzach river, the possible effects on the river bed and the high flood water level due to construction measures along the river can be shown (see: Habersack 2000, this conference).

2. STUDY SITE AND RIVER TYPOLOGY

With a total length of 225 km, the Salzach river drains a catchment of 6734 km². The study site of the 5th – 6th order alpine stream has a total length of 60 km (catchment 1400 km²). About 10 % of the catchment are covered by glaciers. Hydrological classification corresponds with a nivo-glacial discharge regime with a very strong annual character. The heavy modified river within the Oberpinzgau flows through a monotone canal-like channel with a corresponding degraded aquatic community.

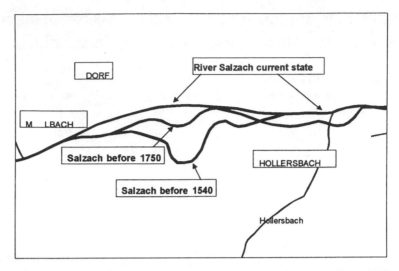

Fig.1: River regulation measures between Dorf and Hollersbach (MADER, EDER 1999)

3. RIVER HISTORY

For the first time in Old-Pliozaen a continuously longitudinal valley of the Salzach river can be discovered. The Salzach valley follows a main alpine fault with the Centralalpine Ingenous Rock formation in the south, and the north alpine schist zone with greywacke formations in the north. The little slope of about 1 ‰ caused by an ongoing drop of the Salzach valley river up the today`s Bruck and a rise downstream.

First clans settled down during the Stone Age. Around 2200 a.Chr. during the Bronze Age first settlements were built along the Salzach river. Activities in mining layed the foundation for an ongoing settlement until 1000 a.Chr.. In the 8th and 9th century, around 300 roman families lived along the Salzach river in the Pinzgau. First documents from 13th century give a report on measures along the Salzach river (weirs, river regulation) for salt-mining and drifting wood.

The clearing of the woodland of the catchment area started before the 15th century. The result was a dramatic loss of the retardation within the forests and the flood retention along the tributaries and the Salzach river. The consequence of this drastic impacts was a significant increase of the surface runoff and the flooding of the Salzach valley. The valley changed into marshland over centuries.

First documents from 1519 give a report on river regulation measures along the Salzach river, were a 2 km long river stretch was shortened over 15 % (see Fig. 1).

War stopped most river regulation works along the Salzach river. During 17th and 18th century no sustainable measures were done for increasing the flood protection of the valley. High flood events had involved a certain risk that the Salzach river could break through the water shed in direction to north into the Saalach river valley.

The start of sustainable river regulation measures (see Fig. 2) was set in 1820 by the emperor Franz I^{st}. The drainage of the sumps and the filling up of the marshlands by the inleading of bed load material from the Salzach river via diversion channels preceded only extremely slow and lasted several decades. By the building of drainage ditches, the groundwater could run off in shortest time. Powerful dams were raised in order to prevent the widening of the water caused by flood events. Several meander were cut off for the increase of the downward gradient. The river was straightened over the whole reach of the study site. The transport capacity of the new river bed was optimized hydraulically. The river width was reduced dramatically. The river bed was lowered around several meters by blasting a rocky ridge at the downstream end of the study site.

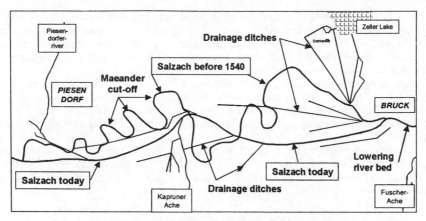

Fig.2: Salzach river1750 and 2000, sustainable measures between
Piesendorf and Bruck, (MADER, EDER 1999)

4. METHODOLOGY

4.1 River morphological Analyzes

Rivers are subject to self-steering dynamic processes in the lapse of time. Therefore single historic inspections only show the situation at a moment within river-history.

By analyzing site descriptions and historic maps, the history of upper Salzach stream course morphology is documented over more than thousand years. The stream course of the Salzach river from 1540 (local maps), 1764 (Josephini land survey), 1830 (Franciscan survey map) and from 1975 has been digitized and combined with the present state. Out of this, the potential stream course is reconstructed within the study site (see fig. 1 & 2).

Because of the lacking of nature like reference reaches on the river Salzach itself, the potential riverbed morphology has to be established by studying and evaluating reference reaches at comparable sections of other rivers of equal type. Further on, results were compared statistically with information out of local historic maps of single river stretches, dated 1795.

In order to classify comparable rivers, the method of cluster analysis has been used for setting up a river typology, based upon morphological, hydrological, biotic and hydraulic parameters covering the complete scaling range. Out of altogether 48 examined Austrian rivers 7 riverswere chosen, which can be compared with the Salzach river were finally determined. In total, 18 nature like reference reaches, which represent 4 different river types, formed the basis for further field data collection and data analysis of the riverbed morphology. In these reference reaches, the current morphologic status corresponds to the potential. Measured physical conditions at reference rivers were statistically compared with analyzed conditions from the current state of river Salzach within a deficit analysis. Data collection is done by random sampling method (≥10 transects, ≥100 single points each reference reach) of following parameters.

- Current near river bed (≥100 points)

- Mean depth, variance of depth, fluctuation coefficient $d_{FC} = \left(\dfrac{d_{max}}{d_{min}} \right)$

- Mean width, variance of width, fluctuation coefficient $w_{FC} = \left(\dfrac{w_{max}}{w_{min}} \right)$

- Moistened surface per 100 m
- Water volume per 100 m
- Mean velocity in transect
- Substrate distribution
- discharge
- river morphology

From the available field data of the comparison streams and the Salzach river, habitat describing bed morphology parameters are analyzed in detail. From this a quantification of available deficits, as well as the classification of the bandwidth for necessary transformations for an approximation to the potential river status is possible. This paper will deal in greater detail primary with the parameters river width and moistened surface.

The necessary area requirements for the realization of measures are determined out of results from the deficit analyses. Scatter plots reveals the dependencies of the parameters from each other or from respective type of stream course. By the confrontation of the results of the Salzach and the comparison waters, as well as with historical data, deficits in the Salzach river become recognizable.

The necessary morphological changes in further consequence serve as input for numerical simulation of the sediment flow (details see HABERSACK this conference). Possible effects on the river bed and on the flood-water protection due to suggested construction measures along the river are shown for a prediction period of 60 years.

4.2 Analyzes of the Sediment input from the Tributaries

The characteristic feature of the tributary bed loads is dominated by instantaneous reaction times, which always lead to a situation characterized by high non-stationarity, time-space variability and fluctuations. The consequence of these extremely non-uniform processes is a lack of fit between computed and observed mean annual bed loads applying deterministic methods. However, a stochastic approach can provide the only effective route towards an description of sediment transport rates in the absence of satisfactory mathematical and physical representations of the laws governing its complexity as well as appropriate field measurements of the apparently random and chaotic processes.

Bayesian techniques are increasingly being used in a variety of highly complex statistical modeling situations. A more recent development has been the recognition that simulation techniques like Monte Carlo Marcov Chain (MCMC) can be applied to get statistical inferences in view of posterior distributions, which may be high-dimensional, unavailable analytically having complex dependence structures. Therefore MCMC methods give a way to carry out Bayesian inference on such a complex problem, like the estimation of bed loads, which could not be easily addressed using any classical approach. Thus stochastic simulations can be used in order to identify a change point in the rate of disasters per year as well as for the computations of the rate of disasters per municipality.

The most commonly used MCMC procedure, the so-called Gibbs sampler, is applied to identify a change point in the rate of disasters following the model of Carlin et al. (1993). In Figure 3 it can be seen that it is almost certain that a change point has been occurred. Therefore further analysis are restricted to the time period from 1898 till 1987.

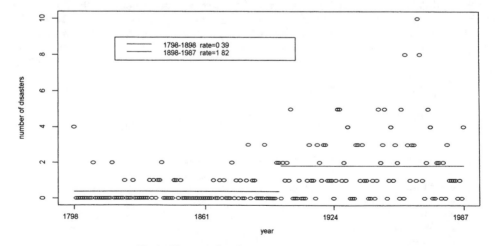

Fig.3: Change point of rate (BOGNER ET AL. 1999)

618

In the following a hierarchical model of George et al. (1993) is used to calculate the rate of disasters for each municipality per year, where the conjugate prior is adopted at the first level, but for any given prior distribution of the hyperparameters, the joint posterior is not of closed form. The number of disasters is assumed to follow a Poisson distribution. The results of this model form the basis for the calculations of annual bed loads of the tributaries. Applying the method of sampling without replacements each time a municipality is "activated" a tributary belonging to this area is drawn from an uniform distribution and its bed load according to the result of the peak value analysis is taken. An uniform distribution is used, because the only available data in a historical chronic is the event time (year of disaster) and the name of the befalling municipality.

5. RESULTS

Speaking about the potential morphologic status of the Salzach River, the specification of one point of reference time is inevitable.

For example: At pre-glacial time, the Saalach river joined the run of Salzach river in the space of the present Zeller lake. At a post-glacial point of view, the drainage of the upper Saalach valley is given according to the today's stream course position. Only by the breakup of the rock threshold near Bruck at 1850, the Zeller lake has been separated from the flooding area of the Salzach.

At the beginning of 20th century, extensive run transfers are implemented and the originally braiding, sinuously, oscillating or meandering river course was stretched due to river regulation measures. Therefore the definition of a potential status of the Salzach must refer either on different points in time for different stretches, or to the time period between the 15th and the 19th century.

The potential stream course was reconstructed as follows:

STREAM COURSE   RIVER STRETCH
straight   Wald – Weyer
sinuously   Weyer – Hollersbach & Mittersill – Niedernsill & Bruck – Högmoos
braiding   Hollersbach – Rettenbach
meandering   Niedernsill – Bruck

Restoration measures have to be coordinated within the guiding view with the base of the potential stream course of the Salzach river. Detailed data about the actual and the potential river bed morphology, specially for the parameters river width and moistened area are shown within Table.1.

Table.1: River width and moistened area, comparison rivers - Salzach 2000 – Salzach 1795

Comparison Rivers	river width (mean)	river width (maximum)	river width (minimum)	moistened area	Historic Salzach	river width (mean)	river width (maximum)	river width (minimum)	moistened area
	[cm]	[cm]	[cm]	[m 100m]		[cm]	[cm]	[cm]	[m 100m]
Mur I	2853	4350	1350	2774	1st peace I	3435	5000	2400	4408
Mur II	3780	4450	3000	3809	1st peace II	4465	5400	4200	3492
Saalach I	3358	4000	2830	3359	2nd peace I	3198	4250	2450	3176
Saalach II	2350	2900	1400	2378	2nd peace II	3380	5300	2050	3261
Enns	4250	6100	3400	4153	3rd peace I	3855	5000	2200	3839
Lech I	2390	3000	1900	2389	3rd peace II	3925	5550	2500	3847
Lech II	2030	2800	1100	2017	9th peace I	3435	4650	2350	3492
Lech III	3760	6400	2000	3656	9th peace II	3525	4200	2550	3542
Lech IV	2890	4000	1630	2887	13th peace I	3110	4050	2100	3044
Lech V	3280	4500	2000	3339	13th peace II	3455	4750	2100	3358
III	1391	1850	950	1435	14th peace I	3290	4100	2200	3278
Isel	1038	1380	770	1050	14th peace II	3690	4500	2400	3811
River Salzach current state									
Wald I	900	930	820	897					
Wald II	882	960	750	897					
Steinach I	1440	1600	1300	1432					
Steinach II	1370	1450	1250	1368					
Uttendorf I	2101	2200	1900	2108					
Uttendorf II	2036	2400	1800	2025					

Comparing the high fluctuation coefficients of the moistened transverse profiles in the comparison rivers with the determined values of the Salzach river, the monotonous conditions in the transverse profile sequence are pointed out clearly. The commonality of the aquatic habitat becomes evident with the comparative analysis of the width-structuring of the river stretches. The values of the variances of the width, taken at the Salzach, are around 10 to 120 times lower than the values within the nature like comparison rivers. The fluctuation coefficients of the width and the middle widths are drastically reduced. The size of the aquatic habitat is reduced clearly.

Comparing the current situation of the moistened surface per 100 m Salzach river, the upper part with stretched river type of run shows very similar, possibly a slightly to low moistened surface. Salzach river downward following, a clear deficit for the potentially sinuously to oscillating run is given. At comparison waters, about 40 % higher moistened surfaces can be determined. Downstream Mittersill, the deficit in opposite to the comparison waters increases up to 70 %. In opposite to the present values of the historical bed morphology from the year 1780 – 1795, the moistened surface was bisected by the adjustment measures over the centuries (see. Fig. 4). Out of this, the necessary increase of the moistened area of the river Salzach and therefore the need of riparian area for river restoration measures is documented.

The very low variability of the monotonous, canal like Salzach is shown clearly by the analysis of the fluctuation coefficient (FC) of the profile widths w_{FC}. For an river stretch within the canalized Salzach, a factor of $w_{FC} = 1,3$ is proven. Comparable nature distances of the same river type indicate fluctuation coefficients between $w_{FC} = 1,4$ and $w_{FC} = 3,2$.

Compared with the current situation of he Salzach, the upper part with stretched run shows clearly to low maximum widths. The range of the variability of widths in current Salzach stretches is very small and illustrates the prevailing deficit of structures. The values of the minimum and maximum widths of the Salzach river are situated within the fluctuation of the values of minimum widths analyzed at comparison waters. Thus, the reduction of the transect by the adjustment measures in direction to the hydraulically necessary minimum profile is documented.

Salzach river downward following, for the potentially oscillating to meandering run a clearly available deficit of the maximum widths is shown. At comparison waters, about 100 % higher maximum width can be determined. The maximum widths of the historical Salzach are about 100 - 120 % higher than the current values of today`s Salzach.

From the confrontation of the current, the historic and the comparison waters values of the minimum and maximum widths, the deficit at heterogeneity within the Salzach river becomes clear (see Fig. 5).

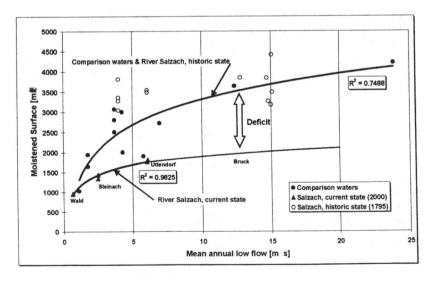

Fig.4: Moistened area, current state and model input

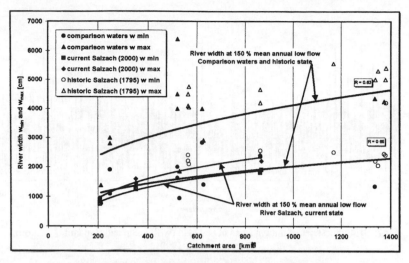

Fig.5: Minimum and maximum river width, confrontation
Salzach 2000 – Salzach 1795 –comparison waters

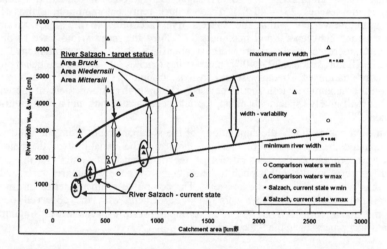

Fig.6: Current and potential width-variability of Salzach river (MADER, EDER 1999)

Out of the stochastic long term simulation of a time period of 500 years, the bed load from peak events of the tributaries has been determined. On the average, per year 29.000 m³ bed load material with a range of 23.800 m³ are transported into the Salzach. The result corresponds very good with the actual quantities, which are taken out of the river Salzach by dredging.

6. DISCUSSION AND RIVER SPECIFIC GUIDING VIEW

The potential morphological status of the river Salzach is described by the selected methodical process of a coordinated and structured investigation of historical data and comparison waters analyses. Out of historical vouchers the potential stream course morphology is derived. Out of analyses from reference distances within type-specific comparison waters the potential bed morphology is shown. Out of the results of the deficit analysis for the present morphological conditions of Salzach river, the target status for further measures is established as a section of the interdisciplinary definition of the guiding view.

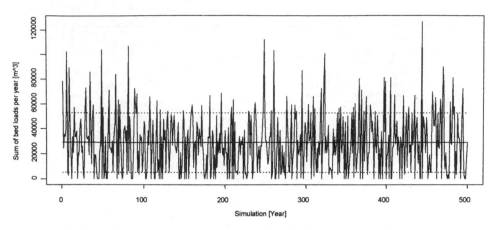

Fig7: Synthetic series of annual sums of bed load (with mean and standard deviation), (BOGNER ET AL. 1999)

For river reaches with a stretched character of the stream course, a running general expansion of the existing river width of about 10 - 15 % is suggested (sinuously & oscillating reaches 15 – 30 %, meandering reaches 25 – 40 %). These values represent minimum claims. For the increase of structure variety within the Salzach river according to the potential situation, punctual expansions between 50 and maximum 120 % of the mean width have to be made. Finally, a mean increase of the river width at about 15 - 20 % for straightened reaches (sinuously & oscillating reaches 20 – 40 %, meandering reaches 40 – 60 %) can be attained (see Fig. 6). A combination of stretch-wise expansions around the minimum and stretch-wise expansions on the maximum width should be used. The mutually arrangement of the widening of the river bed will involve spaciously an appropriate depth structuring and a near natural pool-riffle sequence.

The shown methods of stochastic simulations of the sediment input from the tributaries are capable to manage the complexity of the large area and the variability of the contributing tributaries by incorporating the possibilities of the occurrence of spatial localized events. Exactly these extreme events will happen to be scattered over the whole area restricted to sub catchment areas with small regional extensions. The computed synthetic series of amounts of sediment transport (see Fig.7) show very good agreement with the observed bed loads confirming the above results. They are used as input for further numerical simulations of sediment transport processes in the Salzach river.

7. FURTHER FEASIBILITY STUDIES

The analyzed target status of type specific river morphology and the analyzed sediment input from the tributaries lay the foundation for the simulation of the sediment transport of the Salzach river. The possible effects on the river bed and the water level at high floods are shown for a prediction period of 60 years. The sediment transport analysis was performed with numerical simulations (Habersack 2000, this conference & Habersack et al 1999). Therefore, the feasibility of the required measures can be acknowledged stretch wise.

REFERENCES

BOGNER ET AL. (1999): Upper Salzach stream care plan – sediment situation in the Salzach river tributaries between Wald and Högmoos. Österreichische Wasser- und Abfallwirtschaft, Jahrgang 51, Heft 7/8 (in German).

CARLIN, B. P., A. E. GELFAND, (1993): Parametric likelihood inference for record breaking problems. *Biometrica*, 80, 507-515

GEORGE, E. I. , MC CULLOCH, R. E., (1993): Variable selection via Gibbs sampling. *J. Am. Statist. Ass.*, 85, 398-409

HABERSACK ET AL. (1999): Sediment transport at the upper Salzach river. Engineering concepts and numerical simulation. Österreichische Wasser- und Abfallwirtschaft, Jahrgang 51, Heft 7/8 (in German).

MADER H., B. EDER, (1999): Upper Salzach stream care plan – stream morphology – sectoral deficiency analysis on the basis of historical data and comperative stream investigations. Österreichische Wasser- und Abfallwirtschaft, Jahrgang 51, Heft 7/8 (in German).

Stochastic Hydraulics 2000, Wang & Hu (eds) © 2000 Taylor & Francis, ISBN 90 5809 166 X

Flood estimation in Halil Rood basin of Iran

M. Ghodsian & A. A. Salehi Neyshabouri
Tarbiat Modarres University, Tehran, Iran

M. Molanejad
Tarbiat Modarres University, and Mahab Ghods Consulting Engineers, Tehran, Iran

M. A. Ahmadi Nejad
Tarbiat Modarres University, Tehran, Iran

ABSTRACT: The peak flood discharge from basin is one of important informations required for planning of the water resources projects. In many cases long term data of annual runoff are not available at a particular site. However meteorological data such as the peak rainfall and other informations regarding basin characteristics are generally available.

In this paper the available data from several sub-basins of Halil Rood basin (Iran) have been used to verify the existing empirical relationships for peak flood discharge estimation. Further, using the available data a simple deterministic model for estimation of peak flood discharge for Halil Rood basin has been developed. It was found that important parameters such as basin area, rainfall, slope of basin and return period governs the peak flood discharge, while parameters like percentage of vegetal cover and forest area are of secondary importance.

Key words: Flood estimation, Empirical equation, Halil Rood basin, Kerman, Iran

1. INTRODUCTION

The estimation of flood magnitude and its return period is one of the most important informations required for flood management in a basin. Design of various hydraulic structures such as dam, spillway barrage and levees requires a realistic estimation of flood with a given return period.

When the flood data record of sufficient length are available, it may be useful for estimation of flood runoff. However, historical records of flood are not generally available for all the sites of interest and quite often the record length is not adequate for the single site analysis. The magnitude of a flood peak can be estimated in several manners. Several investigators such as Dickens (1965), Ryves (1986), Inglis (1997), Jallali (1988), Fuller (1992), Ellis Gray (1992), Sangal (1992), Morphy (1992) and Garde and Kothyari (1988) have developed relationships for estimation of annual flood runoff from basins. Because of the limited data on which these relationships are based and due to the fact that they do not include all the variables on which flood discharge of given return period depends, these relationships are not expected to give results with satisfactory accuracy for all the basins.

In this paper, the available data from several sub-basins of Halil Rood basin of Iran have been used to verify the existing empirical relationships for peak flood discharge estimation. Further using the available data, simple deterministic model for estimation of annual peak flood discharge have been developed.

2. DESCRIPTION OF THE STUDY AREA

For the present study, the Halil Rood basin is selected. This basin is situated in the Kerman province of Iran (see Fig. 1). The Halil Rood basin with an area of about $8.3*10^4$ Km^2 lies between north latitude $28° 26'$ to $29° 36'$ and east latitude $56° 17'$ to $57° 43'$. This basin is surrounded by Lalezar and Rastagh mountains in north, by Tange-dolfard and Ghaleh-dokhtar mountains in east, by Khabir, Sange sefeed and Beede shirin mountains in west and by Bezar, Hosein abad, Torangh and Bagh borj monitions in south (1988). The study was carried out for nine sub-basins of Halil Rood. Table 1 shows the characteristics of selected sub-basins (1988).

3. DATA

Rainfall and peak flood data of above-mentioned sub-basins were obtained from water resources research organization and regional water organization of Kerman province. The data used were for sufficient length (data were extended for few of the sub-basins with shorter length of data). Arithmetic average of rainfall values at different stations in the sub-basins were taken to obtain the peak annual rainfall. Thereafter, maximum 24-hr rainfall and annual peak flood discharge, for different return period was computed. The details and ranges of data used are given in reference (1998). Percentage area of forest and agricultural land for each sub-basin were obtained from the land use map of Halil Rood basin.

4. ANALYSIS AND DISCUSION

The above mentioned data for Halil Rood sub-basin are used for the purpose of checking the accuracy of some of the existing relationships for peak flood discharge. As the existing relationships were not found to yield satisfactory results over the whole range of data, the data were then reanalyzed to develop new relationship for estimation of peak flood discharge.

The methods of Fuller (1992), Jallali (1988), SCC (1968), Elis Gray (1992) and Sangal (1992) were checked for accuracy with the above data. Figures 2-5 show typical comparison between actual and computed peak flood discharge using Sangal (1992), Elli Gray (1992), SCS (1968) and Jallali (1988) methods respectively. Figure 2 shows that with Sangal method majority of the data points for Soltani sub-basin lies within ±50% error line. The average percentage error between the actual and computed values due to Sangal method is 38.8. Figure 3 shows the comparison of actual peak flood discharge and computed peak flood discharge using SCS method [10] for all the data of above-mentioned sub-basins. It is obvious that majority of the data points fall within ±200% error line. However, it is clear that SCS method overestimates the peak flood discharge for most of the data. The average percentage error between the actual and computed values of peak flood discharges due to SCS method is 120.8. Figure 4 shows the comparison of actual and computed peak flood discharge using Jallali method (1988). This figure reveals that Jallali equation predicts the peak flood discharge with a maximum error of 70% for most of the data. The average percentage error due to Jallali method is 40.6. Figure 5 shows the comparison of actual and computed peak flood discharge by using Fuller method (1992) for all the data of sub-basins. It is obvious that majority of the data point lies within ±70% error line. However it is clear that Fuller method underestimates the peak flood discharge for most of the data. The average percentage error due to Fuller equation is 49%. Almost similar results were observed while using other methods i.e. Ellis Gray (1992), Garde and Kothyari (1988), etc. For more information reference (1998) can be looked in.

In view of the above discussion and with the purpose of obtaining better relationship of peak flood discharge for Halil Rood basin a new analysis of data was carried out. It was assumed that peak flood discharge Q_p is a function of variables such as maximum 24-hr rainfall P_{24}, general slope of basin S, area of basin A, return period T, forest area A_f and agricultural area A_v, i.e.:

$$Q_p = f(T, P_{24}, S, A, A_f, A_v) \tag{1}$$

Multiple regression analysis of all the data yielded the following relationship for Q_p:

$$Q_p = 10^{-1.76} T^{0.16} P_{24}^{0.77} S^{0.34} A^{0.8} A_v^{0.001} A_f^{-0.002} \tag{2}$$

with regression coefficient of 0.97. It was found that eq. (2) predicts the peak flood discharge with a maximum error of ±50 % for most of the data.

626

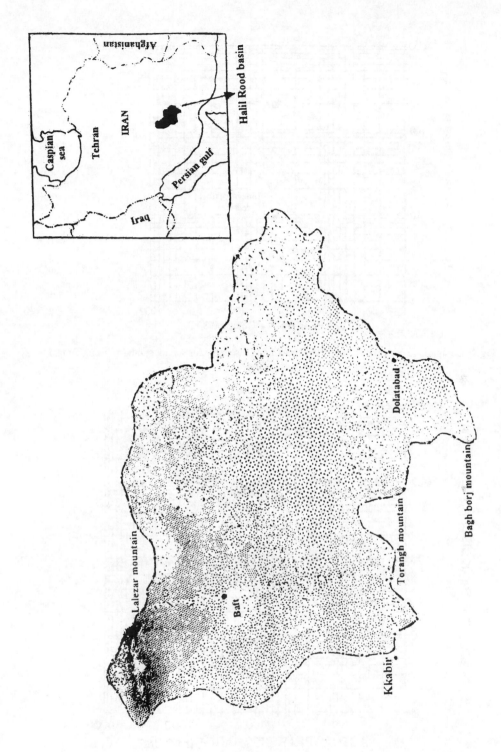

Fig.1: Location of Haiill Rood Basin

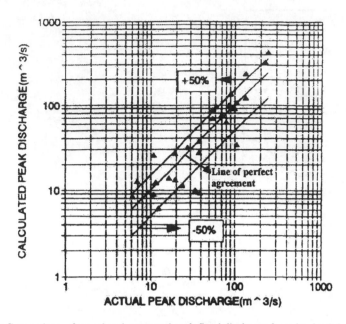

Fig.2: Comparison of actual and computed peak flood discharge for soltani sub-basin using Sangal method

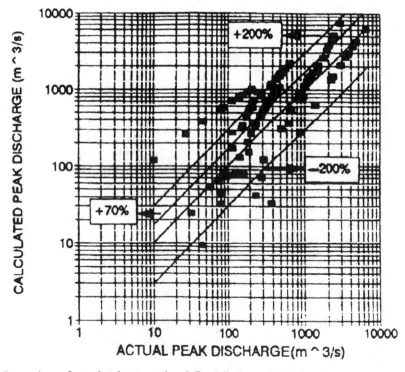

Fig.3: Comparison of actual and computed peak flood discharge for all the sub-basin using SCS method

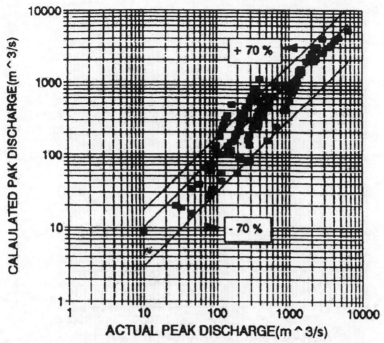

Fig.4: Comparison of actual and computed peak flood discharge for all the sub-basin using Jallali method

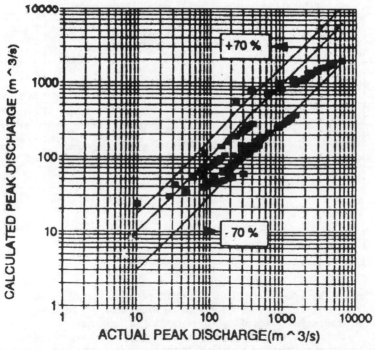

Fig.5: Comparison of actual and computed peak flood discharge for all the sub-basin using Fuller method

629

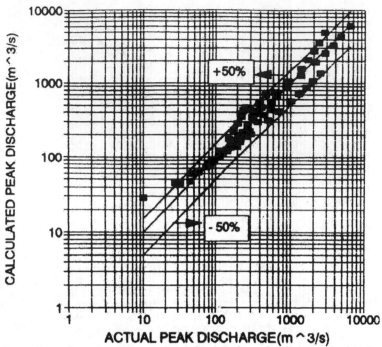

Fig.6: Comparison of actual and computed peak flood discharge for all the sub-basin using equation(3)

Table.1: Characteristic of sub-basins used in this study

Name of sub-basin	Area (Km^2)	Slope (%)	Length of main river (Km)	Forest area (%)	Vegetal cover (%)
Polbaft	235.6	4.3	21.5	1	6.7
Soltane baft	935	2.4	61.5	1	14.3
Cheshme aroos	100.4	8.1	12.5	1	25
Meydan	631.2	7.1	32.5	1	14.3
Zarin	359	7.4	28.8	25	4.0
Goldan	191	8.7	28	7.7	1.3
Dehrood	1365	3.35	42.5	8.3	2.0
Safarzadeh	8420	1.8	92.5	5.0	5.0
Kahange Sheibani	12990	1.2	157.5	10	10

In eq. (2) Q_p is peak flood discharge in m³/s; T is return period in year; P_{24} is maximum 24-hr rainfall in mm; A is total area of basin in Km²; A_f is forest area of basin as percentage of basin area and A_v is agricultural area of basin as percentage of basin area.

In order to study the degree of effectiveness of parameters on right hand side of eq. (2) on Q_p, the β-coefficient of each independent variable with the dependent variable Q_p was obtained and is shown in table 2. Here $β_1 = b_1 δ_1 / δ_1$; in which $δ_1$, b_1 and $β_1$ are standard deviation, regression coefficient and β-coefficient of the ith independent variable respectively and $δ_1$ is standard deviation of dependent variable.

630

Table.2: β-coefficients

Variable	A	P_{24}	T	S	A_v	A_f
β-coefficient	0.96	0.36	0.20	0.16	0.02	0.18

A close study of β-coefficient reveals that the basin area is the most influencing factor on the dependent variable Q_p followed by 24-hr rainfall, basin slope, return period, agricultural land area and forest area. It is realized that land usage has changed during the period of data. But in the absence of any information, the available data was considered as average over the period, which may be the reason that A_v and A_f are of minor importance. Moreover, values of β for A_v and A_f indicate that the effect of these parameters on Q_p for Halil Rood basin can be neglected. Further multiple regression analysis of all the data, after neglecting A_v and A_f, gives:

$$Q_p = 10^{-1.75} T^{0.16} P_{24}^{0.76} S^{0.34} A^{0.34} \tag{3}$$

Figure 6 shows comparison of actual values of Q_p and calculated values of Q_p by using eq. (3) for the data of above mentioned sub-basins. Average percentage error due to eq. (3) for all the data is 22.97 with regression coefficient of 0.98. Comparison of Figures 2-5 with figure 6 and also average percentage error obtained by different methods indicates that the peak flood discharge computed by eq. (3) is more reliable. Hence one can use eq. (3) for estimating Q_p for Halil Rood basin.

In order to verify accuracy of eq. (3), data from another sub-basin of Halil Rood i.e. Ferozabad sub-basin, which were not utilized earlier to establish eq. (3), were used (see table 3). The average percentage error due to eq. (3) for this sub-basin is 26. Therefore one can expect that eq. (3) satisfactorily estimate the peak flood discharge for Halil Rood basin of Iran.

Table.3: Characteristics of Ferozabad sub-basin

General slope (%)	Length of main river course (Km)	Area (Km)
2.3	47.5	1176.2

5. CONCLUSION

The analysis of data for nine sub-basins of Halil Rood basin of Iran indicates that the available methods do not predict amount of peak flood discharge accurately. New relationships have been developed which enables accurate estimation of peak flood discharge.

REFERENCES

Ahmadi Nejad, M. A.,(1998), Rainfall-peak flood discharge relationship for Halil Rood catchment of Kerman, Dissertation presented for master degree in Hydraulic engineering, Civil engineering dept., Tarbiat Modarres University, Tehran- Iran

Dickens (1965), Professional paper on Indian engineering, Volume 2.

Ghosh, S. N. (1992), Flood control and drainage engineering, Oxford and IBH publishing co., India

Garde, R. J. and Kothyari, U. C., (1988)," Flood estimation using regional flood frequency analysis", International seminar on hydrology of extremes, Roorkee, India

Jallali, (1988), Hydrologic report of Sayed Mortaza basin, Soil conservation office, Jahade-sazandegi, kerman- Iran (in Persian)

Mahdavi, M.,(1992), Applied hydrology, Volume 2, Tehran University press, Tehran-Iran (in Persian)

Report of soil conservation of Jeeroft dam -Halil Rood, (1988), soil conservation office and watershed management, forest organization affiliated to ministry of agriculture, Tehran-Iran (in Persian)

Subramanya, K.,(1997),Engineering hydrology,Tata McGraw Hill publishing company limited, New Delhi, India

SCS, (1968), Hydrology suppl. auto. Set. 4, engineering handbook, Washington DC.
Varsheny, R. S. (1986), Engineering hydrology, Nem chand publisher, Roorkee, India

Stochastic Hydraulics 2000, Wang & Hu (eds) © 2000 Taylor & Francis, ISBN 90 5809 166 X

Policy implications of uncertainty integration in design

J.K.Vrijling & P.H.A.J.M.van Gelder
Delft University of Technology, Netherlands

ABSTRACT: In this paper, the implications of integration of uncertainties in the design of flood protecting structures will be analysed. In particular the probability of inundation of the low-lying polders protected by dikes from the Dutch Lake IJssel (1200 km^2) is studied on the basis of a physical model. The uncertainty in the probability of overtopping is analysed with a first order reliability method (FORM). The FORM-calculations showed that the uncertainty in the wind speed and the uncertainty in the water level contributed most in the total uncertainty of the probability of overtopping of a dike with given height. A comparison has been made between the results of the FORM-calculations and the calculations where only intrinsic uncertainty has been assumed in the basic variables. Furthermore, in this paper, the reliability-based optimal design of the dikes along the lake is studied. Finally the policy implications of the uncertainty integration in the design of dikes are discussed.

Key words: Uncertainty integration , First order reliability method, Optimal design.

1. INTRODUCTION

Lake IJssel is situated in the northern part of the Netherlands (Fig. 1). It has an area of approximately 1200km^2. The lake is surrounded by dikes in order to protect the low-lying polders from flooding. The dikes are designed in a probabilistic manner (see for instance CUR, 1990, and Van Gelder et.al., 1995). The required safety against inundation of the polders is 1/4000 yr^{-1}.

In Westphal et.al. (1997), a physical model has been developed for the water levels of Lake IJssel. In this paper, the probability of inundation of the low-lying polders behind the Lake IJssel dikes will be studied on the basis of the physical model of Westphal et.al. (1997). In particular the uncertainty in the probability of inundation will be analysed with FORM. The following variables will be included in the FORM-analysis:
- Water level
- Wind speed
- Model uncertainties in:
  - Water level
  - Wind speed
  - Wind surge
  - Wave height
  - Wave steepness
  - Wave run-up
  - Lake oscillations

The outline of the paper is as follows. First the physical and reliability models will be briefly described. Some aspects of uncertainty modelling and the applied uncertanties in the physical model will be highlighted. The results of the FORM calculations will be discussed. The influences of the uncertainties on the reliability-based optimal design of the dikes at the locations of Enkhuizen and Rotterdamsche Hoek will be investigated. Policy implications and conclusions will end the paper.

2.PHYSICAL AND RELIABILITY MODEL

The physical model is described in Westphal et.al. (1997) and is based on WAQUA (a two-dimensional water flow model) and HISWA (a wave model).

In this paper the following reliability function will be used:

$$Z = K - M - \Delta - z_{2\%} \tag{1}$$

in which:

Z: reliability function [m]
K: the crest height [m]
M: lake level (averaged over four locations on Lake IJssel) [m]
*: wind surge [m]
$z_{2\%}$: 2% wave run-up [m]

The wind surge is based on:

$$\Delta = \alpha \bullet \frac{W^2 \bullet F}{2 \bullet g \bullet D} \tag{2}$$

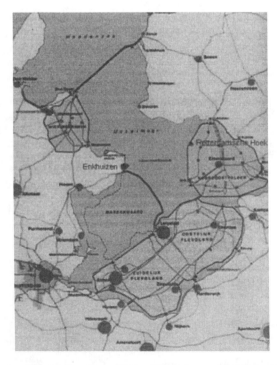

Fig.1: Lake IJssel and the two locations of interest Enkhuizen and Rotterdamsche Hoek

in which:

*   constant; $3.6.10^{-6}$ [-]
W: wind speed [m/s]
F: fetch length [m]
g: gravitation constant; 9.8 [m/s^2]
D: water depth [m]

The wave run-up is modelled with the Van der Meer formula (1997).

3.UNCERTAINTIES

Suppose that the true state of a variable is X. Prediction of X may be modeled by X*. As X* is a model of the variable X, imperfections may be expected; the resulting predictions will therefore contain errors and a multiplicative correction factor N may be applied. Consequentely, the true state of the variable may be represented as stated by Ang, (1973).

$$X = NX^*. \tag{3}$$

If the state of the variable is random, the model X* is also a random variable. The intrinsic variability is described by the coefficient of variation (c.o.v.) of X*, given by (x)/*(x*).The necessary correction N may also be considered as a random variable, of which the mean value *(N) represents the mean correction for systematic error in the predicted mean value, whereas the c.o.v. of N, given by * (N)/ * (N), represents the random error in the predicted mean value.

It is reasonable to assume that N and X* are statistically independent. Therefore we can write the mean value of X as:

$$*(X) = * (N) * (x^*) \tag{4}$$

The total uncertainty in the prediction of X can be written as:

$$C.o.v.(X) = sqrt(C.o.v.^2(N) + C.o.v.^2(x^*) + C.o.v.^2(N)\ C.o.v.^2(x^*)) \tag{5}$$

Beyond a multiplicative uncertainty modelling, also an additive model can be used:

$$X = X^* + A \tag{6}$$

The necessary correction A may also be considered a random variable, of which the mean value *(A) represents the mean correction for systematic error in the predicted mean value, whereas the c.o.v. of A, given by * (A)/ * (A), represents the random error in the predicted mean value.   As in the multiplicative case, it is reasonable to assume that A and X* are statistically independent. Therefore we can write the mean value of X as:

$$*(X) = * (A) + * (X^*) \tag{7}$$

The total uncertainty in the prediction of X is:

$$Var(X) = Var(A) + Var(X^*) \tag{8}$$

Uncertainties in decision and risk analysis can primarily be divided in two categories: uncertainties that stem from variability in known (or observable) populations and therefore represent randomness in samples (inherent or intrinsic uncertainty), and uncertainties that come from basic lack of knowledge of fundamental phenomena (epistemic uncertainty).

635

Inherent uncertainties represent randomness or the variations in nature. For example, even with a long history of data, one cannot predict the maximum water level that will occur in, f.i., the coming year at the North Sea. It is not possible to reduce inherent uncertainties.

Epistemic uncertainties are caused by lack of knowledge of all the causes and effects in physical systems, or by lack of sufficient data. It might be possible to obtain the type of the distribution, or the exact model of a physical system, when enough research could and would be done. Epistemic uncertainties may change as knowledge increases.

The inherent uncertainty and epistemic uncertainty can be subdivided in the following types of uncertainty (Pat*-Cornell 1996): inherent uncertainty in time and in space, parameter uncertainty and distribution type uncertainty (together also known as statistical uncertainty) and finally model uncertainty.

i) Inherent uncertainty in time

Stochastic processes running in time such as the occurrence of water levels and wave heights are examples of the class of inherent uncertainty in time. Unlimited data will not reduce this uncertainty because the realizations of the process in the future stay uncertain.

ii) Inherent uncertainty in space

Random variables that represent the fluctuation in space, such as the dike height. Just as for inherent uncertainty in time it holds that unlimited data (e.g. if the height would be known every centimeter) will not reduce this uncertainty. There will always still be a fluctuation in space.

iii) Parameter uncertainty

This uncertainty occurs when the parameters of a distribution are determined with a limited number of data. The smaller the number of data, the larger the parameter uncertainty.

iv) Distribution type uncertainty

This type represents the uncertainty of the distribution type of the variable. It is for example not clear whether the occurrence of the water level of the North Sea . is exponentially or Gumbel distributed or whether it has a completely different distribution.

Remark: a choice was made to divide statistical uncertainty into parameter- and distribution type uncertainty although it is not always possible to draw the line; in case of unknown parameters (because of lack of observations), the distribution type will be uncertain as well.

Later in this paper we will try to distinguish what type of uncertainty a stochastic parameter represents. Since parameter uncertainty and distribution type uncertainty can not be discerned, another practical -less scientific- division has been chosen. The statistical uncertainty is divided in two parts: 'statistical uncertainty of variations in time' and 'statistical uncertainty of variations in space'.

v) Statistical uncertainty of variations in time

When determining the probability distribution of random variable that represents the variation in time of a process (like the occurrence of a water level), there essentially is a problem of information scarcity. Records are usually too short to ensure reliable estimates of low-exceedance probability quantiles in many practical problems. The uncertainty caused by this shortage of information is the statistical uncertainty of variations in time. This uncertainty can theoretically be reduced by keeping record of the process for the coming centuries.

vi) Statistical uncertainty of variations in space

When determining the probability distribution of random variable that represents the variation in space of a process (like the fluctuation in the height of a dike), there essentially is a problem of shortage of measurements. It is usually too expensive to measure the height or width of a dike in great detail. This statistical uncertainty of variations in space can be reduced by taking more measurements (see also Vrijling and Van Gelder, 1998).

vii) Model uncertainty

Many of the engineering models that describe the natural phenomena like wind and waves are imperfect. They may be imperfect because the physical phenomena are not known (for example when regression models without underlying theory are used), or they can be imperfect because some variables of lesser importance are omitted in the engineering model for reasons of efficiency.

Before was mentioned that inherent uncertainties represent randomness or the variations in nature. Inherent uncertainties cannot be reduced.

Epistemic uncertainties, on the other hand, are caused by lack of knowledge. Epistemic uncertainties may change as knowledge increases. In general there are three ways to increase knowledge:
- Gathering data
- Research
- Expert-judgment

Data can be gathered by taking measurements or by keeping record of a process in time. Research can f.i. be done into the physical model of a phenomenon or into the better use of existing data. By using expert opinions it is possible to acquire the probability distributions of variables that are too expensive or practically impossible to measure.

The goal of all this research obviously is to reduce the uncertainty in the model. Nevertheless it is also thinkable that uncertainty will increase. Research might show that an originally flawless model actually contains a lot of uncertainties. Or after taking some measurements the variations of the dike height can be a lot larger. It is also thinkable that the average value of the variable will change because of the research that has been done.

The consequence is that the calculated probability of failure will be influenced by future research. In order to guarantee a stable and convincing flood defence policy, it is important to understand the extent of this effect.

4 UNCERTAINTIES IN THE PHYSICAL MODEL

4.1 Intrinsic uncertainty in the lake level

The annual maxima of the lake level can satisfactorily be modelled by a Gumbel distribution:

$$F_{M_{1year}}(Mp) = e^{-e^{-\frac{Mp - A_{1year}}{B_{1year}}}} \tag{9}$$

in which:

M_{1year}: the annual maximum of the lake level [m]
Mp: lake level [m]
A_{1year}: 0.02 [m]
B_{1year}: 0.11 [m]

4.2 Intrinsic uncertainty in the wind speed

In The Netherlands the hour-averaged wind speed from a direction sector * can adequately be modelled by a 2-parameter Weibull distribution (Rijkoort, 1983, and Wieringa and Rijkoort, 1983):

$$F_W(w|\phi) = 1 - e^{-\left(\frac{w}{a_\phi}\right)^k} \tag{10}$$

where:
w: the wind speed [m/s]
a_*: a constant dependent of the wind direction

4.3 Statistical uncertainties

The following statistical uncertainties are modelled by normal distributions:

Uncertainty in Lake Level (add)	Mean *	Standard Deviation *
fMp	0.0 m	0.1 m

Uncertainty in Wind (add)	Mean *	Standard Deviation *
fW	0.0 m/s	3 m/s

Uncertainty in Wind (mult)	Mean *	Standard Deviation *
fW	1	0.1

4.4 Model uncertainties

The following model uncertainties are modelled by normal distributions:

Uncertainty in surge (add)	Mean *	Standard Deviation *
fOpw	0.0 m	0.1 m

Uncertainty in Oscillations (add)	Mean *	Standard Deviation *
fOsc	0.1 m	0.05 m

Uncertainty in Significant Wave Height (add)	Mean *	Standard Deviation *
fHs	0.0 m	0.07 m

Uncertainty in Wave Steepness (add)	Mean *	Standard Deviation *
fsop	0	0.005

Uncertainty in Wave Run-up (mult)	Mean *	Standard Deviation *
fOplm	1	0.085

Table.1: FORM results of Rotterdamsche Hoek

K	Mp	$\alpha^2_{\lambda\phi}$	fMp	$\alpha^2_{\beta\phi}$	W	α^2_W	fW	α^2_{fW}	Δ	fOpw	α^2_{fQpw}	fOsc	α^2_{fOsc}
1	2	3	4	5	6	7	8	9	10	11	12	13	14
4.1	-0.20	0.09	0.04	0.04	27.75	0.31	2.97	0.24	1.22	0.05	0.05	0.11	0.01
4.3	-0.19	0.08	0.04	0.03	28.56	0.35	3.39	0.24	1.32	0.05	0.05	0.11	0.01
4.5	-0.18	0.07	0.04	0.03	29.44	0.38	3.76	0.24	1.42	0.05	0.04	0.11	0.01
4.7	-0.18	0.06	0.04	0.02	30.38	0.42	4.10	0.24	1.53	0.05	0.04	0.11	0.01
4.9	-0.18	0.05	0.04	0.02	31.34	0.45	4.40	0.23	1.65	0.05	0.05	0.11	0.01

K	H_s	fHs	α^2_{fHs}	$z_{2\%}$	$fz_{2\%}$	$\alpha^2_{fz_{2\%}}$	$s_{op}*100$	fs_{op}	$\alpha^2_{fs_{op}}$	β	$P_{fl\phi}$	$P_f = P_{fl\phi}*0.101$	T_p	ξ_{op}
15	16	17	18	19	20	21	22	23	24	25	26	27	28	29
4.1	2.12	0.00	0.00	3.04	1.06	0.13	3.18	0.00	0.13	2.00	$2.26*10^{-2}$	$2.291*10^{-3}$	6.54	1.00
4.3	2.16	0.00	0.00	3.13	1.07	0.12	3.15	0.00	0.12	2.29	$1.10*10^{-2}$	$1.111*10^{-3}$	6.63	1.00
4.5	2.20	0.00	0.00	3.22	1.07	0.11	3.12	0.00	0.12	2.56	$5.19*10^{-3}$	$5.245*10^{-4}$	6.72	1.01
4.7	2.24	0.00	0.00	3.30	1.08	0.11	3.10	0.00	0.11	2.82	$2.40*10^{-3}$	$2431*10^{-4}$	6.81	1.02
4.9	2.28	0.00	0.00	3.39	1.08	0.10	3.07	0.00	0.11	3.06	$1.10*10^{-3}$	$1.113*10^{-4}$	6.89	1.02

Table.2: FROM results of Enkhuizen

K	Mp	$\alpha^2_{\lambda\phi}$	fMp	$\alpha^2_{\beta\phi}$	W	α^2_W	fW	α^2_{fW}	Δ	fOpw	α^2_{fQpw}	fOsc	α^2_{fOsc}
1	2	3	4	5	6	7	8	9	10	11	12	13	14
3.1	-0.19	0.12	0.04	0.05	16.38	0.23	1.07	0.15	0.33	0.04	0.05	0.11	0.01
3.3	-0.17	0.13	0.05	0.05	16.68	0.23	1.08	0.14	0.35	0.05	0.05	0.11	0.01
3.5	-0.14	0.14	0.05	0.04	16.98	0.23	1.09	0.13	0.37	0.05	0.05	0.11	0.01
3.7	-0.12	0.14	0.06	0.04	17.27	0.23	1.10	0.12	0.38	0.06	0.04	0.11	0.01
3.9	-0.09	0.15	0.06	0.04	17.54	0.22	1.10	0.11	0.40	0.06	0.04	0.11	0.01

K	H_s	fHs	α^2_{fHs}	$z_{2\%}$	$fz_{2\%}$	$\alpha^2_{fz_{2\%}}$	$s_{op}*100$	fs_{op}	$\alpha^2_{fs_{op}}$	β	$P_{fl\phi}$	$P_f = P_{fl\phi}*0.081$	T_p	ξ_{op}
15	16	17	18	19	20	21	22	23	24	25	26	27	28	29
3.1	1.53	0.03	0.06	2.92	1.06	0.15	2.86	0.00	0.18	1.84	$3.323*10^{-2}$	$2.691*10^{-3}$	5.85	1.44
3.3	1.58	0.04	0.06	3.07	1.07	0.15	2.80	0.00	0.19	2.16	$1.540*10^{-2}$	$1.248*10^{-3}$	6.00	1.45
3.5	1.62	0.04	0.05	3.33	1.08	0.15	2.74	0.00	0.20	2.47	$6.478*10^{-3}$	$5.466*10^{-4}$	6.16	1.47
3.7	1.66	0.04	0.05	3.38	1.09	0.15	2.67	0.00	0.21	2.77	$2.818*10^{-3}$	$2.283*10^{-4}$	6.31	1.49
3.9	1.70	0.05	0.05	3.53	1.10	0.15	2.60	0.00	0.22	3.05	$1.130*10^{-3}$	$9.155*10^{-5}$	6.47	1.51

639

Table.3: Contributions of the uncertainties in the variance
of $P_f \approx 1/4000$ 1/yr at Rotterdamsche Hoek.

Uncertainty: (percentages in variance) Variable	Symbol	Intrinsic	Intrinsic + statististical additive		Intrinsic + statististical multiplicative		Intrinsic + statististical + model additive		Intrinsic + statististical + model multiplicative	
Lake Level	Mp	13%	7%	69%	4.5%	71%	6%	48%	5%	55%
Wind	W	87%	62%		66.5%		42%		50%	
Uncertainty Lake Level	fMp	- -	2%	31%	2%	29%	2%	26%	2%	24%
Uncertainty Wind	fW	- -	29%		27%		24%		22%	
Other	fOpw, fOsc, fHs, $f_{\Omega\gamma}$, fs_φ	- -	- -	- -	- -	- -	26%	26%	21%	21%
Total:		100%	100%	100%	100%	100%	100%	100%	100%	100%

Table.4: Contributions of the uncertainties on the variance of $P_f \approx 1/4000$ 1/yr at Enkhuizen

Uncertainty: (percentages in variance) Variable	Symbol	Intrinsic	Intrinsic + statististical additive		Intrinsic + statististical multiplicative		Intrinsic + statististical + model additive		Intrinsic + statististical + model multiplicative	
Lake Level	Mp	55%	43%	60%	45%	78%	13%	25%	14%	36%
Wind	W	45%	17%		33%		12%		22%	
Uncertainty Lake Level	fMp	- -	6%	40%	6%	22%	4%	31%	4%	16%
Uncertainty Wind	fW	- -	34%		16%		27%		12%	
Other	fOpw, fOsc, fHs, $f_{\Omega\gamma}$, fs_φ	- -	- -	- -	- -	- -	44%	44%	48%	48%
Total:		100%	100%	100%	100%	100%	100%	100%	100%	100%

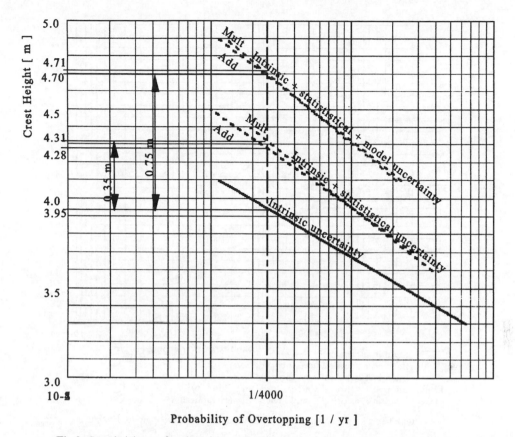

Fig.2: Crest height as a function of the probability of overtopping for Rotterdamsche Hoek.

5.FORM CALCULATIONS

The probability of Z < 0 (overtopping of the Lake IJssel dikes) is calculated by a first order reliability method (FORM). A good overview of FORM is given in Thoft-Christensen and Baker (1982). Given the additive uncertainties in the physical model of section 3, the FORM calculations are presented for various crest heights for the location of Rotterdamsche Hoek in table 1 and for the location of Enkhuizen in table 2. The notation of the 29 variables is summarized in the list of symbols (attached to the last page of this paper).

Rotterdamsche Hoek is situated on the East-side of the lake, "facing" the strong winds from the West (most frequent direction in The Netherlands). Enkhuizen is situated on the West-side of the lake, and therefore more or less protected against the strong winds from the West. This observation also follows from the tables 1 and 2, in which it is seen that the crest heights at Rotterdamsche Hoek are appr. 1 m higher than the crest heights in Enkhuizen, given a fixed probability of overtopping.

In the case of multiplicative uncertainties, FORM calculations have been performed as well. The results were very similar to the results for the additive uncertainties, and therefore not shown in this paper. The reason is that the design wind speed is in the order of 30 m/s for which the multiplicative uncertainty is 30 x 0.1 = 3 m/s (the same as the assumed additive uncertainty).

The results of the FORM calculations can also be presented graphically (Figs. 2 and 3 for Rotterdamsche Hoek and Enkhuizen resp.). Notice the differences between the required crest heights for the three cases: intrinsic uncertainty, instrinsic + statistical uncertainty, and intrinsic + statistical + model uncertainty. These differences can be up to 1 metre. Also notice, that there is not much difference in the results for the choice of the type of uncertainty modelling (additive or multiplicative), for the reason as explained earlier.

641

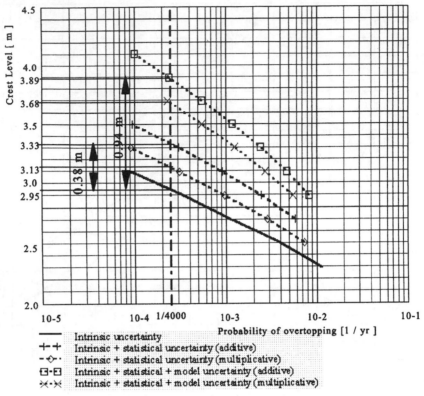

Fig.3: Crest height as a function of the probability of overtopping for Enkhuizen.

Finally, the contributions of the various uncertainties in the overall uncertainty in the probability of overtopping are summarized in tables 3 and 4 (Rotterdamsche Hoek and Enkhuizen resp.). Notice that the ratio $\alpha^2_{fMp} / \alpha^2_{fW}$ is larger for Enkuizen, than for Rotterdamsche Hoek. From this, it follows that Enkhuizen is more "lake level dominant" and Rotterdamsche Hoek is more "wind dominant'.

6.RELIABILITY-BASED OPTIMIZATION OF THE DIKE HEIGHTS

Notice that the probabilities of overtopping are given by straight lines in the semi-logarithmic figures 2 and 3. This means that they can be described by exponential distribution functions: $P(K<k) = 1-\exp(-(k-A)/B)$.
When the following notation is adopted:

P	=	probability of overtopping [1/yr]
i.u.	=	intrinsic uncertainty
s.u.	=	statistical uncertainty
add.	=	additive model
mult.	=	multiplicative model
m.u.	=	model uncertainty
h	=	required crest height
A	=	location parameter
B	=	scale parameter

the table 5 can be derived.

Table.5: Comparison of the distribution parameters and the required crest heights.

Rotterdamse Hoek	A in m 1	ΔA in m 2	B in m 3	ΔB in m 4	h for P = 1/4000 [1/yr] in m 5	Δh in m 6	P [1/yr] for h = 3.95 m 7
i.u.	2.303	- -	0.198	- -	NAP + 3.95	- -	1/4000
i.u. + s.u. (add.)	2.285	-0.018	0.240	0.042	NAP + 4.28	0.33	1/1000
i.u. + s.u. (mult.)	2.113	-0.190	0.267	0.069	NAP + 4.33	0.38	1/1000
i.u + s.u. (add.) + m.u.	2.492	0.089	0.2645	0.0665	NAP + 4.71	0.76	1/250
i.u. + s.u. (mult.) + m.u.	2.357	0.054	0.2832	0.0852	NAP +4.70	0.75	1/275

Enkhuizen	A in m 1	ΔA in m 2	B in m 3	ΔB in m 4	h for P = 1/4000 [1/yr] in m 5	Δh in m 6	P [1/yr] for h = 2.95 m 7
i.u.	1.559	- -	0.166	- -	NAP + 2.95	- -	1/4000
i.u. + s.u. (add.)	1.701	0.142	0.194	0.028	NAP + 3.33	0.38	1/625
i.u. + s.u. (mult.)	1.600	0.041	0.183	0.017	NAP + 3.13	0.18	1/1600
i.u + s.u. (add.) + m.u.	1.782	0.223	0.252	0.086	NAP + 3.89	0.94	1/100
i.u. + s.u. (mult.) + m.u.	1.700	0.141	0.273	0.107	NAP + 3.68	0.73	1/200

The results of this table will serve as input for the analysis of the economic optimal dike height.

In the reliability-based design of hydraulic structures, the idea is to determine the total costs function (Van Gelder et.al., 1997). By assuming the exponential distribution (with parameters A and B) for the probabilities of overtopping, we can write:

$$C_{total} = I_0 + I_1 k + \frac{S}{r}(1 - F(k)) = I_0 + I_1 k + \frac{S}{r} e^{\frac{k-A}{b}} \tag{11}$$

The optimal dike height follows from the minimization of the total costs function and can be expressed by the formula:

$$k_{opt} = A - B \log\left(\frac{I_1 Br}{S}\right) \tag{12}$$

and the optimal probability of failure is given by:

$$p_{opt} = \frac{I_1 Br}{S} \tag{13}$$

It is interesting to notice that the optimal probability of failure is independent of the A parameter of the exponential distribution. The ΔA values from table 1 are therefore neglected in the determination of the optimal failure probability. An increase in the slope of the exponential distribution (i.e. ΔB>0), results in an increase in p_{opt}. From table 5, it was seen that more uncertainty results in a higher ΔB-value. Consequently this leads to a higher optimal probability of failure.

The change in the optimal probability of failure (from p_{opt} to p_{opt}') caused by the increase in uncertainty (from B to B+ΔB) can also be expressed as follows:

$$p_{opt}' = \frac{B + \Delta B}{B} p_{opt} \tag{14}$$

Given an optimal probability of failure of 1/4000 yr^{-1} for Rott.Hoek, the inclusion of all uncertainties (ΔB=0.0852) leads to a new optimal probability of failure of

$$\frac{0.2645}{0.198} \cdot \frac{1}{4000} = \frac{1}{3000} yr^{-1} \qquad (15)$$

Instead of an economic optimal dike height of 4.70 m, a height of 4.60 m is the result. For the location of Enkhuizen, the proposed approach leads to a decrease in the dike height from 3.89 m to 3.77 m.

7. POLICY IMPLICATIONS

The present safety standards of the Dutch dikes and flood defences date to the report of the Deltacommittee (1960) and are expressed as an exceedance frequency of the design water level. The Dutch Ministry of Public Works wants to change this policy to bring it in line with the approach in areas like planning and transport, where failure probabilities are given in a framework of acceptable risk. The transition from water level criteria towards flooding probabilities -and finally to a flood risk approach- requires a model to calculate the probability of a failure of a dike system with several boundary conditions:

1 The model should be widely accepted by the flood defence community and in some sense by the general public.

2 The results of the model should be sufficiently robust. The answers should not vary substantially with slight moderations of the input.

3 The results of the model should not actuate the decision to alter the dike systems. The dike systems are now in full compliance with the Deltacommittee standard and are generally considered to be sufficiently safe.

From the previous sections, it can be observed that there is a seeming difference between the probability of flooding computed with the 'old' (without uncertainties) and the 'new' model (with uncertainties). It is still not decided how to deal with this seeming difference. There are three possible reactions:

1 Accept the difference and do nothing.

2 Heighten the dikes in order to lower the 'new' probability of flooding to the 'old' value.

3 Do research before the transition from 'old' to 'new' model takes place in order to reduce some uncertainties and close the gap between the 'old' and the 'new' probability of flooding.

8. CONCLUSIONS

Beyond intrinsic uncertainty, also statistical uncertainties and model uncertainties should be taken into account and are important quantities in the design philosophy of water defences. In addition to intrinsic uncertainty, the influences of the statistical and model uncertainties have been studied with FORM (tables 2 and 3), and it was observed that these were not negligable. Especially the uncertainty in the wind has a large contribution to the overall uncertainty in the probability of overtopping (about 20%).

Note that in this paper the probabilities of overtopping have been calculated. However, overtopping does not mean failure of the dikes. In order to determine the probability of inundation, the probability of overtopping should be multiplied by a so-called transition probability which describes the probability that a dike fails under the condition of overtopping.

The reliability-based decision-making procedure which has been applied in this paper can succesfully be used in the analysis of the optimal failure probabilities for the dikes along the Lake IJssel. The influence of the uncertainties lead to an increase in the probability of exceedance lines. When the hydraulic boundary conditions are modelled in an exponential way, analytical considerations can be given for the optimal probabilities of failure and the optimal dike heights.

REFERENCES

P.H.A.J.M. van Gelder, J.K. Vrijling, and K.A.H. Slijkhuis, Coping with uncertainty in the economical optimization of a dike design, Proceedings of the 27th IAHR Congress, Water for a Changing Global Community, pp.554-559, San Fransisco, 1997.

CUR, 1990, Probabilistic design of flood defences, Report 141, Gouda, The Netherlands.

Van Gelder, P.H.A.J.M., Roos, A., and Tonneijck, M.R., 1995, On the probabilistic design of dikes in the Netherlands. Applications of Statistics and Probability, ICASP7, Volume 3, pp.1505-1508, A.A. Balkema, July 1995.

Westphal, R., Beyer, D., Blaakman, E., 1997, Backgrounds on hydraulic loads on the dikes of Lake IJssel; Part 8: Reproduction functions (in Dutch), Ministry of Water Management, RIZA, Lelystad.

A.H.S. Ang, Structural risk analysis and reliability-based design, J. of Structural Division, ASCE, Vol.99, No. ST9, Sept. 1973, pp. 1891-1910.

Wieringa, J. en P.J. Rijkoord, Windklimaat van Nederland, Den Haag, Staatsuitgeverij, 1983.

Rijkoord, P.J., A compound Weibull model for the description of surface wind velocity distributions, De Bilt: KNMI, ScientificReport WR83-13,1983.

Bretschneider, Ch. L., Generation of windwaves over shallow bottom, Technical Memorandum no. 51, Beach Erosion Board, Office of the Chief of Engineers, 1954.

P. Thoft-Christensen and M.J. Baker, Structral Reliability Theory and its applications, Springer Verlag, Berlin etc., 1982.

Sanders, J.W., A Growth-Stage Scaling Model for the Wind-Driven Sea, Sonderdruck aus der Deutschen Hydrographischen Zeitschrift, Band 29, 1976, Heft 4.

Van der Meer, J.W., "Golfoploop en golfoverslag bij dijken", juni 1997.

P.H.A.J.M. van Gelder, and J.K. Vrijling, Risk-Averse Reliability-Based optimization of Sea-Defences, Proceedings of the 8th Engineering Foundation Conference on Risk Based Decision Making in Water Resources, pp. 61-76, October 12-17, 1997, Santa Barbara, California.

Pat*-Cornell, M.E. 1996. Uncertainties in risk analysis; Six levels of treatment. Reliability Engineering and System Safety. Nr. 54. Northern Ireland: Elsevier Science Limited.

J.K. Vrijling and P.H.A.J.M. van Gelder, The effect of inherent uncertainty in time and space on the reliability of flood protection, ESREL'98: European Safety and Reliability Conference 1998, pp.451-456, 16 - 19 June 1998, Trondheim, Norway.

LIST OF SYMBOLS

i.u.	=	intrinsic uncertainty
s.u.	=	statistical uncertainty
add.	=	additive model
mult.	=	multiplicative model
m.u.	=	model uncertainty
h	=	required crest height
A	=	location parameter Exponential distribution
B	=	scale parameter Exponential distribution
K	=	crest height (m) (deterministic)
M_p	=	lake level (m)
$\alpha_{M_p}^2$	=	contribution of the lake level in the overall uncertainty
fMp	=	random variable of the uncertainty in the lake level
α_{fMp}^2	=	contribution of the uncertainty in the lake level in the overall uncertainty
W	=	wind speed (m/s)
α_W^2	=	contribution of the wind speed in the overall uncertainty
fW	=	random variable of the uncertainty in the wind speed

α^2_{fW} =		contribution of the uncertainty in the wind speed in the overall uncertainty
Δ =		wind surge (m)
fOpw =		random variable of the model uncertainty in the storm surge
α^2_{fOpw} =		contribution of the model uncertainty in the storm surge in the overall uncertainty
fOsc =		random variable of the model uncertainty in the lake level increase due to oscillations
α^2_{fOsc} =		contribution of the model uncertainty in oscillations in the overall uncertainty
H_s =		significant wave height (m)
fHs =		random variable of the model uncertainty in H_s (m)
α^2_{fHs} =		contribution of the model uncertainty in H_s in the overall uncertainty
$z_{2\%}$ =		2%- wave runup (m)
$fz_{2\%}$ =		random variable of the model uncertainty in the 2%- wave runup (m)
$\alpha^2_{fz2\%}$ =		contribution of the model uncertainty in the 2%- wave runup in the overall uncertainty
$s_{op} * 100$ =		wave steepness $*100$
fs_{op} =		model uncertainty in the wave steepness
$\alpha^2_{fs_{op}}$ =		contribution of the model uncertainty in the wave steepness in the overall uncertainty
β =		reliability index
$P_{f|\phi}$ =		probability of overtopping given wind from direction sector *
P_f =		probability of overtopping
T_p =		peak wave period (s)
ξ_{op} =		surf similarity parameter

Stochastic Hydraulics 2000, Wang & Hu (eds) © 2000 Taylor & Francis, ISBN 90 5809 166 X

Uncertainty analysis of flood wave propagation

Christina Wan-Shan Tsai
V.T. Chow Hydrosystems Laboratory, Department of Civil and Environmental Engineering,
University of Illinois, Urbana-Champaign, Ill., USA

ABSTRACT: Conventionally flood wave propagation has been treated in a deterministic manner, however in reality, the descriptions of the flood waves may be fraught with stochastic variation of the channel geometry, probabilistic nature of roughness coefficients, data sample limitations and errors, model reliability and operational variability. In this study, flood waves propagating in open channels with different geometric cross sections are analyzed. Propagation characteristics such as logarithmic decrement and dimensionless wave celerity are investigated theoretically using the linearized stability analysis, in which the flood wave is treated as a small perturbation to initially uniform flow. Particularly, in light of the uncertainties associated with the probabilistic nature of the channel geometry and roughness coefficient, the first-order second-moment (FOSM) analysis is introduced to theoretically interpret the uncertainties resulting from the randomness of the channel geometry and roughness coefficients for different wave approximations. The overall uncertainties of flood wave propagation can be computed by putting each component uncertainty together using the proposed analytical relationships.

Key Words: Flood wave propagation, Uncertainty analysis, Stochastic hydraulics, Open channel geometric variation, Linear stability analysis

1. INTRODUCTION

Flood wave movement, which can be considered as a shallow water wave propagating along the channel, has been broadly studied for centuries. Lighthill and Whitman (1955) introduced the kinematic wave theory to illustrate the flood wave movement in rivers. Ponce and Simons (1977) used the linear stability analysis to interpret the mechanism of shallow water wave propagation. In their analyses, the channel cross sectional geometry is ideally assumed as hydraulically wide and deterministic. However, the natural channel cross sections may appear in rectangular, triangular, trapezoidal or other irregular shapes and may be subject to different degrees of stochastic variations. Yen and Yen (1992) theoretically investigated the stochastic perspective of open channel equations with most emphasis on the steady uniform flow. It is demonstrated from their study that the open channel geometry parameters such as width, slope, bedforms and roughness can affect the accuracy of open channel flow computation.

This study aims at introducing the stochastic perspectives of the flood wave propagation and discusses the uncertainties due to the probabilistic nature of the channel geometry and roughness coefficient. First, flood waves propagating in open channels with different cross sections are analyzed using the linear stability analysis. As the result, logarithmic decrement and dimensionless wave celerity are expressed in terms of channel geometric parameters, roughness coefficient, dimensionless wave number and initial uniform Froude number. Different levels of wave approximations are presented and analyzed. Next, the first-order second-moment analysis (Yen 1986) is employed to explore the uncertainties of flood wave propagation. This involves the assessment of uncertainties due to the randomness of channel geometry in terms of width, side slope and roughness coefficient and model error. Each contributing variable is evaluated in terms of its mean and coefficient of variation for the given distribution. The overall uncertainties can be computed using the derived close-form relationship from the linear stability analysis.

1.1 Saint-Venant Flow and its Lower-level Approximations

The one-dimensional Saint-Venant equations used to describe the flood wave propagation in open channels are expressed as

$$\frac{\partial y}{\partial t} + u\frac{\partial y}{\partial x} + D_h \frac{\partial u}{\partial x} = 0 \tag{1}$$

$$l\frac{1}{g}\frac{\partial u}{\partial t} + c\frac{u}{g}\frac{\partial u}{\partial x} + p\frac{\partial y}{\partial x} + f(S_f - S_o) = 0 \tag{2}$$

in which u = depth-averaged velocity; y = water depth; D_h = hydraulic depth; g = gravitational constant; S_f = frictional slope; S_o = bed slope; x = spatial coordinate; t = temporal coordinate and l, c, p, f are integer constants of 0 or 1. In the derivation of Eqs. (1) and (2), the flow is considered to be gradually varied, ie., there are no rapid changes in the flow cross section and there is no flow separation. The correction coefficients to modify the nonhydrostatic pressure and nonuniformity of the velocity distribution are assumed constants at a value of one. Though the full Saint-Venant equation is considered relatively complete in describing the wave propagating process, there exist several lower-level wave approximations in the interest of practicability and mathematical tractability. Depending on which terms in the momentum equation are considered important in the practical situation, the full dynamic wave equation can be simplified to different lower-level wave approximations. In Eq.(2), for kinematic wave, $f = 1$ and $l = c = p = 0$; for noninertia wave, $p = f = 1$ and $l = c = 0$; for gravity wave, $l = c = p = 1$ and $f = 0$.

2. LINEAR STABILITY ANALYSIS

The linear stability analysis provides a physical insight into the wave propagating process in the first-order sense. Assuming a first-order disturbance is postulated on the base flow- steady uniform flow, the postulated flow can thus be described as

$$u = u_o + \varepsilon u' \text{ and } y = y_o + \varepsilon y' \tag{3}$$

And the corresponding hydraulic depth and frictional slope are

$$S_f = S_{fo} + \varepsilon S'_f \text{ and } D_h = D_{ho} + \varepsilon D_h' \tag{4}$$

The subscript $_o$ denotes the flow quantities corresponding to the initially steady uniform flow condition and ε represents a variable of smaller order of magnitude. Both Manning's formula $u = K_n R^{2/3} S^{1/2}/n$ and Chezy's formula $u = CR^{1/2}S^{1/2}$ are considered here, in which R= hydraulic radius; K_n =a constant depending on the measurement unit (Yen 1992); n= Manning n and C= Chezy C. Imposing the assumption that the resistance coefficients are unchanged before and after the flow is perturbed, we have the ratio of two slopes as

$$S_f/S_o = (u/u_o)^2 (R/R_o)^{-m} \tag{5}$$

$m = 1$ for Chezy's formula and m=4/3 for Manning's formula. For simplicity, the following dimensionless variables are introduced:

$$(x^*, t^*, u^*, y^*, u'^*, y'^*) = (\frac{x}{L_o}, \frac{t}{L_o/u_o}, \frac{u}{u_o}, \frac{y}{y_o}, \frac{u'}{u_o}, \frac{y'}{y_o}) \tag{6}$$

in which $L_o = y_o / S_o$. Suppose the disturbance postulated on the steady uniform flow is in sinusoidal form as follows:

$$y'^* = \tilde{y} \exp[i(\sigma^* x^* - \beta^* t^*)] \text{ and } u'^* = \tilde{u} \exp[i(\sigma^* x^* - \beta^* t^*)] \tag{7}$$

where σ^* = dimensionless wave number; β^* = dimensionless complex propagation factor; $\tilde{u}$ and $\tilde{y}$ are dimensionless amplitude of flow velocity and depth. Substituting Eqs.(3)-(7) into Eqs.(1) and (2) and expressing the hydraulic radius and depth using the corresponding geometry, the nondimensional governing equation of $O(\varepsilon)$ order is obtained.

$$\mathbf{A}Y'=0 \tag{8}$$

where F_n = initial uniform Froude number, the elements in the coefficient matrix $\mathbf{A}$ are:
$a_{11} = (b/y_o + z)\sigma^* /(b/y_o + 2z)$, $\quad a_{12} = \sigma^* - \beta^*$, $\quad a_{21} = -iF_n^2(l\beta^* - c\sigma^*) + 2f$, and
$a_{22} = pi\sigma^* - fm\{(b/y_o + 2z)/(b/y_o + z) - 2\sqrt{1+z^2}/(b/y_o + 2\sqrt{1+z^2})$. The variable
matrix $Y' = \begin{bmatrix} \tilde{u} & \tilde{y} \end{bmatrix}^T$.

In order to have nontrivial solutions of Y', the coefficient matrix $\mathbf{A}$ must vanish. The characteristic equation describing the flood wave propagation in open channels can thus be expressed as

$$det\mathbf{A} = 0 \tag{9}$$

2.1 Propagation characteristics of Shallow Water Waves in Open Channel Flow

The dimensionless complex propagation number can be expressed as

$$\beta^* = \beta_r^* + i\beta_c^* \tag{10}$$

where β_r^* and β_c^* are real. Note that the amplitude grows exponentially in time if $\beta_c^* > 0$ and decays exponentially if $\beta_c^* < 0$. The dimensionless celerity of the disturbance is defined as

$$c^* = \frac{L/T}{u_o} = \frac{\beta_r^*}{\sigma^*} \tag{11}$$

where L= wavelength of the disturbance and T= wave period. The wave amplification or attenuation mechanism can be quantified by the value of logarithmic decrement

$$\delta^* = 2\pi\beta_r^* / |\beta_c^*| \tag{12}$$

c^* and δ^* are used to characterize and quantify the wave propagation properties in this study.

3. KINEMATIC WAVE APPROXIMATION

In kinematic wave approximation, only friction and gravity terms are considered dominant. Let $f = 1$ and $l = c = p = 0$, the characteristic equation (8) can be simplified. The nondimensional kinematic wave celerity and the logarithmic decrement are obtained as

$$c_k^* = \frac{1}{2}\left(2 + m - m\frac{2\sqrt{1+z^2}}{b/y_o + 2\sqrt{1+z^2}} \frac{b/y_o + z}{b/y_o + 2z} \right) \tag{13}$$

$$\delta_k^* = 0 \tag{14}$$

For rectangular channel, z = 0 and $c_k^* = \{2 + m - 2m/(b/y_o + 2)\}/2$; for triangular channel, b = 0 and $c_k^* = 1 + m/4$; and for wide rectangular channel, z = 0, $b/y_o \to \infty$ and $c_k^* = 1 + m/2$

The propagation of kinematic wave is in one direction, since the dimensionless wave celerity has only one value. Kinematic wave neither attenuates nor amplifies because of zero logarithmic decrement. The propagating characteristics are independent of initial uniform Froude number

649

and the wave number. However, it relies on the geometry factors, namely, side slope z and b/y_o ratio.

4. NONINERTIA WAVE APPROXIMATION

In noninertia wave approximation, inertia terms are considered negligible compared to others. Let $p = f = 1$ and $l = c = 0$, the characteristic equation of the noninertia wave approximation can be simplified to yield the nondimensional noninertia wave celerity and the logarithmic decrement respectively as

$$c_{ni}^* = \frac{1}{2}\left(2 + m - m\frac{2\sqrt{1+z^2}}{b/y_o + 2\sqrt{1+z^2}}\frac{b/y_o + z}{b/y_o + 2z}\right) \tag{15}$$

$$\delta_m^* = -2\pi\frac{\dfrac{b/y_o + z}{b/y_o + 2z}\sigma^*}{2 + m - m\left(\dfrac{2\sqrt{1+z^2}}{b/y_o + 2\sqrt{1+z^2}}\dfrac{b/y_o + z}{b/y_o + 2z}\right)} \tag{16}$$

For rectangular channel, $z = 0$: $c_{ni}^* = (2 + m - 2m/(b/y_o + 2))/2$ and $\delta_m^* = -2\pi\sigma^*/\{2 + m - 2m/(b/y_o + 2)\}$.

For triangular channel, $b = 0$: $c_{ni}^* = 1 + m/4$ and $\delta_{ni}^* = -2\pi\sigma^*/(m+4)$. For wide rectangular channel, $z = 0$, $b/y_o \to \infty$, $c_{ni}^* = 1 + m/2$ and $\delta_m^* = -2\pi\sigma^*/(m+2)$.

Noninertia wave has the same dimensionless wave celerity as that of the kinematic wave. However, the logarithmic decrement of noninertia wave is different from zero and is a function of geometric factors, dimensionless wave number and m value. The noninertia wave in open channels with the aforementioned cross-sectional geometry always attenuates in the propagating process.

5. GRAVITY WAVE APPROXIMATION

Friction and gravity terms are not considered in the gravity approximation. Let $l = c = p = 1$ and $f = 0$ in Eq.(8), the nondimensional gravity wave celerity can be expressed as

$$c_g^* = 1 \pm \sqrt{\frac{b/y_o + z}{b/y_o + 2z}}/F_n \tag{17}$$

and the logarithmic decrement is zero for all kinds of cross sectional shapes

$$\delta_g^* = 0 \tag{18}$$

For rectangular channel, $z = 0$, $c_g^* = 1 \pm 1/F_n$; for triangular channel, $b = 0$, $c_g^* = 1 \pm 1/\sqrt{2}F_n$; for wide rectangular channel, $z = 0$, $b/y_o \to \infty$, $c_g^* = 1 \pm 1/F_n$.

The gravity wave propagates in two directions, at the celerity depending on the geometric factors and the steady uniform Froude number. They both propagate downstream if the flow is supercritical. When the flow is subcritical, the gravity wave propagates in both upstream and downstream directions. The gravity wave neither attenuates nor amplifies since the logarithmic decrement is zero.

6. DYNAMIC WAVE APPROXIMATION

Considering the full Saint-Venant equations and having every term kept in the characteristic equation (8), we obtain the dimensionless wave celerity as

$$c_d^* = 1 \pm \frac{1}{2\sigma^* F_n^{\ 2}} \left(A^2 + B^2\right)^{1/4} \cos\frac{\theta}{2} \qquad (19)$$

and the logarithmic decrement as

$$\delta_d^* = 2\pi \frac{-\dfrac{1}{F_n^{\ 2}} \pm \dfrac{1}{2F_n^{\ 2}}\left(A^2 + B^2\right)^{1/4}\sin\dfrac{\theta}{2}}{\sigma^* \pm \dfrac{1}{2F_n^{\ 2}}\left(A^2 + B^2\right)^{1/4}\cos\dfrac{\theta}{2}} \qquad (20)$$

where $A = -4 + 4F_n^{\ 2}\dfrac{b/y_o + z}{b/y_o + 2z}\sigma^{*2}, B = 4m\left(1 - \dfrac{b/y_o+z}{b/y_o+2z}\dfrac{2\sqrt{1+z^2}}{b/y_o+2\sqrt{1+z^2}}\right)F_n^{\ 2}\sigma^*$

and $\theta = \cos^{-1}[A/(A^2 + B^2)^{1/2}]$.

7. FIRST-ORDER SECOND-MOMENT ANALYSIS

From the derived wave characteristics, which are expressed in terms of channel geometric parameters, roughness coefficient, dimensionless wave number and initial uniform Froude number, first-order second-moment analysis is applied. To account for any error as the result in replacing the full dynamic wave by other lower-level wave approximations, a correction factor λ with mean $\bar{\lambda}$ and coefficient of variation Ω_λ is introduced.

By applying the FOSM analysis for the dimensionless wave celerity and logarithmic decrement, we have the means and coefficients of variation, respectively as

$$\bar{c}^* = \bar{\lambda}f(\bar{z}, F_n, \bar{b}/y_o, \bar{m}) \text{ or } \bar{\delta}^* = \bar{\lambda}g(\bar{z}, F_n, \bar{b}/y_o, \bar{m}) \qquad (21)$$

$$Var(c^*) = (\frac{\partial c^*}{\partial \lambda})_o^2\Omega_\lambda^2 + \sum_i\sum_j (\frac{\partial c^*}{\partial x_i})_o(\frac{\partial c^*}{\partial x_j})_o r_{ij}\sigma_{xi}\sigma_{xj} \qquad (22)$$

$$Var(\delta^*) = (\frac{\partial \delta^*}{\partial \lambda})_o^2\Omega_\lambda^2 + \sum_i\sum_j (\frac{\partial \delta^*}{\partial x_i})_o(\frac{\partial \delta^*}{\partial x_j})_o r_{ij}\sigma_{xi}\sigma_{xj} \qquad (23)$$

in which the bar represents the mean value of the variable; and r_{ij} is the coefficient of correlation between variables x_i and x_j. If x_i and x_j are statistically independent, then only $i=j$ terms survive. Therefore, with the assumption that the random variables are assumed statistically independent in this study, application of Eqs. (21), (22) and (23) yields what follows:

7.1 Kinematic wave

Taking the stochastic aspect of geometric factors b and z, and the roughness coefficient into consideration, the means and coefficients of variation of the dimensionless wave celerity and logarithmic decrement can be deduced, respectively:

$$\bar{c}_k^* = \frac{1}{2}\bar{\lambda}_k\left(2 + \bar{m} - \bar{m}\frac{2\sqrt{1+\bar{z}^2}}{\bar{b}^* + 2\sqrt{1+\bar{z}^2}}\frac{\bar{b}^* + \bar{z}}{\bar{b}^* + 2\bar{z}}\right) \qquad (24)$$

$$\bar{\delta}_k^* = 0 \qquad (25)$$

$$\Omega_{ck}^2 = \Omega_{\lambda k}^2 + \frac{1}{\bar{c}_k^{*2}}\{(\frac{\partial c_k^*}{\partial m})_o^2\bar{m}^2\Omega_m^2 + (\frac{\partial c_k^*}{\partial b})_o^2\bar{b}^{*2}\Omega_b^2 + (\frac{\partial c_k^*}{\partial z})_o^2\bar{z}^2\Omega_z^2\} \qquad (26)$$

$$\Omega_{\delta k}^2 = \Omega_{\lambda k}^2 \qquad (27)$$

where the sensitivity coefficients evaluated at the mean value points are

$$(\partial c_k^* / \partial m)_o = \frac{1}{2}\bar{\lambda}_k\left(1 - \frac{2\sqrt{1+\bar{z}^2}}{\bar{b}^* + 2\sqrt{1+\bar{z}^2}}\frac{\bar{b}^* + \bar{z}}{\bar{b}^* + 2\bar{z}}\right),$$

651

$$(\partial c_k^* / \partial z)_o = \overline{\lambda}_k \overline{b}^* \overline{m} \{ \overline{b}^* - \overline{b}^* \overline{z} - 2\overline{b}^* \overline{z}^2 + 2(-\overline{z}^3 + \sqrt{1 + \overline{z}^2}(1 + \overline{z}^2)) \}$$

$$/ \{ (\overline{b}^* + 2\overline{z})^2 (\overline{b}^* + 2\sqrt{1 + \overline{z}^2})^2 \sqrt{1 + \overline{z}^2} \}$$

$$(\partial c_k^* / \partial b^*)_o = \overline{\lambda}_k \overline{m} \{ -2\overline{z}^3 - 2\overline{z} + (\overline{b}^{*2} + 2\overline{z}^2 + 2\overline{b}^* \overline{z})\sqrt{1 + \overline{z}^2} \} / \{ (\overline{b}^* + 2\overline{z})^2 (\overline{b}^* + 2\sqrt{1 + \overline{z}^2})^2 \}$$

7.2 Noninertia wave

The mean values and coefficients of variation of the dimensionless wave celerity and the logarithmic decrement are

$$\overline{c}_m^* = \frac{1}{2} \overline{\lambda}_{ni} \left(2 + \overline{m} - \overline{m} \frac{2\sqrt{1 + \overline{z}^2}}{\overline{b}^* + 2\sqrt{1 + \overline{z}^2}} \frac{\overline{b}^* + \overline{z}}{\overline{b}^* + 2\overline{z}} \right) \tag{28}$$

$$\overline{\delta}_{ni}^* = -2\pi \overline{\lambda}_{ni} \frac{\dfrac{\overline{b}^* + \overline{z}}{\overline{b}^* + 2\overline{z}} \sigma^*}{2 + \overline{m} - \overline{m} \left(\dfrac{2\sqrt{1 + \overline{z}^2}}{\overline{b}^* + 2\sqrt{1 + \overline{z}^2}} \dfrac{\overline{b}^* + \overline{z}}{\overline{b}^* + 2\overline{z}} \right)} \tag{29}$$

$$\Omega_{cm}^2 = \Omega_{\lambda ni}^2 + \frac{1}{\overline{c}_{ni}^{*2}} \{ (\frac{\partial c_m^*}{\partial m})_o^2 \overline{m}^2 \Omega_m^2 + (\frac{\partial c_m^*}{\partial b^*})_o^2 \overline{b}^{*2} \Omega_b^2 + (\frac{\partial c_m^*}{\partial z})_o^2 \overline{z}^2 \Omega_z^2 \} \tag{30}$$

where the sensitivity coefficients at the mean value points are

$$(\partial c_m^* / \partial m)_o = \frac{1}{2} \overline{\lambda}_{ni} \left(1 - \frac{2\sqrt{1 + \overline{z}^2}}{\overline{b}^* + 2\sqrt{1 + \overline{z}^2}} \frac{\overline{b}^* + \overline{z}}{\overline{b}^* + 2\overline{z}} \right),$$

$$(\partial c_m^* / \partial z)_o = \overline{\lambda}_{ni} \overline{b}^* \overline{m} \{ \overline{b}^* - \overline{b}^* \overline{z} - 2\overline{b}^* \overline{z}^2 + 2(-\overline{z}^3 + \sqrt{1 + \overline{z}^2}(1 + \overline{z}^2)) \}$$

$$\prod / \{ (\overline{b}^* + 2\overline{z})^2 (\overline{b}^* + 2\sqrt{1 + \overline{z}^2})^2 \sqrt{1 + \overline{z}^2} \}$$

$$(\partial c_{ni}^* / \partial b^*)_o = \overline{\lambda}_{ni} \overline{m} \{ -2\overline{z}^3 - 2\overline{z} + (\overline{b}^{*2} + 2\overline{z}^2 + 2\overline{b}^* \overline{z})\sqrt{1 + \overline{z}^2} \} / \{ (\overline{b}^* + 2\overline{z})^2 (\overline{b}^* + 2\sqrt{1 + \overline{z}^2})^2 \}$$

$$\Omega_{\delta m}^2 = \Omega_{\lambda ni}^2 + \frac{1}{\overline{\delta}_m^{*2}} \{ (\frac{\partial \delta_{ni}}{\partial m})_o^2 \overline{m}^2 \Omega_m^2 + (\frac{\partial \delta_m}{\partial b^*})_o^2 \overline{b}^{*2} \Omega_b^2 + (\frac{\partial \delta_m}{\partial z})_o^2 \overline{z}^2 \Omega_z^2 \} \tag{31}$$

where the sensitivity coefficients evaluated at the mean value points are

$$(\partial \delta_{ni}^* / \partial m)_o = -2\pi \overline{\lambda}_{ni} \overline{b}^* \sigma^* (\overline{b}^* + \overline{z})(\overline{b}^* + 2\sqrt{1 + \overline{z}^2})(\overline{b}^{*2} + 2\overline{b}\overline{z} + 2\overline{z}\sqrt{1 + \overline{z}^2})$$

$$/ \{ \overline{b}^{*2}(2 + \overline{m}) + 2(4 + \overline{m})\overline{z}\sqrt{1 + \overline{z}^2} + 2\overline{b}^{*2}[(2 + \overline{m})\overline{z} + 2\sqrt{1 + \overline{z}^2}] \}^2$$

$$(\partial \delta_m^* / \partial z)_o = -2\pi \overline{\lambda}_m \overline{b}^* \sigma^* \{ -\frac{4\overline{b}^*(2 + \overline{m} + 2\overline{z}^2)}{\sqrt{1 + \overline{z}^2}} + \overline{b}^{*2}(-2 + \overline{m}(-1 + \frac{2\overline{z}}{\sqrt{1 + \overline{z}^2}})) + 2(-4(1 + \overline{z}^2)$$

$$+ \overline{m}(-2 - 2\overline{z}^2 + \frac{\overline{z}^3}{\sqrt{1 + \overline{z}^2}})) \} / \{ \overline{b}^{*2}(2 + \overline{m}) + 2(4 + \overline{m})\overline{z}\sqrt{1 + \overline{z}^2} + 2\overline{b}^{*2}[(2 + \overline{m})\overline{z} + 2\sqrt{1 + \overline{z}^2}] \}^2$$

$$(\partial \delta_{ni}^* / \partial b^*)_o = -2\pi \overline{\lambda}_{ni} \overline{b}^* \sigma^* \{ 4(2 + \overline{m})\overline{z}^3 - 2(\overline{b}^{*2} + \overline{z}^2)\overline{m}\sqrt{1 + \overline{z}^2} + \overline{z}(4(2 + \overline{m}) + \overline{b}^{*2}(2 + \overline{m})$$

$$+ 8\overline{b}^{*2}\sqrt{1 + \overline{z}^2}) \} / \{ \overline{b}^{*2}(2 + \overline{m}) + 2(4 + \overline{m})\overline{z}\sqrt{1 + \overline{z}^2} + 2\overline{b}^{*2}[(2 + \overline{m})\overline{z} + 2\sqrt{1 + \overline{z}^2}] \}^2$$

7.3 Gravity wave

The mean values and the coefficients of variation for both dimensionless gravity wave celerity and logarithmic decrement are computed as

$$\overline{c}_g^* = \overline{\lambda}_g (1 \pm \sqrt{\frac{\overline{b}^* + \overline{z}}{\overline{b}^* + 2\overline{z}}} / F_n) \tag{32}$$

652

$$\overline{\delta}_g^* = 0 \tag{33}$$

$$\Omega_{cg}^2 = \Omega_{\lambda g}^2 + \frac{1}{\overline{c}_g^{*2}} \{ (\frac{\partial c_g^*}{\partial b^*})_o^2 \overline{b}^{*2} \Omega_b^2 + (\frac{\partial c_g^*}{\partial z})_o^2 \overline{z}^2 \Omega_z^2 \} \tag{34}$$

$$\Omega_{\delta g}^2 = \Omega_{\lambda g}^2 \tag{35}$$

with the sensitivity coefficients as $(\partial c_g^* / \partial b^*)_o = \overline{\lambda}_g \overline{z} / \{2 F_n (\overline{b}^* + 2\overline{z})^2 \sqrt{(\overline{b}^* + \overline{z})/(\overline{b}^* + 2\overline{z})}\}$,

$(\partial c_g^* / \partial z)_o = -\overline{\lambda}_g \overline{b}^* / \{2 F_n (\overline{b}^* + 2\overline{z})^2 \sqrt{(\overline{b}^* + \overline{z})/(\overline{b}^* + 2\overline{z})}\}$.

7.4 Dynamic wave

The stochastic nature of the dimensionless wave celerity and logarithmic decrement, by assessing each component uncertainty, can be computed as

$$\overline{c}_d^* = 1 + \frac{1}{2\sigma^* F_n^2} \left(\overline{A}^2 + \overline{B}^2 \right)^{1/4} \cos\frac{\overline{\theta} + k\pi}{2} \tag{36}$$

$$\overline{\delta}_d^* = 2\pi \frac{-\dfrac{1}{F_n^2} + \dfrac{1}{2F_n^2} \left(\overline{A}^2 + \overline{B}^2 \right)^{1/4} \sin\dfrac{\overline{\theta} + k\pi}{2}}{\sigma^* + \dfrac{1}{2F_n^2} \left(\overline{A}^2 + \overline{B}^2 \right)^{1/4} \cos\dfrac{\overline{\theta} + k\pi}{2}} \tag{37}$$

$$\Omega_{cd}^2 = \Omega_{\lambda d}^2 + \frac{1}{\overline{c}_d^{*2}} \{ (\frac{\partial c_d^*}{\partial A})_o^2 \overline{A}^2 \Omega_A^2 + (\frac{\partial c_d^*}{\partial B})_o^2 \overline{B}^2 \Omega_B^2 + (\frac{\partial c_d^*}{\partial \theta})_o^2 \overline{\theta}^2 \Omega_\theta^2 \} \tag{38}$$

$$\Omega_{\delta d}^2 = \Omega_{\lambda d}^2 + \frac{1}{\overline{\delta}_d^{*2}} \{ (\frac{\partial \delta_d^*}{\partial A})_o^2 \overline{A}^2 \Omega_A^2 + (\frac{\partial \delta_d^*}{\partial B})_o^2 \overline{B}^2 \Omega_B^2 + (\frac{\partial \delta_d^*}{\partial \theta})_o^2 \overline{\theta}^2 \Omega_\theta^2 \} \tag{39}$$

where

$$\Omega_\theta^2 = \frac{1}{\overline{\theta}^2} \{ (\partial \theta / \partial A)_o^2 \overline{A}^2 \Omega_A^2 + (\partial \theta / \partial B)_o^2 \overline{B}^2 \Omega_B^2 \},$$

$$\Omega_A^2 = \frac{1}{\overline{A}^2} \{ (\partial A / \partial b^*)_o^2 \overline{b}^{*2} \Omega_b^2 + (\partial A / \partial z)_o^2 \overline{z}^2 \Omega_z^2 \},$$

$$\Omega_B^2 = \frac{1}{\overline{B}^2} \{ (\partial B / \partial b^*)_o^2 \overline{b}^{*2} \Omega_b^2 + (\partial B / \partial z)_o^2 \overline{z}^2 \Omega_z^2 + (\partial B / \partial m)_o^2 \overline{m}^2 \Omega_m^2 \}.$$

The sensitivity coefficients at the mean value points are expressed as

$(\partial c_d^* / \partial A)_o = \overline{A} \cos(\overline{\theta} / 2) / \{4(\overline{A}^2 + \overline{B}^2)^{3/4} \sigma^* F_n^2\}$,

$(\partial c_d^* / \partial B)_o = \overline{B} \cos(\overline{\theta} / 2) / \{4(\overline{A}^2 + \overline{B}^2)^{3/4} \sigma^* F_n^2\}$

$(\partial c_d^* / \partial \theta)_o = (\overline{A}^2 + \overline{B}^2)^{1/4} \sin(\overline{\theta} / 2) / (4\sigma^* F_n^2)$

$(\partial \delta_d^* / \partial A)_o = 2\overline{A} \pi \{\cos(\overline{\theta} / 2) + F_n^2 \sigma^* \sin(\overline{\theta} / 2)\}$

$/ \{(\overline{A}^2 + B^2)^{3/4} (2 F_n^2 \sigma^{*2} + (\overline{A}^2 + \overline{B}^2)^{1/4} \cos(\overline{\theta} / 2))^2\}$

$(\partial \delta_d^* / \partial B)_o = 2\overline{B} \pi \{\cos(\overline{\theta} / 2) + F_n^2 \sigma^* \sin(\overline{\theta} / 2)\}$

$/ \{(\overline{A}^2 + \overline{B}^2)^{3/4} (2 F_n^2 \sigma^{*2} + (\overline{A}^2 + \overline{B}^2)^{1/4} \cos(\overline{\theta} / 2))^2\}$

$(\partial \delta_d^* / \partial \theta)_o = \pi \{(\overline{A}^2 + \overline{B}^2)^{1/4} + 2 F_n^2 \sigma^* \cos(\overline{\theta} / 2) - 2\sin(\overline{\theta} / 2)\}$

$/ \{(\overline{A}^2 + \overline{B}^2)^{-1/4} (2 F_n^2 \sigma^{*2} + (\overline{A}^2 + \overline{B}^2)^{1/4} \cos(\overline{\theta} / 2))^2\}$

The above analyses are for trapezoidal channels with stochastic variations on side slope and bottom width, the other cross sections such as wide rectangles, rectangles, and triangles can be

653

treated as special cases of trapezoids and can be computed by defining the values of side slope and bottom width.

8. ASSESSMENT OF COMPONENT UNCERTAINTIES

(1) Model error λ

The model error assigned for each wave approximation depends on the discrepancy resulting from replacing the full Saint-Venant equations by kinematic wave (λ_k), noninertia wave (λ_{ni}) or gravity wave (λ_g). The mean and coefficient of variation of the model error for different levels of wave approximations can be assessed from the theoretical investigation (Ponce et al. 1978, Tsai 1999, Tsai and Yen 2000) or numerical examination (Xia and Yen 1994) on their criteria of applicability. It is expected that adoption of more refined wave approximations will reduce the model error λ.

(2) Side slope z

In flood routing or unsteady flow computation, the side slope of a trapezoidal or triangular channel is ideally assumed constant. While in reality, the side slope may be impaired by natural variability because of erosion, deposition and other environmental impacts. In this study, the side slope with a mean $\bar{z}$ and a coefficient of variation Ω_z is considered.

(3) Bottom width b

Because of vegetation, sedimentation or other factors, the bottom width of the channel is subject to degrees of variation. Since the channel bottom width is stochastic along the channel, its variation can be viewed as fluctuation with a mean b and a coefficient of variation Ω_b in the spatial sense.

(4) Power of roughness coefficient m

In traditional flood wave computation, the resistance coefficient appearing in the Saint-Venant equations or the lower-level approximations is ideally assumed invariant of flow depth. However due to unsteadiness and spatial nonuniformity of the free surface profile, the resistance coefficient is found to be a function of flow depth and therefore the power m indicating the relationship between the roughness coefficient and the flow depth varies. Deterministically, for Chezy's formula $m=1$, and for Manning's formula, $m=4/3$. In this study, the stochastic variation of m due to the unsteadiness or other flow properties is considered. For Chezy's formula, m is assumed with a mean $\bar{m} = 1$ and a coefficient of variation Ω_m, and for Manning's formula, m is assumed with a mean $\bar{m} = 4/3$ and a coefficient of variation Ω_m.

The overall uncertainty of flood wave propagation in open channels due to the randomness of the geometry, roughness coefficient and model error can be assessed using Eqs. (24)-(38), depending on the wave approximate model adopted for computation.

9. CONCLUDING REMARKS

This study aims at introducing the stochastic perspectives of the flood wave propagation and discusses the uncertainties due to the probabilistic nature of the channel geometry, roughness coefficient and model error. From the algebraic relationships derived using the linear stability analysis, the uncertainties resulting from the stochastic variations of the channel geometric factors, the randomness of perturbed roughness coefficient, and the error resulting from replacing the full Saint-Venant equations by lower-level wave approximations are formulated in a first-order manner. This study gives a theoretical framework in which each contributing variable is evaluated in terms of its mean and variance for the given distribution. Further investigation on the correlation among each contributing variable is suggested. More field or experimental data are needed to give the appropriate values of the means and coefficients of variation for each contributing variable.

Stochastic consideration of the flood wave propagation not only provides a valuable insight into the underlying physics of the problem but also makes the engineering designs, flood risk assessment and management more reliable.

REFERENCES

Lighthill, M. J. and Whitham, G. B. (1955). "On kinematic waves: I. Flood movement in long rivers." Proceedings, Royal Society, London, Series A, 229, 281-316.

Ponce, V. M. and Simons, D. B., (1977). "Shallow wave propagation in open channel flow." Journal of the Hydraulics Division, ASCE, 103(HY12), 1461-1475.

Ponce, V.M., Li, R.N. and Simons, D.B., (1978). "Applicability of kinematic and diffusion models." Journal of the Hydraulics Division, ASCE, 104(HY3), 363-360.

Tsai, W.-S. (1999). "Investigation of the effect of channel cross-sectional geometry of shallow water wave propagation." Proceedings, WaterPower 99, ASCE, Las Vegas, Nevada, Sec.H, 1-10.

Tsai, C. W.-S. and Yen, B.C. (2000). "Linear stability of shallow water waves in open channel flows." Submitted to Physics of Fluids

Xia, R. and Yen, B.C., (1994). "Significance of averaging coefficients in open channel flow equations." Journal of Hydraulic Engineering, ASCE, 120(2), 169-180.

Yen, B.C., Cheng S.-T., and Melching, C.S. (1986). "First order reliability analysis." Stochastic and Risk Analysis in Hydraulic Engineering. B.C. Yen, ed. Water Resources Publication, Littleton, Colorado.

Yen, B.C. (1992). "Dimensionally homogeneous Manning's formula." Journal of Hydraulics Engineering, ASCE, 118(9), 1326-1332.

Yen, B.C. and Yen, C.L. (1992). "Stochastic perspective of open channel equations." Proceedings of the 6th IAHR International Symposium on Stochastic Hydraulics, Taipei, Taiwan.

Environmental hydraulics and global climate modeling

Stochastic Hydraulics 2000, Wang & Hu (eds) © 2000 Taylor & Francis, ISBN 90 5809 166 X

Invited lecture: A real time early warning and modelling system for red tides in Hong Kong

J.H.W.Lee, K.T.M.Wong, Y.Huang & A.W.Jayawardena
Department of Civil Engineering, Hong Kong University of Science and Technology, Kowloon, China

ABSTRACT: Harmful algal blooms (HAB) can lead to great economic losses to fisheries and significant adverse impacts on the environment. And yet the onset of HAB, a worldwide problem, is notoriously difficult to predict. This paper describes the design, development, and initial operation of a real time, remotely controlled, early warning system for algal blooms and red tides in a coastal field research station. The system measures solar radiation, wind velocity, tidal level and velocity, dissolved oxygen and chlorophyll fluorescence at three depths, supplemented by regular onsite sampling for subsequent chemical analysis of nutrient and chlorophyll concentrations, and taxonomic examination. An example of successful detection and monitoring of an algal bloom event (and also a red tide) is given. Preliminary results of modelling chlorophyll concentrations using artificial neural networks (ANN) are also presented. Different network structures are trained on a six year biweekly water quality data set, and tested on an independent 3 year data set. Unlike previous work on limnological and riverine applications, the results show that rather good predictions of long term trends in algal biomass can be obtained using only Chlorophyll-*a* and (or without) total inorganic nitrogen (time delayed for one week) as input nodes. On the other hand, the phase error of predictions renders the ANN method unsuitable for short term forecasts of algal blooms - which can take off and collapse in 7-10 days.

Key words: Red tide, Algal bloom, Real time warning system, Red tide modelling, Hongkong

1.INTRODUCTION

In sub-tropical coastal waters around Hong Kong and South China, algal blooms and red tides (due to the rapid growth of microscopic phytoplankton) are often observed. Under the right environmental conditions (favourable temperature, solar radiation, nutrient concentration, predation pressure, wind speed, and tidal flushing) these blooms can occur and subside over rather short time scales - in the order of days to a few weeks. Algal blooms often lead to discoloration of the marine water, which may lead to beach closures, severe dissolved oxygen depletion, fish kills, and shellfish poisoning. Over the past two decades, massive fish kills due to oxygen depletion have been observed in some of the marine fish culture zones in Hong Kong (Lee *et al*: 1991); toxic algal blooms have, however, been relatively rare. Nevertheless, in April 1998, a devastating red tide has resulted in the worst fish kill in Hong Kong's history - it wiped out over 80 percent (3400 tonnes) of fish stocks, with estimated loss of more than HK$312 million (Dickman 1998).

Whereas the general ecological response of phytoplankton to environmental conditions has been extensively studied (e.g. Parsons and Takahashi 1984), the causality and dynamics of algal blooms are extremely complicated and not well-understood. As part of a group research project, two field stations have been developed in the northeastern and southern coastal waters of Hong

Kong to provide an almost continuous database for studying short term algal dynamics and red tides. The field programme is accompanied by various physically-based and data-driven modelling efforts. The main objective is to sharpen our understanding on the causes of HAB, and to develop early warning systems and mitigation measures. In this paper we present initial experiences from a remotely-controlled field research station for studying algal blooms. The system design, instrumentation and key operation details are described. Typical results and recent observations of two algal blooms/red tides successfully "captured" by the early warning system are illustrated. A preliminary study of the use of a data assimilation method to predict algal blooms in coastal waters is also presented.

2. FIELD MEASUREMENT OF ALGAL BLOOM DYNAMICS

2.1 Algal dynamics field monitoring station

The field monitoring station is located in an inner cove at the southern end of Kat O Bay, a pristine tidal inlet in the remote northeastern waters of Hong Kong (Fig.1a). As there is practically very little polluting discharges (Kat O Island is relatively uninhabited) into surrounding waters, the water quality of the bay is generally very good. Nevertheless, every year red tides in Hong Kong are frequently first detected in Kat O. This field monitoring station is also located close to the O Pui Tong marine fish culture zone at the northern end of the bay, and a fisheries research station of the Hong Kong Agriculture, Fisheries and Conservation Department (AFCD). A used 6 m x 9 m fishfarm raft was purchased and refurbished into a secure platform from which the water quality, meteorological and hydrographical measurements can be conveniently made. The raft is located about 100 m offshore; the average depth at the site is around 7 m; tides are semi-diurnal and mixed, with a mean tidal range of 1.7 m and typical current velocities of less than 0.1 m/s. Fig.1a) and Fig.1b) show respectively an aerial view of Kat O Bay and an observed red tide patch close to the field station.

2.2 Telemetry system

Fig.2 shows the instrumentation and data logging system as deployed on site. The telemetry system consists of a data logging system (micrologger and various measuring devices) and a modem which connects the data logger to a telephone line for data transmission back to the University. A Campbell Scientific CR23X micrologger is used for program control and data collection; telecommunication is accomplished by a DC112 modem connecting to the micrologger while the other end is a personal computer with a phone modem. Dissolved oxygen (DO) at surface, mid-depth, and bottom (1m, 3m, 5m) are monitored by positioning tygon tubings at the corresponding depths. With the use of a peristaltic pump, the water at each depth is pumped above board and made to pass through and drain freely from a measuring cell fitted with a YSI membrane type DO probe connected to a YSI 58 DO meter. Four minutes of pumping are required to flush the cell, before the desired DO measurement is made. At intervals of 2 hours, the pump is turned on; the water is sampled from each depth in turn by the use of magnetic valves and the DO measured. The on/off operation of the pump and valves is controlled by relays that in turn are controlled by the micrologger. The DO measured by the telemetry system has been calibrated against laboratory and field tests, and is regularly calibrated on site with air saturation method. Similarly the chlorophyll fluorescence at the corresponding depths is also measured in pump-through mode (see below). Global solar radiation is measured by a Kipp and Zonen double-dome pyranometer. Wind speed and direction are measured by a R.M. Young propeller-vane type anemometer. The water temperature is measured in-situ by YSI thermistors (accuracy 0.2 deg C) at three depths; air temperature is also monitored. Tidal elevation and surface current velocity are measured by a Sontek Acoustic Doppler Current Meter. Details on the DO measurement have previously been reported (Lee and Lee 1995). The pumping system and relay box, and the micrologger and thermistor channels, are powered each with a 12V car battery.

The DO and current data are measured once every two hours. The other variables are sampled every second; wind, solar radiation, and temperature are logged every hour. With the use of a communications software, all data is retrieved every day at the University laboratory. The system deployed on site requires careful maintenance at intervals of about 10-14 days, at which time all instruments and connection tubings are checked and maintenance work carried out. During the site visit, water samples at the three depths are usually taken for subsequent analysis of nutrient, chlorophyll levels, cell counts and species identification.

2.3 Chlorophyll measurement and Alarm system

Chlorophyll concentrations are measured by a Chelsea Minitracka II miniature fluorimeter equipped with a flow-through cell. Water at three depths is pumped consecutively through the flow cell; inside the cell is a blue LED light source emitting irradiation at 470 nm. On irradiation with this blue light, chlorophyll resident in the algal cells fluoresce and emit light in the 650-700 nm range; the measured red emisson of the fluorescent light then provides a sensitive and convenient estimate of chlorophyll biomass. As a control, a second fluorimeter is mounted directly *in-situ* at 0.5 m below the water surface. Continuous one-minute and one-hour average readings of the chlorophyll fluorescence are monitored for the three depths and the surface-mounted fluorimeter respectively.

This method is suitable for *in-situ* remote deployment and is extremely useful in describing temporal variabilities; it is potentially an important component of a red tide early detection and warning system. However, it is recognised that fluorescence is inherently an imprecise measure of chorophyll biomass; the fluorescence yield (ratio of fluorescence to actual chlorophyll concentration) can be influenced by many physiological factors (Cullen *et al.* 1997). For example, it has been observed that the fluoresence yield decreases in strong sunlight, giving apparently depressed surface chlorophyll concentrations during the day. Thus the fluorescence needs to be calibrated against direct measurements of Chlorophyll-*a* on acetone extracts. When left in long term deployment the fluorimeter can also suffer from marine fouling problems.

Fig.3a) shows an example calibration for a sea water sample containing a natural assemblage of diatoms (dominating species *Pseudonizchia seriata, Skeletonema costatum, Leptocylindrus minimus*), with Chlorophyll-*a* concentrations of up to 50 *µg/L*; fig.3b) shows a similar calibration of fluorescence vs Chl-*a* concentration for several diatom and dinoflagellate laboratory cultures. It can be seen the fluorescence varies approximately linearly with chlorophyll-*a* concentration; however, the fluorescence yield (slope) is species and cell-size dependent; however, it appears reasonably consistent calibrations can be generally used for diatom and dinoflagellates respectively.

Fig.1: a) (left) Aerial view of algal dynamics field monitoring station at Kat O, Hong Kong
b) (right) Decaying red tide patch observed on 21 February 2000

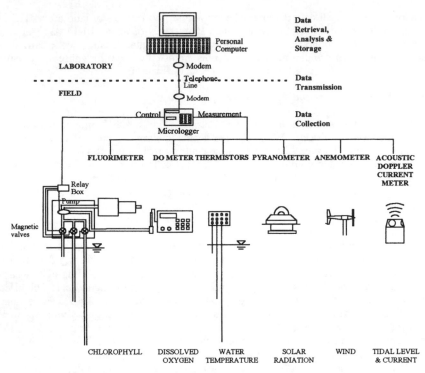

Fig.2: Schematic of telemetry system for study of algal dynamics in Kat O, Hong Kong

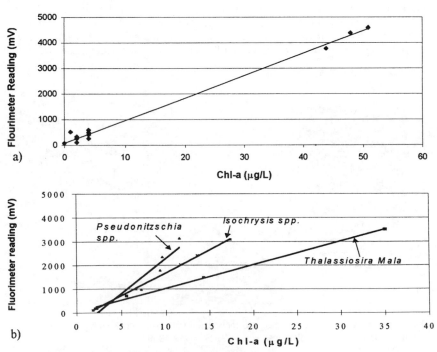

Fig.3: Calibration of fluorescence against chlorophyll-*a* concentrations for
a) natural diatom population;
b) laboratory algal cultures

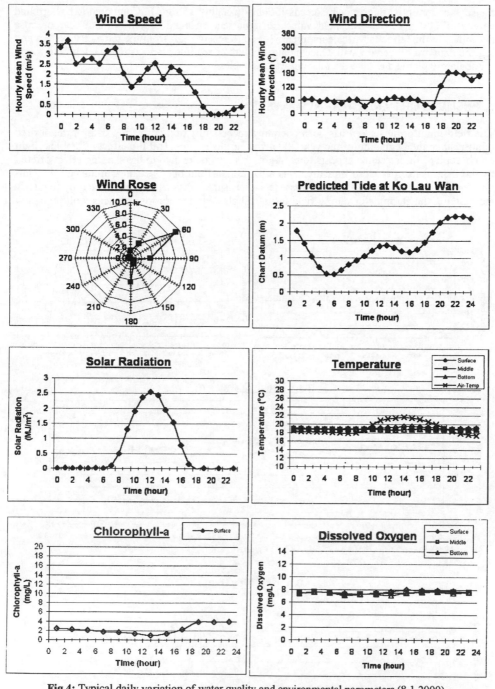

Fig.4: Typical daily variation of water quality and environmental parameters (8.1.2000)

The purpose of the system is to monitor short term algal and dissolved oxygen dynamics. Experience shows that reliance on the daily transmission of water quality data alone is inadequate for detection of algal blooms. An alarm system is set up so that the micrologger will call back the computer at the University when the 1 hour-averaged Chlorophyll fluorescence measured by the surface-mounted fluorimeter exceeds 1500mV (roughly 15 $\mu g/L$). Compared to background values of 200 mV, this threshold value can represent an algal bloom without fear of false alarming. In the laboratory, the computer is set up to receive call from the telemetry system and log the data after receiving call. Once an algal bloom or red tide alarm is received, a field trip is immediately planned to take the relevant water samples.

3. RESULTS OF EARLY WARNING SYSTEM

The field monitoring system has been operating successfully and delivering data of the desired quality since the beginning of the year 2000. Our results suggest that the effect of ambient light on chorophyll fluorescence appears to be slight; this may be related to the shading effect offered by the fishraft itself (i.e. our measurements are different from open-sea measurements). In addition, biofouling (which would lead to erroneous measurements) can be minimized by regularly maintaining the fluorimeter during field visits and cleaning the fluorimeter lens with propanol.

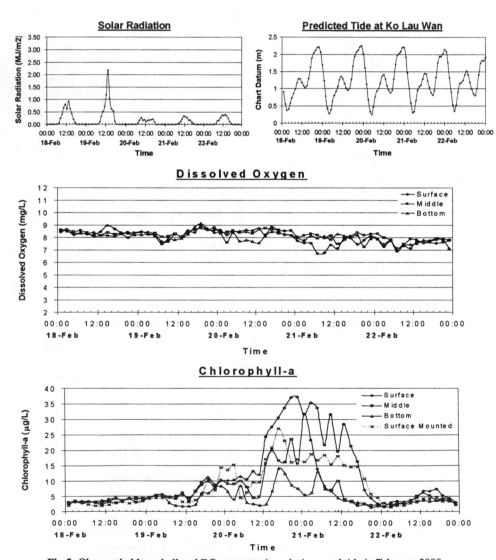

Fig.5: Observed chlorophyll and DO concentrations during a red tide in February 2000

664

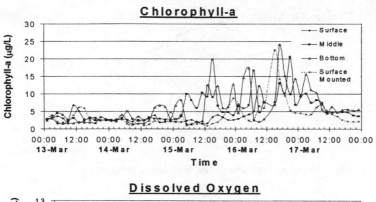

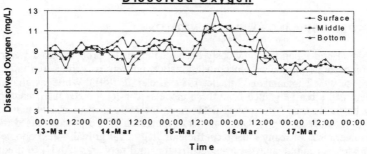

Fig.6: Observed chlorophyll and DO concentration during an algal bloom in March 2000

Provided a smooth water flow is assured through the tygon tubings and peristaltic pumping system, the chlorophyll fluorescence measurement, if properly interpreted, has proved most robust and useful.

Fig.4 shows a typical daily summary of measured water quality and environmental variables in the winter. The Kat O station is characterized by northeasterly winds of typically below 4 m/s; the water is well-mixed, with negligible vertical temperature differentials. With background chlorophyll concentrations of 1-4 $\mu g/L$, the dissolved oxygen variation in the unpolluted bay is rather small, with slight increases of surface DO during the sunlight hours. The predicted tide at the closest station where long term tide records are available, Ko Lau Wan, is also shown.

Fig.5 shows the observed chlorophyll and DO concentrations during a red tide that was detected successfully by the system. An alarm was received over the weekend on February 19. When a field trip was undertaken on February 21, it was observed that the red tide patch already started to decay (Fig.1b). The data shows that the bloom (observed species include *Rhizosolenia spp.* and *Noctilluca Scintillans*) occurred during 18-22 February 2000. The period was preceded by a week of largely clear sunny days; however the weather was cloudy and misty (with trace rainfall) during the bloom event. The temperature and salinity were uniform over the depth, and around 16.8 deg C and 31.5 ppt respectively. It appears that this red tide was transported from the outer bay with a Spring flood tide on 17 February 2000. The chlorophyll measurements of the surface-mounted fluorimeter are consistent with those obtained with the pumping mode from three depths. The somewhat lower values recorded by the fluorimeter mounted 0.5 m below the water surface may reflect the much longer time-averaging interval for the *in-situ* measurement. The data show clearly the greater algal concentrations near the surface, and (consistent with the visual observation) the die off of the bloom (due to presumably nutrient limitation or insufficient solar energy) around February 21, which is associated with large diurnal DO fluctuations (Lee *et al.* 1991). This event illustrates the rather short time scale of an algal bloom in these sub-tropical waters, and demonstrates the ability of the system to detect blooms in spite of their patchiness. A second red tide was also successfully detected during March 13-17, 2000; this red tide was characterised by minor blooms in the preceding week, and significant DO vertical differentials (Fig.5b). During the day chlorophyll fluorescence is clearly higher at the surface; while the situation is reversed at night - suggesting the diel vertical migration of dinoflagellates.

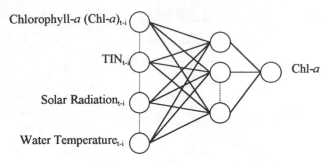

Fig. 7: Neural network structure for algal bloom prediction

4. ALGAL BLOOM MODELLING BY NEURAL NETWORK

Algal dynamics in Hong Kong has previously been studied using process-based water quality models which incorporate phytoplankton growth kinetics, nutrient recycling, and tidal circulation (e.g. Ambrose 1993; Thomann and Mueller 1987; Lee *et al.* 1991). While insights about the ecological system can often be gained from these models, it is recognized the predictions may not necessarily be better than "black-box" data-driven methods such as neural network techniques (e.g. Whitehead *et al.* 1997). One significant limitation of the process-based models is the significant uncertainty of many kinetic coefficients (e.g. algal growth rates) adopted in the water quality model. An initial study of using artificial neural networks (ANN) to predict algal blooms has been carried out. As a first step, while the telemetry system is being evaluated and constantly updated, the analysis is first performed using the comprehensive biweekly water quality data of inner Tolo Harbour, a semi-enclosed bay close to Kat O. The weakly flushed Tolo Harbour is well-known for frequent occurrences of algal blooms and red tides (Hodgkiss and Ho 1991), with a peak at 85 outbreaks in 1988, leveling off to current frequencies of around 20 per year. This data set at inner Tolo (TM3 monitoring station) serves as a natural starting point for testing data assimilation techniques.

4.1 Network structure

The application of neural network techniques to study algal blooms has started to appear only recently (e.g. Recknagel *et al.* 1997). However, the great majority of the work are concerned with limnological or riverine applications. The study of its use for predicting algal blooms in coastal waters has apparently not been reported. Fig.7 illustrates the adopted 3-layer network structure with one input, one hidden and one output layer. The time lags are denoted by the subscripts (t-i). Based on previous field and water quality modelling, the chosen input nodes are chlorophyll-a, total inorganic nitrogen (TIN), solar radiation, and water temperature. As the objective is to study the use of ANN for short-term forecasting, and reliable water quality measurements (especially nutrient and chlorophyll-a) are usually not available more frequently than one-week intervals, all the input nodes are time-lagged. The chlorophyll-a and TIN input nodes refer to measurements from (t-7) to (t-13) days, while for solar radiation time lags of 1-3 days.

Table 1: Root mean square error (RMSE) of different network structures.

Scenarios	Inputs	RMSE	
		Training	Testing
1	chl-a	2.75	2.12
2	chl-a, TIN	2.36	2.37
3	chl-a, TIN, Temperature	2.14	3.06
4	Chl-a, TIN, Solar Radiation	2.19	2.61
5	Chl-a, TIN, Solar Radiation, Temperature	2.17	3.32

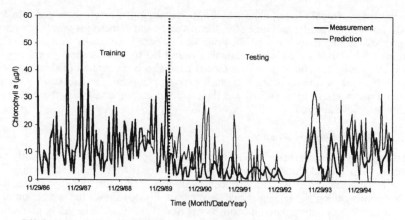

Fig.8: a) Prediction of chlorophyll-*a* in inner Tolo Harbour by neural network model (based on biweekly data), scenario 2.

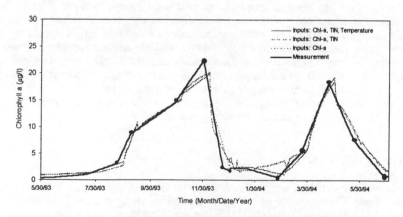

Fig.8: b) Prediction of chlorophyll-*a* in inner Tolo Harbour by ANN using daily interpolated data for network training.

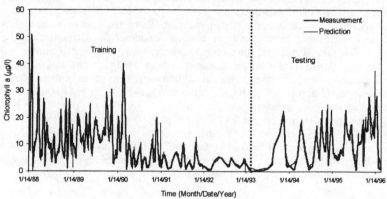

Fig.9: Phase error of ANN prediction in short term algal bloom prediction.

667

The main indicator of algal bloom, chlorophyll-a, is chosen as the network output. In order to apply ANN learning with different time lags, the biweekly and sometimes monthly data is first linearly interpolated into daily data; all the water quality data is vertically averaged. The data from 1988 to 1993 is used for training, while that from 1993 to 1996 is used for testing. Other networks with simpler structures (less parameters) have also been studied (Table 1). For each scenario, the network is trained using a gradient descent algorithm for backward propagation, while the root mean square error (RSME) of the predicted output is taken as the error measure for the steepest gradient search.

4.2 Results of Neural Network Prediction

The general technique was first tried on biweekly data (without interpolation into daily values). Fig.8a) shows the predicted chlorophyll-a concentration from 1986-1994 for scenario 2. For this case the network prediction is in fair agreement with data during the training period, but performs rather poorly during the testing period. In contrast, Fig.8b) shows the prediction of a network trained using daily-interpolated measurements from 1988-1996. The ANN prediction follows the observed trends quite well for both the training and testing periods, and the agreement is quite good - especially considering only Chlorophyll-a and TIN are used as input nodes. As shown in Table 1, the RMSE for different scenarios are not very different from each other. In fact, similar quality of prediction can be achieved using simply the time-lagged chlorophyll-a concentrations - suggesting almost an auto-regressive type of process. This somewhat surprising result is remarkably different from the ANN simulations that have been performed thus far (e.g. Recknagel 1997) which tend to include many more input nodes. The insensitivity of the network to inclusion of TIN may also reflect the fact that nutrients are typically not limiting in the eutrophic bay; this also throws doubt on the advantage of deploying expensive automatic nutrient analysers (for ammonia and nitrate nitrogen) in red tide warning systems in these coastal waters.

Despite the apparent agreement with data on the time scale used in Fig.8, on closer examination the predicted value has a phase error of about 7 days. In Fig.9 the predictions of three scenarios during the testing period are shown for one year, 1993-94. It can be seen the use of Chl-a or Chl-a and TIN gives similar results. Various alternative approaches have also been studied; however all predictions are characterized by this one week time lag which cannot be removed. This implies that until the ANN can be improved by reliable data of higher frequency, the neural network approach cannot be relied upon to give short-term algal bloom predictions.

5. CONCLUDING REMARKS

A telemetry system has been developed to monitor algal blooms and red tides in the northeastern waters of Hong Kong; a similar field station has also been developed in the southern waters of Hong Kong (Lo Dick Wan, Lamma Island) with quite different patterns of water quality changes. The ability of the system to detect algal blooms and red tides is demonstrated. The analysis of a comprehensive set of water quality data in a neighbouring eutrophic bay suggests that the use of artificial neural networks can simulate long term trends of algal biomass reasonably well, but fail to give reliable short term forecasts. In the next phase, the field monitoring system will be further strengthened by a recently installed acoustic current meter and interaction with hydrodynamic and data-driven modelling.

ACKNOWLEDGEMENTS

This study was supported by a Hong Kong Research Grants Council Central Allocation Group Research Project. The assistance of T.T. Chuang and Ironside Lam in the field work and the assistance of the Hong Kong Environmental Protection Department in providing the water quality data are deeply appreciated. The support of the Hong Kong Agriculture, Fisheries and Conservation Department in the fieldwork is gratefully acknowledged.

REFERENCES

Ambrose, R.B. *et al.* (1993). *The Water Quality Analysis Simulation Program, WASP5*. Environmental Research Laboratory, USEPA, Athens, Geogia.

Anderson, D.M. *et al.* (1982). Distribution of the toxic dinoflagellate *Gonyaulax tamarensis* in the Southern New England Region, *Estuarine, Coastal and ShelfScience*, 14, 447-458.

Cullen, J.J., Ciotti, A.M, Davis, R.F. and Lewis, M.R. (1997). "Optical detection and assessment of algal blooms", *Limnology and Oceanography*, 42(5), pp.1223-1239.

Dickman, M.D. (1998). Hong Kong's worst red tide, Proc. Int. Symp. Env. Hydraulics, (Ed. Lee J.H.W. *et al.*), Balkema, pp.641-645.

Hodgkiss, I.J. and Ho, K.C. (1991). Red tides in sub-tropical waters: an overview of their occurrence, *Asian marine biology*, 8, 5-23.

Lee, J.H.W., Wu, R.S.S., Cheung, Y.K., and Wong, P.P.S. (1991). Dissolved oxygen variations in marine fish culture zone. *J. of Env. Engr., ASCE*, 117, 799-815.

Lee,J.H.W., Wu, R.S.S., and Cheung, Y.K. (1991). Forecasting of dissolved oxygen in marine fish culture zone. *J. of Env. Engr. ASCE*, 117, 816-833.

Lee, H.S. and Lee, J.H.W. (1995). Continuous monitoring of short term dissolved oxygen and algal dynamics. *Water Research*, 29, 2789-2796.

Parsons, T.R., Takahashi, M., Hargrave, B. (1984). *Biological oceanographic processes*, Pergamon.

Recknagel, F., French, M., Harkonen, P. and Yabunaka, K. 1997. Artificial neural network approach for modelling and prediction of algal blooms. *Ecological Modelling*, 96, 11-28.

Thomann, R.V. and Mueller, J.A. (1987). *Principles of surface Water quality modeling and control*, Harper and Row, New York.

Whitehead, P.G. and G.M. Hornberger (1984). "Modelling algal behaviour in the River Thames", *Water Research*, 18, 945-953.

Stochastic Hydraulics 2000, Wang & Hu (eds) © 2000 Taylor & Francis, ISBN 90 5809 166 X

Invited lecture: Analysis of instream habitat quality – Preference functions
and fuzzy models

Klaus Jorde, Matthias Schneider & Frank Zöllner
Institute of Hydraulic Engineering, University of Stuttgart, Germany

ABSTRACT: To improve the performance of the habitat simulation model CASIMIR, fuzzy rules are applied to describe habitat preferences of fish species during their different life stages. Fuzzy rules turned out to be more effective to develop because they can be derived from experts' knowledge that is often readily available. Common preference function based approaches often consider parameters seperated from each other or in combination with one or two other parameters. Opposite to that, fuzzy rules allow to include large numbers of combinations of physical paremeters into habitat simulation tools and it is easy to include more parameters if they turn out to be relevant. The results gained with fuzzy rules clearly differ from those gained with traditional preference functions. First comparisons with actual fish findings in several rivers show a higher correlation with habitat prediction for fuzzy based simulations than for those based on preference functions. Methods and results from a river in Germany are presented.

Key words: Habitat simulation, fish habitat, spatial heterogeneity, instream flow, simulation model, river ecology, fish habitat, CASIMIR, preference function, fuzzy model.

1. INTRODUCTION

Rivers and streams throughout the world have undergone and are still undergoing dramatic changes of their physical constitution and their chemical, physiographical and biological state. In highly developped regions, such as central Europe, nearly all rivers have been trained, channelized, straightened, impounded, diverted, etc. Reasons for these measures are human needs and activities like agriculture and irrigation, urbanization of river corridors, hydropower use, navigation, flood protection, and others. Additionally, the hydrological regime of many rivers is influenced by human activitites. These influences can be locally restricted, like in the case of diversion type run-off river hydropower plants. They can also cover large parts of a catchment, like in the case of large alpine storage reservoirs, that collect water from snow melt during spring and summer thus retaining natural floods, and release that water during the next winter due to peak power production during natural low flow season. In this case the flow regime is completely changed in terms of seasonality over large sections of the affected river system. The consequence of these complex physical changes in river systems is a degradation of biological integrity of running waters all over the world.

2. RIVERS AND THEIR HABITATS

The typical course of a river can be divided into three major habitats, all of which are affected by human activities that influence the natural discharge regime or the morphological structure of the river system. The first is the aquatic zone, where fish and macrophytes are living. The second is the river bottom with the hyporheic interstitial where benthic species are responsible for most of the biologic processes, especially in rhitral and epipotamal rivers..The third habitat includes the riparian zone and the floodplain in between the lowest and the highest water table.

The various affects of human activities on these habitats and thus the ecosystem of running waters can be essentially divided into two stages. First, there are changes in the abiotic factors which include the hydrological regime and thus the temporal variability, hydraulic factors such as water depths, flow velocities, shear stresses and turbulence etc., the sediment regime, landscape aesthetics, the ground water situation and certain hydrophysiographic parameters such as the diurnal and seasonal temperature and oxygen regimes. Second, changes in the composition of river bound taxa and communities are resulting. These include the composition of fish and invertebrate biocenoses, age class distributions, dynamics of population and settlement, vegetation communities, food chains, trophic status etc.

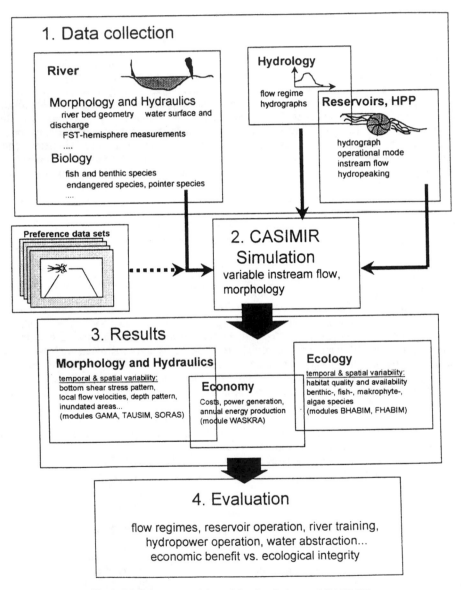

Fig.1: Modular composition of the simulation tool CASIMIR

3. THE SIMULATION MODEL CASIMIR

CASIMIR (Computer Aided Simulation Model for Instream Flow Regulations) is a Toolbox for habitat simulation in rivers. It has been developed at the Stuttgart Institute of Hydraulic Engineering since the beginning of the 1990's (fig. 1). The simulation model comprises modules with individual computing programmes which can be combined to suit a particular case in question. Three main areas of simulation are implied in the current version. The flow regime module includes programs to simulate hydro power plants, including energy generation, reservoir operation and instream flow regulations. The river bed module calculates statistical distributions of near bed flow forces derived from field measurements. The aquatic zone module simulates and analyses hydraulic and morphologic patterns. The latter two modules are complemented with biological components that contain data about habitat preferences and simulation tools for habitat quality and availability.

This modular structure has the advantage that further parameters can be adopted at any time, should they prove to be ecologically relevant. Thus very differing data sets can be evaluated and results gained. The application of this model was originally directed to the optimizing of instream flow regulations. In the meantime CASIMIR has been applied to examine the consequences of various types of human activities or natural processes in river systems.

The basic assumption for ecological evaluation is that hydraulic habitat can be described as a combination of various physical parameters. However it is obvious, that this assumption neglects many important habitat criteria such as food availability, chemical or organic pollution, predator-prey interactions etc. Once these physical patterns have been established they are subsequently compared with the relevant requirements or preferences of animal or plant species or communities. The complexity of the relevant dimensions of influence on the available habitat, particularly when temporal variability is included, calls for computer aided simulation models for its assessment. Whereas there is a long tradition in habitat modelling in North America, especially considering the PHABSIM model (Bovee, 1982), there are now groups all over the world working in this field.

3.1 River bottom

The first stage of development of CASIMIR served to quantify and simulate the hydraulic habitat at the river bottom which is defined mostly by the pattern of substratum and near bottom hydraulic forcees. As measurements of local velocities at the river bottom or local shear stress are difficult and laborious, so called FST-hemispheres have been developped (Statzner & Müller, 1989). FST-hemispheres consist of a set of 24 hemispheres with equal shape (diameter = 7 cm), but different specific weight. In a try-and-error procedure they are placed on a specific ground plate at the river bottom. The heaviest hemisphere that is not taken away from the current serves as a measure for flow force at the river bottom. Sets of 100 measurements at different discharges are taken and the data are transfered into physical data such as bottom shear stresses (Jorde 1997), hemisphere densities or density differences, which is more effective because it requires no calibration. Lognormal or Weibull distributions are then used to formulate the distribution pattern of local hydraulic forces at the river bottom as a function of the discharge.

The combination of these patterns with discharge time series of a river reach and subsequently with the habitat requirements of certain selected benthic organisms permits the incorporation of spatial heterogeneity and temporal dynamics, particularly relevant ecological factors. Here the interfaces to biology are standardized preference curves of benthic organisms (Jorde & Bratrich, 1998).

Preference curves are now available for many species, however their transferability to different rivers has been proved for only about 50 species. This valuable database is being constantly expanded within the framework of ongoing projects.

3.2 Aquatic zone

Further modules within CASIMIR comprise the aquatic volume as habitat for fish and macrophytes. The traditional approach for a qualification of fish habitats are local depths, depth averaged velocities or so called „nose"-velocities (referring to the nose of the fish, Milhous 1999), and substratum, as applied for many years in PHABSIM studies (Jorde & Schneider, 1998).

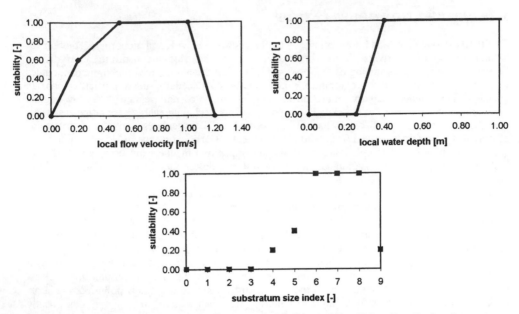

Fig. 2: Depth, velocity and substratum preferences of adult Brown Trout (*Salmo Trutta*), river Lenne

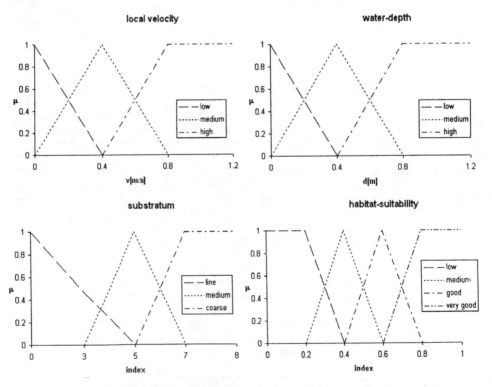

Fig.3: Fuzzy sets describing degrees of membership (membership functions) for input and output

Each of the criteria was treated seperately and the final habitat quality of a certain area was defined by multiplying single criteria results or taking the minimum or average of single criteria results. Preference data as shown in fig.2 are obtained by snorkeling and/or electrofishing or are already available from comparable sites or ecoregions. Newer approaches for preference functions are based on multivariate statistics (Parasiewicz, 1998) and results reveal a higher correlation between fish findings and predicted habitat quality in some cases. This approach takes into account that physical parameters cannot be considered as isolated from each other. However, preference functions generally neglect the plenty of possible physical parameter combinations. Another lack of preference based methods is that spatial connectivity and the networking of habitats is not considered. Furthermore, so called „Individual fish based models" and bioenergetic models (Ludlow & Hardy 1996) were developped and successfully applied for large drift feeding salmonids.

Another new approach to evaluate habitat qualities with CASIMIR are fuzzy models (Schneider & Peter 1999). Fuzzy models allow to work with imprecise or „fuzzy" information. They have the significant advantage that expert knowledge readily available from experienced fish biologists can easily be transferred into preference data sets by setting up check-lists with possible combinations of relevant criteria combinations and let experts define if habitat quality is e.g. good, medium or unsuitable. Table 1 shows an example for combinations of three of the relevant parameters for adult Brown Trout (*Salmo Trutta*) as defined by fish experts for the river Lenne in Germany.

Table.1: Fuzzy rules for habitat suitability for adult Brown Trout (Salmo trutta)

Brown trout *(salmo trutta)*			
velocity	depth	substratum	suitability
H	H	C	+
H	H	M	0
H	H	F	0
H	M	C	+
H	M	M	0
H	M	F	0
H	L	C	0
H	L	M	0
H	L	F	0
M	H	C	++
M	H	M	0
M	H	F	+
M	M	C	++
M	M	M	0
M	M	F	0
M	L	C	0
M	L	M	0
M	L	F	0
L	H	C	++
L	H	M	0
L	H	F	0
L	M	C	++
L	M	M	0
L	M	F	0
L	L	C	0
L	L	M	0
L	L	F	0
H = high	C = coarse	++ = good	
M = medium	M = medium	+ = medium	
L = low	F = fine	0 = unsuitable	

The parameters that are contained in the criteria combinations are linked to physical numbers by sets of fuzzy rules as shown in fig. 3. These rules define the relation between input or explanatory vaiables and consequence, in our case the output or habitat suitability. Not only the input but also the output (habitat suitability) is „fuzzy". The output of the fuzzy model are degrees of fullfillment of habitat suitability. These degrees of fullfillment are used for a „defuzzyfication". Thus the result of the whole fuzzy set of rules is transformed back into a standardized number to describe habitat quality defined between 0 and 1. Mathematical algorithms for fuzzy rule based calculations are described in (Bárdossy & Duckstein, 1995).

It is well known that connectivity is not only a question of longitudinal or lateral direction but connectivity describes a network of microhabitat types. That means for example that good feeding grounds might possibly only be used when suitable refuge is close. Refuges could be for example certain cover types, turbulences or deep pools. The suitability of a location is therefore defined not only by the situation on the spot but also by the situation further or closer around the spot. Such information can be included in complete sets of fuzzy rules and it is easy to consider

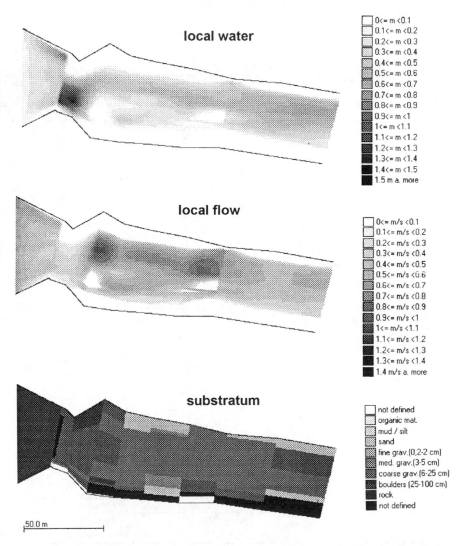

Fig.4: Depth, velocity and substratum patterns in a 200 m reach of the river Lenne at a discharge of 4000 l/s

additional parameters if they turn out to be relevant. Field measurements for the hydraulic and morphologic simulation consist of cross section survey data with additional monitoring of substratum and choriotopes including accessabilities, types and availability of different cover categories, shadowing, pool types, single rocks and bank structure. The data are processed within CASIMIR and a 3-dimensional digital river bed model is generated. Additional informations are stored as index figures with each bottom element. Water tables can be either measured, which is often more effective and accurate especially under low flow conditions, or calculated from a 1-D hydraulic model. CASIMIR utilizes simple rule based methods to derive local depth averaged velocities from mean cross section velocities and cross section geometry. Fig. 4 shows an example of local flow velocities, water depths and substratum of a investigation site at the river Lenne, a tributary to Ruhr and Rhine in Germany, which is directly influenced by water diversions due to hydropower production and additional large reservoir operation upstream.

The results of this abiotic modelling are consequently combined with the preference functions or with fuzzy rules that describe habitat quality. A comparison between both simulation results is given in fig. 5. It is evident that results differ significantly in qualities as well as in locations. First comparisons between predicted habitat quality and actual fish findings are promising for the fuzzy approach.

To include temporal information, hydrographs of the river reach are integrated into the model to indicate seasonal aspects that might be of major importance, such as spawning periods of certain species. This allows to adapt flow regulations or limitate abstraction of water to fullfill certain criteria and thus ensure ecological sustainability of a river reach. Fig. 6 shows the comparative results of integrated habitat suitability for Brown Trout with increasing discharges, for different life stages and with preference versus fuzzy approach in the case of adult Brown Trout.

4 EVALUATION

Different approaches for an evaluation of the simulation results are possible depending on the scope of the study. The simplest approach is based on a comparison of hydraulic and morphological patterns with those of unaffected reference sites. No biologic information is needed then. If biological information is to be considered additionally, there are several possibilities

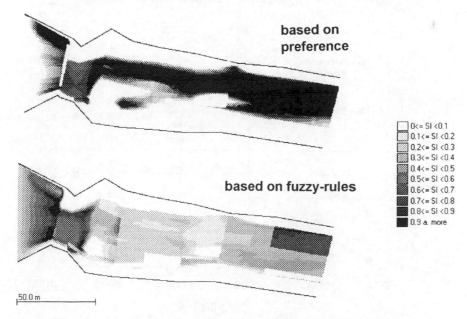

based on preference

based on fuzzy-rules

0<= SI <0.1
0.1<= SI <0.2
0.2<= SI <0.3
0.3<= SI <0.4
0.4<= SI <0.5
0.5<= SI <0.6
0.6<= SI <0.7
0.7<= SI <0.8
0.8<= SI <0.9
0.9 a. more

50.0 m

Fig.5: Comparative habitat simulation results for adult brown trout in a 200 m reach of the river Lenne at a discharge of 4000 l/s

which can be related to species composition, certain target species like the largest potentially occurring predator, or endangered species. It is obvious that an improvement in habitat quality for one species usually means a decrease in habitat quality for other species. Keeping in mind that natural rivers are often inhabitated by many fish species in different life stages with different habitat needs it is clear that evaluation of complex results from habitat simulations (see fig. 6)

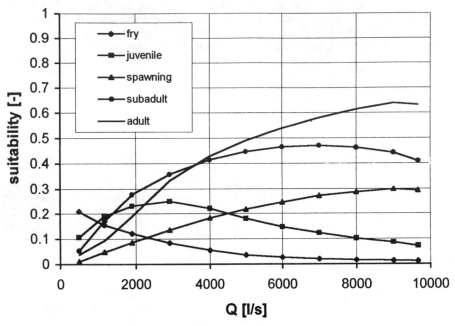

Fig.6a: Integral habitat suitabilities in a 1000 m reach of the river Lenne at 4000 l/s for Brown Trout at different life stages (left) and preference vs. Fuzzy approach for adults (right)

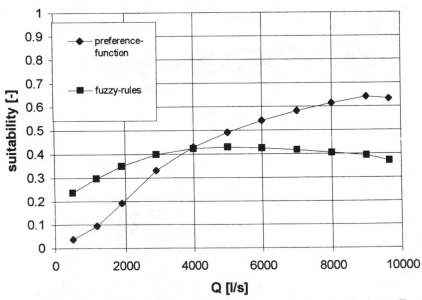

Fig.6b: Integral habitat suitabilities in a 1000 m reach of the river Lenne at 4000 l/s for Brown Trout at different life stages (left) and preference vs. Fuzzy approach for adults (right)

needs experience and clearly defined objectives. Complex multi species evaluations of habitat simulation results have been developped for some cases (Wentzel,1999).

5. VALIDATION OF MODELS

The validation of habitat simulation models is more difficult than the validation of the hydraulic or morphologic part which can easily be verified by field measurements. The problems with biologic information, however, is twofold: First, it is well known, that only a certain share of suitable hydraulic habitat is actually occupied by individual animals. So the availability of suitable hydraulic habitat is necessary, but not sufficient as a prerequisite. Parameters that are not considered in the model investigations, e.g. social pressure or lack of food, are probably relevant to that. Second, it is time and cost intensive, to monitor local species abundance over a longer range of time. At least a period of one lifecycle of target species is necessary for a validation of the simulation results. Additionally, natural abundance of species varies strongly with time and space, even without any human influence. The time scale includes daily and seasonal migrations of highly mobile species, such as fish. Therefore, indexes of biological response caused by habitat improvements or degradation must be selected very carefully in order to yield meaningful results (Karr & Chu, 1999).

6. CONCLUSION

Habitat simulation models provide valuable information to improve the design or operation manner of hydraulic structures or river engineering projects and therefore support the conservation and protection of biological integrity of running water ecosystems. The application of fuzzy logic approaches seems to be a promising and versatile additional tool to be included in simulation models such as CASIMIR.

REFERENCES

BARDOSSY, A. & L. DUCKSTEIN (1995): Fuzzy Rule-Based Modeling with Applications to Geophysical, Biological and Engineering Systems. CRC Press, Boca Raton.

BOVEE, K.D. (1982): A guide to stream habitat analysis using the instream flow incremental methodology. - US Fish and Wildlife Service, Instream Flow Information Paper No. 12, US Fish and Wildlife Servivce Biologic Report 82, 248 pp.

GIESECKE, J., M. SCHNEIDER & K. JORDE (1999): Analysis of Minimum Flow Stretches Based on the Simulation Model CASIMIR, Proceedings 28th IAHR Congress, 22.-27. Aug. 1999, 9 Seiten auf CD-ROM, Graz.

GIESECKE, J. & K. JORDE (1998): Simulation and Assessment of Hydraulic Habiat in Rivers. Proceedings "Modelling, Testing and Monitoring for Hydro Powerplants - III, Aix-en-Provence 1998, The International Journal on Hydropower & Dams, Sutton, UK, pp. 71-82.

JORDE, K. & C. BRATRICH (1998): Influence of River Bed Morphology and Flow Regulations in Diverted Streams on Bottom Shear Stress Pattern and Hydraulic Habitat. In: Bretschko G. & Helešic J. (Eds.), Advances in River Bottom Ecology IV, Backhuys Publishers, 47-63.

JORDE, K. & SCHNEIDER, M. (1998): Einsatz des Simulationsmodells PHABSIM zur Festlegung von Mindestwasserregelungen, Wasser + Boden 50, Heft 4, S. 45- 49.

JORDE, K. (1997): Bottom Shears Stress Pattern and its Ecological Impact. International Journal of Sediment Transport (12), No. 3, Special Issue of Morphological Dynamics and Unsteady Sediment Transport, IRTCES, pp. 369-378

KARR, J.R. & E.W. CHU (1999): Restoring Life in Running Waters – Better Biological Monitoring, Island Press, Washington D.C.

LUDLOW, J.A. & T.B. HARDY (1996): Comparative Evaluation of Suitability Curve Based Habitat Modeling and a Mechanistic Based Bioenergetic Model Using 2-Dimensional Hydraulic Simulations in a Natural River System. Proc. 2nd Int. Symp. Ecohydraulics2000, Quebec City, June 1996, Vol B, p. 519-530.

MILHOUS, R.T. (1999): Nose Velocities in Physical Habitat Simulation. Proceedings 28th IAHR Congress, 22.-27. Aug. 1999, 6 p.on CD-ROM, Graz.

PARASIEWICZ, P. (1998): Computerunterstützte Methoden zur Untersuchung biozönotisch wirksamer hydraulisch-morphologischer Parameter der Fließgewässer. Abt. Hydrobiologie, Universität für Bodenkultur, Vienna.

SCHNEIDER, M. & A. PETER (1999): Ökostrom: Field Study and Use of the Simulation Model CASIMIR for Determination of Fish Habitat in River Brenno. Proc. 3rd Int. Symp. Ecohydraulics, 12.-16. July 1999, 3 pages extended abstract on CD-Rom, Salt Lake City.

STATZNER, B. & R. MÜLLER (1989): Standard hemispheres as indicators of flow characteristics in lothic benthos research. - Freshwater Biology 21: 445-459.

TRUFFER, B., C. BRATRICH & K. JORDE (1999): Ökostrom. Neue Perspektiven der Wasserkraftnutzung. Wasserwirtschaft, 89, Heft 10, S. 488-495.

WENTZEL, M.W. & MATSUMOTO, J.(1999): A genetic algorithm based decision support system for instream flow requirements. Proc. 3rd Int. Symp. Ecohydraulics, 12.-16. July 1999, 6 pages on CD-Rom, Salt Lake City.

Stochastic Hydraulics 2000, Wang & Hu (eds) © 2000 Taylor & Francis, ISBN 90 5809 166 X

Low frequency relationships between Korean precipitation and atmospheric variability

Dae-Hyeong Park & Sung-Hwan Hwang
Water Research Laboratory, Department of Civil Engineering, University of Seoul, Korea

Young-Il Moon
Department of Civil Engineering, University of Seoul, Korea

ABSTRACT: This paper demonstrates the relationships between the time variability of the precipitation data in Seoul, Korea and the Southern Oscillation Index(SOI). Connections between this index and the variability of precipitation data in Seoul are demonstrated. The ability to distinguish a low frequency relationship between the precipitation in Seoul and the atmospheric circulation pattern is important for understanding of the climatic response to the precipitation pattern in Korea. We focus on interannual and interdecadal time scales by using Multi-channel Singular Spectral Analysis(M-SSA). We used M-SSA to identify coherent space-time patterns of low frequency bands between the Seoul precipitation and the Southern Oscillation Index, and cross correlation test was examined to find the lag time between two parameters.

Key words: Seoul precipitation, Southern oscillation index, Multi-channel singular spectrum Analysis

1. INTRODUCTION

The purpose of this paper is to understand the role of atmospheric variability in the time behavior of precipitation pattern in Seoul, Korea, and to examine low frequency relationships between Seoul precipitation and Southern Oscillation Index(SOI). Global climatic variations, such as El Niño and La Niña, lead to hydrologic imbalances with severe damages worldwide. During the last decade, due to sudden climatic imbalances, South Korea had severe flood damages in the year of 1991, 1995, 1996, 1998, and 1999. The total damage cost was 5 billion dollars during that period. Especially, the 1998 flood damage was seemed to have strong relationships with the 1997-98 El Niño effect. The recent studies between Korea climate patterns and El Niño events revealed that the cross correlation coefficient between them was small(Lee, 1998 ; Lee and Kim, 1999). Sin(1998, 1999) investigated the El Niño effects in Korea, and concluded that El Niño events produce lower than average precipitation amount with one year lag time, and has high correlation in the southern part, relative to northern part, of Korea. Lee and Kim(1999) studied the relationships between El Niño and run-off amount in a fixed area (dam), and found the run-off of the next year decreases if an El Niño events are happened. Mechanistic explanations for the low-frequency behavior are being developed through a study of the nature of ocean-atmospheric interactions, specifically in the North Pacific, North Atlantic, and the Tropics(Moon and Lall, 1996). Moon and Lall(1996) presented the low frequency relationships between the Great Salt Lake volume in the United States and several atmospheric flow indices including SOI to understand the impact of climate variability and underlying dynamics of the hydrologic system. Klein and Bloom(1987), Kiladis and Diaz(1989), Cayan and Peterson(1989), and among numerous others investigated and found large scale atmospheric and oceanic conditions exert considerable influences on the low frequency patterns of North America. Also, Lall and Mann(1995) analyze the time series of Great Salt Lake monthly volume change, monthly precipitation, temperature, and stream flow by using Singular Spectrum Analysis(SSA) and Multitaper method(MTM), and found signals which are consistent across the time series. Keppenne and Ghil(1990) used SSA, MTM, and Maximum-entropy method to analyze and predict El Niño event. Keppenne and Ghil(1992) also used SSA to filter out variability unrelated to the SOI, and Maximum-entropy method to ENSO(El Niño and Southern Oscillation) forecast, and predicted 1993-94 La Niña event based on data through Feb. 1992.

In this paper, we applied the Multi-channel Singular Spectrum Analysis(M-SSA) between Seoul precipitation data and the Southern Oscillation Index(SOI). The analysis shows common modes in 0.5, 1, 3.3, 5, and 15 year frequency bands.

2. DATA SETS

Two times series are analyzed here. They are monthly Southern Oscillation Index(SOI) from September 1932 to December 1998, and monthly precipitation data in Seoul which are available after 1908.

Korean peninsula is located approximately 33° 43°N and 124° 132°E which is in the far east of Northern Hemisphere, and the climate depends largely on the Asian Monsoon. By the Monsoon effect, the precipitation pattern in Korea can be characterized as wet summer, and dry winter. Thus, 67% of annual precipitation amount occurs during the rainy summer season from June to September. The average of annual   precipitation amount in Korea is 1,274 mm per year.

Seoul is located in the middle-west part of Korea, 1,327 mm annual mean precipitation, and summer precipitation from June to September accounts 71% of annual precipitation amount.

The Southern Oscillation Index(SOI) series is consist of time series of normalized monthly mean differences in sea-level pressure(SLP) at Tahiti(approximately 150W, 18S) and Darwin (approximately 130E, 13S).

SOI = SLP(Tahiti) - SLP(Darwin)

This paper used the SOI data provided by NOAA(National Oceanic and Atmospheric Administration) in the United States, and Seoul precipitation data provided by Korea Meteorological Administration.

3. MULTI-CHANNEL SINGULAR SPECTRUM ANALYSIS(M-SSA)

Multi-channel SSA(M-SSA) is the Singular Spectrum Analysis(SSA) with dimensions. The SSA is a variant of Principal Component Analysis(PCA) of the delay coordinates for a time series. The basic concept is to filter out nonharmonic element from the original data, and thus helps the easy way to find out a periodicity and trends of the data(Moon and Lall, 1996). By using this method, original time series is divided into several harmonic elements, and after analyzing the elements, choose the dominant elements to figure out the periodicity and trends. The resulted trends can be a helpful tool for forecasting upcoming events. This method was introduced into biological oceanography by Colebrook(1978), into nonlinear dynamics by Broomhead and King(1986), Vautard and Ghil(1989). Then in the 1990's, Vautard et al.(1992), Keppenne and Ghil(1992), Moon and Lall(1996), and Moon et al.(1999) used this method to study the ENSO effect. SSA is a diagnostic method related to empirical orthogonal function(EOF) analysis, but applied in the time domain rather than in the spatial domain. If the given time series is x_i $(1 \leq i \leq N)$, then it can be expressed as follows,

$$x_{i+(j-1)\tau} = \sum_{k=1}^{M} a_i^k E_j^k, \quad 1 \leq j \leq M, \quad 1 \leq i \leq N - M + 1 \tag{1}$$

Here, a_i^k is a projection coefficient called Principal Component(PC), E^k is a Empirical Orthogonal Function(EOF, $1 \leq k \leq M$), M is a Mth dimension(role of smoothing window), and τ is a sampling rate. Thus, kth PC for the kth EOF can be defined as orthogonal projection of time series.

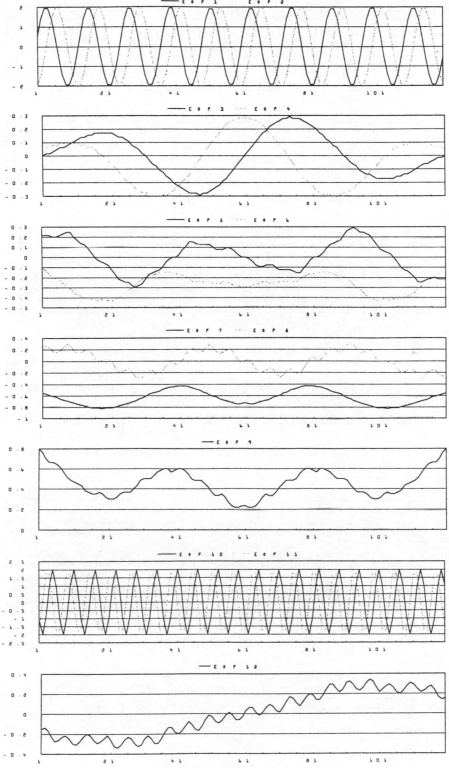

Fig.1: Resulted M-SSA EOFs 1-12

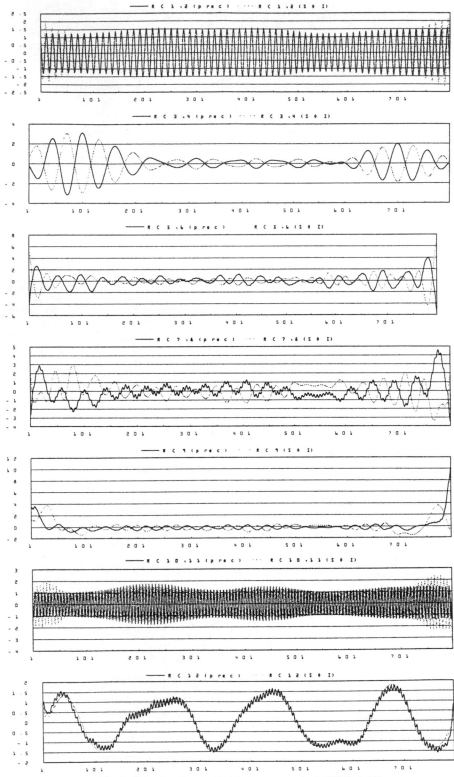

Fig.2: Paired and reconstructed components 1-12(RC 1-12)

$$a_i^k = \sum_{k=1}^{M} x_{i+(j-1)\tau+1} E_j^k, \qquad 0 \leq i \leq N-M \qquad (2)$$

Hence, nonharmonic elements can be separated according to the portions that can express the variances, and original data can be the Reconstructed Componenet(RC) as follows. Summed M number of RCs are the original time series.

$$(R_A x)_i = \frac{1}{i} \sum_{j=1}^{i} \sum_{k \in A} a_{i-j}^k E_j^k, \qquad 1 \leq i \leq M-1 \qquad (3)$$

$$(R_A x)_i = \frac{1}{M} \sum_{j=1}^{M} \sum_{k \in A} a_{i-j}^k E_j^k, \qquad M \leq i \leq N-M+1 \qquad (4)$$

$$(R_A x)_i = \frac{1}{N-i+1} \sum_{j=i-N+M}^{M} \sum_{k \in A} a_{i-j}^k E_j^k, \qquad N-M+2 \leq i \leq N \qquad (5)$$

M-SSA(Moon and Lall, 1996) is a Principal Component Analysis in the space and time domains of delay coordinates for several series, often from several sites. This is similar to extended EOF analysis(Weare and Nasstrom, 1982). M-SSA identifies common and recurring modes of spatial and temporal variation based on the selected embedding dimension M. Here, the "spatial" components are the different time series that are being correlated. For M-SSA, the principal axes(the eigenvectors) of a sequence of M-dimension vectors through an orthogonal basis($E^k, 1 \leq k \leq M$) for SSA

$$X_{ij} = \sum_{k=1}^{M} a_i^k E_j^k, \qquad 1 \leq j \leq M \qquad (6)$$

becomes

$$X_{ij} = \sum_{k=1}^{M \circ L} a_i^k E_j^k, \qquad 1 \leq i \leq N-M+1, \quad 1 \leq j \leq M \circ L \qquad (7)$$

The kth basis vectors is an eigenvector of the block-Toeplitz matrix T_x containing the cross-covariance matrices of the L different channels at lags 0 to M-1. The kth (PCs) and (RCs) are similar to SSA equations (2)-(5) if we consider an EOF length of $M \circ L$.

4. RESULTS

Here, we used M-SSA to identify coherent space-time patterns of low frequency bands between the Seoul precipitation and the Southern Oscillation Index. The embedding dimension 120, and sampling rate, τ, 1 were used. Too small an M leads to smearing of the spectrum and too large an M to splitting of peaks. After figure out the eigenvectors, pairs of nearly equal eigenvectors were reconstructed to identify the oscillatory modes. The EOFs that represent the variances from the data has been identified and summarized in Table 1. The first twelve EOFs were 47.9% of variance. Remaining EOFs were regarded as noises and were not considered.

Table.1: EOFs portion to the total variances

EOFs	1-2	3-4	5-6	7-8	9	10-11	12	13-120
Percent(%)	18.7	8.4	6.4	5.3	2.4	4.6	2.1	52.1

Pairs of nearly equal eigenvectors(empirical orthogonal functions : EOFs) identify oscillatory modes for given embedding dimension, and normalized EOFs are presented in Fig. 1. Vautard and Ghil(1989) observed that the near-equality of a pair of eigenelements were associated with oscillatory phenomena. In Fig. 1, EOFs 1-2(18.7% of total variance) represent annual cycle, while EOFs 3-4(4.26% and 4.17% of total variance) represent 5 year cycles. The EOFs 5-6(3.3% and 3.1% of total variance), EOFs 7-8(2.7% and 2.6% of total variance), and EOF 9(2.4% of total variance) represent 3.3 year cycle. The EOFs 10-11(2.3 and 2.3% of total variance) appears to have semiannual cycle while EOF 12(2.1% of total variance) shows secular mode.

The M-SSA reconstructed components(RCs) are shown in Fig. 2. Vautard et al.(1992) define reconstructed components(RCs) as optimal linear combinations of the corresponding principal components. As similar to Fig. 1, RCs are showing same cycle to corresponding EOFs. That is, 1 year cycle for RC 1-2, 3.3 year cycle for RC 5-6, 7-8, 9, and 5 year cycle for RC 3-4, semiannual cycle for RC 10-11, and secular mode for RC 12. On the other hand, the notice is that every RCs, except RC 1-2 and 12, are showing negative connections between precipitation data and SOI. That can be interpreted as when El Niño happens(SOI with negative value), precipitation in Korea increases based on each low frequency bands.

Finally, the cross correlation test was examined to know the existing "lag time" between SOI and Seoul precipitation. The highest correlation coefficient is approximately -0.72 when lag times is 7 months. That is, the precipitation amount, based on low frequency bands, increases with 7 months lag time if El Niño event happens.

5. CONCLUSION

We investigated to find the low frequency connections between Seoul precipitation data and Southern Oscillation Index(SOI) by using Multi-channel Sigular Spectrum Analysis(M-SSA) method. This analysis revealed common modes in 0.5, 1, 3.3, 5, and 15 years. The annual cycle, caused by seasonal variation, was the most dominant component(18.7% of total variance) of Seoul precipitation. In addition, 3.3-5 year cycle(3.3 year cycle can be interpreted as QBO(Quasi-Biennial Oscillation) mode which is observed in the stratopheric winds, and 5 year cycle as El Niño), was captured with 22.5 % of total variance. Semiannual cycle, which also caused by seasonal variation, was one of the element(4.6% of total variance) captured, and secular mode with longer than 15 year cycle, which could be related with long-term scale climate conditions, with 2.1% of total variance. As a result of cross correlation test, 7 month lag time was the dominant one between Seoul precipitation and El Niño effect. We can conclude that when El Niño happens(SOI with negative values), Seoul precipitation increases with 7 months lag time based on low frequency bands.

REFERENCES

Broomhead, D. S., and G. P. King, Extraction qualitative dynamics from experimental data, Phys. D Amsterdam, 20, 217-236, 1986.

Cayan, D. R., and D. H. Peterson, "the influence of North Pacific atmospheric circulation on streamflow in the west." Aspects of climate variability in the Pacific and Western Americas. American Geophysical Union, Washington, D.C., 1989.

Colebrook, J. M., Continuous plankton records: Zooplankton and environment, North-East Atlantic and North Sea, 1948-1975, Oceanol. Acta, 1, 9-23, 1978.

Keppenne, C. L., and M. Ghil, Adaptive spectral analysis of the Southern Oscillation Index, in Proceedings of the XVth Annual Climate diagnostics Workshop, pp. 30-35, U. S. Department of Commerce, NOAA, Springfield, Va., 1990.

Keppenne, C. L., and M. Ghil, Adaptive filtering and prediction of the Southern Oscillation index, J. Geophysical Res., 97, 20449-20454, 1992.

Kiladis, G. N., and H. F. Diaz, Global climatic anomalies associated with extremes in the Southern Oscillation, J. Climate. 2, 1069-1090, 1987.

Klein, W. H., and H. J. Bloom, Specification of monthly precipitation over the United States from the surrounding 700 mb height field, Monthly Weather Rev., 115, 2118-2132, 1987.

Lall, U., and M. E. Mann, The Great Salt Lake: a barometer of low frequency climatic variability, Water Resour. Res., 31(10), 2503-2516, 1995.

Lee D. R., Relationships between El Niño/La Niña and Korea temperature/precipitation, KWRA papers, vol 31, no. 6, 807-819, 1998.

Lee H. S., and Y. O. Kim, Correlation test between ENSO and Korea run-off amount, KSCE conference 1999, 343-346, 1999.

Moon Y. I., D. H. Park, and S. H. Hwang, El Niño frequency analysis by using prefiltered Singular Spectrum Analysis, KWRA conferences 1999, 171-176, 1999.

Moon Y. I., and U. Lall, Atmospheric flow indices and interannual Great Salt Lake variability, Journal of Hydrologic Engineering, April, 55-61, 1996.

Moon Y. I., Y. I. Cha, and S. H. Hwang, Seoul station missing value expansion by using nonparametric Markov model, KSCE Conference 1999, 269-272, 1999.

Sin H. S., J. H. An, and Y. N. Yoon, Spatial/temporal correlation test between El Niño and Korea precipitation, KWRA conferences 1998, 32-37, 1998.

Sin H. S., Y. S. Kim, J. H. Kim, and Y. S. Kim, El Niño/La Niña effect on Korea weather/hydrologic condition, KWRA conferences 1999, 195-200, 1999.

Vautard, R., and M. Ghil, Singular Spectrum Analysis in nonlinear dynamics, with applications to paleoclimatic time series, Physica D, 35, 395-424, 1989.

Vautard, R., P. Yiou, and M. Ghil, Singular Spectrum Analysis: a toolkit for short, noisy chaotic signals, Physica D, 58, 95-126, 1992.

Weare, B., and J. S. Nasstrom, Examples of extended EOF analysis, Monthly Water Rev., 110, 481-485, 1982.

Stochastic Hydraulics 2000, Wang & Hu (eds) © 2000 Taylor & Francis, ISBN 90 5809 166 X

Effects of onshore-offshore sediment movements on beach profiles

Y.Çelikoğlu, E.Çevik & Y.Yüksel
Faculty of Civil Engineering, Yildiz Technical University, Istanbul, Turkey

ABSTRACT: Change of coastal morphology depends on incident wave conditions, beach slope and sediment characteristics. Onshore-offshore sediment transport rate exposed to waves is very important for coastal morphology for the short term variation of the coasts, because beach profile is formed by the transport of sediment. In this study, a series of experimental investigations were carried out for the determination of volumetric change of beach profiles under three different bed materials and two different beach slopes.

Key words: Onshore-offshore sediment transport, Coastal morphology, Wave, Beach profile

1. INTRODUCTION

The onshore-offshore movement of sediment on beaches has been studying since the beginning of 1950's both in the laboratories and fields (Saville, 1957; Horikawa, 1978; Hallermeier, 1984; Larson et. al., 1989; Dean, 1990; Çevik and Yüksel, 1997). However, the physical processes have not been fully understood.

Johnson (1949), Iwagaki and Noda (1962) and Nayak (1970) have proposed the parameters by which beach profiles could be classified into two types as; bar or step types. These studies did not take the beach gradient into consideration. However, the later sudies have shown that the initial beach slope has a great effect on the final configuration. Sunamura and Horikawa (1974) have proposed a new beach classification based on the displacement of topography from the initial beach slope;

$$C = \left(\frac{H_0}{L_0}\right) \tan\beta^{0.27} \left(\frac{d_{50}}{L_0}\right)^{-0.67} \tag{1}$$

where C is profile parameter, H_0 is deep water wave height, L_0 is deep water wave length, $\tan\beta$ is bottom tlope, d_{50} is grain size. Sunamura and Horikawa (1974) classified the profiles as follows:

C>8   **Profile is Bar Type:** Sand accumulates in offshore zone and shoreline retrogresses. It is also known as winter or storm type profile, because it is an erosional profile.

4<C<8   **Profile is Transition Type:** Shoreline advances and sand piles up offshore.

C<4   **Profile is Step Type:** Sand deposition takes place backshore and shoreline prograds. It is also known as summer or normal type profile, because it is an accretional profile.

As known from previous studies bar type beach profile is produced by waves with relatively large wave steepness, while the step type beach profile is produced with relatively small wave steepness. Profile type depends not only incident wave conditions but also sediment size and beach slope.

Monotonic beach profile has defined as $d(y) = A y^{2/3}$ in which this the water depth at a distance, y, offshore and A is a scale parameter depending on sediment characteristics by Dean (1991).

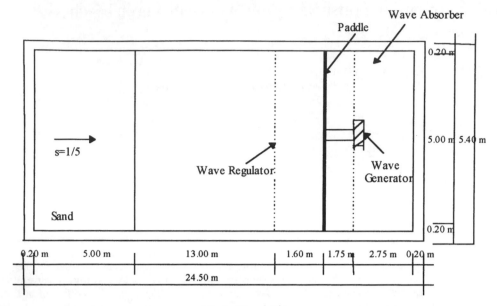

Fig.1: Wave basin.

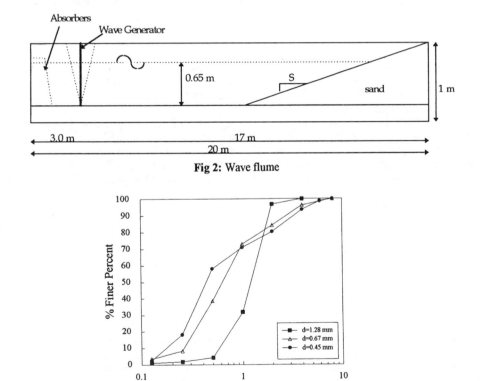

Fig 2: Wave flume

Fig.3: Grain size distributions for three different sand materials

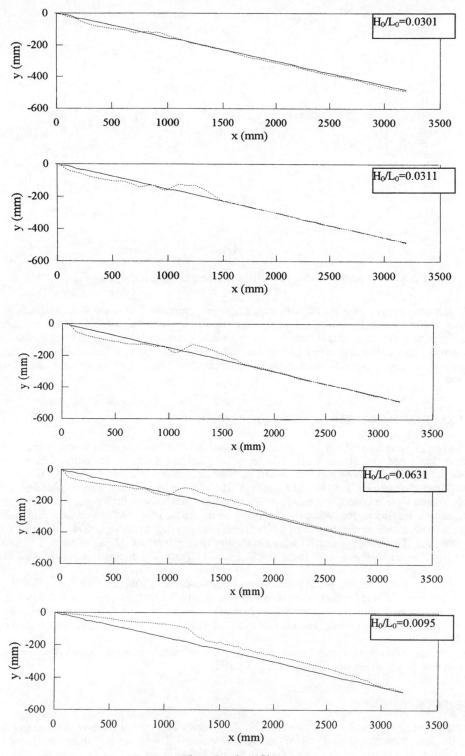

Fig 4: Beach profiles

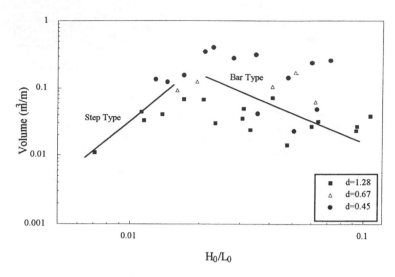

Fig 5: Beach profile volumes versus wave steepness for three different sand sizes

Beach erosion is in progress on many coasts. Seaside resorts such as many Turkish coasts are among those, which have suffered when storms have removed the beaches. However, sand coastlines in many parts of the world have been retreating during the past century. Observations show that many parameters affect the shoreline variation. In this study, it was concluded that the grain size is a significant parameter on the development of profiles as well as the wave height and period.

2. EXPERIMENTAL SET UP AND PROCEDURES

Experiments were carried out in a wave channel and a wave basin in order to avoid the scale effects (Fig.1 and Fig. 2). Sidewalls of the flume are glass and hence the observations have been made during the experiments in Coastal and Harbour Laboratory at Yıldız Technical University. The flume is 20 m long, 1 m wide and 1 m high. The basin is 24.5 m long, 5.40 m wide and 1 m high. Waves were produced as periodic waves by a wave generator in both experiments. A wave generator is located at one end of the flume and basin. At the other end of the flume and basin have a sandy beach model. Two different beach slopes were 1/5 and 1/10 considered during the experiments. Three different sand sizes were used in the experiments. Median diameters (d_{50}) of the materials are 0.45 mm, 1.28 mm and 0.67 mm. The grain size distributions of the sand are shown in Figure 3.

The initial and developed beach profiles were recorded using HR Wallingford Touch Sensitive Two Dimensional Bed Profiler. The profiler consist of a probe that can move up and down and which is mounted on a carriage moving horizontally along a 6 m length supported beam. The profiler has ± 1 mm horizontal and a 0.5 mm vertical accuracies. The bed profile is controlled by a PC.

For each experimental run, the developed profiles were measured after 1000 waves. Experimental conditions were summarized in Table 1.

4. DISCUSSION

Experiments show that a beach subject to breaking waves experiences a return flow across the profile that carries sediment stirred up by the waves offshore in the undertow. Even at equilibrium conditions, when no net change in the profile shape occurs, this transport should take place implying that material has to come onshore above the undertow layer to compensate for the offshore transport (Larson et. al, 1999). When the undertow reaches the break point, the transported sediment has to be resuspended up into the water column and pushed onshore to ensure an equilibrium situation where no material moves offshore. Thus, such a simplified picture yields a surf zone with considerable amounts of sediment moving, but with no net changes in time of the profile depths, and the break point acts almost as a singularity. Fig. 4 shows some examples of the recorded beach profiles in the surf zone at equilibrium conditions.

Table.1: Experimental conditions

Type	Slope ($\tan\beta$)	Wave Height H_0 (m)	Wave Period T (s)	Wave Steepness H_0/L_0 (-)	Grain Size d_{50} (mm)	Remarks
Bar	1/5	0.047	1.0	0.0301	0.67	Basin
Bar	1/5	0.077	1.0	0.0494	0.67	Basin
Bar	1/5	0.097	1.0	0.0622	0.67	Basin
Bar	1/5	0.045	1.2	0.0200	0.67	Basin
Bar	1/5	0.070	1.2	0.0311	0.67	Basin
Bar	1/5	0.092	1.2	0.0409	0.67	Basin
Bar	1/5	0.115	1.2	0.0511	0.67	Basin
Bar	1/5	0.064	1.6	0.0160	0.67	Basin
Bar	1/5	0.078	1.6	0.0195	0.67	Basin
Bar	1/5	0.054	1.0	0.0338	0.45	Basin
Bar	1/5	0.081	1.0	0.0506	0.45	Basin
Bar	1/5	0.101	1.0	0.0631	0.45	Basin
Bar	1/5	0.050	1.2	0.0227	0.45	Basin
Bar	1/5	0.078	1.2	0.0355	0.45	Basin
Bar	1/5	0.105	1.2	0.0477	0.45	Basin
Bar	1/5	0.132	1.2	0.0600	0.45	Basin
Bar	1/5	0.158	1.2	0.0718	0.45	Basin
Bar	1/5	0.058	1.6	0.0145	0.45	Basin
Bar	1/5	0.081	1.6	0.0203	0.45	Basin
Bar	1/5	0.112	1.6	0.0280	0.45	Basin
Bar	1/5	0.139	1.6	0.0348	0.45	Basin
Bar	1/5	0.164	1.6	0.0410	0.45	Basin
Step	1/5	0.059	2.0	0.0095	0.45	Basin
Step	1/5	0.080	2.0	0.0129	0.45	Basin
Bar	1/5	0.106	2.0	0.0171	0.45	Basin
Bar	1/5	0.131	2.0	0.0211	0.45	Basin
Bar	1/5	0.142	2.0	0.0229	0.45	Basin
Bar	1/5	0.094	1.0	0.0600	1.28	Flume
Bar	1/5	0.135	0.9	0.1069	1.28	Flume
Bar	1/5	0.081	1.3	0.0306	1.28	Flume
Bar	1/5	0.104	1.47	0.0310	1.28	Flume
Bar	1/5	0.112	1.86	0.0208	1.28	Flume
Bar	1/10	0.145	1.0	0.0928	1.28	Flume
Bar	1/10	0.099	1.0	0.0640	1.28	Flume
Bar	1/10	0.144	1.05	0.0410	1.28	Flume
Step	1/5	0.077	2.64	0.0071	1.28	Flume
Step	1/5	0.095	2.33	0.0112	1.28	Flume
Step	1/5	0.120	2.58	0.0115	1.28	Flume
Step	1/10	0.107	2.0	0.0171	1.28	Flume
Step	1/10	0.082	1.5	0.0234	1.28	Flume
Step	1/10	0.086	2.0	0.0138	1.28	Flume

When the beach profiles reached at the equilibrium conditions, it is assumed that the change in the sediment transport rate in the undertow is balanced by sedimentation through the water column. This sedimentation represents the net effect from sediment being resuspended locally by the waves and the settling of the material. Thus, the stirring of the sediment at the bottom mobilizes grains that may be brought up into the water column by the turbulence from wave breaking. As it is known that grain size distributions of the bed materials also effect the sediment transport rate. Moreover, sediment with non-uniform grain size distribution under a wave attack of constant intensity will be sorted and armour layers may be established (Asar et. al., 1997). In Fig. 5 the beach profile volumes against deep-water wave steepness for different grain size were plotted. Finer and coarser grain sizes are distinguished in Fig. 5. Beaches with finer sediments have larger profile volume than that of coarser sand. During the transition from step type to bar type beach profile has its biggest volume. The volume of the step type beach profile increases as increasing deep-water wave steepness. But the profile type changes to bar type after a certain values of wave steepness and the volume of the bar decreases as deep water wave steepness increases. However, bar profiles were formed at smaller wave steepness for finer bed material (see Fig. 5 and Table 1). These results were agree with Sunamura and Horikawa (1974).

5.CONCLUSION

Grain size distributions of the bed materials effect the sediment transport rate and hence the form of beach profiles. For finer bed materials, the volume of the beach profiles increase. Armour layers are established over the beach profiles for graded bed materials, and it also effects the equilibrium conditions of the profiles.

ACKNOWLEDGEMENT

The authors wish to thank the research Foundation of Yildiz Technical University for their financial support (project no: 98-A-05-01-02).

REFERENCES

Asar, Y.,Yüksel, Y., Kapdaşlı, S., 1997. "Sorting on a two dimensional beach",Proc. of the 3th Int. Conf. On the Mediterranean Coastal Environment, Medcoast 97, Quawra, Malta, pp. 1003-1010.

Çevik, E, Yüksel, Y., 1997. "Morphological change of beach profiles", Proc. of the 3th Int. Conf. On the Mediterranean Coastal Environment, Medcoast 97, Quawra, Malta, pp. 1021-1027.

Dean, R. G., 1991. "Equilibrium beach profiles: Characteristics and applications", J. of Coastal Research, 7 (1):53-84.

Hallermeier, R. J., 1984. "Wave cuts in sand slopes applied to coastal models", J. of Waterway, Port, Coastal and Ocean Eng., Vol. 110, No 1, pp 34-49.

Horikawa, K., 1978. "Coastal engineering", University of Tokyo Press.

Johnson, J. W., 1949. "Scale effects in hydraulic model involving wave motion", Trans. AGU, Vol. 30, pp 517-525.

Larson, M., Kraus, N. C., Sunamura, T., 1989. "Beach profile change: Morphology, Transport rate, Numerical solution".

Larson, M., Kraus, N. C., Wise R. A., 1999. "Equilibrium beach profiles under breaking and non-breaking waves", Coastal Engineering, 36 (1999), 59-85.

Saville, T. Jr., 1957. "Scale effects in two dimensional beach studies". Proc. of the 7th International Association for Hydraulics Research Congress; Lisbon; Portugal; pp 920-938.

Sunamura, T., Horikawa, K., 1974. "Two dimensional beach transformation due to waves", Proc. of the 14th Coastal Eng Conf., Copenhagenm Denmark, pp 920-938.

Iwagaki, Y., Noda, H., 1962. "Laboratory study of scale effects in two dimensional beach processes", Proc. of the 8th Coastal Eng Conf., Mexico City, Mexico, pp 194-210.

Nayak, I. V., 1970. "Equilibrium profiles of model beaches", Proc. of the 12th Coastal Eng Conf.

Confidence levels of flood and drought forecasts from El Nino/Southern oscillation indexes

Saleh A. Wasimi

Faculty of Informatics and Communication, Central Queensland University, Rockingham, Qld., Australia

ABSTRACT: Ropelewski and Halpert (1996) have identified 19 regions of the world whose precipitation characteristics are related to El Nino/Southern Oscillation phenomena. Although the teleconnections are now extensively described in literature, scientists are still groping to model the nature and strength of the teleconnections. To provide a lumped picture Eltahir (1996) have used the discriminant analysis approach to predict flood and drought conditions in the Nile river. To average out temporal variability Chiew et al. (1998) have proposed contemporaneous averaging of data for a period of up to 36 months. These are useful approaches to provide a broader picture, but a more efficient tool for decision making purposes would be if each forecast is associated with a confidence level. Construction of confidence level has to start from the variability of the input data. As models are developed to process the input data, statistical equations can also be developed which can capture the propagation of the variability of the input data through the models. Wasimi (1994) describes how statistical equations can be applied to capture the propagation of variability in numerical analysis. In this paper methods are developed which can associate confidence levels with climatic forecasts made through remote controls such as El Nino/Southern Oscillation indices.

Key words: Climate forecasting, Confidence interval, ENSO, Regression analysis.

1. INTRODUCTION

Decision making process requires an insight as to how far an actual observation may deviate from the forecasted value. This insight may be provided by the construction of a confidence interval. Frequently confidence intervals are constructed on the assumption of normality. If the input variables have all Gaussian distributions, the output variable may or may not have a Gaussian distribution. The problem of unknown probability distribution of the output variable is usually circumvented by using the Central Limit Theorem. Thus, if X is the forecasted or expected value, the confidence interval for α percent level of significance is

$$\overline{X} \pm Z_{\alpha/2} \frac{S}{\sqrt{n}} \tag{1}$$

where S is sample standard deviation, n is sample size, and Z is corresponding value for α from standard normal table. When the sample size is less than 30 we can replace Z by t provided the parent population is normally distributed. The last assumption is very restrictive because we have to know the probability distribution of the output variable and it has to be Gaussian for the t statistic to be applicable. This aspect is incorrectly ignored in many applications. However, there exists a non-parametric approach for the construction of confidence interval. In that approach the confidence interval is constructed for the median and the concept of ranking is used. For example, the rank of the lower bound for 95 percent confidence interval is given by $(n-1.96\sqrt{n}+1)/2$ and that for the upper bound is given by $(n+1.96\sqrt{n}-1)/2$.

If a system is linear and all input variables are normally distributed, the output variable will always be normally distributed, and thus, the construction of confidence interval is fairly straightforward. If the system is nonlinear but the system dynamics is known, the distribution of the output variable can be determined using stochastic equations. However, the derivation process can be cumbersome, and therefore, oftentimes people resort to Monte Carlo simulation. A large number of simulation runs may provide useful information about the distribution of the output variable. Generally speaking, this approach is fairly reliable and is used in many applications but one has to be aware of possible pitfalls in this approach as discussed in Wasimi (1994). Especially if one is interested in the tails of the distribution, a rigorous theoretical approach is recommended to determine the actual distribution of the output variable.

In this paper, a process of derivation of confidence interval for nested regression models is presented. The method is applied in climate forecasting using El Nino/Southern Oscillation indices.

2. CLIMATE AND El NINO/SOUTHERN OSCILLATION INDICES

Giorgi et al. (1996) noted that the drought and wet climatic conditions of a region do not depend on local factors or meso-scale circulations but rather on some remote controls. One of this remote controls has been found to be the El Nino/Southern Oscillation (ENSO) index . Ropelewski and Halpert (1996) have identified 19 regions of the world where climatic conditions are significantly related to ENSO indices. Many researchers have attempted to model this relationship. Amarasekera et al. (1998) and Whitaker (1999) reported that climatic conditions and ENSO indices are in phase, that is, they occur simultaneously. However, Cane and Zebiak (1985) and Gershunov and Barnett (1998) noted that ENSO itself can be predicted with reasonable accuracy up to 18 months in advance. Thus a bootstrap method of forecasting can be used to forecast flood or drought conditions as suggested by Smith and Ropelewski (1997). Nevertheless, development of such a model is fraught with difficulties. Firstly, there is a controversy as to the definition of ENSO and an informative paper on the subject is by Trenberth (1997). The original definition of ENSO involved the atmospheric pressure difference between Tahiti and Darwin. Later it was not found suitable because there are many small-scale and high frequency phenomena in the atmosphere, such as Madden-Julian Oscillation, that can influence the pressures at these stations but are unrelated to ENSO. Wright (1989) used the sea surface temperature of the equatorial Pacific Ocean as the ENSO index. NOAA currently is using a different region of the Pacific Ocean as the indicator. Secondly, there is no universally accepted criterion to identify and define flood and drought conditions (Chang, 1987; Dracup et al., 1980). Some researchers have used streamflow as an indicator because it is the best integrator of rainfall over a basin. Thirdly, there is so much noise in daily or weekly observations of ENSO index that all attempts to relate it to climatic conditions were unsuccessful and thus various types of smoothing have been done. Chiew et al. (1998) have applied contemporaneous averaging of data for a period of 36 months for the Murray-Darling river basin. At present, three-month contemporaneous averaging is the most popular form of compositing in climate modeling community (Harrison and Larkin, 1998).

Eltahir (1996) has used discriminant analysis approach to predict flood and drought conditions in the Nile river. Despite the controversies and uncertainties he argued that his approach would be useful in the decision making process. In that vain Whitaker (1999) has attempted a more quantitative approach based on regression techniques, and here we look into the confidence levels of his model outputs.

3. CASE STUDY

The Ganges river basin has been identified by Ropelewski and Halpert as one of the 19 regions of the world which is influenced by ENSO, and Whitaker have reported that the annual flow of the Ganges river has a correlation coefficient of -0.54 with the Wright ENSO index of June-July-August (JJA). The autocorrelation for the Ganges annual flow is 0.3 at lag 1. Furthermore, it has been found that the magnitude of flood or drought is significantly related to the gradient of the ENSO index and to some extent the sunspot activity. Whitaker used streamflow data of the Ganges river from 1934 to 1993 (1971 data missing), Wright SST upto 1986 and thereafter NOAA's Nino 3.4 data to build a model that could predict the Ganges annual flow with a lead time of one year based on ENSO index, ENSO gradient, and sunspot data. Because of the complexity of the problem a gradual model building approach was adopted.

As a first step the usefulness of the traditional model was examined. The traditional popular method in annual streamflow modeling is the autoregressive model (Maass et al., 1962). From autocorrelation function, partial autocorrelation function and Akaike information criteria an AR(1) model was found to fit the data with parameter α =0.30. This model for one-year forecasting can thus explain 9 percent (= α^2) of the variability. Whitaker then examined the correlation between the residuals of the AR(1) model with the ENSO signal and obtained a correlation coefficient of -0.44 at lag zero and no other significant correlation at any other lag.

A linear regression analysis of the flow residuals with ENSO index was then performed. This yielded an R-square value of 0.19 and a significant F- statistic of 13.5. Since ENSO index and the Ganges flow variability are simultaneous, the information provided for flow at a given time by the autoregressive model and by the ENSO index are independent. Thus these two models can be combined for an improved forecast, the contribution to the sum of squares by the AR(1) model being 9 percent and that by the ENSO index model 19%. The two together explains 28 percent of the annual variability.

The residuals remaining from the coupled model showed some relationship with the gradient of the ENSO index. An examination of this aspect revealed that there appears to be some sort of inertia to changes in nature. When it begins to shift from wet to dry or from dry to wet, the changing process tends to linger longer with sharper gradient of the ENSO index. Although this was apparent from the visual examination of curves, it did not show up in the statistical analysis. The reason later was found to be drowning of information from large data of average years, which are neither dry nor wet, but where short changes of ENSO occurred with sharp gradients. It then became necessary to isolate dry and wet years from the average years. The next question then arose was should dry and wet years be treated together or they should be treated separately. Hoerling et al. (1997) reported that the relationship between El Nino and La Nina events is non-linear. Since the modeling techniques used are primarily linear, wet and dry episodes were treated separately. The residuals remaining after the two models when regressed with the gradients of El Nino years yielded an R-square value of 0.14 and that for the La Nina years yielded an R- square value of 0.40. Between 1934 and 1993 there were 14 El Nino years and 10 La Nina years. So though the sample size was small, the result was nevertheless significant.

van Loon and Labitzke (1998) have shown that sunspot cycle is related to inter-annual variability in the stratosphere, which in turn is correlated with temperature at 500-hPa height in the troposphere. The relationship they found is purely statistical, but points to the fact that sunspot activity may have some influence on global circulation. Based on this assumption Whitaker related the residuals remaining after the two stages of modeling with the gradients of sunspot data. The regression of the residuals remaining after the two stages of modeling with gradients of sunspot observation yielded an R-square value of 0.02 for the El Nino years and 0.13 for the La Nina years. The variance explained for the El Nino years is thus insignificant, but that for the La Nina years is significant.

The next challenge was then how to construct a confidence interval for such a sequence of models to gain some insight into the reliability of such forecasts.

4. CONFIDENCE INTERVAL FOR NESTED MODELS

Nested regression models become necessary when the length of data of explanatory variables is different. Confidence interval is constructed from the variance of residuals. We know

Variance of Σ(models) = Σ (variance of models) + 2 Σ (covariance between the models)   (2)

Where Σ stands for summation. In this particular case the models are independent thus the covariance terms can be ignored. For the AR(1) model the variance using Bartlett's formula is

$$\frac{1+2\rho^2}{n}(Y_{t-1} - \bar{Y})^2 \tag{3}$$

where ρ is the lag 1 autocorrelation, n is the number of observations, Y_{t-1} is the latest observation, and $\bar{Y}$ is the average flow.

For the nested regression models, let us consider the regression model of the form

$$\hat{Y} = \bar{Y} + \beta_1(X - \bar{X})_. \tag{4}$$

Therefore, for any given value of the explanatory variable X_0

$$
\begin{aligned}
\text{Var}[\hat{Y}_0] &= \text{Var}[\bar{Y} + \hat{\beta}_1(X_0 - \bar{X})] \\
&= \text{Var}[\bar{Y}] + (X_0 - \bar{X})^2 \text{Var}[\hat{\beta}_1] \\
&= \frac{\sigma^2}{n} + \frac{(X_0 - \bar{X})^2 \sigma^2}{S_{xx}} \\
&= \sigma^2 \left[\frac{1}{n} + \frac{(X_0 - \bar{X})^2}{S_{xx}} \right]
\end{aligned}
\tag{5}
$$

where $S_{xx} = \sum_{t=1}^{n}(X_t - \bar{X})^2$. If R_1^2 is the variance explained by the first regression model, then we know $R_1^2 = 1 - \dfrac{\sigma^2(n-2)}{S_{yy}}$. Thus $\sigma^2 = \dfrac{(1-R_1^2)S_{yy}}{n-2}$. The variance unexplained by the first model is $(1-R_1^2)$. The second model explains R_2^2 fraction of this unexplained variance. Therefore, the variance explained by the two regression models in sequence is $R_1^2 + R_2^2(1-R_1^2)$. With similar argument, the third model explains $R_3^2[1-R_1^2-R_2^2(1-R_1^2)]$ of the variance. Thus the three together explains $R^2 = R_1^2+R_2^2+R_3^2-R_1^2R_2^2-R_2^2R_3^2-R_3^2R_1^2+R_1^2R_2^2R_3^2$. A general relationship can be developed following a similar approach but such an exercise can only be a mathematical venture unless the explanatory variables are independent of each other. If not a complex procedure of orthogonalization has to be adopted before coupling the models.

We can now construct a confidence interval for the AR(1) model and three regression models as follows:

$$\hat{Y} \pm t_{\alpha/2, n_1-6} \left[\frac{(1-R^2)S_{yy}}{n_1-6} \left(\frac{1}{n_1} + \frac{1}{n_2} + \frac{1}{n_3} + \frac{(X_0 - \bar{X})^2}{S_{xx}} \right) + \frac{1+2\rho^2}{n_1}(Y_{t-1} - \bar{Y}) \right] \tag{6}$$

5. RESULTS AND DISCUSSION

The modeling exercise mentioned above when applied to the Ganges river annual flow data at Hardinge Bridge yielded the following models:

For El Nino years:

$$\hat{Q}_t = \bar{Q} + 0.30(Q_{t-1} - \bar{Q}) + 551.00 - 15.27S_t - 9.55\Delta S + 10.45\Delta P \tag{7}$$

For La Nina years:

$$\hat{Q}_t = \bar{Q} + 0.30(Q_{t-1} - \bar{Q}) + 1466.80 - 15.27S_t - 12.13\Delta S + 12.03\Delta P \tag{8}$$

where Q_t is the annual flow of the Ganges river for year t (in m^3/sec), S is the SST anamoly as described in Eltahir, P is the sunspot number, and Δ is used to indicate gradient, ie, $\Delta S = S_t - S_{t-1}$. The 95 percent confidence intervals constructed are given in Table 1 for El Nino and La Nina years.

Table 1: 95 percent confidence interval for the Ganges annual flow with one year lead time. (flow in thousands of cubic meters per sec.)

El Nino years	Confidence interval		Observed value	La Nina Years	Confidence interval		Observed value
	Lower	Upper			Lower	Upper	
1939	8.6	12.0	8.5	1938	11.5	13.6	12.0
1941	7.6	11.1	7.7	1950	11.4	14.1	11.6
1951	6.9	10.0	8.2	1955	13.6	16.1	16.1
1953	8.8	11.9	11.2	1956	15.2	18.0	16.7
1957	8.3	11.7	10.2	1964	11.1	13.4	12.4
1963	8.8	12.2	13.8	1970	10.2	13.3	10.3
1965	7.2	10.1	7.6	1971			
1969	8.3	11.5	11.5	1973	9.7	12.4	13.1
1972	6.3	9.5	8.1	1975	11.6	13.8	12.9
1976	8.6	11.3	10.1	1988	10.9	13.1	11.9
1982	6.8	10.2	9.9				
1987			12.3				
1991	8.2	11.6	8.7				
1993	7.8	10.3	8.5				

Considering the fact that out of 22 observations, only two values were outside the confidence interval, such a modeling exercise is useful, and especially important for the Ganges basin where one-sixth of the total world population lives. Their sustenance is largely dependent on the river, and any forewarning can be of immense benefit to the people. However, how much reliance should be given to such forecasts in the planning process needs careful consideration particularly because when we see that 1963 was an El Nino year and yet, the flow was much above average annual flow of 11,200 cubic meters per second and 1970 was a La Nina year and the flow was below average.

REFERENCES

Amarasekera, N. A., Lee, R. F., Williams, E. R., and Eltahir, E. A. B., 1998: ENSO and the natural variability in the flow of tropical rivers. Journal of Hydrology, in press.

Cane, M. A. and Zebiak, S. E., 1985: A theory for El Nino and the Southern Oscillation. Science, 228, 1085-1087.

Chang, T. J., 1987: Drought analysis in the Ohio river basin. Proceedings of Engineering Hydrology, ASCE, 601-609.

Chiew, F. H. S., Piechota, T. C., Dracup, J. A. and McMahon, T. A., 1998: El Nino/Southern Oscillation and Australian rainfall, streamflow and drought: Links and potential for forecasting. Journal of Hydrology, 204, 138-149.

Dracup, J. A., Lee, K. S., and Paulson, E. G. Jr., 1980: On the definition of droughts. Water Resources Research, 16(2), 297-302.

Eltahir, E. A. B., 1996: El Nino and the natural variability in the flow of the Nile river. Water Resources Research, 32(1), 131-137.

Harrison, D.E. and Larkin, N. K., 1998: El Nino-Southern Oscillation sea surface temperature and wind anomalies, 1946-1993. Reviews of Geophysics, 36(3), 353-399.

Hoerling, M. P., Kumar, A., and Zhong, M., 1997: El Nino, La Nina, and the nonlinearity of their teleconnections. Journal of Climate, 10, 1769-1786.

Gershunov, A. and Barnett, T. P., 1998: ENSO influence on intraseasonal extreme rainfall and temperature frequencies in the contiguous United States: Observations and Model results. Journal of Climate, 11(7), 1575-1586.

Giorgi, F., Mearns, L.O., Shields, C., and Mayer, L., 1996: A regional model study of the importance of local versus remote controls of the 1988 drought and the 1993 flood over the central United States. Journal of Climate, 9(5), 1150-1162.

Maass, A., Hufschmidt, M. M., Dorfman, R., Thomas, Jr., H. A., Marglin, S. A., and Fair, G. M., 1962: Design of Water-Resource Systems. Cambridge, Massachusetts: Harvard University Press.

Ropelewski, C. F. and Halpert, M. S., 1996: Quantifying southern oscillation-precipitation relationship. Journal of Climate, 9(5), 1043-1059.

Smith, T. M. and Ropelewski, C. F., 1997: Quantifying southern oscillation-precipitation relationships from an atmospheric GCM. Journal of Climate, 10, 2277-2284.

Trenberth, K. E., 1997: The definition of El Nino. Bulletin of the American Meteorological Society, 78(12), 2771-2777.

van Loon, H. and Labitzke, K., 1998: The global range of the stratospheric decadal wave. Part I: Its association with the sunspot cycle in summer and in the annual mean, and with the troposphere. Journal of Climate, 11(7), 1529-1537.

Wasimi, S. A., 1994: Estimation error of flows computed by hydrodynamic routing models. Proceedings of the Ninth Congress of the Asian and Pacific Division of the International Association for Hydraulic Research, 24-26 August, Singapore. Vol. 1, 172-177.

Whitaker, D. W., 1999: El Nino, Teleconnections, and flow in the Ganges: A new paradigm for streamflow forecasting. Masters Thesis. Department of Environmental Engineering, University of Cincinnati, Ohio.

Wright, P. B., 1989: Homogenized long-period Southern Oscillation indices. International Journal of Climatology, 9, 33-54.

Stochastic Hydraulics 2000, Wang & Hu (eds) © 2000 Taylor & Francis, ISBN 90 5809 166 X

A comparison of parametric and nonparametric estimators for probability precipitation in Korea

Young-Il Cha & Young-Il Moon
Water Research Laboratory, Department of Civil Engineering, University of Seoul, Korea

ABSTRACT: The frequency analyses for the precipitation data in Korea were performed. We used daily maximum series, monthly maximum series, and annual series. In order to select an appropriate distribution, 17 probability density functions were considered for the parametric frequency analyses. They are Gamma II, Gamma III, GEV (Generalized Extreme Value), Gumbel (Extreme Value type I), Log-Gumbel, Log-normal II, Log-normal III, Log-Pearson type III, Weibull II, Weibull III, Exponential, Normal, Pearson type III, Generalized logistic, Generalized Pareto, Kappa and Wakeby distributions. For nonparametric frequency analyses, variable kernel and log-variable kernel estimators were used. Nonparametric methods do not require assumptions about the underlying populations from which the data are obtained. Therefore, they are better suited for multimodal distributions with the advantage of not requiring a distributional assumption. The variable kernel estimates are comparable and are in the middle of the range of the parametric estimates. The variable kernel estimates show a very small probability in extrapolation beyond the largest observed data in the sample. However, the log-variable kernel estimates remedied these defects with the log-transformed data.

Key words: Frequency analysis, Nonparametric, Probability precipitation

1. INTRODUCTION

When designing a hydrosystem to control and use of water resources, a frequency analysis based on hydrological data is one of the most important element for designing and planning an economical hydrosystem. Generally, the rainfall data can be easily observed than the flood data in Korea. Therefore, in this paper a comparison of parametric and nonparametric techniques for probability precipitation in Korea is presented.

A currently used approach to frequency analysis is based on the concept of parametric statistical inference. In these analyses, the assumption is made that the distribution function describing precipitation data is known. Distributions that are often used are Log-normal, Pearson Type III, Gumbel, extreme value distributions, Gamma, and others by using the utliers. The probability method of moment (MOM), maximum likelihood (ML), probability weighted moment (PWM), or L-moment. However, such an assumption is not always justified. Some difficulties associated with parametric estimation are (1) the objective selection of a distribution, (2) the reliability of distributional parameters (especially for skewed data with a short record length), (3) the inability to analyze multimodal distributions, and (4) the treatment of oweight moment (PWM) method (Greenwood et al., 1979; Hosking, 1989, etc.), L-moment method (Hosking, 1990), or others have been complemented the problems of skewed data with a short record length. Nevertheless, in the process of parametric frequency analysis, the choice of best fitted distribution among the other distributions which are passed the goodness-of-fit tests (χ^2 test, Kolmogorov-Smirnov test, Cramer von Mises test, etc) is still not a easy task. Also, the analysis of bimodal probability density function has many complicated problems when the data has a mixed distribution. The assumption of a pre-chosen distribution, which is based on goodness-of-fit tests and selected as the most appropriate distribution, is no longer valid if the size of the data available is increased. Therefore, parametric techniques may be inadequate for reliable frequency estimates.

In recent years, nonparametric kernel density estimation methods have been introduced as viable and flexible alternatives to parametric methods for flood frequency analysis or probability precipitation estimation. Several nonparametric approaches have been introduced by Adamowski (1985, 1989, and 1996), Adamowski and Feluch (1990), Adamowski and Labatiuk (1987), Lall et al. (1993), Moon et al. (1993), Moon and Lall (1994 and 1995), and Moon (1999). Nonparametric methods do not require assumptions about the underlying populations from which the data are obtained. Also, they are better suited for multimodal distributions. Usually, nonparametric kernel density estimator was relatively consistent across the estimation situations considered in terms of bias and root mean squares error (RMSE) with the advantage of not requiring a distributional assumption while providing a uniform procedure (Lall et al., 1993; Moon et al., 1993).

Even though many people have shown that the nonparametric method provides a better fit to the data than the parametric method and gives more reliable flood or precipitation estimates, the nonparametric method implies a very small probability in extrapolation beyond the highest observed data in the sample. In this paper, we tried to show a remedy for these inadequacies by introducing a log-estimator which is a probability density function for log-transformed data, $\ln x_i$ if x_1, x_2, ..., x_n are random variable. The extrapolation is based on the shape of the kernel density function assumed and on the value of bandwidth h. Thus, only a few observations contained in the bandwidth h influence the extrapolation to the tail of the distribution. However, it is possible to remedy these defects by applying the nonparametric kernel estimator to log-transformed data.

2. PARAMETRIC/NONPARAMETRIC FREQUENCY ANALYSES

Which distribution is best for the precipitation data in Korea based on parametric frequency analyses? The question of which distribution gives the best fit may be decided by using the chi-squared statistic and Kolmogorov-Smirnov tests. As shown in Figure 1, the parametric frequency analyses were considered all the distributions available. The parameters of selected distributions were estimated from the method of moments, Maximum likelihood, or L-moments.

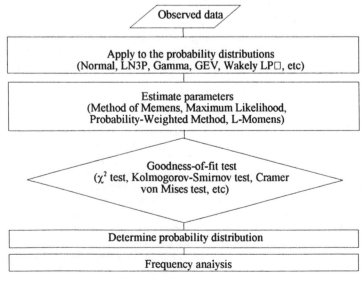

Fig.1: The procedure of parametric frequency analysis

Then, we assumed a particular distribution which was selected from goodness-of-fit tests among the other several distributions to regard it as a population's distribution and progress to analyze. This paper applied Normal distribution, two parameter Log-Normal distribution, three parameter Log-Normal distribution, three parameter Gamma distribution, Log-Pearson Type III distribution and Generalized Extreme Value distribution to precipitation data in Korea. The goodness of fit tests was applied to χ^2 test for probability density functions and Kolmogorov-Smirnov test for cumulative distribution functions with 5% significant level.

Rosenblatt (1956) introduced the nonparametric kernel density estimator, defined for all real x by

$$f(x) = \frac{1}{n}\sum_{i=1}^{n}\frac{1}{h}K\left(\frac{x-X_i}{h}\right) \tag{1}$$

where $x_1, \ldots, x_n$ are independent identically distributed real observations, $K(\cdot)$ is a kernel function, and h is a positive smoothing factor assumed to tend to zero as n tends to infinity. Silverman (1986) explained the basic concept of the nonparametric kernel density estimator. From the definition of a probability density, if the random variable x has density f(x), then

$$f(x) = \lim_{h\to 0}\frac{1}{2h}p(x-h < X < x+h) \tag{2}$$

For any given h, P(x-h < x < x+h) can be estimated by the proportion of the sample falling in the interval (x-h, x+h). Thus, a natural estimator is given by choosing a small number h and setting

$$\hat{f}(x) = \frac{1}{2hn}\ [\#\,of\ X_i, \ldots\ldots, X_n\ falling\,in\,(x-h, x+h)] \tag{3}$$

To express the estimator more transparently, define the weight function w(x) by

$$w(x) = \begin{cases} \frac{1}{2}\ ,\ if\ abs(x) < 1 \\ 0\ ,\ if\ otherwise \end{cases} \tag{4}$$

Then it is easy to see that the estimator can be written as

$$f(x) = \frac{1}{n}\sum_{i=1}^{n}\frac{1}{h}W\left(\frac{x-X_i}{h}\right) \tag{5}$$

It follows from equation (5) that the estimator is constructed by placing a box of width 2h and height (2nh)-1 on each observation and then summing to obtain the estimator. This weight function is the kernel function which satisfies the condition

$$\int K(t)\,dt = 1,\ \ where\ t = \frac{x-X_i}{h} \tag{6}$$

The kernel function is usually required to be unimodal with peak at x = 0, smoothness, and a symmetric function, that is, a density ($\int K(t)\,dt = 1$) with expectation 0 ($\int tK(t)\,dt = 0$) and finite variance ($\int t^2 K(t)\,dt = constant$).

When applying the method in practice, it is necessary to choose a kernel function and a smoothing parameter. Some useful kernel functions are given in Table 1 and Figure 2. Usually, different kernels should be examined depending the objective. For example, if continuity and differentiability of the density is needed, one may choose a kernel with infinite support rather than one with finite support.

While the choice of kernel does not seem to be critical, the choice of smoothing factor is quite a different matter. The value of h is critical and, in practice, not obvious. Too large an h implies large bias, an oversmoothed estimate, and consequent loss of information. Too small an h implies large variance and too rough an estimate (Adamowski and Labatiuk, 1987). Since all error measures depend on the unknown density, generally they cannot be used in deriving analytical expressions for selecting the smoothing factor h. Several measures of performance for using the data to produce suitable values for the smoothing parameter h have been proposed. The smoothing parameter h can be obtained by Maximum Likelihood Cross-Validation (Habbema et al., 1974; Duin, 1976), Least Squares Cross-Validation (Hall, 1983; Hall and Marron, 1987; Stone, 1984), Breiman et al. Method (Breiman et al., 1977), and Adamowski Cross-Validations (Adamowski, 1985).

Table 1: Typical kernel functions

Kernel	K(t)
Rectangular	1/2 for $\|t\| < 1$ 0 if otherwise
Gaussian	$\dfrac{1}{\sqrt{2\pi}}\exp\left(-\dfrac{t^2}{2}\right)$
Epanechnikov	$\dfrac{3}{4}\left(1-\dfrac{1}{5}t^2\right)/\sqrt{5}$ for $\|t\| < \sqrt{5}$
Rajagopalan	$\dfrac{3h}{1-4h^2}(1-t^2)$, where $\|t\| \le 1$
Cauchy	$\dfrac{1}{\pi(1-t^2)}$

Table 2:. Precipitation data of Han River area

Area	Site	Year	Data size	Missing Year
Han River	Seoul	1907~1998	85	1907.1~1097.9, 1950.9~1953.11
	Inchon	1949~1998	50	1950.6~1951.9
	Chungju	1973~1998	26	

Table.3: Probability precipitation of each return period (mm, ▓ : maximum or minimum)

Site	Seoul		Inchon		Chungju	
Return Period (year)	100	200	100	200	100	200
Log-Variable Kernel	2365.6	2423.9	2146.8	2610.4	2005.3	2254.7
Variable Kernel	2353.9	2356.9	2014.6	2032.2	1891.0	1898.5
Exponential					2248.2	2453.6
Gamma	2215.9	2335.8	1792.0	1874.8	1879.9	1970.8
GEV	2330.8	2486.6			2014.2	2151.0
G-Logistic	2442.2	2696.1			2099.4	2315.1
Log-Normal	2330.8	2493.4			2009.7	2148.0
G-Parato					1834.3	1872.5
Gumbel	2401.9	2590.2			2039.8	2188.5
Normal			1713.9	1774.5	1790.7	1856.3
Pearson typeIII	2309.0	2456.5			1988.1	2111.1
Wakeby	2330.3	2458.4				

Lall et al. (1993) demonstrated that one should directly focus on kernel distribution function estimates rather than kernel density estimates. The variable kernel estimate $F_n(x)$ of the cumulative distribution function $F(x)$ is defined as :

$$F_n(x) = \int_{-\infty}^{x} \sum_{i=1}^{n} \frac{1}{nhd_{i,k}} K\left(\frac{t-x_i}{hd_{i,k}}\right) dt$$

$$= \sum_{i=1}^{n} K*\left(\frac{x-x_i}{nd_{i,k}}\right)$$

(7)

where $K(t)$ is a kernel function, h is a bandwidth, $d_{i,k}$ is the distance from x_i to its k th nearest neighbor, and $K^*(t) = \int_{-\infty}^{t} K(u)du$.

Lall et al. (1993) provide a review of this discussion and compare the performance of different kernels and bandwidth selection methods in the flood frequency context. They found that the variable kernel estimator with heavy-tailed kernel (Cauchy) and bandwidth selection based on Adamowski criteria (VK-C-AC) led to the best tail estimates using kernel methods. The Cauchy kernel is a heavy tailed kernel and may have a better capacity for extrapolation, particularly with heavy tailed densities.

3. RESULTS

The frequency analysis for the precipitation of 26 sites in 5 basin areas (i.e. Han River, Nakdong River, Keum River, Sumjin River, Yeongsan River in Korea), which are under control of the Korea Meteorological Office, were performed. We used daily maximum series, monthly maximum series, and annual series. In order to select an appropriate distribution, 17 probability density functions were considered for the parametric method. They are Gamma II, Gamma III, GEV (Generalized Extreme Value), Gumbel (Extreme Value type I), Log-Gumbel, Log-normal II, Log-normal III, Log-Pearson type III, Weibull II, Weibull III, Exponential, Normal, Pearson type III, Generalized logistic, Generalized Pareto, Kappa and Wakeby distributions. The distribution parameters were estimated by method of moments, maximum likelihood, probability weighted moments, or L-moments method. The goodness-of-fit test for the parametric distribution applies Kolmogorov-Smirnov test and χ^2 test with significant level of 5 %.

For nonparametric frequency analyses, variable kernel density function and log-variable kernel density function estimators were used with Cauchy kernel and bandwidth selection Adamowski criteria (VK-C-AC).

The precipitation data of Han River area is shown in Table 2. In this paper, we presented just the data and the results of Han River area to save the space.

The Figures 3~5 represent the probability density functions (PDF) of annual precipitation amounts, monthly maximum precipitations, and daily maximum precipitations for Seoul, Inchon, and Chungju stations respectively. From the Figures 3~5, we observed bimodal distributions in Seoul and Chungju stations, and multimodal one for Inchon station. In those cases, a difficulty associated with parametric approach is the inability to analyze multimodal distributions. Therefore, parametric estimation techniques are inadequate for modelling such an annual maximum process. However, nonparametric methods do not require assumptions about the underlying populations from which the data are obtained. Therefore, they are better suited for multimodal distributions with the advantage of not requiring a distributional assumption.

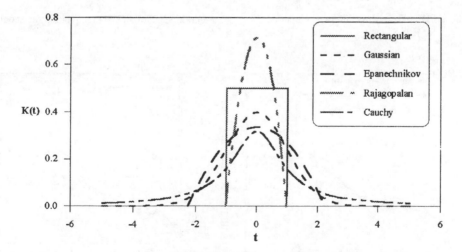

Figure 2. The Shape of Kernel Functions

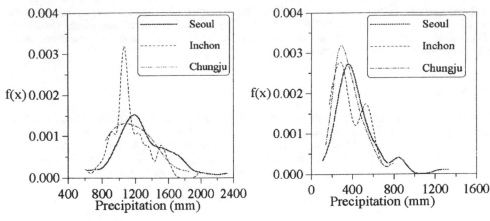

Figure 3. PDF of annual precipitation for Han River

Figure 4. PDF of monthly maximum precipitation for Han River

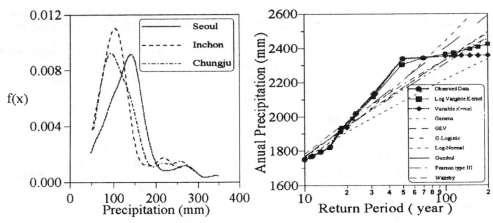

Figure 5. PDF of daily maximum precipitation for Han River

Figure 6. CDF of annual precipitation for Seoul

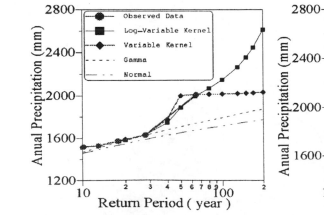

Figure 7. CDF of annual precipitation for Inchon

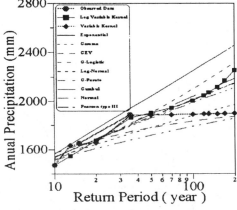

Figure 8. CDF of annual precipitation for Chungju

A comparison among the parametric and the nonparametric distribution estimators of the quantile function for Seoul, Inchon, and Chungju data are shown in Figures 6~8. Here the method of parametric estimation for parametric methods is considered with L-moments. The maximum and minimum recorded precipitations are presented in Table 3. The parametric estimates of the 100-year precipitation range from 2216 to 2442 mm in Seoul, from 1791 to 2248 mm in Chungju per year respectively. The log-variable kernel and variable kernel are comparable and are in the middle of the range of the parametric estimates. As shown Figure 7, only Gamma and Normal distributions were passed the goodness-of-fit test for Inchon station. The variable kernel estimates in Figures 6~8 shows a very small probability in extrapolation beyond the 50-year return period (i.e., the quantiles beyond the 50-year are almost the same). However, the log-variable kernel estimates (i.e., variable kernel estimator applied to log-transformed data) remedied these defects with the log data.

4. CONCLUSION

The frequency analysis for the precipitation data of 26 sites in 5 basin areas in Korea were performed. We applied nonparametric variable kernel estimators, log-variable kernel estimators, and 17 selected parametric distribution estimators to daily maximum series, monthly maximum series, and annual series. Since the results of the parametric estimators varied according to the distributions and the methods of the parametric estimation, it is not easy to say which parametric estimator is the best. However, for each data set, the nonparametric variable kernel estimator with the Cauchy kernel and Adamowski's bandwidth selection is shown to be competitive with any parametric distribution estimators and has the advantage of not requiring a distributional assumption. In particular, the nonparametric kernel estimators (variable and log-variable kernel estimators) worked better than the parametric estimators for multimodal data. This ability to analyze multimodal density by the nonparametric method is particularly useful in hydrology. Even though only a limited data set was available and estimation outside the range of data was wanted, the log-variable kernel estimator provided good results in the upper tail compared with the variable kernel estimator.

REFERENCES

Adamowski, K. 1989. "A Monte Carlo comparison of parametric and nonparametric estimation of flood frequencies." J. of Hydrology, Vol. 108, pp. 295-308.

Adamowski, K. 1996. "Nonparametric Estimation of Low-Flow Frequencies." Journal of Hydraulic Engineering, Vol. 122, No. 1, pp. 46~49.

Adamowski, K. 1985. "Nonparametric kernel estimation of flood frequency." Water Resources Research, Vol. 21, No. 11, pp. 1585-1590.

Adamowski, K., and C. Labatiuk. 1987. "Estimation of flood frequencies b a nonparametric density procedure." Hydrologic Frequency Modeling, pp. 97-106.

Adamowski, K., and W. Feluch. 1990. "Nonparametric flood-frequency analysis with historical inforamtion." J. of Hydraulic Engineering, Vol. 116, No. 8, pp. 1035-1047.

Breiman, L., W. Meisel, and E. Purcell. 1977. "Variable kernel estimates of multivariate densities." Technometrics, Vol. 19, No. 2, pp. 135-144.

Duin, R. P. W. 1976. "On the choice of smoothing parameters for parzen estimators of probability density functions." IEEE Trans. Comput., C-25, pp. 1175-1179.

Greenwood, J. A. Landwehr, J. M. Matalas, N. C., and Wallis, J. R. 1979. "Probability Weighted Moments : Definition and Relation to Parameters of Several Distributions Expressible in Inverse Form." Water Resources Research, Vol. 15, No. 5, pp. 1049~1054.

Habbema, J. D. F., J. Hermans, and V. D. Broek. 1974. "A stepwise discrimination program using density estimation." In G. Bruckman (Ed.). Physical verlag. Compstat Vienna, pp. 100-110.

Hall, P. 1983. "Large sample optimality of least squares cross-validation in density estimation." Ann. Statist., Vol. 11, pp. 1156-1174.

Hall, P., and J. S. Marron. 1987. "Extent to which least-squares cross-validation minimizes integrated square error in nonparametric density estimation." Probability Theory Rel., Fields 74, pp. 567-581.

Hosking, J. R. M. 1989. The theory of probability weighed moments, Research Report, RC 12210. IBM Research Division, T.J. Watson Research Center, New York.

Hosking. J. R. M. 1990. "L-moments Analysis and estimation of distribution using linear combinations of order statistics." Journal of Royal Statistical Society, Vol. 52, No. 1, pp. 105-124.

Lall, U., Young-Il Moon, and K. Bosworth. 1993. "Kernel flood frequency estimators : bandwidth selection and kernel choice." Water Resources Research, Vol. 29, No. 4, pp. 1003-1015.

Moon, Young-Il, Lall, U., and Bosworth, K. (1993). "A comparison of tail probability estimators." Journal of Hydrology, Vol. 151, pp. 343-363.

Moon, Young-Il, and U. Lall. 1995. "Nonparametric flood frequency analysis by a kernel quantile function estimator." European Geology Society.

Moon, Young-Il, and U. Lall. 1994. "Kernel Quantile Function Estimator for Flood Frequency Analysis." Water Resources Research, Vol. 30, No. 11, pp. 3095-3103.

Moon, Young-Il, U. Lall, and K. Bosworth. 1993. "A comparison of tail probability estimators." Journal of Hydrology 151, pp. 343-363.

Moon, Young-Il, Young-Il Cha, and Si-Young Chun. 1999. "The Estimate of Probability Precipitation by Frequency Analysis: Parametric and Nonparametric Methods." Conference Proceedings of Korea Water Resources Association, pp. 117-122.

Rosenblatt, M. 1956. "Remarks on some nonparametric estimates of a density function." Ann. Math. Statist. 27, pp. 832-837.

Silverman, B. W. 1986. Density estimation for statistics and data analysis, New York: Chapman and Hall.

Stone, M. 1984. "An asymptotically optimal window selection rule for kernel density estimates." Ann. Statistics, Vol. 12, pp. 1285-1297.

Stochastic Hydraulics 2000, Wang & Hu (eds) © 2000 Taylor & Francis, ISBN 90 5809 166 X

Shingle beach response to swell wave action

Kaiming She & Paul Canning
The School of the Environment, University of Brighton, UK

Sally Sudworth
The Environment Agency Southern Region, Guildbourne House, Sussex, UK

ABSTRACT : Coastal erosion has long been associated with longshore transport or litteral drift generated by storm waves. At the same time, swell waves have been perceived to have benefiting effects in preserving our beaches. This may not always be the case. Field observations along the southern coastline of England have shown that shingle beaches can suffer large losses of beach material under the action of swell waves. A model study has been carried out to examine the response of a shingle beach to swells of various periods and heights. It will be shown that the onshore or offshore movement of beach material is strongly associated with both the wave period and height. Under certain conditions, swell waves can cause severe damage to the beach by removing the beach crest material and transporting it offshore into deep water where the material becomes irretrievable. The findings of this study suggest that while swells of small amplitude help to maintain the beach material, swell of sufficient heights can cause both short and long term losses of beach material, resulting erosions of beaches. This has significant implications in establishing good coastal protection schemes where needed.

Key words: Sediment transport, Swell waves, Coastal processes

1. INTRODUCTION

Coastal erosion and accretion of sand type beaches has in the past been intensively studied, with a considerable amount of data available from field and laboratory experiments. However, the study of shingle type beaches has been relatively rather limited, despite their existence on many parts of the coast, especially along the coast of the UK. Erosion of beaches is often associated storm wave attacks by way of creating a littoral drift and a sediment movement in the offshore direction. Swell waves, on the other hand, are commonly regarded as having beneficial effects on the beach by transporting sediment material back onshore. While this is generally true, field observations in the south coast of England have seen cases of noticeable loss of beach material without the presence of a storm. The field engineer suggested the possibility of strong swells causing the material loss but the hypothesis could not be verified due to a lack of research in relation to shingle beach processes under swell wave action. In contrast to the large volume of literature in sand transport in coastal waters, investigations of shingle beach response to wave action are very limited and these include laboratory scale models of Powell (1990), van Hijum et al (1982) and field studies of van Wellen et al (1997). This work consists of two series of tests under laboratory conditions and is intended to demonstrate that strong swell waves can cause severe damage to the beachhead.

Table.1: Swell wave test conditions

| Test I.D. | Full scale parameters | | Model scale parameters | | Swell effects |
	Wave Height (m)	Wave period (s)	Wave Height (mm)	Wave period (s)	Beach head loss/gain
f040h20	0.72	15.00	20	2.50	Gain
f040h40	1.44	15.00	40	2.50	Gain
f040h60	2.16	15.00	60	2.50	Gain
f040h80	2.88	15.00	80	2.50	loss
f050h40	1.44	12.00	40	2.00	gain
f050h60	2.16	12.00	60	2.00	loss
f050h80	2.88	12.00	80	2.00	loss
f067h50	1.80	9.00	50	1.50	gain
f067h70	2.52	9.00	70	1.50	loss
f067h90	3.24	9.00	90	1.50	loss
f100h50	1.80	6.00	50	1.00	loss
f100h70	2.52	6.00	70	1.00	loss
F100h100	3.60	6.00	100	1.00	loss

2. EXPERIMENTAL SETUPS AND PROCEDURES

Laboratory modelling of shingle beach processes requires appropriate choice of model sediments. The practice at Hydraulic Research Wallingford, UK (Coates, 1994) employs three scaling criteria for the selection of model sediments and these are related to the permeability, threshold of motion and onshore/offshore transport. There are deep doubts about the use of these criteria (Loveless et al, 1995) and a joint research programme involving the Universities of Brighton, Bristol and Southampton is in progress to investigate the appropriateness of different scaling laws for coastal sediment modelling. It will be some time before concrete conclusions can be reached with respect to the sediment scaling. For the current study, the sediment scaling was based on onshore/offshore criterion (She, 1996). The investigation comprises of two test series under two different offshore beach conditions as detailed below.

The experiment was carried out in 10m long wave flume with a 0.45m × 0.45m cross section. The wave generation was through a piston type random wave generator in a water depth of 0.28 m. The design of the first test series was based on a 1:36 scale. The beach of 1:7 slope was constructed with graded anthracite coal with the toe of the fully mobile beach reaching a water depth of 0.3 m. The median sediment diameter D_{50} was 0.83 mm. This was chosen to give a prototype size of 10 mm, representative of beaches on the South East UK coast (She, 1996). Details of the test parameters can be found in Table 1. The beach head was 83 mm above the still water level. Each test was run for a duration of 1800 s, corresponding to a 3 hour prototype wave action.

3. EXPERIMENTAL RESULTS AND DISCUSSION

3.1. General observation

For all the test cases, rapid transformation of the beach profile took place within the first minutes of each test, after which the profile showed a pseudo-stable state. The general direction of the sediment movement may be in the onshore direction or offshore, depending on the wave frequency and height. Both onshore and offshore sediment movement have been observed with the tested range of wave parameters.

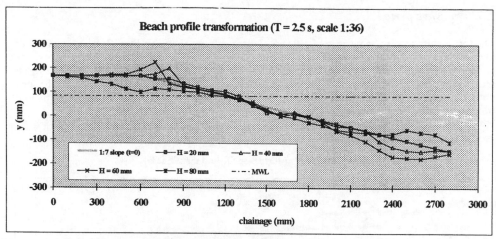

Fig.1: Transformation of beach profiles

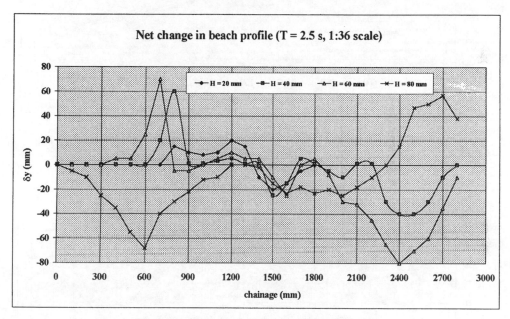

Fig.2: Transformation of beach profiles (net change)

3.2. The wave height

Swell waves are commonly perceived to result in onshore transport of sediment, leading to deposition of material at the beachhead. Swell waves thus act as a natural coastal defence measure. This was shown to be the case under the condition that the wave height did not exceed a certain value. Figure 1 shows the beach profiles at the end of four tests with the same wave period of 2.5 s. As the wave height was increased, the volume of material buildup at the beachhead due to onshore transport was increased. At a wave height of 80 mm, the onshore transport did not take place. The sediment movement took place in the offshore direction instead, resulting in a loss of beachhead material. The net change of the beach profile with respect to the initial profile are shown in Fig. 2, providing a clearer view of the beach material movement.

3.3. *The wave period*

The test results with wave periods of 1.0, 1.5, 2.0 seconds are shown in Figs. 3~8. A summary of the material movement on the beach is shown in Table 1. It can be seen that the wave period has direct influence on the threshold wave height at which the sediment movement changes from onshore to offshore direction. For a wave period of 2.5 s, the offshore transport did not happen until reaching a wave height of 80 mm. In contrast, waves of 1 s period caused offshore transport at only 50 mm height.

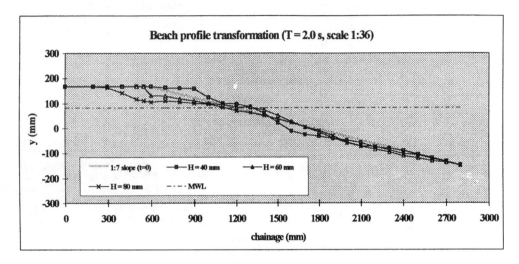

Fig.3: Transformation of beach profiles

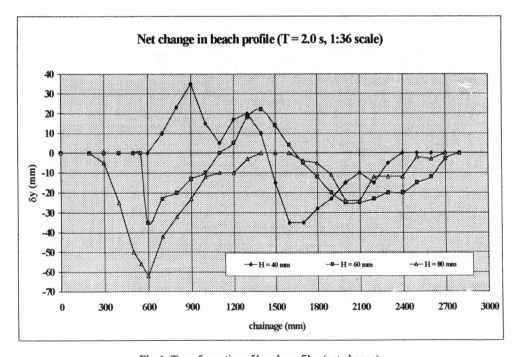

Fig.4: Transformation of beach profiles (net change)

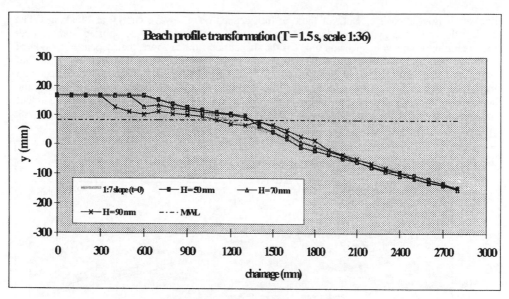

Fig.5: Transformation of beach profiles

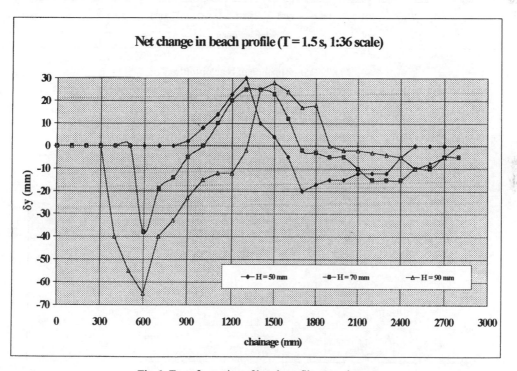

Fig.6: Transformation of beach profiles (net change)

3.4. Discussions

The overall direction of sediment movement are related to two major factors. The first is the suspension time of sediment particles with respect to the wave period and the second is the backwash flow intensity. Both of these in turn depend on the wave energy at breaking. The greater the wave energy, the greater distance the sediment particles will be uplifted from seabed, thus resulting in longer suspension time. If the suspension time is short, the suspended motion of sediment particles will be in the onshore direction. Offshore movement will still take place but under bed load only, which is small compared to the suspended movement. The net sediment movement is therefore in the onshore direction. When the suspension time is sufficiently long the sediment particles will then be carried in suspension by the backwash. The overall sediment movement can therefore take place in the offshore direction.

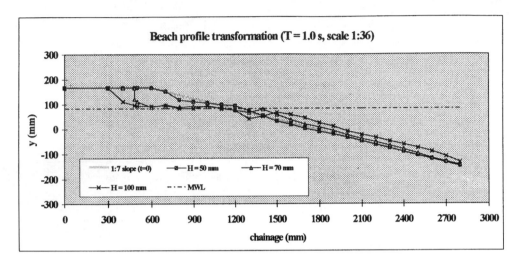

Fig.7: Transformation of beach profiles

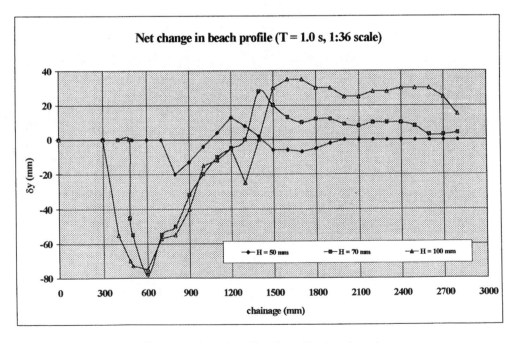

Fig.8: Transformation of beach profiles (net change)

4. CONCLUSIONS

Experiments have been carried out to investigate the beach profile response to swell wave action. The experimental results show that the effect of swell wave attack is not always to re-nourish the beach, as is often supposed. The loss of beach material happens when the swell wave height exceeds a certain value. The threshold wave height at which beachhead depletion takes place decreases with the reduced wave period.

REFERENCES

Coates, T.T., 'Effectiveness of control structures on shingle beaches: Physical model studies', Report SR 387, HR Wallingford, 1994.

Coates, T.T, 'Physical modelling of the response of shingle beaches in the presence of control structures', Proceeedings of Int. Conf. Coastal Dynamics '93, 1993, pp924-937.

Loveless, J.H and Grant, G.T., 'Physical modelling of scour at coastal structures', HYDRA 2000, 1995, vol. 3, pp293-298.

Powell, K.A., 'Predicting short-term profile response for shingle beaches', Report SR 219, HR Wallingford, 1990.

Powell, K.A., Quinn, P.A. and Greated, C.A., 'Shingle beach profiles and wave kinematics', Proceedings of Int. Conf. on Coastal Engineering, 1992, pp2358-2369.

She, K., 'Modelling of shingle beach response to storm and swell wave action', 'Modelling of shingle beach response to storm and swell wave action', Final Report, Environment Agency Shoreham and Lancing Sea Defences project, 1996.

Van Hijum, E. and Pilarczyk, K.W., 'Equilibrium profile and longshore transport of coarse material under regular and irregular wave attack', Rijkswaterstaat, Ministry of Transport and Public Works, pub. No. 274, 1982.

Van Wellen, E., Chadwick, A.J., Bird, P.A.D., Bray, M., Lee, M. and Morfett, J. 'Coastal sediment transport on shingle beaches', Proceedings of Int. Conf. Coastal Dynamics '97, 1997.

Stochastic Hydraulics 2000, Wang & Hu (eds) © 2000 Taylor & Francis, ISBN 90 5809 166 X

Comparison of uncertainty-analysis methods applied to simulation of urban water quality

Charles S. Melching
Department of Civil and Environmental Engineering, Marquette University, Milwaukee, Wis., USA

Willy Bauwens
Laboratory of Hydrology, Vrije Universiteit Brussel, Belgium

ABSTRACT: The Laboratory of Hydrology at the Vrije Universiteit Brussel (VUB) has developed a suite of computer models to simulate water quality in the Zenne River near Brussels, Belgium. The KOSIM model simulates the runoff and pollutant loads to the river, and the SALMON-Q model simulates water quality in the river during unsteady flow. These models have been used to evaluate the effectiveness of various wastewater-treatment-plant (WWTP) alternatives in improving water quality, particularly dissolved oxygen (DO) concentrations, in the river considering conditions throughout a representative one-year period. Because of the many parameters and variables involved in the coupled KOSIM, WWTP, and SALMON-Q models, an evaluation of the uncertainty in the simulated DO concentrations was done to improve confidence in the proposed water-quality-management schemes. Latin Hypercube Simulation (LHS) and the Mean-value First-Order Reliability Method (MFORM) were applied to estimate the uncertainty in simulated DO concentrations throughout the river resulting from uncertainties in 34 parameters and boundary conditions in SALMON-Q, 14 parameters in KOSIM, and 5 parameters in the WWTP model. The uncertainties in these 53 parameters were determined on the basis of measured data and information contained in the literature. The mean values of key statistics of the simulated DO concentrations estimated with the two methods agreed within 5 percent for 57 percent of the output comparisons and within 10 percent for 73 percent of the output comparisons. However, the standard deviation of key statistics of the simulated DO concentrations estimated with MFORM were more than 25 percent different from those estimated with LHS for 50 percent of the output comparisons. Thus, the nonlinearities involved in the simulation of pollutant loads, treatment processes, and instream pollutant transport over an entire year are large enough such that MFORM may not yield reliable results.

Key words: Uncertainty analysis, Water-Quality modeling, Water-Quality management, Dissolved oxygen, Nonpoint source pollution

1. INTRODUCTION

Numerous methods are available to evaluate the uncertainty in estimates made with computer models of environmental systems. Several of these methods and their applications to hydrologic and water-resources models are summarized in Melching (1995). Uncertainty-analysis methods may be grouped into two general classes: simulation-based methods and point-estimation methods. These methods determine the model-output uncertianty from the uncertainty in the basic variables of the model. The basic variables may include model parameters, boundary conditions, forcing functions, input variables, and other factors that are used in the model simulations and are not known with certainty.

In this paper, the results of two uncertainty-analysis methods one simulation based (Latin Hypercube Simulation, LHS) and the other point estimation (Mean value First-Order Reliability

Method, MFORM) are compared for an application to simulation of dissolved oxygen (DO) concentrations in the Zenne River in and near Brussels, Belgium. Water quality in the river is simulated over a one-year planning period (1986) using a coupled system of an urban runoff and pollutant-load model, constant wastewater-treatment plant (WWTP) removal efficiencies, and an unsteady flow and water-quality model. Because of the one-year simulation period and the interaction among tributary loading, treatment, and stream water-quality models more computationally intensive uncertainty-analysis methods, such as Monte Carlo Simulation (MCS), could not be applied.

2. UNCERTAINTY ANALYSIS METHODS

2.1 Mean-Value First-Order Reliability-Analysis Method

In the first-order reliability-analysis methods, a Taylor series expansion of the model-output function is truncated after the first-order term

$$C = g(X_e) + \sum_{i=1}^{p} (x_i - x_{ie})(\frac{\partial g}{\partial x_i}) x_e \qquad (1)$$

where C is the model output of interest, g(.) is a function representing the simulation process, X_e is the vector representing the expansion point, and p is the number of basic variables, x_i. In the Mean-value First-Order Reliability Method, MFORM, the expansion point is at the mean value of the basic variables. Thus, the expected value and variance of the performance function are

$$E(C) \approx g(X_m) \qquad (2)$$

$$Var(C) = \sigma_c^2 \approx \sum_{i=1}^{p} \sum_{j=1}^{p} \left(\frac{\partial g}{\partial x_i} \right) x_m \left(\frac{\partial g}{\partial x_j} \right) x_m E\left[(x_i - x_{mi})(x_j - x_{mj}) \right] \qquad (3)$$

where σ_c is the standard deviation of C and X_m is the vector of mean values of the basic variables. If the basic variables are statistically independent, the variance of C becomes

$$Var(C) = \sigma_c^2 \approx \sum_{i=1}^{p} \left((\frac{\partial g}{\partial x_i}) x_m \sigma_i \right) \left((\frac{\partial g}{\partial x_i}) x_m \sigma_i \right) \qquad (4)$$

where σ_i is the standard deviation of basic variable i. In this study, the basic variables were assumed to be statistically independent and a forward difference was used to compute the partial derivative. A Δx value of 10% was applied to all system-wide basic variables except the removal efficiencies of the activated-sludge WWTP for which a value of 1% was applied and a value of $0.1\sigma_i$ was applied to all variables that change among the drainage basins.

The advantage of MFORM is its simplicity, requiring knowledge of only the first two statistical moments of the basic variables and simple sensitivity calculations about selected central values. However, when applied to engineering design problems, in general, including watershed modeling and design, MFORM has several theoretical and/or conceptual problems (Melching, 1992). The main weakness of MFORM is that a single linearization of the model-output function, g(.), at the central values of the basic variables is assumed to be representative of the statistical properties of system performance over the complete range of basic variables. For nonlinear systems, this assumption becomes more and more inaccurate as the basic variables depart from the central values. As summarized in Melching (1995), several poignant examples are available in the water-resources literature to illustrate the nonlinearity effects on statistical properties of the system. This paper also shows the effect of nonlinearities on the statistical estimates obtained from MFORM.

Latin Hypercube Simulation, LHS, (McKay et al., 1979; McKay, 1988) is a stratified sampling approach that allows efficient estimation of the statistics of output. In LHS, the probability distribution of each basic variable is subdivided into N ranges each with a probability of occurrence equal to 1/N. Random values of the basic variable are simulated such that each range is sampled only once. The order of the selection of the ranges is randomized and the model is run N times with the random combination of basic variable values from each range for each basic variable. The output statistics, distributions, and correlation among output and input variables may then be approximated from the sample of N output values. McKay et al. (1979) showed that the stratified sampling procedure of LHS converges more quickly than the random sampling applied in MCS or other stratified sampling procedures. Accuracy is a function of the number of samples and rules are not available for choosing N. McKay (1988) suggests that N equal to twice the number of suspected important basic variables might provide a good balance of accuracy and economy for models with large numbers of parameters, but ultimately each modeler must check for convergence for the model and problem in question. Iman and Helton (1985) indicated that a choice of N equal to 4/3 times the number of suspected important basic variables usually gives satisfactory results. In this study, the number of simulations was selected as 4/3 times the number of suspected important basic variables.

Melching (1995) applied LHS to a rainfall-runoff model to compare LHS to the MCS, MFORM, and Rosenblueth methods. This comparison indicated that LHS provided the closest match to the results of MCS among the three methods applied. On the basis of the previous studies (summarized in Melching, 1995), the results of LHS are taken as closer to the truth in the method comparison done here.

The UNCSAM computer package developed by Janssen et al. (1992) was used to generate the sets of random basic-variable values corresponding to the LHS procedure. Computer programs were written to read the random parameter sets generated by UNCSAM and then to prepare input data sets for the watershed-load model, the treatment-removal model, and the stream water-quality model. The various programs were run as a batch file wherein the programs preparing model input were first run; the watershed-load models were then run; the output from the watershed-load models were converted to input to the stream water-quality model including the effects of the WWTPs; and, finally, the stream water-quality model was run.

3. A CASE STUDY

Until May 1999, wastewater generated in Brussels, Belgium, and suburban areas including nearly the entire Brussels-Capital Region and a small portion of Flemish Brabant drained untreated to the Zenne River. This area totaling 238 km^2 is drained by a combined-sewer and tributary-river system. Design, construction, and operation of wastewater-treatment plants (WWTPs) to reduce pollution discharged to the Zenne River and, thus, improve water quality in the river have been a priority of the Brussels-Capital Region for more than 15 years. A WWTP in Anderlecht, known as the Brussels South WWTP, that treats the combined-sewer flows originating in the southern suburbs and communes of Brussels (a population equivalent of 360,000) was completed in May 1999 at a cost of 148 million ECUs. Further, 700-800 million ECUs have been dedicated for the design and construction of a WWTP in Neder-Over-Heembeek, known as the Brussels North WWTP and the intercepting sewers to bring combined-sewer and tributary-river flows to the plant. The Brussels North WWTP will be designed to serve a population equivalent of 1.1 million.

When the Brussels North WWTP is completed, combined sewers from the left-bank drainage basins of the Zenne River will flow into a new collector sewer and be brought to the Brussels North WWTP. Overflows from the drainage basins on the left bank will go into the Brussels-Charleroi Canal as they currently do. The drainage to and through the Hoofdmoerriool (right-bank collector sewer) also will be treated at the Brussels North WWTP. Interceptors will be constructed to carry dry-weather flows and runoff from small storms from the Haren and Woluwe Rivers to the Brussels North WWTP. Flows higher than the capacity of the interceptors will directly flow into the Zenne along the current courses of the Haren and

Woluwe Rivers. Thus, the North WWTP will receive flows from the left bank, Hoofdmoerriool, Haren, and Woluwe basins. If the sum of these flows exceeds the WWTP capacity, the flows greater than the WWTP capacity will bypass the WWTP and directly enter the Zenne without treatment. Flows from the Brussels South combined sewer will be treated at the Brussels South WWTP and discharged to the Zenne. Flows greater than the WWTP capacity will bypass the WWTP and directly enter the Zenne without treatment.

4. MODELS APPLIED

To assist in the design of the Brussels North WWTP, the Laboratory of Hydrology of the Vrije Universiteit Brussel developed computer models to simulate pollutant loads and transport to and in the Zenne River. The computer models simulate the hydraulics and pollutant transport in the combined sewers and rivers draining to the Zenne on the basis of the data collected in previous studies, and the effects of loads from these sewers and rivers on the water quality in the Zenne. Demuynck and Bauwens, 1996) describe these models and their calibration and application to the drainage system of Brussels and the Zenne. Melching (1999) describes the assumptions applied in (1) selecting the basic variables for consideration in the uncertainty analysis and (2) determining the probabilistic characteristics of these variables.

4.1 Watershed Pollutant-Load Model

A continuous series of flows and constituent loads from the various drainage basins to the WWTPs and Zenne River were computed with the KOSIM model (Harms and Kenter, 1987). Separate KOSIM models were developed to simulate flows and pollutant loads for five basins draining to the Zenne River—Brussels South (main flow and combined-sewer overflow (CSO)), Left Bank (main flow only), the Hoofdmoerriool (main flow and three CSOs), Haren River (main flow and CSO), and Woluwe River (main flow and CSO). A 10-minute time step is used for simulation of runoff in the Brussels drainage basin using rainfall data from the Royal Meteorological Institute gage at Ukkel for a representative year (1986).

Because a continuous time series of flows and pollutant loads is computed four processes must be simulated: (1) dry-weather flows, (2) dry-weather pollutant loads, (3) storm flows, and (4) storm pollutant loads. Each of these processes has a total-volume and a time-distribution component. Pollutants simulated with KOSIM for the Brussels drainage basin included BOD, suspended solids, ammonia, and phosphorus.

The dry-weather flow per inhabitant-day, the time distribution of flows within a day, and the variability of the dry-weather flow per inhabitant-day were determined on the basis of measured dry-weather flows (Wollast et al., 1992). Pollutant loads in the dry-weather flow are represented in KOSIM as daily loads in grams per liter. These daily loads, the time distribution of the load throughout the day, and the variance in pollutant loads were determined from measured daily-load data (Wollast et al., 1992).

Storm-runoff volumes are computed in KOSIM as the sum of runoff from impervious areas and from pervious areas. These storm-runoff volumes are added to the dry-weather flow to determine the total flow. Because of an inconsistency among the tributary area runoff models for pervious areas the uncertainty in the estimation of storm runoff from pervious areas was ignored. This is reasonable because the amount of runoff from pervious areas is expected to be small compared to the runoff from impervious areas due to the large amount of impervious area in the drainage basins (watershed average 33.5 percent impervious) and the relatively low intensity storms common in Belgium.

Pollutant loads in storm runoff are computed using an average concentration that was determined on the basis of data contained in the literature. The storm loads were then added to the dry-weather loads to determine the total load.

Storm hydrographs and pollutographs for drainage subareas were computed using the Nash unit-hydrograph concept with 3 reservoirs and a storage coefficient that varied among the drainage subareas. The resulting hydrographs and pollutographs then were routed through the combined-sewer system using the hydrograph-translation method with a specified routing delay time that varies among basins. The Brussels South, Haren River, and Woluwe River basins were simulated using single subareas and, thus, routing was not done. The KOSIM model for the left bank was divided into 40 subareas. The KOSIM model of the Hoofdmoerriool was divided into 11 subareas.

In total, 14 parameters in KOSIM were considered in the uncertainty analysis. These are the storage coefficient for the Nash unit hydrograph, watershed area; percentage of pervious land cover; minimum and maximum runoff coefficients, interception loss, and depression storage for impervious areas; BOD, SS, and ammonia concentrations in storm runoff and in dry-weather flow; and the dry-weather flow.

Six of the fourteen parameters were applied system wide, whereas eight of the parameters varied among the different drainage basins. If each of these 8 parameters was considered independent in the uncertainty analysis, more than 200 variables would need to be considered and this would make uncertainty analysis computationally prohibitive. Therefore, these 8 parameters were considered as standardized variables in the uncertainty analysis. A standardized variable is computed as the variable value minus the mean divided by the standard deviation, i.e., $Z_i = (x_i - x_{mi})/\sigma_i$. In this way, the parameters for each drainage basin have a mean and variance appropriate for that basin in the uncertainty analysis. However, in the LHS, the value of the dry-weather flow, for example, will have an identical deviation (relative to its standard deviation) from its mean in each subbasin. Further, in MFORM, the change in the standardized variable is related to the change in the original variable as $\Delta Z_i = \Delta x_i/\sigma_i$. The value of ΔZ_i was taken as 0.1 for all standardized variables and the sensitivity coefficient was computed by uniformly increasing each parameter value 0.1σ, and running the model to determine the change in output resulting from this increase. In reality, the uncertainty in the determination of the dry-weather flow, for example, could result in the value for the Haren basin being higher than the estimated mean and the value for the Woluwe basin being lower than the estimated mean. Thus, some accuracy and reality is lost by the lumping of parameters as standardized variables, but this lumping is necessary to make the uncertainty analysis feasible at a preliminary level.

4.2 Wastewater-Treatment-Plant Model

The Brussels South and Brussels North WWTPs were assumed to use an activated-sludge process with a maximum capacity of 2.5 times the daily mean dry-weather flow (DWF). Flows greater than the plant capacity were assumed to flow through a sedimentation tank before entering the Zenne River. The capacity of this tank also was 2.5 times the DWF. Therefore, flows greater than 5 times DWF discharge directly in the river without any treatment. The removal of BOD, SS, and ammonia in the WWTPs was computed assuming constant removal efficiencies throughout the entire year.

The uncertainty in each pollutant-removal efficiency (5 parameters) was considered in this study. In the uncertainty analysis, the removal efficiency was assumed to be the same for both the South and North WWTPs. This was done because the exact details of the plants were not known because the South WWTP was in construction and the North WWTP has not yet been designed. Thus, the removal efficiencies were more like design targets for water-quality management than actual operational parameters.

4.3 Stream Water-Quality Model

The SALMON-Q model (Hydraulics Research Wallingford, 1993) was used to simulate unsteady flow and pollutant transport in the Zenne River. SALMON-Q simulated flows and water quality in a 20.17-kilometer (km) reach of the Zenne between Paepsem and Eppegem. This segment was divided into four reaches and 21 computational elements (segments) in SALMON-Q. The Zenne River is roofed over for 9.5 km in the modeled reach. The main portion of the covered river begins about 1.15 km upstream of the Hoofdmoerriool discharge point.

Under the future condition, flow will enter the Zenne River at six computational elements in SALMON-Q. Reach 1, element 7 (segment 7) will receive CSO's from the Woluwe basin. Reach 1, element 8 (segment 8) will receive treated wastewater from the Brussels North WWTP, flows by passing the Brussels North WWTP, and CSO flows from the Haren basin. Reach 2, element 4 (segment 12) will receive CSOs from the Hoofdmoerriool at the Nieuwe Maalbeek location. Reach 3, element 2 (segment 14) will receive CSOs from the Hoofdmoerriool at the storm basins location. Reach 4, element 1 (segment 17) will receive CSOs from the Hoofdmoerriool at the Brussels South location. Reach 4, element 4 (segment 21) will receive treated wastewater from the Brussels South WWTP and flows bypassing the WWTP.

Unsteady flow in the Zenne River is computed in SALMON-Q through solution of the full dynamic wave (de Saint Venant) equations. Thus, the flow throughout the system is a function of inputs at the boundaries and from the tributary combined sewers (computed with KOSIM), stream geometry, and flow resistance. Pollutant transport is simulated in SALMON-Q using the advection-dispersion equation with source and sink terms and decay and transformation of pollutants estimated with first-order rate constants. The loss (sink) of oxygen to biological processes in the bed sediment (sediment oxygen demand) is simulated in SALMON-Q as a function of the build up and release of water with low DO and high pollutant concentrations in the pore water as a function of the sediment deposition and erosion processes. SALMON-Q also can simulate the production and consumption of DO by algal photosynthesis and respiration, respectively. However, because the Zenne is roofed over for much of the study area algal growth was not simulated in this study.

Uncertainty in the SALMON-Q parameters of hydraulic resistance, the dispersion model, the sediment-transport model, the first-order rate constants, and the partition coefficients between dissolved and particulate BOD and between rapidly and slowly degraded BOD and organic nitrogen was considered in the uncertainty analysis. In total, 34 basic variables were considered in the uncertainty analysis of SALMON-Q.

5. RESULTS

Fifty-three uncertain basic variables were considered in this analysis, and so 70 simulations were made for LHS and 54 simulations were made to determine the local sensitivity coefficients for MFORM. For these simulations, the output from SALMON-Q includes flow and concentrations of DO, BOD, SS, ammonia, nitrite, and nitrate at the downstream end of each of the 21 computational segments at a time step of 10 minutes for an entire year. Therefore, key output was selected for analysis to keep the comparison manageable. The concentration of DO generally is considered as the primary indicator of ecosystem health. The average DO concentration over the year and the amount of time the DO concentration was less than specified values (1 mg/L, 2 mg/L, and so on) were considered at each location along the stream on the basis of DO concentrations output every 2 hours. One hundred and fifty two DO concentration levels were considered in the comparison of MFORM and LHS. The comparison was done for concentrations of 1-7 mg/L for segments 1-5 and 17-20, 1-6 mg/L for segments 6-8 and 21, 2-13 mg/L for segment 9, 2-9 mg/L for segments 10 and 13, 2-10 mg/L for segment 11, and 2-8 mg/L for segments 12 and 14-16.

The mean of the average annual DO concentration at each segment estimated with MFORM was within 2 percent of the value estimated with LHS for all segments and the average difference between MFORM and LHS was 1.16 percent. Indicating that in terms of estimating the average annual DO concentration the nonlinearities in the modeling system—KOSIM-Treatment Efficiencies-SALMON-Q—are minor. Further, for the DO concentration levels the mean amount of time less than the specified concentration estimated with MFORM was within 5 percent for 87 of 152 cases (57 %). Whereas the error in the MFORM estimate was greater than 10 percent for 40 of 152 cases (26 %). Thirty-eight of these involved DO concentrations that occurred less than 876 hours within (10 % of) the year. Therefore, when estimating the mean of the amount of time that the DO concentration was less than a specified value MFORM estimates agreed well with LHS estimates except at DO concentrations that occurred less than ten percent of the time throughout the year. This illustrates the effect of modeling system nonlinearity at extreme DO concentrations.

The comparison between MFORM and LHS estimates of model-output standard deviations is much less encouraging. The standard deviation of the annual mean DO concentration along the Zenne River estimated by MFORM and LHS is shown in Figure 1. From this figure it can be seen that there are large differences between the MFORM and LHS estimates downstream

722

from river kilometer (RK) 17. For the 21 computational segments the errors range from –2.9 percent (segment 21, RK 19.2) to 51.5 percent (segment 18, RK 16.4) with an average error of 21.5 percent.

The agreement in the estimates of the standard deviation of the amount of time less than specified values is even poorer. The error in the MFORM estimate relative to the LHS estimate is greater than 25 percent for 76 of the 152 cases (50 %) with an average absolute error for these cases of 45.5 percent. Whereas the error in MFORM estimates relative to LHS estimates was less than 10 percent for only 40 cases (26 %).

Figure 1 indicates that the agreement between MFORM and LHS estimates of the standard deviation of the annual mean DO concentration is better downstream of RK 8 (segments 1-8) than between RKs 8 and 17 (segments 9-18). Downstream of RK 8 the error in the MFORM estimate of the standard deviation of the annual mean DO concentration is 14 percent, whereas between RKs 8 and 17 the average error is 33 percent. The agreement in the estimates of the amount of time the DO concentration is less than a specified value also is significantly better downstream of RK 8 than upstream of RK 8. Downstream of RK 8, the MFORM estimates of the standard deviation agreed within 10 percent of the LHS estimates for 23 of 53 (43 %) DO concentration levels considered. Further, the error in the MFORM estimate relative to the LHS estimate was greater than 25 percent for only 16 of 53 (30 %) DO concentration levels considered, and the average absolute error in the standard deviation estimates was 25 percent. Upstream of RK 8, the MFORM estimates of the standard deviation agreed within 10 percent of the LHS estimates for only 17 of 99 (17 %) DO concentration levels considered. Further, the error in the MFORM estimate relative to the LHS estimate was greater than 25 percent for 60 of 99 (61 %) DO concentrations levels considered, and the average absolute error in the standard deviation estimates was 56 percent.

The flow from the Brussels North WWTP enters the Zenne River in segment 8 and the average annual flow in the Zenne nearly doubles from about 4.1 m³/s to about 7.7 m³/s. This large load dominates the water-quality relations in the Zenne downstream from RK 8. Whereas the flow from the Brussels South WWTP (entering in segment 21 around RK 20) only increases the flow 32 percent from about 3.1 m³/s to about 4.1 m³/s. Therefore, the load from the Brussels South WWTP does not dominate downstream water-quality conditions like that from the Brussels North WWTP. For the case, where water quality in the Zenne is dominated by a substantial forcing function the agreement between MFORM and LHS estimates of the standard deviation of model output is good. However, when the water-quality relations are not driven by a strong forcing function, the agreement between MFORM and LHS estimates of the standard deviation of model output is very poor.

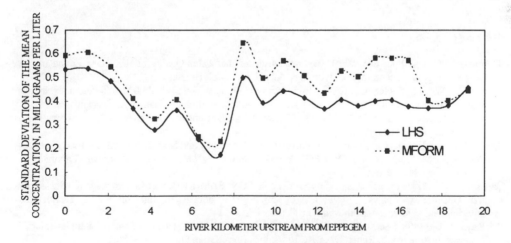

Fig.1:. Standard deviation of the annual mean dissolved oxygen concentration in the Senne River near Brussels, Belgium, simulated over 1986 as estimated with Mean-value First-Order Reliability Analysis (MFORM) and Latin Hypercube Simulation (LHS)

6. CONCLUSIONS

Latin Hydrpercube Simulation (LHS) and the Mean-value First-Order Reliability Method (MFORM) were applied to evaluate the uncertainty in the simulation of dissolved oxygen (DO) concentrations in the Zenne River in and near Brussels, Belgium. A suite of models was applied to estimate the DO concentrations in the Zenne to assist in the design of the Brussels North Wastewater Treatment Plant (WWTP). The KOSIM model was used to simulate flows and pollutant loads reaching the river from the tributary combined sewers and rivers. A simple constant treatment efficiency model was used to simulate pollutant removal at the Brussels South and North WWTPs. The SALMON-Q model was used to simulate flow and water quality in the river under unsteady conditions over a planning year (1986).

The results from LHS were taken as more correct and the results from MFORM were compared to those from LHS for the annual mean DO concentration and the amount of time DO concentrations were less than specified values for each of 21 computational segments (152 DO concentration-location combinations were considered). MFORM estimates of the mean of the model output agreed well (most errors less than 5 percent) with those from LHS except when extreme DO concentrations (resulting less than ten percent of the time) were considered. The deviations for the extreme DO concentrations reflect nonlinearities in the suite of models applied to simulate water-quality in the Zenne River. MFORM estimates of the standard deviation of model output agreed well (most errors less than 15 percent) with those from LHS for locations downstream of the point where flows from the Brussels North WWTP enter the Zenne. MFORM estimates of the standard deviation of model output agreed poorly (most errors greater than 25 percent, and average absolute error equal to 56 percent) with those from LHS for locations upstream of the point where flows from the Brussels North WWTP enter the Zenne River. Thus, MFORM works well when there is a strong forcing function (high loads from the Brussels North WWTP) affecting DO concentrations and making the local sensitivity coefficients used in MFORM reliable and consistent gages of model performance. In the absence of a strong forcing function, the local sensitivity coefficients no longer are reliable and consistent gages of model performance, and the use of LHS is preferred.

ACKNOWLEDGEMENTS

The research described in this paper was supported by the "Research in Brussels" program of the Ministry of Economics of the Brussels Capital Region in Belgium. The support of Dr. C. May, Program Director, and Mr. R. Grijp, Minister of Economics is gratefully acknowledged.

REFERENCES

Demuynck, C. and Bauwens, W., 1996, "Modelling of the Water Quality of the River Zenne in the Brussels Region," Laboratory of Hydrology, Vrije Universiteit Brussel.

Harms, R. W. and Kenter, G., 1987, Mischwasserentlastungen, KOSIM V.3.0, Mikrocomputer in der Stadtenwasserung, Institut fur Technische-Wissenschaftliche Hydrologie, Hannover, Germany.

Hydraulics Research Wallingford, 1993, "SALMON-Q User Documentation, Version 1.0," Howbery Park, Oxfordshire, United Kingdom.

Iman, R. L. and Helton, J. C., 1985, "A Comparison of Uncertainty and Sensitivity Analysis Techniques for Computer Models," *Report NUREGICR-3904, SAND 84-1461*, Sandia National Laboratories, Albuquerque, New Mexico.

Janssen, P. H. M., Heuberger, P. S. C. and Sanders, R., "UNCSAM 1.1: A Software Package for Sensitivity and Uncertainty Analysis," *Report No. 959101004*, National Institute of Public Health and Environmental Protection, Bilthoven, The Netherlands.

McKay, M. D., 1988, "Sensitivity and Uncertainty Analysis Using a Statistical Sample of Input Values," *Uncertainty Analysis*, Y. Ronen, ed., CRC Press, Inc., Boca Raton, Florida, 145-186.

McKay, M. D., Beckman, R. J. and Conover, W. J., 1979, "A Comparison of Three Methods for Selecting Values of Input Variables in the Analysis of Output from a Computer Code," *Technometrics*, 21(2), 239-245.

Melching, C. S., 1992, "An Improved First-Order Reliability Approach for Assessing Uncertainties in Hydrologic Modeling," *Journal of Hydrology*, 132, 157-177.

Melching, C. S., 1995, "Reliability Estimation," Chapter 3 in *Computer Models of Watershed Hydrology*, V.P. Singh, ed., Water Resources Publications, Littleton, Colorado, 69-118.

Melching, C.S., 1999, "Uncertainty Analysis for Holistic River Water-Quality Management Systems," Final Report to the Research in Brussels Action, Ministry of Economics, Brussels Capital Region, Belgium.

Wollast, R., et al., 1992, "Réseau de Surveillance des Écoulements et des Charges Polluantes dans les Collecteurs D'Amenée à la Future Station D'Épuration de Bruxelles – Nord," Laboratorie de Traitement des Eaux et Pollution, Université Libre de Bruxelles, Belgium.

Stochastic Hydraulics 2000, Wang & Hu (eds) © 2000 Taylor & Francis, ISBN 90 5809 166 X

Zoning of the Yellow River basin

Zhang Ouyang & Xu Jiongxin
Institute of Geography, Chinese Academy of Sciences, Beijing, China

ABSTRACT: Based on morphological and hydrological data, this paper presents a zoning of the Yellow River basin, a fluvial system, with Schumm's theory. The results meet well with the idealized fluvial system model, and they also show that the Yellow River system is a hierarchical system. The whole basin, the reaches upstream and downstream Hekouzhen each has its own water production zone, sediment production zone, and transfer zone respectively. Historically, this hierarchical system changes with the landform evolution. The subsystems from the upper to the lower form a cascading system and process-response system. The dividing of the sediment production zone and the transfer zone is important for understanding of the Yellow River basin.

Key words: The Yellow River basin, Fluvial system, System zoning

1. INTRODUCTION

The Yellow River is the Mother River of Chinese people literarily, but it is also a disastrous river due to frequently flooding. It carries heavy sediment load that makes it very active and difficult to harness. With hazards of erosion by water and wind in the middle reaches and channel aggradation and flood menace in the lower reaches, the eco-environment of the Yellow River is so vulnerable that it is always a serious worry of Chinese people. Although the people have successfully controlled the river with levees, hazards of the erosion in the middle and the risks of levee breach in the lower reaches still exist and the eco-environment of the Yellow River still needs great effort to improve and protect (Qian, et al, 1993). To solve the Yellow River problems need systematic consideration. The concept of the fluvial system proposed by Schumm (1977) provides a theoretic base for this issue. The paper employs this concept to solve the problems of the Yellow River system by dividing it into subsystems.

2. METHODS AND DATA

Schumm (1977) divided a fluvial system into water and sediment production zone, transfer zone and deposition zone. Each zone, macroscopically, can be taken as a subsystem that has a dominant way of sediment movement and acting process. The main actions of the three zones are erosion, transfer and deposition respectively. At the same time, a fluvial system has different temporal and spatial scales and is a hierarchical structure. At different temporal and spatial scales, there exists resemblance as well as difference in a fluvial system.

Comparing Schumm's idealized model with the Yellow River basin, a fluvial system, we can easily find that, in the Yellow River basin, upstream Hekouzhen is the production zone, between Hekouzhen and Taohuayu is the transfer zone and downstream Taohuayu is the deposition zone morphologically (Fig.1). And the spatial distribution of the production, transfer and deposition subsystems of the Yellow River system meets well with that of Schumm's. Further, the annual mean water and sediment amount of each station along the stem Yellow River from 1919 to 1979 are calculated to divide quantitatively the Yellow River system. The amount of the annual mean water and sediment production of each zone will get when subtract the annual mean water and sediment amount of the upper station from the amount of the lower station. The percentage of the water and sediment production amount of the very zone occupying that at the controlling outlet station based on the measured data is calculated and used as the main index to divide the Yellow

River system. At the same time, its physic geographical and geomorphic conditions are also in consideration. Using the data before 1980 is mainly because that the environment of the Yellow River was, relatively, slightly impacted at that time by human and thus these data could reflect its real natural conditions. Confined to data resources, the data upstream Hekouzhen is from 1919 to 1967 (before the construction of the Liujiaxia reservoir) (Yang, et al, 1993), and downstream Hekouzhen is from 1919 to 1979 (hydrological bureau of the Ministry of Water Resources of China). These two series of data have minor discrepancy, but it doesn't out of the range of the discussion of this paper. Because the Yellow River basin has the characteristics of water and sediment coming from different zones (Qian, et al, 1993), its water production zone and the sediment production zone are separated.

3. RESULTS

3.1 The Whole Stem System

Using the percentages of water and sediment production amount of each reach of the stem Yellow River occupying that at Lijin station as a controlling criterion, the spatial distribution of the water and sediment production along the whole Yellow River is manifestly shown in the histogram (Fig.2A).

The water production amount upstream Lanzhou is over 78%; the sediment production amount from Hekouzhen to Longmen is about 85%; from Taohuayu to Lijin, on the whole, is at the transferring state; and downstream Lijin, all the water and sediment entered this area. Traditionally, upstream Hekouzhen is the upper reach, between Hekouzhen and Taohuayu is the middle reach and downstream Taohuayu is the lower reach (Qian, et al, 1993). Thus, the water comes mainly from the upper reach; the sediment comes mainly from the middle reach; the lower reach offers a route for water and sediment transporting; and the delta is the depositional site. Simplified Fig.2A to Fig.2B, the subsystems of the whole stem system is manifested. The upper reach upstream Hekouzhen is the water production zone (subsystem); middle reach between Hekouzhen and Taohuayu (Huayuankou station data) is the sediment production zone; the lower reach from Taohuayu to Lijin is the transfer zone; and downstream Lijin is the deposition zone. The general situations of these subsystems are listed in table 1.

Table.1: General situations of the Yellow River basin subsystems

subsystem	Water production zone	sediment production zone	Transfer zone	Deposition zone
Drainage area (km²)	386000	344000	22000	574
River length (km)	3472	1206	682	103
Number of tributaries*	43	30	3	0
Water fall (m)	3496	890	94	7.4
Sediment production (%)	14	103	-17	-100
Water production(%)	62	51	-13	-100

*Drainage area >=1000 km².

The Yellow River basin could be taken as a hierarchical system. Besides the whole basin, within the upper reaches upstream Hekouzhen and the middle and lower reaches downstream Hekozhen have similar distribution of water production zone, sediment production zone, and transfer zone.

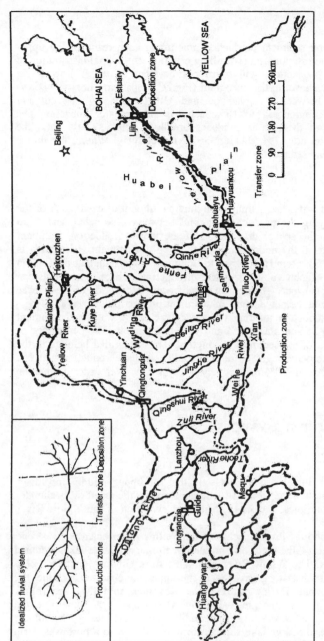

Fig.1: Sketch map of the Yellow River basin (the inset diagram after Schumm, 1977).

729

3.2 Upstream Hekouzhen

Using the percentages of water and sediment production amount of each reach of the upper reaches occupying that at Hekouzhen station as a controlling criterion, the spatial distribution of the water production, sediment production, and transfer subsystem of the Yellow River upstream Hekouzhen is manifestly shown in the histogram (Fig.2C). Upstream Longyangxia is the water production subzone, between Longyangxia and Qingtongxia is the sediment production subzone, and from Qingtongxia to Hekouzhen is the transfer subzone. The percentages of water production of these three subzones are 80%, 39%, -19%, and the percentages of sediment production are 12%, 124%, -36% respectively. Where, the minus mark represents the loss of water and sediment.

3.3 Between Hekouzhen and Lijin

Similarly, using the percentages of water and sediment production amount of each reach of the middle and lower reaches occupying the production amount between Hekouzhen and Lijin station as a controlling criterion, the spatial distribution of the water production, sediment production, and transfer subsystem in the middle and lower reach downstream Hekouzhen is manifested in the histogram (Fig.2D). From Hekouzhen to Longmen is the sediment production subzone, between Longmen and Taohuayu is the water production subzone, and between Taohuayu and Lijin is the transfer subzone. The percentages of water production of these three subzones are 42%, 95%, -36%, and the percentages of sediment production are 98%, 22%, -20% respectively. In the water production subzone, the percentage of the water and sediment production between Longmen and Sanmenxia is 64% and 41% respectively. The percentage of water production is more than that of the sediment production. Thus, this area belongs to the water production subzone. And the percentages of the water and sediment production between Sanmenxia and Taohuayu are 31% and -19% respectively. This area has net water production and net sediment loss.

4. RELATIONSHIPS AMONG THE SUBSYSTEMS

4.1 Dynamic Coupling Relationship

The formation of this hierarchical fluvial system, a system that the upper, middle and lower reaches (subsystems) of the river basin have similar structures, is mainly the result of its climate, soil, and landform evolution. Among these factors the historical evolution of the Yellow River landforms is primarily important.

Historically, the upper, middle and the lower reaches of the Yellow River basin were, at one time, separate fluvial systems. During the early and middle Quaternary, a large lake occupied Qiantao Plain of the Yellow River. The Yellow River originated from Qilian Mountain 1.2Ma BP, and the Huangshui River and the Datong River were the upper regions. At this time, the Qiantao plain was the deposition zone. During the Huanghe Movement in 1.2Ma, the Yellow River cut through the Jishi Gorge and ran into Linxia-Lanzhou Basin, and at the same time, cut through the Sanmenxia Gorge (Li, et al., 1996). From middle and late times of the middle Pleistocene to the late Pleistocene, the large lake shifted north and shrank to a northwest strip lake. Later, the water of the lake effused outside and the ancient lake disappeared gradually. Thus, the Yellow River channel south Baotou formed. But the newly formed channel was unstable and shifted in the Qiantao Plain frequently that, as a result, small lakes developed in the spacious flood-plain sometimes (Min, et al., 1998). After the river cut the Sanmenxia Gorge through, the channels upstream Sanmenxia eroded head-ward and eventually connected with the channels of the Qiantao Plain. Then, the reaches in Qiantao Plain converted to the transfer zone. During the Gonghe Movement in 0.15Ma, the Yellow River eroded head-ward continuously to the Gonghe Basin upstream Longyangxia. At the same time, the Tibet Plateau uplifted to near the modern altitude, which caused the inner Tibet Plateau and west China to become drier (Li, et al., 1996). At this time, the basic modern structure of the upper Yellow River is, by and large, formed. Upstream Longyangxia is the water production zone. From Longyangxia to

Qingtongxia is the sediment production zone in which the river runs through the Loess Plateau, of which the soil can be easily eroded. And between Qingtongxia and Hekouzhen is the transfer zone because it located in arid and semi-arid region that the stem channel lacks tributaries, and at some reaches, the channel is surrounded by desert.

Before the through-cutting of the Sanmenxia Gorge, the channel of the Yellow River upstream Sanmenxia was the deposition zone where the ancient Sanmen Lake once located. With the through-cutting of the Sanmen Gorge, the ancient Sanmen Lake disappeared and the channel of the stem Yellow River upstream Sanmenxia came into being. After the Sanmen Gorge was cut through, the valley slope and the channel gradient upstream Sanmenxia increased rapidly with the strong head-ward erosion. The river in these reaches (especially the reaches between Hekouzhen and Longmen) ran through the Loess Plateau, which can be easily eroded with a large amount of sediment, and therefore this zone converted from deposition zone to the sediment production zone. The sediment from the Loess Plateau ran through the Sanmen Gorge, deposited near the outlet and then transported downward, and eventually formed the alluvial fan of the Yellow River and the alluvial plain—Huabei Plain. At this time, downstream Sanmenxia was the deposition zone. The fore side of this zone extended seaward, and the rear side channeled gradually. Because of the embankment, the channel narrowed and the sediment deposited within the channel bed consequently. So the height of the artificial levees have to be raised now and then to cope with the increasing elevation of the bed. With the raising artificial levees, this reach became the hanging river with few tributaries and therefore became the transfer zone.

After the through-cutting of the Sanmen Gorge, the lower Yellow River breached and shifted frequently and as a result, each breaching and shifting built up a new small fan. Many of these small fans constituted the modern Yellow River fan, of which the building up material was identified as the sediment after the late Holocene (Ye, et al., 1990). When the channel was breached to build up fan, this reach was deposition zone; and when it channeled, this reach became transfer zone to transport the sediment into sea.

The Yellow River delta, the deposition zone, is composed of a serial of small deltas. The constructive process of the delta is very similar to that of the Yellow River fan (Ye, et al., 1990). When the channel breached, it built up small deltas; and when channeled, it transported the sediment to the sea but still with a lot of sediment deposition.

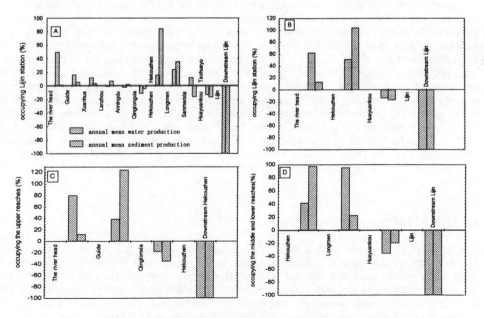

Fig.2: Water and sediment distribution of the Yellow River basin

731

4.2 Process-response Coupling Relationship

Each subsystem (zone or subzone), as stated above, can be considered to be an open system. Each has its own set of morphologic attributes, which can be related to water discharge and sediment movement (Schumm, 1973). The subsystems of the Yellow River from the upper to the lower form a cascading system for the output of the water and sediment production zones is the input of the transfer zone. It is also a process-response system. The production subsystems yield water and sediment discharge and transport them through the transfer system to the deposition system. For the alluvial channel, this discharge acts as a main forcing process on the channels (transfer zone) to change its morphology. The channel system would, inevitably, response to this process by changing its morphology. This process-response relationship can be expressed as:

$$\left.\begin{array}{l} B = F_1(J_0, Q_0, G_0, D_0) \\ h = F_2(J_0, Q_0, G_0, D_0) \\ J = F_3(J_0, Q_0, G_0, D_0) \end{array}\right\} \tag{1}$$

Where, B is the channel width, h is the depth, J is the channel gradient, J_0 is the valley slope, Q_0, G_0, D_0 are the valley slope, water discharge and its process, sediment discharge and its process, sediment composition of the source areas respectively. J_0 represents the energy dissipation of the whole drainage basin. Q_0, G_0 and D_0 reflect the attributes of the transfer zone and have an important impact on the channel morphology of the transfer zone. B, h and J represent the main geomorphic characteristics of the transfer zone and their adjustment processes are the response to the water and sediment input, and, in reverse, influence the processes of the source areas. This process-response relationship of the subsystems stated above is coupled by equation (1). Studies have revealed that the water and sediment production system and the transfer (channel) system of the whole Yellow River basin have a strong coupling relationship (Xu, 1997). This coupling of the sediment production zone (the Loess Plateau) and the transfer zone (the channel of the lower reach) is primarily important for the Yellow River harnessing.

The reaches of the Yellow River from Qingtongxia to Hekouzhen and from Taohuayu to Lijin, the transfer zones, are all alluvial channels and the adjustments of the channel morphologies are all dependent on the water and sediment inputs of the source areas. For the longer time scale, the channels of the two have uplifting tendencies and their plane morphologies are all varied frequently (Qian, et al., 1993). But owing to the difference of the water and sediment attributes input into the channels and the geographical conditions of the two reaches, the siltation process and the channel forming process are also different, and as a result, causing different environment problems. Thus, the coupling relationships of the source systems and the channel systems of the two reaches are different. If the contrast study of the two reaches could carry out with equation (1) to find out the channel pattern change thresholds and the siltation thresholds, the following questions could be answered. Under what water and sediment conditions and combination, the channel will change its pattern? Under what conditions the channel will keep unsiltation? Then the attentional conservation and harnessing of the sediment production zones is in favor of harnessing the channels of the transfer zone and preventing the flood and sediment hazards.

5. CONCLUSIONS

The drainage basin of the Yellow River is a very complex system. It is divided hierarchically based on morphological and hydrological data in this paper with Schumm's fluvial system theory. The results meet well with the idealized fluvial system model. Thus, an example of the Schumm's idealized fluvial system model is provided. On the whole, the upper reach upstream Hekouzhen is the water production zone (subsystem); middle reach between Hekouzhen and Taohuayu is sediment production zone; and downstream Lijin is the deposition zone. Within the upper reaches upstream Hekouzhen and the middle and lower reaches downstream Hekozhen, the river have their own water production zone, sediment production zone, and transfer zone respectively.

This zoning of the Yellow River basin is not exclusive. Indeed, the subsystems are in the change in view of large time scale. For example, the piedmont of the Yellow River is previously deposition zone, but now it is the transfer zone. Each subsystem has its characteristics of geomorphologic process. And these subsystems need to be coupled as a whole. The subsystems from the upper to the lower form a cascading system by material (water and sediment, etc.) movement and energy dissipation. The Yellow River basin is also a process-response system because the subsystems are firmly connected by the input and output of the material and energy. The coupling of the sediment production zones and the transfer zones of the upper reaches and the middle and lower reaches is important for the Yellow River harnessing.

ACKNOWLEDGEMENTS

This study is part of the project (No.: 59890200) supported by the National Natural Science Foundation and the Ministry of Water Resources of China. Thanks are due to Professor Zhaoyin WANG for his constructive comments on this paper.

REFERENCES

DE Boer, D.H., Hierarchies and spatial scale in geomorphology, Geomorphology, 1992, Vol.4, pp.303-318.

Hydrological bureau of the Ministry of Water Resources of China, 1982, Statistics of the attributes of the hydrological data of the main rivers in China (the first part, from 1919 to 1979).

LI, Jijun, Xiaomin Fang, Xiaozhou Ma, et al., Geomorhological process of the upper Yellow River and uplift of Qinghai-Xizang (Tibet) Plateau during Late Cenozoic times, Science in China (series D, 1996, Vol.26, No.4, pp. 316-322. (In Chinese).

MIN, Longrei, Zhenqin Chi & Guanxiang Zhu, Environment change of Qiantao Plain in the Quaternary, In: Zhisheng An (edited), The Loess, the Yellow River and the Yellow River culture, Zhengzhou: The Yellow River water conservancy press,1998, pp.50-54. (In Chinese).

QIAN, Yiyin, et al, 1993, Water and sediment changes and channel process of the stem Yellow River, Beijing: Chinese constructive industrial press, pp.230. (In Chinese).

SCHUMM, S.A., 1977, The fluvial system, New York: John Wiley and Sons, pp.338.

XU, Jiongxin, 1997, A study on the coupling relation between the water and sediment yield sub-system and river channel deposition system: an example from the Yellow River, Acta geographic sinica, Vol.52, No.5, pp.421-429. (In Chinese).

YANG, Laifei, et al, 1993, the prediction of the scour and siltation attributes of the channels from Lanzhou to Hekouzhen after the construction of large reservoirs in the upper Yellow River. see: Foundation Commission of the Yellow River water and sediment change research, proceedings of the Yellow River water and sediment change research (3), pp1-32.

YE, Qingchao, et al., 1990, Fluvial geomorphology of the lower Yellow River, Beijing: Science Press, pp.268. (In Chinese).

Wave and coastal processes

Stochastic Hydraulics 2000, Wang & Hu (eds) © 2000 Taylor & Francis, ISBN 90 5809 166 X

Velocity measurements in an oscillatory boundary layer under irregular waves

Hitoshi Tanaka, Mustafa Ataus Samad & Hiroto Yamaji
Department of Civil Engineering, Tohoku University, Sendai, Japan

ABSTRACT : Present paper describes the experimental system developed in order to measure bottom boundary layer properties under irregular waves. Measured velocity data shows very good comparison with laminar analytical solution and with transitional behavior for sinusoidal waves emphasizing the accuracy of the experimental system. A set of experimental data has also been presented for smooth turbulent flows. Wave irregularity introduces a wide range of pressure gradient conditions which subsequently influences the turbulent behavior in the boundary layer. Existence of logarithmic layer in vertical velocity profiles for many phases facilitated the calculation of bottom shear stress variation. Turbulence is generated under higher waves in the wave terrain and dissipates in the presence of successive smaller waves. This gradual dissipation leaves traces of turbulence in much smaller waves preceded by large turbulent waves, a behavior found to be characteristic under irregular waves.

Key words : Irregular wave, Bottom boundary layer, Turbulence

1. INTRODUCTION

The complex behavior of sediment movement under waves is mainly dependant on the hydrodynamics of shear flows close to the bottom. As waves approach the coast, a thin boundary layer is established at the bottom and fluid-sediment interaction becomes significant with the production and dissipation of turbulent energy. Consequently over the years there have been many studies to investigate bottom boundary layer properties under waves. Most of these studies mainly considered regular sinusoidal wave motion in the free stream (Kamphuis, 1975; Hino et al., 1976; Sleath, 1987; Jensen et al., 1989).

Considering that ocean waves are random in nature and the complexities in determining the governing forcing for sediment transport beneath irregular waves, it is of much practical value that bottom boundary layer properties are investigated under such waves. Although several researchers have considered non-linearity in wave motion to study bottom boundary layer flows (Nadaoka et al., 1994; Tanaka et al., 1998), only very few studies have so far been reported on irregular wave bottom boundary layers. The experimental results from Simons et al. (1994) had been the only notable in recent years. They had measured the time variation of bottom shear stress with a shear plate device in a two-dimensional wave basin and provided some useful information on the variation of bottom shear stress. Other studies on irregular wave bottom boundary layer were mainly analytical and aimed at providing spectral parameterization of boundary layer properties (Madsen et al., 1988; Myrhaug, 1995 etc.). Very recently Samad et al. (1999) had presented some preliminary experimental results measured in a wind tunnel for plane bottom condition. These results show very interesting behavior of turbulence generation and dissipation in the boundary layer.

Present paper describes the experimental system developed in order to measure bottom boundary layer properties under irregular waves. Measured velocity data shows very good comparison with laminar analytical solution and with transitional behavior for sinusoidal waves emphasizing the accuracy of the experimental system. A set of experimental data has also been presented for smooth turbulent flows. Wave irregularity introduces a wide range of pressure gradient conditions which subsequently influences the turbulent behavior in the boundary layer. Existence of logarithmic layer in vertical velocity profiles for many phases facilitated the calculation of bottom shear stress variation. Turbulence is generated under higher waves in the wave terrain and dissipates in the presence of successive smaller waves. This gradual dissipation leaves traces of turbulence in much smaller waves preceded by large turbulent waves, a behavior found to be characteristic under irregular waves.

2. EXPERIMENTAL SET-UP

2.1 Experimental system

The principle of the experimental system has been based on utilizing a servo-motor at the core. The input piston signals, pre-obtained from generated irregular wave free stream velocities, have been applied to drive the servo-motor. Flow measurements then have been performed in a wind tunnel connected through a piston system. As such the experimental system consists of two major components, a) a wave generation unit and b) a flow measuring unit.

a) Wave generation unit: The wave generation unit is made up of signal control and processing components along with piston mechanism. The piston displacement signal has been fed into the instrument through a PC. Input digital signal has been converted to corresponding analog data through a digital-analog converter. The analog signal drives the servo motor connected through a servo motor driver. The piston mechanism has been mounted on a screw bar which again was connected to the servo motor. The feed-back on piston displacement, from one instant to the next, has been obtained through a potentiometer that compared the position of the piston at every instant to that of the input signal, and subsequently adjusted the servo-motor driver for position at the next instant. A smooth piston movement has been ensured repeating the process until the end of the input signal has been reached.

b) Flow measuring unit: The flow measurement unit comprised of a wind tunnel and a one-component laser Doppler velocimeter (LDV) for flow measurement. The wind tunnel is connected to the piston system and has a dimension of 5m in length, 20cm and 10cm in width and height respectively. The wind tunnel was constructed from mirror-plane PVC plates on all four sides. Near the measuring section the side walls were made of transparent fiber-glass sheets to facilitate LDV measurements. A schematic diagram of experimental set-up is shown in Fig.1.

2.2 Accuracy of experimental system

Input piston displacement signal to run the experiment has been derived from irregular wave free stream velocity data generated by using Bretschneider-Mitsuyasu (Mitsuyasu, 1970) spectral density formulation. From a long series of velocity data, a short segment, typically containing about 10 waves, has been selected for the experiments. Selected input signal has then been run cyclically for 50 cycles to facilitate computation of ensemble average quantities. Validation tests for the accuracy of experimental system have been performed at various levels. The primary one being the comparison of input and recorded piston displacement and free stream velocity as shown in Fig.2. It can be seen that the agreement in piston displacement (D_p) is nearly perfect (z_{cl} is depth to the center-line of the wind tunnel). Although sometimes the fine variations in free stream velocity (U, normalized by significant free stream velocity $U_{1/3}$) are not exactly reproduced, as marked with arrows, however, the overall accuracy is generally very high.

The accuracy has been further assessed for laminar and turbulent flow conditions in Fig.3. Considering that irregular waves can be resolved into infinite number of component waves, it is possible to obtain vertical velocity variation in the boundary layer analytically from free stream velocity for laminar motion. Such as given by the following sinu-Fourier expression:

$$u = \sum_i A_{Ui} \left\{ \cos\left(2\pi f_i t + \phi_i\right) - \exp\left(-\beta z\right) \cos\left(2\pi f_i t + \phi_i - \beta z\right) \right\} \tag{1}$$

where u is velocity at a depth z, A_{Ui} are amplitude of component waves, f_i ($=1/T_i$, T_i = component wave periods) are component frequencies, t is time, ϕ_i are component phases and β is a function reciprocal to Stokes layer thickness (δ_i), i.e., $\beta=(\omega_i/2\nu)^{0.5}$, ω_i are component wave frequencies ($=2\pi/T_i$) and ν is kinematic viscosity. For a set of measured data for laminar motion, vertical velocity profiles have been compared with corresponding analytical solution in Fig.3(a). The agreement here is quite satisfactory with the reproduction of even very fine variation in velocity overshooting.

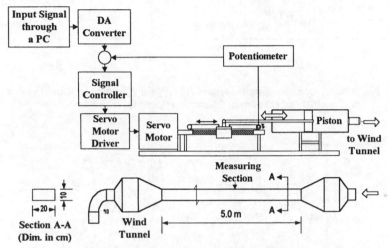

Fig.1: Schematic diagram of experimental set-up

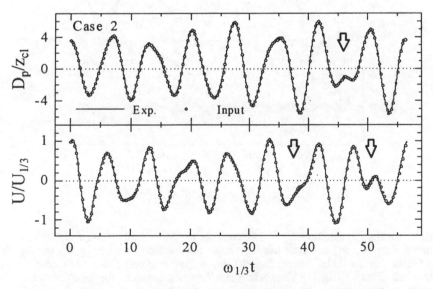

Fig.2: Comparison of measured piston displacement and free stream velocity with input data

739

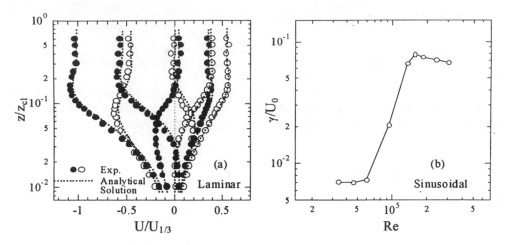

Fig.3: Accuracy of experimental system; (a) irregular wave laminar vertical velocity profiles and (b) average fluctuating velocity for sinusoidal waves

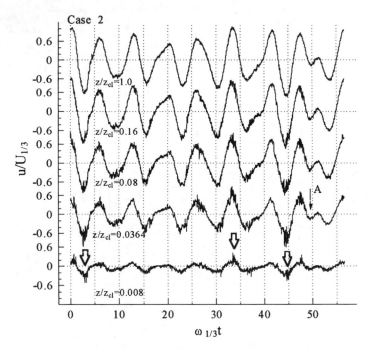

Fig.4: Raw velocity records at selected elevations

It is rather difficult to examine the performance of the system directly under turbulent motion. A qualitative comparison, however, has been made for the transitional Reynolds number from measured data for sinusoidal waves. Fig.3(b) shows the variation of measured time averaged fluctuating velocities (γ) against sinusoidal wave Reynolds numbers, Re ($=U_0^2/\omega\nu$, where U_0 is free stream velocity amplitude). Fluctuating velocities have been measured at a depth of $z/\delta_f\approx1.60$. Measured transitional range is in very good agreement with those observed by, for example, Hino et al. (1976) and Jensen et al. (1989), Re=1.04□10⁵ and 1.60□10⁵ respectively.

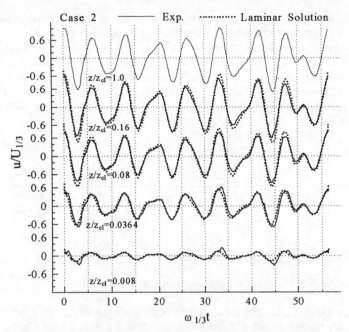

Fig.5: Phase ensemble averaged velocities along with corresponding laminar analytical solutions at selected elevations

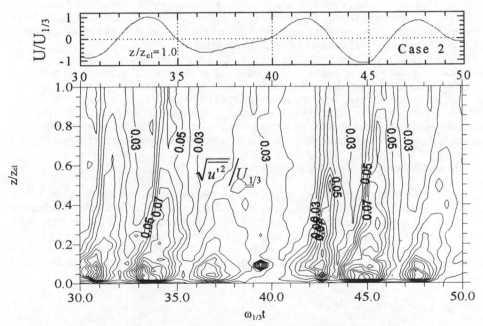

Fig.6: Contour plot of fluctuating velocity for selected time segment, Case 2

2.3 Experimental Conditions

Experiments have been carried out for two turbulent cases. The experimental conditions are presented in Table 1, where $Re_{1/3} = U_{1/3}^2/(\omega_{1/3}\nu)$ is significant wave Reynolds number, $\omega_{1/3}$ is

741

significant wave frequency ($=2\pi/T_{1/3}$), $T_{1/3}$ is significant wave period and ρ is mass density of air. In this paper details of measurement data has mainly been presented from Case 2.

For sinusoidal waves the critical Reynolds number (Re_{cr}) at which transition starts has been proposed by many researchers as mentioned already. The value generally suggested is around $Re_{cr}=1.5\square10^5$. In addition to significant wave Reynolds number, Reynolds number (Re_p) corresponding to half wave period (T_p) and velocity peak (U_p) has also been utilized in this study. The variation of half wave Reynolds numbers within selected wave segments for both the cases is presented later in Fig.7.

Table 1: Experimental conditions

Run	$T_{1/3}$ s	ρ g/cm³	ν cm²/s	$U_{1/3}$ cm/s	$Re_{1/3}$
Case 1	3.0	0.00119	0.152	507	$8.06*10^5$
Case 2	2.5	0.00115	0.160	438	$4.73*10^5$

3. MEASURED VELOCITY TIME VARIATION

3.1 Raw velocity records and ensemble averaged velocities

Velocities have been measured at different elevations in the wind tunnel closely spaced near the bottom. Raw velocity records at selected elevations are presented in Fig.4 for Case 2. A wide range of flow situations can readily be identified from the figure. While waves with relatively higher magnitude show turbulent behavior (marked with arrows), the turbulent energy is dissipated in the presence of successive smaller waves. Also smaller waves those follow large turbulent waves show presence of turbulence indicating gradual energy dissipation process. This is characteristic of irregular wave boundary layer and is particularly significant as it shows turbulent behavior in a wave having wave Reynolds number in laminar region from sinusoidal wave consideration (such as that marked as 'A' in the figure).

Phase ensemble averaged velocities at selected elevations along with corresponding laminar analytical solution are shown in Fig.5. Close to the wall at phases with high turbulent intensities ensemble averaged velocities are larger in magnitude than corresponding laminar solutions, whereas, away from the wall these are much reduced. The behavior is in agreement with those first pointed out by Hino et al. (1976) under sinusoidal waves. Near the wall turbulence production and subsequent momentum exchange causes velocity acceleration, while away from the wall turbulent mixing reduces flow velocities than corresponding laminar solution.

3.2 Fluctuating velocities and level of turbulence

The variation of turbulent fluctuating velocities over depth is shown in Fig.6 for Case 2 for a selected time span. High turbulent intensities are produced near the bottom under decelerating phases and gradually dissipate over the depth. However, under accelerating phases a favorable pressure gradient condition forces a reduction in the fluctuating velocities.

To analyze the presence of turbulence within individual waves, an averaging of measured fluctuating velocities has been made over depth and over time corresponding to half wave periods. Such that:

$$u'_s = \frac{1}{T_{pi}} \int_0^{T_{pi}} \left(\frac{1}{z_{cl}} \int_0^{z_{cl}} \sqrt{\overline{u'^2}}\, dz \right) dt \qquad (2)$$

where u'_s is average fluctuating velocity and $\sqrt{\overline{u'^2}}$ is root mean squared fluctuating velocity.

The variation of normalized u'_s along with those for Re_p for both the cases is presented in Fig.7. These are plotted against the time for corresponding half wave velocity peaks. Re_p values for both the cases indicate flow in turbulent region for most part of the wave terrain. As a result, average half wave fluctuating velocity varies in the same way with Re_p showing higher

742

magnitude of turbulence at higher Reynolds number waves. However, for half waves marked with A18, A23, B12 or B17 a reduction in fluctuating velocities can be observed whereas Re_p shows an increase than those from preceding waves. Such behavior is typical under irregular waves and has been elaborated later in the section.

Figure 8 shows the correlation of u'_s, normalized by corresponding half wave free stream velocity peaks (U_p), with Re_p. In the turbulent region, except for some scatter, variation of average fluctuating velocity is fairly constant. This indicates that average properties of velocity fluctuation, after all, has some regular trend. Further detail examination of the scatter reveal some interesting behavior. Half waves marked as A12 and A23 or B3 and B17 (in Fig.7) have Re_p values in the same range, however, they show a wide variation in normalized average fluctuating velocities, u'_s. Waves A12 or B3 are preceded by higher turbulent waves and the process of gradual turbulence dissipation caused some turbulent energy to be carried into these waves. On the other hand waves A23 or B17 follow smaller laminar waves within which most of the turbulent energy has already been dissipated, therefore, show lower level of turbulence.

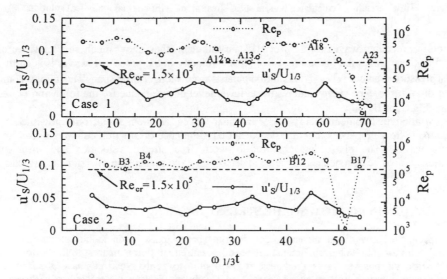

Fig.7: Time variation of average fluctuating velocities and half wave Reynolds numbers

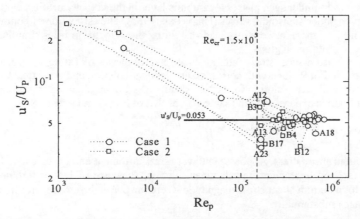

Fig.8: Variation of normalized average fluctuating velocity with half wave Reynolds number

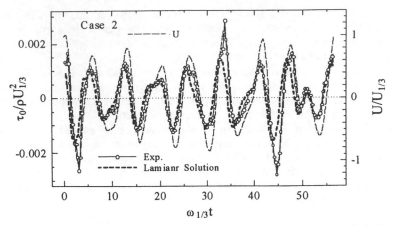

Fig.9: Time variation of calculated bottom shear stress along with laminar analytical solution

Although preceding waves for A18 or B12 are turbulent, they show much smaller value of u'_s/U_p. For these waves the variation of free stream velocity in the preceding waves is such that smaller pressure gradient ($\partial U/\partial t$) condition persists over a longer time span. This allows some dissipation of turbulence energy and results in lower level of turbulence for the following waves (see also Fig.6 for B12).

The variation in turbulent energy under irregular waves, as such, is not only governed by instantaneous flow properties, but also dependant on turbulence dissipation conditions from the previous waves. The phase lag observed between free stream velocities and turbulent fluctuations would be of much significance in determining the suspension of sediment particles.

4. VARIATION IN BOTTOM SHARE STRESS

In the absence of any direct measurement, bottom shear stress needs to be estimated from measured velocity data. When sufficient number of measurement points nearest to the bottom are located inside viscous sub-layer thickness, it is possible to apply Newton's law of viscosity successfully to compute bottom shear stress. Under turbulent motion the thickness of the sub-layer reduces substantially and the nearest measurement point to the wall may not always be located inside. However, under such phases logarithmic layer in the velocity profiles would exist as has been observed under sinusoidal waves (Hino et al., 1983). The estimation of bottom shear stress, therefore, may be most successfully made by fitting the velocity profiles simultaneously in the viscous and logarithmic regions.

Time variation of bottom shear stress determined at selected phases by fitting vertical velocity profiles is presented in Fig.9 along with corresponding laminar solution and free stream velocity for Case 2. Laminar analytical solution for bottom shear stress can be obtained from free stream velocity applying Fourier expansion in the same way as that of Eq.(1), such that:

$$\frac{\tau_0}{\rho} = \sum_i \beta \nu A_{Ui} \left\{ \cos\left(2\pi f_i t + \phi_i\right) - \sin\left(2\pi f_i t + \phi_i\right) \right\} \tag{3}$$

where τ_0 is the bottom shear stress. At phases with very small turbulent energy, calculated bottom shear stress shows good agreement with laminar solution. Presence of higher turbulence are characterized by high bottom shear stress magnitude where the phase difference with free stream velocity also reduces substantially.

CONCLUSIONS

An experimental system has been developed to measure flow velocities in the bottom boundary layer under irregular waves. The system is based on the utilization of a servo-motor as the core. The system shows high accuracy in reproducing input signals and flow conditions for both laminar and turbulent cases. Experiment data for plane bottom condition has also been presented for two turbulent cases.

Wave irregularity introduces a wide range of flow scenarios covering both laminar and turbulent regimes which is also reflected in the velocity records. Average turbulent behavior has been analyzed on a half wave basis. It shows that the level of turbulence under any wave is not only dependant on instantaneous flow properties but also depend on the turbulence conditions in the preceding waves and on the variation in pressure gradient. As a result, much higher turbulent energy can be observed in a relatively smaller wave, whereas, sufficiently larger waves show smaller turbulent intensities. Apart from this effect, the average fluctuating velocities show regular trend, irrespective of half wave Reynolds numbers, when normalized by corresponding half wave free stream velocity peaks.

The bottom shear stress has been successfully calculated by fitting measured vertical velocity profiles simultaneously for viscous and logarithmic layers. Under waves where sufficient turbulent energy is produced the bottom shear stress shows high magnitude and becomes almost simultaneous with free stream velocity.

REFERENCES

Hino, M., Sawamoto, M. and Takasu, S. (1976): Experiments on transition to turbulence in an oscillatory pipe flow. *J. Fluid Mech.*, Vol. 75, pp. 193-207.

Hino, M., Kashiwayanagi, M., Nakayama, A.and Hara, T. (1983): Experiments on turbulence statistics and the structure of a reciprocating oscillatory flow. *J. Fluid Mech.*, Vol. 131, pp. 363-400.

Jensen, B.L., Sumer, B.M. and Fredsøe, J. (1989): Turbulent oscillatory boundary layer at high Reynolds numbers. *J. Fluid Mech.*, Vol. 206, pp. 265-297.

Kamphuis, J.W. (1975): Friction factor under oscillatory waves. *J. Waterways, Harbor and Coastal Engng. Division, ASCE*, Vol. 101, No. WW2, pp.193-203.

Madsen, O.S., Poon, Y.K. and Graber, H.C. (1988): Spectral wave attenuation by bottom friciton: Theory, *Proc. 21st Int. Conf. Coastal Engng.*, pp. 492-504.

Mitsuyasu, H. (1970): On the growth of spectrum of wind-generated waves. *Proc. Coastal Engng., JSCE*, Vol. 17, pp. 1-7. (In Japanese).

Myrhaug, D. (1995): Bottom friction beneath random waves. *Coastal Engng.*, Vol. 24, pp. 259-273.

Nadaoka, K., Yagi, H. Nihei, Y. and Nomoto, K. (1994): Characteristics of turbulent structure in asymmetric oscillatory flow. *Proc. Coastal Engng., JSCE*, Vol. 41, pp.141-145. (In Japanese).

Samad, M.A., Tanaka, H. and Yamaji, H. (1999): Experiments on bottom boundary layer under irregular waves. *Proc. Coastal Engng., JSCE*, Vol. 46, pp. 21-25. (In Japanese).

Simon, R.R., Grass, T.J., Saleh, W.M. and Tehrani, M.M. (1994): Bottom shear stress under random waves with a current superimposed. *Proc. 24th Int. Conf. Coastal Engng.*, Vol. 1, pp. 565-578.

Sleath, J.F.A. (1987): Turbulent oscillatory flow over rough beds. *J. Fluid Mech.*, Vol. 182, pp.369-409.

Tanaka, H., Sumer, B.M. and Fredsøe, J. (1998): Theoretical and experimental investigation on laminar boundary layer under cnoidal wave motion. *Coastal Engng. J.*, Vol. 40, No. 1, pp. 81-98.

Stochastic Hydraulics 2000, Wang & Hu (eds) © 2000 Taylor & Francis, ISBN 90 5809 166 X

Shrinkage of estuarine channels in the Haihe basin and control strategies

Hu Shixiong & Wang Gang
International Research and Training Center on Erosion and Sedimentation, Beijing, China

Wang Zhaoyin
Department of Hydraulic Engineering, Tsinghua University and International Research and Training Center on Erosion and Sedimentation, Beijing, China

ABSTRACT: The Haihe Drainage System is composed of the Luanhe River, Haihe River, Touhai-Majia River and many other small rivers flowing into the Bohai Bay. There are 12 major river mouths in the drainage system. Because of overusing water resources in the upper and middle reaches of the Haihe Drainage System, less and less water flows to the river mouth, and many of the channels have been shrinking quickly. The lower reaches of the river are dried up in spring and early summer. Sediments deposited in the mouth channel are rarely scoured away. The discharge capacity of the channel is consequently reduced greatly, which result in more serious flood hazard. There was a balance between the runoff and tide in sediment carrying capacity decades ago when there was still much water to the river mouths. Quick reduction in runoff resulted in more sediment carried into the river mouth from the silt coast by tidal flow and less out of the mouth. People had to build tide gates to control sedimentation of the estuarine channel. The gates were closed for long time or year round to store fresh water and stop salty water intrusion. The downstream section of the gate silted seriously and people have to dredge the channel every year before the flood season. The source of the sediment and mechanism of estuarine channel shrinkage is analyzed in the paper. Based on the results of scientists and field data, strategies for controlling estuarine channel shrinkage are studied. The Strategies include digger dredging, trailer dredging, scouring with pumping water or stored tidal water, building double guiding dikes and new gates.

Key words: The Haihe river, Shrinkage of estuarine channel, Strategies for controlling estuarine channel shrinkage

1.INTRODUCTION OF THE HAIHE RIVER BASIN

The Haihe River Basin is located in North China with area of 262.6 km^2. It is a quickly developed area with many important cities and industrial hubs, including Beijing, Tianjin, Tangshan, Cangzhou, Dezhou and Huanghua. The area watched fast progresses in urbanization in the past decades, and human activities have resulted in great influences on the environment, river hydrology and sediment budget. The area is projected to be more prosperous with more ocean resources, oil and gas fields, chemical and steel industry bases, and denser railways and express highways in the next century.

Two thirds of the basin are mountainous and hilly area in the north and west (the Yanshan Mountain, the Loess Plateau and Taihang Mountain), and 1/3 is plain in the east. Sediment load comes mainly from the plateau and mountainous areas and is transported to the plain and river mouth to deposit. The drainage system is composed of the Luanhe River, the Haihe River, the Touhai River, the Majia River and many small rivers (Table.1 and Fig.1). These rivers are grouped into north group and south group. The former consist of the Jiyun Canal, Chaobai River, Yongding River and North Canal and the latter consists of the Daqing River, Ziya River, South

Canal and Zhangwei River (Fig.1). There are 12 major river mouths at which floodwater pours into the Bohai Bay. The drainage system is characterized by: a) remarkable discharge difference between dry and flood seasons. There is no flow in most time of the year but flood in summer; b) great influence of human activities. About 600 reservoirs and thousands of wells, numerous dams and tide gates had changed the runoff, sediment load and fluvial process.

Table.1: Major river mouths of the Haihe Drainage System

River	Length(km)	Area (km2)	Main River Mouths
The Haihe River	1090	265,400	Douhe River Mouth, Shahe River Mouth, New Yongding River Mouth, Haihe River Mouth, Duliujian River Mouth, Mapeng River Mouth, Qikou River Mouth, Nanpai River Mouth, Dakou River Mouth
The Luanhe River	877	44,600	Luan River Mouth
The Touhai River	446.5	13136.5	Touhai River Mouth
The Majia River	287.5	10638.4	Majia River Mouth

Table.2: Reduction of annual runoff and sediment into the sea

Period	1950s	1960s	1970s	1980s
Runoff of the Haihe River (billion m^3)	7.3	4.48	1	0.17
Sediment of the Haihe River (million tons)	52.14	1.77	0.1	0.0018
Runoff of the Luanhe River (billion m^3)	5.34		3.79	0.796
Sediment of the Luanhe River(million tons)	24.83		2.4	0.42

Table 3 The reduction of the flood discharge capacity in the river mouths

River Mouth	Period of recurrence	Designed Capacity	Present Capacity	Designed/Present
New Yongding R.	50	1800 m^3/s	460 m^3/s	25.6%
Duliujian R.	20	3200 m^3/s	1500 m^3/s	46.9%
New Ziya Canal	50	6000 m^3/s	3500 m^3/s	58.3%
New Zhangwei R.	50	3500 m^3/s	2000 m^3/s	57.1%
Haihe R.	50	1200 m^3/s	200 m^3/s	16.7%
Total	------	15700 m^3/s	7660 m^3/s	48.8%

Table.4: Dredged and resilted sediment at the Haihe Gate and in the downstream section

Item / Date	Dredged sediment	Annual runoff	Redeposit silt In the next year
1981	0.558 million m^3		
1982	0.527 million m^3	38.8 million m^3	0.599 million m^3
1984	0.411 million m^3	164 million m^3	0.467 million m^3
1985	0.654 million m^3	199 million m^3	0.741 million m^3
1987	0.519 million m^3	251 million m^3	0.548 million m^3
1989	0.326 million m^3	74.1 million m^3	0.482 million m^3
1990	0.423 million m^3	218 million m^3	0.389 million m^3
1992	0.275 million m^3		
1996	0.97 million m^3		
Average* (1981-1996)	0.46 million m^3		0.47 million m^3

* The average sediment in the main river channel.

Many new problems arose from the development of economy among them the shrinkage of the river mouths, high risk of flood and water logging, land subsidence and water shortage are the main challenges to water and river training engineers and scientists (Qiao, et al. 1994; Wang, et al. 1997)

Scientists studied the causes of and patterns of channel siltation (Xu et al, 1983, Zhang X. 1994), analyzed flood risk and proposed control strategies (Zhang H. 1998; Peng, 1997). Others developed computer models and calculated the effects of dredging (Bai et al.1998a, b), conducted physical model study of dredging strategy (Duan, 1997) and made suggestions of engineering methods to reduce sediment siltation (Fang, 1996). Nevertheless, the mechanism of the estuarine channel shrinkage is to be studied and the problem of sedimentation of the river mouth is to be solved yet.

This paper studies the dynamic mechanism of the estuarine channel shrinkage and analyzes the changes in runoff and sediment load. The shrinkage of the river mouth reduces greatly the discharge capacity of the channel and increases the risk of flood and water logging consequently. Different strategies for controlling estuary shrinkage are studied and compared in detail, a major strategy for controlling estuarine channel shrinkage is presented finally.

2. THE MECHANISM OF THE ESTUARY CHANNEL SHRINKAGE AND FLOOD RISK

2.1 The sedimentation and its source

Since 1970 the pressure of population growth and economic development on water resources resulted in over-diversion of river water and great reduction in runoff to the river mouth. Consequently many of the estuarine channels have been shrinking quickly. For examples, the annual runoff to the Yellow River mouth have reduced from 47 billion m3 in 1950s to 17 billion in 1990s; the Haihe River water to the Bohai Bay also reduced from 7.3 billion m^3 in 1950s to 0.17 billion m^3 in the 1980s (Table 2). The Yongding River began to be dry in 1960s and has become an emergency floodway now. The sediment carrying capacity of flow is proportional to the high power of flow discharge. Although the sediment from the river to the estuary are much less than before (Table.2), the sharp reduction in the runoff made the sediment deposited in the mouth channel rarely scoured away, so that the estuarine channel shrinkage become more and more severe.

Almost all estuarine channels in the basin are silted up seriously after the sluice gates were constructed. For example, a volume of 18.62 million m^3 deposited in the 11km channel below the Haihe River Tide Gate from 1958 to 1989. The river was narrowed from 250 m in 1958 to 100m in 1990, the channel bed at the gate was silted up by 6 m. A mouth bar appeared and moved into the river channel from 10 km to 4.8 km downstream the gate. The bed elevation of the mouth bar raised from −3.2m to 0.5m. There is still an apparent channel from the gate to the mouth bar, but the channel disappeared below the mouth bar. The river mouth channel of the Yongding River was silted with sedimentation rate of 3.64 million m^3 per year, which also resulted in the serious estuarine channel shrinkage.

The Yellow River had been entered the Bohai Bay from the the Haihe River Mouth three times in the ancient time (AD1048-1128). The Yongding River, one of main tributaries in the Haihe Basin is also famous for its abundant sediment. The sediment from the rivers formed the river mouth bar and the shallow beach in the past. The data from the Huanghua Harbor shows the modern Yellow River sediment can be transported to the south of the Dakou river mouth, but impossible to the Haihe River Mouth. It is similar the analysis of the Yellow River Commission of China. The sediment source of Haihe River Mouth can not be from the north because of the wave breaker of the Tianjin Harbor. The sediment deposit in the mid of the Bohai Bay can not be initiated and does not affect the river mouths. Therefore, the deposit in the local shallow beach, and sediment from coast erosion are main sources of sediment in the estuarine channel.

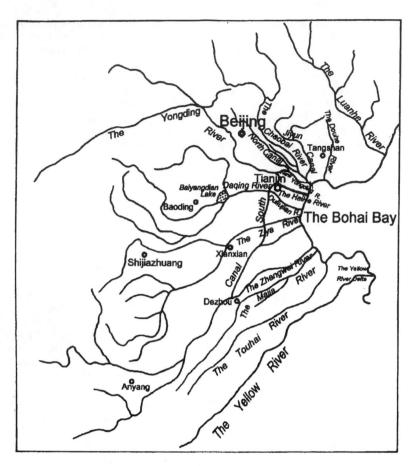

Fig.1: The sketch of the Haihe Drainage System

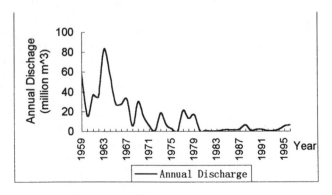

Fig.2: Annual discharge of the Haihe gate

2.2 Dynamic Mechanism of the Estuary Shrinkage

There was a balance between the runoff and tide in sediment carrying capacity decades ago when there was still much water to the river mouths. Quick reduction in runoff resulted in more sediment carried into the river mouth by tidal flow and less out of the mouth. People had to build tide gates to control sedimentation of the estuarine channel. The gates were closed for long time or year round to store fresh water and stop salt water invasion. This results in two changes in dynamic conditions. On the one hand, the tide circle at the gate is deformed, the velocity of the flood tide current became higher than that of the ebb current. On the other hand, runoff discharge through the gate decreased greatly, the scour sediment capacity of the runoff also reduced. All these bring about more sedimentation at the gate and downstream sections because the tidal current carried less sediment out, then the estuarine channel, especially at the gate and downstream sections, shrank. For instances, the ratio of tidal water accommodation capacity in the Haihe River Mouth before and after construction of the gate was 7.5:1 in 1987, the ratio of sediment volume carried by flood tide current to that by ebb current is 6.2:1. The analysis of the sea water samples at the Douhe River Gate in 1998 shows that the sediment concentration in the flood tide is 750mg/l, but that in the ebb current is only 414 mg/l.

The field survey in the Haihe estuarine area shows that the silt carried by flood tidal current is mainly from the right beach. The nearer the estuarine channel is from the gate, the more serious sedimentation happened. It is also found that the stronger the tide current, the more sediment is silted up in the estuarine channel. Harbors and estuaries are often found serious sediment deposition after strong winds. Wind surges scoured sediment from the coastal deposit and carried the sediment into river mouths and resulted in siltation of the mouth channels. If no better measure to mitigate the sedimentation, the river mouth below the gate will become coast finally. The area near the sea will be swamped by flood in the rainy season. The dynamic process of the estuarine channel shrinkage can be illustrated as following (Hu et al.,1999).

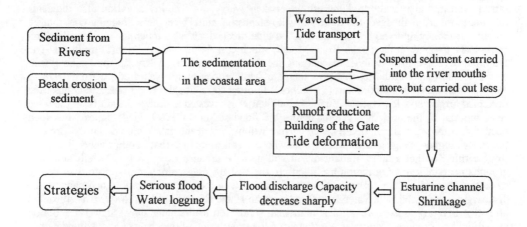

2.3 Flood risk in the Haihe Basin

After the catastrophic flood in 1963, The flood discharge capacity was enhanced by 10 times thanks to the two times river training in the 1960s and 1970s. Unfortunately, sedimentation during the two decades has reduced the capacity again by about 50% and the hazard of flood threatens the local people again. Table 3 shows the reduction of the capacity of the rivers. At the Same time, the flood discharge capacity of the Haihe Gate decreased year by year (Fig. 2)

Before 1980s, the reservoirs and detention basins could be used for controlling flood in the rainy season, which did not cost much because the detention basins were not developed economically. However, it is hard to use the basins for controlling flood. Because population in the basins increased quickly in recent decades, local economy would suffer many times more than ever before, once the detention basins was utilized to control flood. Sedimentation and infrastructure construction also reduced the detention capacity of the reservoirs and basins. Another serious problem in controlling flood is the reduction of flood discharge capacity in the estuarine channels. In 1963, the reservoirs and detention basins played important roles in mitigating the flood hazard by storing 2/3 of the total flood water. However, the total flood discharge capacity of the 5 main river mouths in the basin has decreased to 48.8% of the design capacity as indicated.

No great flood occurred in the north part in the period 1937-1998. The possibility of the catastrophic flood is higher and higher according to the periodical law of the flood in this area. Though floods rarely occurred in both the south and the north parts, it is predicted that the possibility of rainstorms simultaneously occurring in the south and north parts is higher than before because of abnormal global climate change in recent years. The whole basin including Tianjin and Beijing is threatened by such kind of floods.

3. STRATEGIES FOR THE SHRINKAGE OF ESTUARINE CHANNELS

Suggestions are made to solve the problem of estuarine channel shrinkage in the Haihe Basin, for example, digger dredging, trailer dredging, scouring with pumping water or storing tidal water, building double guiding dikes, building a new gate and the non-engineering measures.

3.1 Digger Dredging

Digger dredging is the currently using strategy maintaining the flood discharge capacity of the river mouth channel. To reduce sedimentation at the Haihe Gate and in the downstream section, a total of 9 million m3 silt was dredged from 1973 to 1993 at cost of 40 million Yuan. People have to repeat the effort every year before the flood season because tidal flow carries consecutively sediment from nearby coast and the shallow seabed. Moreover, the dredged channel is often silted up again quickly as given in Table 4. The estuarine channel was narrowed and river bed was raised even the large scale dredging are undertaken every year. The investigation also indicates that the more silt is dredged, the more silt will redeposit again. The endless dredging cost a large amount of money up to now. For example, the total dredged silt is as much as 11.32 million m3 from 1981 to 1997 in the Hiahe River Mouth, which cost 64.62 million Yuan($8 million). The Duliujian River, a main channel discharging the floodwater of Daqing Rive in the Haihe Basin, was silted seriously in the reach below the gate. Million tons of silt was dredged from 1978 to 1993, which cost 10 million Yuan ($1 million). In order to reduce the flood risk and control the predicted large flood, 0.18 billion Yuan ($22 million) was invested to dredge the silt from the three river mouths in the area of Tianjin before the flood season of 1999. Nevertheless, no flood occurred in the year, silt carried by tidal current will deposit at the gate region again. At present, the cost of digger dredging is about $1.5/m3, whose cost is more than that of other ways. The most trouble thing is that a large amount of silt still need be dredged every year. The silt stockpile grounds near the estuarine region are filled up, the cost for digger dredging is more than before since the dredged silt has to be transported far away with long pipes. The designed capacity of discharging flood and rain water is 800 m3/s in the Haihe River Mouth in 2000. It is 400 m^3/s after the dredging in these years. The numerical simulation of dredging shows that 1-1.2 billion m3 silt will be dredged every year. Furthermore, the water stage of the channel above the gate has to be raised and the dikes along the river have to be built higher than before, which will cost a large amount of money and cause many problems in the city.

Nowadays, the main strategy for controling the estuary shrinkage is still digger dredging and transportation of the dredged silt to the stockpile ground with long pipes. A new idea is to use the dredged silt to create new land, which can be developed for industry and other fields. Two problems have to be solved: 1. how the new land is protected from storm wave, which need research on the processes of cohesive sediment movement and coast stability; 2. how to ensure the new land possessed by the dredging group and shared with other departments.

3.2 Trailer Dredging

Experiences and computer simulation find trailer dredging more economical than digger dredging and very effective if it is accompanied with discharged water above the gate, especially in the flood season. Trailer dredging can make better use of ebb current and runoff discharged from the river for carrying sediment away from the gate and therefore save money and time.

Data of the flood season in 1994 illustrate that the dredging with ebb current transported sediment far downstream from the gate as there was a discharged runoff from the gate. The computer simulation also indicates that the silt is moved to the place far away the estuarine channel by the trailer dredger combined with ebb current and discharged runoff. The cost of trailer dredging is about 2 yuan/m³ ($0.25/m³). It needs no silt stockpile ground.

Frankly, the trailer dredging is only a temporary solution because the sediment is continuously transported to the river mouth by the tidal flow. This kind of dredging generates a mud belt, which moves down and up under action of tide. The effectiveness the dredging depends on the runoff discharge. Because the area is plagued by more and more serious water shortage, trailer dredging can not be used as the major strategy controlling river mouth shrinkage.

3.3 Scouring with pumping water

As shown in Fig. 3, the Xingang Harbor is the close neighbor of the Haihe mouth and the gate. Sea water from the harbor is clear (no suspended sediment). From the history of the harbor, the conclusion can be drawn that proper pumping clear water from the harbor to the Haihe river will has little influence on the harbor. In the early period, the harbor was silted seriously. The annual siltation was 5.75 million m³ and the deposit was 3.16m thick per year. There was an apparent floating mud layer in the channel of harbor. However, with the building of the Haihe Gate in 1958, the annual total silt decreased to 4.5 million, even if the depth and width of the harbor channel was doubled, The annual deposit depth reduced to 2.07m. In the 1970s and 1980s, the annual deposit in the harbor was only 1.31m, and the floating mud disappeared.

The strategy suggests to build a pumping station near the gate and the harbor and scour the silt from the Haihe channel with pumping water from the harbor. The main problem is how to harmonize the management of the harbor and the Haihe mouth. In the past, the operation of the Haihe Gate greatly reduced sedimentation of the harbor. Nowadays pumping water from the harbor basin to scour sediment from the river channel may reduce sedimentation at the gate.

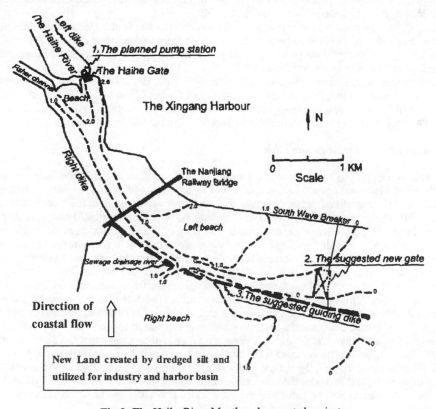

Fig.3: The Haihe River Mouth and suggested projects

Because of siltation below the gate, the bed elevation just below the gate is higher than that above the gate, which leads to the difficulty of discharging water in low flow season. In order to mitigate the severe water logging, the pumping station can also be used to pump water for the river to the mouth channel down the gate, and scour the deposit combining trailer dredging. Although the management and operation cost is high, the efficiencies of flood control, reducing siltation and mitigating water logging in the rive mouth are high, showing the strategy promising and encouraging.

3.4 Scouring with storing tidal water

Understanding the laws of sedimentation below the gate, and learning from the experience of the Aide River in Germany, people suggested to store the flood tidal water and create a flow by releasing the water to flush the silt during the ebb in the river month. (Wu et al. 1992; Wang G. Et al. 1998). Experiments were conducted and the results show the method effective but some problems have to be solved before it is applied as the main strategy. One of the difficulties is that no clear seawater can be stored with the gate because the sediment in the estuary is fine and settles slowly and flood tide carries the suspended sediment back to the gate. Sometimes the stored tidal water contains much sediment and causes the channel above the gate siltating.

3.5 Build new gates

Many gates have been built at the river mouths in the area. They may be divided into short channel gates and the long channel gates. Wherever the gate is located, the severe sedimentation occurs in the estuarine channel below the gate. New gates are requird to be built because of the aged use of the old gates, the ground subsidence of the river mouth, decreasing of the runoff discharge ability, and the shortage of stockpile ground of the dredged silt.

It is suggested to build a new gate close to the mouth sand bar in the Haihe River Mouth as shown in Fig.3 (Fang, 1996). The present gate is 12 km from the mouth and a new gate will be built at 5 km downstream of the present gate. It is estimated that 50% of sediment deposition can be avoided with the new gate operation.

Scientists also suggested to build a new gate near the Tianjin Harbor, and to use the channel as the flood discharge waterway when the catastrophic floods occur (Zhao Y., 1994). The cost of the new gate will be about $500 million. The new gate will be very effective to discharge the flood water. The problems are the huge input and cooperation with the harbor department.

3.6 Building a new double guiding dike

According to the analysis on the sediment source , the sedimentation at the Haihe Gate is mainly from the right beach and coastal sediment disturbed by the wave and carried by tidal current. The calculation of the runoff and ocean dynamics show that the average wave in the river mouth area only can disturb the seabed as deep as 0.65m, the huge wave (once in a month) can work as deep as 4.5m. From the experience of the dredging and building of the wave breaker in the neighboring New Tianjin Harbor, a double guiding dike s are suggested to be built in the important river mouth, which is as long as 9 km and intercept the turbidity belt. From the wave and tide conditions mentioned above, it is concluded that the guiding dike will extend to 9 km into the sea as deep as 4.5 m in order to intercept the high turbidity belt. Then, high turbidity water can not enter into the river mouth and effectively prevent the mouth from siltation. The data also show that the sediment content outside the river mouth 9 km away is less than $0.07kg/m^3$. Therefore, the project will be a permanent way for the serious sedimentation problem in the estuarine channel. The problems are the design of guiding dike and the input. The numerical simulation shows that the direction of the new dike is better to parallel the south wave breaker of the Tianjin Xingang Harbor, the gap of the double dikes is about 1600m, the optical length is about 9 km. A physical experiment needs to be done in laboratory, and provides the verification of the best option of the direction, length and gap of the dikes. The input will be reduced if the silt dredged at present can be made use to build the guiding dike.

In the Haihe River mouth, only the south guiding dike need to be established, another dike can make full use of the south wave breaker of the Tianjin harbor(See Fig.3).The numerical simulation of the Haihe River Mouth shows that the length of the dike should be 9 km, and the best length of the guiding dike in the Duliujian River mouth is about 6km. The cost of the single dike in the Haihe River will cost 0.5-0.6 billion Yuan ($60-70 million). The double dikes in the Duliujianhe River Mouth is about 400 million Yuan ($50 million), however, it can save greatly the dredging fee every year (it is 50 million Yuan in these two river mouths in 1999) and reduce the flood risk in the rainy season. The numerical simulations show that the annual total silt in the river mouth with the dikes is about 0.1million m3, which are only 15% of the silt without the dikes. The sediment carried by the coastal flow will deposit behind the dike (see Fig.3), which will be beneficial for the creating new land in the beach.

According to the new urban plan of the Tianjin City, a industrial zone near the harbor will be built in the right beach (south beach).The total area is as much as 75 km^2, which need a very large scale creation new land from the beach. Building a new double guiding dike will help to realize the plan.

4.THE NEW STRATEGIES FOR THE ESTUARINE CHANNEL SHRINKAGE

4.1Comparison of different strategies

In order to find better ways to solve the estuarine channel shrinkage, it is necessary to estimate the fisiablilty and practice of different strategies mentioned above.

Table.5: Comparison of different strategies for the Haihe estuarine channel shrinkage

Strategy	Investment (Million Yuan)	Difficulties	Management	Output /Input	Influence on environment	Feasibility
Digger dredging	658	Simple; Experienced	Convenient Simple	2.48	Improvement	Feasible
Trailer Dredging	100	Simple; Need better tools	Simple	1.80	Improvement	Practical
Scour with sea water by pumping from the harbor	569	Difficult in land use;	Difficult; High running fee	0.85	No	Unfeasible
Scouring with storing tidal water	550	Difficult High cost	Difficult; Need Dredging silt	1.70	Improvement	Unfeasible
Discharge flood water to The Harbor basin	290	Difficult; Narrow place	Affect the Harbor	3.46	Affect the harbor	Unfeasible
Building new gates	1470	Difficult; Construction in deep water	Difficult; High running fee	0.85	no	Unfeasible
Guiding dike	500	Simple; high cost	Convenient	2.85	Improvement	Feasible

4.2 New strategies for the estuarine channel shrinkage

From the analysis on the different measures, the shortcoming of the dredging way and the source of silt, our new idea on how to mitigate the estuarine channel shrinkage and the flood hazard is given. It shows that a fundamental way to solve the estuarine channel shrinkage should include the following five steps.

1) Building the double guiding dikes to intercept the coastal turbidity belt, and let the flood tidal current carry the clear water (containing less sediment) to the river mouth from the deep sea.

2) Building the new lower gate to scour the silt with storing tidal water. The new gate should be designed as "wide dam and main channel" , in which wide dam in the low bed will discharge the flood by breaking the dam, the main channel with the sluice gate will control the waterlog problem. The local sedimentation law and the runoff and tide data should decide the exact plan of managing and operating the two gates.

3) Building the local port for fishing and transportation, which can mitigate the sedimentation like the trailer dredging, when the boat go out and come back from the sea every day.

4) Creating the new land through combining the dredged silt. The new land can get from the fast sedimentation behind the guiding dike.

5) Developing the new sand for oil and chemical industry and new harbor basin. A large oil field is being under construction, and the new chemical base is planed near the Haihe and Duliujian river mouth, which need much land. These measures should considerate comprehensively the environmental protection, fishery, transportation land developing, urban plan and flood control.

Nowadays, the sediment carried by river runoff decreased sharply, even no sediment is transported into the sea in several years, and the establishment of Huang uh Harbor will cut the transportation of the fine sediment of the Yellow River to the north Bohai Bay. Therefore, it is possible to eradicate the estuarine channel shrinkage in the Haihe Basin. Generally, it is necessary to harmonize the relations among the transportation department, the water conservancy department, the local authority local, troop and central government, abided by the principle of combining the profit of new land development, flood control benefit and investment in the long run. The estuarine channel shrinkage will be overcome by the new steatites through the five steps mentioned above.

5 CONCLUSION

1) For the mechanism of estuarine channel shrinkage, it is concluded that wind surges scoured sediment from the coastal deposit, the flood tide current carried the sediment into river mouths more, and ebb flow carried out less, which resulted in serious siltation of the mouth channels.

2) Physical and mathematical models show that trailer dredging combined with scouring with runoff or tidal water is economical and practical. Guiding dikes at the river mouths to prevent turbidity water entering into the estuarine channel will be a major strategy. Building a new gate combining the harbor may also mitigate the sedimentation of the mouth channels.

3) it is necessary to harmonize the relations among the transportation department, the water conservancy department, the local authority local, troop and central government, abided by the principle of combining the profit of new land development, flood control benefit and investment in the long run. The comparison of different strategies shows that the estuarine channel shrinkage will be overcome by the new steatites through the five steps, which include.

   a) Building the double guiding dikes to intercept the coastal turbidity belt, and let the flood tidal current carry the clear water (containing less sediment) to the river mouth from the deep sea.
   b) Building the new lower gate to scour the silt with storing tidal water.
   c) Building the local port for fishing and transportation;
   d) Creating the new land through combining the dredged silt
   e) Developing the new sand for oil and chemical industry and new harbor basin

ACKNOWLEDGEMENT

Financial support is from National Natural Science Foundation and the Ministry of Water Resource of China under the grant No. 59890200 and the State Key Laboratory of Estuarine and Coastal Research of the East Normal University under the grand No. 99-009

REFERENCE

BAI Yuchuan, 1998a, Study and Application of flow mathematics model of the river channel and flooding area, Water Resource and Hydroelectric Technology, No.7. (in Chinese).

BAI Yuchuan, GU Yuanyan, YU Tianyi, 1998b, Mathematics simulation on sediment movement in the trailer dredging of the Haihe River Mouth, The Theory and Application in the River Simulation, Press of Wuhan University of Hydraulic and Electric Engineering, LI Yitian (ed.), pp146-151.(in Chinese).

DUAN Zhike,1997, Experimental study on the dredging program And discharge capacity in the Haihe River Mouth, Water Resource and Hydroelectric technology,No.11. (in Chinese).

FANG Xiufang, 1996,Suggestion on rebuilding the Haihe River Gate, Water Resource in the Haihe River, No.5.(in Chinese).

HU Shixiong, WANG Zhaoyin and DING Pingxing, 1999,Shrinkage of the Estuarine Channels of the Haihe Drainage System and Its Influences on Flood Hazard, International Journal of Sediment Research, Vol.14, No.2.

QIAO Pengnian, ZHOU Zhide, Zhang Hunan, 1994, Introduction of the estuary evolution in China, Science Press. (in Chinese).

WANG Gang, ZHU Baoliang, YU Qingsong & BIAN Ziran, 1998, Impounding tidal flow for flushing sediment deposits below tidal barriers, River Sedimentation, Theory and Applications, A.W. Jayawardena, J. H. W. Lee & Z.Y. WANG (ed.), A. A. Balkema Publishers, pp917-922.

WANG Zhaoyin, Bingnan LIN and Franz NESTMANN, 1997, Prospects and New Problem of sediment research, International Journal of Sediment Research, Vol.12, No. 1, pp1-15.

WU Deyi, Wang Gang, WANG Jinzhu,1996, Report of the experiment study on the storing tidal water to scour sediment in the river mouths of the Haihe River Basin. (in Chinese).

XU Mingquan, ZHU Zongfa, CHANG Deli, 1983, Impact of sedimentation in the downstream section of the Haihe Gate on the flood discharge capacity, Collected Research Papers of IWHR, No.11, Beijing. (in Chinese).

YUE Shuhong, LI Guangyue, 1997, The characteristics of wave in the shallow sea near Tanggu, The Oceanology of the Yellow and Bohai Sea, Vol. 15, No.1, (in Chinese).

ZHANG Hongxiang, 1998, Report on the flood hazard in Tianjin. (in Chinese).

ZHANG Xiangfeng & Lian Daren,1998, Fluvial process of the Haihe River Estuary, River Sedimentation, Theory and Applications, A.W. Jayawardena, J. H. W. Lee & Z.Y. WANG (ed.), A. A. Balkema Publishers, pp923-928.

ZHANG Xiangfeng, 1994,Analysis on the sedimentation in the Haihe River Mouth, Sediment Research, (4). (in Chinese).

ZHAO Hongkui, 1992, Strategies of flood control in Tianjin, Proceedings of the 8th Sino-Japanese workshop on the river and dam engineering, Tokyo. (in Chinese).

ZHAO Yongpin, 1994,Suggestion on rebuilding the River Gate combining the Harbor and experiment of storing the tidal water to scour sediment for enlarge the discharge capacity, Water Resource in the Haihe River, No.4.(in Chinese).

Stochastic Hydraulics 2000, Wang & Hu (eds) © 2000 Taylor & Francis, ISBN 90 5809 166 X

Seabed shear stresses under random waves plus currents: Predictions vs. laboratory measurements

Dag Myrhaug & Lars Erik Holmedal
Department of Marine Hydrodynamics, Norwegian University of Science and Technology, Trondheim, Norway

Olav H. Slaattelid
Norwegian Marine Technology Research Institute, Valentinlyst, Trondheim, Norway

ABSTRACT: Results of comparison between a parametric model and laboratory measurements of seabed shear stresses under random waves with currents superimposed are presented. The model is based on a modified version of the Myrhaug (1995) approach where the effect of random waves on the bottom friction was studied by assuming that the wave motion is a stationary Gaussian narrow-band random process, and by using the Soulsby et al. (1993a) explicit friction coefficient formula for sinusoidal waves. The data used for comparison are obtained from statistical analysis of Simons et al.'s (1996) and MacIver's (1998) direct measurements of the bottom shear stresses under combined random waves and orthogonal as well as near-orthogonal currents.

Key words: Seabed shear stresses; Random waves plus current; Statistical analysis.

1. INTRODUCTION

Along the coast, at intermediate and shallow water depths, the particle movements induced by the surface waves have a strong effect in the entire flow region from the surface to the bottom of the ocean. The total flow in this region arises from two major effects; the surface waves and the currents in the ocean. At the seabed there is a thin flow region, the bottom boundary layer, that is dominated by friction arising from bottom roughness. The boundary layer flow determines the bottom shear stresses, which are of fundamental interest for sediment transport and thereby the evolution of coastal morphology. A number of models have been developed in order to calculate shear stresses under regular waves, most of them assuming the current to be slowly varying over a wave length, see e.g. Soulsby et al. (1993b). Recently, studies on the effect of the randomness of the wave motion on the bottom friction have been made. Among these are Zhao and Anastasiou (1993), Ockenden and Soulsby (1994), Simons et al. (1994, 1996), Madsen (1994), Myrhaug (1995) (hereafter denoted as M95), Myrhaug and Slaattelid (1996), Myrhaug and Hansen (1997), MacIver (1998), Myrhaug et al. (1998) and Holmedal et al. (1999).

The purpose of this paper is to compare a modified version of the M95 approach with statistical analysis of the Simons et al. (1996) and MacIver (1998) data from laboratory experiments of direct measurements of the bottom shear stresses under combined random waves and orthogonal as well as near-orthogonal currents. The probability distributions of the maximum bottom shear stress together with some characteristic statistical values of the maximum bottom shear stress are presented. An acceptable agreement is found between measurements and predictions.

2. THEORETICAL BACKGROUND

Myrhaug (1995) found the dimensionless maximum bottom shear stress $(\hat{\tau})$ for rough turbulent flow to be Weibull distributed, i.e.,

$$P(\hat{\tau}) = 1 - \exp(-\hat{\tau}^{\beta}) \qquad\qquad \hat{\tau} = \frac{\tau_m}{\rho U_*^2} \geq 0 \tag{1}$$

where

$$U_*^2 = \frac{1}{2} c \left(\frac{A_{rms}}{z_0}\right)^{-d} U_{rms}^2 \tag{2}$$

$$\beta = \frac{2}{2-d} \tag{3}$$

The root-mean-square (rms) values A_{rms} and U_{rms} are defined below, ρ is the density of the fluid, and z_0 is the seabed roughness parameter. τ_m is the maximum bottom shear stress induced by individual random wave calculated by

$$\frac{\tau_m}{\rho} = \frac{1}{2} f_w U^2 \tag{4}$$

where f_w is the friction coefficient taken as

$$f_w = c \left(\frac{A}{z_0}\right)^{-d} \tag{5}$$

with

$$c = 1.39 \ , \ d = 0.52 \tag{6}$$

A is the orbital displacement amplitude at the seabed, U is the orbital velocity amplitude at the seabed, and ω is the angular wave frequency. One should notice that Eqs. (5) and (6) were proposed by Soulsby et al. (1993a) and are valid for sinusoidal waves and rough turbulent flow, obtained as best fit to data in the range $10 \ . \ A/z_0 \ . \ 10^5$ (see Soulsby et al. (1993a, Fig. 1)). This distribution of $\hat{\tau}$ is based upon assuming that (1) the free surface elevation is a stationary Gaussian narrow-band random process with zero expectation, and (2) the friction coefficient formula for sinusoidal waves are valid for random waves as well.

Use of Eqs. (4) and (5) implies that each wave is treated individually, and consequently that the friction coefficient is taken to be constant for a given wave situation without including the "memory" in the turbulence in the wave boundary layer. The accuracy of this assumption should be validated by using a full boundary layer model to calculate the shear stress under random waves. However, results from some preliminary studies have been presented by Penny et al. (1994) as well as Tran Thu et al. (1995), and overall the results suggest that the M95 approach, which applies c and d in Eq. (6), is adequate as a first approximation and can be used to predict e.g. integrated effects such as bedload sediment transport with a reasonable degree of accuracy. Further details are given in M95 and Myrhaug and Hansen (1997).

One should notice that M95 applied c and d in Eq. (6), while here a modified version of M95 with other values of c and d will be used. Further details will be given in the forthcoming.

Based on the present assumptions, the bed orbital displacement $a(t)$ as well as the bed orbital velocity $u(t)$ will be stationary Gaussian narrow-band processes with zero expectations.

Now A and U will both be Rayleigh-distributed, i.e.,

$$P(\hat{x}) = 1 - \exp(-\hat{x}^2) \ ; \ \hat{x} = x/x_{rms} \geq 0 \tag{7}$$

where x represents A or U, and x_{rms} represents A_{rms} and U_{rms}. The mean zero-crossing wave frequency is obtained as

$$\omega_z = \frac{U_{rms}}{A_{rms}} \tag{8}$$

One should notice that Eqs. (1) to (3) are obtained by using Eqs. (4) and (5) with $A = U/\omega$ and ω replaced by ω_z from Eq. (8); then it follows that τ_m is distributed as U^{2-d}, from which Eqs. (1) to (3) follow by transformation of random variables by using Eq. (7) with $x = U$ and $x_{rms} = U_{rms}$. This model using c and d in Eq. (6) has only been compared with estimates of seabed shear stresses under random waves from field measurements. In that case good agreement was found between predictions and data from the Strait of Juan de Fuca, Washington State (Myrhaug et al., 1998).

The characteristic statistical values of the maximum bottom shear stress in random waves can be obtained when the probability distribution is known for given values of A_{rms} and U_{rms}. The quantities considered here are:

The expected value of $\hat{\tau}$, $E[\hat{\tau}]$, and the standard deviation of $\hat{\tau}$, $\sigma_{\hat{\tau}}$, given by, respectively (see e.g. Bury, 1975)

$$E[\hat{\tau}] = \Gamma(1 + \frac{1}{\beta}) \tag{9}$$

$$\sigma_{\hat{\tau}} = [\ \Gamma(1 + \frac{2}{\beta}) - \Gamma^2(1 + \frac{1}{\beta})\ ]^{1/2} \tag{10}$$

where Γ is the gamma function.

The value of $\hat{\tau}$ which is exceeded by the probability $1/n$, $\hat{\tau}_{1/n}$, and the expected value of the $1/n$ largest values of $\hat{\tau}$, $E[\hat{\tau}_{1/n}]$, given by, respectively (M95)

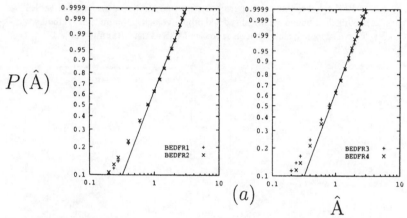

Fig.1: a) Probability distribution of normalized bed orbital displacement amplitude in Weibull scale: --Rayleigh distribution; other symbols represent LUCIO data. See also Table 1.

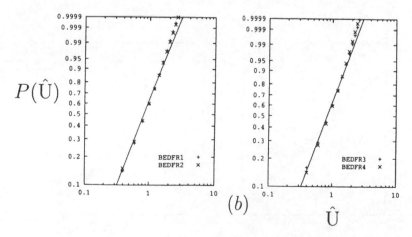

Fig.1: b) Probability distribution of normalized bed orbital velocity amplitude in Weibull scale: --Rayleigh distribution; other symbols represent LUCIO data. See also Table 1.

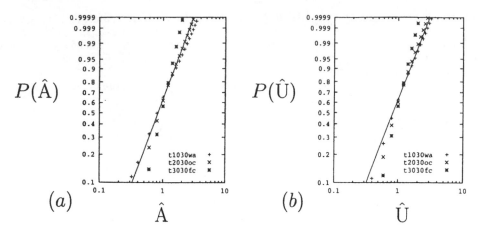

Fig.2: a) Probability distribution of normalized bed orbital displacement amplitude in Wiebull scale:--Rayleigh disribution; other symbols represent MAST3 data. See also Table1.
b) Probability distribution of normalized bed orbital velocity amplitude in Weibull scale:--Rayleigh distribuion; other symbols represent MAST3 data. See also Table 1

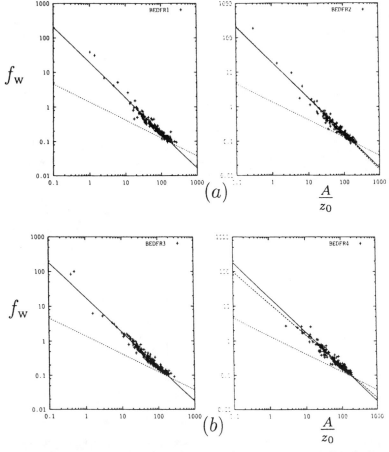

Fig.3: Friction coefficient versus amplitude to roughness ratio for the LUCIO half-cycle data:a) −best fit of Eq.(5) to BEDFR1 data;--best fit of Eq.(5) to BEDFR2 data; b) −best fit of Eq.(5) to BEDFR3 data; --best fit of Eq(5) to BEDFR4 data; Soulsby et al.(1993a); other symbols represent·LUCIO data. See also Table 1.

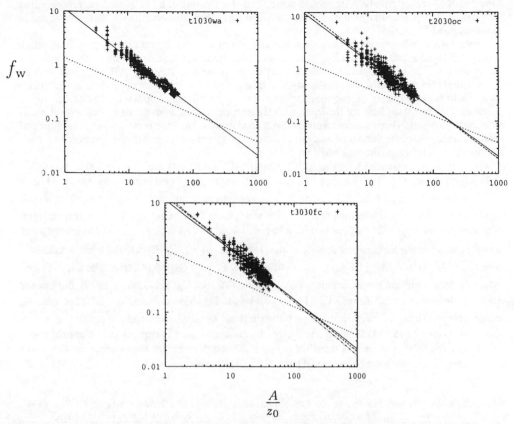

Fig.4: Friction coefficient versus amplitude to roughness ratio for the MAST3 half-cycle data:--best fit of Eq.(5) to t1030WA data; --best fit of Eq.(5) to t2030OC data; -.- best fit of Eq.(5) to t3030FC data;Soulsby et al.(1993a); other symbols represent MAST3 data. See also Talbe 1.

$$\hat{\tau}_{1/10} = (\ln n)^{1/\beta} \tag{11}$$

$$E[\hat{\tau}_{1/n}] = n\, \Gamma(1+\frac{1}{\beta}, \ln n) \tag{12}$$

where $\Gamma(\cdot,\cdot)$ is the incomplete gamma function.

The expected largest value among N values given by (see e.g. Bury, 1975)

$$E[\hat{\tau}_N] = (\ln N)^{1/\beta}\, (1+\frac{0.5772}{\beta \ln N}) \tag{13}$$

The first term in Eq. (13) can be interpreted as the Acharacteristic largest value$\cong$, $\hat{\tau}_N$, which has, on the average, only one exceedance in a sample of size N, i.e., $1-P(\hat{\tau}_N) = 1/N$, giving

$$\hat{\tau}_N = (\ln N)^{1/\beta} \tag{14}$$

3. RESULTS AND DISCUSSION OF PREDICTIONS VS MEASUREMENTS

Comparisons between predictions and data from measurements will now be presented and discussed. Here the Simons et al. (1996) and MacIver (1998) data, which are from laboratory measurements of bottom shear stresses under random waves plus current for fully rough turbulent flow conditions, will be used for comparison. Hereafter these data sets will be referred to as LUCIO

763

and MAST3, respectively. One should notice that the Simons et al. (1994) data, representing similar data, but for intermediate smooth to rough turbulent flow conditions, were discussed in Myrhaug and Slaattelid (1996).

Both data sets represent bottom shear stresses which were measured directly at a fixed rough bed by using a shear plate device together with simultaneous measurements of three velocity components. The LUCIO data include two sequences of random waves in still water and with an orthogonal current superimposed. The MAST3 data include one sequence of random waves in still water and with two near-orthogonal currents superimposed; one opposing and one following current. The data used here are the friction coefficients calculated from the half-cycle amplitude of the shear stress τ_m (between consecutive maxima and minima) and the corresponding amplitude of wave-induced velocity. Statistical analysis of the data has been performed in order to make a proper comparison between predictions and data.

The actual test conditions for the LUCIO and MAST3 data sets are given in Table 1 together with some analysis results which will be discussed subsequently. Here $T_z = 2\pi / \omega_z = 2\pi A_{rms} / U_{rms}$ is the mean zero-crossing wave period, and $\overline{U}$ is the depth-averaged current velocity over the water depth $h = 0.50$ m and 0.49 m for LUCIO and MAST3, respectively. The roughness Reynolds number is defined as $Re = k_N u_* / \nu$, where k_N is the Nikuradse sand roughness, $<$ is the kinematic viscosity of the fluid, and $u_* = (\tau_m / \rho)^{1/2}$ is the friction velocity. For the LUCIO and MAST3 data $k_N = 1.87$ cm and $z_0 = k_N / 30 = 0.0623$ cm. The LUCIO and MAST3 data represent rough turbulent flow (i.e., $Re > 70$, see e.g. Schlichting, 1979), but appear to be in the lower A / z_0 range, i.e., $1 . z_0 . 300$ (see Fig. 3) and $3 . A / z_0 . 60$ (See Fig. 4), respectively. Thus, a direct comparison between these two data sets and the results based on Soulsby et al.=s (1993a) friction coefficient formula in Eqs. (5) and (6) is not appropriate. Therefore the M95 approach is modified to cover the rough turbulent flow regime for the lower A / z_0 range by making use of the LUCIO and MAST3 data sets.

Table.1: Main flow variables and results for : Simons et al. (1996) LUCIO data, where $271 \leq Re \leq 1580$; MacIver (1998) MAST3 data, where $230 \leq Re \leq 1142$. For both data sets $z_0 = 0.0623$ cm.

Data set	Record	N	T_z (s)	θ_{wc} (deg)	$\overline{U}$ (cm/s)	U_{rms} (cm/s)	$\dfrac{A_{rms}}{z_0}$	c	d	β
LUCIO	BEDFR1	202	1.91	-	0	19.70	105.5	19.3	1.02	2.04
	BEDFR2	176	1.83	90	12.5	20.62	106.2	19.6	1.04	2.08
	BEDFR3	221	2.03	-	0	17.66	100.4	18.0	1.00	2.00
	BEDFR4	196	1.92	90	12.6	17.78	95.7	12.1	0.91	1.84
MAST3	t1030W	783	1.22	-	0.6	8.28	25.87	11.39	0.91	1.83
	t2030OC	695	1.22	112.6	13.5	8.20	25.62	13.36	0.95	1.90
	t3030FC	617	1.21	76.3	15.2	8.47	26.18	14.63	0.99	1.98

A consequence of the narrow-band assumption is that A and U both are Rayleigh-distributed with the distribution functions given in Eq. (7), and thus it is of interest to compare the data of A and U with the narrow-band assumption. Figures 1a and 1b show $P(\hat{A})$ and $P(\hat{U})$ according to Eq. (7) with $\hat{x} = A / A_{rms}$ and $\hat{x} = U / U_{rms}$, respectively, together with the LUCIO data in Weibull scale. Figures 2a and 2b show similar results for the MAST3 data. For the LUCIO data it appears to be differences between the Rayleigh distribution and the data for $P(\hat{A})$ for lower values of $\hat{A}$ (Fig. 1a). It is also noticed to be significant differences between the Rayleigh distribution and the data for $P(\hat{A})$ and $P(\hat{U})$ for the MAST3 following current condition (Fig. 2). However, overall it appears that the Rayleigh distribution gives an adequate representation of most of the data for waves alone as well as waves plus current.

The present approach for comparison with the LUCIO and MAST3 data sets is as follows. First, Eq. (5) is fitted to the half-cycle data for random waves as well as random waves with a current superimposed. Thus c and d are determined for each data set. Second, the probability distribution function of the normalized bottom shear stress is obtained by using Eqs. (1) to (3). Finally, the characteristic statistical values given in Eqs. (9) to (14) are calculated.

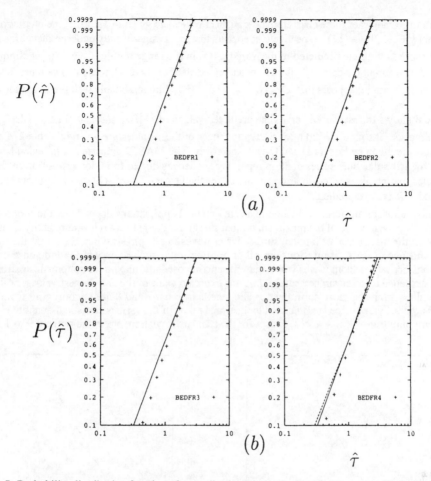

Fig.5: Probability distribution function of normalized maximum bottom shear stress in Weibull scale:a) – Eqs.(1) to (3) with c, d and β values corresponding to BEDFR1; --Eqs.(1) to (3) with c.d. and β values correspinding to BEDFR2; b)—Eqs.(1)to (3) with c,d and β values correspinding to BEDFR3; --Eq.(1) to (3) with c,d and β values correspinding to BEDFR4; other symbols represent LUCIO data. See also Table 1.

Figures 3 and 4 show the results of fitting Eq. (5) to the half-cycle data for the LUCIO and MAST3 data set records, respectively. The Soulsby et al. (1993a) formula is included for comparison. One should note that Eq. (5) represents a straight line in log-log scale. As observed in Figs. 3 and 4 the effect of the current is insignificant, as should be expected for orthogonal as well as near-orthogonal currents superimposed on waves, as long as the current is not too dominant (see e.g. Soulsby et al., 1993b). This is also demonstrated in the subsequent discussion of the results.

Figures 5 and 6 show the probability distribution function of the normalized bottom shear stress in Weibull scale for the LUCIO and MAST3 data set records, respectively. Both the model, i.e., Eqs. (1) to (3) with the β values given in Table 1, and the data are shown. Overall it appears that the model gives an adequate representation of most of the data for larger values of $\hat{\tau}$, i.e., for $\hat{\tau}$ / 1. However, differences between the model and the data are noticed for the MAST3 record representing the following current condition (Fig. 6). For this record (t3030FC) significant differences between the Rayleigh distribution and the data for $P(\hat{A})$ and $P(\hat{U})$ were noted in Fig. 2. In this case the model gives a smaller probability than the data for larger values of $\hat{\tau}$.

Figure 7 shows the measured versus the predicted values of $E[\hat{\tau}]$, $\sigma_{\hat{\tau}}$ as well as $\hat{\tau}_{1/n}$ and $E[\hat{\tau}_{1/n}]$ for $n = 3$ and $n = 10$ for the LUCIO and MAST3 data set records. The predicted to

765

measured ratio ranges of these characteristic statistical values are given in Table 2. The predictions are given by Eqs. (9) to (12), respectively. It appears that $\sigma_{\hat\tau}$ is significantly underpredicted for all the data records with the predicted to measured ratios in the range 0.45 to 0.66. $E[\hat\tau]$ is slightly underpredicted for all the data. $\hat\tau_{1/3}$ is slightly underpredicted for LUCIO, while it is well predicted for MAST3. Overall it appears that $E[\hat\tau_{1/3}]$, $\hat\tau_{1/10}$ and $E[\hat\tau_{1/10}]$ are reasonably well predicted for all he data.

Figure 8 shows the measured versus the predicted values of $E[\hat\tau_N]$ and $\hat\tau_N$ for the LUCIO data set records. The predicted to measured ratio ranges of these values are given in Table 2. The predictions are given by Eqs. (13) and (14), respectively. The MAST3 data are not included here because the irregular time series contain repeating sequences due to too poor resolution in the frequencies, which gives interpretation problems for the largest values. Overall it appears that $E[\hat\tau_N]$ and $\hat\tau_N$ are overpredicted.

The most appropriate statistical value to use in practical applications depends on the problem dealt with. The mean value of the maximum bottom shear stress might be a relevant quantity to use to represent the dissipation of random surface water waves in e.g. physical models for predicting coastal and ocean flow circulations. In other applications, such as in suspended sediment calculations beneath random waves with a current superimposed, the maximum bottom shear stress which is exceeded by a certain percentage $1/n$, the expected value of the $1/n$ largest values, or the largest value, might be more appropriate. The results in Figs. 7 and 8 are encouraging from a practical point of view, since the predicted to measured ratios of the various statistical quantities of the maximum bottom shear stresses, except for the standard deviation, are in the range 0.9 to 1.2.

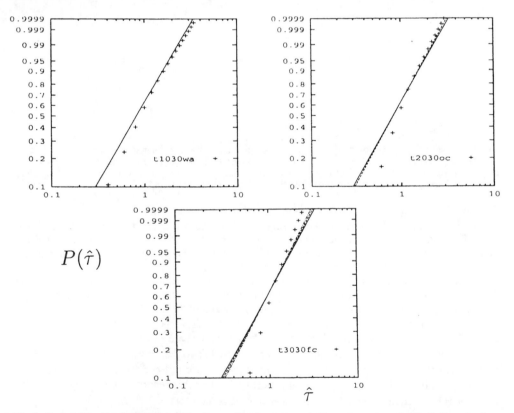

Fig.6: Probability distribution function of normalized maximum bottom shear stress in Weibull scale:a) – Eqs.(1) to (3) with c, d and β values corresponding to t1030WA; --Eqs.(1) to (3) with c.d. and β values corresponding to 2000C;-.- Eqs.(1)to (3) with c,d and β values corresponding to 3030FC; other symbols represent LUCIO data. See also Table 1.

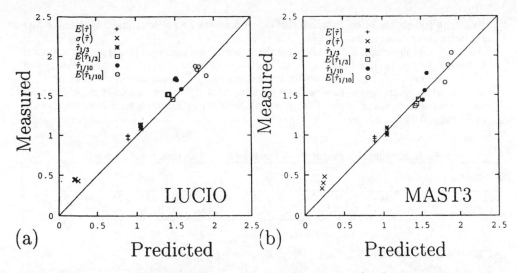

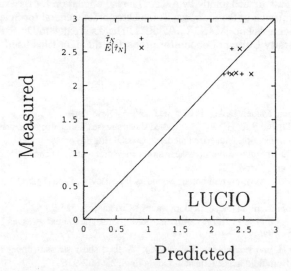

Fig.7: Measured versus predicted values of $\sigma_{\hat{\tau}}$, $E[\hat{\tau}]$, $\hat{\tau}_{1/n}$ and $E[\hat{\tau}_{1/n}]$ for n=3,10

Fig.8: Measured versus predicted values of the largest normalized maximum
bottom shear stress for the LUCIO data.

The results presented here suggest that the present approach can be used as a first approximation to represent the seabed shear stress under random waves plus orthogonal and near-orthogonal currents.

CONCLUSIONS

Overall the parametric model based on a modified version of the Myrhaug (1995) approach gives an adequate representation of the Simons et al. (1996) and MacIver (1998) data, showing that:
1. The probability distribution function of the maximum seabed shear stress is reasonably well represented by the Weibull distribution for larger shear stress values for most of the data. The best agreement is found between the model and the Simons et al. (1996) data. There are differences between the model and the MacIver (1998) record representing waves plus the

767

following current condition, in which case the model gives a smaller probability than the data for larger shear stress values.

2. An acceptable agreement is found between the measured and predicted characteristic statistical values of the normalized maximum seabed shear stresses $E[\hat\tau]$, $\hat\tau_{1/3}$, $\hat\tau_{1/10}$, $E[\hat\tau_{1/3}]$ and $E[\hat\tau_{1/10}]$ for both data sets. The predicted to measured ratios for these quantities are in the range 0.87 to 1.15. $\sigma_{\hat\tau}$ appears to be significantly underpredicted. For the Simons et al. (1996) data, the predicted to measured ratios of the largest values of the maximum seabed shear stresses are in the range 0.91 to 1.21.

Table.2: Predicted to measured ratio ranges for the characteristic statistical values.

Data set	$\sigma_{\hat\tau}$	$E[\hat\tau]$	$\hat\tau_{1/3}$	$E[\hat\tau_{1/3}]$	$\hat\tau_{1/10}$	$E[\hat\tau_{1/10}]$	$\hat\tau_N$	$E[\hat\tau_N]$
LUCIO	0.45-0.59	0.89-0.93	0.92-0.97	0.92-1.01	0.87-0.99	0.94-1.08	0.91-1.14	0.96-1.21
MAST3	0.53-0.66	0.90-0.96	0.96-1.05	1.01-1.04	0.89-1.06	0.93-1.09	-	-

ACKNOWLEDGEMENTS

This work was undertaken as part of the MAST project "The Kinematics and Dynamics of Wave-Current Interactions". It was funded jointly by Anders Jahres Foundation for the Advancement of Science and by the Commission of the European Union Directorate General for Science, Research and Development under contract no. MAS3-CT95-0011. Thanks are due to Dr. R.R. Simons and R.D. MacIver at University College London for preparing the data files used in this study.

REFERENCES

Bury, K.V. (1975). "Statistical Models in Applied Science", Wiley, New York.

Holmedal, L.E., Myrhaug, D. and Rue, H. (2000). "Seabed shear stresses under irregular waves plus current from Monte Carlo simulations of parameterized models", Coastal Eng., in press.

MacIver, R.D. (1998). "3-D irregular wave bed shear stress measurements", Technical Report HYD98-2, University College London, London, UK.

Madsen, O.S. (1994). "Spectral wave-current bottom boundary layer flows", Proc. 24th Conf. on Coastal Eng., ASCE, Kobe, Japan, Vol. 1, pp. 384-398.

Myrhaug, D. (1995). "Bottom friction beneath random waves", Coastal Eng., 24:259-273.

Myrhaug, D. and Hansen, E.H. (1997). "Long-term distribution of seabed shear stresses beneath random waves", Coastal Eng., 31: 327-337.

Myrhaug, D., Slaattelid, O.H. and Lambrakos, K.F. (1998). "Seabed shear stresses under random waves: predictions vs estimates from field measurements", Ocean Eng., 25: 907-916.

Myrhaug, D. and Slaattelid, O.H. (1996). "Comparison between a parametric model and measurements of seabed shear stresses under random waves with a current superimposed", In: K.S. Tickle et al. (Editors), Proc. 7th IAHR Int. Symp. on Stochastic Hydraulics, Balkema, Rotterdam, pp. 291-299.

Ockenden, M.C. and Soulsby, R.L. (1994). "Sediment transport by currents plus irregular waves", Report SR 376, HR Wallingford, Wallingford, UK.

Penny, M.D., Davies, A.G. and Soulsby, R.L. (1994). "Sediment transport under irregular waves plus currents", Book of Abstracts, MAST-2, G8M Coastal Morphodynamics, Overall Workshop, Gregynog, Wales, pp. 3.15-1/3.15-4.

Schlichting, H. (1979). "Boundary-Layer Theory", 7th Ed., McGraw-Hill, New York, N.Y.Simons, R.R., Grass, T.J., Saleh, W.M. and Tehrani, M.M. (1994). "Bottom shear stresses under random waves with a current superimposed", Proc. 24th Conf. on Coastal Eng., ASCE, Kobe, Japan, Vol. 1, pp. 565-578.

Simons, R.R., MacIver, R.D. and Saleh, W.M. (1996). "Kinematics and shear stresses from combined waves and longshore currents in the UK Coastal Research Facility", Proc. 25th Conf. on Coastal Eng., ASCE, Orlando.Soulsby, R.L., Davies, A.G., Fredsøe, J., Huntley, D.A., Jonsson, I.G., Myrhaug, D., Simons, R.R., Temperville, A. and Zitman, T. (1993a). "Bed shear-stresses due to combined waves and currents", Book of Abstracts, MAST-2, G8M Coastal Morphodynamics, Overall Workshop, Grenoble, pp. 2.1-1/2.1-4.

Soulsby, R.L., Hamm, L., Klopman, G., Myrhaug, D., Simons, R.R. and Thomas, G.P. (1993b). "Wave-current interaction within and outside the bottom boundary layer", Coastal Eng., 21:41-69.

Tran Thu, T., Carmo, J. and Temperville, A. (1995). "Bed shear stresses due to combined waves and currents on flat bottom", In: M.J.F. Stive et al. (Editors), Advances in Coastal Morphodynamics, Delft Hydraulics, Delft, pp. 4-84/4-88.

Zhao, Y. and Anastasiou, K. (1993). "Bottom friction effects in the combined flow field of random waves and currents",Coastal Eng., 19:223-243.

Stochastic Hydraulics 2000, Wang & Hu (eds) © 2000 Taylor & Francis, ISBN 90 5809 166 X

Wave group statistics of northern Taiwan

John Z.Yim, Chung-Ren Chou & Jaw-Guei Lin
*Department of Harbour River Engineering, National Taiwan Ocean University, Keelung, Taiwan,
China*

ABSTRACT: Analysis of wave group statistics has been carried out using data measured in
northern Taiwan. Five statistical models were applied to study the statistical distribution of total
run and run length of the wave groups. It was found that the Weibull distribution could be used
to describe the statistics of run length and total run. When the restriction of more than two
higher waves must exist in a wave group is adopted, the statistics of the group run length and
the group interval can be modeled with the exponential distribution as a first approximation.

Key words: Wave group statistics, Weibull distribution, Exponential distribution.

1. INTRODUCTION

Wave groups may have important effects on the stability of coastal structures (Medina et al.,
1994). Their existence is often the main reason for the tranquility problems of a harbour
(Ouellet & Theriault, 1989). Due to the existence of wave groups, waves with long-periods can
be induced (Schäffer, 1993, Zuo & Yao, 1991). These long waves can then cause serious
problems in the laboratory when generating waves, which are to be used for physical model
testing (Sand, 1982). It is therefore desirable to acquire further knowledge of the nature of wave
groups.

Wave groups are waves high waves running in sequence. Generally speaking, there are two
ways in treating wave groups. The first one considers that the waves are basically linear. All the
waves in a wave field have independent frequencies and wave numbers, and therefore each
wave travels freely on the ocean surface. High waves are accumulated by small amplitude
waves through a random process (Goda, 1973, 1983; Kimura, 1980; Longuet-Higgins, 1984).
The second way of dealing with wave groups is to consider that the waves, as they are subjected
to the nonlinear boundary conditions, are nonlinear. One of the most predominant effects of
nonlinearity is to generate higher harmonics of the basic, sinusoidal waves. The other effect is
the exchange of energy among wave components through nonlinear wave-wave interactions
(Phillips, 1981; Hasselmann, 1961). Due to these reasons, researchers have argued that, in
studying the characteristics of random wave groups, the effect of nonlinearity should not be
neglected (Mase & Iwagaki, 1987). These researchers have used the nonlinear Schrödinger
equation, or its modified version, to study the phenomenon of ocean wave grouping.

When all wave components are free traveling on the sea surface, it is nature to assume that
their phases have equal chances to have values in the interval of $[0, 2\pi]$. Under these
circumstances, as pointed out by Benoit et al. (1997), all wave components will be independent
of each other. And, as a consequence, the probability distribution of the water surface
fluctuations will be Gaussian. In fact, except in nearshore regions, researchers do find that the
Gaussian model is rather satisfactory.

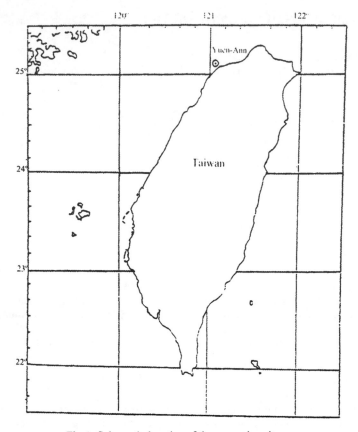

Fig.1: Schematic location of the measuring site.

It must be pointed out that, when the waves are really free and independent, consecutive wave heights should have zero correlation. Since the correlation between consecutive heights for a realistic sea state is generally in the range of 0.2 to 0.5, real sea waves are therefore not independent (Lamberti & Rossi, 1994). Nonlinearities, albeit small, must have some affects on the wave characteristics.

To study the effect of wave grouping of an actual sea, wave records measured at the northern part of Taiwan were used. In the following, we further divide this article into three parts. In Section II we describe shortly the measuring site and the data acquisition procedures, as well as the models used. In Section III we present the results of our analyses, associated with discussions. And, with a brief conclusion in Section IV we then close this paper.

2. DATA ACQUISITION AND MODELS USED

Wave data consisting of 20 (twenty) months of field measurements were used to study wave group characteristics. The data were measured in the northern part of Taiwan. Water depth at the measuring site is 25 meters. The sampling frequency of the data was 2.56 Hz ($\Delta t = 0.390625$ sec) for a period of 20 minutes every hour. Each wave record thus has a total of 3072 data points. A total of 10,867 wave records were used for the analyses. Figure 1 shows the measuring site schematically.

Surface elevations were first mean-removed and de-trended according to the usual procedure (Bendat & Piersol, 1986). A least square technique was used to determine the possible trend, which was assumed to be polynomial with an unknown order of n, where $n \leq 5$ was assumed. Quite often, a linear trend, $n = 1$, was found to satisfy the purpose.

Wave heights were calculated from the records using the usual zero-downcrossing method. Expressing the surface displacement of a narrow-banded sea surface as:

$$\varsigma(t) = \mathrm{Re}\left\{ A(t)e^{i\phi(t)} \right\}$$

(1)

then

$$A(t) = \sqrt{\varsigma^2(t) + \xi^2(t)}$$

(2)

and

$$\phi(t) = \tan^{-1}\left[\frac{\xi(t)}{\varsigma(t)} \right]$$

(3)

where $A(t)$ and $\phi(t)$ are real functions of time t, representing, respectively, the envelope and the phase of $\varsigma(t)$, and $\xi(t)$ is the Hilbert transform of $\varsigma(t)$, defined as:

$$\xi(t) = \frac{1}{\pi} P \int_{-\infty}^{\infty} \frac{\varsigma(\tau)}{t - \tau} d\tau$$

(4)

where P is the Cauchy principle value of the integral evaluated at $\tau = t$. $\xi(t)$, and therefore $A(t)$ and $\phi(t)$, can be calculated rather easily from $\varsigma(t)$ through Fourier transformation.

Spectral analyses for carried out for both the wave records and the envelope. Assemble average were first performed by dividing the records into 12 segments, each having 256 data points. The 12 rough estimates of the spectral densities were averaged to yield a smoothed estimate. A Hanning window was applied in the frequency domain to increase the degrees of freedom (DOF) of the spectral estimated. The final spectra for both surface elevation and envelope have a resolution of $\Delta f = 0.01$ Hz, with 24 DOF.

The statistical properties of the surface elevations, of the wave heights, and of the wave group properties, were tested using probability models found in the literature. Due to space limitations, only a brief description of these models will be given in the following:

1. The Gaussian distribution

$$p(x, \mu, \sigma) = \frac{1}{\sigma\sqrt{2\pi}} \exp\left[-\frac{(x - \mu)^2}{2\sigma^2} \right]$$

(5)

where x is the normalized quantity, $x = \dfrac{x_i}{x_{ref}}$, with x_i the respective variables, i.e., surface fluctuations, wave heights etc, x_{ref} is the normalization factor, when the distributions of the surface fluctuations is considered, $x_{ref} = \xi_{rms}$, i.e., the root-mean-square values of ξ, otherwise $x_{ref} = x_{mean}$. μ is the mean value, and σ is the standard deviation.

2. The Rayleigh distribution

$$p(H') = \begin{cases} \dfrac{\pi}{2} H' \exp\left(-\dfrac{\pi}{4} H'^2 \right) & H' \geq 0 \\ 0 & \text{otherwise} \end{cases}$$

(6)

3. The Weibull distribution

$$p(H', \alpha, \beta) = \alpha\beta H'^{\alpha-1} \exp\left(-\beta H'^{\alpha} \right) \qquad \alpha, \beta \geq 0; H' > 0$$

(7)

where α and β are, respectively, the shape and scaling parameters of the distribution.

The above three models were used for the distributions of wave heights and wave groups statistics. For the surface distribution, besides the Gaussian model, the Gram-Charlier series expansion was also used. Since the mathematical expressions for this model is rather lengthy, the interested reader is referred to the monograph of Ochi (1992). For the statistical distribution of the parameters for wave groups, the exponential distribution is also applied. The equation for this distribution was also omitted for space limitations. All the statistical fits were verified by the χ^2- and/or the Kolmogorov-Smirnov goodness-of-fit tests.

Following Xu et al. (1993) we have also calculated the so-called group height factor (GHF), which is defined as:

$$GHF = \frac{m_{A,0}}{m_{\zeta,0}} \tag{8}$$

This parameter characterizes the relative height of the wave groups. As pointed out by Xu et al. (1993) if the process $\xi(t)$ under consideration is normal, one would have:

$$GHF = 2 - \frac{2}{\pi} \approx 0.43 \tag{9}$$

This parameter can thus be used as an indicator for the normality of the process. Another parameter used by Xu et al. is the Group Length Factor (GLF), defined as:

$$GLF = \frac{f_p}{f_{A,p}} \tag{10}$$

where f_p and $f_{A,p}$ are, respectively, the peak frequencies of the wave and the envelope spectra. This parameter can be considered as an indicator for the relative length (or period) of wave groups (Xu et al., 1993). All calculations were carried out on an IBM compatible PC with a Pentium II processor.

4. RESULTS AND DISCUSSION

Figure 2 shows the spectral densities measured on different dates. This figure is used here to denote that the spectral densities are rather different, representing various sea states. In general, it was found that, irrespective of the sea states considered, the distribution of water surface elevations is well represented by the Gaussian model. This is demonstrated in Figure 3, where three curves representing, respectively, the Gaussian, the Gram-Charlier series expansion, and that of the Stokes finite amplitude theory due to Huang et al. (1984) are shown. This is probably because that the water depth at the measuring site, 25 meters, is rather deep, and nonlinear effects due to shoaling are not very pronounced.

Figure 4 shows the probability distribution of wave heights fitted with the three models mentioned earlier. It has been shown (Plate, 1978; Piroth, 1994; Yim, 1997) that for measurements conducted in wind-wave flumes, the Gaussian distribution is more superior in describing the wave height distributions. The exact reason for this is not clear at present. However, judging from Figure 4, it is seen that all the three models can be used to describe the distribution of wave heights rather satisfactorily.

The total number of successive high waves exceeding a prescribed threshold, e.g., $H_{1/3}$, is called the *run length*, and the repetition of such high waves is the *total run* (Goda, 1985). Figure 5 shows the distribution of run length for the data measured on February 1, 1993. It can be seen that both the Weibull and the exponential models can be used to approximate the distribution of run length. Dawson et al. (1991) argued that a group must have two or more waves. They used the terms of *group length* and *group interval* instead. When this restriction is adopted, the number of wave groups contained in the individual records becomes rather sparse. We have summarized the number of group lengths and group intervals in a whole month to test their possible statistical distributions. Figures 6 and 7 show, respectively, the distributions of group lengths and group intervals for the month December 1992. In these two figures, the Weibull distribution is seen to describe the trends of the distributions better.

Finally, in Figure 8 we present the values of group length factor (GLF) with group height factor (GHF) discussed by Xu et al. (1993). It can be seen that the values of the group height factors are centered around the theoretical value of 0.43 for normal process, with, however, a tendency to have larger values.

774

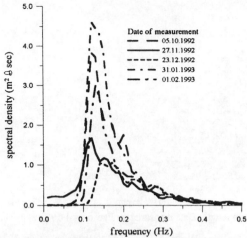

Fig.2: Spectral densities measured on different daytime.

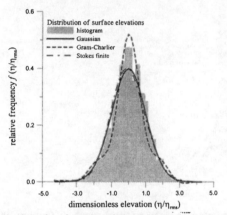

Fig.3: Distribution of surface elevations. Data measured on 27 November 1992, at 8:00 a.m.

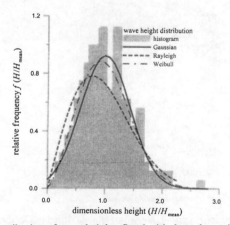

Fig.4: Distribution of wave heights fitted with three theoretical models. Data measured on October 5 1992, at 6:00 a.m

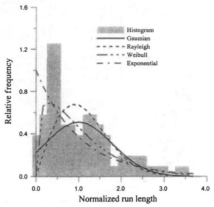

Fig.5: Distribution of run length. Data measured on February 1 1993 at 3:00 a.m.

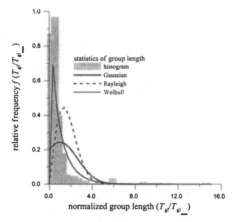

Fig.6: Statistical distribution of group lengths. Data measured in October 1992.

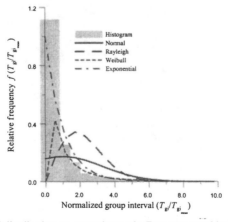

Fig.7: Statistical distribution on group intervals. Data measured in October 1992

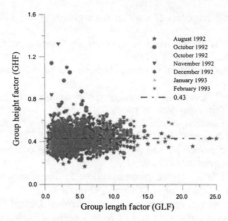

Fig.8: Comparison of group length factor (GLF) with group height factor (GHF).

5. CONCLUSIONS.

Studies of wave field characteristics have been carried out using data measured at the northern Taiwan. Possible distributions of the surface fluctuations, the wave heights, the group lengths, and the group intervals are examined. It was found that:

1. The distribution of the surface elevations can be approximated with the Gaussian model rather satisfactorily. The curve of the Gram-Charlier series expansion does not necessarily represent a better fit for records measured on this site.

2. The distribution of the wave heights can be modelled with the Gaussian, the Rayleigh, and the Weibull distribution. However, the Weibull distribution, with its two parameters, is more adjustable in its form, is preferred.

3. Both the group lengths and the group intervals can be approximated with either the exponential or the Weibull distributions. The exponential function can fit the peak values centered at the origin more adequately.

ACKNOWLEDGEMENT

The paper is partly supported by the National Science Council, Taiwan China Contract Nos. NSC-89-2611-E-019-024 (JZY), and NSC-89-2611-E-019-027 (CRC).

REFERENCE

Bendat, J. S. & A. G. Piersol 1986 *"Random data. Analysis and Measurement procedures"* 2nd ed. John Wiley, New York, N.Y., pp. 566

Benoit, M., P. Frigaard and H. A. Schaeffer 1997, "Analysing multidirectional wave spectra: A tentative classification of available methods" in *"IAHR Seminar Multidirectional waves and their interaction with structures"* 27th IAHR Congress, San Francisco, 10-15 August 1997, pp. 131-158

Dawson, T. H., D. L. Kriebel & L. A. Wallendorf 1991 "Experimental study of wave groups in deep-water random waves" Appl. Ocean Res., Vol. 13 pp. 116-131

Goda, Y. 1976 "On wave groups" in "Behaviour of Off-Shore Structures" (BOSS '76) pp. 1-13

Goda, Y. 1983 "Analysis of wave grouping and spectra of long-travelled swell" Rep. Port & Harbour Res. Inst. Japan vol. 22 No. 1 pp. 3-41

Hasselmann, K. 1962 "On the non-linear energy transfer in a gravity-wave spectrum. Part 1. General theory" J. Fluid Mech., Vol. 12, pp. 481-500

Huang, N. E., S. R. Long, C. C. Tung & L. F. Bliven 1984 "The non-Gaussian joint probability density function of slope and elevation for a nonlinear gravity field" J. Geophys. Res., Vol. C89, pp. 1961-1972

Kimura, A. 1980 "Statistical properties of random wave groups" Proc. 17th Int'l Conf. Coastal Engng. ASCE, pp. 2955-2973

Lamberti, A. & V. Rossi 1994 "Wave grouping and sequential correlation, field studies in Italian coastal areas" Coastal Engng., Vol. 21 pp. 271-300

Longuet-Higgins, M. S. 1984 "Statistical properties of wave groups in a random sea state" Phil. Trans. R. Soc. Lond., Vol. A312 pp. 219-250

Mase, H. & Y. Iwagaki 1987 "Evolution of wave groups in shallow water and wave group properties of random waves" Coastal Eng. Japan Vol. 30 pp. 19-32

Medina, J. R., R. T. Hudspeth & C. Fassardi 1994 "Breakwater armor damage due to wave groups" J. Waterway, Port, Coastal & Ocean Eng. ASCE vol. 120 pp. 179-198

Ochi, M. K. 1992 "Applied probability and stochastic processes in engineering and physical sciences" John Wiley & Sons, Singapore, pp. 499

Ouellet, Y. & I. Thériault 1989 "Wave grouping effect in irregular wave agitation in harbors" J. Waterway, Port, Coastal & Ocean Eng. ASCE vol. 115 pp. 363-383

Phillips, O. M. 1981 "Wave interactions – the evolution of an idea" J. Fluid Mech., Vol. 106, pp. 215-227

Piroth, K. 1994 "*Modellierung von Wellen und welleninduzierten Kraeften im Windwellenkanal*", Ph.D. Dissertation, University Karlsruhe, Institut für Hydrologie und Wasserwirtschaft, pp. 210 (in German)

Plate, E. J. 1978 "Wind-generated water surface waves: The laboratory evidence" in "Turbulent Fluxes Through the Sea Surface, Wave Dynamics, and Prediction" A. Favre and K. Hasselmann (eds.), Plenum Press, New York, pp. 385-401

Sand, S. E. 1982 "Long wave problems in laboratory models" J. Waterway, Port, Coastal & Ocean Eng. ASCE vol. 108 pp. 492-503

Schäffer, H. A. 1993 "Infragravity waves induced by short-wave groups" J. Fluid Mech., Vol. 247 pp. 551-588

Xu, D.-L., W. Hou, M. Zhao & J. Wu 1993 "Statistical simulation of wave group" Appl. Ocean Res., Vol. 15 pp. 217-226

Yim, J. Z. 1997 "A comparative study of the statistical distributions of wave heights" China Ocean Engineering, Vol. 11, pp. 285-304

Zuo, Q.-H. & G.-Q. Yao 1991 "Long waves induced by wave groups over a trench" China Ocean Engng., Vol. 5 pp. 191-202

Stochastic Hydraulics 2000, Wang & Hu (eds) © 2000 Taylor & Francis, ISBN 90 5809 166 X

An improved definition of random wave breaking on a beach of constant slopes

Paul Canning & Kaiming She
School of the Environment, University of Brighton, UK

ABSTRACT: A Video Image Processing (VIP) technique has been developed at Brighton to study water wave evolution in coastal waters. The VIP technique allows detailed examination of a continuous image sequence of a progressing wave, providing a complete and precise parametric description for processes such as development of a breaking wave and post-breaking mixing. Simultaneous analysis of sediment processes can also be carried out.

The definition of wave breaking on sloping beaches has been a subject of great interest in recent years. Although significant progress has been made in the prediction of breaking of monochromatic waves there has been very little progress in the prediction of random wave breaking due to the extreme uncertainties involved in measurements of this nature. With the new VIP technique, we have successfully carried out a series of wave flume experiments under storm wave conditions. The complete automation of the VIP technique has made it possible to analyse very large number of breakers with great accuracy. A large number of individual breakers have been examined, resulting in a well defined correlation between the non-dimensional breaking wave height H_b/gT^2 and breaking depth d_b/gT^2, where H_b and d_b are breaking height and breaking depth, respectively. g is the acceleration due to gravity. The effects of the beach slope and bed roughness on the relation between H_b/gT^2 and d_b/gT^2 have also been investigated.

Key words: Wave breaking, Video imaging, Breaking criterion.

1 INTRODUCTION

Experimental studies of wave breaking all involve acquisition of wave data through one or more techniques. Commonly used techniques include wave gauge arrays, Laser Doppler Anenometry (LDA) and Particle Image Velocimetry (PIV). LDA allows measurement of particle velocity of a single point while PIV achieves 2D full field velocity maps. Traditional PIV technique involves taking and analysing photographic images of the flow field (Skyner et al, 1998; Greated et al, 1992) but the latest development simply uses high resolution video cameras (Nadaoka et al, 1997). A Video Image Processing (VIP) technique has been developed at Brighton to study water wave evolution in the near-shore zone. This allows detailed investigation of continuous image sequences of progressing waves, providing complete and precise parametric description of breaking wave development and post-breaking behaviour.

The shoaling and breaking of waves onto beaches, and the resultant change in the wave geometry is of interest in the area of coastal defence, and in relation to the understanding of sediment transport on beaches. Random waves throughout their evolution have wave-wave interaction, but when shoaling becomes strongly non-linear and subsequently the wave surface near breaking cannot be adequately measured with time series of the wave surface. Therefore, experimental measurement of the wave surface during shoaling and breaking has been performed using the VIP technique, allowing spatial and temporal measurement of the wave evolution in this region. The present work focuses particularly on the wave breaking point, although experimental measurements of pre- and post-breaking wave behaviour have been collected.

1.1 Previous random wave studies

Random waves provide great complexity in the shoaling and breaking zones, with difficulty in defining the parameters in a similar fashion to the monochromatic wave condition. The majority of random wave parameter studies involve probabilistic treatment of measurements from field work or laboratory experiments (Baldock et al, 1998; Dally et al, 1986; Wesson et al, 1998). Additionally, much of the field data has been collected in deep water (Gentile, 1998; Zhang et al, 1992) for its value in offshore structure design, with relatively fewer studies in the near-shore region (Chadwick et al, 1995; Wessen et al, 1998). Although the shoaling process tends to separate a wave group into its constituent wave frequencies, it has been noted that on steeper beaches, such as the 1/7 to 1/10 slopes included here, wave groupiness can increase towards the shore (Kobayashi et al, 1989), causing non-linearities to be even greater.

In order to predict the random wave behaviour, two types of technique are generally used; (1) The decomposition of a random wave into the component waves, with subsequent super-position, or (2) substituting in a monochromatic wave which is deemed representative of a wave field. However, with the first method, as the component waves are modelled separately, no wave-wave interactions are incorporated in any form. The second technique models the representative monochromatic wave using high-order numerical schemes, which does involve non-linear effects, but in an essentially different sense (see Zhang et al, 1992, for details). Parameterisation of random waves is mainly statistical, often using representative wave parameters. As the majority of random wave records are in time series format, the distance between adjacent zero-upcrossings or zero-downcrossings gives the wave period, and the wave height is defined within these two points. Furthermore, these parameters are then treated statistically, to produce representative parameters depending on the preferred usage (see Goda, 1992, for details). In the near-shore zone, where non-linear behaviour is inherent in wave propagation, and the approach of breaking involves rapid changes in the wave geometry, solely temporal measurements are no longer sufficiently accurate (for example, fixed wave gauges). Alternately, spatial visualisation techniques are well suited to this application (Bonmarin, 1989), allowing direct definition and calculation of wave parameters. Recently, some progress has been achieved in deterministic measurement (Spell et al, 1996; Zhang et al, 1999) of random wave fields, although again focusing on offshore wave fields rather than the near-shore zone. Furthermore, the effects of wave breaking, viscosity and high-order non-linearities are still to be included in the numerical wave model of Zhang et al (1999). A summary of random wave breaking criteria is given in Southgate (1993).

1.2 The effects of bed friction on wave breaking

Previous experimental and numerical work has often used a smooth, impermeable plane bed as a simplified representation of a beach, which is applicable to sandy beaches (Griffiths et al, 1992). The shallowest slope in the experimental work included here, 1/30, pertains to this description, and so only a smooth bed was used. However, on steeper, shingle beaches, the roughness of the beach, however it is defined, will be greater than for sandy beaches. This is reflected in the literature; steeper bed slopes are roughened to simulate the greater particle size on steep, shingle beaches (Kobayashi et al, 1990). The roughness particles used in the Kobayashi et al (1990) study were well sorted, with median diameter of 21mm. The 21mm median diameter corresponds to large gravel size (and therefore not at model scale), whereas the random wave conditions are obviously at a model scale, highlighting the complicated and poorly understood scaling requirements for bed particles, even when non-mobile. The issues relating to scaling of sediment particles are extremely complex and extensive research into this area is ongoing.

It has been noted previously (Baldock et al, 1998) that bed friction effects are minimal in the surf zone; although this region has been observed, the breaking and pre-breaking zone are investigated herein, where bed friction will possibly have an effect on the wave geometry or parameters. The inclusion of the roughened beds in this study will also give insight into the important variables associated with random wave breaking on mobile beds; an extension of this work into mobile bed interaction with wave behaviour in the near-shore zone is currently underway.

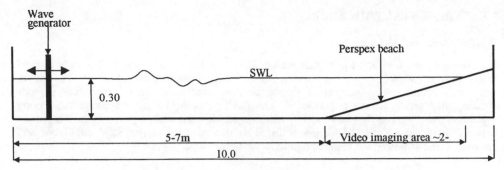

Fig.1: Wave flume setup

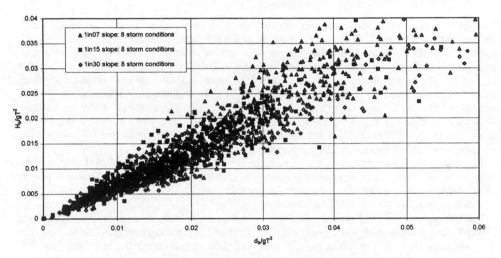

Fig.2:. Random wave breaking, smooth bed.

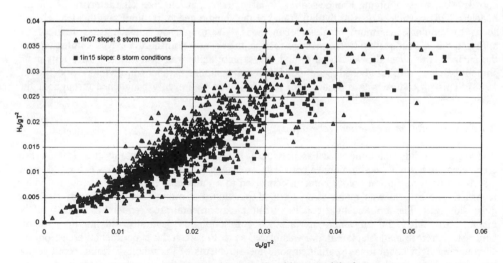

Fig.3:. Random wave breaking, rough bed.

781

2 EXPERIMENTAL PROCEDURE

2.1 Wave tank and beach models

Experiments were conducted in a glass walled, 10m long, 0.50m deep and 0.45m wide channel fitted with a piston-type wave generator, controlled by computer (see Fig. 1). The waves generated shoaled over a number of sloping beaches constructed from 10mm thick perspex sheeting, stiffened by struts along its length. This sheeting was held in place by concrete supports underneath, and fixed to the glass walls by a thin bead of silicone sealant (a 1mm gap was present between the perspex sheet and the side walls). This allowed relatively practical installation and removal whilst achieving the required rigidity of the beach slope.

2.2 Experimental setup

Two main experimental conditions were carried out. These were variations on the model beach surface using the smooth surface of the perspex and a roughened surface using anthracite coal, with 1.18mm>D>1.00mm. The random waves were generated in deep to transition depth water, and allowed to shoal and break due to the presence of impermeable, sloping beaches (Canning et al, 1998). The beach slopes of 1/7, 1/15 and 1/30 were used to represent different types of beaches: shingle beaches (~1/7), and sand beaches (~1/30). The roughened bed experiment was carried out for the 1/7 and 1/15 slopes, which can be seen as representative of shingle beaches. This allowed comparison between the respective types of beaches. The shallowest slope had a steeper ramping 1/10 slope of 1.4m length, which then intersected with the 1/30 slope. The ramping slope was added solely due to wave tank length limitations.

The random waves generated were based on the JONSWAP spectrum, based on the storm wave data recorded during the Environmental Agency: Lancing and Shoreham Sea Defences project (She, 1996). Eight wave climates were generated, described in Table 1, two each from representative 1, 10, 50 and 100 year storms.

The smooth surface used was simply the untreated perspex surface. The smooth surface was used to allow the possibility of investigating the effect of bottom roughness, by comparison with the roughened surface. The model material used was crushed anthracite, due to its previous well documented use for similar studies (Powell et al, 1992; She, 1996), and its relative cheapness. The size of particle applied had a median diameter between 1.00mm and 1.18mm. This corresponds, when applying Froude scaling, to large gravel particle size, characteristic of shingle beaches. This particle size is at the larger end of the shingle beach grading, and this was chosen so as to simulate maximum roughness conditions. The roughened surface was produced by applying a thin skin of glue to the perspex surface, then gently pouring the sieved anthracite over the surface- this was allowed to cure, and excess anthracite was brushed off. Any remaining smooth patches were roughened again individually, in a similar manner to previous work (Sleath, 1988). This procedure was repeated until the whole bed was evenly roughened throughout the whole surface.

2.3 Experimental measurement technique

The video imaging technique used has previously been described in Canning et al (1998). For completeness this will be summarised here. The wave evolution over the beaches is imaged using a JAI-204 high resolution colour camera, connected to a Panasonic SVHS high resolution video recorder. The camera was mounted on a tripod, positioned approximately 1-2m away from the glass side wall. The ambient light conditions in the experimental area reduced the clarity of the image, particularly the air-water interface. Use of floodlights generally did not improve the contrast; therefore matt black material was placed so as to screen the experimental area from the surrounding light conditions. Spatial/temporal measurements of the whole variable depth region have been taken for the steeper slope of 1/7. However, due to the requirement for high spatial resolution, preferably around 1-2mm, the shallower slopes have measurements taken with the breaking point in a central position, but did not include imaging the complete variable depth region.

2.4 Post-experimental measurement procedure

The analogue recording of the wave shoaling and breaking process was then digitised using a Kramer RGB Decoder connected to a dedicated Pentium computer. The computer is fitted with a DataCell SnapperTool v2.02 real time frame grabber card. The digital image sequences are captured in ImagePro Plus v3.0, using the SnapperTool v2.02 capture modules. This achieves a temporal resolution of 20ms, with a spatial resolution of 1-2mm depending upon the experimental setup. The digital sequences are then processed and analysed using Wipa32NT, a (W)ave (I)mage (P)rocessing and (A)nalysis programme written at the University of Brighton particularly for this use. This automatically processes the images to define the wave surface and bed surface, and from this calculates wave parameters throughout the wave evolution. Further details are given in Canning et al (1998).

3. EXPERIMENTAL RESULTS AND DISCUSSION

Random wave data (normally from field studies) is often non-dimensionalised using the apparent period calculated from time series. This is a somewhat simplified approach used due to data limitations. The subject of random wave parameter definition has been investigated previously (Seyama et al, 1988), with some success in reducing the data scatter inherent in field and experimental measurements. In the paper mentioned above the definition of water depth was studied, respective to SWL, MWL, and other datums (see also Weishar et al, 1978).

As the experimental measurements collected in this series of experiments has a spatial as well as temporal dimension, the standard non-dimensional approach is followed, but by calculating the wave period at breaking by a different method. The crest speed as the wave crest approaches breaking is calculated using the horizontal crest movement between the successive frames. As the wavelength at breaking is known by direct measurement from the image frame, the period is then calculated by dividing the wavelength by the crest speed.

The random wave parameter relationships are presented in Fig. 2 for the smooth bed experiment. For the first time in the experimental studies of random wave breaking, we are able to establish a relatively well defined relationship between H_b/gT^2 and d_b/gT^2. The beach slope does not appear to have significant effects on the height to depth relationship. The same observations can be made for wave breaking on roughened beds, as shown in Fig. 3. There is increase in the data scatter compared to the smooth bed measurements. For the 1/7 bed slope, the introduction of the bed roughness resulted in a slight increase of breaking wave height (~3%) compared to the smooth bed results. For the 1/30 slope, however, no difference can be observed between the rough and smooth bed results. This means that the bed friction causes loss of wave energy but the point of breaking is still primarily governed by the non-dimensional water depth.

4 CONCLUSIONS

A series of random wave breaking experiments have been carried out using a newly developed laboratory measurement technique. This technique has allowed accurate measurement of random wave breaking on a number of slopes, under smooth and rough bed conditions. The experimental results show the following:

- A definitive relationship exists between H_b/gT^2 and d_b/gT^2, where H_b and d_b are the height and depth at breaking, respectively.
- The beach slope and the bed roughness have a very small influence on the above relationship.
- The relationship between H_b/gT^2 and d_b/gT^2 is of the same order of accuracy for both random waves and monochromatic waves.

5 FURTHER WORK

The present experiments have highlighted the random wave breaking behaviour on impermeable, immobile beach slopes. Experimental measurements of the shoaling and post-breaking processes have also been collected, and are currently being analysed. The limited investigation of variable bed conditions will be extended so as to study permeable, mobile bed response to wave breaking. This will allow a detailed understanding of the interaction between the wave breaking events and sediment motion.

Table 1: Experimental Random Wave Conditions (1:50 scale)

Test Case	T_s (s)	H_s (mm)	Return period (years)
IR1	0.66	35.20	1
IR2	0.86	68.40	10
IR3	0.89	74.00	1
IR4	0.93	82.00	50
IR5	0.98	90.80	100
IR6	1.00	95.20	10
IR7	1.05	107.60	50
IR8	1.07	113.40	100

ACKNOWLEDGEMENTS

The first author was in receipt of an EPSRC studentship.

REFERENCES

Baldock, T.E., Holmes, P., Bunker, S. and Van Weert, P. 'Cross-shore hydrodynamics within an unsaturated surf zone', Coastal Engineering, Vol.34, 1998, pp173-196.

Bonmarin, P. 'Geometric properties of deep-water breaking waves', J. Geophysical Research, Vol.209, 1989, pp405-433.

Canning, P., She, K. and Morfett, J. 'Application of video imaging to water wave analysis', Proc. 3rd Int. Conf. HydroScience and Engineering, 1998, p55, (paper on CD-ROM).

Chadwick, A.J., Pope, D.J., Borges, J. and Ilic, S. 'Shoreline directional wave spectra Part 2. Instrumentation and field measurements', Proc. Instn Civ. Engrs Wat. Marit. And Energy, Vol.112, 1995, pp209-214.

Dally, W.R. and Dean, R.G. 'Transformation of random breaking waves on surf beat', Proc. 17th Int. Conf. Coastal Eng., 1986, pp109-123.

Greated, C.A., Skyner, D.J and Bruce, T. 'Particle Image Velocimetry (PIV) in the Coastal Engineering Laboratory', Proc. 23rd Int. Conf. Coastal Eng., 1992, pp212-225.

Griffiths, M.W., Easson, W.J., Greated, C.A. 'Measured internal kinematics for shoaling waves with theoretical comparisons', J. Waterway, Port, Coastal, and Ocean Eng., Vol.118, No 3, 1992, pp280-299.

Gentile, R. 'Statistical properties of random waves in deep water directional seas involving f^4 frequency spectra', Proc. 8th Offshore and Polar Eng. Conf., Vol.3, 1998, pp203-211.

Goda, Y. 'Random seas and design of maritime structures', University of Tokyo Press, 1985.

Kobayashi, N., DeSilva, G.S. and Watson, K.D. 'Wave transformation and swash oscillation on gentle and steep slopes', J. Geophysical Research, Vol.94, 1989, pp951-966.

Kobayashi, N., Cox, D.T. and Wurjanto, A. 'Irregular wave reflection and run-up on rough impermeable slopes', J. Water, Port, Coastal and Ocean Eng., Vol.116, No.6, 1990, pp708-726.

Nadaoka, K., Ono, O. and Kurihara, H. 'Near crest pressure gradient of irregular water waves approaching to break', Proc. Int. Conf. Coastal Dynamics, 1997.

Powell, K.A., Quinn, P.A. and Greated, C.A. 'Shingle beach profiles and wave kinematics', Proc. 23rd Int. Conf. Coastal Eng., 1992, pp2358-2369.

Seyama, A. and Kimura, A. 'The measured properties of irregular wave breaking and wave height change after breaking on the slope', Proc. 15th Int. Con. Coastal Eng. 1988, pp419-432.

She, K. 'Modelling of shingle beach response to storm and swell wave action', Final Report, University of Brighton, Dept Civ. Eng., August 1996.

Skyner, D.J. and Easson, W.J. 'Wave kinematics and surface parameters of steep waves travelling on sheared currents', J. Waterway, Port, Coastal and Ocean Eng., Vol.124, No.1, 1998, pp1-6.

Sleath, J.F.A. 'Transition in oscillatory flow over rough beds', J. Waterway, Port, Coastal and Ocean Eng., Vol.114, 1988, pp18-33.

Southgate, H.N. 'Review of wave breaking in shallow water', Advances in Underwater Technology, Ocean Science and Offshore Engineering, Vol.29, 1993, pp251-275.

Spell, C.A., Zhang, J. and Randall, R.E. 'Hybrid wave model for unidirectional irregular waves- part 2. Comparison with laboratory measurements', Applied Ocean Research, Vol.18, 1996, pp93-110.

Weishar, L.L. and Byrne, R.J. 'Field study of breaking wave characteristics', Proc. 9th Int. Conf. Coastal Eng., 1978, pp487-506.

Wessen, J.C, Su, M-Y. and Burge, R.E. 'Some near shore breaking wave statistics outside the surf zone', Proc. 8th Int. Offshore and Polar Eng. Conf., Vol.3, 1998, pp144-151.

Zhang, J., Randall, R.E. and Spell, C.A. 'Component wave interactions and irregular wave kinematics', J. Waterway, Port, Coastal and Ocean Eng., Vol.118, No.4, 1992, pp401-415.

Zhang, J., Prislin, I., Yang, J. and Wen, J. 'Deterministic wave model for short-crested ocean waves: Part 2. Comparison with laboratory and field measurements', Applied Ocean Research, Vol.21, 1999, pp189-206.

Stochastic Hydraulics 2000, Wang & Hu (eds) © 2000 Taylor & Francis, ISBN 90 5809 166 X

Spectral analysis of group waves run-up

Zheng Jinhai, Yan Yixin & Qu Yonggang
Research Institute of Coastal and Ocean Engineering, Hokai University, Nanjing, China

ABSTRACT: The run-up spectrum of group waves on smooth slopes ranging from 1/1 to 1/10 is explored through laboratory investigations, in which the group waves consist of two wave series with equal wave height and frequency interval 0.02 *Hz*. The spectral estimation is conducted by using the fast Fourier transformation approach and the smooth technique provided by the VAX system. It is found that the run-up spectrum depends on the incident wave parameters, the water depth, the frequency interval of component waves and the slopes. Moreover, a significant influence of the nonlinear terms of group waves, including the self-interaction term, the cross-interaction term and the wave set-down term, on the spectral form is demonstrated. The experimental results show a phenomenon of energy saturation at three frequency regions, which are named as the zero-order frequency, the first-order frequency and the second-order frequency, and corresponding to the wave set-down term, the linear term, and the interaction terms of group waves, respectively. The Irribarren number ξ is employed to detect the run-up spectrum. The run-up spectrum with the maximum energy density at the zero-order frequency region emerges when ξ varies from 1.5 to 2.5. The run-up spectrum with the maximum energy density at the first-order frequency region appears while ξ changes from 3.5 to 5.1. The run-up spectrum with the maximum energy density at the second-order frequency region arises as ξ becoming more than 5.2.

Key words: Group waves, Run-up spectrum, Nonlinear term, Spectral analysis, Laboratory investigation

1.INTRODUCTION

Wave run-up on coastal structures, such as seawalls, dykes, surge barriers and so on, is an important factor in the design of structures. Besides, the height of wave run-up is the limit of on-shore side for on-offshre and littoral sand transports. There are mainly two ways to analyze the characteristics of wave run-up. One is the individual wave run-up analysis, the other is the spectral analysis. From the engineering viewpoint, such as to determine the heights of coastal structures, the individual run-up wave analysis is preferable, because frequency distributions or extreme statistics of individual run-up wave heights are required. The spectral analysis is employed to investigate the dynamic response between the incident waves and the run-up variations and the spectral characteristics of run-up variations themselves (Mase H.,1988). Compared with studies and conclusions on the run-up of monochromatic waves (Xue Hongchao et al., 1991), there are few studies which discussed group waves and little understandings about their run-up properties. However, model tests of three different wave patterns to investigate their run-up and run-down on the permeable and impermeable slopes and the stability of dolos armour have demonstrated a significant influence of the succession of waves on coastal structures. Also a conclusion of strongly grouped waves are more critical than regular wave was reached (Burcharth H.,1979).

According to the theoretical analysis of group waves by Yan and Zheng (1997), the impact of the nonlinear terms of group waves on the water surface is remarkable, especially in the shallow water. Therefore, the nonlinear terms should be taken into account while the spectral characteristics of group waves run-up are studied. The purpose of present paper is to examine the spectral properties of group waves run-up on smooth slopes experimentally by considering two wave series with equal wave height and the frequency interval is 0.02 Hz.

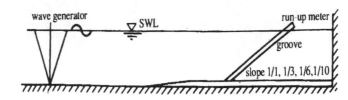

Fig.1: Experimental Apparatus

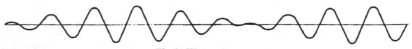

Fig.2: Wave pattern

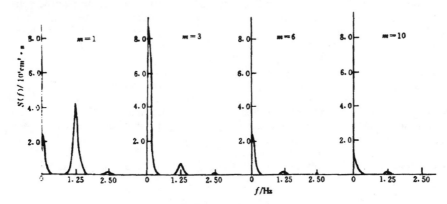

(a) D/H=6.4, f_0=1.25Hz

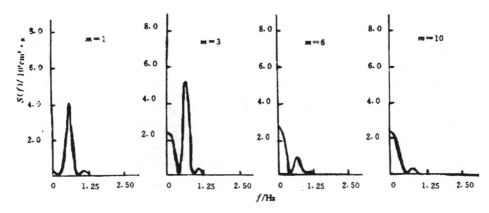

(b) D/H=6.4, f_0=0.52Hz

788

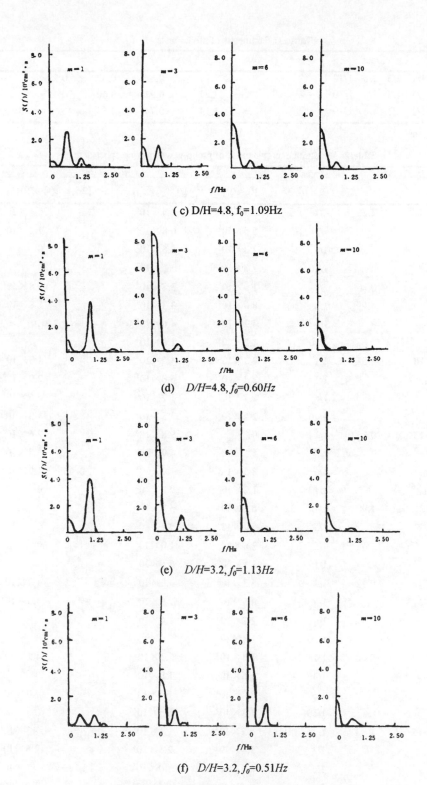

(c) D/H=4.8, f_0=1.09Hz

(d) D/H=4.8, f_0=0.60Hz

(e) D/H=3.2, f_0=1.13Hz

(f) D/H=3.2, f_0=0.51Hz

Fig.3: Runup energy spectra for each beach slopes
duc to six kinds of incident waves with different water depth

789

Table.1: Parameters of model tests

D/H	slopes				f_0			
6.4	1/1	1/3	1/6	1/10	1.25	0.80	0.52	0.35
4.8	1/1	1/3	1/6	1/10	1.29	0.92	0.60	0.41
3.2	1/1	1/3	1/6	1/10	1.13	0.74	0.51	

Table.2: Measured energy density of group waves run-up spectra

D/H	L_0/D	f_0	slopes	$S(f)$ /cm^{2}s		
				$[0, \triangle f]$	$[f_0, f_0+\triangle f]$	$[2f_0, 2(f_0+\triangle f)]$
6.4	3	1.25	1/1	2.3×10^4	4.4×10^4	6.5×10^2
			1/3	8.5×10^4	6.8×10^3	1.2×10^2
			1/6	2.3×10^4	6.5×10^2	0
			1/10	7.8×10^3	1.2×10^2	0
	6	0.80	1/1	1.6×10^3	3.8×10^4	0.8×10^2
			1/3	8.0×10^4	3.3×10^4	1.8×10^3
			1/6	4.2×10^4	4.6×10^3	1.2×10^2
			1/10	1.5×10^4	1.0×10^3	0
	10	0.52	1/1	4.3×10^2	4.5×10^4	1.3×10^3
			1/3	2.3×10^4	6.8×10^4	2.9×10^3
			1/6	2.7×10^4	1.2×10^4	6.5×10^2
			1/10	2.2×10^4	3.8×10^3	1.2×10^2
	15	0.35	1/1	0	9.8×10^3	2.6×10^3
			1/3	5.1×10^3	4.7×10^4	3.7×10^3
			1/6	8.5×10^3	1.7×10^4	1.2×10^2
			1/10	1.2×10^4	9.8×10^3	0
4.8	3	1.90	1/1	8.1×10^3	3.7×10^4	0
			1/3	6.7×10^4	9.1×10^3	6.5×10^2
			1/6	2.6×10^4	1.0×10^3	0
			1/10	1.1×10^4	6.5×10^2	0
	6	0.92	1/1	5.2×10^3	3.0×10^4	1.3×10^3
			1/3	8.6×10^4	1.5×10^4	4.0×10^2
			1/6	2.7×10^4	1.4×10^3	0
			1/10	1.3×10^4	6.5×10^2	0
	10	0.60	1/1	1.4×10^3	2.2×10^4	4.2×10^3
			1/3	1.6×10^4	1.8×10^4	0
			1/6	2.7×10^4	1.4×10^3	4.0×10^2
			1/10	1.3×10^4	6.5×10^2	0
	15	0.41	1/1	5.2×10^3	1.1×10^4	7.3×10^3
			1/3	1.9×10^4	2.5×10^4	2.5×10^3
			1/6	1.5×10^4	1.0×10^4	9.1×10^2
			1/10	2.1×10^4	6.5×10^3	1.2×10^2

3.2	6	1.13	1/1	6.5×10^3	3.8×10^4	6.5×10^2
			1/3	10.0×10^4	4.0×10^3	1.2×10^2
			1/6	3.3×10^4	1.0×10^3	0
			1/10	1.4×10^4	1.2×10^2	0
	10	0.74	1/1	1.8×10^3	2.4×10^4	3.3×10^3
			1/3	13.0×10^4	1.4×10^4	1.4×10^3
			1/6	5.8×10^4	6.5×10^3	1.2×10^2
			1/10	2.8×10^4	1.4×10^3	0
	15	0.51	1/1	0	6.9×10^3	4.8×10^3
			1/3	3.6×10^4	1.3×10^3	1.3×10^4
			1/6	4.9×10^4	1.7×10^4	0
			1/10	1.9×10^4	5.2×10^3	0

2.LABORATORY INVESTIGATIONS

A series of experiments are conducted in a 1.50m wide, 30.00m long and 0.75m deep wave flume in the Estuarine and Coastal Engineering Laboratory of Hohai University. At one of the ends of the wave flume, a random wave generator is installed. Beach slopes are selected as 1/1,1/3,1/6, and 1/10, and the water depths are 0.24m, 0.32m, and 0.64m, respectively. The outline of the experimental apparatus is illustrated in Fig.1.

The parameters of model tests are described in Tab.1, where f_0 represents the typical frequency of group waves, and D/H is the ratio of water depth to the maximum wave height which is defined as the maximum difference in water level between two successive zero-crossings in the down-ward direction.

The interval frequency of component waves, Δf, equals to 0.02Hz. The wave pattern used here is shown in Fig.2, where the vertical scale is amplified.

A capacitance type wave gauge is used as a run-up meter, which is calibrated by moving every 5cm along the beach surface. The calibration curve is almost approximated by a straight line.

In the spectral estimation, the fast Fourier transformation approach and the smooth technique provided by the VAX system are employed. The number of data point is 1024, the Nyquist frequency is 1.25 Hz, the number of degrees of freedom is 42, and the bandwidth of spectral resolution is 0.064 Hz.

3.EXPERIMENTAL RESULTS

Table 2 shows the measured energy density of group waves run-up spectra on smooth slopes.

It can be seen from Tab.2 that the run-up spectrum of group waves depends on the parameters of incident waves, such as the maximum wave height and the wave period which is the time elapsing between two successive zero-crossings in the down-ward direction, the water depth, the frequency of component waves and the slopes. It is also found that the maximum energy densities in the low frequency region for different kinds of relative water depth and different frequencies of component waves appear when the slope is 1/3. Another interesting result can be obtained is that the energy densities in the high frequency region for different kinds of relative water depth and different slopes become larger with the increasing of L_0/D, where L_0 is deep-water wave length.

Fig.3 shows the run-up energy spectra for each beach slopes due to six kinds of incident waves with different water depth.

An important conclusion that might be drawn from the graphs is that a phenomenon of energy saturation at three frequency regions, which are named as the zero-order frequency, the first-order frequency and the second-order frequency, and corresponding to the wave set-down term, the linear term, and the interaction terms of group waves, respectively. It is illustrated from those figures that the energy densities in the second-order frequency region become larger when the frequencies of component waves get smaller, which demonstrates that the smaller the frequencies of component waves, the much more important influence of the interaction terms of group waves.

By applying the surf similarity parameter $\xi = \left(H / L_0 \right)^{-\frac{1}{2}} \tan \alpha$, where α is the angle of the slope, the run-up spectrum is detected. The run-up spectrum with the maximum energy density at the zero-order frequency region emerges when ξ varies from 1.5 to 2.5. The run-up spectrum with the maximum energy density at the first-order frequency region appears while ξ changes from 3.5 to 5.1. The run-up spectrum with the maximum energy density at the second-order frequency region arises as ξ becoming more than 5.2.

4 CONCLUSIONS

On the basis of experimental study on group waves run-up on smooth slopes, a significant influence of the nonlinear terms of group waves on the spectral form is demonstrated. The measured run-up energy spectra of group waves indicate a phenomenon of energy saturation at three frequency regions, which are named as the zero-order frequency, the first-order frequency and the second-order frequency. The run-up spectrum with the maximum energy density at the zero-order frequency region emerges when ξ varies from 1.5 to 2.5. The run-up spectrum with the maximum energy density at the first-order frequency region appears while ξ changes from 3.5 to 5.1. The run-up spectrum with the maximum energy density at the second-order frequency region arises as ξ becoming more than 5.2.

REFERENCES

Burcharth H. F., 1979. The effect of grouping waves on on-shore structure. *Coastal Engineering*, 2(3):189~199

Mase H., 1988. Spectral characteristics of random wave run-up. *Coastal Engineering*, 12(2):175~189

Xue Hongchao, Guo Da, Zhong Husui, Pan Shaohua, 1991. Wave runup-rundown amplitude on slopes. *China Ocean Engineering*, 5(1):39~50

Yan Yixin, Zheng Jinhai, 1997. Runup and rundown of group waves on smooth slopes. *Journal of Hohai University*, 25(5):90~95 (in Chinese)

Probabilistic control of reservoir management

Stochastic Hydraulics 2000, Wang & Hu (eds) © 2000 Taylor & Francis, ISBN 90 5809 166 X

Invited lecture: Optimisation of hydropower schemes: A case study in Ethiopia

C.Jokiel
Lahmeyer International GmbH, Bad Vilbel, Germany

ABSTRACT: The main aspect related to the planning and design of hydropower schemes besides purely technical and socio-environmental issues is the economic evaluation. In case a project or several candidate projects are technically feasible the decision as to which project will be implemented first will be based on key economic parameters, e.g. the specific generation costs or the net benefits of a project. In order to maximize the benefits of a project, optimization procedures have to be incorporated into the planning process. This paper presents some of the optimisation procedures performed during the feasibility study of the Chemoga-Yeda-Sens Hydropower Scheme in Ethiopia, a scheme consisting of three interconnected reservoirs.

Key words: Hydropower schemes, Optimisation, Ethiopia, Interconnected reservoirs

1.THE PROJECT

In January 1998, an international Joint Venture lead by Lahmeyer International was contracted by the Ethiopian Electric Power Corporation (EEPCO) to perform the feasibility study of three high head hydropower schemes in the Ethiopian highlands, comparing them with a fourth scheme to establish which scheme should be implemented first. One of these 4 schemes is the Chemoga-Yeda-Sens scheme located in the basin of the Abay River (Blue Nile) close to Debre Markos some 200 km north of Addis Ababa (see Figure 1).

The main features of the three reservoirs and the hydropower component for the Stage I development at Sens are summarised in Table 1.

Table.1: Main Features of the Project

	Chemoga	Yeda	Sens
Dam Height	36 m	40.5 m	28.5 m
Dam Volume	2.25 MCM	1.5 MCM	0.33 MCM
Active Storage	220 MCM	215 MCM	8.13 MCM
FSL	2428 m asl	2290 m asl	2290 m asl
Hydropower Components Stage I at Sens Headpond			
Length of Headrace Tunnel			4300 m
Length of Penstock			3500 m
Rated Discharge			37.9 m³/s
Rated Head			790 m
Installed Capacity			245 MW
Firm Annual Energy			589 GWh/y

The scheme consist of two interconnected reservoirs, Chemoga and Yeda, transferring water to the headpond on the Sens River from where the water drops into the Abbay Gorge utilising a head difference of some 1400 m. The scheme shall be developed in 2 stages to more closely match the demand growth of the Ethiopian power system and to ease financing. Figure 2 gives an overview of the scheme.

2. LEVELS OF OPTIMIZATION

Within the planning process of large hydropower schemes optimisation procedures have to be performed at different levels and in different contexts. In the following, three typical levels of optimisation will be presented. These levels cannot be seen in isolation but are typically interdependent. Furthermore, the three levels should be regarded as rough categories which are not necessarily comprehensive but serve as examples. As each project has its individual characteristics the optimisation procedures applied have to be selected and adopted accordingly.

Level 1: Optimisation in System Context
Large hydropower projects need to be optimised as part of an overall system expansion plan. This includes the comparison with candidate power plants (hydro, thermal, nuclear) to identify the most attractive scheme(s) and the most favourable sequence of implementation to best meet the predicted future demand at the least cost and in an environmentally and socially acceptable manner. The specific generation costs [US$/kWh] are most commonly used as the key economic parameter for this evaluation.

Level 2: Optimisation in Project Context
For a specific project, alternative layouts will be compared and the main characteristics of the scheme, e.g. storage volume, installed capacity, mode of operation, will be optimised by means of reservoir operation and energy generation studies. The depth of these studies depends on the stage of the planning process, i.e. prefeasibility or feasibility level. The key economic parameters used for this optimisation are the least present costs [US$] or the maximum net benefits [US$].

Level 3: Optimisation of Project Components
For the selected layout, the dimensions of individual structures of a scheme will be optimised. Examples include the diameter of power waterways or the length of a spillway crest. In general, such optimisations will be performed at a feasibility level. The key economic parameters are the same as those for the project optimisation or minimum construction costs.

3. OPTIMIZATION OF THE CHEMOGA-YEDA-SENS SCHEME

3.1 *General*

As mentioned before, four different schemes were investigated within the feasibility study. Out of these four schemes the Chemoga-Yeda-Sens scheme proved to be the most attractive with a specific installed capacity cost of 1420 US$/kW. During feasibility design various optimisation procedures were performed for the Chemoga-Yeda-Sens scheme, of which the following will be presented:

(1) Optimisation of the Chemoga and Yeda reservoir capacities (Level 2 Optimisation),
(2) Optimisation of the spillweir crest lengths of Chemoga and Yeda reservoirs (Level 3 Optimisation),
(3) Optimisation of the power waterways at Sens reservoir (Level 3 Optimisation).

3.2 Optimisation of reservoir capacities

The capacities of the Chemoga and Yeda reservoirs were optimised to define the firm flow as a function of the active reservoir capacities and to define the active storage capacity which results in the least present cost discounted throughout the life of the project.

To calculate the firm flow ("firm flow" defined as the yield with 98 % reliability) of the two reservoirs, detailed reservoir simulations had to be performed using the long-term discharge records developed during the hydrological studies. The simulations runs were carried out using Lahmeyer International's proprietary computer program APROS. This program solves, subject to operating rules, the storage continuity equation for reservoirs. The computations were carried out with a monthly time step over the hydrological base period (1960 – 1996) taking into account inflows, flow releases, spillages and net losses.

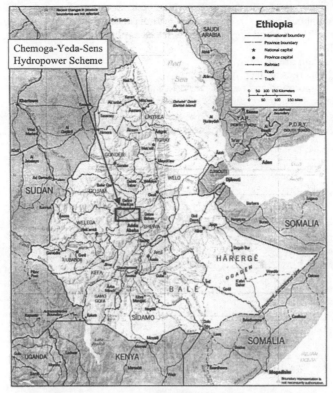

Fig.1: Location of the Chemoga-Yeda-Sens Hydropower Scheme

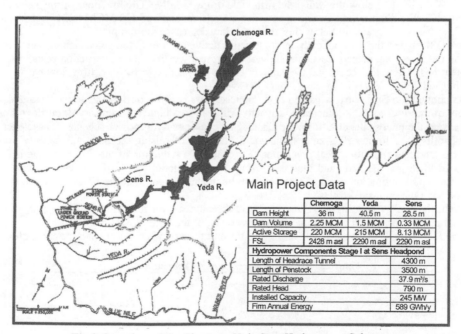

Main Project Data

	Chemoga	Yeda	Sens
Dam Height	36 m	40.5 m	28.5 m
Dam Volume	2.25 MCM	1.5 MCM	0.33 MCM
Active Storage	220 MCM	215 MCM	8.13 MCM
FSL	2426 m asl	2290 m asl	2290 m asl
Hydropower Components Stage I at Sens Headpond			
Length of Headrace Tunnel			4300 m
Length of Penstock			3500 m
Rated Discharge			37.9 m³/s
Rated Head			790 m
Installed Capacity			245 MW
Firm Annual Energy			589 GWh/y

Fig.2: Overview of the Chemoga-Yeda-Sens Hydropower Scheme

797

The operation is driven by operating rules which seek to meet a target continuous flow demand or target continuous power in each month without having the reservoir drop below minimum operating level (MOL) and they seek to minimise spillage in any month, thus maximising the reservoir outflow.

Inputs to the program are as follows:
-reconstituted discharges at the project site,
-net losses (i.e. evaporation and rainfall on surface area) of the reservoir, and
-the elevation–area–capacity relationship of the reservoir.

Simulation runs to determine the firm flow were carried out for a range of active reservoir storages. The simulation results for both reservoirs are shown in Figure 3.

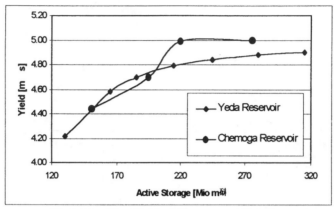

Fig.3: Firm Flow Yields of Chemoga and Yeda Reservoirs

For the Chemoga reservoir there is almost no change in the firm yield for reservoir capacities above 220 MCM. Below this value the firm yield drops rapidly. The function for the Yeda reservoir has two distinct ranges; above an active storage capacity of 200 MCM to 215 MCM the function has a low gradient; below this, the yield flow decreases more rapidly.

In the next step the economically most attractive active reservoir storage capacities which result in the least total present cost consisting of the present worth of the construction costs, resettlement costs, land loss costs and the value of energy losses due to reduced firm flow has to be established.

As shown, the firm outflow gained from the reservoirs is a function of the active storage capacity of the reservoirs. An increased capacity will lead to a higher firm reservoir outflow thus increasing the energy benefits from the Chemoga-Yeda-Sens hydropower scheme. These energy benefits have to be offset against the additional costs of raising the dam, submerging productive land and resettling local population. For example a reduction of the active storage is related to lowering the FSL and thus decreasing the construction costs of the dam as well as environmental mitigation costs as less area will be inudated. An increase of the active storage would have the opposite effects.

Both, the energy benefits and the costs, were discounted over the life time of the project to the reference year, which is the year of commissioning of the hydropower plant.

The following Table 2 gives an overview of the unit prices and economic parameters applied:

Table .2. Unit Prices and Economic Parameters

Item	Chemoga	Yeda
Construction Costs	1.14 MUS$/m	1.03 MUS$/m
Resettlement Costs	0.19 MUS$/km²	1.03 MUS$/km²
Costs of Energy	0.048 US$/kWh	0.048 US$/kWh
Surcharges	55 %	55 %
Interest Rate	12 % p.a.	12 % p.a.
Construction Time	3 years	3 years
Life Time of Project	50 years	50 years

The average construction costs per meter of dam height were calculated by applying unit costs for excavation (overburden and rock) and placing of dam fill material (rockfill and clay) and multiplying those with the respective quantities. Also included are the costs for the river diversion which increases with increasing dam height. Surcharges of 55 % for contingencies, engineering and administration were added to the construction costs.

The results of the optimisation of the reservoir capacities are presented in Figures 4 and 5. They show the total present costs vs. the active storage of the reservoir. Both the costs as well as the energy benefits are expressed as costs relative to a reference project layout. This reference layout comprises an active storage of 275 MCM and 215 MCM for the Chemoga and Yeda reservoir, respectively, as determined during an earlier phase of the project.

At Chemoga for reservoir capacities below 220 MCM the rapid decrease of the firm yields (see Figure 3) leads to an increase of the total present costs. As there is no increase in energy benefits above this value the total present costs increase again due to the increasing construction costs. Thus, the most attractive reservoir capacity for Chemoga is 220 MCM, which is some 55 MCM less than estimated at an earlier stage.

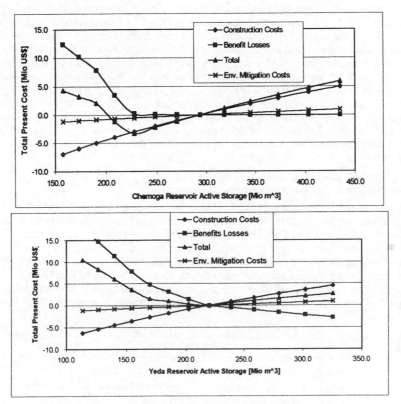

Fig.4 and 5: Results of Reservoir Capacity Optimisation
for Chemoga and Yeda Reservoirs

The reduction of the active storage capacity from 275 MCM to 220 MCM corresponds to a lowering of the FSL by 1.5 m. This leads to a reduction of some 2.7 km2 of submerged land of which 54 % are cultivated and 45 % are grazing areas. The population effected by resettlement is reduced by 420 people. While the lowering of the FSL reduces the environmental impacts, the firm yield of the reservoir is not compromised.

As can bee seen from Figure 5 there is a minimum total present cost for a reservoir capacity at Yeda of 215 MCM. Above this value the total present cost increases due to the increasing construction and environmental mitigation costs. Below 215 MCM the steep gradient of the energy benefits lead to an increase of total present cost. Thus, the most attractive reservoir storage capacity is 215 MCM. The optimisation resulted in the following main project data:

Table.3: Results of Reservoir Optimisation

	Chemoga	Yeda
Active Storage	220 MCM	215 MCM
Reservoir Area at FSL	28.8 km²	29.8 km²
Flow Yield	4.99 m³/s	4.78 m³/s

3.3 *Optimisation of the spillweir crest lengths of Chemoga and Yeda Reservoirs*

The objective of the optimisation is to establish that spillweir crest length which results in the least overall cost of construction of the dam, the spillweir and the diversion conduit. For a given FSL the height of the dam is a function of the spillweir crest length: an increased crest length will result in a lower surcharge above the FSL thus reducing the final dam height. As this will also lead to a narrower dam foundation the length of the diversion conduit will decrease too. Thus, the reduced spillweir construction costs have to be offset against the additional costs of heightening the dam and the additional costs for extending the diversion conduit.

Due to the prevailing topographic and geologic condition an ungated side channel spillweir had been selected for the Chemoga reservoir and an ungated bellmouth type spillweir for the Yeda reservoir. For a range of spillweir crest lengths, outline designs and cost estimates were made for the dam, the spillweir, and the diversion conduit.

To calculate the reservoir surcharges and outflows the relevant design discharges were routed through the reservoirs. The method of flood routing solves the simple water continuity equation within the reservoir as the flood enters and passes through the reservoir. Inputs for the calculations are as follows:

Hydrographs of design discharges (HQ10000), and
the elevation–area–capacity relationships of the reservoirs

In addition to the flood surcharge the final dam height has to provide sufficient freeboard to prevent overtopping by wave run-up. The results of routing the design floods through the two reservoirs are shown in Figure 6. These functions define the required dam heights (flood surcharge + wave run-up) for a range of spillweir crest lengths.

In order to establish the least cost combination of dam height, diversion conduit length and spillweir crest length, the construction cost of each structure has to be related to the spillweir crest length. The construction cost of each alternative is expressed as a cost relative to a reference layout. This reference layout comprised a spillweir with a crest length of 20 m. The

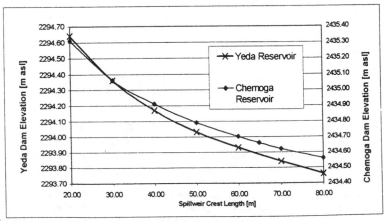

Fig.6: Flood Routing Results for Chemoga and Yeda Reservoirs

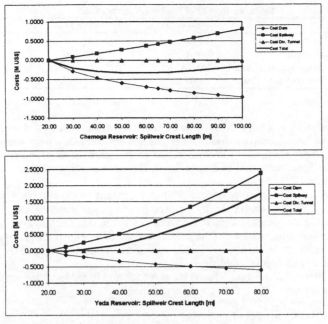

Fig.7: Results of Spillweir Optimisation for a) Chemoga
and b) Yeda Reservoir

changes in dam height and diversion conduit length associated with changes in the spillweir dimensions relative to the reference layout were costed by applying unit rates for the civil works.

The results of the optimisation are presented in Figure 7 for a) the Chemoga and b) the Yeda reservoir. They indicate that there is a least cost alternative with spillweir crest lengths of 65 m and 25 m, respectively. At Yeda the steep gradient of the spillweir construction costs governs the optimisation resulting in a shorter spillweir crest than at Chemoga. This is due to the fact that a bellmouth type spillweir has to be extended in both the lateral and longitudinal directions to increase the spillweir crest length. The optimisation results in the following dimensions:

3.4 Optimisation of the power waterways at Sens Reservoir

In the case of the Chemoga-Yeda-Sens hydropower scheme the power waterways consist of the power intake and the power conduit (headrace tunnel and surface penstock) of which only the latter has been optimized.

Again the objective of the optimisation is to establish the least total present cost which consists of the present worth of the construction costs and energy losses due to hydraulic friction. A larger diameter conduit has the advantage of reducing the friction losses along the waterway thus increasing the energy benefits from the scheme. These benefits have to be offset against the additional costs of excavating and lining the larger diameter conduit. As construction costs are incurred during the construction period whereas the energy losses accumulate over the life time of the project the costs and benefits have to be transformed to a reference year, which is in general the year of commissioning of the hydropower plant.

At rated discharge, the hydraulic losses through the entire power conduit from inlet to tailwater are as follows:

Penstock: hv = 24.289 m
Headrace Tunnel: hv = 3.988 m
Total   28.277 m

The losses total 28.3 m and represent some 3.6 % of the gross head.

Table. 4: Results of Spillweir Optimisation

	Chemoga	Yeda
Spillw. Crest Lenght	65 m	25 m
Dam Height	36 m	40.5 m
Diversion Conduit Length	200 m	220 m
Estimated Constr. Costs	21.65 MU$	13.69 MUS$

The power conduit consists of a concrete lined headrace tunnel and a steel surface penstock pipe. For a range of tunnel diameters, outline designs and cost estimates were made for both the headrace tunnel and the penstock. The construction costs include costs of excavation and concrete lining for the headrace tunnel and costs for steel pipes including adequate support structures for the penstock.

The friction losses in the conduits were quantified by means of the average value of energy and considered as benefit losses (i.e. costs) incurred by the power plant. These friction losses were calculated using the Manning equation. Surcharges to the construction costs as well as interest rate, construction time, life time of the project and the energy costs are given in Table 1.

The optimisation has to be carried out for a range of tunnel diameters until the diameter resulting in the least present cost is established.

The results are presented in Figure 9 for a) the penstock and b) the headrace tunnel. They indicate the least present cost (PC) for the conduits as follows:

Penstock : D = 3.0 m   PC = 99.2 MUS$
Headrace T. :   D = 4.5 m   PC = 23.3 MUS$

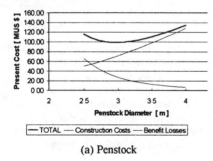

(a) Penstock

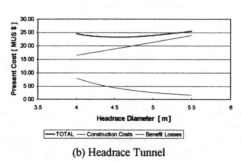

(b) Headrace Tunnel

Fig.8: Results of Power Waterways Optimisation

As can be seen from Figures 9 and 10 the accumulated present cost is quite sensitive to the chosen diameter, e.g. a reduction of the penstock diameter by 0.5 m leads to a total present cost of close to 120 MUS$ which is an increase of 20 % or 20 MUS$ compared to the optimised solution.

4 CONCLUSION

Three typical levels of optimisation were introduced – system context, project context, project components – with the first having the main objective to best meet the energy demand side. The aim of the latter two levels of optimisation, project context and project components, is to maximise the net benefits of a specific project.

Most optimisation procedure include an offset of benefits gained from a project against the project costs. Both, benefits and costs, have to be accumulated over the life time of the project and discounted to a reference year to allow an economic evaluation.

It has been shown that the net benefits of a project can be very sensitive to the project layout or the design of the project components. By applying adequate optimisation procedures the most attractive layout can be elaborated and the net benefits of a project can be increased significantly.

Stochastic Hydraulics 2000, Wang & Hu (eds) © 2000 Taylor & Francis, ISBN 90 5809 166 X

Hydrologic budget of the Shihmen reservoir watershed

Chin Yu Lee
Department of Soil and Water Conservation, National Pingtung University of Science and Technology, Taiwan, China

ABSTRACT: The main purpose of this study is to indicate the factors affecting hydrologic budget of precipitation, evaporation and discharge of Shihmen reservoir watershed in Taiwan. Four phenomena have been developed: 1. downpour after drought, 2. Precipitation in excess of 10mm over three days, 3. Drought period (Nov.-Apr.) and flood period(May-Oct.) , i.e. Flood/Drought index analyses, and 4. Applying the hydrologic statistics methods, should involve: (1) ARMA stochastic model (2) least square method and (3) Fourier series method. The results can accurately describe the hydrologic characteristics of a watershed in details, flood hazards prediction on the different return periods also present.

Key words: Hydrologic budget, ARMA, Flood/Drought index, Least square method, Fourier series

1. INTRODUCTION

In recent years, the increasing imbalance between water supply and water demand has give rise to great attention from both the relevant authorities and the general public on water resources planning programs, in which the long-term forecasting of the water cycle and its distribution is one of the very important topics (Xiong and Guo, 1999). The planning, design, operation, and maintenance of water resources developments require statistical information in the form of various hydrologic series. A thorough understanding of the structure of hydrologic time series is a prerequisite for any reliable input data in the planning and operation of water resource projects. Apart from the stochastic variation of hydrologic quantities with time, diverse sources of inconsistency and non-homogeneity superpose their changes to the stationary stochastic and non-stationary deterministic variations. Therefore, the study of the effects of non-homogeneity and inconsistency of data on the properties of a hydrologic time series is a very important subject for practical application. The essential practice in the water resource field is to use the statistical data of past observations, make an inference about the population of a hydrologic variable, and expect that the basic properties will hold true in future samples. However, if the past data show an inconsistency and non-homogeneity, the statistical inference about a unique population may not correspond to future samples. Future samples may not have non-homogeneity or they may experience another type (Yevjevich and Jeng, 1969).

This study was developed to estimate the hydrologic budget relationships for the Shihmen reservoir watershed. To provide the hydrologic budget information it was necessary to develop estimates of precipitation, evaporation and discharge. It should be recognized that the true hydrologic budget of Shihmen reservoir watershed is complex, and extremely variable (Wymore, 1974). This analysis represents a long-term average situation for some of the variables, and is based on a large extent on data from regional weather stations. Two examples, the Kaoyi and Yufon station, are discussed have for a better illustration of inconsistency and non-homogeneity in order to show both their importance and ways of treating them.

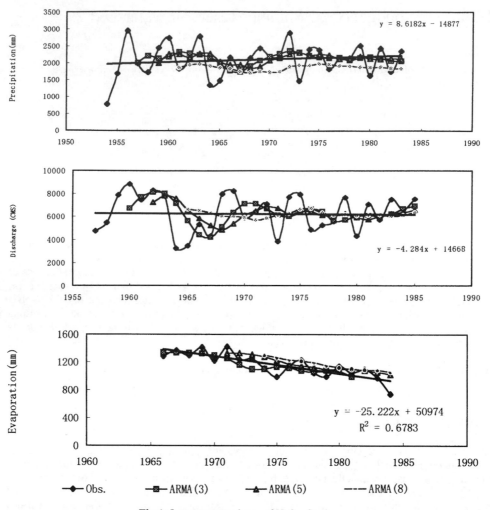

$y = 8.6182x - 14877$

$y = -4.284x + 14668$

$y = -25.222x + 50974$
$R^2 = 0.6783$

—◆—Obs.   —⊞—ARMA(3)   —▲—ARMA(5)   ---ARMA(8)

Fig.1: Long-term tredency of Yufon Station.

2. METHODOLOGY

1. For each simulation scenario, numerous traces of monthly precipitation, evaporation and discharge can be using an autoregressive moving average(ARMA) stochastic model (Box and Jenkins, 1976). The monthly hydrologic data sets ARMA model has parameters, which can be assumed to be, vary on yearly basis (Shen and Guillermo, 1992).

2. To find "good" estimators of the regression parameters β_o and β_1, we shall employ the method of least squares. For each sample observation (X_i, Y_i), the method of least squares considers the deviation of Y_i from its expected value:

$$Y_i - (\beta_o + \beta_1 X_i) \tag{1}$$

In particular, the method of least squares requires that we consider the sum of the n squared deviations. This criterion is denoted by Q:

$$Q = \sum_{i=1}^{n} (Y_i - \beta_o - \beta_1 X_i)^2 \tag{2}$$

804

According to the method of least squares, the estimators of β_0 and β_1 are those values b_0 and b_1, respectively, that minimize the criterion Q for the given sample observations (X_i, Y_i).

3.   During the past years much controversy has been raised among hydrologists about whether to use or not to use Fourier series analysis for estimating the periodic parameters of periodic stochastic models. Therefore, it is fair to put forth some arguments and criteria about why and in what cases the use of Fourier series is justified. Consider the periodic time series $X_{\upsilon,\tau}$, where υ denotes the year and τ denotes the time interval within the year. The parametric or Fourier series representation can be obtained by Salas *et. al.* (1980).

3. ANALYSIS OF RESEARCH WATERSHED DATA

The Shihmen reservoir watershed is a hydraulic engineering multiple purposes, the benefits of the project are irrigation, power-generation public water supply, flood control and tourism. It has been 36 years since the construction work was accomplished in 1964 and began with its operation. For all these years, Shihmen reservoir has contributed a great deal to the Northern part of Taiwan,

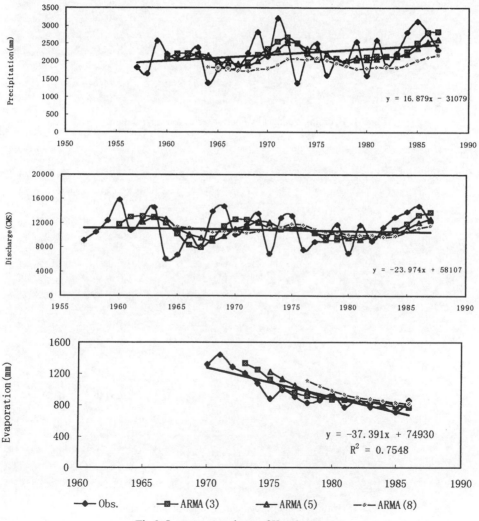

Fig.2: Long-term tredency of Kaoyi station.

805

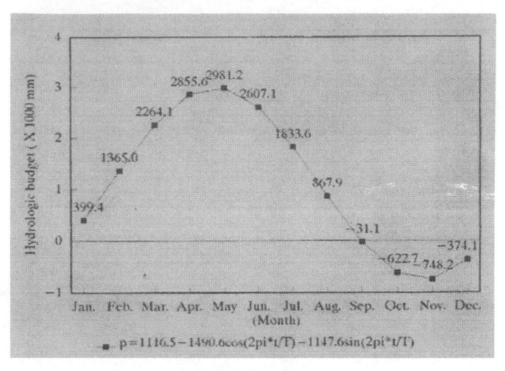

Fig.3-1: Hydrologic budget by Fourier series of Kaoyi station.

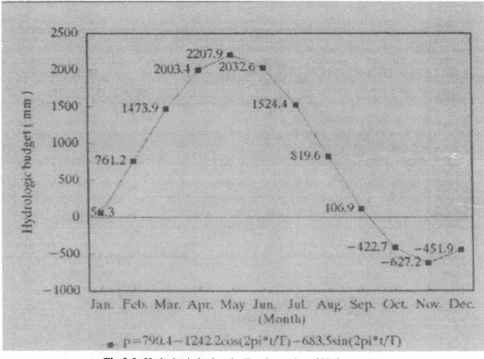

Fig.3-2: Hydrologic budget by Fourier serier of Yufon station.

806

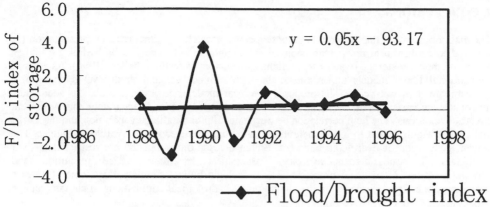

y = 0.05x − 93.17

Fig.4-1: Flood/Drought index of Yufon station.

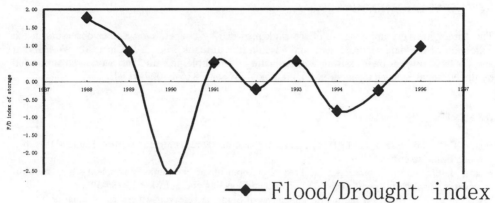

Fig.4-2: Flood/Drought index of kaoyi station.

including the agricultural improvement, industry development, enhancement of people's living standard, increase in the opportunity of employment, and prevention of flood and drought. The basic precipitation, evaporation and discharge data were provided to us by the Shihmen reservoir Commission of Taiwan, R.O.C. A total of 44 years(1954-1997) of observed these datum were used in this study (Lee, 1994).

The performance of the hydrologic budget in simulating these historic data sets was evaluated using several criteria:(1) downpour after drought, (2) precipitation in excess of 10mm over three days, (3) flood/drought index analyses. From the stochastic simulation analysis by use of the historical records., it reveals the contribution to the increase of water supply ability of the Shihmen reservoir watershed.Several ARMA model orders were fitted to the transformed data and the characteristics of the hydrologic budget were obtained from the observed data collected at Kaoyi and Yufon station located near the central part of the reservoir are partly shown in Fig. 1 through Fig. 2. And Fig. 3 shows the hydrologic budget was calculated by Fourier series. In the condition that Kaotai reservoir can join the operation of Shihmen system after year 2000, the results of stochastic simulation show that the only observed improvement is the reduction from 10% to 15% in vulnerability for both users. More details can be found in Wan and Guo (1992). The index of flood/drought in Fig. 4 shows that precipitation and discharge under the decreasing tendency, but the evaporation and storage will be in the state of increases by time. However, selection of the most appropriate parameterization in the future will require more extensive testing on longer data sets and across different hydroclimatologists (Evans and Jakeman, 1998).

807

4. CONCLUSIONS

The main objective of this paper is to propose a new approach of mathematical representation (as opposed to conventional linear regression) of the climatic effects, specifically hydrologic budget effects on monthly storage (water use) in Shihmen reservoir watershed (Miaou, 1990). As a result, conventional linear models underestimate the high summer demand in dry years while they overestimate it in wet years. The proposed nonlinear models, however, are able to adequately reflect storage variations and estimate the high summer demand (flood) accurately under different climatic conditions. The high correlation between annual precipitation, evaporation and discharge makes it straightforward to use climate model to predict changes in the hydrologic budget. The current generation of hydrologic budget of Shihmen reservoir watershed are restricted in their usefulness for subgrid-scale hydrologic assessment by their coarse resolution and parameterization of terrestrial processes. The hydrologic budget simulation by some stochastic analyses is one such approach that has found direct hydrological applications in climate change impact assessments (Wilby,1994).

ACKNOWLEDGMENTS

The author are very appreciative of the work provided by cooperative weather observers in the collection of the data reported here, and I would like to thank Ms. Y.-J. Chen, Mr. W.-M.Tsai and C.-T. Chow, for their assistance in evaluating the hydrologic data. This work was supported by the National Science Council(NSC), grant no. NSC-88-2815-C-020-011-E.

REFERENCES

Box, G.E.P., and Jenkins, G.M., 1976. Time series analysis: Forecasting and control, Holden-Day, Inc., San Francisco.

Evans, J. P., and Jakeman, A. J., 1998. Development of a simple, catchment-scale, rainfall-evapotranspiration-runoff model. Environmental Modeling and Software,13:385-393.

Lee, C.-Y., 1994. Study on the hydrologic budget of Shihmen reservoir watershed. Annual of Taiwan Museum, 27:1-22 (in Chenese).

Miaou, S.P., 1990. A class of time series urban water demand models with nonlinear climatic effects. Water Resour. Res.,26(2):169-178.

Salas, J.D. *et. al.*, 1980. Applied modeling of hydrologic time series. Water Resources Publications, Littleton, Colorado.

Shen, H. W., and Guillermo, Q., 1992. Risk assessment of reservoir sedimentation accumulation. Proceedings of the 6th IAHR Int. Symp. on Stochastic Hydraulics(ISSH), Taipei, TAIWAN,139-146.

Wan, S., and Guo, J. T., 1992. Reliability, resiliency and vulnerability analysis of a reservoir system by stochastic simulation. Proceedings of the 6th IAHR Int. Symp. on Stochastic Hydraulics(ISSH), Taipei, TAIWAN, 147-154.

Wily, R. L., 1994. Stochastic weather type simulation for regional climate change impact assessment. Water Resour. Res.,30(12):3395-3403.

Wymore, I. F., 1974. Estimated average annual water balance for Piceance and Yellow Creek watersheds. Technical Report Series No.2, Fort Collins, Colorado State University.

Xiong, L., and Guo, S., 1999. A two-parameter monthly water balance model and its application. J. Hydrol.,216:111-123.

Yevjevich, V., and Jeng, R. I., 1969. Properties of non-homogeneous hydrologic series. Hydrology paper No.32, Fort Collins, Colorado State University.

Stochastic Hydraulics 2000, Wang & Hu (eds) © 2000 Taylor & Francis, ISBN 90 5809 166 X

The application of probabilistic calculus method in the management of reservoirs

Ma Rongyong

Institute of Water Conservancy and Hydropower Science, Guangxi University, Nanjing, China

ABSTRACT: It takes several years for long-period storage reservoirs to be filled or emptied. The income flow of the reservoirs is random. Therefore the process of impounding and emptying, as a matter of fact, is stochastic, which has brought great difficulty in the management of the reservoir. This article studies the storage law of the long-period storage reservoirs to perfect the management.

According to the stochastic nature and periodicity of the hydrological phenomenon, the probabilistic calculus method is adopted to deduct the principle and method of the storage law for long-period storage reservoirs. The calculus equation of the process of the calculation and finally an example of Bo Se Reservoir are presented in the paper. By using this method the designed flood level for one thousand years is lower than the original one by 0.8 meters. Thus we can cut down the construction cost for a new project or improve the benefit of a built project on the basis of not lowering the flood control standard.

Key words: Management of reservoir, Probabilistic calculations method, Long term storage, Bo se reservoir

1. DEDUCTION OF THE PROBABILITY CALCULUS EQUATION

Basing on the periodic principle of hydrological phenomenon, and supposing that the random process of the inflow of the same month in different years has a similar but independent probability distribution, we set V_t and V_{t+1} as the impounded volume at the beginning and the end of t time-interval which is calculated monthly, with Q_t as the natural inflow discharge and Q_{ft} as the most optimizational power generation discharge of hydro- power station. From the water conservation law we get the equation of the impounding state

$$V_{t+1} = V_t + Q_t - Q_{ft} \tag{1}$$

A simple method is applied to show the relation between the values of Q_t and V_{t+1}. That is to calculate on the known V_t and Q_{ft} and make a matrix based on the calculation results, with V_t value as the column vector, and V_{t+1} value as the row vector. The elements of the matrix are made up of the frequency value corresponding to Q_t.

The state transition equation (1) can be written into $V_{t+1} = V_t + Y_t$, in which $Y_t = Q_t - Q_{ft}$ represents the net water impounding into the reservoir (positive value) or the net water discharging from the reservoir (negative value). Because V_t depends on Q_{t-1} while Y_t only has the relation with Q_t and Q_{ft}, V_t doesn't rely on Y_t. Suppose s as the initial impounding in the t time-interval and r as the final impounding, then the frequency relation should be:

$$
\begin{aligned}
P\{V_{t+1} = r\} &= P\{V_t + Y_t = r\} \\
&= \sum_s P\{V_t = s\} \bullet P\{Y_t = r - s\}
\end{aligned}
\tag{2}
$$

$$= \sum_{s} a_{rs} \bullet P\{V_t = s\}$$

Here, $a_{rs} = P\{Y_t = r - s\} = P\{Q_t = r - s + Q_{ft}\}$ can be derived from the Qt frequency curve. That is to say, every Q_t value marked in the matrix can be substituted by the frequency value a_{rs} corresponding to Q_t. On account that the initial and final impounding state at every time-interval are firmly graded and Q_{ft} is only connected to V_t, so in regard to every time-interval a_{rs} only has the relation with frequency curve of natural discharge. Therefore the initial impounding state frequency of every grade at the end of the time-interval can be calculated by using the following probabilistic calculus matrix equation:

$$\begin{bmatrix} P_{r_1} \\ P_{r_2} \\ \vdots \\ P_{r_k} \end{bmatrix} = \begin{bmatrix} P_{r_1 s_1} & P_{r_1 s_2} & \cdots & P_{r_1 s_k} \\ P_{r_2 s_1} & P_{r_2 s_2} & \cdots & P_{r_2 s_k} \\ \vdots & \vdots & \cdots & \vdots \\ P_{r_k s_1} & P_{r_k s_2} & \cdots & P_{r_k s_k} \end{bmatrix} \begin{bmatrix} P_{s_1} \\ P_{s_2} \\ \vdots \\ P_{s_k} \end{bmatrix}$$

(3)

Here, P_{s_j} is the impounding probability when the initial impounding state is s_i; P_{r_j} is the impounding probability when the final impounding state is r_j; $P_{r_j s_i}$ is the state transition probability when s_i(initial impounding state) transfers into r_j(final impounding state) (i,j=1,2,3,....k).

2. PROBABILITY CALCULUS PROCESS

The derivation of the matrix equation of the probabilistic calculus above has applied the additive and multiplicative principle. The impounding probability of the reservoir at the end of time-interval has relations with that of probability at the beginning of the same time-interval as well as the probability of natural inflow. The initial impounding and the natural inflow in this time-interval are two simple and independent events. Their probabilities happening at the same time are an multiplication of probabilities respectively. In addition, as for the final impounding state of the same reservoir, every initial impounding state is a non- compatible thing, or at least one of the state probabilities is the sum of every initial impounding state probability. Therefore the probability in a certain final impounding state is equal to the sum of partial probability of each initial impounding state corresponding to this final impounding state at the same time-interval.

We can divide the initial and final impounding state in the every time-interval into k grades, and then regulate them from the empty state of the reservoir. Suppose t=1,and the initial impounding state is $V_{t=1} = S_k$, in this case the corresponding frequency vector is:

$$P_0 = \begin{bmatrix} P_{s_1} \\ P_{s_2} \\ \vdots \\ P_{s_k} \end{bmatrix} = \begin{bmatrix} P\{V_{t=1} = s_1\} = 0 \\ P\{V_{t=1} = s_2\} = 0 \\ \vdots \\ P\{V_{t=1} = s_k\} = 1 \end{bmatrix}$$

(4)

Putting P_o into the probability calculus equation (3), we can obtain the column vector P_1 of the final impounding frequency in the first time-interval. Then putting P_1 which is taken as column vector of the initial impounding frequency in the second time-interval into equation, we can get the column vector of the final impounding frequency P_2. In like manner, we can get the column vector $P_1, P_2, ... P_{12}$ of each month final impounding of the year. In this way, 12 impounding frequency curves at each end of the month can be obtained. For the reason that the initial state of the regulation and its frequency column vector are assumed, it is required to alternate and deduct till each value of the frequency column vector is stable and then we can get the stable impounding frequency curve.

3. MONTHLY PROBABILITY CURVE OF THE WATER LEVEL IN FLOOD SEASON

The first step is to substitute the stable impounding probability curve of a month in flood period for six graded impounded levels $Z_i=Z_0,Z_1\cdots\cdots Z_5$, in which the corresponding probability is Δ $P_i=\Delta P_0, \Delta P_1\cdots\cdots \Delta P_5$, and $\sum_{i=0}^{5}\Delta P_i = 1$. The second step is to take different impounded levels Z_i as the initial adjustment water level, and to calculate the flood regulation at different flood frequency, for the purpose of getting the flood levels corresponding to different frequency and drawing probability curves $Z_m\sim P$ of flood levels with initial adjustment water level Z_i as the parameter. If regarding Z_i and the flood as two independent things, then according to the association probability principle, we can obtain the probability curve $Z_m\sim P$ of the flood level in that month.

The probability curves of flood level in each month in flood period can be obtained according to the ways above. Finally on the basis of the designed frequency value of the reservoir flood control standard, we can check the probability curves of each month's flood level to select the maximum as the designed flood level for the reservoir.

4. CALCULATION EXAMPLE

The calculation of the normal storage level is programmed to be 233 meters for Bo Se Reservoir.

4.1 Theoretical frequency curves of monthly discharge

We can input the measured monthly discharge in 47 years and dispersion mean coefficient value Φ, to calculate the statistic parameter X, and C_v of monthly measured discharge according to the formula of method of moment. Suppose $C_s= (1.5\sim4)C_v$, on the basis of the least squares techniques, we apply golden section method (0.618 method) to adjust the initial parameter C_s, C_v and $\bar{x}$ in given scope, and to conduct a interation gradually until the sum of squares of deviation between the theoretical frequency curve and measured point vertical coordinates reach to the minimum. In this way, we can obtain the theoretical frequency curve and the statistic parameters..

4.2 Relationship between Vt and Qft of the reservoir

On considering of the random influence on the inflow of the reservoir, the random dynamic programming method has been adopted to calculate the optimizing adjustment and to deduct the relational curve between the monthly initial state and the optimizing generating discharge of the hydro-power station . Based on the optimum dewatering decision, we can make further adjustment and calculation.

According to the optimizing principle of random dynamic programming, the recurrence formula of setting up optimizing adjustment model for the reservoir of hydro-power station is

$$FN_t^1(V(t))^* = \max_{Q_L(t)\in\Omega}\left\{\sum_{m=1}^{m}P_m(t)\big[FN_t(V(t),V(t+1),Q_L(t))+FN_{t+1}^1(V(t+1))^*\big]\right\}$$

(5)

Here, V(t) refers to the initial storage of the t time-interval; V(t+1) refers to the storage at the end of the t time-interval that is the initial storage of the t+1 time-interval; $Q_L(t)$ refers to dischargement of the adoption of L decision at t time-interval; $FN_t(V(t),V(t+1),Q_L(t))$ refers to the product value that the hydro-power station has produced from the initial state V(t) to the final state V(t+1), on condition that the inflow is $QT_m(t)$ and the decision dischargement is $Q_L(t)$ at t time-interval; $P_m(t)$ refers to the segment frequency at t time-interval corresponding to the inflow $QT_m(t)$ of the reservoir; M refers to the numbers of frequency segment of the inflow; $FN_{t+1}^1(V(t+1))^*$ refers to when the impoundage of the initial state is V(t+1) in t+1 time-interval,

811

the maximal expectation value of the general product for the hydro-power station from t+1 to the first time-interval or the final period of remains, which can be obtained by V(t+1) from the recursive curve in t+1 time-interval; $FN^1{}_t (V(t))^*$ refers to the maximal expectation value of the general product from t time-interval to the first time-interval when the impoundage of the initial state in t time-interval of the reservoir is V(t); Ω refers to the feasible zone of the decision dischargement $Q_L(t)$.

The constraint condition is

$$Q(N_{min}(t)) \le Q_L(t) \le Q(N_{max}(t)); N_{min}(t) \le FN_t(V(t),V(t+1),Q_L(t)) \le N_{max}(t) \Big\}$$
$$V_{dead} \le V(t) \le V_{normal}; \quad V_{dead} \le V(t+1) \le V_{normal} \qquad \Big\}$$

(6)

Here, $N_{min}(t)$, $N_{max}(t)$ refers to the minimum technology product and the potential product of the hydro-power station at t time-interval; V_{dead}, V_{normal} refers to the dead storage and the proper storage corresponding to normal impounded level.

The state transition equation is

$$V(t+1) = V(t) + QTm(t) - QL(t) \qquad (7)$$

4.3   To deduct the monthly stable frequency curves for final impounding of the reservoir

Basing on the monthly natural inflow frequency curve, the relation curve between monthly initial impounded state and the optimizing generating discharge of the hydro-power station, as well as the reservoir's state transition equation and probability routing matrix equation, we can work out the monthly stable frequency curves of the final impounding by using the probability routing method.

4.4   To work out the monthly flood level probability curves for the reservoir in flood period

Divide each month's stable impounded probability curves in flood period into 6 grade storage levels; Take every grade as the initial adjustment level; Then conduct a flood regulation calculus on this month's different flood frequency respectively. In this way, the probability curve of the flood level can be obtained with the initial adjustment level Z_i as the parameter. Finally, we can get each month's probability curve of flood level by combined probability calculation.

5. CONCLUSION

(1) The result of this method reflects the impounded rule of long-period storage reservoir on the condition of the optimizing adjustment for the hydro-power station. It shows the possibility of every different impound state at the end of every month in the form of probability.

(2) The previous flood adjustment calculus for flood control design of reaervoir began from the fixed flood restricted level. That is to say, it was considered right in the time that the impounded level was at the flood-restricted level before the coming of the flood period. The flood level frequency gained from the adjustment is equal to the flood frequency. In fact, the real impounded level of the reservoir for long-period storage before the fixed flood period is not surely equal to the flood restricted level. Therefrom, the floods control storage for the reservoir, which is based on the designed flood restricted level, is usually a bit larger.

(3) The result shows that the maximum possibility of the normal impounded level, which is 233 meters high for Bo Se Reservoir in flood period, is only 35.7%. By adjustment calculus and probability combined calculus, the designed flood level happening in 1000 years is lower than the original value by 0.8 meters, and the check flood level in 5000 years by 1.25 meters. On these grounds, under the condition of not lowering the flood control standard for those reservoirs being designed, the flood control storage can be properly reduced to reach the purpose of reducing the fabrication cost. For those constructed reservoirs, according to their real impounded conditions, we can adopt methods to increase the rate of full storage of the reservoir to gain the purpose of increasing the generating capacity of the hydro-power station and other integrated benefits.

(4) It is estimated that by using the method in this article to manage Bo Se Reservoir, the guaranteed output can be increased by 4,100kW and the annual generating capacity 58,000,000kW.h.

(5) The calculus method in this article actually combines the benefit calculation and flood control calculation. In this way, the management of the reservoir can be more reasonable. It also provides a feasible way for improving the economic benefits of the reservoir project. This method is right and feasible which can be referred to use by manufacturing departments.

REFERENCES

Bras N., The Scientific Allocation of Water Resource, Pejing Water Conservancy and Electric Power Publishing House

Liu Jiyin, Ma Rongyong, The Calculus Method on Reservoir Optimizing Adjustment and Flood Control Storage, The Journal of He Hai University, Dec.1990 (in chinese)

Ma Rongyong, The Application of Probabilistic Calculus Method on Research of Impounded Law of Reservoir for Long-period Storage, Red Water River 1990(4) (in chinese)

The Mathematics Department of Zhe Jiang University, Probability Theory and Mathematical Statistics, People's Education Publishing House, Sept.1979 (in chinese)

Stochastic Hydraulics 2000, Wang & Hu (eds) © 2000 Taylor & Francis, ISBN 90 5809 166 X

A new approach to estimation of inflow into a dam reservoir based on filtering theory

Hiroyuki Suzuki, Kazuyoshi Hasegawa & Mutsuhiro Fujita
Division of Civil and Environmental Engineering, Hokkaido University, Sapporo, Japan

ABSTRACT: A new approach to estimation of inflow into a dam reservoir based on filtering theory is proposed in the paper The smoothing filter consisted of a notch filter (NF) and low-pass filter (LPF) is designed considering the seiche period. Water inflow into a dam reservoir can be estimated by using the proposed smoothing filter, which eliminates influence of noise in time series data of water level.

Key Word: Inflow estimation into dam reservoir, Seiche, Notch filter, Low-Pass filter, Filter performance

1. INTRODUCTION

Data of water inflow into a dam reservoir is very important for efficient dam operation. In Japan, the inflow is estimated from change in the hydrostatic water level measured at constant time intervals. However, the signal of hydrostatic water level data is disturbed by many kinds of noise that originate from water surface fluctuations, such as wind wave, seiche and water surface set-up caused by wind. It is necessary to eliminate these noises included in time series data of water level, for exact estimation of inflow into a dam reservoir. Although estimation of water level is important for dam control, little attention has been given to the problem of inflow estimation. Almost all dam control offices in Japan employ the moving average method for smoothing the time series data of water level. However, the moving average method is not effective for real time estimation, because, this process requires long time calculations to eliminate long-term periodic noise such a seiche.

This paper will explain about a new method for estimating hydrostatic water level using a digital smoothing filter that was designed on the basis of theoretical values of seiche periods derived from our seiche model. The performance of the smoothing filter will be discussed from the viewpoint of error characteristics.

2. WATER LEVEL MEASUREMENT AND PREDICTION OF THE SEICHE PERIOD USING A SEICHE MODEL

Pressure gauges for measurement of water level were installed at 4 different sites in Kanayama Dam Reservoir (Fig. 1). The measuring time interval is 5 (sec.). Kanayama Dam Reservoir, which is located in Hokkaido, is a relatively large-scale dam reservoir for Japan. The water surface area of this reservoir is 9.2km^2, when the water level is FWL(=345m).

The results of water level measurement showed that the seiches occur in this reservoir with a long period such as 2500 (sec.) of 1st mode and 1400 (sec.) of 2nd mode (Hasegawa et al.,1998) The seiche period can be calculated using Eq.(1), derived from the solution of a 1-dimensional seiche model (Suzukiet al.1999)expressed by the type of wave equation:

$$T_i = \frac{4\pi L}{j_{1,i}\sqrt{gh_0}} \tag{1}$$

where, T_i is the i-th mode seiche period, L is the length of the reservoir, g is gravity acceleration, h_0 is water depth near the dam body, and $j_{1,i}$ is an eigenvalue defined by the i-th zero point of the Bessel function of order 1. Eq.(1) is obtained for a reservoir with a rectangular plane view and with a water triangular longitudinal profile of depth that becomes shallow toward the upstream. If the plane view of reservoir is triangular shape, the eigenvalue in Eq.(1) changes to $j_{2,i}$ defined by the i-th zero point of the Bessel function of order 2 (Hasegawa et al.,1998). Fig. 2 shows a comparison of the theoretical seiche period calculated by Eq.(1) and the seiche period obtained by spectral analysis of water level data from simultaneous water level measurement. For calculating the theoretical seiche period in Fig. 2, 10000 (m) and 17 (m) were used for the value of L and h_0 in Eq.(1), respectively. In Fig. 2, it is clear that the theoretical seiche period coincides well with the observed seiche period. Thus, if the seiche period is not known, it can be calculated using Eq.(1).

3. SMOOTHING FILTER FOR TIME SERIES DATA OF WATER LEVEL

3.1 Structure and Design of the Smoothing Filter

The smoothing filter is consists of a notch filter (NF) and a low-pass filter (LPF), whose digital transfer functions have a 2nd-order polynomial of z (variable of z-transform) in the numerator and denominator. NF is used for eliminating a seiche component, and LPF is used for eliminating other noise from the time series data of water level. For a system using these filters, an Infinite Impulse Response (IIR) system was adopted, because an IIR filter enables the output to be obtained by calculating only a few inputs, and this is an advantageous property for real time estimation of water inflow.

These filters were designed by the frequency conversion method following the flowchart shown in Fig. 3. A fundamental LPF that has a Butterworth property as a gain characteristic in the pass band was used for the design of each filter, because, this property does not have a ripple in the pass band of gain characteristic, and this is an advantage for eliminating noise. The transfer function of a fundamental LPF for designing IIR NF and IIR LPF is expressed by

$$H_{A-LPF1}(s) = \frac{1}{s+1} \tag{2}$$

$$H_{A-LPF2}(s) = \frac{1}{(s/\omega_c)^2 + \sqrt{2}(s/\omega_c)+1} \tag{3}$$

where H_{A-LPF1} and H_{A-LPF2} are fundamental LPF for designing IIR NF and IIR LPF, respectively; s is a variable of Laplace transform; and ω_c is the cut-off angular frequency of obtaining −3dB gain. Eq.(4) is an analog transfer function of NF obtained by frequency conversion of Eq.(2) using Eq.(a) in Fig. 3.

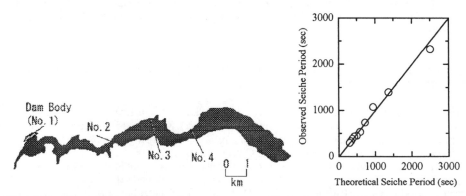

Fig.1: Plane view of Kanayama Dam Reservoir and each gauging station.

Fig.2: Comparison of theoretical seiche period and observed seiche period

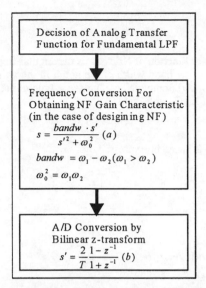

Fig.3: Flowchart of designing IIR NF and IIR LPF

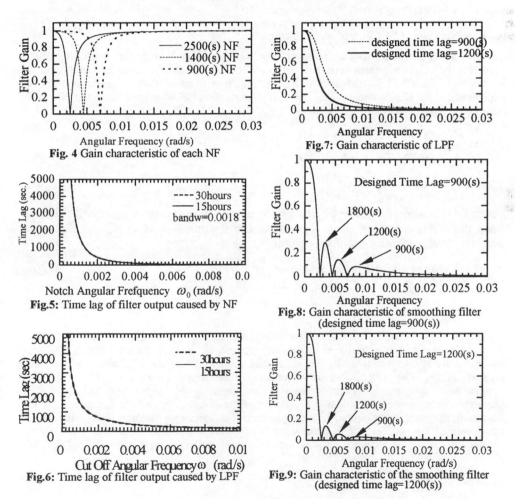

Fig. 4 Gain characteristic of each NF

Fig.7: Gain characteristic of LPF

Fig.5: Time lag of filter output caused by NF

Fig.8: Gain characteristic of smoothing filter (designed time lag=900(s))

Fig.6: Time lag of filter output caused by LPF

Fig.9: Gain characteristic of the smoothing filter (designed time lag=1200(s))

$$H_{A-NF}(s) = \frac{s'^2 + \omega_0^2}{s'^2 + bandw \cdot s' + \omega_0^2} \tag{4}$$

where, H_{A-NF} is an analog transfer function of NF, ω_0=is the angular frequency of the component eliminated by NF, and $bandw$ is the band width of the notch in the gain characteristic of NF. A bilinear z-transform is used for A/D conversion. The same type digital transfer function H_D is obtained by A/D conversion of Eq.(3) and Eq.(4).

$$H_D(z) = \frac{a_2 z^{-2} + a_1 z^{-1} + a_0}{b_2 z^{-2} + b_1 z^{-1} + b_0} \tag{5}$$

Then, filter factor a_i and b_i (i=1,2,3) can be expressed by
=LPF=

$$\begin{cases} a_2 = \omega_c^2 T^2 & b_2 = 4 - 2\sqrt{2}\omega_c T + \omega_c^2 T^2 \\ a_1 = 2\omega_c^2 T^2 & b_1 = -8 + 2\omega_c^2 T^2 \\ a_0 = \omega_c^2 T^2 & b_0 = 4 + 2\sqrt{2}\omega_c T + \omega_c^2 T^2 \end{cases} \tag{6}$$

=NF=

$$\begin{cases} a_2 = 4 + \omega_0^2 T^2 & b_2 = 4 - 2bandwT + \omega_0^2 T^2 \\ a_1 = -8 + 2\omega_0^2 T^2 & b_1 = -8 + 2\omega_0^2 T^2 \\ a_0 = 4 + \omega_0^2 T^2 & b_0 = 4 + 2bandwT + \omega_0^2 T^2 \end{cases} \tag{7}$$

where T is the sampling period.

3.2 Consideration of the Characteristics of Output

The time lag of the filter output is a problem for inflow estimation. Furthermore, each filter needs to remain the trend component in the time series data of water level. The necessary conditions for a smoothing filter will be discussed in this section. It is assumed that a flood wave can be expressed by the Cosine wave. When $x_1 = Cos(\theta t)$ or $x_2 = \alpha t + \beta$ is inputted into the NF, each output is given by the following equations.

$$Y_1(s) = \frac{s^2 + \omega_0^2}{s^2 + bandw \cdot s + \omega_0^2} \left(\frac{s}{s^2 + \theta^2} \right) \tag{8}$$

$$Y_2(s) = \frac{s^2 + \omega_0^2}{s^2 + bandw \cdot s + \omega_0^2} \left(\frac{\alpha}{s^2} + \frac{\beta}{s} \right) \tag{9}$$

where Y_1 is the output for x_1, Y_2 is the output for x_2, and θ is the angular frequency of the Cosine wave that indicates a flood wave. The inverse Laplace transform of Eq.(8) and Eq.(9) are expressed by the following equations, respectively.

$$y_1(t) = F_1 Cos(\theta t - \phi_1) + F_2 e^{-\alpha t} Cos(K_1 t - \phi_2) \tag{10}$$

$$y_2(t) = \alpha t + \beta - F_3 \left\{ 1 - e^{-\alpha t} Cos(K_2 t) \right\} + F_4 e^{-\alpha t} Sin(K_2 t) - F_5 e^{-\alpha t} Sin(K_2 t)$$
$$\alpha = \frac{bandw}{2} \tag{11}$$

Here, y_1 and y_2 are the inverse Laplace transform of Y_1 and Y_2, respectively, and F_1, F_2, F_3, F_4, F_5, K_1 and K_2 are constants determined by $bandw$, ω_0, and θ. Eq.(11) indicates that the NF can retain the trend component in the output. It was confirmed that the LPF also has the same property as that of NF with respect to the remainder of the trend component. Eq.(10) and Eq.(11) indicate that damping oscillation appears at the beginning of output from the filter. The parameter α represents the damping coefficient of the damping oscillation that appears at the beginning of output. In the case of LPF, the parameter that represents the damping coefficient can be calculated by $(\omega_c^2/2)^{1/2}$. The parameters $bandw$ and ω_c, which determine the damping coefficient, must be determined considering the control method of each dam. In Eq.(10) , the phase shift of output ϕ_1 is expressed by

$$\phi_1 = Tan^{-1}\left(\frac{bandw \cdot \theta}{\omega_0^2 - \theta^2} \right) \tag{12}$$

In the case of LPF, it is confirmed that the phase shift of output ϕ_L is expressed by

$$\phi_L = Tan^{-1}\left(\frac{\sqrt{2}\omega_c\theta}{\omega_c^2 - \theta^2}\right) \tag{13}$$

The time lag caused by NF and LPF can be calculated using Eq.(12) and Eq.(13), respectively, if the duration of the flood can be expressed using the period of the Cosine wave.

4. APPLICATION OF THE FILTERS AND INFLOW ESTIMATION

4.1 Application of the Smoothing Filter to Water Level Data

The best way to eliminate a seiche component is to remove the signal of the seiche by using only NF. This way make the minimal loss of information included in the time series data of water level. However, the seiche period has many values that is close to the value of the theoretical seiche period, because the complex topography of a real reservoir has various reflection points for surface waves. Therefore, it is difficult to eliminate a seiche component by using only NF. A smoothing filter consisting of NF and LPF is therefore used for eliminating the seiche component from the time series data of water level.

An amplitude of seiche decreases with a decreases in mode number of seiche. Therefore, a seiche with 1st, 2nd and 3rd mode is eliminated using NF. The value of ω_0 in Eq.(7) is given by the theoretical seiche period calculated by Eq.(1). In the case of Kanayama Dam Reservoir, the theoretical seiche periods are about 2500 (sec.), 1400 (sec.) and 900 (sec.), and the sampling period T is 5 (sec.). The value of 0.0018 is used for *bandw* in Eq.(7). The amplitude of damped oscillation that appears at the beginning of output decreases to 10% in the time same as 1st mode seiche period by using this value for *bandw*. Fig. 4 shows the 3 kind of gain characteristic corresponding to each NF. Fig. 5 and Fig. 6 show the time lags caused by the NF and LPF, respectively. The time lags are calculated from Eq.(12) and Eq.(13), in which θ is given as the angular frequency corresponding to a 30-hour or 15-hour period. There is not a noticeable difference in time lag depending on the value of θ, because, θ is a very small value compared with the value of the other parameters in Eq.(12) and Eq.(13). It is confirmed from Fig. 5 that the total time lag caused by 3 kinds of NF is about 400 (sec.). When total time lags of 1200 (sec.) and 900 (sec.) can be allowed in smoothing filter for dam control, the time lag that can be allowed in the LPF is 800 (sec.) and 500 (sec.), respectively. From Fig. 6, $\omega_c=1.77\times10^{-3}$(rad/s) and $\omega_c=2.83\times10^{-3}$(rad/s) can be obtained as the cut-off angular frequencies corresponding to time lags of 800 (sec.) and 500 (sec.), respectively. Gain characteristics in the case of $\omega_c=1.77\times10^{-3}$ and $\omega_c=2.83\times10^{-3}$(rad/s) are shown in Fig. 7. Fig. 8 and Fig. 9 show gain characteristics of the smoothing filter expressed as the product of NF gain characteristics and LPF gain characteristic. From Fig. 8 and Fig. 9, it is clear that the smoothing filter has gain characteristic that can eliminate a periodic component whose period is shorter than the longest seiche period.

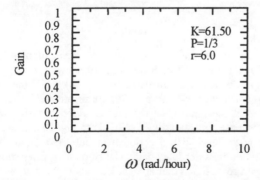

Fig.10: Gain characteristic of the flood wave for rainfall input

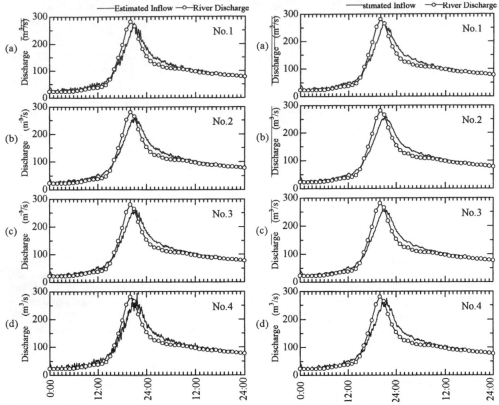

Fig.11: Inflow calculated using time series data of water level processed by the smoothing filter. (1998.9.16 0:00-9.17 24:00: designed time lag=900(s))

Fig.12: Inflow calculated using time series data of water level processed by the smoothing filter. (1998.9.16 0:00-9.17 24:00:designed time lag=1200(s))

When the seiche period is very long, as is the case in Kanayama Dam Reservoir, it is possible that a smoothing filter might eliminate a signal of true water level fluctuation. However, there is no method to know the real fluctuation of hydrostatic water level. Therefore, assumptions about the true fluctuation of hydrostatic water level have to be given. The gain characteristic of flood wave for rainfall input (Tanaka et al.,1999), which is calculated from the storage function, is given by Eq.(14):

$$G = \frac{1}{\sqrt{1+(K \cdot P \cdot r^{P-1} \cdot \omega)^2}} \qquad (14)$$

where, G is the Gain, K is the storage coefficient, P is the storage exponent, ω is the angular frequency (rad/hour), and r is the rainfall intensity. Fig. 10 shows the gain characteristic of a flood wave for rainfall input, which was calculated by Eq.(14). Then the values of K and P, which were used for runoff calculation by Tateya (1999) are substituted into Eq.(14). The rainfall data used for r is recorded at the time when the biggest flood occurred in Kanayama Dam Reservoir in 1998. If the response corresponding to $G=0.05$ in Fig. 10 appears in the flood wave, fluctuation of hydrostatic water level with a period over 5600 (sec.) ($\omega=4$ rad/hour) can be regarded as the true fluctuation in water level data. The smoothing filter should be designed to retain over 80% of filter gain for a signal that is regarded as true fluctuation in water level. The gain of the smoothing filter for fluctuation with a period of 5600 (sec.) is 0.852 (1200sec. designed time lag) and 0.906 (900sec. designed time lag).

4.2 Inflow Estimation

Generally, inflow is estimated by the following equation:

$$Q_I = \frac{\partial V(H)}{\partial t} + Q_O \tag{15}$$

where Q_I is the estimated inflow, H is the smoothed water level, $V(H)$ is a function that expresses the H-V(storage) relation, and Q_O is outflow. If the change in water level in short time is linear, H is expressed by the following equation.

$$H = at + b \quad , \quad \frac{\partial H}{\partial t} = a \tag{16}$$

Eq.(17) is obtained by substituting Eq.(16) into Eq.(15).

$$Q_I = \frac{\partial V(H)}{\partial H} \frac{\partial H}{\partial t} + Q_O = \frac{\partial V(H)}{\partial H} \cdot a + Q_O \tag{17}$$

Inflow can be estimated by Eq.(17) using water level data processed by the smoothing filter. Fig. 11 shows the result of estimation of inflow in the case of designed time lag of 900 (sec.), and Fig. 12 shows the result of estimation in the case of designed time lag of 1200 (sec.). The inflow shown in Fig. 11 and Fig. 12 were estimated from the time series data of water level measured at each gauging station and processed by the smoothing filter. In Fig. 11 and Fig. 12, the dotted lines indicate river discharge measured at the inflow point of the biggest river connected to the reservoir. The flood peak of the estimated inflow is smaller than the river discharge. An error in observed discharge may have caused this result. The amplitude of oscillation in the estimated inflow, which was calculated from the observed data at No.2 station and No.3 station, is smaller than that of the estimated inflow calculated from the No.1 and No.4 water level data, because the locations of No.2 and No.3 gauging stations are close to a node position of 1st mode and 2nd mode seiches. These node positions were confirmed in our previous study[2]. Therefore, the water level data measured at No.2 and No.3 stations will be useful for dam control. In Fig. 11 (d) and in Fig. 12 (d), amplitude of oscillation in the estimated inflow is larger than that of other estimated inflows, because, the locations of No.1 and No.4 gauging stations are close to the loop position of the seiche. The periods of these oscillations coincides with the pass band in Fig. 8 and Fig. 9, such as 1800 (sec.), 1200 (sec.) and 700 (sec.). This means that seiches with various periods occur in a real reservoir. This is one of the reasons why inflow estimation is difficult.

The results of estimation shown in Fig. 12 are better than those shown in Fig. 11 in the sense that the amplitude of oscillation in the results of estimation is small. Kanayama Dam Reservoir is a large-scale dam reservoir in Japan, and the seiche period in Kanayama Dam Reservoir is very long. From Fig. 5 and Fig. 6, it is clear that this is a disadvantageous condition for designing a smoothing filter from the viewpoint of time lag. However, Kanayama Dam Reservoir has a large flood control capacity sometimes enabling floods to be controlled without operating the dam gate. Therefore, in Kanayama Dam Reservoir, time lag of 1200 (sec.) is not problem for dam control. The seiche periods in other reservoirs in Japan are shorter than that in Kanayama Dam Reservoir because of the smaller size of other reservoirs. Therefore, a smoothing filter can be designed easily from the viewpoint of time lag. Therefore, a smoothing filter can be used in almost all dam reservoir in Japan.

5. FILTER PERFORMANCE

It was confirmed that a smoothing filter can estimate the hydrostatic water level in which the influence of noise is restrained. However, the true hydrostatic water level or true water inflow cannot be measured. Therefore, the accuracy of the estimated hydrostatic water level that obtained using the smoothing filter cannot be evaluated by comparison between the output and any data.

Error characteristic of the filter will be discussed in this section to solve this problem.

5.1 Statistical Characteristic of Filter Input

The statistical characteristic of filter input is determined by referring to the water level data from simultaneous water level measurement. Fig. 13 show a time series of water surface fluctuations measured at each gauging station. The trend component of the data has been eliminated in Fig. 13. Fig. 14 shows the auto-correlation coefficient for the water surface fluctuations shown in Fig.

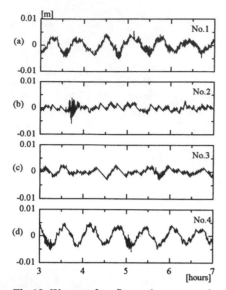

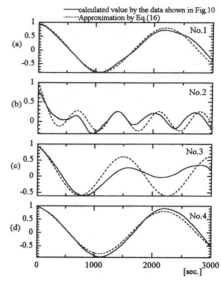
calculated value by the data shown in Fig.10
Approximation by Eq.(16)

Fig.13: Water surface fluctuation measured at each gauging station(1998.10.13 3:00-6:00)

Fig.14: Auto-correlation of water surface fluctuation and results approximated by Eq.(18)

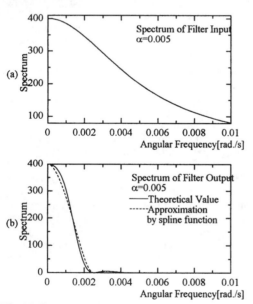

Fig.15: Spectrums of output and input (α=0.005)

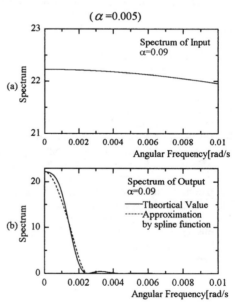

(α=0.005)

Fig.16: Spectrums of output and input (α=0.09)

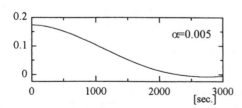

Fig.17: Auto-correlation of filter output

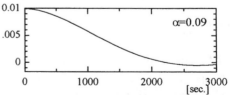

Fig.18: Auto-correlation of filter output (α=0.09)

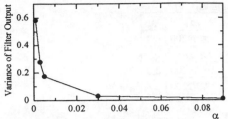

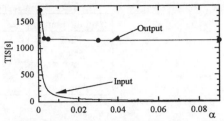

Fig.19: Relation between α and variance of output

Fig.20: Relation between α and time integral scale

13. Oscillation with periods of 2200 (sec.), 1500 (sec.) and 700(sec) appear in Fig. 14. This means that seiches with these periods occur in this time. From the figure, the auto-correlation coefficient could be approximated by the following equation:

$$R(\tau) = ACos(\omega_k \tau) + \sigma e^{-\alpha \tau} \tag{18}$$

where R is the auto-correlation coefficient, τ is the lag time, and α, A, σ., ω_k are constant parameters. In Eq.(18), the 1st and 2nd terms represent the auto-correlation coefficients derived from fluctuation of seiche and inflow, respectively. The broken line in Fig. 14 shows the results of approximation using Eq.(18). Table 1 shows the values of each parameter in Eq.(18) for this approximation. In another case of water surface fluctuation, it was confirmed that the auto-correlation can be approximated using $\alpha = 0.09$.

5.2 Statistical Characteristic of Filter Output

Auto-correlation of filter output can be calculated by a mathematical method of Fourier analysis

if the auto-correlation of input is known. The smoothing filter can eliminate the seiche component. Therefore, the 1st term of Eq.(18) can be ignored. The auto-correlation of filter input R_{in} and the spectrum of filter input S_{in} can then be expressed by Eq.(19) and Eq.(20), respectively.

$$R_{in}(\tau) = e^{-\alpha \tau} \tag{19}$$

$$S_{in}(\omega) = \frac{2\alpha}{\alpha^2 + \omega^2} \tag{20}$$

Then Eq.(19) is normalized by σ. Gain characteristics of LPF and NF are obtained from Eq.(4) and Eq.(5), respectively:

$$G_L(\omega) = \frac{\omega_c^4}{\left(\omega_c^2 - \omega^2\right)^2 + 2\omega_c^2\omega^2} \tag{21}$$

$$G_{Ni(i=1,2,3)} = \frac{\left(\omega_{0i}^2 - \omega^2\right)^2}{\left(\omega_{0i}^2 - \omega^2\right)^2 + bandw^2\omega^2} \tag{22}$$

where G_L is the gain characteristic of LPF, and $G_{Ni(i=1,2,3)}$ is the gain characteristic of NF (i corresponding to 3 kinds of NF). The gain characteristic of the smoothing filter is expressed by the product of all gain characteristics. The spectrum of smoothing filter output can be calculated from Eq.(23):

$$S_{OUT}(\omega) = G_L(\omega)G_{N1}(\omega)G_{N2}(\omega)G_{N3}(\omega)S_{in}(\omega) \tag{23}$$

where, S_{OUT} is the spectrum of output from the smoothing filter. The auto-correlation coefficient of filter output can be obtained by the inverse Fourier transform of Eq.(23), and it is clear that Eq.(23) is even function from the order of ω in Eq.(21) and Eq.(20). Therefore, the inverse Laplace transform of Eq.(23) is given by Eq.(24):

$$R_{OUT}(\tau) = \frac{1}{\pi}\int_0^\infty S_{OUT}(\omega)Cos(\omega \tau)d\omega \tag{24}$$

823

Table 1: Each parameter in Eq.(18)

Gauging Station	α	σ	ω_k	A
No.1	0.005	0.2	2π /2200	0.8
No.2	0.005	0.75	2π /700	0.25
No.3	0.005	0.4	2π /1500	0.6
No.4	0.005	0.2	2π /2200	0.8

However, it is difficult to calculate an integral in Eq.(24) theoretically, because Eq.(23) is very high order function of ω. Therefore, Eq.(23) is approximated by using a 3-order spline function. Fig. 15 (a) and (b) show spectrums of input and output of the smoothing filter, calculated by Eq.(20) and Eq.(23), respectively, in the case of α=0.005. Fig. 16 (a) and (b) also show spectrums of input and output in the case of α=0.09. The spectrums of output and input vary depending on the value of α. Fig. 17 and Fig. 18 show the auto-correlation of output calculated from Eq.(24), whose integration is obtained approximately by using the spline function in the cases of α=0.005 and α=0.09, respectively. From Fig. 17 and Fig. 18, it is confirmed that the variance of output is smaller than that of input from 1 order to 2 orders. This is an advantageous function of the smoothing filter.

The variance of output is influenced by the value of α. The length of an interval with a high correlation is also influenced by the value of α. Fig. 19 shows the relation between α and variance of output from the smoothing filter. Although the variance of output does not exceed the variance of input, the variance of output increases rapidly with decreases in the value of α. It is possible that sufficient accuracy of output cannot be obtained, depending on the necessary accuracy for dam control if the variance of output becomes large.

The duration in which a strong auto-correlation is maintained can be evaluated by a time integral scale calculated by Eq.(25):

$$TIS = \frac{S}{R_{OUT}(0)} \tag{25}$$

$$S = \int_0^{t_z} R_{OUT}(\tau)d\tau$$

where, TIS is the time integral scale, and t_z is the time when $R_{OUT}(\tau)$ becomes 0 at first. Fig. 20 shows the time integral scales of output and input. Then time integral scale of input is calculated by the following equation:

$$TIS = \int_0^\infty e^{-\alpha\tau}d\tau = \frac{1}{\alpha} \tag{26}$$

From Fig. 20, it is clear that the time integral scale of output is larger than that of input. This is caused by the filter characteristic that gives a time lag to output. In fact, the difference between time integral scales of input and output is about 1200 (sec.). This difference coincides with the designed time lag of the smoothing filter.

6. SUMMARY

A new method for estimating water inflow into a dam reservoir based on the filtering theory was proposed in this paper. The smoothing filter consisted of a notch filter (NF) and low-pass filter (LPF) is designed considering the seiche period. Water inflow into a dam reservoir can be estimated by using the proposed smoothing filter, which eliminates influence of noise in time series data of water level.

The performance of filter was evaluated in this study. It was confirmed that the variance of output from the smoothing filter is smaller than the variance of input, whose auto-correlation remains strong for a long time as seems in the auto-correlation obtained from fluctuation of inflow. This is advantageous for the smoothing filter.

The smoothing filter has a characteristic that gives time lag to output. The smoothing filter is needed to incorporate the function of prediction filter for solving problem of time lag.

REFERENCES

Hasegawa, K., K. Ishida and H. Suzuki: Estimation of Hydrostatic Water Levels in A Dam Reservoir on Windy Days Using Multiple Water Gauges Data, Advanced in Hydro-Science and –Engineering, Vol.3, P.92, 1998. (full paper on CD-ROM).

Suzuki, H., K. Hasegawa, M. Nakatsugawa, and M. Iwasaki: Characteristics of Water Surface Oscillations in Kanayama Dam Reservoir, Proceedings of the 54th Annual Conference of The Japan Society of Civil Engineering 2, pp.506-507, 1999. (in Japanese)

Tanaka, G., M. Fujita, M. Kudo and K. Uchijima: Comparison Between The Kinematic Wave Model And The Storage Function Runoff Model, Journal of Hydraulic Coastal and Environmental Engineering, pp.21-36, 1999, 2. (in Japanese)

Tateya, K.: Thesis for Doctor's Degree, Hokkaido University, pp.195-220, (in Japanese)

Stochastic Hydraulics 2000, Wang & Hu (eds) © 2000 Taylor & Francis, ISBN 90 5809 166 X

Energy-reliability tradeoff analysis for multipurpose reservoir operation using LP-SDP

Wang Jinwen, Zhang Yongchuan & Yuan Xiaohui
Department of Hydroelectric Engineering, Huazhong University of Science and Technology, Wuhan, China

Zhang Youquan
School of Electric Power Engineering, Yunnan Polytechnic University, Kunming, China

ABSTRACT: Few existing reservoir operation models are oriented toward generating operational trade-offs. Regarding the Markovian inflow, this paper adopted stochastic dynamic programming (SDP) with punish-variables as adjustable weights, and varying them, a great lot of efficient long-term reservoir operation policies and the associated performance indices (PIs) could be derived for tradeoff analysis between hydropower generation and PIs of water supply reliability. Proposed model herein conceptually consists of two modules. Module 1 solves Linear Programming (LP) for water allocations optimally for different objectives within the current period for given state of the system and total release from reservoir, and module 2 uses Stochastic Dynamic Programming (SDP) for deriving the steady state reservoir operating policy. Finally, A case study on Yudong Reservoir in Yunnan province is given.

Key word: Multipurpose reservoir, Trade-off analysis, Stochastic, Dynamic programming, Linear programming.

1. INTRODUCTION

Evaluation of the trade-off between economic cost and reliability of achieving the desired goals remains an active research field(*Takyi and Lence*, 1999). In many multi-objective cases, it seem more sensible to obtain the trade-offs between different performance indictors (PIs) for the decision-makers. Unfortunately, with few exceptions(*Zhenpeng and Shangyou*, 1989; *Georgakakos*, 1993), existing reservoir operation models are not oriented toward generating operational trade-offs. In this paper, the trade-offs between expected yearly energy and various reliability goals can be derived for decision-makers.

Mentioned as *Cohon and Marks* (1975) and *Piccardi and Soncini-Sessa* (1991), a multi-objective analysis can be conducted by using as a tool the optimal control problem and the related statistics. The similar idea to the weighting method(*Zaden*, 1963), the set of efficient policies can be derived by repeatedly solving the optimal control problem with different weights in the cost function. Regarding the Markovian inflow, this paper adopted stochastic dynamic programming (SDP) with punish-variables as adjustable weights, and varying them, lots of efficient long-term reservoir operation policies could be derived. Different from the early works, however, the reliability PIs are related more implicitly to the efficient operation policy in this study. The simplest way to determine those PIs is by statistics after the operation simulation with the historical inflow series as input(*Ruxiang*, 1987). However, the solutions by this method exist large deviation from the real ones because of the inflow sampling error, which can be decreased through increasing the input samples through inflow simulation model. Another typical way is by the probability formula (*Piccardi and Soncini-Sessa*, 1991), which needs to derive the steady state probability through the transition probability matrixes multiplying until steady. When there are many periods, state variables or high number of discretization intervals for each state variable, to derive the PIs by the probability formula is very difficult. In this paper, the second approach, operation simulations with vast man-made runoff series, is adopted.

The objective function of SDP is the generated energy amount minus several penalty terms, which penalize failures to meet the water supply targets. Early literatures typically use three forms of penalty functions, linear one(*Shiqian*, 1988), quadratic one(*Ponnamb-alam and Adams*, 1996), and exponential one(*Georg-akakos et al.*, 1997), respectively. The last two, in the price of more difficulty, have the capabilities to take into account the severity of larger supply deficits in the manner of imposing an increasingly higher penalty as water supply falls short of the target demands. In present case study, we select the linear penalty terms to produce our desired PIs, such as the expected yearly energy over long-term reservoir operation and the probabilities of meeting the demands from power generation, irrigation, industrial water supply and so on.

The present model conceptually consists of two modules. Module 1 solves LP for water allocations optimally for different objectives within the current period for given state of the system and total release from reservoir, resulting in the maximized immediate reward. And the requirements of reservoir water balance in each period form the constraints. Module 2 uses stochastic dynamic programming (SDP) for deriving the steady state operating policy for reservoir, resulting in the maximization of expected annual system performance. In the model, the results of module 1 and transition probabilities of reservoir inflow form the input to SDP. We note that such consideration is not new, especially in irrigation studies, where multiple crops comprise the competitive water-consumers. In those studies, they are mainly three forms of combinations from linear programming (LP), dynamic programming (DP) and stochastic dynamic programming (SDP), for examples, DP-SDP by *Vedula and Mujumdar*(1992), DP-DP by *Mujumdar and Remesh*(1997),and LP-SDP by *Vedula and Kumar*(1996). Here, we adopt LP-SDP, which effectively avoids the descretization of decision variables. Rather than the multi-crop cases, however, the competitive water-use in this study comes from multiple demands for power generation, irrigation, domestic and industrial water supply.

For the convenience of description, the schematic layout of a case reservoir is shown in Fig.1.

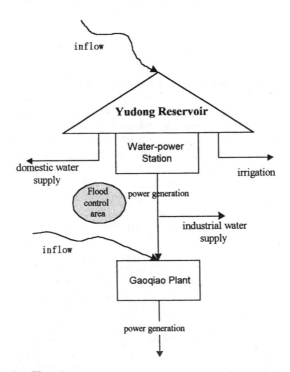

Fig. 1:   The schematic layout of Yudong Reservoir for multiple supplies.

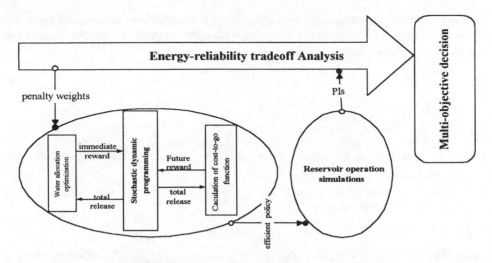

Fig.2: The framework of the energy-reliability analysis system.

2. THE MODEL STRUCTURE

Given a set of penalty weights, solving SDP can obtain the optimal reservoir operation policy, under which the associated PIs can be determined by simple statistics after vast simulations of reservoir operation with man-made runoff series as input. Thus, varying the penalty weight set, lots of such optimal policies can be derived. Those policies consist of the efficient policies. Here, the word *efficient* means that there does not exist another policy that improves one of the competitive objectives without worsening the other one(*Piccardi and Soncini-Sessa*, 1991). This means that, corresponding to the *efficient* policies, lots of the *efficient* sets of PIs are obtained. As each of the PIs represents one objective, the energy-reliability tradeoff analysis is practically a multi-objective analysis. The analysis result is delivered for decision-makers in the way of tradeoff graphs, through which the decision-maker can select the most satisfactory point. The model structure is shown in Fig. 2.

3. THE MODEL OF SDP

When the state variables in SDP are V_t (reservoir storage at the beginning of period t) and Q_t (average inflow during period t), which is assumed known, then the recursive equation can be expressed as:

$$M_t(V_t, Q_t) = \max_{U_t \in \Omega_t} \{B_t(V_t, Q_t, U_t) + E[M_{t+1}(V_{t+1}, Q_{t+1})]\} \tag{1}$$

where E is expectation operator, M_t is the maximized future return from operating the reservoir system from the beginning of period t, u_t is the control variable and Ω_t its feasible space. And B_t is the immediate reward.

There are two state transition equations, runoff model (2) and water balance equation (3):

$$Z_{t+1} = r_1 Z_t + \varepsilon_{t+1} \tag{2}$$

$$V_{t+1} = V_t + (Q_t - u_t)T_m \tag{3}$$

where, $Z_t = \frac{Q_t - \mu_t}{\sigma_t}$ is the normalized inflow of Q_t (here, μ_t and σ_t are respectively the mean monthly flow of period t and the associated standard deviation.), r_1 the first-order auto-regression coefficient (reflecting the correlation degree between inflows of contiguous periods), T_m the time length of each period, and ε_{t+1} an independent random variable. In addition, the constraints on reservoir storage and discharge must be satisfied, i.e.,

$$V_{\min} \le V_t \le V_{\max} \tag{4}$$

$$\max\{V_{\min}, V_t + (Q_t - u_{t,\max})T_m\} \le V_{t+1} \le \min(V_{\max}, V_t + Q_t T_m) \tag{5}$$

$$u_t T_m = V_t + Q_t T_m - V_{t+1} \tag{6}$$

where, $V_{\min}$ and $V_{\max}$ are respectively the minimum and maximum allowable storage, and $u_{t,\max}$ is the maximum attainable release.

In practice, a termination criterion of expression (1) must be added. The following convergence test has been adopted:

$$\max_{(V_t, Q_t) \in \Sigma_t} \left| [M_{n+1}(V_t, Q_t) - M_n(V_t, Q_t)] - [M_n(V_t, Q_t) - M_{n-1}(V_t, Q_t)] \right| \le \Delta \tag{7}$$

where $\Delta > 0$ is selected depended on the desired precision, $M_n(V_t, Q_t)$ is the maximized value of cost-to-go function from operating the reservoir system from the beginning of period t of year n when reservoir state is (V_t, Q_t), and Σ_t the feasible space of reservoir state.

Furthermore, the solving of SDP needs the state discretization. Since the probability distribution of average flow of each period is different according to different period, the discrete interval of runoff state Q_t should be different from each other. In this paper, in light of the principle of probability statistics, the flow discretization space of each period is treated differently. Concretely, the flow state Q_t is split equably between 0 and $Q_{t,P\%}$, and the flow of each period is represented by its midpoint value. Here, $Q_{t,P\%}$ denotes the value which the probability of the average flow of period t is more than or equal to is a very small percentile $P\%$. In addition, the storage state V_t is split equally between $V_{\min}$ and $V_{\max}$.

4. THE MODEL OF LINEAR PROGRAMMING

When the linear penalty terms, which penalize failures to meet the water supply targets, are adopted, the objective function of LP can be expressed as:

$$B_t(V_t, Q_t, u_t) = \max_{D_t \in \Lambda_t} \left\{ \xi d_1(r_{t,ir} - R_{t,ir}^d) + \xi d_2(r_{t,in} - R_{t,in}^d) + \beta d_3(E_t - E_{t,d}) + E_t \right\} \tag{8}$$

subject to the constraints:

$$r_{t,ir} + r_{t,ye} \le u_t - R_{t,dom} \tag{9}$$

$$r_{t,ir} + r_{t,in} + r_{t,ge} \le u_t - R_{t,dom} \tag{10}$$

$$r_{t,ir} \ge (1 - p_{ir}) \cdot R_{t,ir}^d \tag{11}$$

$$r_{t,in} \ge (1 - p_{in}) \cdot R_{t,in}^d \tag{12}$$

$$r_{t,ye} \le R_{t,ye}^{\max} \tag{13}$$

$$r_{t,ge} \le R_{t,ge}^{\max} - Q_{t,L} \tag{14}$$

in which, $r_{t,ir}$, $r_{t,in}$, $r_{t,ye}$ and $r_{t,ge}$ are decision variables of decision vector D_t, denoting the release supply from Yudong Reservoir for irrigation, industry, power generation for Yudong Plant and Gaoqiao Plant in period t, respectively. $R_{t,do}$ is water supply for domestic use, which is directly drawn from reservoir regarding it is very small but demanded with high guarantee percentile (100%). E_t is the total generation by Yudong and Gaoqiao. Subscription d denotes the demanded supply or energy. d_t is the penalty weight and δ 、 ξ and β are penalty coefficients, which equal to 0 or 1 depending upon whether the demanded supply is met or not. $p_{ir}(\%)$ and $p_{in}(\%)$ are respectively the maximum allowable failures to meet the demand of irrigation and industry. $R_{t,ye}^{max}$ and $R_{t,ge}^{max}$ are respectively the maximum flow capacity through turbines of Yudong Plant and Gaoqiao Plant. And $Q_{t,L}$ is the local runoff between Yudong and Gaoqiao.

It is noted that the generation discharge of Yudong Plant can be reused for downstream industrial supply and by Gaoqiao Plant for power generation. When such discharge is too much to be fully utilized by Yudong Plant, certain spillage is possibly favorable for power generation of downstream Gaoqiao Plant. The constraint (10) can take account this situation.

Notice that since there exist three discontinuous penalty coefficients, the objective function of LP is obviously discontinuous, too. In order to solve LP, its objective function needs to be continuous. The simple way to treat it is to enumerate all possible values of penalty coefficients, needing 8 times solving of LP in total and adding three corresponding constraints in each time. Such treatment is not sensible. Again, notice that when the water supply for irrigation or industry use is more than the demanded amount, the surplus will not bring any benefit. Thus, regarding that the irrigation or industry supply is less than or equal to the demanded amount, two constraints can be added. And the optimal solution will not be omit, but the model can be simplified greatly since the penalty coefficients δ and ξ is constantly set to be 1. As for generation discharge, adding the corresponding constraints depending on the two different value of β, the optimal solution can be obtained by comparison between two results from two runs of LP.

5. CALCULATION OF THE COST-TO-GO FUNCTION

Calculation of the cost-to-go function is in fact a expect operation, which can be expressed as:

$$E[M_{t+1}(V_{t+1},Q_{t+1})] = \sum_{(V_{t+1},Q_{t+1})} M_{t+1}(V_{t+1},Q_{t+1}) \cdot p\{(V_{t+1},Q_{t+1})|(V_t,Q_t),u_t\} \tag{15}$$

According to the stage-to-stage transformation equa-tion (2), when stage vector (V_t,Q_t) and total release u_t are known, from water balance equation (3), V_{t+1} the storage at beginning of period t is known, too. Thus,

$$E[M_{t+1}(V_{t+1},Q_{t+1})] = \sum_{Q_{t+1}} M_{t+1}(V_{t+1},Q_{t+1}) \cdot p\{Q_{t+1}|Q_t\} \tag{16}$$

The key to determine the above term is to derive the transition probability $p\{Q_{t+1}|Q_t$, which, since Q_t is the representing value of discrete interval, is considered to be the conditional probability for Q_{t+1} to belong to the interval $[Q_{t+1}-\frac{1}{2}\Delta_{t+1},Q_{t+1}+\frac{1}{2}\Delta_{t+1})$ when the mean flow of period t is Q_t. Here, Δ_{t+1} is the interval length of Q_{t+1}.

Based on the runoff model (2) and the parameter estimation of seasonal AR(1), the expression can be further developed as:

$$Z_{t+1} = r_t Z_t + \sigma_{\varepsilon,t+1}\Phi_{t+1} \tag{17}$$

where, Φ_{t+1} is a random variable, which accords with standard P-III distribution and whose mean, standard deviation and skewness are respectively 0, 1 and $C_{s\Phi,t+1}$. The Z_t and the estimation of $\sigma_{\varepsilon,t+1}$ are respectively:

$$Z_t = \frac{Q_t - \mu_t}{\sigma_t} \tag{18}$$

and $\hat{\sigma}_{\varepsilon,t+1} = \sqrt{1 - r_{t+1}^2}$ \hfill (19)

Given Q_t is known, from (17), the Φ_{t+1} is a function of Q_{t+1}:

$$\Phi_{t+1}(Q_{t+1}) = \frac{\frac{Q_{t+1}-\mu_{t+1}}{\sigma_{t+1}} - r_t \frac{Q_t - \mu_t}{\sigma_t}}{\sqrt{1 - r_{t+1}^2}} \tag{20}$$

Set the probability distribution function of
Φ_{t+1} to be: $F_{t+1}(x) = P\{\Phi_{t+1} \geq x\}$ \hfill (21)

Then,

$$p\{Q_{t+1}|Q_t\} = F_{t+1}\left[\Phi_{t+1}(Q_{t+1} - \tfrac{1}{2}\Delta_{t+1})\right] - F_{t+1}\left[\Phi_{t+1}(Q_{t+1} + \tfrac{1}{2}\Delta_{t+1})\right] \tag{22}$$

In addition, set the boundary value to be:

$$F_{t+1}\left[\Phi_{t+1}(Q_{t+1})\big| Q_{t+1} \geq Q_{t+1,N}\right] = 0\% \tag{23}$$

$$F_{t+1}[\Phi(0)] = 100\% \tag{24}$$

where $Q_{t+1,N}$ is the final discrete value of Q_{t+1} with the interval as Δ_{t+1}.

6. CASE STUDY: YUDONG RESERVOIR IN YUNAN PROVINCE OF CHINA.

Yudong reservoir, located in Jule River (exactly, the right upstream of Zhaotong, a city of Yunnan province of China), is designed as an over-year reservoir with multipurpose mainly for flood control, irrigation, water supply, and power generation. Recently finished of Yudong project, a detailed study in reservoir operation is urgently desired. In the process of construction, Gaoqiao, a daily hydropower station with much high head drop of about 550 m, is located in the downstream from Yudong. And it mainly utilizes the discharge from Yudong for power generation during refill period. In this paper, a joint consideration of Yudong and Gaoqiao has been taken (Fig.1).

In the *Preliminary Report*(1991), Yudong reservoir operation is designed to satisfy following requirements: demand-met probability for irrigation 81%, demand-met probability for industrial supply 100%. Here, the flood control elevation is set to be 1983m from June to September and downstream crop risk from flood is not taken into account. Furthermore, through running model, energy-met probability is found usually having direct ratio with energy, as well as no special requirement for power generation exists, the penalty weight for energy-met failure is set to be 0.

In application, inputting the 42 years of historical inflow data (1954-1995) and the 100 man-made runoff samples that are simulated through seasonal AR(2)(in detail by *Jinwen et al.*, 1999)and repeatedly running LP-SDP with different set of penalty weights, lots of efficient policies and the associated long-run PIs by operation simulations then can be obtained. It is observed that, except for few points, each run of LP- SDP with different set of punish-weights can produce a trade-off point. The resulting graphs are shown in figure 3. Apparently, among the three PIs, one of their improvements will cause at least one of the other two to be worse. In this study, we noticed that, when either the demand-met probability for irrigation or that for industrial supply is very large (say, more than 99%), its tiny increase will result in great reduction in the expected yearly energy. Thereby, in this stochastic study, it is not only uneconomical but also impossible to meet the demands with 100% probability. This study does find this problem when handling the domestic demand as a strict one. But such impact is neglectable since this demand is very small.

The sensitivity analysis for demand-met probability of industrial supply and irrigation are shown in Fig4 and Fig. 5, respectively.

In *Fig 4*, when setting the demand-met probability for irrigation to be 81%, the Pareto Boundary between expected yearly energy versus the demand-met probability for industrial supply can be gotten. Trans-parently, as the demand-met probability of industry is less than 95%, its fluctuation has small impact upon the expected yearly energy; rather, when it is more than 97%, its enhance will cause the reduction of expected yearly energy greatly. Under such consideration and also the requirements from various reservoir-users, this study suggests that maybe it is favorable to choice the industrial demand-met probability as 97%.

In Fig.4 provided the industrial demand-met probability are set to be 97%, the trade-off relationship between expected yearly energy and irrigation demand-met probability can be obtained. From the graph, the probabilities about 88% are the critical range of the curve. Before this range, the impact of expected yearly energy from the change of irrigation reliability is much less, but after it, the situation is reversed. Thereby, it is recommended that it had better not to selected the probability of meeting the irrigation more than 90%.

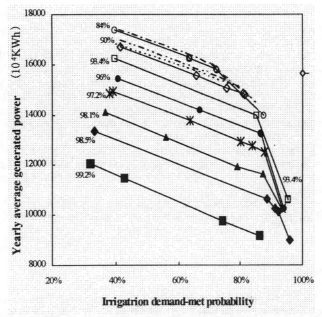

Fig. 3: The trade-off relationships between irrigation reliability level and power generation with various demand-met probabilities for industrial water supply.

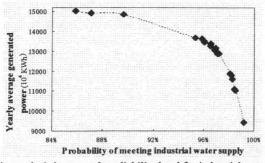

Fig. 4: The sensitivity analysis between the reliability level for industrial water supply and power generation. Here, demand-met probability for irrigation is set to be 81%..

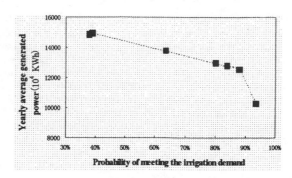

Fig. 5: The sensitivity analysis between irrigation reliability level and power generation. Here, demand-met probability for industrial supply is set to be 97%.

Finally, giving a demonstration, this study wants to decide a more satisfactory trade-off point and its corresponding policy. As mentioned earlier, the industry and irrigation reliability is respectively required to be 100% and 81%. Upon the above trade-off analysis, suitably receding the industrial reliability level is much favorable for power generation. At the same time, the industry department demands the water supplying with much high dependability. Accordingly, this study suggests the demand-met probability for industrial supply to be 97% or so. As for irrigation reliability, the requirement for it is not so high, but since there exist no severe conflict between it and the power generation when it is about 81%, maybe it is considerable to suitably increase the irrigation-met probability, which we feel may reach 88% or so. Based upon those considerations, this study has derived one such tradeoff point: expected yearly generation 12877.07 (10^4 *kilowatt*), irrigation reliability 87.12%, industrial supply reliability 96.94%. Other PIs can be also determined by operation simulation, e.g., met-probability of firm energy 93.62%, mean supply for domestic use 0.57 *cubic meter per second*, max deficit for irrigation 30%, max deficit for industry 10%.

It is worth noting that each trade-off point costs about 15 minutes in PC P II -200 by running LP-SDP. In addition, since each run can be carried on independently, there exists no computer burden.

7. CONCLUSIONS

In this paper, for the purpose of deriving various trade-off points, the LP-SDP with adjustable penalty weights is put forward. Thereinto, LP is solved for water allocations optimally for different objectives within the current period for given state of the system and total release from reservoir, resulting in the maximized immediate reward. And SDP is used to derive the steady state operating policy for reservoir, resulting in the maximization of expected annual system performance. In the model, the results of LP and transition probabilities of reservoir inflow form the input to SDP.

Finally, with the example study of Yudong Reservoir, this study concludes:

(1) The SDP with linear penalty weights is enough to produce the trade-off points of reliability PIs like various demand-met probabilities. Further research may orient to take into account the other risk PIs, e.g., resiliency and vulnerability, etc.

(2) It is not easy to derive a desired tradeoff point by varying the penalty weights in SDP. This makes a sensitivity analysis an even more difficult task when several PIs are set constant. Since Artificial Neural Networks (ANNs) boasts the special merits in set mapping, further research may orient to approximate the tradeoff curves (e.g., Fig.3) by training an ANNs, which may make the sensitivity analysis easier.

(3) In theory, there exists no computation problem for SDP to solve the presented operation optimization. However, the computation burden cannot be neglected when many intervals for state variable are used, many elements are taken into account, or other optimization technique is combined into.

REFERENCES

Cohon, J. L., and D. H. Marks, A review and evaluation of multiobjective programming techniques, *Water Resour. Res.,* 11(2), 208-220, 1975.

Georgakakos, A. P., Operational trade-offs in reservoir control, *Water Resour. Res.,* 29(11), 3801-3819, 1993.

Georgakakos, A. P., HuaMing Yao, and Yongqing Y., A control model for dependable hydropowerr capacity optimization, *Water Resour. Res.,* 33(10), 2349-2365, 1997.

Jinwen, W., C. Si, and Youquan Z.,Random simulation of Monthly runoff process for Yudong Reservoir, Yunnan Water Power, 15(1), 8-13,1999.

Mujumdar, P. P., and T. S. V. Ramesh, Reel-time reservoir operation for irrigation, *Water Resour. Res.,* 33(5), 1157-1164, 1997.

Piccardi, C., and R. Soncini-Sessa, Stochastic dynamic programming for reservoir optimal control: Dense discretization and inflow correlation assumption made possible by parallel computing, *Water Resour. Res.,* 27(5), 729-741, 1991.

Ponnambalam, K., and B. J. Adame, Stochastic optimization of multireservoir systems using a heuristic algorithm: Case study for India, *Water Resour. Res.,* 32(3), 733-741, 1996.

Preliminary Report on the Feasibility of Yudong Project, *Yunnan Water Cons. and Hydroelec. Reconnaissance, Design institution,* Oct., 1991.

Shiquan, H., Guide for system analysis in water resources, *Water Conser. and elec. Power Pub. Company,* 1988.

Ruxiang, Y., Analysis and Applications of Water Resources system, StingHua Uni. Press., 1987.

Takyi, A. K., and B. J. Lence, Surface water quality management using a multiple-realization chance constraint method, *Water Resour. Res.,* 35(5), 1657-1670, 1999.

Vedula S., and D. N. Kumar, An integrated model for optimal reservoir operation for irrigation of multiple crops, *Water Resour. Res.,* 32(4), 1101-1108, 1996.

Vedula S., and P. P. Mujumdar, Optimal reservoir operation for irrigation of multiple crops, *Water Resour. Res.,* 28(1), 1-9, 1992.

Zaden, L. A., Optimality and no-scarlar-valued performance criteria, *IEEE Trans. Autom. Control,* Ac-8, 59-60,1963.

Zhenpeng, H., and F. Shangyou, Multiobjective risk analysis for the contradiction between flood control and benefit promotion in reservoir operation, *J. Of Wuhan Univ. Of Hydr. and Elec. Eng.,* China, 22(1), 71-79, 1989.

Stochastic Hydraulics 2000, Wang & Hu (eds) © 2000 Taylor & Francis, ISBN 90 5809 166 X

Stochastic analysis of water hammer in a Reservoir-Pipe-Valve system

Zhang Qinfen & Suo Lisheng
Hokai University, Nanjing, China

Wenzhu Guo
Bureau of Power Industry of Anhui Province, Hofei, China

ABSTRACT: The randomness of conditions under which water hammer occurs and factors that influence the magnitudes of water hammer pressures results in the stochasticism of water hammer. The statistical characteristics and probability distributions of such influential factors as boundary conditions, initial states and parameters of hydraulic systems are analyzed, and the analytic probability functions of the maximum water hammer pressure in a reservoir-pipe-valve system are derived, based on simplified water hammer formulas. A stochastic numerical simulation is then proposed. Computational examples show that the results from both the analytic analysis and the numerical simulation agree well with each other, but differ from those by traditional definite methods.

Key words: Single pipe; Water hammer; Stochastic analysis; Numerical simulation

1. INTRODUCTION

The water hammer is stodiasic because many factors that influence the magnitudes of water hammer pressures results in the stochasticism of water hammer. As one of main loads in structural design of penstocks or pressure tunnels, the stochastical characteristics of water hammer is important for reliability analysis in structural design. But so far, such random influential factors as boundary conditions, initial states and parameters of hydraulic systems are still considered as definite parameters, and the most unfavorable combinations of all factors are selected in analysis of water hammer to obtain the maximum magnitudes of water hammer pressures. These traditional definite models can not provide the probability distributions of water hammer, and no longer fit in with the needs of reliability analysis in structural design.

The water hammer in a frictionless reservoir-pipe-valve system shown in Fig.1 with linear valve closure can be divided into two types. The maximum of First-type water hammer, which usually takes place in high-head hydropower plants, occurs at the end of first-interval. And the maximum of Second-type, which usually takes place in low-head hydropower plants, occurs at the end of valve-closure.

The stochastic analysis of water hammer in a reservoir-pipe-valve system is performed in this paper. First, the statistical characteristics and probability distributions of such influential factors as boundary conditions, initial states and parameters of hydraulic systems are analyzed based on a large number of observed data from existing hydropower plants in China. Then, the analytic probability functions of the maximum water hammer pressure in a reservoir-pipe-valve system are derived based on simplified formulas associated with the two types of water hammer, respectively, and a stochastic numerical simulation is proposed. Computational examples are presented, and the probability distribution of the annual maximum value of total hydraulic pressures (the static plus the dynamic, namely water hammer) and that of the dynamic pressure are all provided.

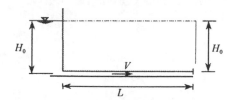

Fig.1: Reservoir-pipeline-valve system

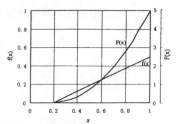

Fig.2: Probability Distribution of m_0

2. PROBABILITY DISTRIBUTIONS OF MAIN RANDOM VARIABLES

2.1 Probability distributions of initial discharge and valve opening

Assume that N_f is the full load produced by the governed turbine, N_0 the actual load before the load rejection. The ratio $m_0=N_0/N_f$ is defined as the load degree coefficient. Since N_0 depends on both the demand of electrical system and the amount of available hydropower, m_0 is a random variable with $0<m_0\leq1$. Statistical analysis of the observed data from 323 load-rejection accidents in 11 hydropower plants shows that, for hydraulic unit with single turbine-generator undertaking peak load, m_0 is subject to increasing triangular probability distribution between [0.2, 1.0] as shown in Fig 2. The probability density function (PDF) can be expressed as

$$f(x) = \frac{2(x-0.2)}{(1-0.2)^2}, \qquad x \in [0.2, \ 1.0] \tag{1}$$

With the turbine simplified as a valve of definite-valued discharge coefficient, the following relations can be derived:

$$m'_0 = \frac{Q_0}{Q_f} = \frac{V_0}{V_f} = m_0\frac{H_d}{H_0} \tag{2}$$

$$\tau_0 = m'_0\sqrt{\frac{H_d}{H_0}} = m_0(\frac{H_d}{H_0})^{1.5} \tag{3}$$

in which Q_0, V_0 and H_0 are initial values of flow, velocity and head, respectively. m'_0 is called as the initial discharge degree coefficient, τ_0 the initial dimensionless valve opening, H_d the designing head of hydropower plant, Q_f and V_f the full discharge and velocity with head H_d and maximum valve opening $\tau_f=1.0$. H_d, N_f and τ_f are all definite-valued constants.

It is obvious that both m'_0 and τ_0 are random variables related to random variables m_0 and H_0. The probability distributions of m'_0 and τ_0 can be obtained from Eqs. (2) and (3) as long as the probability distributions of m_0 and H_0 are known.

2.2 Probability distributions of valve closure time Tc

For a given τ_0, designed valve closure time T_{c0} controlled by the governor is a definite-valued constant. Particularly, for linear valve closure, $T_{c0}=C\tau_0$, in which C indicates the designed closure time with valve opening changing from 1.0 to 0. However, the sensitivity of governed devices, the instability of oil pressure and the ways of operating in every accident lead to a discrepancy ΔT_c between the actual closure time T_c and T_{c0}, i.e. $T_c = T_{c0} + \Delta T_c$. Experiences show that the discrepancy is subject to Gauss distribution, so it can be assumed that $T_c \propto N(T_{c0}, 0.01T_{c0}^2)$, indicating that T_{c0}, associated with random variable τ_0, can be taken as mean value of T_c, and the discrepancy is usually within 10%. In short, T_c is subject to Gauss distribution with respect to τ_0.

2.3 Probability distributions of the head H0

According to Ref. [1], the annual maximum value of reservoir surface level is a random variable subject to Gauss distribution, and there is a statistical relation between its mean value μ_{H_G} and the normal reservoir surface level H_n

$$\mu_{H_G} = 0.935 H_n - 0.33 \tag{4}$$

where the base of dam is taken as the elevation datum. The variance C_V is approximately 0.1. These statistical characteristics have been verified by the authors, too. For the simplified system shown in Fig.1, the static head H_0 can be calculated as the annual maximum reservoir elevation H_G, and its PDF may be written as

$$f(x) = \frac{1}{\sigma_{H_G} \sqrt{2\pi}} \exp\left[-\frac{(x - \mu_{H_G})^2}{2\sigma_{H_G}^2} \right] \tag{5}$$

3. ANALYTIC PROBABILITY FUNCTIONS OF THE MAXIMUM WATER HAMMER PRESSURE

In a frictionless reservoir-pipe-valve system as shown in Fig.1 and with linear valve closure from initial opening τ_0 to finial opening 0, the resulted water hammer can be divided into two types and the simplified formula to calculate their maximum value are, respectively,

For First - type $\qquad \Delta H_m = \xi_m \cdot H_0 = \dfrac{2\sigma}{2 - \sigma} \cdot H_0 \qquad$ when $\rho\tau_0 > 1$ $\tag{6}$

For Second - type $\qquad \Delta H_1 = \xi_1 \cdot H_0 = \dfrac{2\sigma}{1 + \rho\tau_0 - \sigma} \cdot H_0 \qquad$ when $\rho\tau_0 \le 1$ $\tag{7}$

where water hammer constants σ, ρ are defined, respectively, as $\sigma = \dfrac{LV_0}{gH_0 T_c}$, $\rho = \dfrac{aV_0}{2gH_0}$,

with L the length of the pipeline, a the wavespeed.

From the definition of σ and Eq. (6), the following equation can be developed:

$$\Delta H_m = \frac{2LV_f H_d H_0 m_0}{2gH_0^2 T_c - LV_f H_d m_0} \tag{8}$$

Represent the three random variables H_0, m_0 and T_c by three random variables X_1, X_2 and X_3, respectively, and constitut a random rector $X = [X_1, X_2, X_3]^T$. Furthermore, represent $[\Delta H_m, X_2, X_3]$ by another random rector $Y = [Y_1, Y_2, Y_3]^T$. Applying the transformation formula of random rector and the relation of marginal PDF and joint PDF leads to the PDF of ΔH_m [2]:

$$f_{Y_1}(y_1) = \int_{-\infty}^{+\infty} \int_{-\infty}^{+\infty} \Phi(y_1, y_2, y_3) \frac{y_2 y_3}{c_1 y_1^2} \, dy_2 \, dy_3 \tag{9}$$

in which $\qquad c_1 = \dfrac{g}{LH_d V_f}$, function $\Phi(y_1, y_2, y_3)$ can be derived from joint PDF of rector

X. The integrating range of y_3 may be [0, 1], and that of y_2 may be from H_{min} to H_{max}, depending on actual conditions.

Similarly, the PDF of ΔH_1 and total pressure H_1, H_2 can be developed as well.

4. NUMERICAL SIMULATION BY MONTE CARLO METHOD

The inherent stochasticism of the system results in the indefiniteness of water hammer value. If the initial parameters were given, the transient process would be definite. For different initial parameters, a series of definite transient process can be simulated by use of Monte Carlo method. As long as the initial random parameters are subject to their inherent distribution laws, the stochastic characteristics of the transient can be resulted.

Monte Carlo method is a random sampling technology and each simulated result represents a sample. With the number of samples increasing, statistical characteristic of object would approach the actual solution. The procedure of numerical simulation for the transient process by Monte Carlo method is:

(1) Generating the random series of H_0, m_0 by transferring random values of uniform distribution between (0, 1) to random series which are, respectively, subject to their distributions described as Eqs. (1) and (5). Then, producing the random series of m'_0, τ_0 and T_c etc. according to their relations with H_0, m_0;

(2) Calculating each value of water hammer from Eq. (6) or Eq. (7) corresponding to each series of sample H_0, m_0;

(3) Performing statistical analysis of the simulated results of water hammer to obtain the stochastic characteristics and probability distributions.

5. EXAMPLE AND ANALYSIS

Example 1   In the system shown in Fig.1, static head H_0 is subject to Gaussian distribution $H_0 \propto N(100, 10^2)$.   Pipe length $L=500$m. Wavespeed $a=1000$m/s.   With $H_d=100$m and valve opening $\tau_0=1.0$, the velocity in the pipeline reaches its maximum $V_f=5$m/s.   The time duration from $\tau_0=1.0$ to 0 (full linear closure) is 10s. The analyses of the maximum of water hammer at the end of the pipe are presented as following.

(1) Definite Calculation

Taking the most unfavorable values for all indefinite parameters, $m_0=1.0$, $H_0=110$m, it is known that $\rho\tau_0>1$, so this is a First-type of water hammer problem. From Eq. (6) the head rise of water hammer $\Delta H_m=30.42$m, the total head of water hammer $H_m=H_0+\Delta H_m=140.42$m.

(2) Stochastic Analysis

Numerical results from Monte Carlo simulation are obtained and compared with the solutions by use of analytic method, as shown in Fig.3. The solid line represents the PDF of pressure head at the end of the pipeline from Monte Carlo program, while the dashed line refers to the analytic results. In application of Monte Carlo method, it is noted that $\rho\tau_0>1$ should be satisfied.

(3) Discussion

With respect to water hammer pressure, the results of definite calculation $\Delta H_m=30.42$m is smaller than the 0.95-point-value of probability distribution from the stochastic analysis, while for the total head, $H_m=H_0+\Delta H_m= 140.42$m from definite calculation is very close to the corresponding 0.95-point-value.   The reason of the former is that the value of ΔH_m is very sensitive to the closure time T_c, but the error of T_c has not been taken into account in definite calculation.   Fortunately, the traditional design load of total head H_m, the summation of the maximum static head and the head rise of water hammer, seems acceptable in magnitudes.

Example 2   In the system shown in Fig.1, static head H_0 is subject to Gaussian distribution $H_0 \propto N(500,50^2)$. Pipe length $L=800$m. Wavespeed $a=1000$m/s. With $H_d=500$m and valve opening $\tau_0=1.0$, the velocity in the pipeline reaches its maximum $V_f=6$m/s.   The time duration from $\tau_0=1.0$ to 0 (full linear closure) is 5s.   The analyses of the maximum of water hammer at the end of pipe are presented as following.

(1) Definite Calculation

With $m_0=0.369$ and $H_0=550$m, $\rho\tau_0<1$ and this is a Second-type of water hammer problem. Since $T_c=2L/a$, the situation is the most serious, in fact, this is a direct water hammer. The head rise of water hammer is $\Delta H_1=204.60$m and the total head $H_1=H_0+\Delta H_1=754.60$-m.

(2) Stochastic Analysis

Numerical results from Monte Carlo simulation are obtained and compared with the solutions

840

by use of analytic method, as illustrated in Fig.4. The solid line represents the PDF of pressures head at the end of the pipeline from Monte Carlo program; while the dashed line refers to the analytic results. In application of Monte Carlo method, $\rho\tau_0 \leq 1$ is maintained.

(3) Discussion

The results of definite calculation are all very close to the corresponding 0.95-point-value from stochastic analysis. It should be noted that severe water hammer of the Second type takes place when head is higher and discharge is smaller.

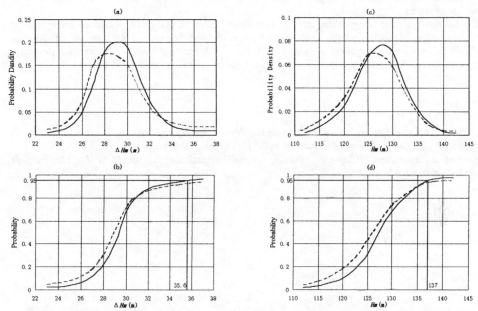

Fig.3: Comparison of numerical and analytical results of the distribution First-type water hammer

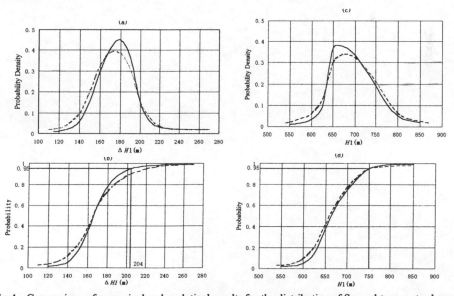

Fig.4 : Comparison of numerical and analytical results for the distribution of Second-type water hammer

841

6. CONCLUSIONS

Compared with traditional definite models, the stochastic analysis of water hammer may take various indefinite parameters and comprehensive conditions into account, thus provide more exact hydraulic pressure of probability information for the reliability analysis of structural design of penstocks or pressure tunnels.

The reasonableness of Monte Carlo method has been verified by the results of analytical analyses in computational examples. This method can be easily applied to more general or complex systems.

Compared with the design value at 0.95-propability-point in results of stochastic analysis, the total hydraulic pressure and the dynamic pressure of First-type water hammer from definite analysis is somehow acceptable in magnitudes, but the dynamic pressure of Second-type water hammer is less. The latter is worth to be noted by hydraulic engineers.

REFERENCE

Wu Shiwei, Zhang Sijun and Yu Qiang, Statistical Values for Annual Maximum of Water Level at Upstream of Dams, Journal of Hohai University (in Chinese), 1984(4).

Mathematics Department of Zejiang University, Probability and Statistics (in Chinese), The People's Education Press, March 1982.

Stochastic modeling

Stochastic Hydraulics 2000, Wang & Hu (eds) © 2000 Taylor & Francis, ISBN 90 5809 166 X

Invited lecture: A new empirical equation of longitudinal dispersion coefficient

Huang Kezhong & Hou Yu
Department of Geography, Zhongshan (Sun Yatsen) University, Guangzhou, China

ABSTRACT: A new empirical equation for predicting longitudinal dispersion coefficient in natural streams by using hydraulic and geometric data is presented. The genetic algorithm that can search out the global optimum is used in this study. The new empirical dispersion coefficient equation is proven to be superior to some existing empirical dispersion coefficient equations by comparison with seventeen measured data taken from Jialing River and Three Gorges Reservoir of China.

Key words: Dispersion coefficient, Empirical equation, Genetic algorithm, Natural stream, Global optimum

1 INTRODUCTION

The one-dimension dispersion equation

$$\frac{\partial C}{\partial t} + V \frac{\partial C}{\partial x} = \frac{1}{A} \frac{\partial}{\partial x} \left(KA \frac{\partial C}{\partial x} \right) \tag{1}$$

derived by Taylor (1954) has been widely used to get a reasonable estimate of the cross-sectional average concentration of pollutants in natural streams, in which A is the cross-sectional area, C is the cross-sectional average concentration, V is the cross-sectional average velocity, K is the dispersion coefficient, t is the time, and x is the direction of the mean flow. Numerical solution of Eq.(1) are easily obtained with given initial condition, boundary condition and channel geometry when the river flow is steady and the dispersion coefficient is given (Huang, 1997). However, to elect the value of dispersion coefficient is an important and difficult task.

Many empirical equations for predicting the longitudinal dispersion coefficient in natural streams have been proposed by various investigators, e.g. Fischer (1975), Liu (1977), and Seo and Cheong (1996). However, their predictions were often overestimating or underestimating significantly compared to the measured data. About the occurrence causes of these situations, the selections of main factors that influence the dispersion characteristics of pollutants and the quality and quantity of measured data sets were important causes, but the regression method for driving the empirical equations also played the part. The regression coefficients given by the traditional regression methods were probably not the global optimum because the optimum results of the regression coefficient given by the traditional regression methods in complex multimodal spaces may be the local optimum results. Therefore, the genetic algorithm (GA) which can search out the global optimum (Holland, 1975, and Goldberg, 1989) is used in this study.

By using the dimensional analysis (Iwasa and Aya, 1980 and Seo and Cheong,1996) or by taking reasonable approximation of the integral relation for the dispersion coefficient in natural streams (Fischer, 1975), the functional relationship pertinent to the dispersion coefficient K can be obtained as follows:

$$\frac{K}{hu_*} = a\left(\frac{W}{h}\right)^b\left(\frac{V}{u_*}\right)^c$$

(2)

In which the factors are the channel width W, the depth h, the average velocity V, and the shear velocity u_*; and the constants a, b, and c are the regression coefficients. We are sure that Eq.(2) used by Fischer (1975), Liu (1977), Magazine et al. (1988), Iwasa and Aya (1991), and Seo and Cheong (1996) is reasonable. The values of regression coefficients in their equations are shown in Tab.1.

Tab.1: Regression coefficients of existing dispersion coefficient equations

Authors	a	b	c
Fischer	0.011	2	2
Liu	0.18	2	0.5
Magazine et al.	338	0	- 0.632
Iwasa and Aya	2.0	1.5	0
Seo and Cheong	5.915	0.620	1.428

GA developed by Holland(1975) has been used in some domains of water resources engineering (e.g. Yeh ,1986, Goldberg et al.,1987, Suzuki et al.,1996, Savic et al.,1996, Liu et al.,1996, Wang et al. 1997, and Jin, 1999). GA has been theoretically and empirically proven to provide robust search in complex spaces. It can search out the global optimum solutions in the optimization problems. On the technology process of GA, it is an optimization program based on the mechanism of natural selection and natural genetics. Natural selection is implemented through the selection and the recombination operators. A population of candidate solutions, usually coded as bit strings, is modified from one generation to the next by the probabilistic application of the genetic operators.

2. REGRESSION COEFFICIENT IDENTIFICATION BY GA

The criterion for the regression coefficient identification may be stated as to minimize the objective function

$$f(a,b,c) = \sum_{i=1}^{n}\sqrt{\left(K_{c,i} - K_{m,i}\right)^2} \rightarrow \min$$

(3)

In which $K_{c,i}$ and $K_{m,i}$ are the calculated and measured dispersion coefficients, respectively, and n is the number of measured data sets for deriving the empirical equation. Tab.2 shows fifty-nine hydraulic and dispersion data measured at 26 streams in United States. Seo and Cheong (1996) used these data partly for deriving and partly for verifying the dispersion coefficient equation. Now we apply all the data to derive the dispersion coefficient equation.

The flowchart illustrating the implement processes of GA in this study is shown in Fig.1. The main processes of the GA are the production of the initial population of bit, the reproduction by selection, crossover, and mutation operators, the decoding, and the evaluation.

The parameters of GA are adopted in this study as follows: The population size is 100, the length of binary string is 6, and the election probability, crossover probability, and mutation probability are 0.1, 0.2, and 0.1, respectively.

The following two conditions are used at a time as the terminating conditions: (1) the minimum value of objective function is unchanged among 10 successive generations, and (2) the repeating generation times is 500.

By using the GA, we obtain the regression coefficients $a = 3.5$, $b = 1.125$, and $c = 0.25$. Therefore, the new dispersion coefficient equation is

$$\frac{K}{hu_*} = 3.5\left(\frac{W}{h}\right)^{1.125}\left(\frac{V}{u_*}\right)^{0.25}$$

(4)

or

$$K = 3.5 \frac{W^{1.125} V^{0.25} u_*^{0.75}}{h^{0.125}}$$

(5)

Tab.2: Hydraulic and dispersion data measured at 26 streams in United States

Streams	h (m)	W (m)	V (m/s)	u*(m/s)	K (m2/s)	References
Antietam Creek, MD	0.30	12.80	0.42	0.057	17.50	Nordin et al. (1974)
	0.98	24.08	0.59	0.098	101.50	
	0.66	11.89	0.43	0.085	20.90	
	0.48	21.03	0.62	0.069	25.90	
Monocacy River, MD	0.55	48.70	0.26	0.052	37.80	
	0.71	92.96	0.16	0.046	41.40	
	0.65	51.21	0.62	0.044	29.60	
	1.15	97.54	0.32	0.058	119.80	
	0.41	40.54	0.23	0.040	66.50	
Conococheague Creek, MD	0.69	42.21	0.23	0.064	40.80	
	0.41	49.68	0.15	0.081	29.30	
	1.13	42.98	0.63	0.081	53.30	
Chattahoochee River, GA	1.95	75.59	0.74	0.138	88.90	
	2.44	91.90	0.52	0.094	166.90	
Salt Creek, NE	0.50	32.00	0.24	0.038	52.20	
Difficult Run, VA	0.31	14.48	0.25	0.062	1.90	
Bear Creek, CO	0.85	13.72	1.29	0.553	2.90	
Little Pincy Creek	0.22	15.85	0.39	0.053	7.10	
Bayou Anacoco, LA	0.45	17.53	0.32	0.024	5.80	
Comete River, LA	0.23	15.70	0.36	0.039	69.00	
Bayou Bartholomew ,LA	1.40	33.38	0.20	0.031	54.70	
Amite River, LA	0.52	21.34	0.54	0.027	501.40	
Tickfau River, LA	0.59	14.94	0.27	0.080	10.30	
Tangipahoa River, LA	0.81	31.39	0.48	0.072	45.10	
	0.40	29.87	0.34	0.020	44.00	
Red River, LA	1.62	253.59	0.61	0.032	143.80	
	3.96	161.54	0.29	0.060	130.50	
	3.66	152.40	0.45	0.057	227.60	
	1.74	155.14	0.47	0.036	177.70	
Sabine River, LA	1.65	116.43	0.58	0.054	131.30	
	2.32	160.32	1.06	0.054	308.90	
Sabine River, TX	0.50	14.17	0.13	0.037	12.80	
	0.51	12.19	0.23	0.030	14.70	
	0.93	21.34	0.36	0.035	24.20	
Mississippi River, LA	19.94	711.20	0.56	0.041	237.20	
Mississippi River, MO	4.94	533.40	1.05	0.069	457.70	
	8.90	537.38	1.51	0.097	374.10	
Wind/Bighorn River, WY	1.37	44.20	0.99	0.142	184.60	
	2.38	85.34	1.74	0.153	464.60	
Copper Creek, VA	0.49	16.66	0.20	0.080	16.84	Godfrey et al.(1970)
Clinch River, VA	1.16	48.46	0.21	0.069	14.76	
Copper Creek, VA	0.38	18.29	0.15	0.116	20.71	
Powell River, TN	0.87	16.78	0.13	0.054	15.50	
Clinch River, VA	0.61	28.65	0.35	0.069	10.70	
Copper Creek, VA	0.84	19.61	0.49	0.101	20.82	
Clinch RIVER, VA	2.45	57.91	0.75	0.104	40.49	
COACHELLA Canal, CA	1.58	24.69	0.66	0.041	5.92	
Clinch River, VA	2.41	53.24	0.66	0.107	36.93	

Copper Creek, VA	0.47	16.76	0.24	0.080	24.62	
Missouri River, IA	3.28	180.59	1.62	0.078	1486.45	Yotsukura et al.(1970)
Bayou Anacoco, LA	0.94	25.91	0.34	0.067	32.52	McQuivey et al.(1974)
	0.91	36.58	0.40	0.067	39.48	
Nooksack River, WA	0.76	64.01	0.67	0.268	34.84	
Wind/Bighorn River, WY	1.10	59.44	0.88	0.119	41.81	
	2.16	68.58	1.55	0.168	162.58	
John Day River, OR	0.58	24.99	1.01	0.140	13.94	
	2.47	34.14	0.82	0.180	65.03	
Yadkin River, NC	2.35	70.10	0.43	0.101	111.48	
	3.84	71.63	0.76	0.128	260.13	

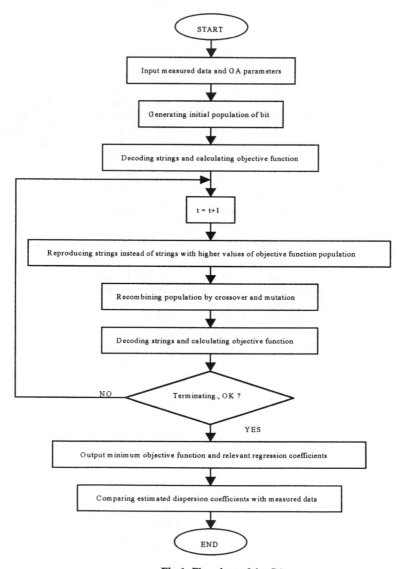

Fig.1: Flowchart of the GA

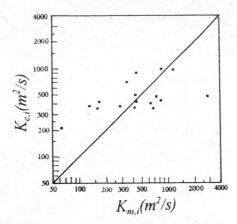

Fig.2: Comparison of dispersion coefficientspredictedby Eq.(5) with 17 measured data

3. VERIFICATION

Tab.3 shows that seventeen dispersion coefficients measured at the reaches of Jialing River and Three Gorges Reservoir in China (Huang et al. 1998). These data are used to verify Eq.(5).

Tab.3: Hydraulic and dispersion measure data at Jialing Riverand Three Gorges Reservoir in China

Streams	h (m)	W (m)	V (m/s)	u⁎ (m/s)	K (m²/s)
Jialing River	11.0	194	1.00	0.13	61.2
	7.4	374	0.90	0.12	163.4
	6.0	330	1.09	0.10	415.6
	6.8	301	1.56	0.10	671.3
	5.5	340	1.03	0.10	154.9
	9.8	380	1.22	0.12	737.7
	6.6	363	1.23	0.10	640.9
	7.0	386	1.25	0.10	821.4
	7.7	315	1.89	0.10	279.3
	5.8	352	1.17	0.10	126.1
Three Gorges Reservoir	23.3	319	0.55	0.21	434.3
	10.9	398	0.95	0.14	419.3
	6.3	349	2.09	0.11	2883.5
	22.3	510	1.96	0.20	1150.1
	23.2	520	2.23	0.20	839.1
	22.2	499	1.82	0.20	427.7
	25.1	403	1.50	0.21	333.6

For the verification of Eq.(5), the dispersion coefficients predicted by Eq.(5) are compared with the measured dispersion coefficients, and the comparison results are shown in Fig.2. In the meantime, the comparisons of three existing dispersion coefficient equations with the measured dispersion coefficient are shown in Fig.3. These results show that the predicted results of the three existing equations are overestimating significantly in most cases, while the precited results of Eq.(5) is better than them.

In order to evaluate the difference between measured and predicted values of the dispersion coefficient more quantitatively, we define a coefficient of deviation C_d as follows:

$$C_d = \frac{\sum_{i=1}^{n}\sqrt{(K_{c,i} - K_{m,i})^2}}{\sum_{i=1}^{n}K_{m,i}} \tag{6}$$

849

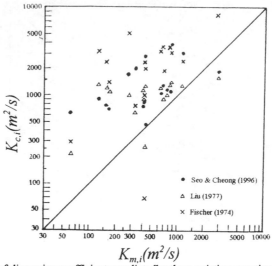

Fig.3: Comparison of dispersion coefficients predictedby three existing equations with 17 data

in which n ($=17$) is the number of measured data set for verification. We obtain $C_d = 0.0116$, 0.00335, 0.00946, and 0.00196 for Fischer's, Liu's, Seo and Cheong's equation, and Eq.(5), respectively. The value of C_d for Eq.(5) is smallest among these values. It obviously demonstrates that Eq.(5) is superior to the existing equations.

4. CONCLUSIONS

When the one-dimensional dispersion equation is applied to predict concentration variations of pollutants in natural stream, selection of proper dispersion coefficient is an important and difficult task. It is a main way to predict the dispersion coefficient by using empirical equation . The traditional regression methods can probably not give the global optimum, while the GA that can search out the global optimum may be used to derive the regression coefficients of the empirical equation. Eq. (5) given in this study is superior compared to the existing empirical equations. It can be used to estimate the longitudinal dispersion coefficient in natural stream when the measured dispersion data could not be collected.

Various complex optimization problems arise in water resources engineering. The GA is a probability method for solving optimization problems. We ought to extend the applications of GA in this area, since the GA is not only effective, but also easily realizable due to the conceptual simplicity of the basic mechanisms.

REFERENCES

Fischer, B.H. (1975), Discussion of "Simple Method for Predicting Dispersion in Streams", by McQuivey, R. S., and Keefer, T. N., Journal of Environmental Engineering Division, ASCE, 101(3), 453-455.

Godfrey, R, G,, and Frederick, B.J.(1970),Stream Dispersion at Selected Sites, United States Geological Survey Professional Paper 433-K, Washington, D.C.

Goldberg, D.E., and Kno, C.H.(1987), Genetic Algorithms in Pipeline Optimization, Journal of Computing in Civil Engineering, 1(2), 128-141.

Goldberg, D.E. (1989), Genetic Algorithms in Search, optimization and Machine Learning, Addison Wesley Inc., USA.

Holland, J.H. (1975), Adaptation in Natural and Artificial Systems, University of Michigan Press.

Huang, K.Z. (1997), Environmental Hydraulics, Zhongzhan University Press. (in Chinese)

Huang, Z.L., Li, J.X., and Huang, J.C. (1998), Preliminary Study on Longitudinal Dispersion Coefficient for the Three Gorges Reservoir, Proceeding of the Second International Symposium on Environmental Hydraulics, "Environmental Hydraulics", Edited by Lee, J.H.W., Jayawardena, A.M., and Wang, Z.Y., Hong Kong, 669-674.

Iwasa, Y., and Aya, S. (1980), Dispersion Coefficient of Natural Streams, Proceedings of the Third International Symposium on Stochastic Hydraulics, Japan, 527-538.

Iwasa, Y., and Aya, S. (1991), Predicting Longitudinal Dispersion Coefficient in Open-Channel Flows, Proceedings of the International Symposium on Environmental Hydraulics, Hong Kong, 505-510.

Jin , J.(1999), Theorem and Application Research in Optimization of Water Problem Using GA, Systems Engineering, 17(3), 77-80.(in Chinese)

Liu, S.W., and Feng, S.Y.(1996), Genetic Algorithm and Its Application in Water Pollution Control System Planning, Journal of Wuhan University of Hydraulic and Electric Engineering, 29(4), 95-99.(in Chinese)

Liu, H.(1977), Predicting Dispersion Coefficient of Stream, Journal of Environmental Engineering Division, ASCE, 103(1), 59-69.

Magazine,M.K., Pathak,S.K., and Pande, P.K. (1988), Effect of Bed and Side Roughness on Dispersion in Open-Channels, Journal of Hydraulic Engineering, ASCE, 114(7), 766-782.

McQuivey, R.S., and Keefer, T.N.(1974), Simple Method for Predicting Dispersion in Streams, Journal of Environmental Engineering Division, 100(4), 997-1011.

Nordin, C,F., and Sabol, G.V.(1974), Empirical Data on Longitudinal Dispersion in River, United States Geological Survey Water Resources Investigation 20-74, Washington, D.C.

Savic, D.A., and Walters, G.A.(1966), Stochastic Optimization Techniques in Hydraulic Engineering and Management, Proceeding of 7th IHAR International Stochastic Hydraulics Symposium, "Stochastic Hydraulics'96", Edited by Tickle,k., Goulter,I., Xu Chengchao, Wasimi, S.A, and Bouchart,F., Australia, 65-72.

Seo, I.W., and Cheong, T.S.(1996), Prediction of Longitudinal Dispersion Coefficient in Natural Streams, Proceeding of 7th IHAR International Stochastic Hydraulics Symposium, "Stochastic Hydraulics'96", Edited by Tickle,k., Goulter,I., Xu Chengchao, Wasimi, S.A, and Bouchart,F., Australia, 459-466.

Suzuki, T., Harada, M., Hammad, A. And Itoh, Y. (1996), An Inverse Analysis of Model Parameters for heterogeneous Aquifer Based on Genetic Algorithm, Proceeding of 7th IHAR International Stochastic Hydraulics Symposium, "Stochastic Hydraulics'96", Edited by Tickle, k., Goulter, I ., Xu Chengchao, Wasimi, S.A, and Bouchart,F., Australia, 617-624.

Taylor, G.I.(1954), Dispersion Matter in Turbulent Flow through a Pipe, Journal of the Royal Society of London, Series A., 223, 446-468.

Wang, L., and Ma, G.W.(1997), A Genetic Algorithm for Parameter Optimization of Streeter-Phelps Model, Advances in Water Science, 8(1), 32-37.(in Chinese)

Yeh, W.W-G. (1986), Review of Parameter Identification Procedure in Groundwater Hydrology: the Inverse Problem, Water Resources Research, 22(2), 95-108.

Yotsukura, N., Fischer, H.B., AND Sayre,W.W.(1970), Measurement of Mixing Characteristics of the Missouri River between Sioux City, Iowa, and Plattsmouth, Nebraska, United States Geological Survey Water-Supply Paper 1899-G, Washington D.C.

Stochastic Hydraulics 2000, Wang & Hu (eds) © 2000 Taylor & Francis, ISBN 90 5809 166 X

A fuzzy logic approach to river flow modelling

D. Han, I. D. Cluckie & D. Karbassioun
Water and Environmental Management Research Centre, Department of Civil Engineering, University of Bristol, UK

J. Lowry
Department of Engineering Mathematics, University of Bristol, UK

ABSTRACT: This paper describes the attempt of using Fuzzy Logic approach for river flow modelling from rainfall data. Firstly, both rainfall and stream flow are converted into representing fuzzy sets. A special format for the fuzzy sets has been established to cope with the characteristics of rainfall and flow records. The process adopted for the modelling is one that generates sets of qualified linguistic conditionals using an ID3 type algorithm proposed by the Artificial Intelligence group at the University of Bristol. A powerful fuzzy programming language FRIL was adopted in this research.

Key word: Fuzzy set, River flow, Artificial intelligence, Fuzzy modelling

1. INTRODUCTION

The first publication in area of fuzzy set theory appeared in 1965 [Zadeh 1965] and the development of this theory for almost 20 years remained in the academic realm. From a methodological point of view, the application areas for FT (Fuzzy Technology) can be classified into the following categories [Zimmermann 1996]:

1) Algorithmic Applications, e.g.
 Fuzzy Mathematical Programming; Fuzzy Planning Methods (CPM, Graphs);
 Fuzzy Petri Nets; Fuzzy Clustering, etc.
2) Information Processing
 Fuzzy Data Bank System; Fuzzy Query Languages; Fuzzy Languages, etc.
3) Knowledge Based Applications
 Expert Systems; Fuzzy Control; Knowledge Based Diagnosis, etc.
4) Hybrid Application Areas
 Fuzzy Data Analysis; Fuzzy Supervisory Control, etc.

Nowadays, combination of FT with neural nets and genetic algorithms is rapidly increasing. Attempts have been made in the past to introduce this new technology into water engineering. It has been found that modelling water movement in the unsaturated soil matrix was suitable for adopting fuzzy rules and in comparison with the numerical solution from the Richards equation, the fuzzy model performed quite well [Bardossy, Bronstrert and Merz 1995]. Fuzzy algorithm has also been applied in urban drainage system modelling that was able to incorporate the GIS and remote sensed thermal map to estimate the runoff potentials [Campana, Mendiondo and Tucci , 1995]. Bardossy [1996] tried to use the fuzzy rules to model infiltration, surface runoff and unsaturated flow. It was found that fuzzy systems provide a robust tool which can handle non-linearities, without requiring a prescribed functional structure. Furthermore the fuzzy rule-based models can easily be coupled; for example, a model for flow in porous media may be coupled with a bacteriological growth model. They are capable of combining physical laws, expert knowledge and measurement data. Efforts were spent on combining the fuzzy theory with Genetic Algorithm to simulate infiltration process and it was found more accurate simulation could be achieved than Newton-Raphson method [Zeigler, Moon, Lopes and Kim 1996]. On flood control, Cheng [1999] proposed a fuzzy optimal model for the flood control system of the upper and middle reaches of the Yangtze River and found the model effective and flexible when it was validated with three typical historical floods.

It is clear that fuzzy technology is playing a more and more important role in modern day water engineering. One area deserving further investigation is real time flood forecasting. Stuber and Gemmar [1997] proposed an approach for data analysis and forecasting with neural fuzzy systems. In their paper, two different system approaches are discussed: a) a neural network for supervised learning of the functional behaviour of time series data and its approximation, and b) a fuzzy system for modelling of the system behaviour with possibilities to exploit expert information and for systematic optimisation. Recently, See and Openshaw [1999] also combined fuzzy logic model with neural networks. They split the forecasting data set into subsets before training with a series of neural networks. These networks are then recombined via a rule-based fuzzy logic model that has been optimized using a genetic algorithm. The overall results indicate that this methodology may provide a well performing, low-cost solution, which may be readily integrated into existing operational flood forecasting and warning systems.

This paper will adopt a different approach using a special artificial intelligent programming language, named FRIL, to model river flows. It is known that a modern real time flood forecasting system demands its mathematical model(s) to handle highly complex rainfall runoff processes. Uncertainty in real time flood forecasting will involve a variety of components such as measurement noises from telemetry systems, inadequacy of the models, insufficiency of catchement conditions, etc. Fuzzy technology has a great potential to tackle the uncertainty problems in this field.

2. THE CATCHMENT

The data used for this paper were collected in a region called Bird Creek in the USA. The data set is divided into two parts: a calibration (training) period and a verification (testing) period. The daily rainfall values were derived from 12 rain gauges situated in/near the catchment area. The river flow values were obtained from a continuous stage recorder. The period used for model calibration spanned some eight years from October 1955 to September 1963, and the verification period ranged from November 1972 to November 1974. During the calibration period the discharge at the basin outlet ranged from 0 to 2540m³/s and rainfall up to 153.8mm/day. The highest recorded discharge during the verification period was 1506m³/s [Hajjam 1997].

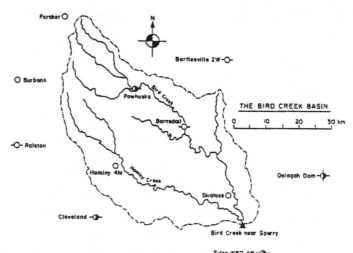

Fig.1: The Bird Creak drainage base (Source Georgakakos et al 1988)

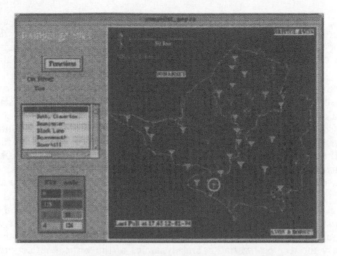

Fig.2: A real time rain gauge telemetry systrem in SW England [Han et al 2000]

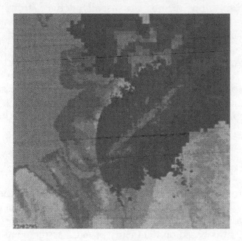

Fig.3: Rainfall observation by weather Radars [Han et al 2000]

The bird Creek catchment covers an area of 2344km^2 and located in Oklahoma close to the northern state border with Kansas. The outlet of the basin is near Sperry about ten kilometers north of Tulsa. The catchment is relatively low lying with altitudes ranging from 175m up to 390m above the mean sea level. There are no mountains or large water surfaces to influence local climatic conditions. Some twenty percent of the catchment surface is covered by the forest while the main vegetative cover is grassland. The storage capacity of the soil is very high [Georgakakos et al 1988]. The river basin and the stream network are shown in figure 1. The catchment receives significant rainfall in most years, and the catchment can be classified as humid although extended periods with very low rainfall can occur.

3. FUZZINESS IN HYDROLOGICAL SYSTEMS

Hydrological system is dynamic and fuzzy. The contributing factors are difficult to evaluate and measure. Traditionally, rainfall records were obtained by raingauges that could produce up to 10% measurement errors due to a variety of causes, such as wind, splashing, trees, surrounding areas, etc. The catchment average rainfall is usually calculated by Thiessen

855

polygon method, or more complicated mathematical processes (such as polynomial/Spline surface fitting or Kriging, etc). In the real time flood forecasting situations, the telemetry raingauge network is usually quite sparse (about 1 in 250 sq km^2 in SW England, see Figure 2) and the derived catchment average rainfall is an uncertain number due to spatial sampling problem.

To overcome the spatial sampling problems caused by sparse raingauge networks, more gauges are required but in reality, due to economic reasons, it is impractical to set up a dense raingauge network for operational real time flood forecasting purposes. One solution is to use remote sensing technology such as weather radars to provide high resolution rainfall measurements in space and in time. Some modern hydrological radars can achieve high definition rainfall measurements up to 250 by 250 m in space over a large area (50km to 76 km in radius) and one of the systems developed at Bristol is illustrated in Figure 3.

However, radar measurement of rainfall is not perfect and could still suffer from a variety of detrimental factors, such as variations in the relationship between back scattered energy and rainfall rate, effects of variation in precipitation with height, anomalous propagation of the beam, etc. Figure 4 illustrates a vertical profile of precipitation measured in Manchester. The data is in the form of a height time image (HTI) with a resolution of 7.5 m in the vertical and 2 seconds. This shows clearly the temporal variation and dynamics of the vertical reflectivity profile in general during this event thus emphasising the difficulties in determining what a scanning radar is actually 'seeing' at any given time or place [Tilford, Han and Cluckie 1995].

On the other hand, stream flow records are usually derived from hydraulic measuring structures, such as weirs. These flow records are more accurate than rainfall measurements, however discrepancies still exist and uncertainties in the records are inevitable.

Traditional hydrological modelling tends to treat the input and output as definite values and lacks of effective means to tackle the uncertainties in the rainfall and flow measurements. Nowadays more and more attentions have been focused on uncertainty issue in real time flood forecasting. Fuzzy technology could take a more active role in this field and in this research, a powerful fuzzy programming tool called FRIL is adopted to explore the potential benefits of using fuzzy theory to improve the performance of operational real time flood forecasting systems.

4. FUZZY PROGRAMMING LANGUAGE FRIL

Fril was originally an acronym for "Fuzzy Relational Inference Langauge", which was the precursor of the present Fril language, which was developed in the late 1970's following Jim Baldwin's work on fuzzy relations in the Artificial Intelligence group of the university of Bristol [Baldwin 1995]. Fril is based on the theory of mass assignments that express families of probability distributions and generalise the representation of both probabilistic uncertainty and fuzzy sets. The mass assignment is related to the basic probability assignment of the Shafer Dempster theory, but the algebra of mass assignments is quite different to the Shafer Dempster method of combining evidences. A special case of the mass assignment is the support pair which defines an interval containing the unknown probability. The support pair is a fundamental form of representation of uncertainty in Fril. A fuzzy set is interpreted as a possibility distribution which is equivalent to a family of probability distributions and hence to a mass assignment. Both continuous and discrete fuzzy sets can be used in Fril.

Fril can deal with uncertainty in data, facts and rules using fuzzy sets and support pairs. In addition to the Prolog rule, there are four different types of uncertainty rule as follows:

Basic Fril rule; Extended Fril rule; Evidential logic rule; and Causal Rule.

The extended rule is important for causal net type applications, and the evidential logic rule is relevant to case-based and analogical reasoning. Each rule can have associated conditional support pairs, and the method of inference from such rules is based on Jeffrey's rule which is related to the theorem of total probability. Fuzzy sets can be used to represent semantic terms in Fril clauses, and support for Fril goals can be obtained by a process of partial matching of such fuzzy terms called semantic unification. Fril implements a calculus of support logic programming, which defines the method of computing support pair inferences. Fril rules can also implement Fuzzy Control knowledge simply and directly.

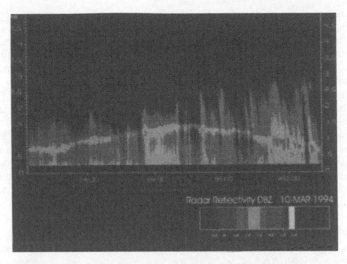

Fig.4: A Typical Height Time Image Obtained from a Vertically Pointing Radar Located in Manchester [Tilford, Han and Cluckie 1995]

Fril Language Components are: Support Logic/Prolog combination language, Declarative programming style; List-based syntax; Portable 'C' implementation; High performance Prolog; Incremental compilation of clauses; Debugging tools; On-line help; Uncertainty rules at the object level of the language; Easy development of expert systems; Modular code development and optimisation; Sophisticated window and menu interface, including dialogue boxes; Fril can call functions written in procedural languages; Fril can be called from procedural code (e.g. a 'C' program); Fril can generate stand-alone applications.

5. FUZZY MODELLING

The mass assignment interpretation [Lawry, Baldwin and Martin 1996] is used to classify the semantic function by a voting model. The membership value is then taken to be the proportion of the voters who include each word as a subset from the finite set of the term set. This concept is used to classify the data of both rainfall and river flow into representing fuzzy sets. For example, rainfall values are split up into five overlapping fuzzy sets described by labels such as: very low, low, medium, high and very high. Both the number of fuzzy sets used and the degree of overlap can be varied noting that an increase generally affects the level of interpolation (usually increasing complexity). The fuzzy sets used take the shape of trapezoid but can be formatted to preference.

The modelling process is based on ID3 type algorithm as proposed by Lawry et al [1996]. This process offers a method which generates a decision tree based on conditional probabilities. The conditional probabilities are relative to each particular branch of the decision trees, which are linked through IF-THEN type statements: i.e.

IF the rainfall today is *medium*
    AND IF the rainfall yesterday was *very high*
        THEN it is *highly likely* that the river flow today will be *high*

The above expression may be concluded from a branch of the decision tree with a conditional probability like:

Pr(River flow today *high*| Rainfall today *medium* AND Rainfall yesterday *very high*)=0.95
Essentially both the river flow output and the conditional probabilities are fuzzy concepts where a probability of 0.95 has a high membership value in the 'highly likely' fuzzy set.
Since the universe of the predicted output (river flow) is fuzzy, it is necessary to aggregate the results based on the relative conditional probabilities since any one output may fall in a number

857

or even all of the fuzzy set. This aspect is also a component of the approach adopted, which offers to weigh values by proportioning the probability that the particular output is chosen multiplied by the probability that the respective fuzzy set labels are appropriate [Baldwin, Lawry and Martin 1998]. For the prediction to produce an output, a process of 'defuzzification' is necessary. This is done simply by weighting the river flow values according to their respective probabilities and the average fuzzy set value for each distinctive fuzzy set (i.e. the centroidal value which best suits the linguistic description).

In this research, the ID3 algorithm [Lawry et al 1996] is adopted to generate a set of classification rules from the data, based on Shannon's measure of entropy, in the form of a traditional decision tree for a collection of data related to a number of classes. Features are selected on the basis of maximising the expected information gain of evaluating them as quantified by Shannon's measure. The decision tree is a combination of nodes (corresponding to features) and branches (corresponding to particular feature values). The prime aim of the ID3 algorithm is to construct a tree compromising the shortest route for checking features for classification. An iterative method is conducted ranking the features according to their effectiveness in partitioning the set of defined classes.

Therefore, the fuzzy sets for rainfall and stream flow are

For Rainfall: (Over a domain from 0 to 50)
F(very low) = [10:1, 16.5:0]
F(low) = [3.5:0,10:1,20:1,26.5:0]
F(medium) = [13.5:0, 20:1, 30:1, 36.5:0]
F(high) = [23.5:0,30:1, 40:1, 46.5:0]
F(very high) = [33.5:0, 40:1]
For streamflow: (over a domain from 0 to 6500)
F(very low) = [73:1, 105.15:0]
F(low) = [0:0, 73:1, 137.3:1, 189.05:0]
F(medium) = [105.15:0, 137.3:1, 240.8:1, 580.25:0]
F(high) = [189.05:0, 240.8:1, 919.7:1, 6500:0]
F(very high) = [580.25:0, 919.71]

Further actual and predicted values are compared quantitatively and qualitatively to demonstrate the power of the fuzzy system as a prediction tool.

Three formats were chosen for the fuzzy modelling structures [Karbassioun and McQueen, 1999]:

1) Exponential rainfall time series;
2) 10 consecutive rainfall time series with self-selection;
3) 10 consecutive, including feedback.

The simulation results are summarised in table 1. The analysis shows that the best accuracy based on the flood peak and timing is achieved when the system is given a time series of ten consecutive preceding days. Utilising the ID3 algorithm, the system selects the four rainfall values sought most significant on building a powerful decision tree. Hydrologically it is noted that when the program selects the four most influential inputs from ten preceding rainfall values, the model regards those values four, five or six days as providing more information than the more recent rainfall values. This feature illustrates the glass box nature of fuzzy logic methods where the model provides feedback on what method is undertaken in classification. This information can thus lead to highlight the values relevant to an ideal training set to optimise further accuracy on prediction.

The above comparison shows that introduction of a feedback function does increase accuracy but consequently reduces the warning time offered by the prediction. One solution may be to incorporate a feedback term for stream flow two or three days before rather than the day before. The extra stream flow information gives the system the classification attributes it needs to cope with an abundance of zero rainfall data.

Table 1: Statistical comparison of the three main approaches

Bird Creek Data Training: 700day Testing: 150 days	Format 1 Exponential time steps	Format 2 10 consecutive steps	Format 3 Incorporation of a feedback term
Approximate prediction time interval	1 day	2-4 days	1 day
Percentage error of output universe on training	12.3%	5.74%	4.55%
Percentage error of output universe on testing	13.8%	9.7%	7.51%

6. CONCLUSIONS

This paper explores the possibility of using fuzzy technology to model river flows for real time flood forecasting. There are a variety of uncertainties in rainfall and stream flow measurements, in spite of the advances of modern telemetry technology, and it is difficulty to treat these uncertainties using traditional deterministic methods. Potentially, fuzzy theory could be very useful to tackle the uncertainty problems in this field. It was found that the fuzzy programming language FRIL is a powerful tool in river flow modelling and further research is needed to combine it with other tools, such as neural networks, support vector machines and genetic algorithms to form an integrated artificial intelligent river flow modelling sytem.

REFERENCES

Baldwin, J.F , T.P. Martin and B.W. Pilsworth , Fril : fuzzy and evidential reasoning in artificial intelligence, Taunton : Research Studies Press , c1995

Baldwin, J.F., Lawry, J. and Martin, T.P. (1998) A note on the conditional probability of fuzzy subsets of a continuous domain, Fuzzy Sets and Systems 96 211-222

Bardossy A, Bronstrert A, Merz B (1995) 1 Dimensional, 2 Dimensional and 3 Dimensional Modelling of Water Movement in the Unsaturated Soil Matrix using a Fuzzy Approach, Advances in Water 18: (4) 237-&

Bardossy A (1996) The use of fuzzy rules for the description of elements of the hydrological cycle, Ecological Modelling, 85: (1) 59-65 FEB

Campana NA, Mendiondo EM, Tucci CEM (1995) A multisource approach to hydrologic parameter estimation in Urban basins, Water Science and Technology, 32: (1) 233-239 Campana NA, Mendiondo EM, Tucci CEM (1995) A multisource approach to hydrologic parameter estimation in Urban basins, Water Science and Technology, 32: (1) 233-239

Cheng CT (1999) Fuzzy optimal model for the flood control system of the upper and middle reaches of the Yangtze River, Hydrological Sciences Journal, 44: (4) 573-582 AUG

Karbassioun, D and McQueen, M (1999), Artificial Intelligence Application to Flood Forecasting,April 1999, Dept of Civil Engineering, University of Bristol

Georgakakos K.P., Rajaram, H. and Li, S.G. (1988) On improved operational hydrological forecasting of stream flows. IAHR report No. 325

Hajjam, S. (1997) Real time flood forecasting model intercomparison and parameter updating using rain gauge and weather radar data, PhD thesis, Telford Research Institute, University of Salford, UK

Han, D. et al , Rainfall measurement over urban catchment using weather radars, Journal of Urban Technology, April 2000

Lawry, J., Baldwin, J.F. and Martin, T.P. (1998) Mass Assignment Induction of Decision Trees on words, AI Group, University of Bristol, UK

See L, Openshaw S (1999) Applying soft computing approaches to river level forecasting, Hydrological Science Journal, 44: (5) 763-778 OCT

Stuber M, Gemmar P (1997) An approach for data analysis and forecasting with neuro fuzzy systems - Demonstrated on flood events at River Mosel, Computational Intelligence, 1226: 468-477

Tilford KA, Han D and Cluckie, I.D.(1995) Vertically Pointing and Urban Weather Radars, In Tilford, K.A. (ed.) *Hydrological Uses of Weather Radar*, British Hydrological Society Occasional Paper No. 5, pp.147-164, (Wallingford, Institute of Hydrology)

Yager RR and L.A. Zadeh, Fuzzy sets, neural networks, and soft computing Imprint, New York : Van Nostrand Reinhold , c1994

Zadeh L.A. (1965) Fuzzy Sets. Information and Control 8:338-353

Zeigler BP, Moon Y, Lopes VL, Kim J (1996) DEVS approximation of infiltration using genetic algorithm optimization of a fuzzy system, Mathematical and Computer Modelling, 23: (11-12) 215-228 JUN

Zimmermann H. –J. (1996) Recent Developments in Fuzzy Logic and Intelligent Technology,Fuzzy Logic, Editor J.F. Baldwin, John Wilwy & Sons Ltd

Stochastic Hydraulics 2000, Wang & Hu (eds) © 2000 Taylor & Francis, ISBN 90 5809 166 X

Two-dimensional hydrodynamic modelling of Dongting Lake

Hua Zhang, Wes Dick & Yujuin Yang
AGRA Earth and Environmental Limited, Calgary, Alb., Canada

Raymond S. Pentland
Water Resource Consultants, Regina, Sask., Canada

Zhou Beida, Gu Qingfu & Liu Dongrun
Hunan Hydrology and Water Resources Bureau, Changsha, China

Zhang Zhenquian
Hunan Hydro and Power Department, Changsha, China

ABSTRACT: This study was part of the Dongting Lake Flood Management Project. The objective of the project was to reduce the economic and social cost of flooding damage in the Dongting Lake area by developing a real-time flood forecasting system as a basis for improved flood warnings. The system predicts tributary flows and lake levels in order to provide flood warnings. The components of the project consist of rainfall-runoff modeling, hydrodynamic modeling, reservoir operation modeling, development of a decision support system, and the use of Geographic Information Systems (GIS). Dongting Lake is hydraulically very complex with lakes, polders and network of rivers all located in a wide alluvial plain. The Lake has four major tributaries and is connected to Yangtze River by five main channels. A two-dimensional hydrodynamic model was developed to study the Dongting Lake system. The model used one-dimensional simulation in the river reaches and two-dimensional simulation in the lake. It was determined that the model can accurately reproduce historical floods. It reproduced water levels and flows at twelve key stations in the project area. Some of the problems encountered in applying the model to this project are discussed. The calibrated model was judged to provide a reliable basis for future Dongting Lake flood forecasting.

Keywords: Hydraulic model; Yangtze River; Dongting Lake; lood forecast; Flood control

1. INTRODUCTION

Dongting Lake is the second largest inland lake in China. It lies adjacent to the Yangtze River in Hunan Province. The lake has four major tributaries and is connected to the Yangtze River by a network of channels. Floods occur frequently, either because of flood flows in the Yangtze River or because of rainfall in the Dongting catchment. During a flood event, the lake water level may rise by 10 meters or more. There is considerable development in polders around the lake, protected by dykes of various levels. During flood events, some dykes may be overtopped or breached. A hydraulic forecast model is necessary to predict lake levels in order to manage the lake and upstream reservoirs and to minimize flood damages and loss of life.

This study is part of Dongting Lake Area Flood Management Project which was carried out by AGRA Earth & Environmental Limited of Calgary, Canada in conjunction with Hunan Hydro & Power Department (HHPD) of China. The work was financed by the World Bank. The goal of the project is to reduce flood damages through better forecasting and operation based on numerical modelling.

2. DONGTING LAKE SYSTEM

2.1 Geography

The Dongting Lake area is located south of the Yangtze River in the northeast portion of Hunan Province of China. Dongting Lake has a surface area of 4109 km^2 (2691 km^2 lake area and 1418 km^2

river network area). It is a very complex hydraulic system with a network of rivers, polders and lakes located in a wide alluvial plain. The lake is fed primarily by five river channels from Yangtze River (the Three Mouths) and by the Four Rivers from the south and west in Hunan Province (Xiangshui, Zhishui, Yuanshui, and Lishui Rivers).

The main river channels in the project area are the Yangtze River, the Three Mouths and the Four Rivers. Dongting Lake itself consists of three major lakes identified as the West Lake, the South Lake and the East Lake. The lakes are connected by channels which constrain the flow, resulting in significant attenuation of flood peaks.

2.2 Hydrology

Damaging floods have occurred frequently in the Dongting Lake area. Recent major floods occurred in 1986, 1988, 1991, 1996, 1998 and 1999. Flooding at Dongting Lake can result from floods on the Yangtze River which arise from precipitation events beyond Hunan Province, from floods on the four major tributaries within Hunan or from a combination of these two sources. The Dongting flood season often starts in June and lasts until September.

The typical distribution of flows in the Dongting Lake system based on recent flood years is shown on Figure 1. During these flood years, 22% of the Yangtze River discharge upstream of Dongting Lake entered the lake through the Three Mouths. Of the total inflow to Dongting Lake in these six flood seasons, about 40% came from the Yangtze River and about 60% from the Dongting Lake basin. The Dongting Lake outflow amounted to 43 % of the Yangtze River discharge downstream of the lake at Luoshan.

3. PREVIOUS STUDIES

Starting early in 1990, China launched a series of major studies of midstream and downstream flood forecasts of the Yangtze River under the supervision of the Ministry of Water Resources (MWR) and the Changjiang Water Resources Commission (CWRC).

During 1993 and 1995, Nanjing Hydrological Research Institute (NHRI) developed a numerical model of the Dongting Lake system (NHRI, 1995). The model included separate one-dimensional (1-D) river and two-dimensional (2-D) lake components. Output of the 1-D component was used as input to the 2-D component. The upstream boundaries of the model were located at the Three Mouths and the Four Rivers. The downstream boundary was at Luoshan.

The Danish Hydraulic Institute (DHI) and CWRC developed a hydraulic model of the Yangtze River in 1994 using DHI's MIKE-11 (1-D) hydraulic model (CWRC, 1994). The model extended from Yichang upstream to Wuhan downstream. Dongting Lake was simulated as a 1-D water body.

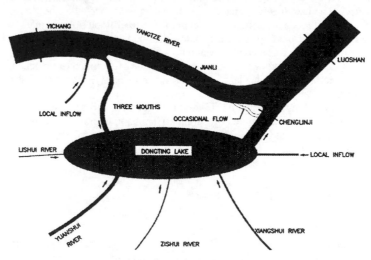

Fig.1: Dongting Lake Water Schematic Diagram

Fig.2: Modelling area of the Hydraulic Model

In 1997, Wuhan Hydroelectric University (WHU) conducted 1-D hydrodynamic modeling of polder dyke breaching operations (WHU, 1997). The model boundaries were set the same as in the NHRI model. Dongting Lake was again simulated as a series of 1-D channel cross-sections. The above models were calibrated but have not been applied to real-time flood forecasting in Dongting Lake.

4. MODEL DESIGN

The hydraulic model simulates the movement of water through the network of channels and lakes which make up the Dongting Lake system. Most of the channel cross-sections have become very regular after years of dyke construction along the channels. They can be appropriately modelled as 1-D reaches. However, a 1-D model cannot accurately model the hydraulics of the three interconnected lakes which make up Dongting Lake, nor the complex connections between the lakes and the tributary channels. Therefore, a 2-D model was used to model the lake and some of the channel linkages. The SMS-RMA2 hydrodynamic model was selected based on its ability to combine 1-D and 2-D hydraulics in a single model.

The boundaries of the hydraulic model are shown on Figure 2. The upstream boundaries are Yichang on the Yangtze River, and Xiangtang, Taojiang, Taoyuan and Shimen on the Four Rivers. The downstream boundary is Luoshan on the Yangtze River. The local inflows, including local catchments, local tributary rivers and polders, were assigned to the rivers and lakes using 40 local inflow boundaries. The Dongting Lake model includes 1-D river channels, 2-D lake areas and features for planned and unplanned polder dyke breaching. The 1-D channels were simplified as trapezoidal cross-sections. The 2-D mesh was constructed based on the 1995 Dongting Lake survey bathymetric data. The average 2-D element size is about 3 km^2. In addition, special units were designed to link river reaches and local inflow boundaries to 1-D rivers, to connect the 1-D river reaches with the 2-D mesh, and to link polders to rivers or lakes.

The model was designed to forecast water levels at 12 key locations as shown on Figure 2: Chenglinji, Nanzui, Xiaohezui, Yuanjiang and Xiangying in the lake area, Changsha, Yiyang, Changde and Jinshi on the Four Rivers; and Shigueshan, Anxiang on Nanxian at the Three Mouths channels. These are locations of active hydrometric gauging stations.

5. MODEL CALIBRATION

The Dongting hydraulic model was calibrated by adjusting the roughness, specified in terms of Manning's n value, for each cross section in the 1-D reaches and each element in the 2-D areas. The 1999 flood data was used for calibration because it represents a high flood condition while there was only one case of polder dyke breaching (comparing to massive breaching in 1996 and 1998) and that case was well documented. To calibrate the model, recorded inflows were applied at the upstream boundaries and recorded water levels at the downstream boundary. Inflows from 27 local basins were estimated using the SSARR hydrological model. The model calibration was separated into several parts, as discussed in the following sections.

5.1 Four Rivers Calibration

The 1-D reaches of the Four Rivers were calibrated independently of the rest of the Dongting Lake system. The calibration was based on the flood data of 1986, 1988, 1991, 1996 and 1998. The reaches were simulated using the observed inflows to the upstream ends of the reaches and the observed water levels at the downstream ends.

In RMA2, 1-D channels are input as trapezoidal sections. The simplified cross sections developed from the surveyed sections were found to simulate the real channel accurately at high water levels. At very low water levels some problems occurred because the bed elevation of a simplified cross section could be different from the actual bed elevation. In general, however, the simplified cross sections provided good results for flood conditions, which are the focus of the model.

The initial model predictions for the Yuanshui and Zhishui Rivers were found to be better than for the Xiangshui and Lishui Rivers. On the Xiangshui and Lishui Rivers the model tended to overestimate water levels during the ascending limbs and underestimate water levels during the descending limbs. This discrepancy is believed to be caused by offstream storage in the channels of several major unmodelled tributaries. Offstream storage was therefore added to the model to account or the tributary channels, resulting in a significant improvement in the simulation.

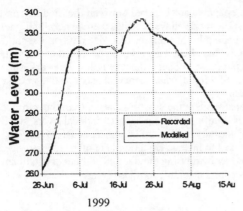

Fig.3: Chenglinji Water Level Calibration

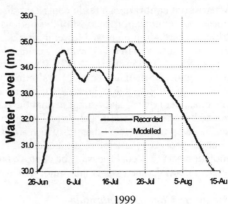

Fig.4: Nanzui Water Level Calibration

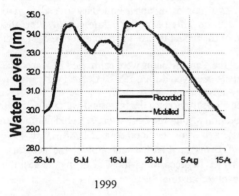

Fig.5: Xiaohezui Water Level

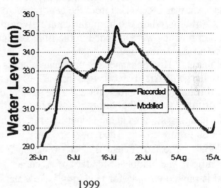

Fig.6: Changsha Water Level Calibtarion

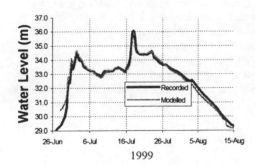

Fig.7: Yiyang Water Level Calibration

Fig.8: Changde Water Level Calibration

5.2 Dongting Lake and Three Mouths Calibration

Dongting Lake and the Three Mouths channels were calibrated in two stages. In the first stage, the upstream boundary of the model was set at Xinjiangke, Shadaoguan, Mituosi, Kangjiagang, and Guanjiapu on the Three Mouths and at Jianli on the Yangtze River. In the second stage, the upstream boundary was extended to Yichang on the Yangtze River. The lake system was calibrated using 1999 flood data and verified with 1988, 1991 and 1996 flood data.

865

After initial calibration, a fairly consistent discrepancy was observed between the modeled and recorded water elevations at most of the stations. The modeled water levels were generally higher during the ascending limbs and lower during descending limbs than the recorded data, in the same way as was observed in the Four Rivers calibration. This effect has also been observed in previous studies (NHRI, 1995a; WHU, 1997). The discrepancy was observed not only for the years when massive polder breaching occurred, but also for the 1999 flood, when there was only one polder dyke breach, suggesting that dyke breaching is not the cause.

The consistent discrepancy suggests that the real storage in the system is larger than the storage simulated in the numerical model. In reality, there is significant storage capacity in local rivers, small lakes, polders and diversion channels which are outside the extent of the lake surveys used to develop the model geometry. Therefore the model was revised to include Dongting lake offstream storage. Simulation accuracy could possibly be improved further by revising the size and number of offstream storage units if detailed topographic information becomes available for those areas.

5.3 Yangtze Channel Calibration

There was less data available for the Yangtze River channel and the upstream ends of the Three Mouths than for the reaches included in the first stage. The cross section spacing was greater, and the survey data was older. Therefore the calibration results after moving upper boundary to Yichang are not as good as when the boundaries are at Jianli and on Three Mouth channels. However, the upstream reaches were included in the final version of the model to provide a complete model with an increased warning time.

The Yangtze River channel and upstream reaches of the Three Mouths were calibrated to reproduce the discharge distribution between the various channels observed during 1999. Correct distribution of the discharge is important for accurate simulations downstream.

The primary goal of the calibration was to have the model reproduce the historical water levels at the twelve gauging stations in the Dongting area. Calibration results for 1999 at six selected stations are shown on Figures 3 - 8. The model accuracy was assessed using the flood data of 1988, 1991, 1996 and 1999. The Coefficient of Determination (COD) was used as the indicator of the accuracy. According to the National Standard of Accuracy for Flood Forecasting Model of China (1985), the model is ranked Class A for Dongting Lake water level forecasting.

5.4 Forecasting Accuracy

The model was tested in forecast mode for lead times of one, two and three days. The difference between the calibration mode and forecast mode was that the downstream water level is unknown in forecast mode. The downstream water level is therefore estimated using a variable stage-discharge rating curve. A variable curve is necessary because of significant hysteresis in the observed stage-discharge relationship for the Yangtze River downstream of Dongting Lake.

The model accuracy in forecast mode was assessed according to the national standard of China (Ministry of Water Resources and Electrical Power of People's Republic of China, 1985). The standard identifies three classes of accuracy, depending on the value of the coefficient of determination, which is defined as:

$$d_y = 1 - \frac{\Sigma(Y_i - Y)^2}{\Sigma(Y_i - \bar{Y})^2} \qquad (1)$$

where: Y_i is the observed value,
  Y is the forecast value, and
  Y is the mean of the observed values

The overall coefficient of determination in terms of water levels was over 0.95 for all three lead times, making it a Class A model.

6. CONCLUSIONS

The RMA2 two-dimensional hydrodynamic model is an appropriate tool for simulating the complex one- and two-dimensional hydraulics of the Dongting Lake system. RMA2, as implemented in the Surface-water Modeling System (SMS) was applied to the lake system with excellent results.

In forecast mode, the model accuracy in terms of water levels is Class A according to the Chinese national standard, indicating that it is acceptable for use in real-time modeling.

The model will be useful for real-time flood forecasting and operation, and also provides a valuable tool for planning and optimizing flood management measures, and assessing the benefits of flood control strategies such as polder dyke breaching.

ACKNOWLEDGMENTS

This project was commissioned by the Hunan Hydro and Power Department, and was financed by the World Bank.

REFERENCES

Brigham Young University (BYU), 1997. SMS Surface-Water Modeling System Reference Manual, Version 5.0. March, 1997. Engineering Computer Graphics Laboratory, BYU.

Ministry of Water Resources and Electrical Power of People's Republic of China, 1985. Guidelines for Hydrological Information Forecasting. SD-138-85.

Nanjing Hydrological Research Institute (NHRI), 1995a. Yangtze Protection System Modeling (Zicheng to Luoshan)

Nanjing Hydrological Research Institute, 1995b. Yangtze River Flood Forecast Decision Support System.

WES, 1996a. Users Guide to RMA2 - Version 4.3, February, 1996, US Army Corps of Engineers, Waterway Experiment Station - Hydraulic laboratory.

Wuhan Hydroelectric University (WHU), 1997. Numerical Modeling of Dongting Lake Flood Polder Operation

Changjiang Water Resources Commission, 1994, Numerical Modeling of Yangtze Midstream Flood Forecast and Operation (Report).

Stochastic Hydraulics 2000, Wang & Hu (eds) © 2000 Taylor & Francis, ISBN 90 5809 166 X

Embedding deterministic models into a stochastic environment: A discussion on the numerical accuracy

J.W. Stijnen, M. Verlaan & A.W. Heemink
Delft University of Technology, Netherlands

ABSTRACT: Monte Carlo simulation is becoming increasingly popular in areas with oceonographic and atmospheric applications. Examples are risk and uncertainty analysis. Usually what is done, is that a stochastic part is added to an already existing, deterministic model, providing the user with error bounds of the model outcome. The underlying deterministic model is often a discretization of a set of partial differential equations describing physical processes such as transport, turbulence, buoyancy effects, and continuity. A lot of effort is also put into numerically efficient schemes for the time integration. The resulting model is most of the time quite large and complex. In sharp contrast with this, the stochastic extension used for Monte Carlo experiments is usually achieved by adding white noise. Unfortunately the order of time integration for the complete stochastic model is reduced, because white noise is not smooth enough. Instead of completely replacing the old numerical scheme and implementing a higher order scheme for stochastic differential equations, we suggest a different approach that uses the existing scheme. The method uses a smooth colored noise process to obtain a higher order of convergence. The above idea was successfully tried on a stochastic version of the highly non-linear Lorenz model, where we performed an uncertainty analysis to gain insight in the accuracy of the model predictions. In the near future the method will be applied to real-life problems.

Key words: Stochastic differential equation, White noise, Colored noise, Convergence

1. INTRODUCTION

White noise is an important process in many applications involving some form of error prediction. In these types of applications the deterministic model is often already available. That includes the numerical scheme, which most likely will have second, or even higher order of convergence. When extending the existing models for some form of error estimation, often only a minor modification of the existing code is required. In many cases only white noise is added. Unfortunately the addition of (linear) white noise to the numerical model reduces the order of convergence for the system as a whole. The impact of the addition of white noise is often underestimated.

By using colored noise instead of white noise, the order of convergence can be restored. The additional computations are most of the time negligable compared to the total amount. Some care needs to be taken, however, when using this method, because in some cases the approximating sequence does not converge to the correct solution.

When using a linear model, there is no difference between the Itô and Stratonovich interpretations. However, since Monte Carlo simulation is specifically used in combination with strongly non-linear models, this difference in interpretation is important. By introducing a small correlation in the noise process, both interpretations coincide. This would seem to solve the problems on both mathematical and physical accounts.

In the next section we will discuss the idea in greater detail while we continue afterwards with a more specific example. Next, we apply the beforementioned idea to a stochastic version of the Lorenz equations. Results will be discussed and we will end with some concluding remarks.

2. THE CONCEPT

Suppose that we have available some deterministic model (either linear or non-linear) described by the following set of equations:

$$\frac{d\mathbf{X}_t}{dt} = f(\mathbf{X}_t, t) \tag{1}$$

where $\mathbf{X}_t$ is the state vector, and $f(\mathbf{X}_t,t)$ is a function completely describing the dynamics of the system. When the model is extended by adding white noise, the equations change to

$$\frac{d\mathbf{X}_t}{dt} = f(\mathbf{X}_t, t) + g(\mathbf{X}_t, t)\mathbf{W}_t \tag{2}$$

here $g(\mathbf{X}_t,t)$ is a function specifying the amount of noise, and $dW_t = W_T dt$ is a Brownian motion.

Assume next, that the original system (1) was solved using a kth order scheme. Unfortunately, after the addition of the white noise term to this equation, the order of convergence for the complete system will be reduced to first order.

White noise works well theoretically, because with white noise the Markov property of the system is maintained. Furthermore it is quite easy to use as forcing in discrete time systems. Looking a bit further, however, we realize that the noise term that we added is nowhere differentiable (even worse: it is not even continuous). This is the core of the problem, since the original scheme that we built into our model (of order k) needs functions to be at least k+1 times continuously differentiable. If this would not be the case, the original scheme, might not have that kth-order convergence. Because we added a non-differentiable noise term to our deterministic system, our higher order convergence is lost and reduced to first order, which is in general the best a stochastic numerical scheme can do with linear white noise.

The overall order of convergence will be reduced to that of the lowest order. The problem obviously lies in the additon of the white noise term. The most common way to work about this problem would be to implement a higher order stochastic scheme. This is not as trivial as it seems, because of the difference in interpretation between different stochastic differential equations and the physical world. The two most common interpretations come from Itô (1942) and Stratonovich (1950's). The two interpretations of Itô and Stratonovich have been extensively researched for different higher order schemes, the relationship between them and their connection to the ' physcical world (see for example Kloeden 1995, Jazwinsky 1970 or Milstein 1995). These two different interpretations are of no concern as long the noise term is linear, because then they are equivalent. Since many applications are specifically used in combination with a strongly non-linear model, this difference in interpretation becomes important. Another problem of interest is that the solution of a stochastic differential equation (SDE) does not always converge to the correct physical solution (see for example Bagchi(1993)).

The problem with higher order schemes for SDE's is that the schemes get quite complicated (especially in more dimensions), and Rümelin1982 has shown that generally the maximum order of convergence concerning the quadratic mean of the one step error is of order O(h3).

A completely different point of view is adopted when we take a closer look at what seems to lie at the base of the problem: white noise. Is it possible to use some form of time-correlated noise perhaps? If so, it should be smooth enough so that the order of convergence for the system as a whole does not change of course. Additionally, it should preferably have the same behaviour as white noise. When performing time-integration on white noise, we obtain the well known Brownian motion process. This process is continuous, but still not differentiable. In order to get a k-times differentiable process we would need to integrate the white noise process k+1 times. For example a special case of an Ornstein-Uhlenbeck process (Gardiner 1983) can

870

accomplish this. This process, which we shall call for now, is the solution of a scalar Langevin equation:

$$\dot{U}(t) = -\frac{1}{a}u(t) + \frac{1}{a}W(t) \tag{3}$$

where a is a characteristic timescale specifying the time over which the process stays significantly correlated, and W(t) is a Gaussian white noise process. The SDE (3) generates a zero-mean, stationary, exponentially correlated, Gaussian process, often called colored noise for short. Since both terms contain the scaling parameter and it is part of the family of first order autoregressive models, we shall call it a scaled autoregressive model, or SAR(1) for short. The SAR(1) process has the pleasant property (in contrast with white noise) that it is continuous, although it is still not differentiable. The idea now, is to couple k+1 different SAR(1) processes as follows:

$$\dot{U}(t) = -\frac{1}{a}u_1(t) + \frac{1}{a}W(t) \tag{4a}$$

$$\dot{U}_2(t) = -\frac{1}{a}u_2(t) + \frac{1}{a}u_1(t) \tag{4b}$$

$$\vdots = \vdots \quad + \quad \vdots$$

$$\dot{U}_{k+1}(t) = -\frac{1}{a}u_{k+1}(t) + \frac{1}{a}u_k(t) \tag{4c}$$

where $u_i(t)$ is some process driven by the Gaussian white noise process W(t) and a as above. Notice that in the limit where the time-correlation approaches zero, i.e. $a \to 0$, all $u_i(t)$ equal white noise processes.

Let us take a look at a more specific case. Assume our model is for now a linear one, described by the following process:

$$dX = -\frac{1}{m}Xdt + dW, \tag{5}$$

where X is some stochastic process, m is the characteristic time-scale of the model, dt is the timestep, and W is a Gaussian white noise process with variance equal to s2. The goal is to obtain a second order of convergence for this stochastic equation. Instead of worrying about higher order stochastic schemes, we make the model deterministic so we can solve it with an ordinary scheme. We use correlated noise and extend the model with the following set of SAR(3) equations

$$dY = -\frac{1}{m}Ydt + u_3dt, \tag{6}$$

$$du_1 = -\frac{1}{a}u_1dt + \frac{1}{a}dw \tag{7a}$$

$$du_2 = -\frac{1}{a}u_2dt + \frac{1}{a}u_1dt \tag{7b}$$

$$du_3 = -\frac{1}{a}u_3dt \quad \frac{1}{a}u_2dt \tag{7c}$$

Generally speaking, the model equation will be very large compared to the additional set of SAR-equations. Therefore the additional computational costs will be minimal. The good news is that equation (6) is now indeed a deterministic one. Which means that this equation can be solved using any conventional second order numerical scheme such as the Heun scheme. An interesting aspect comes up when we notice that as a approaches zero, the time-correlated noise becomes white. This in turn means that in the limit the deterministic equation 6 changes to the stochastic one (equation 5). However, the second order convergence is lost when applying a second order scheme on equation 5. To conclude: somewhere along the way when taking the limit for we loose an order of convergence.

871

3. CALCULATING THE ERROR

When we want to get some margin for the error we make when using the expanded time-correlated system (6) instead of the white noise one (5), we can break down that error into the summation of three parts: firstly the discretization of the system equation, secondly the discretization of the SAR(3) process and thirdly the approximation of the white noise with colored noise.

Note that when using numerical techniques on the SAR process, it is necessary to use a timestep ds that is much smaller than the timestep dt used in the model equation itself. The first order term in the error will be then be small in comparison with the second order term, thus governing the desired second order behaviour. Realize also that the timesteps should fulfill the following relation with respect to the characteristic timescales: a>ds and m>dt.

The first part will be something of the form C1dt2, while the second one will behave as C2ds, with C1, C2 constants. The part we are most interested in however, is the third part: in what way does the solution of equation 6 converge to equation 5?

In order to find out how this happens, we found that probably the simplest way was to look at the power spectral density functions. From the convolution property of the Fourier transform it follows that for a SAR(3) process the Fourier transform is equal to

$$U(w) = H(\omega)W(\omega) = ((\frac{1}{ia\omega + 1})^3 W(\omega) \tag{8}$$

where H(w) is called the frequency response, and i2=-1. The Fourier transforms of the two different systems can now be described as follows:

$$X(\omega) = \left(\frac{m}{im\omega + 1}\right)W(\omega) \tag{9}$$

$$Y(\omega) = \left(\frac{m}{im\omega + 1}\right)U(\omega) \tag{10}$$

where U(w) is given in equation (8). We can now write down the Fourier transform for the error R(w):

$$R(\omega) = X(\omega) - Y(\omega) = \left(\frac{m}{im\omega + 1}\right)V(\omega) \tag{11}$$

$$\text{With } V(\omega) = (1 - (\frac{1}{ia\omega + 1})^3)W(\omega) \tag{12}$$

The power spectral density fRR(w) of the error is the Fourier transform of the autocovariance function CRR(t) of that error (see for example Mortensen (1987)). The autocovariance function is defined as

$$C_{RR}(\tau) = E\{R(t + \tau)R(t)\} \tag{13}$$

The inverse relation between fRR(w) and CRR(t) is characterized as follows:

$$C_{RR}(\tau) = \frac{1}{2\pi} \int_{-\infty}^{\infty} \phi_{RR}(\omega) e^{i\omega\tau} d\omega \tag{14}$$

The variance of the error (t=0) can therefore be found from

$$E\{R^2(t)\} = \frac{1}{2\pi} \int_{-\infty}^{\infty} \phi_{RR}(\omega) dt \tag{15}$$

In this case the driving process of the error is a N(0,s2) distributed white noise process, and therefore the power spectral density of the output satisfies

$$\phi_{RR}(\omega) = H_R(\omega)H_R(-\omega)\sigma^2 = |H_R(\omega)|\sigma^2 \tag{16}$$

where HR(w) is the frequency response function of the error. Combining this relation with the result obtained in (12), we see that the power spectral density function is

872

$$\phi_{RR}(\omega) = \left| (\frac{m}{im\omega + 1})(1 - (\frac{1}{ia\omega + 1})^3 \right| \sigma^2 \qquad (17)$$

Substituting this into (15) and solving the integral results in

$$VAR_{RR} = ((\frac{33}{16}a + \frac{9}{2}\frac{a^2}{T} + o(a^3))\sigma^2 \qquad (18)$$

We see that the standard deviation of the error depends on the square root of the characteristic timescale of the SAR process. In other words this tells us that when we want to halve the error, we would need to take the square root of the timecorrelation factor a to accomplish this. Practically, this means that the timestep ds used in the numerical solution procedure of the SAR process must be very small, since ds<a. This in turn means, that when we make the assumption of white noise in our model, and we try to approximate this with colored noise hoping to obtain a faster convergence, we will be a bit disappointed. In most cases it will probably be more useful to use white noise with a first order approximation and a finer timestep, than it would be to use colored noise with a second order deterministic scheme.

4. AN EXAMPLE: THE LORENZ EQUATIONS

In order to get more familiar with the idea mentioned above, we will apply it to the Lorenz equations. These equations are highly non-linear and given by

$$dX = \sigma (Y\text{-}X) dt$$
$$dY = (\rho X - Y - XZ) dt \qquad (19)$$
$$dZ = (XY - \beta Z) dt$$

where the parameters s=10, r=2, and b=8/3 are chosen to obtain a chaotic solution set. The initial condition is given by (-5.91652,-5.52332,24.5723). Since very small deviations in the parameters or initial condition can have dramatic effects after a few jumps, we limit our solution procedure to one jump. The end time for the simulation is T=0.8. Since the system of equations (20) cannot be solved analytically, we solved it using the second order Heun scheme for a different number of timesteps (ranging from 2 to 216). The solution we found using the maximum number of timesteps is assumed to be the "exact" solution. A measure for the error was based on comparing the numerical solution for different timesteps with the "exact" one. Not surprisingly, when plotting the results of the log error against the log error of the stepsize we get a straight line with slope minus two.

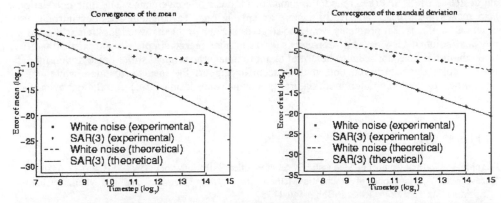

Fig. 1: The absolute error between the "exact stochastic" solution (216 timesteps) and the subsequent numerical solutions with less timesteps (27-215). The endtime was set at T=0.8 (a) Convergence of the mean in case of white noise (slope of the dashed line equals minus one), and with the SAR(3) process (slope of the dashed line equals minus two). (b) Convergence of the standard deviation, again for both white noise and the SAR(3) process.

We would now like to do some error prediction, so we add a standard Gaussian white noise process to obtain the stochastic set of equations

$$dX = \sigma \, (Y\text{-}X) \, dt + DW_1$$
$$dY = (\rho X - Y - XZ) \, dt + Dw_2 \qquad\qquad (20)$$
$$dZ = (XY - \beta Z) \, dt + dW_3,$$

where $(W1.W2,W3) \sim N(0,1)$. We again solve the system with the Heun scheme and proceed the same way as in the deterministic case. An "exact" solution is generated using 216 timesteps. The solutions generated with larger timesteps are based upon the same set of increments as the "exact" solution. Since the noise term is linear we now expect a first order convergence. The results are plotted in figure (1a) for the mean. Indeed, the slope of the dashed line through the dots is equal to one. In figure (1b) the results for the variance are shown.

Next, we extend the model with a SAR(3) process as introduced before and use that as the driving process instead of white noise. The solution obtained with white noise and 216 timesteps is used as the "exact" reference solution. When varying the amount of time-correlation (parameter a), we are able to obtain a second order of convergence for a small, and this gradually moves towards a first order convergence when a gets larger. The first order behaviour should disappear when the stepsize gets smaller. Unfortunately as we have seen, this goes with a factor . Figure (1a) again shows the results for the mean. The dashed line through the dots now has a slope equal to minus two, as does the variance (1b). Clearly, the use of the SAR(3) process results in faster convergence of both the mean and the variance of the solution, than with white noise.

5. DISCUSSION

Although white noise is a widespread and common phenomenon in many engineering applications, it is not always the easiest to work with. Mostly it is added to an already existing model when some form of error estimation is required.

White noise is physically speaking an approximation of a process with extremely short correlation times. A side-effect of adding white noise to an already existing model is, unfortunately, that the carefully constructed higher order scheme of the original model has lost its higher order convergence rate. To circumvent this problem it is suggested to use a wide-band process that approximates white noise. This process is correlated and smooth enough so that the original higher order scheme keeps its higher order convergence. More specifically, it is suggested to use a SAR(k+1) model to maintain a k order of convergence. The time-correlation also makes the process physically a more acceptable one. To conclude: besides introducing practical mathematical problems, white noise gives serious problems when looking at it from a physical point of view as well. Concepts such as infinite energy simply do not exist in nature. Even though it does not seem to be useful to approximate the original white noise driven model with a colored noise driven one, it does seem to support the idea to abandon white noise alltogether. It would be much more natural to assume some form of time-correlation, even if it were small.

REFERENCES

Bagchi, A., Optimal Control of Stochastic Systems Prentice Hall, New York, 1993, ISBN: 0-13-6386105.

Gardiner, C.W., Handbook of Stochastic Methods for Physics, Chemistry and the Natural Sciences Springer-Verlag, Berlin, 1983, ISBN: 3-540-11357-6.

Jazwinsky, A.H. Stochastic Processes and Filtering Theory Academic Press, New York, 1970, ISBN: 0-12-381550-9.

Kloeden, P.E., Platen, E., Numerical Solution of Stochastic Differential Equations Springer-Verlag, New York, 1995, ISBN: 0-387-54062-8.

Milstein, G.N., Numerical Integration of Stochastic Differential Equations Kluwer Academic Publishers, Dordrecht, 1995, ISBN: 0-7923-3213-X.

Mortensen, R.E., Random Signals and Systems John Wiley & Sons, New York, 1987, ISBN: 0-471-634956.

Rümelin, W. Numerical Treatment of Stochastic Differential Equations Siam J. Numer. Anal., Vol. 19, No. 3, June 1982, pp. 604-613.

Stochastic Hydraulics 2000, Wang & Hu (eds) © 2000 Taylor & Francis, ISBN 90 5809 166 X

Monte Carlo simulations in hydrodynamics

Bangmin Zheng & Qingfeng Ji
Wuhan University of Hydraulic and Electrical Engineering, China

Hongwei Fang
Department of Hydraulic Engineering, Tsinghua University and International Research and Training Center on Erosion and Sedimentation, Beijing, China

ABSTRACT: With advance of powerful computers, it becomes more and more available to develop the stochastic method to discrete the convection and diffusion equation and to predict particle motion in hydrodynamics. In this paper some testing cases by using Monte Carlo method to simulate potential flow in irregular grid, free and submerged jet and sediment random movement in the upper pool of pump station have been presented. As well the truncation error and accuracy of this method are analyzed, in which the grid size and the step numbers of random walk are involved. Results show that the Monte Carlo method agrees well with other numerical methods, moreover this method can save the memory of the computer and be easy to extend to two- and three-dimensional calculations.

Key Words: Monte Carlo method, Convection and diffusion equation, Particle motion

1.INTRODUCTION

In general, Monte Carlo Method (MCM) consists of a large number of deterministic simulations where the input series are constructed on the basis of a number of randomly generated parameters with a prescribed probability distribution. The set of results from these simulations is used to determine the statistical properties of the predictions. Both of the determined problems (such as integration equation, algebraic series equation and partial differential equation) and undetermined problems (such as particle random motion) can be solved numerically by MCM. In the research on hydrodynamics the convection and diffusion equation captures most of the transport phenomena of flow and sediment. The governing equation and specified boundary conditions in Cartesian coordinates system read:

$$\frac{\partial \varphi}{\partial t} + u_j \frac{\partial \varphi}{\partial x_j} = a \frac{\partial^2 \varphi}{\partial x_j^2} + S \tag{1}$$

$$\varphi(Q) = f(Q) \tag{2}$$

where $u_j = u_j(P)$; $a = a(P) > 0$; $S = S(P)$; P and Q are the points in the calculation domain Ω and on the boundary respectively.

Equation (1) can be discreted by central difference, upwind and time difference scheme, then a discretization of the convection and diffusion equation is obtained (Chiu,1971;Zheng, 1989):

$$\varphi(P) = [1 + P_0(P)]^{-1} [\sum_{j=1}^{nb} P_{ij} \varphi(P_j) + P_s(P)S(P)] \tag{3}$$

where P_{ij} is the transport probability from point $P = P_i$ to the neighbor nodes P_j and can be evaluated from random walk model; $P_0(P)$ is the probability induced by the time difference term.

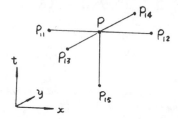

Fig.1: Random walk in time space domain

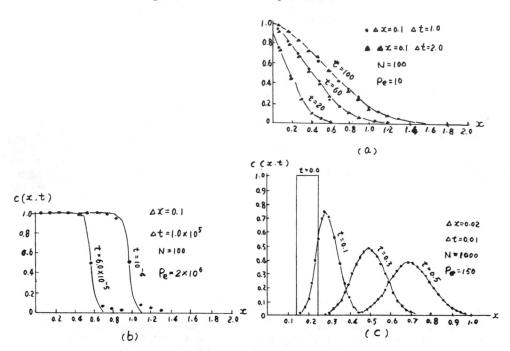

(a)

(b)

(c)

Fig.2: Compute examples for convection and diffusion equation
with different time and space step

The random model can be described as follows: Suppose that a particle starts from point $P=P_0$ in the calculation domain and walks stochastically to its neighbor point every step continuously, as shown in figure 1. Therefore the path of the random walk is $P_0 \rightarrow P_1 \rightarrow P_2 \rightarrow \ldots\ldots \rightarrow P_k = Q$, the model provides that the particle stop until it reaches the boundary. Due to the stochastic movement, the path of the particle is a random variable and can be written as:

$$\xi = g(v_P) = \sum_{m=1}^{k-1} (\prod_{i=1}^{m} W_i) Z_m + (\prod_{i=1}^{k-1} W_i) f(Q) \tag{4}$$

Let $W(P)=[1+P_0(P)]^{-1}$ and $Z(P)=P_s(P)S(P)$, substitute them into equation (3) then yields:

$$\varphi(P) = W(P)[\sum_{j=1}^{nb} P_{ij}\varphi(P) + Z(P)] \tag{5}$$

The relationship of equation (4) and (5) produces the mathematical expectation of the random walk model (Zheng et.al., 1989; Zheng, 1992)

$$E[g(v_P)] = W_0 Z_0 + W_0 \sum_{j=1}^{nb} P_{ij} E[g(v_{pij})] = \varphi(P) \tag{6}$$

878

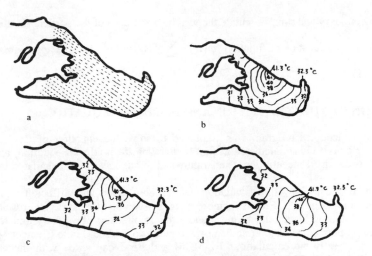

Fig.3: comparison of velocity and temperature fields in cooling pool
of heat power station by Monte Carlo method
(a) Velocity field by MCM   (b) Temperature field by MCM
(c) Temperature by F.E.M.   (d) Temperature by experiment

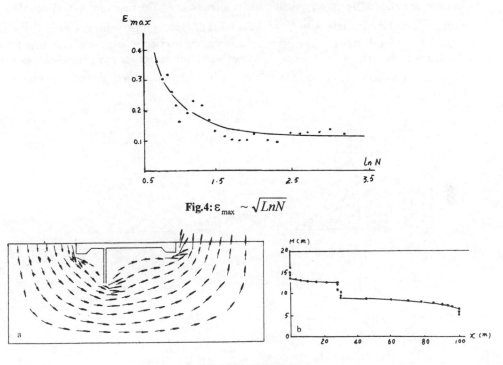

Fig.4: $\varepsilon_{max} \sim \sqrt{LnN}$

Fig.5: MCM in seepage flow
(a) Velocity distribution
(b) Pressure distribution

The above equation shows that the mathematical exception of the random model is equal to the solution of the convection and diffusion equation (1). When the model is performed on the computer, the first step is to produce a series pseudo random number between 0 to1. Then let the particle randomly walk from point P_i to P_j as the transport probability P_{ij}, Finally the particle reaches the boundary and stop. Calculate the weighted averages of the particle on the boundary, we can obtain the evaluation of function $\varphi(P) = \sum_{i=1}^{N} f(Q)n_i / N$. It is the approximate solution of equation (1).

2.SOLUTION OF THE CONVECTION AND DIFFUSION EQUATION

Generally the uncoupled program was employed to solve the equation of flow and scalar, namely the velocity field is calculated at first by iterating the nonlinear equation, and then the scalar (such as density, concentration, temperature and so on) equation is calculated by using the velocity field.

Figure 2a, b and c show the calculations of the convection and diffusion equation with different time and space step. From the figure it can be seen that the numerical results coincide with analytic solution. Figure 3a, b and c show the comparison of velocity and temperature fields in cooling pool of heat power station by Monte Carlo method. Finite Element Method and experiment. Generally the calculations by MCM and FEM can agree with the experiment. However the predicted figure also indicates that the results of MCM agree better than FEM. Moreover we can learn from the calculations that MCM can easily get a stable convergence solution even if the Peclet number Pe= 10^6 or Reynolds number Re= 10^8.

Analyze the calculations, the error and accuracy not only depend on the mesh size (Δx and Δy), but also are affected by the step numbers of random walk N. The relationship between the max error $\varepsilon_{\max}$ and the variable $\sqrt{\ln N}$ of the number of random walk is illustrated in figure 4, in which Δx=0.5 and Δt=1.0. The figure appears that the max error $\varepsilon_{\max}$ gradually tend to a fixed value with increase of the number of random walk. Similar results can be also obtained from reference (Chiu, 1971), but the variable are $\sqrt{N}$ and variance σ instead of variable $\sqrt{\ln N}$.

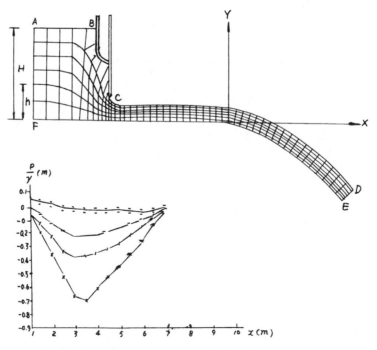

Fig.6: MCM in free surface flow   (a)Grid   (b)Pressure along the Spillway.

0	0	0	0	0	0	0	0	0	0	0
0	0	0	0	0	0	0	0	0	0	0
0	0	0	0	0	0	0	0	0	0	0
0	0	0	0	0	0	0	0	1	0	0
0	0	0	0	0	0	0	0	0	0	0
0	0	1	0	0	0	0	0	0	0	0
0	0	8	4	3	3	1	0	0	0	0
0	1	75	58	24	41	25	3	0	1	1
1000	935	297	304	293	205	140	32	26	6	2
1000	989	521	390	312	241	174	93	41	4	2
1000	950	501	361	290	207	148	27	29	4	1
1	2	11	62	47	35	21	1	1	0	0
0	0	6	5	3	1	0	0	0	0	0
0	0	2	1	0	0	0	0	0	0	0
0	0	0	0	0	0	0	0	0	0	0
0	0	0	0	0	0	0	0	0	0	0
0	0	0	0	0	0	0	0	0	0	0
0	0	0	0	0	0	0	0	0	0	0
0	0	0	0	0	0	0	0	0	0	0

t=5

0	0	0	0	0	0	0	0	0	0	0
0	0	0	0	0	0	0	0	0	0	0
0	0	0	0	0	0	0	0	0	1	1
0	0	0	0	0	0	0	1	1	3	2
0	0	0	1	1	2	1	2	5	9	11
0	1	2	4	3	4	2	9	12	14	21
0	5	6	11	10	21	8	40	49	75	20
0	47	11	39	111	115	121	140	155	141	140
1000	933	350	796	669	630	501	473	416	417	410
1000	944	893	821	709	664	572	532	501	495	415
1000	938	941	805	661	635	501	411	411	419	211
0	51	75	92	105	112	126	136	153	142	135
0	3	5	7	18	24	33	43	32	71	80
0	1	1	3	4	5	7	11	15	10	25
0	0	0	1	1	0	1	2	2	1	1
0	0	0	0	0	0	0	1	1	2	3
0	0	0	0	0	0	0	0	0	0	1
0	0	0	0	0	0	0	0	0	0	0
0	0	0	0	0	0	0	0	0	0	0

t=15

0	0	0	0	0	0	0	0	0	1	1
0	0	0	0	0	0	0	1	1	3	4
0	0	0	0	0	2	1	3	5	9	11
0	0	0	0	2	2	2	6	12	24	35
0	0	0	1	5	3	10	14	35	50	73
0	1	1	2	13	25	38	51	58	92	105
0	3	5	10	70	91	113	123	155	166	174
0	46	74	85	154	175	210	221	235	219	291
1000	945	860	810	680	632	658	622	610	570	293
1000	957	905	832	761	733	699	670	653	614	782
1000	946	858	806	683	669	656	623	607	575	491
0	48	76	90	157	174	201	225	241	277	304
0	2	5	2	73	89	117	126	155	163	173
0	1	2	3	12	23	36	55	39	95	101
0	0	0	1	4	5	9	8	33	49	65
0	0	0	0	1	2	3	5	14	22	34
0	0	0	0	0	1	1	2	6	9	12
0	0	0	0	0	0	0	1	1	4	3
0	0	0	0	0	0	0	0	0	1	1

t=30

Fig.7: (a) Particle motion from a outlet and its distribution

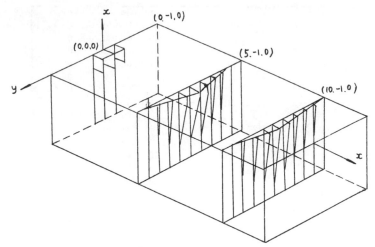

Fig.7: (b) MCM in 3-D jet

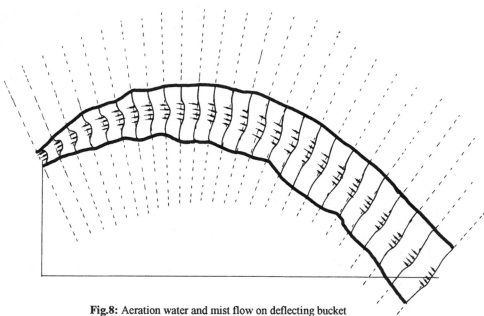

Fig.8: Aeration water and mist flow on deflecting bucket

There are also two tests calculating seepage flow at a gate of underground water (as shown in figure 5) and free surface flow along the spillway (as shown in figure 6). The above two cases are linear problem (linear equation and nonlinear boundary condition) that we just only solve the local grid point near the surface, which people pay close attention to, instead of the whole calculation domain.

3.LAGRANGE PARTICLE MOTION BY MONTE CARLO SIMULATIONS

The application of Monte Carlo method in Lagrange particle motion concerns on the problem of water environment and sediment transport. The fundamental idea is the same to solve the convection and diffusion equation, namely suppose to release a large number of particles (10^3 -

10^6) on a point in calculation domain then follow the trail of particles' motion. Figure 8a and b show the two- and three-dimensional concentration distribution that usually occurs in the outlet of the jet. Reference [5] has also proposed the calculations of sediment transport in sedimentation. The distribution of the aeration water and mist flow on deflecting bucket is illustrated in figure 9. The predicted results follow the law of scalar distribution in the aeration and mist flow.

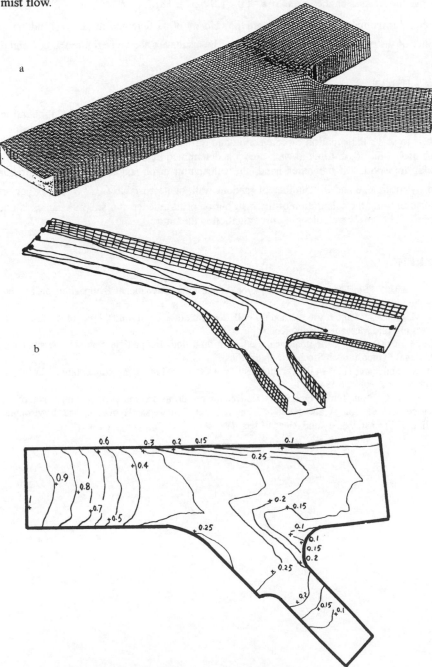

Fig.9: Sediment particles settling in pumping station fore-bay
(a) Computation grid   (b) Path of sediment motion
(c) Computated concentration distribution by using MCM

883

The further development for sediment particle settling and force acting on the particle is two phase flow theory (Zheng and Fang, 1992):

$$\frac{du_{pi}}{dt} = \frac{1}{\tau}(u_{pi} - u_i) + C_i \tag{7}$$

where τ is the stresses on the particle $\tau = (\rho_p + 0.5\rho_f)/18\mu f$; $f = C_D R_{ep}/24$; C_i is the coefficients for the model; u_{pi} and u_i are the velocity of particle and fluid; ρ_p and ρ_f are the density of the particle and fluid respectively. The results are shown in figure 9a, b, c and d.

4.CONCLUSION

Monte Carlo simulation is an important and prospective numerical method in the application of hydrodynamics. In this paper, some testing cases by using the random walk model are performed to solve the equation of convection and diffusion, Laplace and two-phase flow. The transport probability P_{ij} of random walk model is determined by the physical mechanism of flow and scalar transport. On the other hand, the calculation error and accuracy is significantly affected by mesh size and the number of random walk or its variable ($\sqrt{N}$, $\sqrt{\ln N}$ or σ). Effort of improving the calculation of the distribution of turbulence statistic mechanics is under way. This will be surely a challenging investigation in the future.

5.REFERENCES

Chiu C. L. (1971), "Stochastic Hydraulics", Proc, of the First International Symposium on Stochastic Hydraulic, Pittsburgh. U.S.A

Zheng B. M. (1989), "Monte Carlo method for pollution and diffusion in fluid", Proc. of the fourth Asian Congress of Fluid Mechanics, Hong Kong.

Zheng B. M. (1992), "Stochastic numerical method in fluid flow and particle motion", the sixth IAHR. International Symposium on Stochastic Hydraulics.

Zheng B. M. and Fang H. W. (1992) "Stochastic process of sediment in sedimentation Basin", J. of Hydrodynamics, Vol. 4, No. 4.

Zheng B. M., Ji. Q. F. and Lin Q. Q. (1989), "Convection and diffusion problems: A comparison of Finite Difference and Random Walk methods", International Conference Hydraulic and Environmental Modeling of Coastal, Estuary and River Waters. 1989.9.

Stochastic Hydraulics 2000, Wang & Hu (eds) © 2000 Taylor & Francis, ISBN 90 5809 166 X

Stochastic modeling for flood emergency management in urban areas

Marco Ferrante
Department of Water and Structural Engineering, University of Perugia, Italy

Luigi Natale
Department of Hydraulic and Environmental Engineering, University of Pavia, Italy

Fabrizio Savi & Lucio Ubertini
Department of Hydraulics, Transportation and Highways, University of Rome 'La Sapienza', Italy

ABSTRACT: The management of emergency during urban flooding requires different numerical models and risk analysis methodologies, compared to those used for non urban areas. From a hydraulic point of view, the flow propagation in a street network involves uncommon aspects in modeling a two-dimensional phenomenon. A complete risk analysis can involve many buildings of remarkable social and cultural interest, (e.g., hospitals, monuments, schools, etc.), and also facilities and transportation networks reliability analysis. A statistical analysis concerning casualties occurred in flood events during the last century in Italy (Guzzetti *et al.*, 1996) has shown that most casualties occurred in the streets, crossing bridges or driving cars. So, knowledge of the flow characteristics in the street network seems to be necessary for a reliable flood risk analysis in urban areas. In this paper a hydrological and hydraulic models are used to simulate the propagation of flood waves along the Tiber river valley, including tributaries, and to simulate the inundation of the urban area of Rome. As a result, the flow characteristics in the street network are evaluated. Because of the randomness of rainfall, of the uncertainties in the flood hydrograph evaluation and propagation, and hence in the risk analysis, the described process required a stochastic approach.

Key words: Stochastic modelling, Urban flood, Risk analysis, Statistical analysis.

1. INTRODUCTION

In order to map flood-prone areas, the usual practice involves the following steps (Weiss and Midgeley, 1977):
- evaluation of the design peak discharges, for defined values of the return period;
- computation of water stages by integrating the steady, 1-D flow equations;
- individuation of critical sections where flow overtops the levees;
- mapping of flood-prone areas where the elevation of the ground is lower than the computed water stages.

This procedure can not be applied to map the flood-prone areas in Rome when extreme floods propagates along the Tiber river because:
- the volumes of the overtopping waves are of the same order of magnitude of the volume of the flood-waves in the river; therefore, the simulation must be carried out in unsteady-state conditions;
- the number of the stream gauging stations along the Tiber river does not allow to evaluate the spatial variation of the peak discharge;
- Rome can be inundated due to the overtopping of levees upstream the center of the city. Rome is protected by the floods by walls, called "muraglioni", built at the end of 1800. These walls are located along both sides of Tiber river in the center of the city, but they were built more or less 500 m downstream an historical bridge in Rome, "Ponte Milvio", which is first bridge, starting from upstream, in the urbanized area;

- the flood waves in Rome depend on the superimposition of the flood waves released by the regulated Corbara lake and that coming from the tributaries of Tiber river.

As a result, the evaluation of the Rome flooding risk may be carried out considering many scenarios with different flood waves. In this study the flood-prone areas in Rome are mapped following a Monte Carlo simulation which allows to analyse the randomness of the distribution of rainfall in space and in time and the uncertainties in the estimation of the values of the parameters of the mathematical models.

The mapping of the flood prone areas involves the following steps:
- rainfall-runoff analysis, to compute the discharge hydrographs of the Tiber river into the Corbara lake and of all the 33 tributaries of Tiber river;
- simulation of the propagation of the flood waves along the Tiber river, taking into account the inflows due to the tributaries and the outflows due the overtopping of the levees;
- simulation of the flooding of the urban area.

2. DESCRIPTION OF THE WATERSHED AND OF THE HYDROGRAPHIC NETWORK

In this study the watershed of Tiber river, from the Corbara lake to the Tirrenian sea is considered. The catchment area of Tiber river at Ripetta gauging station, in Rome, is 16620 km². Fig. 1 shows the river network downstream Corbara, a reservoir encompassing a volume of 190×10^6 m³. The catchment area at the Corbara dam is 6075 km².

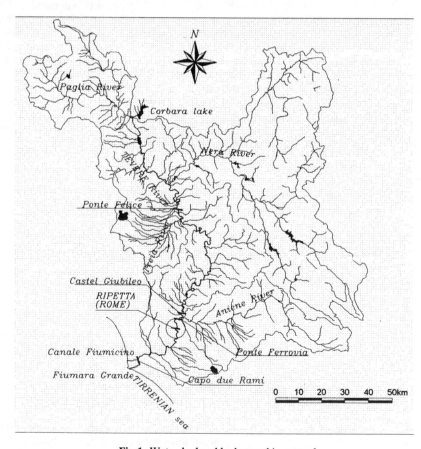

Fig.1: Watershed and hydrographic network

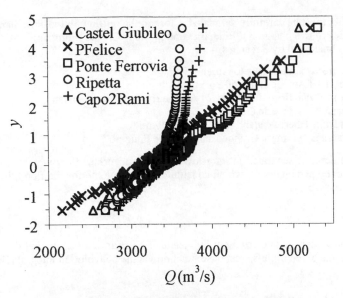

Fig.2: Frequency distribution of the computed peak discharge

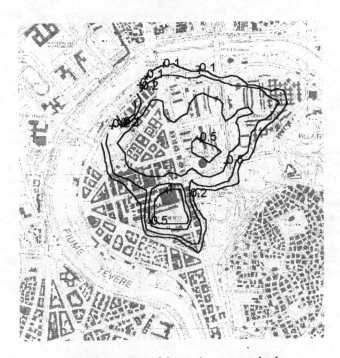

Fig.3: Mean values of the maximum water depths

Between Corbara lake and Rome, 33 tributaries flow in Tiber river; the four most important are shown in Fig. 1. For these tributaries (Paglia, Nera, Treia and Aniene rivers) discharges data are available at gauging stations, located close to their inflow in Tiber river. In Rome, discharge measurements have been carried out at Ripetta gauging station starting from 1922.

In the centre of Rome, the Tiberina island separates the Tiber river in two channels, less than 1 km long. In these channels 4 bridges and 2 sills are located, which strongly influence the flow conditions and that must be carefully simulated. Close to the outflow in the Tirrenian sea, Tiber river also separates in two channels, "Fiumara Grande" and "Canale Fiumicino". We consider a looped network, constituted by 8 reaches:

-     Tiber river from Corbara lake to Aniene river,
-     the final reach of Aniene river (60 km),
-     Tiber river in Rome from the inflow of Aniene river to Tiberina island,
-     the two channels of Tiberina island,
-     Tiber river from Tiberina island to "Capo due Rami"
-     the two channels "Fiumara Grande" and "Canale Fiuicino".

The infiltrated depth is separated in interflow r_i, rapid subsurface flow which contributes to flood wave, and deep percolation r_p which constitutes a lost volume for the flood, by means of the following relationship:

$$r_i = C_s [r(t) - r_d(t)] \tag{1}$$

If the rainfall-runoff analysis is carried out by means of a linear, lumped model, the runoff can be expressed in terms of the impulse response function via the convolution integral (Bedient and Huber, 1988):

$$Q(t) = \int_0^t u(\tau) i_n (t - \tau) d\tau \tag{2}$$

where i_n is the net rainfall intensity. For both surface and subsurface runoff, the watershed is schematised by means of a cascade of linear reservoirs and a linear channel, in series:

$$u(t) = \frac{1}{k_i \Gamma(n_i)} \left(\frac{t - C_i}{k_i} \right)^{n_i - 1} e^{-(t - C_i)/k_i} \tag{3}$$

where k is the storage delay time of the reservoirs, n is the number of reservoirs and C is the translation time in the channel. The subscript i indicates that different values of the parameters k and n are used for surface and subsurface flow.

The hydrological module was calibrated on the basis of the measured hourly discharge hydrographs and hourly rainfall ietographs for historical floods. More details can be found in the paper by Natale et al (1999).

The density of the rainfall gauging stations does not allow to correctly simulate the rainfall volumes distribution, due to the lack of stations located in altitude. Therefore, in the calibration of the hydrological module a further parameter C_p was introduced in order to take into account distribution of rainfall in space and time. This parameter was varied in time during the storm.

It follows that for each of the 5 watersheds, including Tiber river at Corbara lake, 10 parameters were calibrated: $f_i, f_f, C_s, k_i n_i, C_i$ (for surface and subsurface flow), C_p.

The calibrated values of k, n and C for subsurface flow resulted very similar for each watershed. For surface flow the calibrated values of k and n were related to the area of the gauged watersheds and regression relationships were found. These relationships were used to estimate the values of the parameters k and n for the surface flow for the ungauged watersheds.
In Tab. 1 are shown the average values and the standard deviation of the calibrated parameters. The values estimated for Nera river were not reported because the discharges are strongly influenced by many regulated lakes.

3. PROPAGATION OF THE FLOOD WAVES IN THE HYDROGRAPHIC NETWORK

The flow in each stream was simulated by integrating the 1-D equations of flow for free-surface, gradually varied flow (Cunge et al, 1980):

$$\frac{\partial Y}{\partial t} + \frac{1}{B} \frac{\partial Q}{\partial x} + q = 0 \tag{4a}$$

$$\frac{\partial Q}{\partial t} + \frac{\partial}{\partial x}(\frac{Q^2}{A}) + gA \frac{\partial Y}{\partial x} + gAS_f = 0 \tag{4b}$$

where: x is the spatial coordinate along the reach; Y is the water stage; Q is the discharge; B is the surface width; A is the wetted cross-section; g is the gravitational acceleration; S_f is the slope friction, computed according to Manninf formula; and q is the outflow unit discharge due to the levee overtop $q = \mu\sqrt{2g}(Y - Z_a)^{1.5}$, where Z_a is the elevation of the levee and μ is a discharge coefficient.

For the generic node, where M reaches are connected, the following conditions were imposed:

mass balance equation: $\displaystyle\sum_{k=1}^{M} Q_k = 0$ (5)

conservation of the total heads: $H_1 + \Delta H_1 = H_2 + \Delta H_2 = \ldots = H_M + \Delta H_M$ (6)

where Q_k and H_k are discharge and total heads in the boundary sections of the reach connected with the node and $\Delta H = \alpha_k \dfrac{Q_k |Q_k|}{2g A_k^2}$ is the localised head loss, where α is a calibration parameter.

Equations (4a,b) are discretized according to the implicit finite difference scheme proposed by Preissmann (Cunge et al, 1980). The resulting system of equations, including 4 boundary conditions (discharge in the upstream nodes and stages in downstream nodes) was solved by means of the condensation method proposed by Uan (1984).

The model simulates the flow through bridges, sills and drop structures. The flow through a singularity is simulated by imposing an internal boundary condition, i.e. rating curves, computed by imposing the momentum balance equations. The model allows to evaluate the head losses through bridges, to simulate the submergence of bridges, and the structure overtopping.

The values of the Manning roughness coefficient, considering compound channels, were calibrated on the basis of discharge measurements carried out for different flow conditions, and vary from 0.036 m$^{-1/3}$s to 0.055 m$^{-1/3}$s. The 8 reaches of the hydrographic network were discretized with 557 cross sections, including 65 bridges, 12 sills and drop structures, and 4 dams upstream Rome.

4. PROPAGATION IN THE URBAN ENVIRONMENT

Hydraulic models for flood routing are usually based on the de Saint Venant partial differential equations (4a,b).

During a flood propagation in urban areas, flow enters the street network on the floodplains and it divides into a system of channels, linked to the main river channel. In each street/channel the flow is mostly one-dimensional and only in small areas, mainly at crossroads and in squares, two-dimensional effects prevail. This suggests that a significant simplification may be introduced considering a model based on a one-dimensional open-channel network, provided that some hydraulic conditions are specified at the junctions (Cunge, 1975; Braschi et al, 1991). Some models have been developed, that can be used also for open channel looped systems (Fread, 1993); because of the large number of loops to be considered, still their use can result in a cumbersome computational burden and a large computation time for urban areas flooding applications.

Further simplifications may be introduced. The first three terms in (4b) may be neglected for gradually varying floods, approximating the flow equations to that for steady-state conditions. Considering the channel linking nodes i and j, the previous assumptions allow to explicitate the dependence of the discharge, Q_{ij}, on the water levels at the nodes, z_i and z_j, by

$$Q_{ij} = \frac{A_{ij} R_{ij}^{2/3}}{L_{ij} n_{ij}} \frac{(z_i - z_j)}{\sqrt{|z_i - z_j|}}$$ (7)

where A_{ij} and hydraulic radius R_{ij} depend on z_i and z_j and L_{ij} is the channel length.

Eq. (4a) may be simplified too, considering the storage capacity of the network as allocated in the nodes:

$$a_i \frac{\partial z_i}{\partial t} - \sum_{j=1}^{k} Q_{ij}\left(z_i, z_j\right) - Q_{ie} = 0 \tag{8}$$

where the sum is extended to each j of the k nodes connected by a channel to node i, Q_{ie} is the flow rate exchanged at node i with outside elements; z_i is the water level at node i. The free surface elevation at endpoints of each channel converging to node i, is assumed to be the same as that of the node, z_i, reducing the number of the unknowns to N.

Equations (8) my be discretized in time, yielding the non linear system of algebraic equations:

$$a_i \frac{z_i^n - z_i^{n-1}}{\Delta t} - \sum_{j=1}^{k} \left[\Theta Q_{ij}\left(z_i^n, z_j^n\right)\right] - \sum_{j=1}^{k} \left[(1-\Theta)Q_{ij}\left(z_i^{n-1}, z_j^{n-1}\right)\right] - Q_{ie} = 0 \tag{9}$$

where: Θ is a temporal weighting coefficient, $1 \geq \Theta \geq 0$; Δt is the time step; and the n index denotes the time $n\,\Delta t$. The Newton-Raphson method is used to solve the non linear equations system given by (9).

5. DEFINITION OF THE SCENARIOS

Some preliminary computations showed that the concentration time for Tiber watershed in Rome is about 7 days. For the return period T=200 years, the daily intensity-duration frequency (IDF) functions were estimated for 3 main subbasins: Tiber at Corbara lake, catchment from Corbara lake to Passo S. Francesco, catchment downstream Passo S. Francesco. The 7 days rainfall for T=200 years vary from 215 mm, to 261mm, to 249 mm starting from the upstream watershed. For these computations the daily rainfalls recorded in 17 gauging stations were used.
-       the rainfall of each watershed must be equal to the value of the corresponding main subbasin;
-       the rainfall intensity was limited by the values of the corresponding IDF function.
The generated 100 realizations of daily rainfall, for 7 days and for each watershed, are consistent with a constant rainfall intensity during each day.

For each watershed and for each realization, the hydrological module was applied. In order to take into account the uncertainties in the parameter estimations, the values of the parameters were randomly extracted from a gaussian distribution with the mean values and standard deviation reported in Tab. 1.

The outflow from Corbara lake was simulated by means of a simplified operating rule: it was assumed that available storage in the lake was 105×10^6 m^3, more or less the 50% of the maximum storage. The outflow was equal to 200 m^3/s until the storage was completely filled and after the outflow was assumed equal to the inflow.

Table.1: Calibrated parameters of the hydrological module

Parameters	Watersheds			
	Corbara lake	Paglia river	Treia river	Aniene river
C_p	2 ± 2	1.1 ± 0.3	1.1 ± 0.2	1.0 ± 0.0
f_i (mm/h)	2.5 ± 2.2	3.1 ± 1.9	4.6 ± 2.2	4.3 ± 0.4
f_f (mm/h)	0.7 ± 0.6	1.5 ± 1.6	2.8 ± 2.5	4.0 ± 0.0
C_s	0.00	0.09 ± 0.02	0.08 ± 0.03	0.08 ± 0.04
$C_{surface}$	1.5 ± 1.5	1.5 ± 2.1	4.5 ± 0.7	0.0
K_{sur}(h)	12.5 ± 0.7	6.0 ± 0.0	4.8 ± 0.0	14.0
N_{sur}	2.3 ± 0.1	1.5 ± 0.0	1.5 ± 0.0	2.5
$C_{subsurface}$	0.0 ± 0.0	0.0 ± 0.0	0.0 ± 0.0	0.0
K_{sub}(h)	20.0 ± 0.0	20.0 ± 0.0	20.0 ± 0.0	20.0
N_{sub}	2.2 ± 0.0	2.2 ± 0.0	2.2 ± 0.0	2.2

The inflow of Nera river in Tiber was assumed equal to 350 m³/s for each scenario.

6. ANALYSIS OF THE RESULTS

The simulation of the propagation of the 100 flood waves from Corbara lake to the Tirrenian sea, taking into account the inflow of the tributaries, allowed to compute the stages and discharge hydrographs in every sections of network.

In Fig. 2 are reported, on Gumbel probability paper, the frequency distribution of the peak discharge computed for T=200 years in the 5 sections reported in Fig. 1.

The outflows due to the levees overtopping strongly deform the frequency distribution of the peak discharge at Rome. It seems that the extreme discharges tend to an asymptotic value of about 3700 m³/s for Ripetta station. In this station the expected value for T=200 years was 3327 m³/s; the 60% of the realizations is included in the range 3126-3512 m³/s, and the 90% of the realizations lie in the range 2833-3584 m³/s. Those values are very similar with those that can computed using the Gumbel probability distribution for the historical peak discharge measured at Ripetta.

For each realization, the hydrographs of the overtopping waves are computed. Significant inundation of the urbanized areas occur in both left and right side, with median of outflow volumes 21 and 53 10⁶ m³, respectively; the median of outflow peak discharges are 116 and 207 m³/s, respectively.

The inundation of the historical centre of Rome in the left side was simulated by means of the model described in section 5; 443 elements and 269 nodes has been used to simulate the urban environment. In Fig.3 the inundated area and the mean values of the maximum water depths computed for each realisation are shown. It clearly emerges that the riverside is inundated with significant water depths, greater than 1 m.

7. CONCLUSIONS

The mapping of the flood-prone areas in Rome can not be carried out with the common procedures based on steady state computations, but an unsteady state model is required. In this study the flood-prone areas for a return period T=200 years in Rome are mapped following a Montecarlo simulation, which allow to define the confidence interval of the peak discharge in Rome and the outflow volumes.

These preliminary computations, which refer to a portion of the left riverside, indicates that the historical center of Rome can be flooded for levee overtopping which occur in the upstream part of the urbanized area.

Further developments of the research will concern the simulation of both riversides, extending the computational domain to the whole urbanised area, in order to evaluate the risk of vulnerable elements within an urban area and to manage the civil protection actions.

REFERENCES

P.B. Bedient, W.C. Huber, *Hydrology and Floodplain Analysis*, Addison-Wesley, New York, 1988

G. Braschi, M. Gallati., L. Natale: *Modelling Floods in Urban Areas*, ASFPM 15th Annual Conference, Denver, Co., June 1991.

J. Cunge, F.M. Holly, A. Verwey, A., *Practical aspects of computational river hydraulics*, Pittman Publ, Boston, 1980.

J. Cunge: *Two Dimensional Modelling of Flood Plains*, Unsteady Flow in Open Channels, Mahmood e Yevjevich Editors, Water Resources Publications, Vol.II, Fort Collins, 1975.

D.L. Fread, *Flow routing*, Handbook of Hydrology, D.R. Maidment ed., Mc Graw-Hill, 1993

F. Guzzetti, M. Cardinali, P. Reichenbach, *Map of sites affected by landslides and floods in Italy - The AVI project*, National Research Council (CNR) National Group for Prevention of Hydrogeological Hazards (GNDCI) Publication n. 1356, map at 1:1,200,000 scale, 1996

L. Natale, F. Savi, L. Ubertini, *Probability of inundation of Rome*, Proceedings of the IASTED International Conference on Modelling and Simulation, Philadeplhia, 1999.

M. Uan, *Principe de resolution de modelisation fileaires des ecoulements a surface libre: application a un reseu maille'*, EDF Report HE/43-80-40, 1984.

H.W. Weiss, D.C. Midgeley, *Suite of mathematical flood plain models*, Journal of Hydraulic Division, ASCE, 104 (3), 1977.

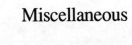

Stochastic Hydraulics 2000, Wang & Hu (eds) © 2000 Taylor & Francis, ISBN 90 5809 166 X

Subject index

Stochastic Hydraulics 2000, Wang & Hu (eds) © 2000 Taylor & Francis, ISBN 90 5809 166 X

Author index

Printed and bound by CPI Group (UK) Ltd, Croydon, CR0 4YY

23/10/2024

01777707-0003